Cloning, Gene Expression, and Protein Purification

Experimental Procedures and Process Rationale

Charles Hardin
Jennifer Pinczes
Andrew Riell
David Presutti
William Miller
Dominique Robertson

North Carolina State University

New York Oxford
OXFORD UNIVERSITY PRESS
2001

Oxford University Press

Oxford New York
Athens Auckland Bangkok Bogotá Buenos Aires Calcutta
Cape Town Chennai Dar es Salaam Delhi Florence Hong Kong Istanbul
Karachi Kuala Lumpur Madrid Melbourne Mexico City Mumbai
Nairobi Paris São Paulo Shanghai Singapore Taipei Tokyo Toronto Warsaw

and associated companies in
Berlin Ibadan

Published by Oxford University Press, Inc.
198 Madison Avenue, New York, New York 10016
http://www.oup-usa.org

Library of Congress Cataloging-in-Publication Data

Cloning, gene expression, and protein purification : experimental procedures and process rationale / Charles C. Hardin... [et al.].
p. cm.
Includes bibliographical references and index.
ISBN 0-19-513294-7 (alk. paper)
1. Molecular cloning—Laboratory manuals. 2. Gene expression—Laboratory manuals.
3. Proteins—Purification—Laboratory manuals. I. Hardin, Charles C.
QH442.2.C567 2001
572.8—dc21 00-045312

Printing (last digit): 9 8 7 6 5 4 3 2 1

Printed in the United States of America
on acid-free paper

CONTENTS

APPENDICES

PREFACE

This manual evolved to train advanced undergraduate and introductory graduate students to enter high-tech jobs in health care diagnoses and treatments, pharmaceutical industries, and graduate studies across a broad set of disciplines. Our goal was to produce a broadly encompassing tool that could help these students develop first-rate laboratory skills. Since we focus on quantitative analyses of biological macromolecules and reactions, students should have taken at least one year of quantitative and organic chemistry courses prior to this training. Additional experiences with common laboratory tools (pipets, quantitative spectroscopy, electrophoresis, centrifugation) and some microbiology training are all preferable, but not strictly essential if appropriately trained instructors work interactively with the students to correct the deficiencies. This book cannot teach the information alone, but it attempts to do the next best thing, teaching students how to teach themselves. Students who acquire these skills should be able to compete successfully for jobs in biochemistry and biotechnology laboratories.

The laboratory exercises and supplementary information were designed to peel back the surface of the experiments and expose their inner workings. They are also meant to be informative and yet be relatively easy to read by students from a variety of scientific backgrounds. The desired outcome is a fully trained student ready to perform truly independent research, understanding what it means to pursue a fully integrated program of laboratory techniques, experimental planning, and both pre- and post-process troubleshooting. Concepts and terms are readily accessed in an extensive index. The purposes of techniques are usually described in detail, as are the consequences of the various types of manipulations. Supplemental literature explains a variety of theoretical explanations designed to foster a better intuitive sense of the experimental outcome as a result of the molecular behavior of the participants.

The approach used in the manual is distinctive in the following ways.

1. The plasmid, cDNAs, and antibody probes are all well characterized in the research literature and are available from an easily accessible source, the American Type Culture Collection (ATCC). Instructors are generally familiar with the techniques used in this manual; we facilitate their expertise by providing these tested materials on a worldwide basis.
2. Information is provided to accompany the lectures and labs from the product and research literature, including primary sources and reviews of both practical and theoretical utility.
3. The relevant parts of original methods and reagent literature are reproduced as seen in the original sources as a means of introducing students to the interwoven nature of the linkage between literature sources and laboratory use.
4. An alphabetized list of defined terms is provided for each of the two parts, allowing the student to rapidly define unfamiliar phraseology or discipline-specific jargon. New terminology and important summary statements are italicized in the text to help the students identify important phrases or conclusions.
5. An alphabetized reagents list is provided in the appendices, allowing the student to rapidly reference common reagents used in the laboratory protocols.
6. All topics and subtopics are organized in a format consistent throughout the manual to assist ease of reading and organizing thought processes. Main topics are numbered and in bold, subtopics are italicized and numbered with numerical extensions, and subtopics are subdivided into categories that are italicized and numbered with two number extensions.
7. A flow chart accompanies each part to facilitate understanding the order and progress of the course experiments.
8. Five types of articles (Innovation/Insight, Theory/Principles/Principles, Process Rationale, Vendor Literature, and Alternative Approach) are included in this manual. These titles have been chosen to accompany each one of the articles to clarify the type of supplemental literature the section provides.
9. Symbols are located at the top outer corner of each new section to indicate the section type (Introductory Material - Compass rose, Flow Charts - Direction coordinates, Appendices - Reference books, Labs - Pipet in Eppendorf tube, lectures (Part I) – Vector, lectures (Part II) - Folded protein, Original literature (Part I) - DNA helix, Original literature (Part II) - ELISA schematic, and Index - Lighthouse).
10. This laboratory manual contains an introductory unit and two main parts. The introductory unit provides practice with many basic skills that are used in the biotechnology laboratory. Part 1 focuses on nucleic acids, including DNA and RNA. Part 2 focuses on proteins, including antibodies and combinatorial technology.
11. Students are encouraged to prepare laboratory writeups in the same format as primary literature manuscripts. This helps the student learn how to clearly and succinctly define the immediate and overall problems, the approach to solving the problem, the nature of the results, and the importance of the results to the overall goal of the project.

The ultimate goal is to make students aware of how to approach multistep projects by ushering them through the learning stages involved in proceeding from the initial cloning experiments, through the gene expression procedures, and finally to the protein purification and analysis techniques.

NOTE TO INSTRUCTORS

The plasmids and antibody used in this manual are available from Dominique Robertson at NCSU (niki_robertson@ncsu.edu). Although the plasmids are available at no cost, there is a small processing charge for the monoclonal antibody.

ACKNOWLEDGMENTS

Special thanks to Richard Guenther, Jane Petitte, Matthew and Bruce Corregan, Bernard A. Brown II, David Lieberman, Kristin Sullivan, Nathan Nicely, Rebecca Gunn, and Erin Warren. We also gratefully acknowledge the contributions of Dean James Oblinger, Professors James Moyer, Dennis Brown, Paul Agris, James Knopp, Cynthia Hemenway, Scott Shore, David Miller, Peter Bann, and Bob Rogers.

We acknowledge grants from the Howard Hughes Educational Fund and the North Carolina Biotechnology Center. We acknowledge the ongoing support from the Office of Academic Affairs, the College of Agriculture and Life Sciences, and the Department of Biochemistry. This manual was produced at and fostered by North Carolina State University.

We dedicate this book to Derek, Hong, and my parents (by Chuck), Steven, my husband (by Jennifer), Mo, and my family (by Andy), Elizabeth, Madelene, and my parents (by Dave), Laura, my wife (by Bill), Robert and Monica (by Niki).

C. H., J. P., A. R., D. P., W. M., and D. R.

NCSU
Raleigh, NC
April 7, 2000

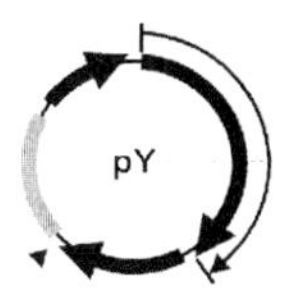

INTRODUCTORY UNIT: INTRODUCTION TO THE BIOCHEMICAL LABORATORY

1. Pipets

The laboratories, lectures, and appendices in this section emphasize the critical nature of proper pipeting. They provide practice in preparing nucleotide and amino acid samples and then characterizing them using ultraviolet absorbance spectrophotometry. Pipeting errors are characterized and the concept of propagated error is explained and illustrated. Protein concentration is measured next, illustrating a simple assay of the total peptide group concentration using a coupled redox-activated dye technique, the BCA assay. All enzyme activity assays (see section 2) require determining protein concentrations, so accomplishing the key goal of monitoring improvements in protein activities requires multiple applications of this fundamental measurement.

2. Introduction/Overview

A majority of the scientific activities done to date under the heading "molecular biology" have involved studying (*1*) production of proteins, (*2*) functions catalyzed by them, or (*3*) controls mediated by them. Most typical projects proceed by the fundamental steps: (*1*) the gene that encodes the protein is captured in an expression plasmid, (*2*) the DNA sequences are determined, and (*3*) the protein is produced via transcription and translation by a virus, bacterium, yeast, or higher eukaryotic cell. Thus, we will focus on the pathway of processes that lead from the DNA to its RNA transcript to the translated protein, with emphasis on how biomolecular structures and the conditions of the surrounding medium work together to produce functionality.

The laboratories and lectures in this section are intended to teach the points to consider in the course of this type of research. Doing reliable experiments is necessary to obtain patents, move a product through early R & D stages, on to clinical trials and finally to sale. This requires careful use of analytical techniques and statistical analyses of the data. Reproducibility is the central tenet of responsible science. Doing otherwise can constitute production of unreliable results and possibly committing fraud!

A remarkably diverse set of factors can affect biomolecular functions. The dependence of functional characterization protocols on isolation, storage, reconstitution, and function assay conditions must be characterized and standardized in order to claim that one has developed a trustable material. This book is designed to explain how one thinks when working to purify, preserve, and assay DNAs, RNAs, and proteins.

3. Sample Preparation

Molecular biological experiments typically involve dissociating one or a few of the macromolecular components of a particular functional assembly, recovering the individual components in as pure a form as possible, then reconstituting the system under conditions "simple enough" to control variables sufficiently well so one can claim reproducibility. For many multistep experimental schemes, this involves obtaining pure components and one or a set of enzymes, DNAs, or RNAs, for use as analytical reagents (e.g., catalysts, redox reagents, labeling or linking materials, internal standards, control substrates). As a result, a huge "biotechnology" industry has evolved to supply verified supporting reagents, thereby reducing the difficulty of assembling the required reagents; the key problem becomes obtaining the biomolecules and specialized small molecules that aren't commercially available. The labs, lectures, and appendices in this book are primarily concerned with preparing pure samples and understanding the factors that will allow one to reproducibly maintain their functions. Mastering these skills, and understanding how to generalize the concepts and apply them to a previously uninvestigated problem, is the key to becoming a working experimental molecular biologist.

A tip. Try to avoid the very worrisome situation we call "kit mentality." This is practiced among students who have not been adequately taught or don't care to make the effort to understand how the chemistry of the situation determines the chances for success, especially when troubleshooting is required. The "kit mentality" practitioner treats the experimental protocol like a "black box" recipe that need not be understood explicitly. For example, they'll transfer "2 μl from the reagent in the green tube" while having no idea what the material is or does.

4. Buffers and Other Variables Considered to Preserve Functional Activities

Proper use of buffers is essential if a researcher is to claim that he/she has developed a reproducible molecular biological method. Since reproducibility is the hallmark of conscientious science, it is imperative that we discuss detailed analytical chemical aspects of *buffers* and *factors one must control* to ensure that they function properly in the experimental milieu of the reaction studied. Norman Good elaborated this subject in his classical papers.

Consider the difference between the chemical *concentrations* and the *activities* of the relevant functional groups on all of the chemical species. Molar specific activities of components in a sample might change with variations in any or all of the following factors: (*i*) pH, (*ii*) one or more buffer components, (*iii*) binding and/or enzymatic substrates, (*iv*) activity-modifying "spectator" molecules, (*v*) mono-, di- and polyatomic ions, (*v*) the enzyme(s), protein(s), nucleic acid(s), carbohydrate(s), and/or polysaccharide(s), (*vi*) lipids, (*vii*) other nonpolar or amphipathic molecules, and many other possibilities, depending on the situation under study.

Temperature changes usually affect the activities of buffer components as well as the other factors listed above. Some buffers (e.g., phosphate) have pK_as that are relatively stable over a range of temperatures. In contrast, the pH of the ubiquitous buffer Tris(hydroxymethyl)aminomethane ("Tris") changes by several tenths of a unit as the temperature is varied across the ranges encountered in typical protocols. This "pH drift" can dramatically affect biochemical activities. For example, preserving the activity of a protein in a particular DNA binding and cleavage reaction sometimes requires that one maintain the pH in a particular range. Moving outside that range can decrease the activity of the protein, sometimes irreversibly. Using a protocol that requires changing temperature, which shifts the pH of Tris, might fail to yield products because the pH change inactivates DNA recognition.

Troubleshooting problems with your reactions can require some subtle considerations. One must sometimes untangle the chemical consequences of interactions between synergistically linked chemical species, energy sources, forces, constraints, entropies, etc. In the case cited above, one should look up the temperature-dependence pK_a coefficient of the pK_a of Tris and other buffers, and select one with a less temperature-dependent pK_a. This consideration is reasonably subtle to the novice troubleshooter.

Our hope is that students will read such analyses and try to learn how to generalize the learned lesson(s) to analogous situations. More importantly, this approach will benefit diligent and mastery-minded students in future pursuits, as they work to develop new protocols and use those obtained from others in the "real world," both privately communicated and commercial. All seasoned molecular biologists have spent time working through "snags" that developed because they don't understand how the chemistry of the situation leads to products, or why the process is "screwed up" by unsuspected interactions between components, constraints, etc. "Paying one's dues" involves undergoing enough such encounters to learn how to minimize one's time in failed processes. Success involves learning how to preserve sufficient materials and activities from one reaction for the next stage of the process.

Reduction of theoretical principles to practical demonstrations is always necessary to prove a new fact. The truly capable experimentalist produces more materials and facts (i.e., they're more successful) because they spend less time repeating flawed protocols. This is often accomplished by taking the time and effort to think about the problem carefully before merely "trying again because I must have screwed something up last time." Maybe the process is not working because some subtle problem(s) lurks beneath the "recipe" of the protocol. This is one potential danger of "recipe mentality," a cousin of the "kit mentality" concept described above. We discourage student and practicing researchers from making up 'recipe cards' with buffer preparation protocols for general distribution to novices.

Make your solutions and samples according to the general procedure summarized below. (Standard brackets notation indicates concentrations throughout this book.) A specific example is used to illustrate.

Note that the units and decimal places are aligned carefully. It may seem overbearing to ask that the information be provided this way. However, when troubleshooting, it can be very enlightening to have this clear exposition of all of the components,

Targeted [Component] in Final Sample/Solution	[Stock Solution]	Volume of Stock Solution Added	Fold Dilution
20 mM Tris-HCl, pH 7.4	200.0 mM	2.0 µl	10.0x
100 mM NaCl	2.0 M	2.0 µl	10.0x
4 mM $MgCl_2$	0.1 mM	0.5 µl	25.0x
10 % ethanol	95 %	2.1 µl	9.5x
H_2O		+13.4 µl	
	Total volume =	20 µl	

stocks intended for use, and final intended concentrations, along with intended volume changes to be used to achieve the final concentrations (typically dilutions). Having the student's (or employee's) notebook with this information for use as the basis of discussions between professor (manager) and student (employee) provides good discipline and can pay off when troubleshooting a problematic experiment. This level of organization will also help when preparing to expand a "pilot protocol" to a more fully realized experiment, in which a range of solutions must be prepared in parallel.

When making stock solutions, be certain that they've been adjusted to the proper pH and volume and kept clean and "microbe free." Most labs are equipped with H_2O "polishing" facilities, which pass the solvent through several stages of ion exchange and organic chemical-removing resin materials, achieving H_2O that is generally clean enough to use in molecular biological bench experiments. However, if organisms are to be grown, e.g., in food-source containing media, solutions should be treated by passage through syringe filtration devices, such as those sold by Amicon (Centricon®), or by autoclaving. Solutions will not 'keep' forever on the shelf in a lab at ca. 22°C. Withholding Mg^{2+} for storage, and only combining it with the other components for the reaction or process when needed, can help limit bacterial contamination. Going one step further, placing the chelator *ethylenediamine-tetraacetic acid* (really Na_2EDTA^{2-}; usually called "EDTA") in the solution will ensure that trace metal or cation contaminants are scavenged. Conversely, it makes little sense to add a required metal (e.g., Mg^{2+}) to a reaction mixture that contains a lot of EDTA, since it will render the metal unavailable for its role in the reaction.

5. pH Meters and Adjustment Techniques

A pH meter is used to determine and adjust the pH of a solution. Two general approaches exist. In the first, one places the calibrated pH probe directly in the buffer solution *while* one adjusts the pH. Of course, one must be particularly careful to avoid transferring contaminants, including chemicals, ions, or microbes, along with the pH probe. In the second approach, one makes concentrated stock solutions, composed of pure protonated and deprotonated buffer species whose pK_a values bracket the targeted pH. One then makes the target buffer by mixing appropriate amounts of the buffer components (taken from a previously developed table) and adjusting the volume to the final value by adding H_2O.

The first technique is superior to the second because it is less restrictive. Adjusting the pH, while monitoring the approach to the target value directly, ensures that the proper buffer is produced. In contrast, the stock-mixing procedure requires the table or doing calculations using the Henderson-Hasselbach equation

$$\mathrm{pH} = \mathrm{pK_a} + \log\frac{[\mathrm{A^-}]}{[\mathrm{HA}]} \qquad (1)$$

where A^- is the less protonated conjugate base species and HA is the more protonated conjugate acid, the two species surrounding the target pH.

Four considerations will help ensure that the buffer is prepared correctly. First, the pK_a must be within one unit of the target pH in order to ensure that the buffer has sufficient buffer capacity to truly limit pH changes. Second, solutions must be equilibrated *at* the temperature used *in* the experiment (see above). Third, one must fine-tune the pH when large volumes of H_2O are added to dilute stock components to their final concentrations. Lastly, when preparing stock solutions of pH-dependent solution components (e.g., EDTA) one must adjust the pH using an immersed pH probe *while* dissolving the minimally soluble tetraprotonated powder. It will dissolve only as it becomes partly deprotonated. If not careful, one risks large errors in achieving the desired concentration.

6. Biotechnology Notes

It is instructive to place the concept of pH measurements in a broader context. It can be very important to determine the concentrations of other types of cations *in vivo* (e.g., Ca^{2+}). Some clever methods have been developed to accomplish this goal, e.g., the now classical technology developed by Roger Tsien and others during the 1980s. When a Ca^{2+} chelator, which is molecularly tethered to a partially inactivated fluorophore, binds cation, the fluorescent molecule becomes activated to fluoresce. As a result, the fluorescence is proportional to the $[Ca^{2+}]$ and one can study their fluctuations. These reagents have been used to follow progressions in cell cycle stages, development, diseased states, etc. Similar reagents have also been developed to determine $[K^+]$, $[Na^+]$, $[H^+]$ and others.

A related set of techniques has also been developed to study enzymatic reactions. These reagents consist of a molecular "cage" that carries an enzymatic reactant or ion cofactor and can be triggered with a flash of light of the appropriate wavelength to release the contents of the cage, thereby initiating the reaction to be studied.

7. Spectroscopy

The *absorbance spectrum* is used in many, many protocols to determine concentrations of nucleic acids, proteins, and many important substrates and cofactors. Spectroscopic approaches have one key advantage over wet lab techniques: they are relatively noninvasive. In fact, materials sometimes emerge from the analysis sufficiently unscathed for further use.

Analytes, the subject of the spectroscopic experiment, are typically scanned in scaled-down reactions in quartz cuvettes. Spectra are also obtained using diluted analytical samples that have been prepared from "stock samples." This stock is also used to execute several reactions, in parallel, allowing use of the same characterized reactant, without requiring analysis of each individual sample.

The *spectrum* presents a two-dimensional plot of absorbance as a function of wavelength, usually scanned across the ultraviolet (UV) range (200 to 350 nm). Proteins containing one or more of the aromatic chromophores – phenylalanine, tyrosine, and tryptophan – will have reasonably intense absorbance. For example, 1 mg of an immunoglobulin G (M_r ca. 150 kDa) produces, very roughly, an absorbance at 280 nm (A_{280}) of 1.0. Nucleic acids always contain a large complement of their chromophoric nucleobases, producing more intense net absorbance than proteins. For example, a ca. 25 μg/ml transfer RNA solution (M_r 25 kDa) produces an A_{260} of about 1.0. These figures are approximations. The exact absorbance at a stipulated wavelength per mole of material is embedded in the "molar extinction coefficient" (ε_{280}), the key to the "Beer-Lambert equation" (Beer's Law).

Introductory lab 2 demonstrates the use of Beer's Law and how to acquire absorbance spectra with amino acids and nucleotide solutions. The student is exposed only to a very small sampling of the range of incredibly powerful spectroscopic techniques. Fluorescence, circular dichroism, Raman spectroscopy, and nuclear magnetic resonance are not covered in detail here, but are very important methods in biochemistry. Consult a text such as *Physical Biochemistry* by van Holde, Johnson, and Ho (Prentice-Hall, 1998) for detailed descriptions. Any current issue of *Biochemistry*, *Journal of Molecular Biology*, or even *Cell* will give illustrations of relevant.

BCA redox-coupled reaction scheme:

Step 1 - Biuret reaction:

$1\ Cu^{2+}$ + 1 protein (reduced peptide bond) $\xrightarrow[OH^-]{\text{BCA reagent}}$ $1\ Cu^{1+}$ (pale blue) + 1 protein (oxidized)

Step 2 - $BCA_2\ Cu^{1+}$ complex formation:

$1\ Cu^{1+}$ + 2 BCA $\longrightarrow$ BCA_2Cu^{1+} complex (dark purple)

8. Protein Concentration Determination Assays

The BCA method is used in Introductory lab 2. The basis of the technique builds on its popular predecessors the Bradford and Lowry methods. The reagents in the BCA reaction include bicinchonic acid and copper hydroxide. This technique begins with the Biuret reaction, which involves protein-induced reduction of Cu^{2+} to Cu^{1+}. In this reaction, Cu^{2+} atoms are reduced to Cu^{1+} by peptide groups in the protein backbone in a redox-coupled reaction. Second, a redox-active dye reacts with the Cu^{1+} that was produced by the peptide-Cu^{1+} redox reaction. Then the dye bicinchonic acid (BCA) is added, which produces an intense purple color by binding to and amplifying the Cu^{1+} signal. The intense purple color is proportional to the amount of peptide groups in the protein. The technique is preferred for two reasons, because it is less sensitive to inhibitory effects caused by other components in the samples (e.g., buffers, cations, substrates) and because all proteins react in proportion to the number of peptide groups. In contrast, the inferior A_{280} method requires knowledge of the number of aromatic amino acids per mole of protein, which differs from protein to protein, resulting in a semiquantitative analysis at best.

Bovine serum albumin (BSA) concentration standards are typically used to calibrate the measurements. A large range of BSA products are listed in a typical source catalog (e.g., Sigma), and products differ in lipid content, purity, and other factors. The researcher must make their choice carefully since assay standardization depends critically on the assumption that BSA acts analogously to the protein of interest. One should generalize this mindset to include all comparison-oriented experiments. It's very common to do so in molecular biology. For example, one typically assumes proteins electrophorese in linear proportionality to log M_r in sodium dodecyl sufate (SDS) gel experiments. Unfortunately, this is not always the case in certain applications (e.g., when they are glycosylated).

9. An Aside

The point of view of the physicist typically limits study to a very small number of interacting species; simplicity rules. This is done to create a controlled situation that can be modeled mathematically and where statistical variations can be characterized thoroughly. A biological cell is exceedingly complicated, whether studied intact, as chemically treated ("fixed") preparations, as excised tissues, or as soluble biomolecules immersed in a reaction mixture. Problems associated with controlling variables increase dramatically as the size, number of interacting components, and complexity of relations increase. While our ability to probe these factors *in vivo* has advanced dramatically during our lives, especially with the advent of isotopic, tomographic, and nuclear magnetic resonance imaging methods, in most cases we're far from being able to pass a Star Trek-like medical instrument over a diseased or flawed body part in order to decipher detailed genetic problems or even fix them.

THEORY/PRINCIPLES
COURSE DESCRIPTION

1. Overview

The sets of techniques in Parts 1 and 2 are performed as part of a multistepped experimental chain designed to "subclone" a fragment of DNA and then verify that each substep was successful. This overall process involves purifying a defined sequence from a biological source, preparing its ends enzymatically, and then splicing it to a second larger fragment to create a replaceable circular plasmid. The location of this inserted fragment is designed to create a defined circular DNA sequence that supports the RNA synthesis process (transcription). The RNA product is designed to support use by the ribosome to make a specified protein (via translation).

In Part 1, students perform a project composed of restriction enzyme digestion, DNA ligation, DNA separation by cloning procedures such as restriction enzyme digestion, DNA ligation, DNA separation by agarose gel electrophoresis, plasmid isolation, and Southern blot analyses of DNA. A flowchart that shows the detailed connection between the subprojects is shown at the beginning of Unit 1.

In Part 2, students learn how to design expression vector constructs, and perform bacterial transformation, polyacrylamide electrophoresis, and immunoblot analysis of proteins, respectively. A flowchart is shown at the beginning of Unit 7. Researchers often resolve biomolecules that differ with respect to size, charge, polarity (electroneutrality), solubility (hydrophobicity), or in terms of substrate- or ligand-binding properties, using an appropriately selected chromatography method. Structural and solution factors control protein purification by affecting its binding capacity and catalytic function(s). Utilities that have been created by taking advantage of these functional groups are described, from a variety of viewpoints, in accompanying literature supplements. Procedures used to make plasmid derivatives that encode fused protein sequences are described. Ligand affinity chromatography is used to specifically purify the "fusion protein." Purification fractionates are analyzed to determine how quantitative functional activities distribute across the chromatography profile. As a general lesson, students learn how to make "tailored" proteins with the possibility of producing dual functions, or even introducing brand new biochemical capabilities.

An interconnected set of chromatography experiments is performed to help the students learn how to optimize their ability to confront less successful experimental outcomes – a common real-life situation. This challenge helps students develop an understanding of the types of questions to ask when you have to weigh the relative utilities of two different approaches, each of which produces marginal success. This touch of realism simulates a typical juncture in decision making. It is often necessary to proceed down both roads and then assess the relative merits of each path. The best path is chosen after measuring quantitative specific activities of samples from the four types of chromatography and the initial cellular extracts, and making direct comparisons. More knowledge regarding the specific identity of the proteins is obtained using protein immunoblotting experiments, after resolving them according to size by gel electrophoresis. We emphasize the utility and generalizability of protein- and antibody-enzyme conjugate technologies.

Students learn to think about exactly why molecules bind a set of surfaces with distinctly different compositions, including carbohydrate-based gel filtration resins and matrices containing either cation-exchange sites, alkane groups, or specified bioligands. The results show how solution conditions affect biomolecule-resin binding interactions and thereby the relative purities of the biochemical products

Putting this set of processes in terms that are defined in the glossary:

a. Subclone a DNA fragment of the *myo*-3 gene and express the encoded protein as a β-galactosidase-*myo*-3 fusion protein. *Myo*-3 is a gene that encodes one of four myosin heavy chain isoforms in the nematode *Caenorhabditis elegans*. The gene was obtained by heterologous hybridization selection from a genomic DNA library prepared from *C. elegans*.
b. Use antibodies generated to bind a β-galactosidase-*myo*-3 fusion protein to identify *E. coli* cells that translate protein from the cloned *myo*-3 expression vector.
c. Purify and analyze the purity of the fusion protein obtained from this expression.

2. Procedures

2.*1 Part 1.*

1. Prepare master plates of the cloned *myo*-3 gene vector and "empty" protein expression vector.
2. Large scale "plasmid preps" of the *myo*-3 gene-containing vectors (pW and p2D) and the β-galactosidase expression plasmid (pUR288).
3. Restriction digestion and gel electrophoresis of the *myo*-3 cloning vectors and expression vector with *Hin*dIII.
4. Isolate purified *myo*-3 DNA fragment and digested expression vector from excised agarose gel pieces.
5. Clone the 480-bp *myo*-3 *Hin*dIII fragment into pUR288.

6. Polymerase chain reaction amplification of *myo*-3 fragment from the expression vector to test for correct insert orientation and reading frame.
7. Prepare competent *E. coli* cells and transform them with the recombinant expression vector.
8. Identify, by immunoassay, colonies that were transformed with the *myo*-3-containing protein expression vector and translate the protein encoded by *myo*-3 correctly.
9. Isolate genomic DNA from 2 strains of *C. elegans* (CB1407 and N2) and perform *Hin*dIII restriction digestion.
10. Construct a *myo*-3 digoxigenin-labeled probe from the cloned *myo*-3 gene fragment.
11. Southern blot the *Hin*dIII-digested *C. elegans* genomic DNA and probe it with the digoxigenin-labeled DNA to verify the duplication of the *myo*-3 gene in one strain of genomic DNA.

2.2 Part 2.

12. Isolate colonies derived from a single *E. coli* colony and induce cultures with IPTG to stimulate the β-gal-*myo*-3 fusion protein promoter.
13. Harvest and disrupt *E. coli* cell walls to release the expressed fusion protein.
14. Purify fusion proteins by gel filtration column chromatography and quantitate protein concentration and perform β-galactosidase enzyme activity assays with eluted fractions.
16. Purify the fusion protein by ion exchange column chromatography and quantitate the protein concentration and perform β-galactosidase enzyme activity assays with eluted fractions.
17. Purify the fusion protein by substrate (galactose) affinity column chromatography and quantitate the protein concentration and perform β-galactosidase enzyme activity assays with eluted fractions.
18. Quantitate the protein concentration and perform β-galactosidase enzyme assays of eluted affinity chromatography fractions.
19. Construct an enzyme assay activity purification table.
20. Perform discontinuous SDS polyacrylamide gel electrophoresis (PAGE) to characterize the size and purity of the purified β-gal-*myo*-3 fusion protein.
21. Perform 'Western' immunoblot assay to verify correct protein folding and the extent of purification of materials at different stages of purification.

3. Useful Literature

3.1 Cloning.

1. Kricka, L. J. (1992) *Nonisotopic DNA Probe Techniques*, Academic Press, Inc., San Diego, CA.
2. Titus, D. (1991) *Protocols and Applications Guide*, 2nd Ed., Promega Corporation, Madison, WI.
3. Brown, T. A. (1990) *Gene Cloning: A Guide to Molecular Biology*, 2nd Ed., Chapman and Hall, N.Y.
4. Old, R. W. and Primrose, S. B. (1989) *Principles of Gene Manipulation: An Introduction to Genetic Engineering*, 4th Ed., Blackwell Scientific Publications, Osney Mead, Oxford.
5. Sambrook, J., Fritsch, E. F. and Maniatis, T. (1989) *Molecular Cloning*, 2nd Ed., Cold Spring Harbor; a *three volume set*, excellent resource.
6. Ausabel F., *et al.* (1988) *Current Protocols in Molecular Biology*, Green and Wiley, N.Y. Excellent resource in a convenient notebook format.

3.2 Biochemistry.

1. Piszkiewicz, D. (1997) *Kinetics of Chemical and Enzyme-Catalyzed Reactions*, Oxford University Press, N.Y.
2. Burden, D., and Whitney, D. (1995) *Biotechnology: Proteins to PCR, a Course in Strategies and Lab Techniques*. Birkhauser, Boston, MA.
3. Creighton, T. E. (1993) *Proteins: Structures and Molecular Properties*, 2nd Ed., W. H. Freeman and Company, N.Y.
4. Boyer, R. F. (1993) *Modern Experimental Biochemistry*, 2nd Ed., Benjamin-Cummings Pub. Co., Redwood City, CA.
5. Robyt, J. F. and White, B. J. (1987) *Biochemical Techniques: Theory and Practice*, Waveland Press, Prospect Heights, IL.
6. Scopes, R. K. (1987) *Protein Purification*, Springer-Verlag, N.Y.
7. Fersht, A. (1999) *Structure and Mechanism in Protein Science*, W. H. Freeman and Company, N.Y.
8. van Holde, K. E., *et al.* (1998) *Principles of Physical Biochemistry*, Prentice Hall, Upper Saddle River, N.J.

3.3 Overview and References.

1. Lewin, B. (1999) *Genes VII*, Oxford/Cell Press.
2. Kendrew, J., and Lawrence, E. (1994) *Encyclopedia of Molecular Biology*, Blackwell Science Ltd., Osney Mead, Oxford.
3. Alberts, B. *et al.* (1989) *Molecular Biology of the Cell*, Garland Press, N.Y.

3.4 Ancillary Subjects.

1. Jack, R. C. (1995) *Basic Biochemical Laboratory Procedures and Computing*, Oxford University Press, N.Y.
2. Ravel, R. (1995) *Clinical Laboratory Medicine*, 6th Ed., Mosby-Year Book, Inc., Saint Louis, MO.
3. Hamkalo, B. A. and Elgin, S. (1991) *Functional Organization of the Nucleus. A Laboratory Guide*, Academic Press, Inc., San Diego, CA.

THEORY/PRINCIPLES

"CENTRAL DOGMA OF MOLECULAR BIOLOGY"

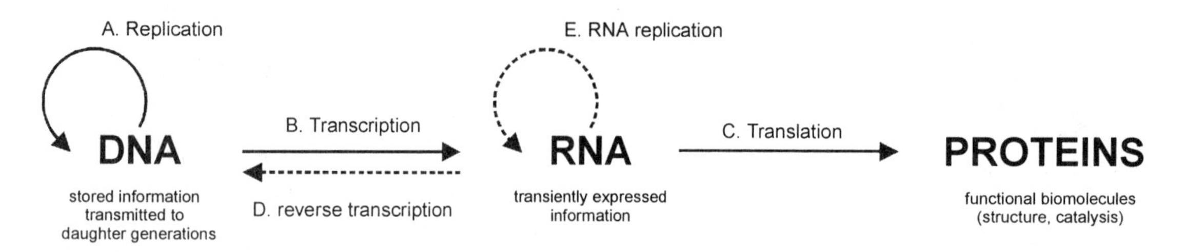

This key statement encompasses the essence of the field that has been named—for better or worse—"molecular biology." It outlines mechanisms, essential requirements, and information flow used to regenerate and propagate the genetic material of a cell. Pivotal roles are played by the following key biomolecules:

(A) DNA polymerase, template DNA, protein "factors" (helicases, single-strand binding proteins)

(B) RNA polymerase, mRNA transcription factors, repressors, activators

(C) ribosomes, aminoacyl tRNA synthetases, transfer RNA, amino acids, translational factors

Some life forms reproduce via two other important processes that illustrate nucleic acid metabolism that falls outside the limits of the traditional dogma. Nature is not dogmatic.

(D) reverse transcriptase – transcription catalyzed by this enzyme proceeds in the reverse direction, from RNA template to produce DNA. RNA tumor viruses (retroviruses) produce their DNA life-cycle form this way.

(E) Qβ replicase – with some viruses (e.g., Qβ) the genome is a RNA and replication is catalyzed by and RNA-dependent RNA; RNA is used to make replicated RNA.

THEORY/PRINCIPLES
LABORATORY SAFETY

Safety is important. Researchers and students must have a heightened awareness of each and every possibility of hazard in a biochemical lab situation. The experiments described here can be (and have been) performed by both college and high school students. High school students are in a much more stringent category in terms of adult supervision. Every hazard must be discussed with the students, at every stage in every process. An adult teacher/teaching technician who has been specifically assigned responsibility for the students who are performing the experiments must be present at all times. To do otherwise is dangerous and disrespectful to the student. Furthermore, by leaving the students unattended the appointed supervisor is increasing the risk of danger, injury, and financial loss to themselves, the students, their parents, and the institution. This subject is so important that we begin the manual with it.

1. Introduction

The **O**ccupational **S**afety and **H**ealth **A**ct Safety Plan (OSHA) directs that each operation which *uses potentially hazardous chemicals* shall have a *safety plan.* The plan also covers the hazards that might be encountered with equipment in the lab such as pumps, hoods, electrical components, etc.

This plan must be signed by all lab personnel (users) attesting that they've read the safety plan and understand how to use it.

1. *Material Safety Data Sheets* (MSDS) – A very important part of the safety plan; it contains a set of specifications for the hazardous chemicals that are used.
2. *Emergency exit routes* – The lab is identified on a map of the building, which indicates the floor and room(s) of interest.
3. *Chemical waste disposal* – there are three types of waste streams: solid, liquid, and mixed; kept them separate; organic wastes are stored in labeled containers.
4. *Biological waste* – describes definitions and hazards; considerations regarding bacteria, vectors, enzymes; autoclaving.
5. *Radioactive waste* – These compounds are carefully monitored because they can be digested or absorbed into biochemical pathways. The result is radioactivity within tissues, an event that could have toxic consequences. These compounds have been used in many key experiments, and, when properly contained, provide powerful tools. In recent years, they've been actively phased out in favor of fluorescent or other spectroscopically tagged substrates.
6. *Financial considerations* – The OSHA office can and will cite us for any violations. The fines can range up to $7,000 for nonwillful violations, and up to $70,000 for willful violations. For example, if they found a lunch in a chemical refrigerator and could prove that the person knowingly violated the code, a $70,000 fine would be levied!

2. Rules

You will be exposed to potential hazards during the course of the laboratory exercises that include toxic compounds, UV radiation, and radioactivity. Special precautions must be taken when working with recombinant DNA. The following rules MUST be observed at all times in the laboratory:

1. No smoking, drinking, or eating in the laboratory. Do not store food in the laboratory freezers or refrigerators.
2. Wear a lab coat, gloves, and safety glasses when working in the fume hood or in the presence of dangerous or potentially dangerous substances. No sandals or open-toed shoes are allowed in the lab at any time.
3. Always wear gloves when working with ethidium bromide, phenol, or any hazardous or potentially hazardous substance. Remove your gloves before leaving the laboratory and wash the outsides of your gloves frequently to prevent contamination of equipment, etc. with caustic agents. Two pairs of gloves are advisable when working with phenol or radioactivity. When applicable, check your gloves with a Geiger counter to monitor for radioactivity before leaving the area.
4. Tie back long hair and loose clothing before working with open flames, sterile cultures, and in the fume hood.
5. Dispose of microorganisms, including the tubes used for their growth, in orange bags marked "BIOHAZARD" for autoclaving. Liquid medium and plastic pipets used to transfer microorganisms will be collected in specially marked flasks containing bleach.
6. Dispose of glass, including culture tubes, pipets and broken labware, ONLY in the designated marked boxes. DO NOT put glass or any sharp object in the orange autoclavable bags marked "BIOHAZARD!"
7. Keep your lab bench free of unnecessary clutter.
8. Wear ear protection when working with the

sonicator.

9. Always wear UV protective goggles when using the transilluminator. Do not try to analyze your gel pattern on the transilluminator; Take a picture and analyze the picture. If you need to excise a band from the gel, make sure you're wearing a UV protective face shield.
10. Wash your hands thoroughly before you leave the laboratory.
11. Spills should be cleaned up immediately. Make sure you have an adequate supply of paper towels at your station. Notify the instructor if you spill a chemical that is labeled poison or caustic. If you spill liquid containing live microorganisms, notify the instructor and pour disinfectant on it.
12. Report all accidents (such as spills, cuts, burns, or other injuries) to the instructor immediately.
13. Know the location of the fire extinguisher, eye wash station, and emergency shower.
14. If you have trouble with a power supply or the leads to a gel, report it to the instructor immediately. If you see someone receiving an electrical shock, use a nonconducting object to break the circuit or you may also get shocked. The most convenient nonconducting object may be the sole of your shoe.
15. No pets are allowed in the laboratory.
16. Discard hazardous chemicals only in designated containers; DO NOT pour them down the sink. Ethidium bromide will be collected and decontaminated before pouring the liquid down the sink. Phenol and chloroform waste will be collected and stored in labeled containers.

THEORY/PRINCIPLES

THE SCIENTIFIC METHOD: SURVIVING RECIPE MENTALITY

1. The Scientific Method

This technique provides an approach to follow when attempting to answer unsolved questions about a process of interest in a scientific problem area. The following steps outline the procedure and some key considerations.

1. Identify the specific question to be investigated.
2. Study the scientific literature, and any other resources available (e.g., the notebooks of previous students in the lab), to determine everything you possibly can about your research problem.
3. Develop a hypothesis that frames your question in terms of a predicted outcome. If your prediction is in fact true then you can test it by performing a specific experiment.
4. Design an experiment to either verify your hypothesis or demonstrate that it is inadequate or incorrect. It is crucial that your tests lead to unambiguous answers.
5. Perform the experiment and acquire the appropriate data to test your hypothesis.
6. Analyze your results. Use an acceptable number of replications to ensure the statistical viability of your results. This is crucial. The ultimate trustworthiness of a conclusion often depends on how certain one is of specific numerical results. Large intrinsic uncertainties can completely undermine the proof of a fact.
7. Reach your conclusion(s). Did your experiment confirm your hypothesis, demonstrate that your hypothesis is never true, or produce a more complicated conclusion.

For example, the result might confirm your hypothesis for a narrow set of circumstances and yet demonstrate that it is not correct for others. To the extent that your hypothesis stands proven, you have confirmed the fact in question.

2. Hypothesis, Theory, Fact, and Law

The scientific quest essentially always involves developing a prediction then designing an appropriate experiment to confirm or deny its accuracy. One seeks to assemble specified results to test a predefined property, fact, or relation and then fit them successfully to a designed logical construct composed of connected concepts, mathematical equations, or both.

The terms postulate, hypothesis, and prediction are often used interchangably. A theory is usually thought of as being more solidly established than this first group of terms. Facts are the goal of the pursuit, in one way or another, depending upon how the scientific analysis and product are defined. A law constitutes a generalized statement of a group of principles or facts.

1. postulate – assumption that an unproven fact is true.
2. hypothesis – a supposition or conjecture put forward to account for certain facts; used as the basis for further investigation(s) aimed at proving or disproving its correctness (truth); a theory which has not been fully proven by experiment.
3. prediction – prophesy; product of a forecast.
4. theory – a set of ideas formulated by reasoning through from known facts to explain an observation; general ideas or suppositions; statement of principles on which a subject is based; the reduction of data or facts to a principle, and the demonstration of their interrelations.
5. fact – proven property, concept, or relation, or existence of a logical connection between or among these items.
6. law – a generalized statement of facts or principles; a factual statement describing events that always occur in certain circumstances.

3. Kit Mentality and Developing a Scientific Mentality

We've found it useful to define the way experiments are performed in terms of three levels of sophistication. One overarching goal of this manual is to help students mature into fully functioning scientists. These definitions help define the characteristics exhibited by experimentalists as they work toward the level of a truly competent scientific thinker. The levels are defined from least to most sophisticated.

1. *Kit mentality* – Experimentalists do not even know exactly what the sample contains. Solutions were made without any attention to how the various components function. While this seems somewhat ludicrous, modern biotechnology kits often contain materials whose identity is kept secret on purpose, to protect against competitors and to discourage the researcher from making the components themselves. As a result, reactions are commonly assembled from tubes that were labeled with meaningless colors, letters, or number codes that do not convey any functional information.
2. *Recipe mentality* – these fledgling researchers know what is in the sample but do not understand how the components work to

accomplish the goal of the reaction or experiment.

3. *Scientist mentality* - these researchers know what's in the sample and how the components work. They have developed a mental database that allows them to manipulate the components to achieve a specific predetermined result. In advanced cases they can even predict a pathway to make related target materials then perform the steps to produce them. This constitutes the ability to develop a hypothesis then carry out the procedures necessary to confirm that it is correct.

How does the laboratory teacher differentiate the students and assign them to one of these groups? The key involves explicitly observing and probing their ability to solve problems. It is difficult to tell the three types of students apart when everything works properly. Researchers have achieved "scientist mentality" when they've learned how to develop a hypothesis, how to design an appropriate experimental protocol to test it, how to perform this protocol, and how to react to mistakes should they occur.

Most real scientific pursuits involve using chemical products or conclusions from one experiment (or series of steps) as the reactant materials or input information for a subsequent experiment. The process of working through a research project and reaching an "ultimate" goal determines whether one is practicing "scientist mentality." Students must think through a series of experiments and design an approach which will lead to their goal. By doing so, they gain an overview perspective of the research problem and an integrated understanding of the experimental steps to the net process. This perspective is crucial if one is to anticipate difficulties that might be encountered along the way.

Mistake-ridden experiments are not included in any laboratory teaching manual we've encountered. Instead, all the protocols "work" and have been optimized for student use. Students gain confidence in their abilities to learn and execute protocols, however, working through unexpected problems is an important aspect of scientific trainings. Detecting troublesome procedures and fixing mistakes train students to think proactively about the detailed mechanisms of the substeps and role of each reagent in the overall experiment.

4. Teaching Responsibilities: Roles of Teachers and Students

Teachers are charged in part with providing information to students. However, experience has indicated that many of our students are not retaining key information from their training. Retention is crucial if they are to advance to the next level, learning how to manipulate the concepts they've already absorbed so they can successfully pose questions in new uncharted problems.

The "complete teaching experience" involves challenging learners to determine for themselves whether the information is absorbed and whether they can apply the lessons to investigate newly encountered circumstances. Students sometimes misconstrue meanings or interpretations, especially when the information is laced with complicated logic, has counterintuitive ideas at crucial junctures, or is not presented with sufficient clarity. The "complete teacher" reiterates or rephrases the lessons from an alternative viewpoint to highlight and reinforce subtle aspects of the lesson's objective(s).

The "complete student" demands clarification or reiteration. The first step is to carry out an investigation. Students then develop questions that have been carefully phrased to clearly define for themselves and the instructor why they do not understand the subject. If they cannot reach clarification, it is the student's responsibility to discuss the matter with their teacher.

The teacher cannot read a student's mind, however, the responsible teacher will take the time to probe the student's level of understanding. If this is not your experience, the teacher's department head and Dean should be informed. It is your right to be given your moneys worth. The pressures on professors to carry out research, be on committees, write grants to improve your school, multitask, etc. in many modern university situations is enormous. This situation can lead to short-changed students. Administrators are becoming aware of this, but the situation is with us, like it or not. You – the student and training researcher – are in the best position to prevent such problems from becoming permanently embedded.

THEORY/PRINCIPLES
PROACTIVE TROUBLESHOOTING

1. Definitions (Random House American College Dictionary)

1. *proactive* – seeking to cause change, exerting influence.
2. *troubleshooter* – an expert in discovering and eliminating the cause of trouble in the operation of a device, process, or logical construct.

2. Troubleshooting Tips

This subject can be approached by listing a set of generalizable tips or by explaining exactly how to think about some specific situations and hope that by analogy the student can learn how to generalize the approach to newly encountered scenarios. The problem with the generalization approach is that it can be somewhat vague and therefore unclear how to apply it to specific problems. The specific scenario approach often works, but sometimes lessons learned in one given context cannot be transferred and applied to analyze a different problem. This often occurs because of differences in the two situations that are relatively subtle. As is so often the case, the answers are hidden among the details. We need a general approach to any problem.

3. Figuring Out Why and Where Errors Occur: The Importance of Troubleshooting

Why do problems occur when attempting to carry out biochemical reactions? The typical source of most errors or screwups is a failure to *take into account how the mechanism of each subprocess in the project occurs*. A key consideration is the physical and chemical state of the biochemical reagents used in the processes. Careful control of the experimental materials, including their preparation, storage, and handling, is crucial to success. The key is to prevent degradation and thereby preserve the activity of the biomolecules. The terms "degradation" and "preservation" certainly encompass the obvious aspects of biomolecular breakdown as a result of chemical, protease, or nuclease action. However, the savvy researcher/student also knows to investigate more subtle ways in which the various components might interact with each other and thereby lead to activity losses.

Sometimes a problem can be reversed and "reactivated" material can be recovered. For example, the *dialysis* technique can be essential to "transfer" a biomolecule from one solution composition, which has inhibitor present, to another, which is compatible with activity. Since so many reactions involve charged functional groups of the molecules, salt concentrations that are too high can surround these groups and shield them from others with which they must interact to accomplish the mechanistic steps necessary for the reaction to proceed. Learn how to think about the effects of the reaction components on the mechanistic functionality of the active components on the biomolecules and on the substrates and other factors used by them.

Look for the existence of dated, incompatible, or badly made reagents. You must read the product literature accompanying the biomolecules and other reagents to be used to be certain that the entire system you have designed is going to work when combined together. This brings us to the concept of a "*control reaction*." Many biomolecules are supplied with instructions for a procedure designed to verify its activity in a predesigned *assay* with verified substrates. You must know for certain whether an enzyme (for example) works in the context of the milieu that you intend to achieve success with your own research substrate. If not, the solution components somehow inhibit the reaction. Sometimes this type of problem can be "fixed" by adjusting the concentrations of the salts, polyamines, buffer, or other components in such a way as to regain activity. Sometimes your substrate is different from the "control" substrate in some subtle way. For example, if an enzyme requires a single-stranded DNA substrate but your DNA folds back upon itself and forms a hairpin duplex structure, the enzyme will not work. Lowering the salt concentration can decrease the stability of such a duplex structure, allowing it to disassemble and thereby making the correct unfettered substrate available for the enzyme of interest.

One must verify these details at each step in the process. Troubleshooting involves narrowing the range of possible difficulties to specific steps and then to specific substances or interactions between substances at that step. Following this approach at every step that a problem might occur constitutes completely troubleshooting the experiment.

The believability of a particular result is expressed in quantitative terms in the form of measurement statistics, the result of a proper *error analysis*. Most often one assesses the range of results possible based on the standard error (deviation) in a small number of real measurements. Wide variation in a measured activity implies a wide range of behaviors or that some component is subject to a wide range of possible functional capabilities. For example, if a particular functional group protonates/deprotonates with a specific pK_a and the buffer pH is adjusted to that value, or varied across a range of values in that vicinity one might expect to observe a wide range of activities. This implies that both inactive and active forms of the functional group contribute to the net result. The solution to the difficulty is to restrict the operational pH to a range that is unaffected by this phenomenon. One can understand the problem and develop a solution only by accounting for the molecular aspects of the catalytic event(s).

Ask teachers how things work if you don't understand. Good proactive troubleshooters become good by learning how to question themselves and others regarding the inner workings of the process. Listen to other students, no matter how "stupid" their question might seem at first consideration. Questions are often an indication that the student has focused on something important. However, the specific question may or may not accurately address the most important issue. It is often just an indication that some issue exists. The exact nature of the actual problem can then be uncovered by the more experienced troubleshooter. This course of events, while seeming indirect, is typical and provides a great opportunity to train in the art and science of troubleshooting. "Stupid" questions are usually not stupid, they merely require redirecting and refinement. An experienced teacher encourages all questions; if the answer is truly obvious, the respectful teacher says so and might reasonably defer the answer. However, "obvious" is different for students at different levels of development. Such questions should sometimes be answered in the presence of the entire class. Several students will probably benefit from the "remedial lesson."

Ask your colleagues questions. They've either worked on related problems or analogous situations, often with many of the same reaction components. Becoming a capable scientist requires that you develop your own mental database in which you've stored and integrated information that explains how the typical reagents work, alone and together in the context of reaction component mixtures of varying complexity. Look at past experiments in your own repertoire with similar characteristics or courses of events. It may also be necessary to contact the researchers who developed the method you're attempting to use or trying to extend for use in your previously uninvestigated situation. Their publication usually does not impart a database of information about situations that do not work or specific manipulations that must be performed in a very specific way to achieve success.

Errors can propagate along with the damaged materials and then present themselves in a form that is detectable one or more steps further along in the procedure (or multistepped process). As a result, it is often necessary to look one or more steps up- or downstream in the process for the solution to a problem that becomes detectable as a result of an unexpected result at a particular substep. Each juncture in a process must be interrogated ruthlessly when the cold gloom of experimental failure alights upon your work.

4. Interrogation 101

Three examples are presented below to demonstrate some ways of thinking about troubleshooting. They discuss explicit details and demonstrate typical types of chemical concepts one must consider to understand the causes and devise solutions. They are complicated, but chosen to illustrate the range of problem types you will encounter in the areas addressed in this manual. The general philosophy is discussed to help the reader learn to generalize these approaches—in the context of any investigation in any area they might be trying to troubleshoot.

Try to find weak links in your protocols. Consider how steps in the reaction mechanisms might be inhibited. Scrutinize all situations in which a seemingly innocuous reagent might compete with an intended reactant for a functional surface or group and thereby prevent the normal course of events from transpiring. Be thorough. Focus on (*i*) charge-charge and nonpolar-nonpolar interactions, (*ii*) ligand-binding site and substrate-enzyme recognition events, and (*iii*) irreversible chemical reactions.

1. The DNA fragment-to-fragment ligation reaction.

This example illustrates a basic incompatibility between two commonly used reagents. Divalent cations such as Mg^{2+} are usually included in reactions that involve nucleic acids and proteins. They help stabilize certain structures and make them more compatible for binding to the active site of the enzyme in question. Many types of biochemical reactions include the reagent EDTA. This reagent binds unwanted divalent cations and thereby prevents them from inhibiting reactions that are poisoned by their presence. However, you do not want to include EDTA in an Mg^{2+}-requiring nucleic acid modification reaction. The EDTA will bind the cation and thereby make it unavailable for the reaction. While this seems like an obvious mistake, many students have fallen prey to it over the years.

2. Casting an SDS electrophoresis gel.

Polyacrylamide gel electrophoresis is commonly used to assess the success of a protein purification protocol, to follow the fate of nuclease reactions, and to determine the sizes of both types of macromolecules. Gel formation involves adding a catalyst that initiates a free radical-dependent polymerization chain reaction. Oxygen is an excellent terminator (poison) of free-radical-induced polymerization. Merely swirling the gel solution prior to gel formation can introduce sufficient oxygen from the air to prevent the process. Purging inhibitory oxygen from the solution by bubbling argon gas through it for a minute prior to catalyst addition helps ensure rapid catalysis and trouble-free gel formation.

3. Western blot immunodetection of proteins resolved on a protein gel.

This is a classic application of an "immunoconjugate" – an antibody that has been covalently tethered to an enzyme, in this case alkaline phosphatase (AP). The overall process involves separating proteins embedded within a polyacrylamide slab gel and then transferring them laterally out of the gel to a membrane that retains the same (separated) pattern. The membrane with the attached proteins is subsequently *washed* and exposed to a "primary antibody" that binds the protein of interest, blotted to the membrane from the gel. In a typical example, the primary antibody is from mouse. Its presence is detected using a "secondary antibody AP immunoconjugate," which specifically recognizes mouse primary antibodies. If the primary antibody detects and binds the protein of interest on the gel (i.e., because it was present in the electrophoresis sample), the secondary immunoconjugate binds the primary antibody-protein complex. If the protein was not in the initial sample, neither primary antibody nor secondary antibody immunoconjugate remains bound to the Western blot membrane. (Got all that?)

Now the screwup and its anatomy. A researcher accidentally used a *sodium phosphate buffer* to *wash* the Western blot membrane prior to the detection stage of the process. To detect membrane-bound protein of interest, the researcher exposes the membrane to an AP enzymatic substrate that turns black at the location of the AP immunoconjugate, if it remains bound to the membrane. The erroneous finding was that no black color formed on the membrane – *no signal was detected.* This is an erroneous result because the protein of interest was in fact present on the electrophoretic gel. It *did* transfer satisfactorily to the blot membrane, it *was* bound properly by primary antibody, and the secondary antibody immunoconjugate *did* bind the primary antibody correctly. The lack of black color occurred because phosphate is an active site-directed inhibitor of the AP reaction. It clogs up the phospate-binding site, which receives phosphate after its removal from the substrate to create the black product compound. Normally this site would then allow the phosphate to diffuse from the enzyme (eject it). Phosphate buffer provides a sort of chemical back pressure (excessive external concentration) that inhibits ejection of the phosphate product fragment from the enzyme. Everything in the protein detection process chain was in place but the AP detection reaction was foiled. Phosphate buffer inhibits AP so its presence is incompatible with detection using an AP immunoconjugate reaction. The wash step screwed things up!

Don't be overly daunted by this last example. We include it at the outset so you can see how complicated a process logic chain in biotechnology methodology can get. This is kind of a worst-case situation. Most of your troubleshooting will be less tricky. Notice that in this example even the *buffer* is a chemical that can doom your experiment. Biochemistry, molecular biology, chemical biology, biology ... chemistry – same difference. Happy hunting!

THEORY/PRINCIPLES

INTRODUCTION TO THE BIOTECHNOLOGY LABORATORY

Burden, D. W., and Whitney, D. B. (1995) *Biotechnology: Protein to PCR, a Course in Strategies and Lab Techniques*, Birkhauser, Bonston,, MA (Adapted)

1. Notes, Records and Labels

It is the duty of the student and researcher to keep an orderly laboratory notebook that saves and reports their activities while preparing, executing and analyzing all of their experiments, even those that "didn't work." The notebook is a permanent record of laboratory protocols and experiments. It should state how and why an experiment was performed, with sufficient detail, so that you or someone else can reproduce the results of that experiment in the future. The organization of the lab notebook should be meticulous, and include references to other protocols and information as necessary.

To illustrate the value of good notes, an analogy can be made to searching for a lost item. While traveling from your bedroom to school you unknowingly dropped your favorite ring, and the only way to find the ring is to retrace your steps. If you immediately retrace, then your short-term memory will suffice. However, if you attempt to retrace your steps two years later, finding the exact path will be at best a guess. Trying to remember the protocols of an experiment is the same as retracing your steps. In the laboratory, using information recorded in a notebook limits the guesswork associated with experimentation. You limit the variables.

In industry, documentation is more than simply a means of minimizing guessing in the lab. Rigid protocols are often adopted so that products, especially pharmaceuticals, will receive approval for use and sale from regulatory agencies (e.g., the FDA). Researchers must adopt GLPs (good laboratory practices) while scientists in production follow GMPs (good manufacturing practices). If approved protocols are violated during the design or production of a drug, then the product may be removed from the market. Carelessness and error are not tolerated when as much as $350 million may be spent developing and testing a drug. This is especially important because a complete notebook is essential for getting your research ideas patented.

Notebooks should be kept as instructed. Some institutions require specific books which are countersigned by supervisors, while others use computer-based notebooks. Records of experiments, such as photographs, printouts, chromatograph, etc., should be taped directly into notebooks. Pertinent information (date, experiment, trial) should be written directly on the record. Scanners can be used to incorporate records directly into computers. Never copy data on a paper towel and incorporate it into the notebook later. This is all too common.

Keeping bottles and tubes accurately labeled can be just as important as an organized notebook. Most people working in a laboratory, at one time or another, have placed an un*labeled* tube in the freezer or refrigerator, knowing they would return to it shortly. When that tube reemerges several weeks, months, or years later, the contents are unknown. It could be water or botulism toxin but would be useless because you wouldn't knows which it was.

Tubes, organisms, media, and reagents must be adequately *labeled* to avoid such confusion. Not only is it important to keep tabs on what is what, but federal and state laws can mandate that all containers in a laboratory be adequately labeled.

2. Handling and Storing Chemicals

It is important to establish a clear protocol for the handling and storage of stock chemicals, (i.e., purchased chemicals). Since most stock chemicals are dry salts, they can be stored at room temperature. Many reagents, such as acids, bases, and solvents should be stored in vented storage cabinets. Flammable solvents must be stored in explosion proof storage cabinets. When using these stocks, only clean spatulas and pipets should enter the containers, and any unused reagent taken out of its container should not be replaced (due to the risk of contamination). It is a good idea to separate these stocks from the solutions they are used to formulate. Think about every action. For example, could the metal in your spatula act as a catalyst when you stick it in a chemical you've never worked with before and cause an explosion? Can long-term storage of a chemical make it become explosive? (Yes, ethers can accumulate peroxides powerful enough to blow a metal door off its hinges and four inches into the cinder block wall across the hall!)

Many solutions can be prepared in advance of an experiment and stored. Inert solutions, such as 5 M NaCl, will not spoil and can remain indefinitely on a shelf as long as it is sealed well. However, other solutions will become inactive, contaminated, or otherwise not usable, especially those that will support microbial growth such as glycine and citrate. Take the time to differentiate between stable solutions and those that will degrade or support microbial growth. Autoclaving a solution before storage will often prevent its spoiling.

3. Equipment Maintenance

Nothing is more frustrating than working in a lab filled with disabled equipment. When instruments have multiple users, it is common for equipment to be abused and break. Maintaining functional equipment can only be accomplished if individuals take responsibility for instrumentation they use and if supervisors ensure that all users are adequately trained. When equipment does break (which even well maintained equipment will do), then a mechanism should be in place for its rapid repair. Inoperable equipment results in the loss of productivity and money. Research labs usually appoint specific students to specific instruments, apparatus, or bench spaces/areas to assure that one person is responsible for its maintenance and repair.

4. Preparation and Storage of Buffers

Biological materials require buffered solutions. Consequently the expiration and storage of buffers and other solutions (salt, organic) are fundamentally important. Several conventions are used to prepare buffers, for instance:

a. Molar solutions (M) – number of moles of a substance in a total of 1 liter of solvent,
b. % Weight – number of grams of a substance per 100 ml of solvent,
c. % Volume – number of milliliters of a solution in a combined total of 100 ml of solvent.

It is important to realize that researchers will define these conventions many different ways. Furthermore, to confuse the issue, more than one of these measurements can be found in a single buffer. For example, the solution 50 mM Tris, pH 7.5, 0.1% SDS, 0.05% Tween 20, contains Tris (M), SDS (w/v), and Tween 20 (v/v). The ability to decipher the components of a solution is gained with experience.

Many buffers are inert, but some may degrade or support microbial growth. If a buffer is susceptible, it should be made as needed. However, solutions which can be prepared in bulk or as concentrates are often convenient. Certain buffers, such as those containing the amino acid glycine or the carboxylic acid citrate, are wonderful sources of energy for microbial contaminants and should be sterilized prior to storage.

5. Care and Maintenance of Cultures

Microorganisms that are continuously used for research need to be maintained (kept fresh). Maintenance can be accomplished by subculturing organisms onto fresh medium on a routine basis once or twice a week.

The techniques used for culturing are dependent upon the organism. Bacteria and yeast are easily cultured by streak plating. Animal cell culture requires subculturing dissociating cells from the culture flask with trypsin, diluting them with a serum based medium, and transferring them to new flasks. Due to the complex and expensive nature of cell culture, this manual will not include these techniques.

The most common method of maintaining microbes is to streak them onto solidified agar media. Prior to working with any microorganism, the cultures must be grown and pure. Streaking microbes onto agar plates allows one to separate individual cells and develop them into isolated colonies.

Between each streak, the loop is sterilized with a flame, allowed to cool (it may be touched to a sterile area of the agar), and then used to spread or dilute the microorganisms over the surface of the plate.

Streaked plates are inverted during incubation so that condensation that forms on the lid doesn't drip onto the surface of the plate. Throughout this procedure, aseptic technique is used.

Aseptic technique prevents contaminating cultures and sterilizes items. Though it sounds like a protocol, aseptic technique is not a specific set of memorizable steps, but rather a general modus operandi and state of mind. It is important to realize that all equipment, solutions, gases, and other items (especially hands) are laden with microorganisms and enzymes that are potential contaminants. Aseptic technique recognizes these risks and avoids exposing pure cultures or sterile systems to items which are not sterile. Some simple guidelines for aseptic technique include:

a. Never needlessly expose a sterile object to a nonsterile object, especially air.
b. All work surfaces should be cleaned and sanitized. Never place caps from culture flasks or media bottles onto a surface. Containers should be opened and closed quickly.
c. All solutions used to culture cells must be sterile. Sterilization can be accomplished by filtration, autoclaving, or irradiation.
d. All cultureware, e.g., pipettes, flasks, dishes, must be sterile. It is advisable to use disposable materials when available.
e. Gloves should always be worn to protect the cultures from microbes present on hands.
f. Don't breathe on open culture flasks and dishes!

The number of organisms used in recombinant DNA research is continuously increasing, with species ranging from phages to yeast and geraniums to mice. *Escherichia coli,* the molecular biologist's workhorse, itself has numerous variants and strains. Each variant possesses a genotype that makes it valuable for a particular line of experimentation.

Deciphering the genetic nomenclature used to describe organisms can be complicated. Organisms that are regularly manipulated, e.g., *E. coli* and *S. cerevisiae,* have been extensively mutated for both biological control and experimental purposes.

When discussing the genetic characteristics of

these organisms, it is important to differentiate between phenotype and genotype. Phenotype is the visible characteristics of a gene, e.g., resulting color, enzyme activity, growth characteristics, sugar assimilation or fermentation, etc. Genotype refers to the actual genetic make-up of an organism, and such characteristics as dominant, recessive, wild type, and mutant.

Handling and storing microbial strains correctly are extremely important in molecular biology (as they generally are in microbiology). *E. coli* strains are genetically modified to be sickly (a form of biological control) and tend to die easily. Even *E. coli* streaked on LB agar and stored at 4°C can die within several weeks. Yeasts, though generally more hearty, are also much the same. Their care includes the following techniques.

6. Freezing

Most cultures are easily preserved frozen, either in liquid nitrogen or in a -80°C freezer. However, the freezing of microorganisms tends to cause cell lysis due to the intercellular formation of ice crystals. The addition of glycerol to 10% (v/v) prior to freezing cells helps to prevent cell lysis and thus preserves the cultures.

A simple method of preserving both yeast and *E. coli* involves preparing sterile cryogenic vials (e.g., Corning Costar 2.0 ml screw capped tubes) with small glass beads (Fisher Scientific 2 mm glass beads) and a 20% solution of glycerol (i.e., 1/3 full of beads and with 250 μl of 20% glycerol). When cells are cultured, strains can be maintained by simply adding 250 μl of culture broth to a prepared vial, vortexing briefly, labeling, and freezing at -80°C. The importance of establishing a simple practice such as this for maintaining commonly used cultures cannot be emphasized strongly enough. To reactivate a culture, simply remove the vial from the freezer (keep on ice for a minimum of time), and with a sterile inoculating needle, break away several glass beads, pour the freed beads onto an agar plate, roll the beads around on the agar surface, and incubate the plate. A smear of cells usually develops which provides an inoculum for subsequent streaking.

7. General Rules for Protein Handling

Proteins are large, complex biomolecules that are extremely sensitive to denaturation (i.e., loss of native conformation and, consequently, loss of biological activity) under typical laboratory conditions. The loss of biological activity can be partial or complete depending on the extent of the denaturation. Manipulation of proteins in the laboratory requires the researcher to follow special rules to avoid contaminating and denaturing samples. The following rules also apply to handling enzymes. The first rule of protein handling is to wear gloves. Proteases from the skin can degrade the sample as well as contaminate an otherwise pure protein solution.

Mixing and dissolving proteins should be done under mild conditions. Generally, protein solutions should not be vigorously stirred because the shear forces that would be generated can degrade the integrity of the protein molecule. Additionally, protein solutions tend to foam if stirred or vortexed too vigorously.

Generally protein solutions should be maintained at high concentration (ideally > 1 mg/ml) as the protein is usually more stable. Proteins tend to stick to most surfaces, and this can be a serious source of protein loss during manipulation and handling, especially with dilute solutions. Loss of protein in this way can be minimized by maintaining a high concentration or by adding a second inert protein, such as albumin, which increases the overall protein concentration. The inert protein should be one that can be easily removed later on.

Glassware or plasticware used in protein studies should be meticulously washed and rinsed with deionized water. Frequently, treatment with EDTA before the water rinse will lower the chances of contamination of the protein by metal ions. Silanization of glassware helps to prevent protein adsorption to the glass.

Long and short term storage of proteins present many potential stability problems. Proteins can generally be stored either frozen, at 4°C, or lyophilized. Ultralow temperatures, e.g., -80°C, can irreversibly denature some proteins. When working with protein solutions in the laboratory for any appreciable length of time, the solutions should be kept cold. This can be done by keeping tubes and bottles on ice.

Proteins can be stored short term in a typical 4°C refrigerator providing that the buffer conditions, i.e., salt concentration, pH, reducing environment, etc., support protein stability. Sometimes addition of a protease inhibitor is required to minimize degradation. Addition of stabilizers such as glycerol may also be necessary to maintain protein activity (maintenance of activity is usually a good measure of protein stability). The presence of glycerol (typically < 20%) generally promotes protein stability and will prevent the protein from freezing, even at temperatures below 0°C.

Addition of glycerol, 50% v/v, is useful for long term storage but may present problems in any subsequent purification procedures (e.g., chromatography). The use of antibacterial agents, such as sodium azide, is appropriate in order to limit microbial growth during storage at higher temperatures.

Although many proteins are quite stable when

stored frozen, repeated freezing and thawing can result in loss of activity. During the freezing process the protein may be subjected to extremes of pH and/or buffer salt concentrations which may result in denaturation. This effect can be minimized by rapidly freezing the protein solution, usually by submersion in a dry ice bath of either ethanol or acetone. The frozen protein solution can then be safely stored at -20°C. Storage of proteins by freezing should be considered a long term storage solution and repeated freezing/thawing should be avoided. One approach to avoid freeze/thawing is to divide the protein sample into aliquots. In this way, each aliquot has to be thawed only once, thus preventing serious loss of activity. This technique can also be used for other unstable nonprotein molecules, such as ammonium persulfate solutions.

Lyophilization is a very effective method of protein storage. The protein solution is frozen (rapidly!) and then placed in a lyophilizer. Under a vacuum, the frozen liquid sublimes leaving the protein behind, usually as a fluffy white solid. The protein to be lyophilized must be dissolved in either water or a buffer solution containing volatile buffer salts that also sublime under the lyophilization conditions. The drawback to lyophilization is the occasional difficulty in redissolving the protein or in recovering full activity.

8. Storage of Nucleic Acids

DNA is generally a very stable molecule and can be stored for years either desiccated (dry) or in solution (frozen). Environmental nucleases can destroy DNA rapidly, however, and care must be exercised not to contaminate solutions. These nucleases require Mg^{2+} for activity; thus dissolving DNA in solutions containing EDTA chelates the Mg^{2+} and helps to preserve the sample. DNA that is contaminated with nucleases either disappears from solution, or, when assessed on an agarose gel, appears smeared or with an altered electrophoretic pattern (as compared to the original pattern). Autoclaving solutions will destroy most, but not all nucleases. DNA solutions can be dried in a Speedvac® which remove liquid and yields crystalline DNA. Tubes can be capped and stored at room temperature, or more preferably at 4°C or 20°C.

Due to extremely resilient RNases, it is more difficult to isolate and store RNA than DNA. RNases tend to be thermostable and are more difficult to destroy. These enzymes are present in cells and ubiquitously on glassware instruments, in water, and in solutions. Several precautions can be followed to prevent ribonuclease contamination, such as baking glassware, treating solutions with diethyl pyrocarbamate (DEPC), and working in conditions that are free of ribonuclease contaminants. (Caution: DEPC is a carcinogen!) RNases arising from cells can be controlled by adding ribonuclease inhibitors, such as vanadyl complexes or RNasin.

THEORY/PRINCIPLES
ERROR ANALYSIS AND ASSAY SENSITIVITY

1. Introduction

We must adhere to a strict set of practices to ensure that *data are properly analyzed* and *conclusions are statistically significant*. The first and most important rule is that everything must be reproduced at least twice to be trusted. Science is only science if it is reproducible. You should at least consider whether you can replicate every aspect of the work that could lead to experimental or statistical variation, including using more than one protein preparation (or other biomolecule) or different batches of substrate, reagents, etc.

Two different concepts, *accuracy* and *precision*, are frequently confused by novices. A result might be repeatable but inaccurate because some intrinsic factor in the measurement keeps causing the same error. For example, if one does not subtract a blank value (e.g., the cuvette plus background buffer), the reading from a measured absorbance is wrong. This is *inaccuracy*. In contrast, if we subtract the absorbance of the blank properly but the cuvette wobbles in the holder, which changes the pathlength by ±12%, the measurement is inherently *imprecise*. We are less certain of the correct answer relative to some correct mean cuvette position and absorbance value. The correct value is not obtained consistently.

We often make quantitative *comparisons*. We often must determine whether two numbers are significantly different or the same. It is not unusual during an academic seminar to hear a sincere critic from the audience discuss how *errors* or *propagated errors* disallow the speaker's alleged conclusion (or render it ambiguous).

Basic derivations and proofs usually begin with an analysis of *how a series of forces or properties are related to or affect each other*. When the analysis yields an equation that can be used to obtain meaningful quantities, the variables are sometimes related to each other by factor-root, or logarithm-exponential relationships. The study of how statistics propagate through one or more such relationships and affect derived results is called *dimensional analysis*.

The *slope of the best-fit line* for a set of data changes when individual data points change. The statistical spread of significant lines comprises a field on the graph that is bracketed by the correlated statistics of the points, which in turn determine whether the results properly conform to a line or curve. This statistical spread can, and often does, get wider at the edges of a graph. This occurs because our knowledge of the experimental values in the numerator or denominator becomes less precise.

Sometimes one must decide whether a point must be discarded from the data set because it is not consistent with the general trend. These problems are discussed in the sections concerning the 'Q-test' and "Student's *t*-test."

One can use statistics to *determine when a line deviates from statistical linearity*. First, one calculates the standard deviation at a given level of statistical certainty for the obviously linear data points. One then adds the next point to the data set and recalculates the statistics. If the newly calculated slope differs from the previous slope by less than about 5%, they're within the 1 σ_i. If the slopes differ by <35%, they are within 2 σ_i confidence. One progressively adds single points to the end of the line then calculates the new slope and statistics. When the new slope is out of the range of the previously calculated value of σ_i (or 2 σ_i to stretch the definition) the 'line' is no longer statistically linear.

2. Assay Sensitivity

1. One considers the following factors to determine how many *significant figures* one can use in determining a *dependent quantity*:
   a. Estimated uncertainties of *independent quantities* can be used to determine effects that get propagated into dependent values (like the absorbance of a weighed/dissolved/volume-adjusted sample).
   b. In weighing a 1 g sample on a balance which is built and optimized to weigh 10 g, the uncertainty also changes by 10-fold. Use an instrument that is optimized for a particular sample size.
   c. *Limiting reagents* might artificially narrow the detection range by tying up excess substrate that should be present to guarantee the large population needed to assume that the solution acts like an equilibrium thermodynamic system.
   d. Error(s) associated with replicate determinations.

2. Concentrations are reported in units of moles per liter. The number of *micrograms* of a DNA fragment depends upon its *size and molecular weight*. One typically determines the concentration of a protein using a 1-μg protein sample in a 50-μl aliquot from a stock solution. Let's assume that the protein molecular weight is 20,000 g/mole (Daltons), or 20 kDa. The protein concentration in the stock solution is calculated as follows:

$$\text{Concentration} = \frac{1\,\mu\text{g protein}}{50\,\mu\text{l}} \times \frac{10^{-6}\,\text{g}}{1\,\mu\text{g}} \times \frac{1\,\text{mole}}{20{,}000\,\text{g}} \times \frac{10^{-6}\,\mu\text{l}}{1\,\text{L}}$$

$$= 10^{-6}\,\text{M} = 1\,\mu\text{M}$$

This **"conversion factor" presentation** helps to show the canceling scheme that converts the units from one frame of reference to a different one. The final units are **mole/L (molar, M)**. This explicit presentation helps to simplify interpretations and can give useful insights into the relationships between the experimental variables.

THEORY/PRINCIPLES

TREATMENT OF ANALYTICAL DATA

Fritz, J. S. and Schenk, G. H. (1974) *Quantitative Analytical Chemistry*, 3rd Ed., Allyn & Bacon, Inc., Boston, MA. (Adapted)

1. Introduction

The task of the analytical biochemist goes beyond that of correctly performing the manipulations and readings required in a procedure. To make their work meaningful, they must also do the following.

1. Properly record and correctly calculate the results of each analysis.
2. Since the analyses are done in replicate (usually triplicate), the analyst must determine the best value to report. Although the best value is usually the arithmetic mean, or average, of the individual results, there is often the question of whether to include a result that seems out of line with the others.
3. Finally, the analyst must evaluate the results obtained and establish the probable limits of error that can be placed on the final result. A simple statistical treatment can be very helpful in this evaluation. A clear concept of the terms "accuracy" and "precision" will aid in understanding the methods of treating analytical data in this chapter. *Accuracy* is the nearness of a result (X_i) or of the arithmetic mean ($\bar{X}$) of a set of results to the true value (μ). Accuracy is usually expressed in terms of error, $X_i - \mu$ (or $\bar{X} - \mu$). *Precision* is the agreement of a set of results among themselves. Precision is usually expressed in terms of the deviation of a set of experimental results from the arithmetic mean of the set. Precision is really a measure of the ability to reproduce a result. Although good precision *usually* is an indication of good accuracy, it is entirely possible to obtain good precision but poor accuracy, and vice versa.

Special mention should be made of these terms as they refer to a single result. It is not always meaningful to speak of the accuracy of a single result, because it represents something that has not been checked or reproduced. By definition there is no such thing as precision for a single result, but it is proper to speak of the uncertainty of a single result.

2. Error and Deviation

2.1 Systematic and Random Errors. Errors in analysis may be classified as systematic (determinate) or as random (indeterminate). *Systematic errors* are caused by a defect in the analytical method or by an improperly functioning instrument or analyst. For example, if the indicator used in a titration changes before the equivalence point, a systematic error will result. Similarly, a titration with a dirty burette will cause a somewhat systematic error (high results). The only way to deal with this type of error is to rectify the cause.

2.2 Random Errors. Random errors are unavoidable because there is some uncertainty in every physical measurement. Even the most careful analyst can read a 50 ml burette accurately only to the nearest 0.01 or 0.02 ml, for example. However, a truly random error is just as likely to be positive as negative. This fact makes the average of several replicate measurements more reliable than any individual measurement. Unfortunately, random errors do set a definite limit on accuracy, even when the measurement is repeated many times.

Relative weighing error:

$$\text{ppt} = \frac{(0.1001\,\text{g} - 0.100\,\text{g})}{0.1000\,\text{g}}(1000) = 1.0$$

2.3 Absolute and Relative Deviation. The precision of a set of results may be expressed in terms of absolute or relative deviation. Like absolute error, absolute deviation is expressed in the dimensions of the results: grams, milliliters, etc. Relative deviation is equal to the absolute deviation divided by the mean $\bar{X}$. Relative deviation has no dimensions, and is also expressed in terms of parts per hundred or parts per thousand.

Although by definition there is no precision of a single result, it is proper to calculate the uncertainty of a single result. In the absence of qualifying information, it should be assumed that the last digit is uncertain by ±1. Suppose the length is 0.1 cm. The relative uncertainty, in parts per hundred, is

$$\frac{0.1\,\text{cm}}{10.0\,\text{cm}}(100) = 1\,\text{pph}$$

The relative uncertainty of a single result permits the significance of various numbers to be compared.

3. Significant Figures

Only significant figures should be used in recording analytical results. *Significant figures* by definition are those digits in a number that are known with certainty, plus the first uncertain digit. The last digit of a number is generally considered uncertain by ±1 in the absence of qualifying information.

To illustrate the concept of significant figures, consider the following numbers, all of which have three significant figures: 0.104, 1.04, 104, and 1.04×10^4. The 1 and the middle 0 are certain, and the 4 is uncertain, but significant. Note that an exponential number has no effect on the number of significant figures.

3.1 Treatment of Zeroes. Special attention should be paid to whether zeroes are significant or not. Zeroes, which appear between other digits, as in 10.04, are of course significant. Initial zeroes, such as in 0.104 or 0.0014, are never significant since they serve only to locate the decimal point and are not "certain" digits. For example, the weight of an object can be expressed as 0.0105 g or as 10.5 mg without changing the uncertainty of the weight or the number of significant figures. The weight is still uncertain by ±1 in the last digit; this can be expressed as ±0.0001 g or as ±0.1 mg. Terminal zeroes are generally considered to be significant.

3.2 Absolute Error. The accuracy of a result may be expressed in terms of absolute or relative error. The absolute error of a result X_t, is equal to $X_t - \mu$ and is expressed in the dimension of the two numbers, such as grams, milliliters, percentage, etc. The subscript t is a summation index. For instance, the absolute error in the weighing of a 0.1000 g (μ) object whose weight on the balance is 0.1001 g (X_t) is 0.0001 g.

3.3 Relative Error. The *relative error* of a result X_t is equal to the *absolute error* ($X_t - \mu$) divided by the *true value* (μ). Relative error, therefore, has no dimensions. Usually relative error is expressed in terms of parts per hundred (pph) or parts per thousand (ppt). This may be done by multiplying the relative error by 100 to give pph, or by 1000 to give ppt. This is illustrated below for relative error in general and for the relative error in weighing the 0.1000 g object where the weight on the balance was found to be 0.1001.

Relative error:

$$\text{pph} = \frac{(X_t - \mu)}{\mu}(100)\;;\; \text{pph} = \frac{(X_t - \mu)}{\mu}(1000)$$

Relative weighing error:

$$\text{pph} = \frac{(0.1001\,\text{g} - 0.1000\,\text{g})}{0.1000\,\text{g}}(100) = 0.10$$

As written, 10,100 is considered to have five significant figures. If only three digits are meant to be significant, then it should be written as 1.10×10^4. The absolute uncertainty of the number written the first way is considered to be ±1; the absolute uncertainty of the number written the second way is $\pm 1 \times 10^2$ or ±100.

3.4 Rounding Off. Although calculations can be made with numbers containing figures which are not significant, the result of the calculations should be reported with only as many figures as are significant. It therefore may be necessary to eliminate digits from the answer that are not significant. In general, digits that are not significant should be eliminated by rounding off. Increase the last retained figure by 1 if the next discarded digit is 5 or greater. If it is less than 5, do not change the last retained figure.

3.5 Addition and Subtraction. The proper use of significant figures in addition and subtraction involves a comparison of only the *absolute* uncertainties of the numbers. This usually means you should retain only as many digits to the right of the decimal in the answer as the number with the fewest digits to the right of the decimal. This number of course has the largest absolute uncertainty. It will be necessary to round the last retained digit up if the next discarded digit is 5 or greater. An alternate method is to adjust the number of significant figures in each number before adding or subtracting.

If numbers with positive or negative exponents are involved, adjust the exponents so that they are all the same before adding or subtracting, as shown in this example.

3.6 Multiplication and Division. The proper use of significant figures in multiplication and division involves a comparison of only the relative uncertainties of the numbers. The rule is that the relative uncertainty of the answer should be of the same order of magnitude as the number with the largest relative uncertainty. In practice this means it should fall between 0.2 to 2 times the largest relative uncertainty in the data.

3.7 Log Terms. Quantities such as pH, etc. should be expressed with the same number of significant figures to the right of the decimal as the *total* number of significant figures in the nonexponential numbers used to calculate them. This is because the digits to the left of the decimal come from the exponent, not the nonexponential number.

4. The Central Tendency of a Set of Results

Once a set of results has been calculated with the proper number of significant figures, the best value must be chosen for a comparison with μ, to evaluate the accuracy of the results. Choosing the best value is done by calculating the central tendency of the set of results. The central tendency may be measured by finding the arithmetic mean $\bar{X}$, or less commonly the *median* (M), or the mode of the set of results or sample.

4.1 Sample. The term sample refers to a small practical number of measurements that forms part of the population. The term *population* embraces an infinitely large number of analytical measurements obtained from operations involving the same material. The sample mean $\bar{X}$ is an estimate of the *population mean* (μ). As long as there is no systematic error, μ should approach the true value quite closely. Hence, the symbol μ will be used interchangeably for the population mean and the true value.

Although the arithmetic mean is commonly chosen as the central tendency of a sample, it should be emphasized that there is no *statistical law* that says the mean must always be chosen. Statisticians recognize the validity of all three measures of the central tendency and recommend that the circumstances dictate which should be used.

4.2 Mean. The arithmetic mean, or average, of a sample is calculated as follows:

$$\bar{X} = \frac{X_1 + X_2 + \ldots + X_n}{n} = \frac{\sum X_i}{n} \quad (1)$$

The summation sign (sigma): $\sum_{i=1}^{n}$ indicates that all values of X_i from X_1 to X_n are summed up. The term n is the number of results in the sample.

The chief advantage of the mean compared to the other measures of the central tendency is that each analytical result in a sample contributes its share to the value of the mean and has the same importance as every other result. It thus has the highest precision, or is the most effective, of all the measures of central tendency.

Statistical theory indicates that the mean of n results is $\sqrt{n}$ times as reliable as any X_i. Thus $\bar{X}$ of four results is twice as reliable as one result; $\bar{X}$ of six results is 2.45 times as reliable as one result; and $\bar{X}$ of eight results is 2.83 times as reliable as one result. Thus if the number of results obtained is limited by time or sample size, it may be deemed inefficient to obtain more than four results.

The main disadvantage of the mean is that for small samples, the mean is more affected than the median by the presence of a result having a large error. This is generally true when n is less than 10, and is especially true where $n = 3$. For three measurements where one is suspected to contain a large error, it is better to use the median than to take the mean of the two closest results.

4.3 Median. The central tendency of a sample may also be calculated by finding the *median* (*M*), which is defined as the middle value of a sample of results arranged in order of magnitude. If n is even, M is found by averaging the two middle results of such a sample. In a truly symmetrical distribution of results, M is the same as $\bar{X}$, but if the results are not symmetrically distributed about the mean, M is not the same.

In general, the M is less precise than the $\bar{X}$ except if $n = 2$, when the results are symmetrically distributed about the mean. The precision (1) of M relative to the mean for $n = 3$ is 0.74; for $n = 4$ it is 0.84; for $n = 5$, it is 0.69; for $n = 6$, it is 0.78. The median is always more efficient when n is even because the two middle values are averaged.

The advantage of M over $\bar{X}$ is that a gross error in one result in a small sample will cause a large error in the $\bar{X}$, but not in M. The same would be true for several such results if they were all higher or all lower than the central tendency.

4.4 Mode. The mode is the simplest measure of the central tendency. It is defined as the value that occurs most frequently in a sample. In a truly symmetrical distribution of results, the mode is the same as $\bar{X}$ (and M). The mode is not very useful for small analytical samples because the chances of having identical results with four significant figures are small. However, where n is at least 10 or more and where the results contain only two or three significant figures, a mode can easily exist. The advantage of using the mode in this situation is that it can be found very quickly by visual inspection.

5. Precision

The precision of a set of analyses is a measure of the magnitude of indeterminate error associated with the analytical method. The quantities most commonly used to measure precision are the *average deviation*, *standard deviation*, and *range*. Some uses of precision are as follows:

1. By establishing the precision of various analytical procedures we can compare them with regard to attainable precision.
2. Precision serves as a check on the analyst. If The analyst's precision is as good as that expected from a given procedure, the analyst has more confidence that the analysis has been done correctly.
3. The *Q test*, which is based on precision, can be used to decide whether a result that differs appreciably from the others should be rejected or should be included in computation of the mean.
4. The *t test*, which is based partly on precision, establishes the confidence limits for the mean.

5.1 Average Deviation. The average deviation (*d*) is a simple and fairly useful concept. It is the average of the deviation of the individual results from the mean *without regard to sign.* The average deviation is calculated as follows:

$$d = \frac{1}{n}\sum\left|X_i - \bar{X}\right| \quad (2)$$

Example. Calculate the average deviation of the following sample: 9.8, 10.6, 10.8, and 10.4.

X_t	$\lvert X_t - X \rvert$
9.8	0.6
10.6	0.2
10.8	0.4
10.4	0.0

First calculate the mean, which in this case is 10.4. Now calculate the absolute values of the individual deviations and sum them. None of the individual deviations will be negative because the absolute values are used.

$$\sum\left|X_t - \bar{X}\right| = 1.2\,;\, d = \frac{1}{4}(1.2) = 0.3$$

5.2 Standard Deviation. A more valid and useful measure of precision is σ. It is the square root of the average of the square of the individual deviations from the mean. It is preferred to average deviation because it has a theoretical basis and because it is more sensitive to gross divergent values. The standard deviation of a sample is symbolized as σ and is always calculated using n - 1 results. The standard deviation of *a population* is symbolized as σ and is calculated using n results. The standard deviation of a sample is an estimate of σ.

The standard deviation of a sample of n results is calculated as follows:

$$\sigma = \left(\frac{1}{n-1}\sum\left(X_t - \bar{X}\right)^2\right)^{\frac{1}{2}} \tag{3}$$

If $\bar{X}$ is not known, the following identity permits more rapid machine calculation:

$$\sum\left(X_t - \bar{X}\right)^2 = \sum X_t^2 - \frac{1}{n}\left(\sum X_t\right)^2 \tag{4}$$

If $\bar{X}$ is known, a still more convenient formula is easily derived from Eq. 4 and is as follows:

$$\sum(X_t - \bar{X})^2 = \sum X_t^2 - n(\bar{X})^2 \tag{5}$$

The standard deviation and the average deviation are expressed in the same dimensions or units as the data; both can also be expressed as relative deviation when divided by $\bar{X}$ and multiplied by 100 (pph) or by 1000 (ppt).

Example. Calculate the standard deviation of the following sample: 10.1, 10.5, 9.9, 9.5, 10.6, 9.4, 11.5, 9.5, 9.5, 10, 9.5, $\bar{X}$ = 10.

$$\sum\left(X_t - \bar{X}\right)^2 = 4.24$$
$$\sum X_t^2 = 1104.24$$

X_t	$X_t - X$	$(X_t - X)$	X_t^2
10.1	0.1	0.01	102.1
10.5	0.5	0.25	110.25
9.9	-0.1	0.01	98.01
9.5	-0.5	0.25	90.25
10.6	0.6	0.36	112.36
9.4	-0.6	0.36	88.36
11.5	1.5	2.25	132.25
9.5	-0.5	0.25	90.25
9.5	-0.5	0.25	90.25
10.0	0	0	100.00
9.5	-0.5	0.25	90.25

$$\sigma = \left[\left(\frac{1}{10}\right)(4.24)\right]^{\frac{1}{2}} = 0.65$$

$$\sigma = \left[\left(\frac{1}{10}\right)\left(1104.24 - 11(10.0)^2\right)\right]^{\frac{1}{2}} = 0.65$$

$$\text{Relative standard deviation} = \frac{0.65}{10.00}(100) = 6.5 \text{ pph}$$

5.3 Range. The *range* is the difference between the largest and smallest results in a sample. If the sample is arranged in order of increasing magnitude, the range is symbolized as $X_n - X_l$. The range is the most rapidly calculated measure of precision because the mean need not be calculated first. For samples where n = 3 to 10, it is recommended that the range be used instead of σ for the following reasons.[1] For samples where n is 10 or less the standard deviation, as well as the average deviation, is not a good estimate of precision. For n = 2 to 4, the range is almost as good an estimate of precision as σ. In summary, there is little to be gained in spending time calculating σ for a small sample when the range is almost as good. When the sample contains more than 10 results, the time spent is justified because the standard deviation is so much better than the range of measuring precision.

Estimating the standard deviation from the range. It was stated above that calculation of σ is more involved than the simple subtraction needed to find the range. However, it is often useful to know the value of σ especially for a small sample, to compare it with the standard deviation obtained in the past for similar samples.

Table 1 - Estimation of Standard Deviation from the Range: Formula: σ = k (range)

N	k
2	0.89
3	0.59
4	0.49
5	0.43
6	0.40
7	0.37
8	0.35
9	0.34
10	0.33

Dean and Dixon[2] suggested a simple method for estimating the standard deviation of a small number of results from the range; simply multiply the range by the *deviation factor* (k) in Table 1.

Example. Estimate the standard deviation of the following data: 19.13, 19.25, and 19.30. The range is 0.17 and the deviation factor is 0.59; so $\sigma = 0.59 \times 0.17 = 0.10$. In this case the standard deviation calculated by the conventional way is also 0.10.

5.4 Normal Distribution Curve. The discussion above was concerned only with the calculation of the standard deviation of a sample, not of an infinitely large number of results. For the latter, the theory of measurements having only random error must be used. Suppose that on an analytical sample an infinite number of measurements were taken. This infinite number would form a population, not a sample, of measurements. The arithmetic mean could be calculated as $\bar{X}$, but it would be symbolized as μ because it would be the mean of a population, not a sample. If the standard deviation of the population were calculated, it could be used to illustrate graphically how closely all the measurements would cluster about the mean, multiples (1, 2, 3, etc.) of the standard deviation being used. Such a graph, called the normal distribution curve or normal error curve, is shown in Figure 1. It is a "normal" (symmetrical) curve because only random errors are present.

The vertical axis of Figure 1 represents the relative frequency of occurrence of a measurement X_i or its error $X_i - \mu$. The horizontal axis is divided into units of standard deviation σ of the population. Note three properties of the curve. (*a*) Because the curve is not symmetrical, there tends to be a negative error for every positive error of the same absolute value. (*b*) The relative frequency of measurements having a small error is very great. Over 68% of the measurements fall within plus or minus one σ unit of

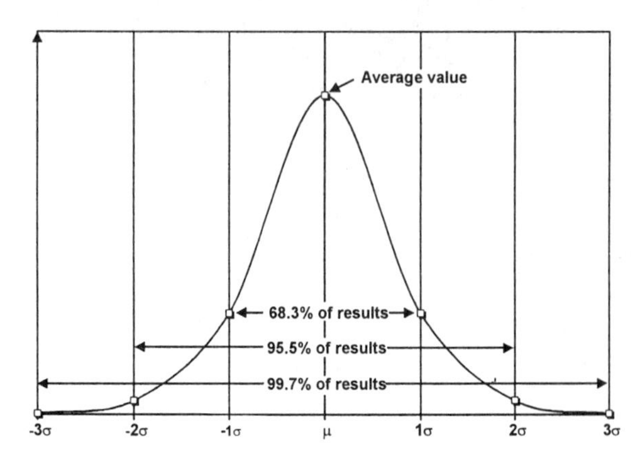

Figure 1 - Relative Frequency of occurrence of a measurement or its error.

Table 2 - Values of t for Calculating Confidence Limits

Number of Measure-ments (n)	Degrees of freedom (n - 1)	Risk and Probability Level		
		0.10 90%	0.05 95%	0.01 99%
2	1	6.314	12.706	63.657
3	2	2.920	4.303	9.925
4	3	2.353	3.182	5.841
5	4	2.132	2.776	4.604
6	5	2.015	2.571	4.032
7	6	1.943	2.447	3.707
8	7	1.895	2.365	3.499
9	8	1.860	2.306	3.355
10	9	1.833	2.262	3.250
11	10	1.812	2.228	3.169
12	11	1.796	2.201	3.106
13	12	1.782	2.179	3.055
14	13	1.771	2.160	3.012
15	14	1.761	2.145	2.977
16	15	1.753	2.131	2.947
21	20	1.725	2.086	2.845
26	25	1.708	2.060	2.787
31	30	1.697	2.042	2.750
41	40	1.684	2.021	2.704
61	60	1.671	2.000	2.660
∞ + 1	∞	1.645	1.960	2.576

the mean. (*c*) The relative frequency of measurements having a large error is very small. Because 99.74% of the measurements fall within 3σ units of the mean, only 0.26% of the measurements will fall outside these limits. These latter measurements may then be said to have a large error, and there are obviously only a few of them.

5.5 Uses of the Normal Distribution Curve. For a small sample, the curve can be used to give an estimate of how good the mean of the results is. For example, if $\bar{X}$ = 10.0% and σ = 0.65% for a sample, then one can estimate that about two thirds (68%) of the results should fall in the range of 10.0 ±0.65%. Thus, in general two thirds of the results should fall within plus or minus one σ of the mean, assuming a normal population.

6. Accuracy of an Analysis: Confidence Limits

The previous section was concerned with the precision of a series of replicate analyses, something that is easy to measure. However, the mean, or best value, is the value that will be reported for an analytical determination. Now the important question is: How closely does the mean or best value agree with the true value μ. In other words, how accurate is the analysis? This job is as easy to measure as the precision. Experimentally, an analytical procedure is often tested on a standard sample that has been carefully prepared so that its composition is known to a high degree of accuracy. Then the standard sample is analyzed by the analytical procedure to be tested, and the analytical results compared with the stated value for each constituent. In favorable cases, this gives a direct indication of the accuracy of which the analytical procedure is capable. The difficulty with this approach is that actual samples have somewhat different composition from the standard sample, and this can affect the accuracy likely to be obtained. Another approach is to use the precision of the analysis, as outlined below.

6.1 Confidence Limits: the T Distribution. If a valid analytical procedure having no systematic error is used to obtain a sample mean, then statistical theory may be used to describe the accuracy of the procedure. Specifically, the theory is used to predict within what *limits* the mean is likely to agree with μ. The theory will not enable this prediction to be made with 100% probability. There is always some *fraction of risk* (α) or percentage probability ($100 - 100\alpha$) involved in such a prediction. The limits predicted for a certain risk or probability are called *confidence limits*. Instead of depending on the normal distribution curve, the limits depend on the t, or "Student's t" distribution curve. This distribution yields values for a constant called t.

To formulate the confidence limits, the standard deviation σ of the sample is calculated, and a value for t is found by consulting Table 2. The constant t depends on α and n, the number of measurements in the sample. (Many statisticians prefer n - 1, which is termed the *degrees of freedom*, and the probability level; both approaches are given in the table.)

The confidence limits for $\bar{X}$ are calculated as follows:

$$\bar{X} \pm \frac{t\sigma}{\sqrt{n}} \qquad (6)$$

It may be said that the fraction of risk that μ lies outside these confidence limits is α. Obviously, $\alpha/2$ is the fraction of risk that μ is *either* larger *or* smaller than the confidence limits. It may also be said that the probability that μ lies inside these limits is $100 - 100\alpha$.

Example. A common fraction of risk is $\alpha = 0.100$. Suppose 10 results have a mean of 56.06% and σ is calculated to be 0.21%. Using $t = 1.833$ for $n = 10$ from Table 2, the confidence limits are calculated to be:

$$\bar{X} \pm \frac{t\sigma}{\sqrt{n}} = 56.06\% \pm 0.12\%$$

The fraction of risk is 0.100 times that if an infinite number of measurements were made, μ would fall outside the confidence interval of 55.94 to 56.18%. Conversely, the probability is 90% that μ would fall inside this interval. This is illustrated in Figure 2, which shows the t distribution curve for a sample of 10 measurements. Reducing the fraction of risk means increasing the probability that μ will fall within a given interval, and it means increasing the size of this interval. For example, the confidence limits for $\alpha = 0.010$ for the above example would be 56.06% ± 0.21%. The confidence interval would be 55.85% to 56.27%.

7. Handling Small Sets of Data

The discussion of precision was mainly concerned with the treatment of large sets of data where n was 10 or greater. However, the usual analysis is performed in triplicate on three weighed samples, or on three aliquots of a solution of a single carefully weighed sample. The analyst must of course calculate correctly the results of each determination, but in addition the analyst must arrive at a single value to report for the triplicate analysis. Using knowledge of the usual precision of the analytical method, the experienced analyst will also be able to decide whether the precision of the result is satisfactory.

Usually the best value to report for an analysis is the mean, or average, of the individual results. Where it appears that one or more of the results contains a large error, it is preferable to report the median instead of the mean. However, this requires some experience, and in the laboratory the beginner is advised to report the mean, and use the median only after consultation with the instructor.

Frequently two results will agree rather closely and a third value will be appreciably higher or lower. In this case, there is a natural tendency to reject the third value and to report the mean of the remaining two values. Unless there is a good experimental reason to do so, the questionable value should not be rejected because it often happens that the close agreement of the two results is fortuitous; the third result should then be used to calculate the mean. A method to test whether such a result should be rejected will now be described.

7.1 The Q Test. The Q test uses the range to determine whether a questionable result should be rejected when n = 3 to 10.[1] This test is most commonly conducted at the 90% confidence level, which means that the questionable result may be rejected with 90% statistical confidence that it is significantly different from the other results. The *rejection quotient* is labeled $Q_{0.90}$; at the 96% confidence level; it is designated $Q_{0.96}$, etc. Values for rejection quotients have been compiled by Dixon[3] and are given in Table 3.

Table 3 - The Q Test

N	$Q_{0.90}$	$Q_{0.96}$	$Q_{0.99}$
3	0.94	0.98	0.99
4	0.76	0.85	0.93
5	0.64	0.73	0.82
6	0.56	0.64	0.74
7	0.51	0.59	0.68
8	0.47	0.54	0.63
9	0.44	0.51	0.60
10	0.41	0.48	0.57

* If $Q_{0.90}$ reject X_n. Arrange sample in order of increasing magnitude.

To perform the Q test, the results are arranged in increasing order of magnitude and are labeled X_1, X_2, X_n. The difference between a questionable result and its nearest neighbor is divided by the range to obtain a quotient Q. If this Q is equal to or greater than the quotient given in Table 3 for n results, the questionable one is rejected.

Example 1. An analysis in triplicate gives the following results: 30.13%, 30.20%, and 31.23%. Should any of the results be rejected?

Questionable result	*Formula for testing*
Smallest value:	$Q(X_1) = \dfrac{X_2 - X_1}{X_n - X_1}$
Largest value:	$Q(X_n) = \dfrac{X_n - X_{n-1}}{X_n - X_1}$

Using the Q test, the smallest result is retained. The calculation of the Q for the largest result of 31.23% is as follows:

$$Q = \frac{X_3 - X_2}{X_3 - X_1} = \frac{1.03}{1.10} = 0.94 \quad (7)$$

Since Q is equal to $Q_{0.90}$ (Table 3) the questionable result may be rejected, but just barely. If possible, it would be better to analyze an additional sample than to report the average of the two remaining results. The skeptic may doubt the necessity for applying the Q test in the previous example because it is "obvious" that the third result is too far off. Let us therefore consider another example.

Example 2. Use the Q test to determine whether any of the following results should be rejected: 40.12, 40.1, and 40.55.

The close agreement of the first two results makes it tempting to discard the third result. However, the Q test will not quite permit this.

$$Q = \frac{X_3 - X_2}{X_3 - X_1} = \frac{0.40}{0.43} = 0.93$$

(0.94 needed for rejection)

Unconvinced by the failure of the Q test to reject a result the analyst "knew" was not as good as the first two, the analyst decided to do two additional analyses. The new results were: 40.20, 40.28. The average of the five results is 40.26, which is much closer to the average of the original three results (40.27) than to the average of the two lower original results (40.14).

In this case, the additional analyses supported retention of the first three results. Of course, the results of additional analyses could have been lower and showed that the 40.14 average was more correct. But unless additional analyses are made it is better statistically to average all three (unless one can be rejected by the Q test).

When there are more than three results, it may be necessary to test more than one questionable result by the Q test. The smallest value is tested first. If it is rejected, the largest value of the remaining set is then tested, etc.

Example 3. Use the Q test on the seven results in this sample: 5.12, 6.82, 6.12, 6.32, 6.22, 6.32, 6.02.

When should precision be considered "good"? This can only be answered from past experience. If we know that a given analytical procedure should give a relative standard deviation of 5 ppt or less and that this precision has been obtained in the past on similar samples, then it is reasonable to assume that a single analysis in triplicate should give precision of this order. By estimating the standard deviation from the range using the deviation factor (Table 1), it is

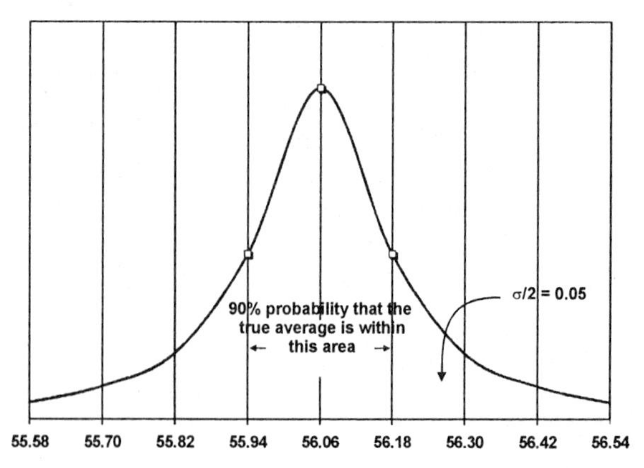

Figure 2 - A graphical representation of the *t* function for 10 results, where ($\bar{X}$) = 56.06, σ = 0.21, and α = 0.10. Note that the risk that μ will be higher than 56.18 is half that of α: the same risk holds for μ being lower than 55.94. The probability is 90% that μ is within the confidence limits of 55.94 to 56.18.

convenient and easy to evaluate the results of an analysis by comparing the precision with that which should be expected. (The Q test should be applied before estimating σ.)

Example 4. A certain analytical method should give a relative standard deviation of 5 ppt or better. A sample is first analyzed three times using this method to give the following set of results: 40.12%, 40.15%, and 40.55%. Because the 40.55% result appears questionable, two additional results are obtained as follows: 40.20% and 40.39%. What is the standard deviation of the first set (n = 3) and of the complete set (n = 5)? The range of the first set is 0.43% (note that the Q test does not reject 40.55%); the range of the second set is 0.43%. The absolute value of each standard deviation is calculated as follows:

$$\sigma(\text{first set}) = 0.59 \times 0.43 = 0.25$$
$$(\text{relative } \sigma = 6.3 \text{ ppt})$$

$$\sigma(\text{full set}) = 0.43 \times 0.43 = 0.18$$
$$(\text{relative } \sigma = 4.6 \text{ ppt})$$

Thus, the first set, which contains the questionable results, does not meet the precision requirements, but the full set does have a relative standard deviation within the expected limits for the method used.

References:

1. Blaedel W. J., Meloche V. W., and Ramsay J. A. (1951) J. Chem. Edu. 28, 643.
2. Dean, R. B. and Dixon, W. J. (1951) Anal. Chem. 23, #636.
3. Dixon, W. J. (1951) Ann. Math. Stat. 22, 68.

THEORY/PRINCIPLES

CONCENTRATION AND TEMPERATURE EFFECTS ON pK_a

Buffers, CalBiochem, San Diego, CA, ph. (800) 854-3417. (Adapted)

1. Introduction

Buffers cannot always be treated as ideal solutions (in which components do not affect each others functions). As a result, the equations which have been developed to illustrate the basic buffer principles must be modified to accommodate these deviations from the ideal. The equilibrium constant (K_a) was defined as the product of the concentration of the dissociated species divided by the concentration of the undissociated species. Thus, for the dissociation of HA the following expression holds when HA forms ideal solutions.

$$K_{HA} = \frac{[H^+][A^-]}{[HA]}$$

In the general case, where the dissociation is not ideal, the term *activity* (a_i), is substituted for concentration giving:

$$K_{HA} = \frac{a_{H^+} a_{HA^-}}{a_{HA}}$$

The use of the activities of the interacting species permits the accommodation of effects of temperature, concentration and other properties of the solute and solvent upon the equilibrium.

While the mathematical relationship of activity and temperature is quite complicated, the actual change of pK_a with temperature ([$\Delta pK_a/\Delta T(°C)$]) approximates a linear relationship, eliminating calculations concerning activity.

The interactions of solute species vary with concentration. At infinite dilution, activity equals concentration (C). At finite dilutions, the deviation between activity and concentration is expressed as the *activity coefficient* (γ_i).

$$a_i = \gamma_i C$$

Substituting this equation into the equilibrium expression gives:

$$K_{HA} = \frac{\gamma_{H^+}\gamma_{A^-}[H^+][A^-]}{\gamma_{HA}[HA]}$$

The γ_i values vary with concentrations and species, so the H^+ concentration varies with changing A^- and HA concentrations. Even though the ratio $[A^-]/[HA]$ is held constant, the equilibrium constant changes because the γ_i values change.

As the concentrations of the buffer species approach zero, activity coefficients approach unity and the equilibrium expression again approaches that of an ideal solution. Table 1 lists the activity coefficients of some common ions found in biological buffers.

The activity coefficients in the table illustrate that some buffer systems will show only a slight dependence of pK_a on a changing concentration (i.e., acetate) whereas systems containing the species PO_4^- or citrate$^-$ show large deviations from ideality. Zwitterionic buffer materials, because of their zwitterionic character, show a slight pK_a with changing concentration.

2. Selection of a Buffer

Most biological reactions occur in the pH range between 6 and 8, yet many commonly used buffers have serious functional drawbacks in this range. Phosphate is a feeble buffer above pH 7.5, precipitates or binds many polyvalent cations, and is frequently a participating or inhibiting metabolite in the reaction of interest or competing reactions with "spectator" species that are not acting innocuously. TRIS is a poor buffer below pH 7.5, a potentially reactive primary amine, and an active participant in various reactions, e.g., alkaline phosphatase.

Carbonate is a poor buffer because dissolved carbon dioxide has limited solubility. Borate binds to a wide range of organic compounds including important respiratory metabolites. To quote Norman Good *et al.*, "It is impossible to even guess how many exploratory experiments have failed, how many reaction rates have been depressed and how many processes have been distorted because of imperfections of the buffers employed." Because of these imperfections, Good and his co-workers designed buffer materials toward the following criteria:

Table 1 - Activity Coefficient of Some Ions in Aqueous Solution

	Buffer Concentration		
Ion	0.001 M	0.01 M	0.1 M
H^+	0.98	0.93	0.86
OH^-	0.98	0.93	0.81
OAc^-	0.98	0.93	0.82
$H_2PO_4^-$	0.98	0.93	0.74
HPO_4^-	0.90	0.74	0.45
PO_4^{2-}	0.80	0.51	0.16
H_2Cit^-	0.98	0.93	0.81
$HCit^-$	0.90	0.74	0.45
Cit^-	0.80	0.51	0.18

1. pK_a between 6 and 8
2. High solubility in aqueous systems
3. Exclusion by biological membranes
4. Minimal salt effects
5. Minimal effects on dissociation due to concentration, temperature, and ionic composition
6. Well defined or nonexistent interactions with mineral cations.
7. Chemical stability
8. Insignificant light absorption between 240 and 700 nanometers
9. Easily available in pure form

These criteria, for the most part, are self explanatory, however, a short discussion of some of them is useful. Good recognized that the majority of biochemical experiments require a pH range of 6-8. While a buffer with a pH of 7 would satisfy this narrow range, a series of buffers with pK_a from 6-8 would give an effectively buffered working range of 5-9 and allow selection of a buffer with a pK_a very near the working pH, permitting one to reduce the buffer concentration substantially.

Solubility in aqueous solutions is an obvious criteria for the selection of a biological buffer. If a buffer has a high ratio of solubility in water to solubility in relatively nonpolar solvents, the biological phase in particular systems is more likely to exclude the buffers from the particle.

In many experiments, exclusion by a biological membrane is of little importance since all participating molecules are in solution. However, the scientist working with bioparticles is acutely aware of the problems created by a buffer system capable of passing through biological membranes. Good and his coworkers demonstrated that, unlike TRIS and phosphate, zwitterionic buffers were not able to pass through biological membranes and gave clearly superior results in the Hill reaction with spinach chloroplasts in succinate oxidation by bean mitochondria, and in the incorporation of carbon-14 labeled leucine into TCA-precipitate protein by *E. coli* cells. Zwitterionic buffers, therefore, manage the pH of the suspending medium without interfering with the internal pH of the bioparticle.

A buffer should be selected so that its effect on other participating ions is nonexistent or at least negligible. If a reaction is dependent on a number of ionic participants, an ionic buffer system can and frequently does interact with other ions. Phosphate would certainly be a poor choice as a buffer in a biochemical process that required or consumed phosphate and might also affect a process requiring sulfate. A buffer which complexes and precipitates cations changes biochemical processes that depend on them.

Ideally, the dissociation constant of the buffer system should not shift as concentrations change, although this is rarely the case. If the shift in dissociation of a buffer material is small, a stock solution can be conveniently prepared and diluted as required without readjustment of the pH. When diluted to 0.05 M, the pH of a 0.5 M phosphate buffer at pH 6.6 rises to 6.9. Another 10-fold dilution increases the pH to 7.1. TRIS, on the other hand, decreases approximately 0.1 pH unit per 10-fold dilution. A dissociation constant shift with dilution must be recognized, especially near the working limits of the buffering range.

Temperature changes markedly affect dissociation constants of buffers. TRIS is particularly notorious for its large dissociation constant shifts with temperature. Expressed as a change in pK_a per degree centigrade, $\Delta pK_a/\Delta T = +0.031$. If you happen to use pH 7 TRIS, made up in the cold room at 4°C, when you incubate at 37°C, the pH will have dropped 1.05 units to 5.95. If the effective buffer range is taken as 1 pH unit on either side of the pK_a, TRIS at 20°C with pK_a 8.3 would be an effective buffer from pH 7.3 to 9.3. The same TRIS used at 4°C becomes a very poor buffer at 7.3 since the pK_a shifts up to 8.8. Thus, two important points have been illustrated: First, always prepare your working buffer at conditions under which you plan to use it; second, choose a buffer with a pK_a that will allow for any pK_a that might occur due to temperature changes.

If complexing occurs between the buffer and cations in the biological system, two problems arise: First, the formation of a complex between the buffer and the cation lowers the buffer capacity for hydrogen ions. Thus, if a significant amount of the buffer is complexed with cations, the hydrogen ion concentration increases.

Second, if a required cation has complexed with the buffer, it may no longer be available for participation in the biochemical reaction. Metal binding by a buffer should be at least well defined and preferably nonexistent. The buffer should be stable under working conditions to oxidation and light and should be unaffected by the biochemical system. Difficulties may arise if a buffer is also a metabolite. Buffer components, which will not adsorb UV or visible light, are convenient in photometric experiments. Table 2 shows the properties of the zwitterionic buffers currently in general usage and how closely they meet Good's criteria.

3. Helpful Hints

3.1 Recognize the Importance of the pK_a. Select a buffer material with a pK_a close to the desired working pH. If you expect the pH to drop during the experiment, choose a buffer with a pK_a slightly lower than the working pH. This will permit the buffering action to become more resistant to changes in hydrogen ion concentration as hydrogen ion is liberated. Conversely, if you expect the pH to rise during the experiment, choose a buffer with a pK_a slightly higher than the working pK_a.

3.2 Prepare Buffers at Working Conditions. Always try to prepare your buffer solution at the temperature and concentration you plan to use during the experiment. Stock solutions are acceptable as long as pH adjustment is made after temperature and concentration adjustment.

3.3 Buffer Materials and Their Salts for Convenient Buffer Preparation. Many buffer materials are supplied both as a free acid (or base) and its corresponding salt. This provides a convenient method for making a series of buffers with varying pH. For example, solutions of 0.1 M HEPES and 0.1 M HEPES sodium salt can be mixed in infinite number of ratios between 10/1 and 1/10 to provide 0.1 M HEPES buffer in a continuum of pH values from 6.55 to 8.55.

3.4 Phosphate Buffer. Mixing precalculated amounts of monobasic and dibasic sodium phosphates has long been established as the method of choice for preparing phosphate buffer. Stock solution A is prepared by dissolving 27.6 g (0.2 moles monobasic sodium phosphate, monohydrate) in deionized water to a total volume of 1000 ml. Stock solution B is prepared by dissolving 28.4 g (0.2 moles) of dibasic sodium phosphate in deionized water to a total volume of 1000 ml. By mixing the appropriate milliliters of A and B as shown in Table 3 and diluting to a total volume of 200 ml, a 0.1 M phosphate buffer of the required pH can be prepared.

3.5 Buffer Materials Must Be Adjusted to the Working pH. Many buffers, especially the Good buffers, are supplied as crystalline acids or bases. The pH of these buffers materials in solution will not be near the pK_a and the materials will not become buffers until the pH is adjusted. In practice, one selects a buffer material with a pK_a near the desired working pH. If this buffer material is a free acid, it is adjusted to the working pH with sodium hydroxide, potassium hydroxide, tetramethylammonium hydroxide or other appropriate base. Buffer materials obtained as free bases must be adjusted by addition of a suitable acid.

3.6 Buffers Without Mineral Cations. Frequently there is a need to prepare buffers without mineral cations. Tetramethylammonium hydroxide solves this problem. The basicity of this organic quaternary amine is equivalent to that of sodium or potassium hydroxide. Buffers prepared with this base can be supplemented at will with various inorganic cations during the evaluation of mineral-ion effects upon enzymes or other bioparticulate activities.

3.7 Buffer Composition from a Graph. Figure 1

Table 3 - Dilutions to achieve proper pH

ml A	ml B	pH	ml A	ml B	pH
92.0	8.0	5.8	45.0	55.0	6.9
90.0	10.0	5.9	39.0	61.0	7.0
87.7	12.3	6.0	33.0	67.0	7.1
85.0	15.0	6.1	28.0	72.0	7.2
81.5	19.5	6.2	23.0	77.0	7.3
77.5	22.5	6.3	19.0	81.0	7.4
73.5	26.5	6.4	16.0	84.0	7.5
68.5	31.5	6.5	13.0	87.0	7.6
62.5	37.5	6.6	10.5	89.5	7.7
56.5	43.5	6.7	8.5	91.5	7.8
51.0	49.0	6.8			

Table 2 - Physical Properties of Some Zwitterionic Buffers

Buffer	pK_a (20°C)	ΔpK_a/ΔT	MW	sat. soln. M at 0°C	Molality of Metal Binding Constants (log K)			
					Mg^{2+}	Ca^{2+}	Mn^{2+}	Cu^{2+}
MES	6.15	-0.011	195.23	0.65	0.8	0.7	0.7	negl
ADA	6.60	-0.011	212.15	-	2.5	4.0	4.9	9.7
BIS-TRIS PROPANE	6.80	-0.016	282.35	2.3	-	-	-	-
PIPES	6.80	-0.009	342.26	1.4	negl	negl	negl	negl
ACES	6.90	-0.020	182.20	0.22	0.4	0.4	negl	0.4
MOPS	7.20	-0.006	209.26	3.0	-	-	-	-
TES	7.50	-0.020	229.25	2.6	negl	negl	negl	3.2
HEPES	7.55	-0.014	238.31	2.3	negl	negl	negl	negl
HEPPS	8.00	-0.007	252.33	2.5	-	-	-	-
TRICINE	8.15	-0.021	179.18	0.8	1.2	2.4	2.7	7.3
Glycine Amide HCl*	8.20	-0.029	110.56	4.6	-	-	-	-
TRIS*	8.30	-0.031	121.13	2.4	negl	negl	negl	-
BICINE	8.30	-0.018	163.17	1.1	1.5	2.8	3.1	8.1
Glycylglycine	8.40	-0.028	132.13	1.1	0.8	0.8	1.7	5.8
CHES	9.50	-0.009	207.30	0.85	-	-	-	-
CAPS	10.40	-0.009	221.32	0.85	-	-	-	-

shows the theoretical plot of ΔpH versus $[A^-]/[HA]$ on two-cycle semilog paper. Since the majority of buffer materials in common usage show little deviation from theory in the pH range about their pK_a values, when the pK_a is accurately known this plot can be of enormous value in calculating the relative amounts of buffer components required for a particular pH. Suppose that you need pH 7.6 MOPS buffer, 0.1 M at 20°C. At 20°C, the pK_a for MOPS is 7.2. Thus, the working pH is 0.4 pH units above the reported pK_a.

Using the chart (Figure 1), this pH corresponds to a MOPS sodium/MOPS ratio of 2.5 and 0.1 M solutions of MOPS and MOPS sodium mixed in this ratio will give the required pH. Check the value with your pH meter if you wish, but if there are deviations, especially when using Good buffers, suspect your meter, not the buffer.

The graph can also be used to calculate the amount of acid (or base) required to adjust a free base buffer material (or free acid buffer material) to the working pH required. 0.1 M TRIS has a pK_a of 8.8 at 4°C. How much TRIS and hydrochloric acid are required to prepare a 0.1 M TRIS buffer at pH 8.2?

Using the graph, pH 8.2 requires a ΔpH of -0.6 and thus a $[A^-]/[HA]$ ratio of 0.25. Calculating the required H^+, the unknown value x:

$$0.25 = \frac{[A^-]}{[HA]} \qquad 0.25 = \frac{\text{Moles TRIS} - \text{Moles H}^+}{\text{Moles H}^+},$$

$$0.25 = \frac{0.1 - x}{x}$$

Solving for x: $1.25x = 0.1$: $x = 0.08$ moles of H^+ required. Thus, 0.1 moles of TRIS and 0.08 moles of hydrochloric acid will yield 1.0 liter of pH 8.2 TRIS buffer at 4°C. When using a free acid buffer material and a base:

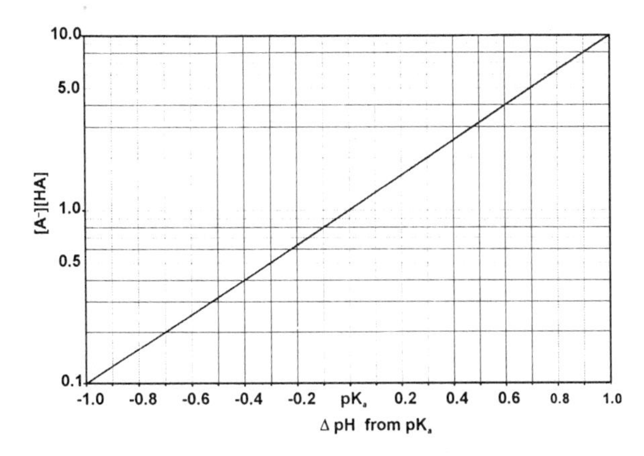

Figure 1 - Theoretical plot of ΔpH versus $[A^-]/[HA]$ on two-cycle semilog.

$$\frac{[A^-]}{[AH]} = \frac{\text{Moles OH}^-}{\text{Moles Buffer} - \text{Moles OH}^-}$$

If you required pH 7.7, 0.5 M TES buffer (pK_a = 7.5 at 20°C), ΔpH = 0.2 and from the graph $[A^-]/[HA] = 1.6$. Calculating:

$$1.6 = \frac{\text{Moles OH}^-}{\text{Moles TES} - \text{Moles OH}^-}$$

$$= \frac{\text{Moles OH}^-}{0.1 - \text{Moles OH}^-}$$

$$0.16 = 2.6 \text{ moles OH}^-$$

$$\text{Moles OH}^- = 0.061$$

Combining 0.1 moles of TES with 0.061 moles of sodium hydroxide in 1.0 liter of water at 20°C will give the required TES buffer.

4. Buffers Listing (Calbiochem™ Acronyms)

1. ACES (*N*-2-acetamido-2-aminoethane sulf-onic acid), pK_a 6.90 at 20°C, MW 182.2
2. ADA (*N*-2-acetamidoiminodiacetic acid, mono-sodium salt), pK_a 6.60 at 20°C, MW 212.2
3. BICINE [*N,N*-bis(2-hydroxyethyl)glycine], pK_a 8.35 at 20°C, MW 163.2
4. BIS-TRIS PROPANE {1,3-bis[tris(hy-droxy-methyl)methylamino]propane}, pK_a 6.80 at 20°C, MW 282.4
5. Citric Acid, monohydrate, MW 210.1
6. CAPS [3-(cyclohexylamino)propanesulfonic acid], pK_a 10.40 at 20°C, MW 221.3
7. CHES (cyclohexylaminoethanesulfonic acid), pK_a 9.50 at 25°C, MW 207.3
8. Glycine Amide, hydrochloride (glycin-amide), pK_a 8.20 at 20°C, MW 110.6
9. Glycylglycine, pK_a 8.40 at 20°C, MW 132.1
10. Glycylglycine, hydrochloride; pK_a 8.40 at 20°C, MW 168.6
11. HEPES (*N*-2-hydroxyethylpiperazine-*N'*-2-ethanesulfonic acid), pK_a 7.55 at 20°C, MW 238.3
12. HEPES, SODIUM SALT (*N*-2-hydroxy-ethyl-piperazine-*N'*-2-ethanesulfonic acid, sodium salt), pK_a 7.55 at 20°C, MW 260.3
13. HEPPS (*N*-2-hydroxyethylpiperazine-*N'*-3-pro-panesulfonic acid), pK_a 8.00 at 20°C, MW 252.3
14. IMIDAZOLE, pK_a 7.0 at 20°C, MW 68.1
15. MES, [(2-(*N*-morpholino)ethanesulfonic acid], pK_a 6.15 at 20°C, MW 195.2
16. METES, SODIUM SALT [2-(*N*-morpho-lino)ethanesulfonic acid, sodium salt], pK_a 6.15 at 20°C, MW 217.2

17. MOPS (3-(*N*-morpholino)propanesulfonic acid) , pK_a 7.20 at 20°C, MW 209.3
18. MOPS, SODIUM SALT [3-(*N*-morpho-lino) propanesulfonic acid, sodium salt], pK_a 7.20 at 20°C, MW 231.2
19. Phosphate, Potassium, dibasic, powdered and sieved for easy solubility, A grade, MW 174.2
20. Phosphate, Potassium, monobasic, powdered and sieved for easy solubility, MW 136.1
21. Phosphate, Sodium, dibasic, powdered and sieved for easy solubility, MW 142.0
22. Phosphate, Sodium, monobasic, mono-hydrate, powdered and sieved for easy solubility, MW 137.99
23. PIPES, SODIUM SALT [piperazine-*N*, *N'*-bis(2-ethanesulfonic acid), monosodium salt, mono-hydrate], pK_a 6.80 at 20°C, MW 342.3
24. Succinic Acid, Disodium salt, hexahydrate, MW 270.2
25. TAPS [*N*-tris-(hydroxymethyl)methyl-3-amino-propanesulfonic acid], pK_a 8.40 at 20°C, MW 243.2
26. TES [*N*-tris-(hydroxymethyl)methyl-2-amino-ethanesulfonic acid), pK_a 7.50 at 20°C, MW 229.3
27. TES, SODIUM SALT [*N*-tris(hydroxy methyl) methyl-2-aminoethanesulfonic acid, sodium salt), pK_a 7.50 at 20°C, MW 251.2
28. Tricine [*N*-tris-(hydroxymethyl) methyl-glycine], pK_a 8.15 at 20°C, MW 179.2
29. TRIS [tris-(hydroxymethyl)aminomethane], heavy-metal free, pK_a 8.30 at 20°C, MW 121.1

INTRODUCTORY LAB 1:
BASIC BIOCHEMICAL TECHNIQUES I: PIPET CALIBRATION AND SOLUTION PREPARATION

Overview:

Accuracy and reproducibility of results in many biochemical experiments are highly dependent on the use of pipets; thus it is absolutely necessary that they function properly. Inaccurate results can occur when pipet parts wear out or corrode. In this lab, you will calibrate your assigned pipets. By doing so, we'll be more certain that volumes delivered in future experiments are accurate. Then, you will use the calibrated pipets to prepare a commonly used buffer stock, adenosine-5'-monophosphate (AMP) stock solution, and tryptophan (trp) stock solution. These stock solutions will then be used to make buffer/AMP and buffer/trp samples.

Accurate solution preparation and correct micropipet use are probably the most important basic biochemical techniques. Buffers are commonly used in biochemical laboratories, so in this lab we practice preparing them. Buffer solutions contain both the acidic and basic forms of a weak acid. They are used to minimize the change in pH of a solution when strong acids or bases are added. Buffers contain acids to react and minimize the effect of added base to react with added acid. The ability to keep solutions at a particular pH is necessary to maintain functional conditions in almost all biochemical reactions.

Experimental Outline:

1. Calibration of micropipets.
2. Preparation of buffer and absorbance spectroscopy samples.

Required Materials:

pipets: 1 ml, 200 µl, 100 µl and 10 µl; pipet tips; solid potassium phosphate, monobasic; distilled H_2O; concentrated HCl; solid KOH; solid AMP; solid trp; top-loading balance; pH meter; magnetic stir-plate; stir-bars; weighing paper or boats; beakers; volumetric flasks; screw-capped 50-ml tubes; 2-ml Eppendorf-capped microfuge tubes

Procedures:

1. Calibrating Micropipets

1. Pipet the maximum deliverable volume of each pipet into tared (preweighed) weigh boats and record the weights to the closest mg in the spaces provided below.
2. Calculate the *averages* (means) as follows:

$$\text{average} = \sum\left(\frac{X_i}{t}\right)$$

The X_i are measured values and t is the number of measurements; record the values below.

3. Calculate the standard deviations for the set of samples as follows:

$$\sigma = \left[\frac{1}{(n-1)}\sum_{i=1}^{t}\left(X_i - \overline{X}\right)^2\right]^{\frac{1}{2}}$$

where t indicates the total number of samples; i is the counting variable; i.e., $t = 6$ for six samples and $i = 1$, 2, 3, 4, 5, and 6 for individual samples.

4. Calculate the *standard mean errors* for the set of samples as follows:

$$\text{standard mean error} = \frac{\text{standard deviation}}{\sqrt{t}}$$

NOTE: Since H_2O has a density of 1.0 g/ml at 25°C, the weight in mg is approximately equal to the volume in µl.

a. 1000-μl pipet:

i.	vi.
ii.	vii.
iii.	viii.
iv.	ix.
v.	x.

Avg. = ______________ σ = ______________ $\sigma/\sqrt{t}$ = ____________

b. 200-μl pipet:

i.	vi.
ii.	vii.
iii.	viii.
iv.	ix.
v.	x.

Avg. = ______________ σ = ______________ $\sigma/\sqrt{t}$ = ____________

c. 100-μl pipet:

i.	vi.
ii.	vii.
iii.	viii.
iv.	ix.
v.	x.

Avg. = ______________ σ = ______________ $\sigma/\sqrt{t}$ = ____________

d. 10-μl pipet:

i.	vi.
ii.	vii.
iii.	viii.
iv.	ix.
v.	x.

Avg. = ______________ σ = ______________ $\sigma/\sqrt{t}$ = ____________

2. Buffer Preparation; Absorbance Spectroscopy

1. Prepare 100 ml of 8 M KOH; F.W. = 56.11 g/mole.

2. Prepare 500 ml of 1 M potassium phosphate (pH 7.5); F.W. = 136.09 g/mole.

NOTE: 136.09 g/mole x 1 mole/l x 0.5 l = 68.045 g.

a. Add the solid potassium phosphate and 400 ml H_2O to a 1-liter beaker.
b. Place the beaker on a stir-plate, then put a small stir-bar and a pH electrode (H_2O rinsed and prestandardized) in the solution.
c. Slowly add the required number of drops of 8 M KOH while stirring and monitoring the pH on the meter as you adjust the pH to 7.5.
d. Transfer the solution to a clean 500-ml volumetric flask and add H_2O to adjust the volume to 500 ml.

3. Prepare a solution of 100 mg trp/250 ml 200 mM potassium phosphate (pH 7.5); F.W. = 204.2 g/mole. Make three solutions.
   a. Dissolve 100 mg of tryptophan in 50 ml of 1 M potassium phosphate (pH 7.5).
   b. Add 200 ml H_2O to each solution in a volumetric flask to give the final solution.
4. Prepare a solution of 100 mg AMP/250 ml 200 mM potassium phosphate (pH 7.5); F.W. = 433.2 g/mole. Make three solutions.
   a. Dissolve 100 mg of AMP in 50 ml 1 M potassium phosphate (pH 7.5).
   b. Add 200 ml of H_2O to each solution in a volumetric flask.
5. Save these solutions for the next period to measure absorbance spectra.

Notes:

This subsection is included to provide details, additional considerations, and associated information about processes or functional activities of components of the associated processes.

1. *If you suspect faulty pipet delivery*, repeat the described measurements. Recalibration is necessary if readings are different by more than 5% from the correct values. Do not adjust the calibration screws unless prior consent is obtained.
2. *Standard Deviation.* When measurements are made using a population that follows a Gaussian distribution relative to the average value ($\overline{X}$), σ the signifies the fact that 68% of the total population will fall statistically within the range $\overline{X} - \sigma$ to $\overline{X} + \sigma$. The population ranging from $\overline{X} - 2\sigma$ to $\overline{X} + 2\sigma$ includes 95% of the total.

Additional Report Requirements:

This subsection is included to ensure that the student recognizes the specific information to be included in the laboratory reports. All of the techniques and experiments should be properly introduced, including any equations and explanations regarding how they relate to the experimental results. The results should be described. An assessment should be made regarding their use in drawing the conclusion(s). Address the purpose of the experiment. Merely performing the physical manipulations associated with testing the hypothesis (or verifying a material) does not constitute completing the experiment. One must then describe the logical use of the data to reach the conclusion(s) and assess your statistical uncertainties. How well has the hypothesis been confirmed? Finally, your must consider carefully whether the results can be interpreted in any other way. Is it possible that the most obvious (expected) conclusion is not the correct one?

Do ambiguities in the logic of the method used to test the hypothesis, or in the results, undermine the degree of confidence with which one views the conclusion? Scientific facts are completely unambiguous. Less firm information might be useful or useless, depending on whether one can make the necessary comparisons with adequate certainty. When the error bars of two numbers overlap, no matter how different their mean values, the numbers are for comparative purposes the same. Drawing conclusions based on a trend may be very tenuous or not even acceptable. Be careful, this is a common subject in seminar discussions and can make or break a "proof."

Report on the following information for this lab.

1. Pipet calibration tables
2. Solution preparation calculations

PROCESS RATIONALE

PIPETS

Oxford-Labware, Sherwood Medical, St. Louis, MO. (Adapted)

1. Volume Adjustment

To select the desired volume, loosen the lock nut mechanism by turning it counterclockwise. To reduce the volume, turn the thumb knob clockwise. Turning the thumb knob counterclockwise will increase the volume. Set the desired volume on the digital display to correspond to the arrow mark molded on the base of the window frame. The selected volume is fixed by turning the lock nut clockwise and can be confirmed on the digital display as in the following examples:

2. Operating Instructions

1. Attach a clean tip firmly to the instrument.
2. Before entry into the sample solution, depress the thumb knob to the "First Stop."
3. Now immerse the tip approximately 3 mm into the sample solution. (Step 1).
4. Smoothly return the plunger knob to the release position, allowing sample to enter tip. (Step 2) Do not allow the knob to "snap" back to release position.
5. Withdraw the tip from the sample solution. Do not wipe the tip.
6. Place the tip against the side wall of the receiving vessel. (Step 3)
7. Smoothly depress the plunger knob to the first stop (Step 4), pause; then depress the knob to the second stop. (Step 5)

NOTE: When dispensing serum and other viscous fluids, it is necessary to pause about two seconds before moving to the secondary stop.

8. With the knob still held in its lowest position, slowly withdraw the tip while sliding it against the wall of the receiving vessel.
9. Return the knob to the release position. Do not allow the knob to "snap" back.
10. Remove the disposable tip by firmly depressing the tip ejector button. (Step 6)

3. Tips for Better Use

1. To improve sampling precision try to effect the same speed for both the intake and delivery of a sample. Smooth depression and release of the plunger knob will give the most consistent result. Never allow the plunger to "snap" back. Consistent technique is a key to precision.
2. Always depress the plunger knob to the proper stop before inserting the tip into the solution. Depressing the plunger knob after insertion may cause the formation of an air bubble in the tip and result in a filling error.
3. Try to insert the tip to approximately the same depth in the sample each time, never going deeper than 3 mm. Hold the instrument as vertically as possible (10° maximum from vertical).
4. When sampling hot or cold material, the temperature of the tip should be equalized to that of the solution to prevent contraction or expansion of sample.

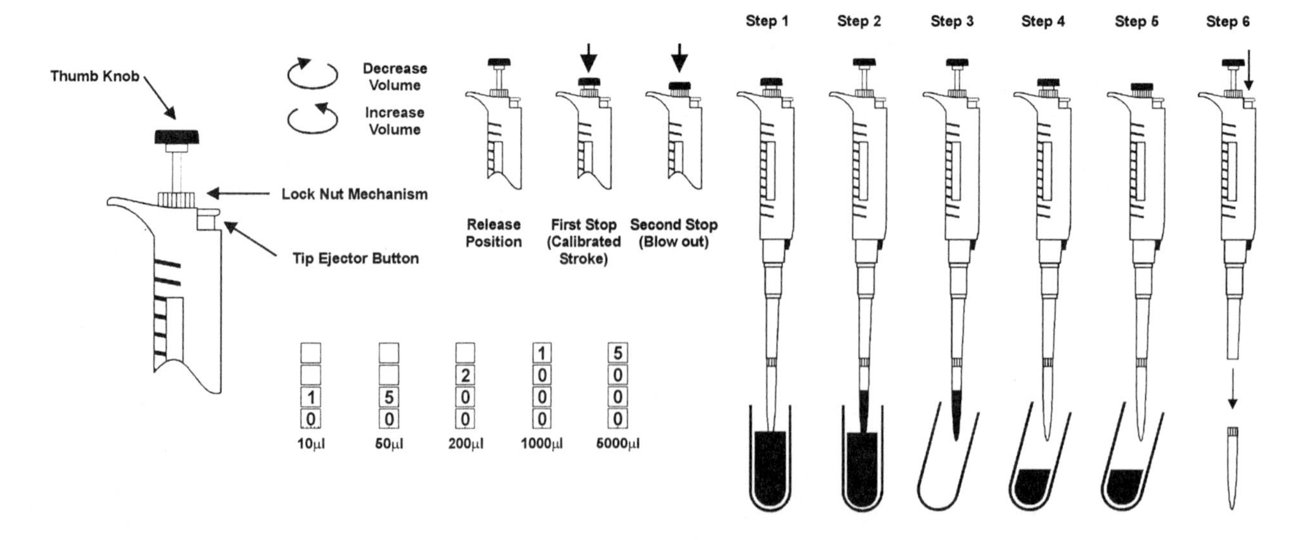

INTRODUCTORY LAB 2:
BASIC BIOCHEMICAL TECHNIQUES II: ABSORBANCE SPECTROSCOPY AND PROTEIN CONCENTRATION DETERMINATIONS

Overview:

In the first part of the lab, you will obtain UV absorbance spectra using the adenosine-5'-monophosphate (AMP) and tryptophan (trp) solutions that were prepared in Introductory lab 1. You will then determine the concentration of your samples using the measured absorbance spectroscopy data and published "molar extinction coefficients." These values will be compared with the expected values calculated from the known (weighed) masses and assumed molecular weights. Comparisons will allow us to assess the accuracy of your absorbance readings and your dilution calculations.

In the second part of the lab, you will perform the BCA protein concentration assay. This colorimetric assay is used because it is faster and easier than other methods (e.g., Lowry, Bradford). First, a bovine serum albumin (BSA) standard curve will be constructed. The BSA standard curve is a plot of absorbance at 562 nm versus protein concentration. BSA is commonly used because of its low cost and availability. Then, the spectroscopic readings of the BCA assay reaction products will be used to determine the concentration of an unknown protein solution from the standard curve.

Experimental Outline:

1. Solution preparation and absorbance spectroscopy of tryptophan and adenosine monophosphate solutions.
2. Protein concentration determination (BCA method).

Required Materials:

pipets; pipet tips; 2-ml Eppendorf capped microfuge tubes; trp and AMP samples from Introductory lab 1; UV-visible spectrophotometer; distilled H_2O; 10 mM sodium phosphate (pH 7); bovine serum albumin standard solution (2 mg/ml); BCA kit (Pierce) reagent A (contains sodium carbonate, sodium bicarbonate, BCA detection reagent, and sodium tartate in 0.1 M NaOH); BCA reagent B (contains 4% $CuSO_4$); 60°C water bath.

Procedures:

1. Solution Preparation and Absorbance Spectroscopy: Tryptophan and Adenosine-5'-Monophosphate

1. Using the samples in the last lab: 100 mg trp in 250 ml of 200 mM of potassium phosphate (pH 7.5) [F.W. = 204.2 g/mole] and 100 mg of AMP in 250 ml of 200 mM potassium phosphate (pH 7.5) [F.W. = 433.2 g/mole], calculate the molar concentrations of the trp and AMP solutions and the dilutions required to give 0.02 mM trp and 67 μM AMP samples. This dilution is necessary for an optimized spectroscopy reading. Calculate the resultant buffer concentration.

 a. 100 mg L-tryptophan in 250 ml of 200 mM potassium phosphate (pH 7.5) [F.W. = 204.2 g/mole].

  i. Calculate the molar concentration of the trp solution using the provided formula weight:
  initial conc. = 1 mole/204.2 g x 100 mg/250 ml x 10^{-3} g/1iter mg x 1 ml/10^{-3} liter = 1.96 x 10^{-3} mole/l
  = 1.96 mM.

  Conversion factors are included and canceled to produce explicit yet efficient units (mM).

  ii. Calculate the dilution required to give a 0.02 mM trp sample:
  stock dilution factor = 0.02 mM/1.96 mM = 0.0102
  So, to make a 2 ml absorbance sample:
  2 ml x 0.0102 = 0.0204 ml = 20 μl + H_2O (2000 μl - 20 μl) = 1980 μl

  iii. The final potassium phosphate concentration is 200 mM x 0.102 = 20.4 mM

 b. 100 mg adenosine-5'-monophosphate in 250 ml of 200 mM potassium phosphate (pH 7.5) [F.W. = 433.2 g/mole].

  i. Calculate the molar concentration of the AMP solution using the provided F.W as described for the trp sample.

ii. Calculate the dilution required to give a 67 μM AMP sample.

iii. Make a 2-ml absorbance sample (your turn to set it up):

Notes Regarding Sample Preparation.

1. Dilute the samples to give the appropriate concentrations. Always use the pipet with the smallest possible delivery volume, to minimize errors introduced by the measurement itself.
2. Incomplete mixing when making solutions or adding new components midway through a timed protocol can result in devastating errors. Mix stock samples well before removing the aliquot that is to be diluted. But be careful, some biomolecules are prone to mechanical denaturation. When a protein solution foams due to mixing, the foam is actually denatured protein. The protein concentration of the solution must have decreased due to lost sample (going from dissolved to a constituent of the foam). Therefore, the stock concentration is no longer correct.
3. When making a biochemical reaction mixture, the order of addition of the reagents can also affect the structural and functional state of the biomolecule(s). One typically adds reagents in the following order: H_2O, buffer, salts, EDTA, β-mercaptoethanol (or dithiothreitol), cofactors (including polyamines), nucleic acids, enzyme(s). Reagents will be listed in protocols by order of addition when possible.

2. Absorbance Spectroscopy

Beer's Law defines the relationship between concentration and measured *absorbance*:

$$A_\lambda = \varepsilon_\lambda bc$$

where A_λ is the measured absorbance at wavelength λ, ε_λ is the *molar extinction coefficient* in units of liter-mole^{-1}cm^{-1} at wavelength λ, b is the pathlength of the cuvette in cm, and c is the molar concentration.

1. Calculate the theoretical and expected absorbance values of the trp and AMP samples at the given wavelengths. Extinction coefficients at the subscripted wavelengths are listed for each compound. Note that they apply only at the subscripted wavelengths.

   a. trp: ε_{219} = 35,000 M^{-1} cm^{-1}; ε_{280} = 5,600 M^{-1} cm^{-1}; $A_{219,\ blank}$ = __________

   b. AMP: ε_{259} = 15,400 M^{-1} cm^{-1}; ε_{240} = 6,000 M^{-1} cm^{-1}; $A_{240,\ blank}$ = __________

2. Measure absorbances of each of the three solutions at both wavelengths.

NOTE: Instrumental description and procedural comments:

1. Two light sources are used, a tungsten lamp for measurements in the visible wavelength range (300 to 800 nm) and a hydrogen lamp that emits UV light in the 340-200 nm range.
2. It is not advisable to read samples that have absorbances greater than 2 because the amount of transmitted light is < 1%. Under these circumstances the total signal is very small, so the signal to noise ratio (S/N) will be unacceptable.

NOTE: General spectrophotometer operation information:

a. Turn on the spectrophotometer, and allow it to "warm up" for 5 minutes.
b. Zero the spectrophotometer with buffer at the appropriate wavelength in a cuvette.
c. Measure the absorbance of the sample in a cuvette at the appropriate dilution and wavelength.

NOTE: Zero the machine with buffer prior to changing samples or wavelengths.

3. Calculate the mean and standard deviation in each case.
   a. trp
      i. trp (219 nm): ________, ______________, ______________
         $\overline{X}$ = ______; σ = ___________

      ii. trp (280 nm): ________, ______________, ______________,
         $\overline{X}$ = ______; σ = ___________

   b. AMP
      i. AMP (240 nm):________, ______________, ______________,
         $\overline{X}$ = ______; σ = ___________

      ii. AMP (259 nm):________, ______________, ______________,
         $\overline{X}$ = ______; σ = ___________

4. Compare these observed values with the calculated values. Was your original calculated value within the experimentally determined range ($\overline{X} \pm \sigma$)?
5. Determine the measured extinction coefficients for trp and AMP using Beer's Law. These determinations require using the measured weights, pipeted volumes, and experimental absorbance values.

NOTE: The uncertainty characterized by the standard deviation affects uncertainty only in the measured "dependent" variable. The absorbance depends on many factors, including pipeting errors, weighing errors (including sample hydration; especially in humid months), volumetric flask calibration, spectrophotometer calibration, and variation in σ due to solvent effects.

3. Protein Concentration Determination: BCA Reagent (Pierce)

1. Prepare a stock protein solution, then dilute it with buffer for the protein determination assay.
   a. Prepare 10 mM sodium phosphate (pH 7) for dilutions:
      i. for 50 ml, use 0.069 g sodium phosphate, monobasic
      ii. adjust to pH 7 and adjust the volume to 50 ml
   b. Prepare 5 ml bovine serum albumin (BSA) standard at a concentration of 2 mg/ml:
      i. use 0.01 g BSA
      ii. Adjust the volume to 5 ml
   c. Prepare "working reagent" (2 ml per sample: a, b, c):
      i. For 17 ml "working reagent": 50 parts BCA reagent A (16.667 ml) per 1 part BCA reagent B (333 μl)
         a. 1 blank (working reagent only)
         b. 5 standards (BSA and working reagent)
         c. 2 unknowns (unknown protein and working reagent)
   d. Prepare one set of standards for the lab in duplicate:

Vol. of albumin standard	*Vol. of sodium phosphate*	*Final concentration*
50 μl	1.95 ml	50 μg/ml
100 μl	1.90 ml	100 μg/ml
150 μl	1.85 ml	150 μg/ml
200 μl	1.80 ml	200 μg/ml
250 μl	1.75 ml	250 μg/ml

   e. Prepare unknown:
      i. Use 10 μl of unknown stock protein and 340 μl of 10 mM sodium phosphate buffer for a total volume of 350 μl.
      ii. Prepare duplicate unknown samples for comparison.

2. Protocol
   a. Pipet 100 μl of each standard or unknown protein sample into labeled tubes. For a blank, use 100 μl of 10 mM sodium phosphate.
   b. Add 2.0 ml of "working reagent" to each tube, then mix well.

c. Incubate all tubes at 60°C for 30 minutes.
d. Cool all tubes to room temperature.
e. Quickly measure absorbance values at 562 nm for each tube. Use the "reagent blank" to set the zero value on the spectrophotometer.

Standard curve values:

50 μg/ml	__________		
100 μg ml	__________	unknowns	__________
150 μg/ml	__________		__________
200 μg/ml	__________		
250 μg/ml	__________		

4. Data Analysis

1. Plot the absorbance values obtained with the BSA standards (*y*-axis) as a function of the final calculated concentrations (*x*-axis).
2. Using a *graphical statistics program*, determine the linear regression fitting line for the standard curve data. What is the standard deviation? Do any of the values deviate wildly from the fit line? If so, delete the point from the file and redetermine the linear regression fit.
3. Using this calculated line, find the concentration that corresponds to the absorbance values obtained with the unknown samples. Did the standard deviation of the fit line convincingly encompass all of the values? Are any of them obviously bad data?

Report Requirements:

1. All absorbance spectroscopy measurements and calculations
2. Protein absorbance table
3. Standard curve and answers to the questions
4. Extrapolated protein concentrations and answers to the question

References:

1. Smith et al. (1985) Anal. Biochem. 150: 76-85.
2. Phone re. BCA kit: Pierce (800-874-3723).

PROCESS RATIONALE

AMP AND TRYPTOPHAN ABSORBANCE SPECTRA; SAMPLE CALCULATIONS

1. AMP

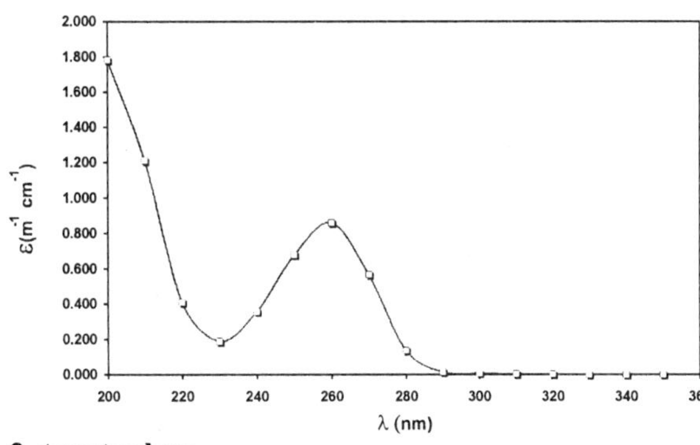

λ	A_λ	λ	A_λ
350	-0.001	340	0.000
330	0.001	320	0.004
310	0.005	300	0.009
290	0.016	280	0.138
270	0.566	260	0.861
250	0.680	240	0.358
230	0.190	220	0.410
210	1.212	200	1.783

2. tryptophan

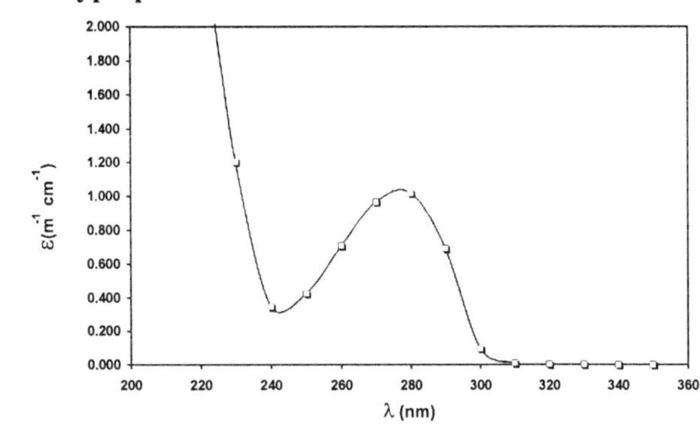

λ	A_λ	λ	A_λ
350	0.001	340	0.001
330	0.003	320	0.004
310	0.011	300	0.097
290	0.692	280	1.019
270	0.966	260	0.708
250	0.426	240	0.347
230	1.204	220	2.495
210	2.495	200	2.495

3. DNA Absorbance Concentration Calculation:

i. Assume a measured A_{260} value of 0.100 and that you have 300 μl of stock solution.

ii. Dilution factor = 200 fold (i.e., you diluted 5 ml of sample with 995 μl buffer)
Absorbance of the stock solution = 0.100 A_{260} ml x 200 = 20 A_{260}/ml
Total amount of sample = 20 A_{260}/ml x 300 μl x 1 ml/1000 μl = 6 A_{260} units

iii. A typically used equality is 20 μg per 1 A_{260} unit (in a 1-cm cuvette)
total amount of DNA (approximate) = 6 A_{260} units x 20 μg/1 A_{260} unit = 120 μg

THEORY/PRINCIPLES

ABSORPTION DATA FOR THE NUCLEOSIDE MONOPHOSPHATES

1. Adenosine 5'-phosphate

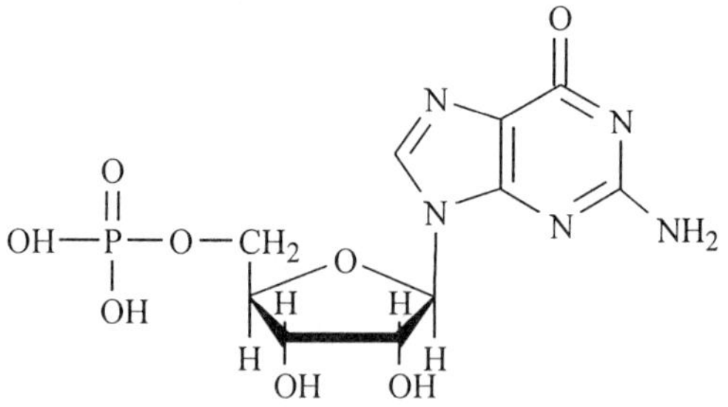

1.1 Abbreviation
Ado-5'-P, pA, AMP

1.2 Formula (MW)
$C_{10}H_{14}N_5O_7P$ (347.22)

1.3 Melting Point
192°C

1.4 pK_a
Basic Acidic
3.8 -

pH	λ_{max}	ε_{max} (x 10^{-3})	λ_{min}	ε (x 10^{-3})					
1	257	15,0	230	230 nm	240 nm	250 nm	270 nm	280 nm	290 nm
7	259	15.4	227	0.23	0.43	0.84	0.68	0.22	0.04
11	259	15.4	227	0.18	0.39	0.79	0.68	0.16	0.01
				-	-	0.79	-	0.16	-

2. Cytidine 5'-phosphate

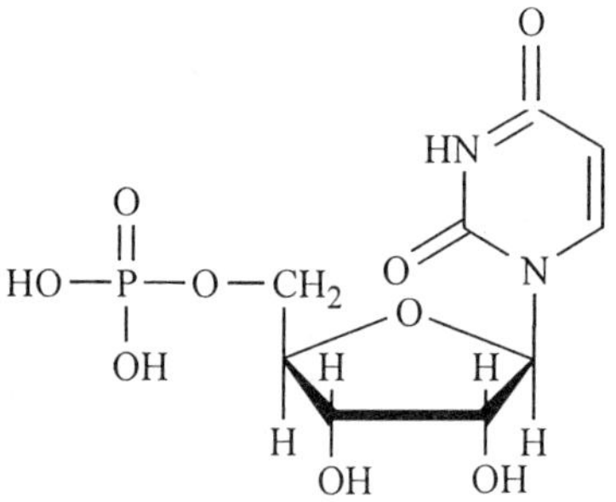

1.1 Abbreviation
Cyd-5'-P, pC, CMP

1.2 Formula (MW)
$C_9H_{14}N_3O_8P$ (323.21)

1.3 pK_a
Basic Acidic
4.5 -

pH	λ_{max}	ε_{max} (x 10^{-3})	λ_{min}	ε (x 10^{-3})					
2	280	13.2	241	230 nm	240 nm	250 nm	270 nm	280 nm	290 nm
7	271	9.1	249	0.56	0.25	0.44	1.73	2.09	1.55
11	271	9.1	249	1.07	0.92	0.84	1.21	0.98	0.33
				-	-	0.84	-	0.98	0.33

3. Guanosine 5'-phosphate

1.1 Abbreviation
Guo-5'-P, pG, 5'-GMP

1.2 Formula (MW)
$C_{10}H_{14}N_5O_8P$ (363.22)

1.3 Melting Point
190-200°C

1.4 pK$_a$

Basic	Acidic
2.4	9.4

pH	λ_{max}	ε_{max} (x 10^{-3})	λ_{min}
1	257	12.2	228
7	252	13.7	224
11	258	11.6	230

ε (x 10^{-3})					
230 nm	240 nm	250 nm	270 nm	280 nm	290 nm
0.22	0.55	0.96	0.74	0.67	0.29
0.36	0.81	1.16	0.81	0.66	0.29
0.38	0.82	0.90	0.97	0.61	0.29

4. Uridine 5'-phosphate

1.1 Abbreviation
Urd-5'-P, pU, UMP

1.2 Formula (MW)
$C_9H_{13}N_2O_9P$ (324.18)

1.3 pK$_a$

Basic	Acidic
-	9.5

pH	λ_{max}	ε_{max} (x 10^{-3})	λ_{min}
1	262	10.0	230
7	262	10.0	230
11	261	7.8	241

ε (x 10^{-3})					
230 nm	240 nm	250 nm	270 nm	280 nm	290 nm
-	-	0.73	-	0.39	0.03
0.21	0.38	0.73	0.87	0.39	0.03
0.79	0.50	0.80	-	0.31	0.02

PROCESS RATIONALE

ABSORPTION SPECTRA DATA FOR AROMATIC AMINO ACIDS - PH 6

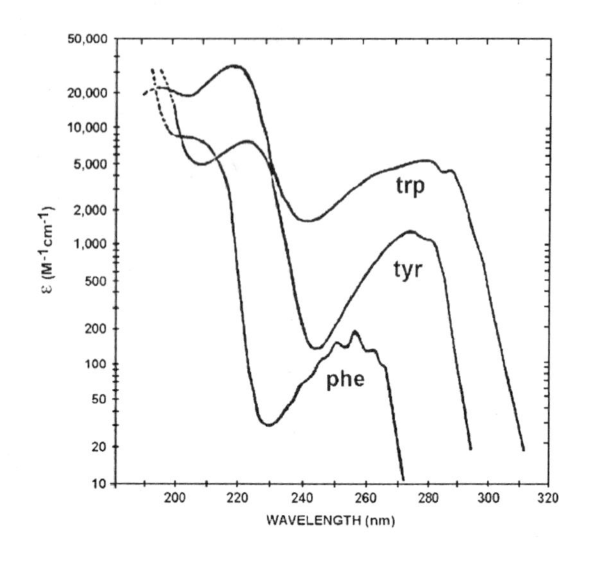

COO^- / $H_3N^+—C—H$ / CH_2 / N−H

tryptophan

COO^- / $H_3N^+—C—H$ / CH_2 / OH

tyrosine

COO^- / $H_3N^+—C—H$ / CH_2

phenylalanine

UV ABSORPTION CHARACTERISTICS OF THE AROMATIC AMINO ACIDS: SELECTED EXTINCTION COEFFICIENTS

1. Phenylalanine:

	Water		Ethanol	
	λ	ε (x 10^{-3})	λ	ε (x 10^{-3})
Inflection	208	10.20	208	10.40
Inflection	217	5.00	217	5.30
Minimum	240	0.080	244	0.088
Maximum	241.2	0.086	242.0	0.093
Maximum	246.5	0.115	247.3	0.114
Maximum	251.5	0.197	258.3	0.195
Inflection	260.7	-	261.2	-
Maximum	267.1	0.091	267.8	0.096

2. Tryptophan:

	Water		Ethanol	
	λ	ε (x 10^{-3})	λ	ε (x 10^{-3})
Minimum	205	21.40	206	21.30
Maximum	219	35.00	221	37.20
Minimum	245	1.900	245	1.560
Maximum	279.8	5.600	282.0	6.170
Maximum	288.5	4.750	290.6	5.330

3. Tyrosine:

	Water		Ethanol	
	λ	ε (x 10^{-3})	λ	ε (x 10^{-3})
Maximum	193	51.70	-	-
Minimum	212	7.00	212	6.20
Maximum	224	8.80	227	10.20
Minimum	247	0.176	246	0.174
Maximum	274.6	1.420	278.4	1.790
Inflection	281.9	-	285.7	-

4. Cysteine:

	Water		Ethanol	
	λ	ε (x 10^{-3})	λ	ε (x 10^{-3})
Inflection	250	0.360	253	0.372
Inflection	260	0.280	260	0.320
Inflection	280	0.110	280	0.135
Inflection	300	0.025	300	0.035
Inflection	320	0.006	320	0.007

5. *N*-Acetylcysteine:

	Water		Ethanol	
	λ	ε (x 10^{-3})	λ	ε (x 10^{-3})
Inflection	250	0.015	250	0.020
Inflection	280	0.005	280	0.005
Inflection	320	-	320	-

PROCESS RATIONALE

BCA ASSAY SAMPLE DATA

Copper I (Cu^{+}) binds to protein backbone amide groups, which become oxidized to Cu^{2+} in a process that is limited by the protein oxidant concentration. When reduced bicinchonic acid (BCA) is added to this mixture, it becomes oxidized by Cu^{2+}, which reverts to Cu^{1+}. We measure the purplish-blue oxidized BCA spectroscopically. The amount of bovine serum albumin (BSA) is determined by the absorbance at 480 nm. When the BSA absorbance vs. [protein] standard curve is constructed, the full data set might include an outlying data point (a point that does not conform to the clearly established trend). If one measurement distorts the slope of the standard best fit curve, one is justified to delete the outlying point. Deletion of the outlier results in a more accurate standard curve interpolate of an unknown protein concentration.

1. Full data set. $y = 3.0032 \times 10^{-3} + 0.0595$ $r^2 = 0.823$

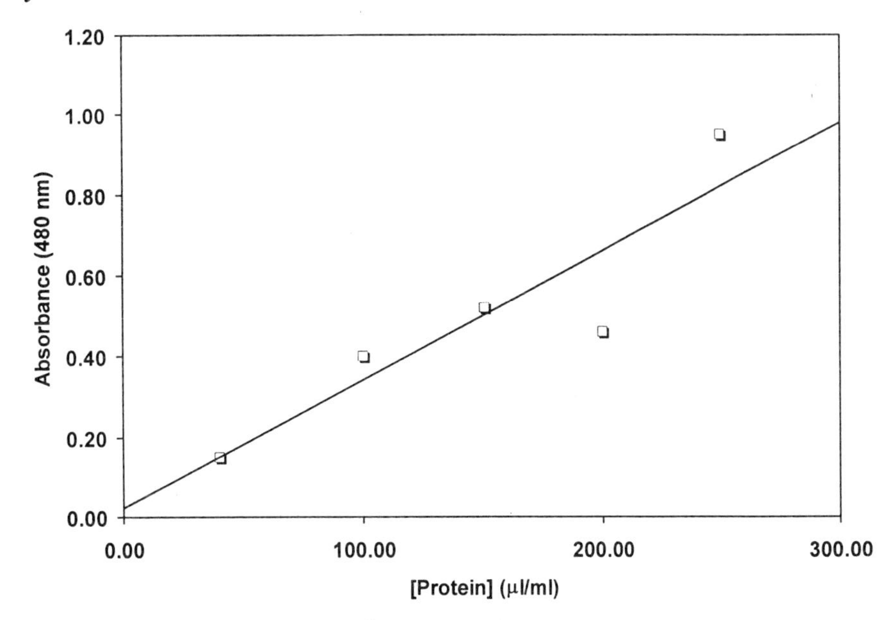

2. Outlying data point deleted. $y = 4.1139 \times 10^{-3} + 0.0041$ $r^2 = 0.998$

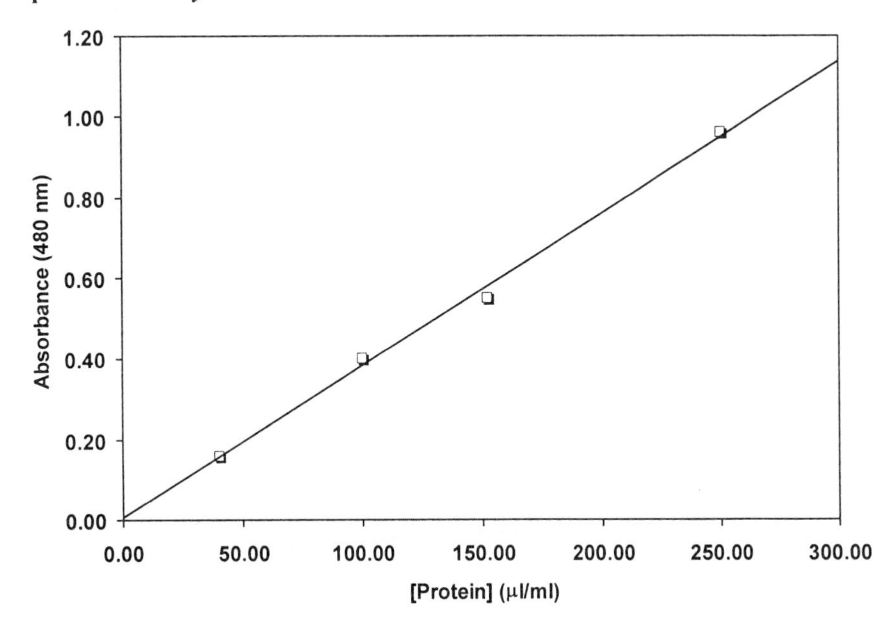

Innovation/Insight

Measurement of Protein in 20 Seconds

Akins, R. E. and Tuan, R. S. (1992) Biotechniques 12: 496-499. (Adapted)

Assay kits available from Pierce Chemical™ (Rockford, IL) take advantage of a sensitive and highly specific interaction between cuprous (Cu^{1+}) ions and the sodium salt of bicinchonic acid (BCA) for the determination of protein concentrations in solution.[5] At alkaline pH, proteins will reduce Cu^{2+} to Cu^{1+}, and in the presence of BCA, this reaction forms a product that absorbs strongly at 562 nm. Since the production of Cu^{1+} in the BCA assay is a function of protein concentration and incubation time, the protein content of unknown samples may be determined spectrophotometrically by comparison to known protein standards. The ease with which an accurate BCA assay may be executed has made it preferable to and of greater general utility than the Lowry *et al.*[3] or Bradford[2] techniques, especially because of the low levels of interference caused by reagents commonly used in protein preparations, such as detergents, certain buffers and salts, and some reducing agents;[4,5] a particular concern in protein assays involves instances where interfering compounds accumulate or partition differentially among samples.

By two tailed Student's *t*-tests, sodium dodecyl sulfate (SDS) (1%), Triton X-100 (1%), HEPES buffer (0.1 M, pH 7.2) and dithiothreitol (DTT, 1 mM) did not interfere with the assay; however, Tris-HCl buffer (0.1 M, pH 8 or 11), glycine (1 M, pH 11) and β-mercaptoethanol (β–ME, 1%) all substantially affected the region. We found that for solutions containing Tris-HCl or glycine, linear graphs of BSA concentration vs. A_{562} could be obtained by including these compounds in the standard and all samples. On the other hand, β-ME interfered to such a large extent that the assay was not usable in its presence. Any compound added to the microwave BCA assay can be easily checked for interference in a similar manner to assess effects on both sensitivity and accuracy (See Figure 1).

References:

1. Boon, M. E. and L. P. Kok. 1989. *Microwave Cookbook: The Art of Microscopic Visualization.* Coulomb Press, Leyden, Leiden.
2. Bradford, M. M. 1976. A rapid and sensitive method for the quantitation of microgram quantities of protein utilizing the principle of protein-dye binding. Anal. Biochem. 72: 248-254.
3. Lowry, O. H., N. J. Rosebrough, A. L. Farr and R. J. Randall. 1951. Protein measurement with the folin phenol reagent. J. Biol. Chem. 193: 265-275.
4. Pierce BCA Protein Assay Reagent Protocol and Information Manual 23220/23225. Pierce Chemical Co., Rockford, IL.
5. Smith, P. K., R. I. Krohn, G. T. Hermanson, A. K. Mallia, F. H. Gartner, M. D. Provenzano, E. K. Fujimoto, N. M. Goeke, B. J. Olson and D. C. Klenk. 1985. Measurement of protein using bicinchonic acid. Anal. Biochem. 150: 76-8.

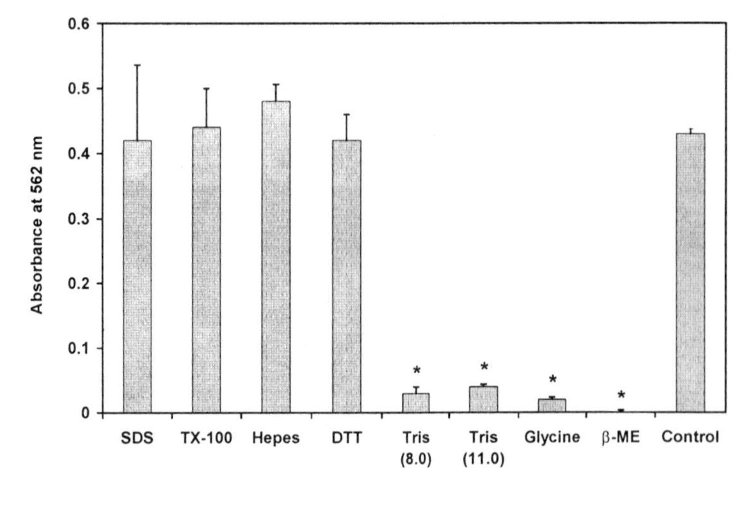

Figure 1 - Effect of common reagents on microwave BCA assay. Stock preparation of BSA was Triton-X 100 (TX-100), HEPES-NaOH, DTT, Tris-HCl (pH 8 and 11), glycine and β-ME at the concentrations indicated in the text. Irradiation was carried out in a General Electric microwave oven. Data presented are mean + or − SD (*n* = 7). Student's t-test was carried out to compare treated samples to control (BSA in water only), and statistically significant differences are indicated with asterisks.

PART 1

Nucleic Acids and Cloning

INTRODUCTION

This set of experiments is organized as a set of interconnected processes. The initial exercises in Part 1 (Lab 1.1) involve growing *E. coli* cells and subsequent isolation of plasmid DNA containing the *myo*-3 gene fragment from the nematode *C. elegans*. During the same lab, in a parallel procedure, the plasmid DNA that will receive the *myo*-3 fragment (via ligation) and then be used for protein expression is isolated. The plasmid strands are cleaved with a sequence-specific restriction endonuclease (*Hin*dIII), then the desired fragment is resolved from the remaining DNA by electrophoresis on a polyacrylamide gel (Lab 1.2). The amount of DNA fragment is quantitated, then to obtain pure DNA fragment the band is excised from the gel (Lab 2.1). In Lab 2.2, the *myo*-3 fragment is ligated into the expression plasmid, which is equipped with sequences that direct transcription of the appropriate messenger (m)RNA. This mRNA is capable of being bound by the ribosome and decoded to make *myo*-3 linked to the protein β-galactosidase, which comes connected to the fusion protein expression vector. In Lab 3.1, a DNA fragment is produced in large quantities from the expression plasmid by amplification using the polymerase chain reaction. Gel electrophoresis allows verification of the correct *myo*-3 insert orientation. The expression plasmid is inserted into *E. coli* cells using the transformation procedure (Lab 5.1), allowing the researcher to both propagate it and store it for future use. The ability of the *E. coli* cells to make the protein is assayed by colony immunoblotting (Lab 5.2). This protein is revisited in Part 2. Note that no lab is included in Unit 4.

In the next line of experimentation, the *myo*-3 gene-containing plasmid is used to make DNA by replication and thereby label it with the molecular tag digoxigenin (DIG), producing a DNA "probe" (Lab 6.1). In a parallel procedure (Lab 6.2), genomic DNA from *C. elegans* cells is isolated and size fractionated by gel electrophoresis. In Lab 6.3, the genomic DNA is transferred laterally out of the gel surface onto a Southern blot membrane, retaining the same DNA pattern. The DIG-labeled *myo*-3 DNA probe is then used to verify that the *C. elegans* DNA that is bound to the blot membrane contains the *myo*-3 gene. Detection involves forming a double helix between the denatured DNA probe and the unlabeled *myo*-3 sequence in unlabeled, denatured, membrane-bound genomic DNA. If no *myo*-3 DNA is in the *C. elegans* DNA, there will be no labeled band on the appropriately treated membrane. If a band appears, the DNA and its flanking sequences in authentic genomic chromatin can be isolated and deduced.

In summary, these experiments demonstrate how to isolate a DNA and place it into a plasmid that produces an mRNA and ultimately a *myo-3-β-gal* fusion protein. The product is verified using an immunochemical technique. In the Southern blot experiments, the student isolates genomic DNA and (separately) makes a DIG-labeled DNA probe using an authentic *myo*-3 DNA in a plasmid. The genomic DNA is fractionated and transferred to the blot membrane. Labeled DNA is used to probe the genomic DNA for the presence of the *myo*-3 gene.

FLOWCHART PART 1

Unit 1: Lab 1.1

1. Growth of *E. coli* bacterial cultures containing the *C. elegans* gene in a cloning vector.
2. Plasmid miniprep to isolate cloning vectors from *E. coli*.
3. Restriction endonuclease catalyzed cleavage of vector (*Hin*dIII).

Unit 1: Lab 1.2

1. Concentrate digested DNA.
2. Gel electrophoresis separation of products to verify correct digestion.

Unit 2: Lab 2.1

1. Excise, extract, and purify digested DNA from agarose gel.
2. Quantitate the yield of *myo*-3 DNA fragment by comparison to known amounts of molecular weight marker DNAs, based on relative gel band intensities.

Unit 2: Lab 2.2

1. Ligate *myo*-3 DNA insert into fusion protein expression vector.

Unit 6: Lab 6.1

1. Nucleotide precursors to label *myo*-3 DIG probe.
2. Quantitation of DIG-labeled

Unit 6: Lab 6.2

1. Isolate *C. Elegans* genomic DNA.
2. Quantitate DNA yield spectroscopically at 260 nm.

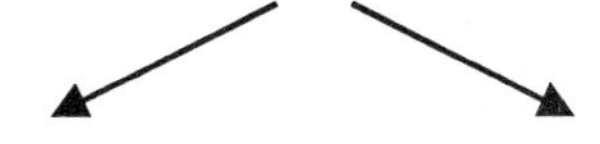

Unit 3: Lab 3.1

1. PCR amplification of expression vector to verify correct insert orientation.
2. Gel electrophoresis to separate and verify successful ligation products.

Unit 5: Lab 5.1

1. Culture and transform competent *E. coli* cells.

Unit 5: Lab 5.2

1. Induce fusion protein translation.
2. Colony immunoblot expressed recombinant protein to test for correct insert orientation and translational reading frame.

Unit 6: Lab 6.3

1. Southern blot of *myo*-3 gene excised from *C. elegans* genome.

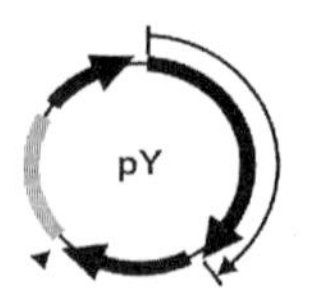

UNIT 1: DNA ISOLATION

1. Media Preparation and Growing Bacterial Cultures

A *culture medium* is an aqueous solution containing the appropriate nutrients necessary to sustain bacterial growth. Liquid media or solid media containing agar may be prepared. Culture media may be either chemically defined or undefined (complex). Chemically defined media contain exact amounts of specific chemicals. Complex media usually contain crude, complex mixtures of nutrients, such as enzymatic digests of animal tissues, yeast cells, or protein. In general, preparation of complex media is much easier and very good growth of bacterial cultures is obtained using them. For most of the techniques used in molecular biology, *E. coli* cultures are grown in complex media, since precise knowledge of the exact chemical make-up is not required.

Since contaminating bacteria are always present – on glassware, in media nutrients, and on human skin and other surfaces – culture media must be sterilized soon after preparation. This is normally done using a large pressurized device called an autoclave. In order to ensure sterility, liquids must be maintained at a temperature of 121°C for 15 to 20 minutes.

After the media are sterilized, care must be taken to prevent contamination that may occur during handling and inoculation of bacterial cultures. The method of handling culture media and pure bacterial cultures in order to prevent contamination is called *aseptic technique.*

2. Plasmid Miniprep

E. coli strains containing plasmids are grown in rich, complex media such as Luria-Bertani (LB) medium, containing an appropriate antibiotic. Cells from an individual colony growing on an agar plate are inoculated into the medium using aseptic technique and the culture is grown overnight at 37°C. Additional treatments (i.e., chloramphenicol treatment) that increase plasmid yield can be employed, however, most plasmids currently in use are produced in large amounts without additional treatment.

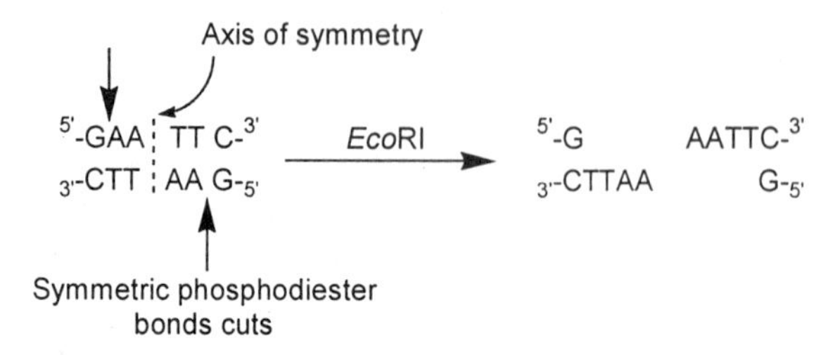

Figure 1 - *Eco*RI recognition sequence before and after enzyme action.

Many plasmid miniprep procedures (small-scale plasmid preparations) are currently in use. They all take advantage of certain physical properties of plasmids (i.e., small size, covalently closed), which allow them to be separated from the other cellular constituents. Most commonly used plasmid miniprep procedures are variations on either the alkaline extraction method or the boiling method.

3. Concentration of Nucleic Acids

Nucleic acid solutions frequently require intentional concentration increases during a purification procedure or after enzyme manipulations. This is typically done by precipitating the nucleic acid with either ethanol or isopropanol in the presence of monovalent cations, which shield repulsive anionic charges. Several different salts may be used for this purpose, e.g., sodium acetate, sodium chloride, ammonium acetate, and lithium chloride. Precipitation of nucleic acids is relatively independent of temperature and can be performed in a few minutes. Following precipitation, the nucleic acids must be centrifuged, dried, and dissolved in a small volume of buffer or water.

4. Restriction Endonucleases

One of the most important advances in the development of modern molecular biology techniques was the discovery of Type II restriction endonucleases. These enzymes recognize 4- to 8-bp duplex DNA sequences with two-fold symmetry axes (Figure 1). After binding to their specific recognition sequence, they cleave each strand in the duplex DNA. Usually, the cleavage sites are positioned such that the enzyme generates ends with single-stranded overhangs; however, some enzymes produce blunt-ended fragments.

Restriction enzymes protect prokaryotic cells from foreign DNA by cutting or *restricting* at specific sites. Cellular genomic DNA is protected by companion enzymes (*methylases*) that methylate adenine or cytosine residues within specific sites. A restriction enzyme and its corresponding methylase recognize and bind to the same specific recognition site. Nucleic acids research has been revolutionized by the discovery of these enzymes. The overhanging strand of duplex DNA produced by a particular enzyme can hybridize to (form a self-complementary helix with) and thereby join to the sequence in the receded space produced by the same enzyme (e.g., *Hin*dIII) on the other strand. These hybridized DNA fragments can be covalently joined by formation of phosphodiester bonds catalyzed by DNA ligase.

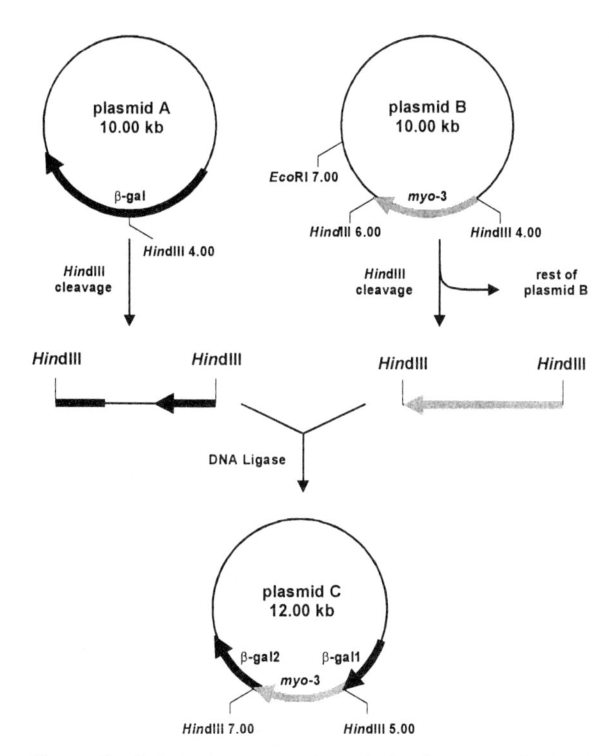

Figure 2 - Subcloning a *myo*-3 restriction fragment obtained from plasmid B into the *Hind*III site created by cleavage at the site containing the β-gal gene in plasmid A.

Since all *Hind*III cohesive ends are equivalent (Figure 2), the small fragment from plasmid B can ligate into plasmid A in two different orientations. The orientation of an insert can be directed by cutting with two different restriction enzymes, which produce noncomplementary cohesive ends (Figure 3).

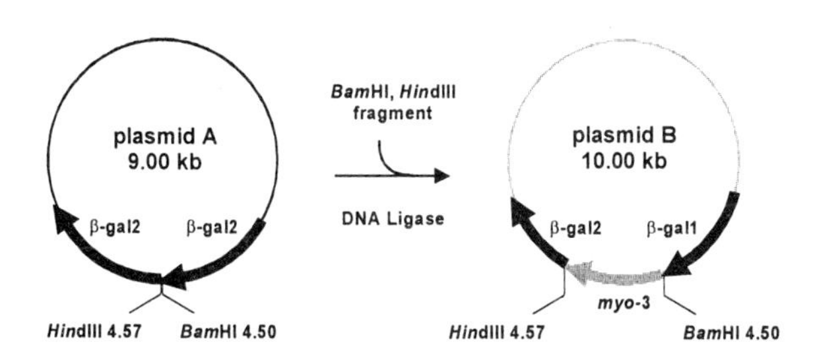

Figure 3 - "Directionally cloning" a *Bam*HI, *Hind*III *myo*-3 fragment in the single allowed orientation in plasmid B.

5. Electrophoresis (Hoeffer)

In electrophoresis, charged molecules in solution, chiefly proteins and nucleic acids, migrate in response to an electrical field. Their *rate of migration*, their *mobility*, through the electrical field depends upon the strength of the field, the net charge, size and shape of the molecules, the ionic strength, viscosity, and temperature of the medium in which the molecules are moving. As an analytical tool, electrophoresis is simple, rapid, and highly sensitive. It is generally used to determine the size of a single charged species; it is used less often as a preparative separation technique.

The method is usually carried out in gels formed in tubes, slabs, or on a flat bed. In a *tube gel* unit, the gel is formed in a glass tube, usually about 12 cm long with 3 to 5 mm inner diameter. A *slab gel* is formed in a glass sandwich made by clamping together two flat glass plates separated by two spacer strips at the edges to make a water-tight seal. Both tube and slab gels are mounted vertically. The gel in a flat bed unit is poured on a horizontal surface and has no cover plate in it. In most electrophoresis units, the gel is mounted between two buffer chambers containing separate electrodes so that the only electrical connection between the two chambers is through the gel.

5.1 Interrelation between Resistance, Voltage, Current, and Power. Two basic electrical equations are important in electrophoresis. The first is *Ohm's Law*:

$$E = IR \tag{1}$$

which states that *electrical current* (I, in amperes) is directly proportional to *voltage* (E, in volts) and inversely proportional to *resistance* (R, in ohms). The second is:

$$P = EI \tag{2}$$

which states that *power* (P, watts), a measure of the amount of heat produced, is the product of voltage and current. This can also be expressed as $P = I^2R$.

In electrophoresis, one electrical parameter, either current, voltage, or power, is held constant. According to the equations, if resistance increases during the run, the consequences will differ in the following ways, depending on which parameter is held constant.

1. Under constant current conditions (molecular mobility is directly proportional to current), the velocity is maintained but heat is generated.
2. In constant voltage, the velocity slows, but no additional heat is generated during the course of the run.
3. In constant power, the velocity slows but heating is constant.

The important consideration, regardless of which mode is selected, is that losses in sample activity due to heating be minimized.

5.2 The Net Charge of Proteins Is Determined by the pH of the Medium. Proteins are *amphoteric* compounds, i.e., they contain both acidic and basic residues. Their net charge is determined by the pH of the medium. Most of the charge of a protein comes from the pH-dependent ionization of carboxyl and amino groups.

$$(X\text{-}COOH \rightarrow X\text{-}COO^- + H^+ ; Y\text{-}NH_2 + H^+ \rightarrow Y\text{-}NH_3^+)$$

Each protein has its own characteristic charge properties, depending on the number and kinds of amino acids carrying amino or carboxyl groups. Unlike proteins, normal duplex nucleic acids are not amphoteric. They remain negative in the pH range normally used for electrophoresis.

The pH at which a protein carries no net charge is called the *isoelectric point* or p*I*. Each protein has its characteristic p*I*. For example, the p*I* of human hemoglobin is 7.07 and that of β-lactoglobulin is 5.34. In a solution with a pH above a protein's isoelectric point, it has a net negative charge and migrates toward the positive electrode (anode) of the electrical field. Below its isoelectric point, the protein is positive and migrates toward the negative electrode (cathode). The pH of a solution in an electrophoresis system must be kept constant to maintain the charge, and, hence, the mobilities of the proteins. For this reason, solutions used in electrophoresis must be buffered.

5.3 Maintaining Constant Temperature. Temperature regulation is important at every stage of the electrophoretic process. Polymerization of gels is an exothermic reaction. If not regulated, convection currents can produce irregularities in the pore size of the gel. To keep pore size uniform and reproducible, the gelation reaction should be carried out at a constant temperature. For this reason, gels should be surrounded by a liquid coolant that is held at a constant temperature during polymerization.

During electrophoresis, constant temperature is essential to avoid denaturing heat-labile proteins. When running slab gels, heat is not conducted away from the slab. Even in an SDS gel, where denaturation is not a consideration, the samples will move faster in the center of the gel than at the edges. This produces a "smile" effect and makes the comparison of band mobilities difficult. When temperature is held constant across the gel, separated samples across a series of parallel lanes will maintain side-to-side flatness, moving downward evenly.

Most buffer systems have been developed at either 0 or 25°C. To maintain an acceptable temperature throughout the run, the lower buffer should cover the part of the tube or slab that contains the gel. For precise regulation, the electrophoresis apparatus must include a system for cooling and circulating buffer in the tank. The cooled buffer will then circulate around the gels and dissipate whatever heat is generated during the run. It is important to use electrophoresis tanks with lower buffer chambers that are large enough to immerse the gel tubes or sandwiches in temperature-regulated buffer.

5.4 Support Matrix. Samples are generally run in a support matrix, such as paper, cellulose acetate, starch gel, agarose, or polyacrylamide gel. The matrix inhibits convections caused by heating. At the end of the run, the matrix can be stained and used for densitometric scanning, autoradiography, or storage. It provides a hard record of the electrophoretic run. The most commonly used support matrices, *agarose* and *polyacrylamide*, provide the operator with an added separating technique. Although agarose and polyacrylamide differ greatly in their physical and chemical structures, they both make porous gels. A porous gel acts as a sieve by retarding or completely obstructing the movement of larger macromolecules while allowing smaller molecules to migrate more freely. Small molecules can also be restricted as a result of the time delay required to undergo one or a series of incursions into the matrix particles themselves. In this case, larger molecules elute faster. These comments apply to nonelectrophoretic separations, e.g., filtration gel chromatography. In contrast, small molecules migrate faster than larger ones in most electrophoretic techniques.

By preparing a gel with different restrictive pore sizes, the operator can take advantage of molecular size differences among proteins. For example, hemoglobin weighs 66 kDa and β-lactoglobulin weighs 38 kDa. It is possible to prepare a gel with a pore size that will be more restrictive to hemoglobin than to β-lactoglobin. At pHs above the isoelectric points of both proteins, where both are anionic, the two will separate not only by charge but by molecular size. Because the pores of an agarose gel are relatively large, agarose is used to separate macromolecules such as high-molecular-weight nucleic acids, large proteins, and protein and nucleic complexes. Polyacrylamide has smaller pores and is used to separate most proteins and small oligonucleotides.

It is important that the matrix be electrically neutral to prevent the flow of solvent toward the electrode (usually the cathode) due to the presence of immobile charged groups in the matrix. This phenomenon, electroosmosis, decreases the resolution of the sample.

5.5 Agarose Gels. Agarose is a highly purified polysaccharide derived from agar. Unlike agar, it is not heavily contaminated with charged material. Most preparations contain some anions, such as pyruvate and sulfate, which may cause some electroosmosis. The matrix comes in powder form and dissolves when added to boiling liquid. It remains in a liquid state until the temperature is lowered to about 40°C, at which point it gels. The gel

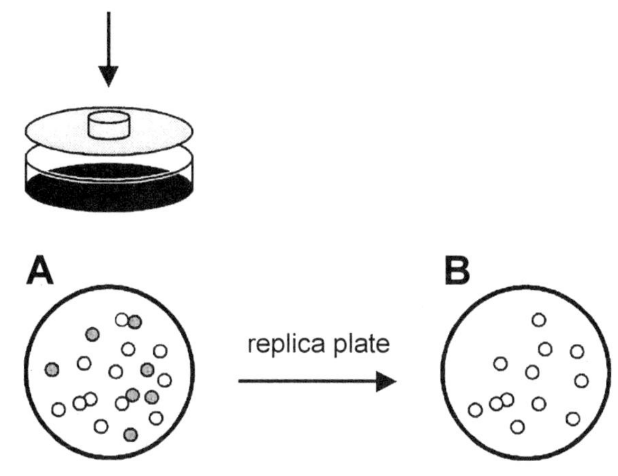

Figure 4 - Replica plating. The filter is stamped on plate A, lifting all of the colonies from its surface. This pattern is transferred to plate B to produce a replica of plate A. Colonies that have not grown have not survived the selection inhibitor (gray): survivors contain the plasmid (white).

is stable; it will not dissolve again until the temperature is raised back to about 100°C. Its pore size may be predetermined by adjusting the concentration of agarose in the gel: the higher the concentration, the smaller the pore size. Working concentrations are frequently in the range of 0.4 to 2% (w/v).

Agarose gels are fragile; they are actually hydrocolloids, and are held together by formation of weak hydrogen and hydrophobic bonds. For this reason, an agarose gel should be handled with special care. If the gel bends, it will break. It should not be picked up without the use of a tray or a special tool.

5.6 Polyacrylamide Gels. Polyacrylamide gels are tougher than agarose gels. In forming the gel, acrylamide monomers polymerize into long chains that are covalently linked by a crosslinker. It is the crosslinker that actually holds the structure together. The most common crosslinker is *N,N'*-methylene-bis-acrylamide, or "bis" for short. Other crosslinkers whose special properties aid in solubilizing polyacrylamide are also used.

Because *oxygen inhibits polymerization*, the monomer mixture must be deaerated. This may be done by purging the mixture with an inert gas or by evacuating the mixture with a vacuum. When the gel is poured, the top of the gel forms a meniscus. If ignored, this curved top on the gel will cause distortion of the banding pattern. To eliminate the meniscus, a thin layer of water or water-saturated *n*-butanol is carefully floated on the surface of the gel mixture before it polymerizes. After polymerization, the water or butanol layer is poured off, leaving a flat upper surface on the gel. The layer of water or water-saturated butanol not only eliminates the meniscus but excludes oxygen – which would inhibit polymerization on the gel surface.

5.7 Determining the Pore Size. The pore size in a polyacrylamide gel may be predetermined in either of two ways. One way is to adjust the total percentage of acrylamide, i.e., the sum of the weights of the acrylamide monomer and the crosslinker. For example, a 20% gel would contain 20% w/v acrylamide plus 5% bis. As the percentage of acrylamide increases, the pore size decreases. The other way to adjust pore size is to vary the amount of crosslinker expressed as a percent of the sum of monomer and crosslinker or %C. Thus, a 20% acrylamide/5% bis gel would have 20% w/v acrylamide plus bis, and the bis would account for 5% of the total weight of the acrylamide. It has been found that at any single percentage of acrylamide, 5% crosslinking creates the smallest pores in a gel. Above and below 5%, the pore size increases.

When there is a wide range in the molecular weights of the material under study, the researcher may prepare a *pore gradient gel*. The pore size in a gradient gel is larger at the top of the gel than at the bottom; the gel becomes more restrictive as the run progresses.

5.8 Polymerization. Polymerization of a polyacrylamide gel is accomplished either by a chemical or photochemical method. In the most common chemical method, *ammonium persulfate* and the quaternary amine, *N,N,N',N'*-tetramethyl-ethylenediamine or *TEMED*, are used as the initiator and catalyst, respectively. In photochemical crosslinking, riboflavin and TEMED are used. The photochemical reaction is started by shining long wavelength ultraviolet light, usually from a fluorescent light, on the gel mixture (e.g., between facing fluorescent lamps).

Since only a minute amount of riboflavin is required, photochemical polymerization is used when a low ionic strength is to be maintained in the gel. It is also used if the protein studied is sensitive to ammonium persulfate or the by-products of chemical polymerization. Outside of these special conditions, however, the most popular electrophoretic systems use chemical polymerization.

Polymerization generates heat. If too much heat is generated, convection currents will form in the unpolymerized gel, resulting in inconsistencies in the gel structure. To prevent excessive heating, the concentration of initiator-catalyst chemicals should be adjusted to complete polymerization in 20 to 60 minutes.

6. Analysis of the Gel

After the electrophoresis run is over, the gel is usually analyzed by one or more of the following procedures:

1. Staining or autoradiography followed by densitometry.

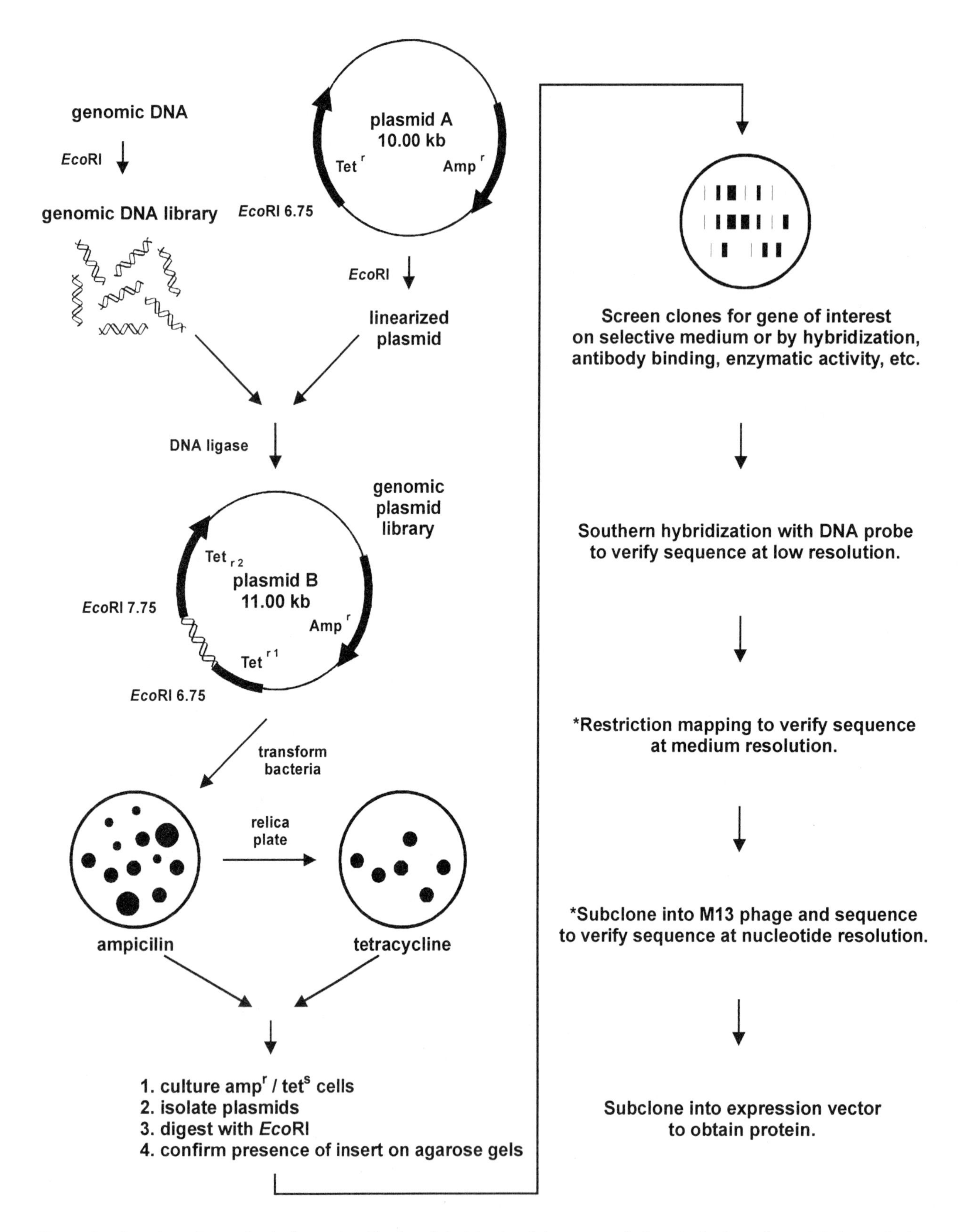
genomic DNA
EcoRI
genomic DNA library
EcoRI 6.75
plasmid A
10.00 kb
Tet r
Amp r
EcoRI
linearized
plasmid
DNA ligase
genomic
plasmid
library
Tet r2
plasmid B
11.00 kb
EcoRI 7.75
Amp r
Tet r1
EcoRI 6.75
transform
bacteria
relica
plate
ampicilin
tetracycline
1. culture amp r / tet s cells
2. isolate plasmids
3. digest with EcoRI
4. confirm presence of insert on agarose gels
Screen clones for gene of interest
on selective medium or by hybridization,
antibody binding, enzymatic activity, etc.
Southern hybridization with DNA probe
to verify sequence at low resolution.
*Restriction mapping to verify sequence
at medium resolution.
*Subclone into M13 phage and sequence
to verify sequence at nucleotide resolution.
Subclone into expression vector
to obtain protein.

Figure 7 – Overview of steps in cloning and verification. Asterisks denote processes that are typically performed during a cloning protocol to characterize the product DNA sequence. We do not describe or include the protocols for these techniques in this manual. Such characterization is commonly performed in current laboratory practice in collaboration with a commercial or university sequence facility.

2. Blotting to a membrane, either by capillarity or by electrophoresis, followed by one of the following:
   a. Nucleic acid hybridization.
   b. Direct autoradiography.
   c. Immunodetection.

The most common postelectrophoresis procedure is chemical staining. Protein gels are most frequently stained with *Coomassie Blue.* Photographic amplification systems containing *silver* or other first row transition metals are also in common use. Coomassie Blue staining is only sensitive to about 100 ng of protein, whereas the photographic amplification systems are sensitive to about 10 ng of protein. Once the gel is stained, it can be photographed or scanned by densitometry to produce a record of the position and intensity of each band.

Nucleic acid bands are often stained with *ethidium bromide*, a fluorescent dye that fluoresces pinkish-orange when bound to nucleic acids and excited by UV light. About 10-50 ng of DNA can be detected with ethidium bromide. Gels are usually photographed to retain a record of the run and avoid unnecessary exposure to the potentially carcinogenic UV radiation.

A second common analytical procedure is *autoradiography*. It is used to detect radioactive samples separated on a slab gel. This procedure frequently requires that the gel be first dried to the consistency of a sheet of paper then placed in contact with X-ray film. The film will be exposed only in the locations of radioactive bands or spots. The resulting autoradiogram is usually photographed or scanned by *densitometry*.

The highly sensitive technique, blotting, is used to transfer proteins or nucleic acids from a slab gel to a membrane such as nitrocellulose, nylon, DEAE, or CM paper. The transfer of the sample can be done by capillarity (Southern blotting) for nucleic acids, or by electrophoresis for proteins or nucleic acids. Southern blotting draws the buffer and sample, usually DNA, out of an agarose gel by placing the slab in contact with blotter paper. A nitrocellulose or nylon membrane, layered between the gel and the blotter paper, binds the nucleic acids as they flow out of the gel. Since the membrane binds the DNA or RNA in the same pattern as on the original gel, it comes out as a faithful copy of the original. A disadvantage of Southern blotting is that it usually takes a long time to prepare, frequently 10-20 hours.

Electrophoretic transfers are much faster than Southern blots, taking from one-half to 2 hours to perform. The gel, containing protein or nucleic acids, is placed next to a membrane in a gel-holding cassette and then placed in a tank filled with buffer. An electric field is applied perpendicular to the plane of the gel cassette and the sample *electroelutes* out of the gel and onto the membrane. The result is a copy of the original gel.

Another nucleic acid transfer technique that is faster than traditional Southern blots is *vacuum blotting*. In this technique, a nylon or nitrocellulose membrane is placed on a porous support covering a vacuum chamber. The gel is placed on the membrane. By applying a slight vacuum to the chamber, the DNA or RNA in the gel moves out of the gel and onto the membrane. The time required for a vacuum transfer is usually about 1 hour.

In all of these methods the sample moves from the gel matrix to the surface of the membrane. Once the sample is bound to the membrane, it is detected by one of several methods. If the sample is radioactive, the membrane can be subjected to autoradiography. If it is not, radioactive probes, such as complementary DNA or specific antibodies, can be bound to either specific nucleic acids sequences or protein antigens. To detect the positions of the probes, the membrane is subjected to autoradiography. Very sensitive techniques have now been developed, which involve detection of fluorescence, eliminating the use of radioactive reagents.

7. Plasmids and Molecular Cloning

Plasmids are small, circular, extrachromosomal duplex DNAs typically found in bacterial cells. They can also be made compatible with both prokaryotic and eukaryotic systems if proper sequences are present. They are important because they can self-replicate within the cell and thus propagate a copy of their genetic information. Plasmids contain genes for nonessential functions not encoded by the chromosome, such as *antibiotic resistance*, *bacteriocin production,* and *sexual fertility*. Many of these plasmid-borne genes can be used as *growth-selectable genetic markers*. Selective media only allow growth of cells that contain a particular gene, carried on a plasmid. For example, the ampicillin resistance gene is carried by plasmids such as pBR322. This gene encodes a β-lactamase enzyme that degrades the antibiotic ampicillin. Thus, if ampicillin is included in the culture medium, only cells containing the plasmid will grow. Alternatively, a particular *phenotype* may be selected by using a medium containing a chemical that changes color when protein is produced from a particular plasmid-borne gene. Growth-selectable markers are essential characteristics of plasmids that are used as *cloning vectors*.

Cloning of a gene requires a vector (usually a plasmid or bacteriophage) that can replicate within the host cell. Growth-selectable markers are required so that cells containing the vector may be selected or identified. Another desirable quality of cloning vectors is that they contain several unique restriction

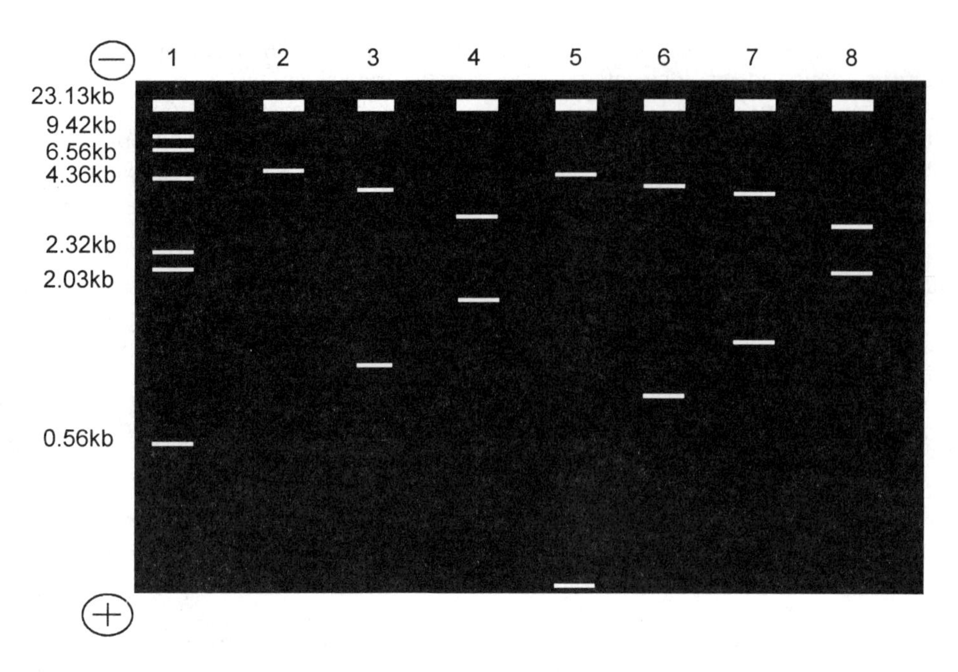

Figure 5 - Restriction analysis on an agarose gel. Lane 1, lambda DNA *Hin*dIII digest; Lane 2, pCH2 following *Eco*RI treatment; Lane 3, pCH2 after *Eco*RI/*Kpn* I; Lane 4, pCH2 after *Eco*RI/*Bam*HI; Lane 5, pCH2 after *Eco*RI/*Hin*dIII; Lane 6, pCH2 after *Kpn*I/*Bam*HI; Lane 7, pCH2 after *Kpn*I/*Hin*dIII; Lane 8, pCH2 after *Bam* HI/*Hin*dIII. The fragments migrate from top (-) to bottom (+).

enzyme recognition sites, usually in a cluster. This facilitates insertion of a fragment and subsequent covalent linkage to the cloning vector. When the gene or DNA fragment is inserted into the vector, it can be *transformed* into *competent* cells. It is important to determine if the transformed cells contain a vector with an inserted DNA or an "empty" vector. The most common method for accomplishing this is called *insertional inactivation.* Plasmids such as pBR322 or the plasmids shown in Figure 2 contain an antibiotic resistance gene, which can be used as a selectable marker, and another gene (often an antibiotic resistance gene), which can be insertionally inactivated. If ampicillin resistance is being used as the selectable marker, the DNA fragment to be cloned can be inserted into a unique restriction enzyme site, or pair of sites, located within the coding region of the tetracycline resistance gene. Cells that contain only the unmodified vector will be resistant to both ampicillin and tetracycline. In contrast, cells that contain plasmids with inserted DNA fragments will be *resistant to ampicillin* (the gene still works) and *susceptible to tetracycline* (the resultant tetracycline resistance protein is rendered useless as a result of the nonsense sequence encoded by the DNA insert).

In the example previously described, the process of *replica plating* must be performed in order to determine tetracycline sensitivity (Figure 4). Carefully isolated colonies, grown on ampicillin plates, are transferred using sterile filters to tetracycline- and ampicillin-containing plates, making exact replicas of the original plates. This process yields colonies that grow on ampicillin and not on tetracycline. Alternatively, colonies can be picked from plates with sterile toothpicks and scored onto marked plates that contain different selective antibiotics. When insert-bearing plasmids are identified, screening methods to identify specific genes are necessary. Specific screens typically involve enzymatic plate assays, hybridizations, or specific antibody reactions.

Isolated genes may be further characterized by *restriction mapping*. This technique allows one to determine the order of specific restriction enzyme recognition sites along a chromosome or DNA fragment. This is done by using single digestions, simultaneous digestions with two or more restriction enzymes, or sequential digestions with multiple enzymes. One determines the exact ordering of the

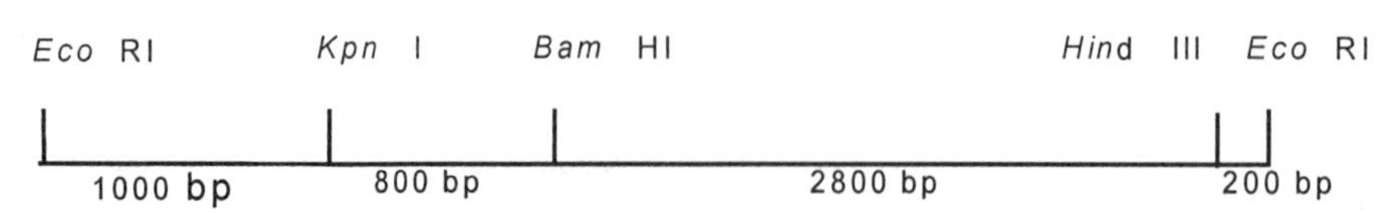

Figure 6 - Order of restriction sites in the DNA fragment whose fragmentation patterns are shown in Figure 5.

sites. The sizes of the restriction fragments are determined by agarose gel electrophoresis. This is done by comparing the mobility of the digested DNA product with those of *DNA standard fragments* of known molecular weights. Size standards are prepared by restriction digestion of well-characterized (completely sequenced) plasmids or viruses with an enzyme or enzymes that create a set of known restriction fragments with known lengths. It is important to understand that the ordering of a set of restriction fragments by size, as seen on a gel, has no relationship to their positions or orientations within the initial uncut fragment. Fragments are connected at the same sticky ends; fragment lengths are determined by the distance between sites that produce the two ends, or between a subterminal site and the terminus.

The order of the restriction sites, deduced from the patterns in Figure 5, is shown in Figure 6. An overview of the procedures involved in cloning DNA in a plasmid then, identifying and analyzing such clones is outlined in Figure 7.

8. RFLP Mapping

A restriction mapping technique can be used with the DNA of higher eukaryotes to diagnose genetic diseases (Figure 8). This method, *RFLP mapping,* is used to detect differences in restriction fragment lengthspolymorphisms within a defined genetic locus. These polymorphisms are most useful

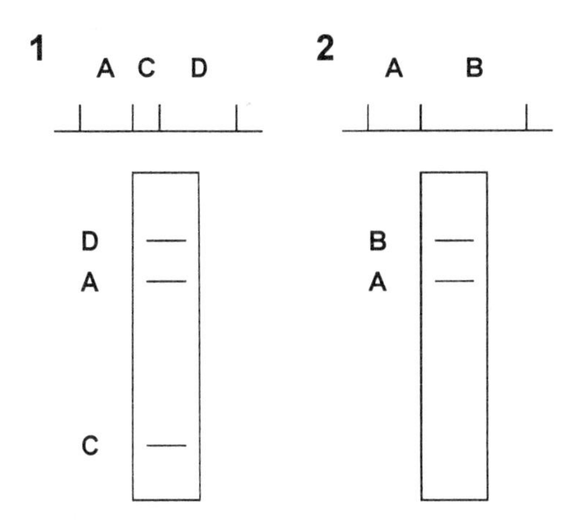

Figure 8 - RFLP analysis. DNA fragment 1 contains 4 target sites. A mutation in DNA 2 (designated as x) has eliminated the third target site found in DNA 1 (between fragments C and D). This mutation can be detected as a change in restriction fragments following the same digestion steps.

when associated with mutant *genotypes* that lead to genetic diseases, such as cystic fibrosis and Huntington's chorea. If a mutation causes a change in the target site for a particular restriction enzyme, the pattern of restriction fragments obtained (detected by specific hybridization probes) will be different. This technique has also become instrumental in law enforcement.

THEORY/PRINCIPLES
SUBCLONING PROCEDURE

"Subcloning" is a somewhat misleading term for the process of removing a DNA fragment from one plasmid (the donor, top right) then inserting it into a desired location within the sequence of another plasmid (the recipient, top left). The project described in this manual involves removing *myo*-3 gene DNA from the donor plasmid and, in parallel, "linearizing" the recipient plasmid to prepare it for the *myo*-3 fragment insertion procedure. To complete the overall process, the ends of the *myo*-3 fragment must be ligated to the ends of the recipient expression plasmid. They must be in the proper orientation so the leading portion of the *myo*-3 gene is adjacent to ("downstream from") the "start sequences" and aligned properly to support the intended functions of the control sequences and the decoded codons. If all is well, the product plasmid is capable of producing viable mRNA, which can be used in protein synthesis to generate the desired protein.

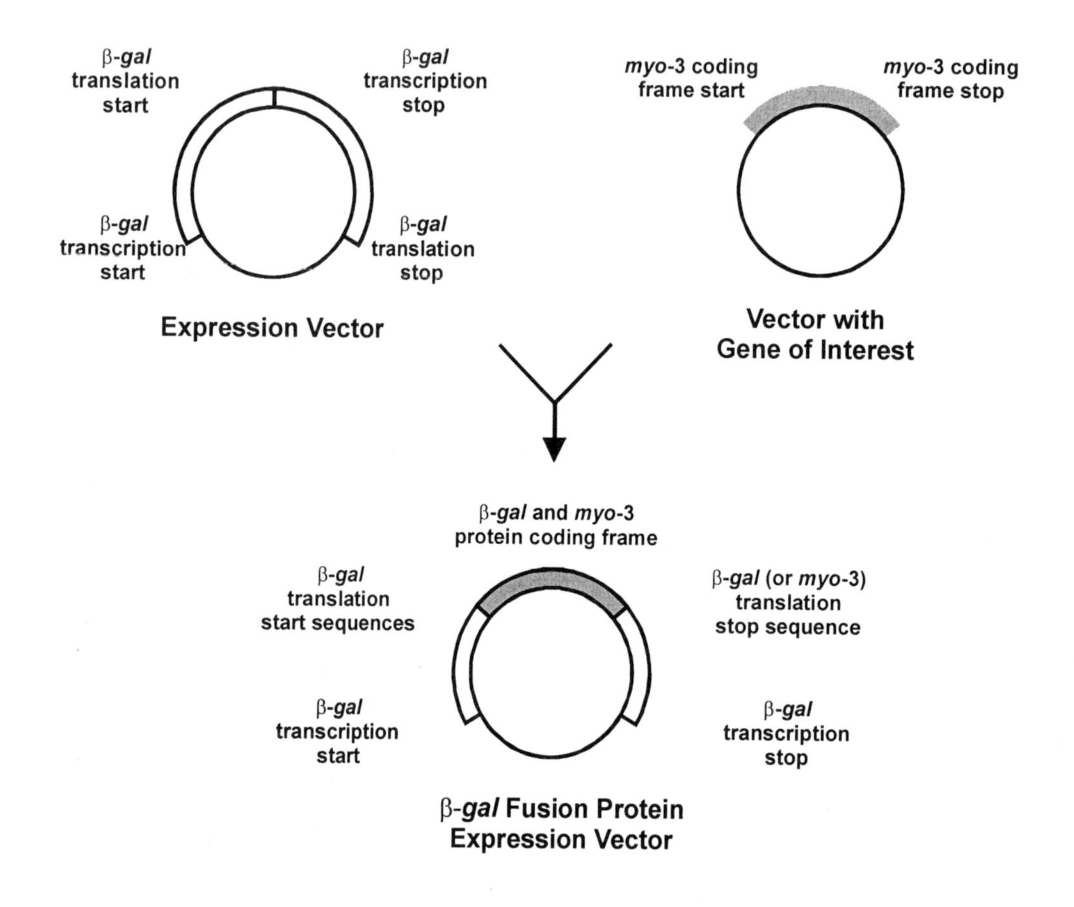

INNOVATION/INSIGHT
THE pET BACTERIAL PLASMID SYSTEM (NOVAGEN)

1. Introduction

The pET System is the most powerful system yet developed for the cloning and expression of recombinant proteins in *E. coli.* Target genes are cloned in pET plasmids under control of *strong bacteriophage T7 transcription and translation signals*; expression is induced by providing a source of T7 RNA polymerase in the host cell. T7 RNA polymerase is so selective and active that *almost all of the cell's resources are converted to target gene expression*; the desired product can comprise more than 50% of the total cell protein after a few hours of induction. Perhaps just as important as the strength of the T7 promoter is the ability of the system to maintain transcriptionally silent target genes in the uninduced state. Target genes are initially cloned using hosts that do not contain the T7 RNA polymerase gene, so they are virtually "off" and cannot cause plasmid instability due to the production of proteins potentially toxic to the host cell. Once established, plasmids are transferred into expression hosts containing a chromosomal copy of the T7 RNA polymerase gene under *lacUV5* control, and expression is induced by the addition of IPTG. Several hosts that differ in their stringency of suppressing basal expression levels are available. These controls provide the greatest amount of flexibility to optimize the expression of a wide variety of target genes.

Since introducing the original vectors, Novagen™ has developed new versions with enhanced features designed for increased versatility, ease of use, and ability to purify expressed proteins in one step at low cost. All of the pET vectors and companion products are available as kits designed for convenient cloning, expression, detection, and purification of target proteins.

Several features of the system help to ensure plasmid stability by maintaining *transcriptionally silent* target genes in the uninduced state. Tight control over basal expression levels is extremely important, since many foreign proteins are toxic when expressed in *E. coli* and thus can be difficult or impossible to clone in many other systems. The pET system has a unique advantage for the establishment of *stable recombinants*, since target genes are initially cloned using hosts that do not contain the T7 RNA polymerase gene. Background expression is minimal in the absence of T7 polymerase because the host enzyme does not initiate from T7 promoters and the cloning sites in pET plasmids are in regions weakly transcribed (if at all) by *read-through activity* of bacterial RNA polymerase. Strains recommended for this purpose are NovaBlue or HMS174, which are *recA*⁻ and produce good yields of plasmid monomer.

It should be noted that several popular commercial vectors carry *T7 promoters* and in principle could be used with the pET expression hosts. However, vectors that carry the *lac* promoter without an additional source of *lac repressor* are inappropriate because multiple copies of the *operator will titrate repressor* and partially induce the gene for T7 RNA polymerase, which is also controlled by *lac* repressor. As a result, basal polymerase activity becomes high enough that many target genes cannot be stably maintained.

For protein production, the recombinant plasmid is transferred to a host containing a chromosomal copy of the T7 RNA polymerase gene. The *hosts are lysogens* of bacteriophage λDE3, which contains the polymerase gene under the control of the inducible *lacUV5* promoter. *Addition of IPTG to a growing culture induces the polymerase, which in turn transcribes the target DNA in the plasmid.* Either of two lysogenic strains, BL21 (DE3) and HMS174 (DE3), may be used. BL21 (DE3) has the advantage that it lacks the *ompT outer membrane protease* that can degrade some proteins during purification. As an *E. coli* B strain, BL21 also lacks the *lon* protease. However, some target genes are more stable in HMS174 (DE3), possibly because this strain is *recA*⁻ and therefore cannot support the induction of DE3 prophage by certain proteins. BL21 expression strains are supplied with the pET Systems and the HMS174 equivalents are available separately.

Even in the absence of IPTG there is some expression of T7 RNA polymerase from the *lacUV5* promoter. If target gene products are sufficiently toxic to *E. coli*, this basal level can be enough to prevent the establishment of plasmids in BL21 (DE3) or HMS174 (DE3). There are several ways to reduce basal expression with the pET System.

The first approach is to use vectors that contain what is termed a T7*lac* promoter [(i.e., pET-11 series, pET-15b, pET-16b, pET-19b, pET-21 series, pET-22b(+), pET-24 series, pET-25b(+)]. These plasmids contain a *lac* operator sequence just downstream from the T7 promoter.

2. Advantages

2.1 Versatile.

1. Choice of N-terminal and C-terminal peptide tags, for protein export, immunoassay and affinity purification.
2. Expanded polylinker containing cloning sites in all three *reading frames*.

3. Choice of expression controls and host strains to optimize expression levels.
4. phage fl *origin of replication* for *mutagenesis* and *sequencing* with single-stranded templates.

2.2 Rapid.

1. *E. coli*-based system for fastest results.
2. Convenient restriction sites for subcloning from other vectors.
3. One-step purification of target proteins without antibodies.

These plasmids also carry the natural promoter and coding sequence for the lac repressor (*lacI*), oriented so that the T7*lac* and *lacI* promoters diverge. When this type of vector is used in DE3 lysogens to express target genes, the lac repressor acts both at the *lacUV5* promoter to repress transcription of the T7 RNA polymerase gene by the host polymerase and at the T7 *lac* promoter to block transcription of the target gene by any T7 RNA polymerase that is made. Only a few target genes have been encountered that are too toxic to be stable in these vectors in BL21(DE3) or HMS174 (DE3).

A second way to stabilize target genes is to express them in host strains containing a compatible plasmid (pLysS) that provides a small amount of T7 lysozyme. The *T7 lysozyme* has two functions. As a natural inhibitor of T7 RNA polymerase, it *stabilizes* target *plasmids by decreasing the basal activity of T7 RNA polymerase*, but it does not prevent induction of high levels of target proteins. Second, it *cleaves a bond in the peptidoglycan layer of the E. coli cell wall* and *allows cells to be lysed under mild conditions*, such as by *freeze-thaw* or the addition of 0.1% Triton X-100. A plasmid that provides a higher level of lysozyme (pLysE) reduces the induced expression to a point where innocuous target genes can be expressed continuously without killing the cell. The combination of a pET-3 type vector and pLysS has been successful in expressing a wide variety of proteins.

The combination of a T7*lac* promoter-containing vector and pLys5 allows target genes having even greater toxicity to be maintained and expressed. A final alternative is to introduce the T7 RNA polymerase by infection with bacteriophage CE6, a lambda recombinant that carries the cloned polymerase gene. Although the method is less convenient than induction of DE3 lysogens, it can be used if target gene products are too toxic to be maintained any other way.

A wide variety of pET vectors is available. All are *derived from pBR322* and vary in leader sequences, expression signals, relevant restriction sites, and other features. There are two major categories of pET plasmids known as *transcription vectors* and *translation vectors*. The transcription vectors (including pET-2l, pET-23, and pET-24) express target RNA but do not provide translation signals. They are useful for proteins from target genes that carry their own bacterial translation signals. Note that the transcription vectors can be identified by lack of a letter suffix after the name.

The translation vectors contain efficient translation initiation signals and are designed for protein expression. In many cases it is advisable to test several vectors having different combinations of promoters and upstream leaders, as well as different host strains, to optimize yields of the target protein.

3. Basic Considerations for Choosing a pET Vector

1. *Intended purification method*: Many of the newer pET vectors contain a His-Tag® sequence, which is a consecutive stretch of either 6 or 10 histidine residues that can be expressed at the amino or carboxy terminus of the target protein. The His-Tag® sequence allows the protein to be purified in one step by metal chelation chromatography using Novagen's His-Tag® Resin and Buffer Kit. This method is rapid and inexpensive and can be used under native or denaturing conditions.
2. *Promoter*: Vectors with the "plain" T7 promoter are suitable for cloning and expression of most target proteins, while those having the T7*lac* promoter are better in most cases when expressing toxic target genes.
3. *Upstream leader*: Most proteins appear to be expressed at equally high levels whether they have T7-Tag, His-Tag®, *ompT*, or *pelB* leaders. The His-Tag® sequence allows one-step purification of the target protein whether it is soluble or insoluble in the host cell, and can be removed if necessary with site-specific proteases. The 12 amino acids and 260 amino acid T7-Tag™ leaders allow immunoassay of any target protein using the T7-Tag Antibody. The longer 260 amino acid T7-Tag leader stabilizes the expression of small peptides and eliminates the need for carrier conjugation before immunization. The *ompT* and *pelB* leaders are designed to direct the target protein to the periplasm, which may promote proper folding and protect against proteolytic breakdown. The *HSV-Tag® sequence* is a *C-terminal epitope* that can be used with the HSV-Tag® monoclonal antibody for immunodetection of proteins expressed in the pET-25b(+) vector.

4. Cl*oning sites*: Many pET vectors contain multiple cloning sites in all three reading frames for convenient cloning. In some cases engineering of the target sequence is needed to create compatible ends, proper reading frames, or to minimize extraneous sequences. This can usually be accomplished most readily by designing appropriate PCR primers having the desired features and 10-12 extra bases at the end to allow for efficient cutting. Many of the cloning sites are compatible between the various vector types so that target genes can be easily transferred between them.

4. pET His-Tag® System

1. Simple, one-step affinity purification without antibodies (Figure 1).
2. Gentle elution conditions to maintain protein activity.
3. Can also be used under denaturing conditions (urea, guanidine, SDS).
4. Protease cleavage sites: factor Xa, thrombin, or enterokinase.
5. Same high expression and tight control as the original pET vectors.
6. Removal of His-Tag® sequence generally not necessary for biological activity.

Figure 1 - The soluble protein fraction prepared from a β-galactosidase recombinant in pET-14b was purified using Navogen's His-Bind™ Resin and Buffer Kit. The indicated fractions were analyzed by SDS-PAGE and Coomassie blue staining. Lane 1, starting material; lane 2, column flow-trough; lane 3, wash; lane 4, wash #2; and lane 5, elution.

LAB 1.1:
MEDIA PREPARATION; BACTERIAL GROWTHS; PLASMID MINIPREPS; *HIND*III DIGESTION OF DNA, COMMERCIAL BACTERIOPHAGE λ DNA *BST*EII DIGEST SIZE STANDARDS

Overview:

In this segment of the course, you will clone the DNA fragment *myo*-3. The theory behind DNA fragment cloning is that most foreign DNA fragments cannot self-replicate; therefore, they must be joined to a vector that can autonomously replicate in a cell. Typically, each vector joins a single fragment of foreign DNA. Cloning allows the purification of large amounts of vector DNA. To clone a plasmid vector, the plasmid and foreign DNA are cut at the same restriction enzyme site and mixed together. Then the ends of each DNA are reannealed by ligation. The resulting recombinant DNA molecules are introduced into bacterial host cells. Often, the vector contains an antibiotic-resistant gene, which results in the production of a protein that renders the host cells insensitive to these antibiotics. Thus, plating the host cells on nutrient agar containing the relevant antibiotic will allow only those cells containing plasmid DNAs to grow.

Prior to this lab period, the TA will digest and ligate plasmids together with *C. elegans* DNA fragments and prepare 'overnight cultures' of XL1-Blue *E. coli* host cells that carry plasmids. Cells harbor one of the following plasmids: pUR288, expression vector; p2D, containing a 3.8 kb fragment of the *C. elegans* genome that contains the *myo*-3 gene; and pW, a control plasmid containing the 488 bp *myo*-3 insert. An "overnight culture" refers to plasmid-containing host cells grown in nutrient rich broth, containing the appropriate antibiotic selection marker. During the overnight culture, the bacterial cells will grow and divide, during which time the recombinant plasmids will replicate many times within the cells. You will isolate the plasmids from these bacteria cultures using a procedure called a "plasmid miniprep." This involves removal and purification of plasmid DNA from a host cell. Then you will carry out restriction endonuclease reactions with the plasmids, enabling you to achieve two goals. First, this will allow you to be certain that the bacteria contain the correct plasmids. Second, this will allow you to excise the cloned *myo*-3 gene from your cloning vector. Restriction endonuclease reactions will be performed with the expression vector, pUR288, so that it can be used in ligation experiments in Lab 2.1. Finally, this vector will be dephosphorylated at its 5' termini to prevent self-ligation.

Experimental Outline:

1. Media preparation.
2. Bacterial growths.
3. Plasmid minipreps.
4. Restriction endonuclease cleavage reactions with *Hin*dIII.
5. Dephosphorylate *Hin*dIII-digested pUR288.

Required Materials:

10 ml, 100 ml, and 1 L flasks; stir bars; stir plate; "NZCYM" media (recipe below); bacterial slants/plates (*E. coli* XL1-Blue) containing the 3 different types of plasmids; 1.5-ml microfuge tubes; microfuge; pipets; 95% ethanol - stored at -20°C prior to use; 3 M sodium acetate, pH 5.2; "TE buffer": 10 mM Tris-HCl (pH 7.5), 1 mM EDTA; "TENS buffer": 60 mg 0.1 N NaOH and 0.5% sodium dodecyl sulfate (0.375 ml of a 20% stock) dissolved and adjusted to a final volume of 15 ml with TE buffer; 0.6-ml microfuge tubes; *Hin*dIII, 10 U/μl [1 unit (U) will cut 1 μg of DNA once in 1 hour]; 10x buffer (specific for the particular restriction enzyme); calf-intestinal alkaline phosphatase (CIP)

Prelab Preparations:

1. Media Preparation

1. Prepare "NZCYM" media.
   a. Place the following in a clean flask:

NZ amine	10 g	NaCl	5 g
Yeast extract	5 g	$MgSO_4 \cdot 7\,H_2O$	2 g

   b. Dilute to 1 liter with deionized H_2O.
   c. Autoclave.

2. Bacterial Growths

1. Pick (remove specks with tips of individual toothpicks) single colonies of XL1-Blue bacteria from the agar streak plates, containing the different plasmids (pUR288, pW, p2D) using a sterile pipet tip and sterile technique.
2. Suspend the picked colony in a sterilized culture flask containing 3 ml of NZCYM media.
3. Incubate these "overnight cultures" at 37°C while shaking to aereate.

Procedures:

1. Plasmid Minipreps

1. Obtain the overnight cultures prepared prior to lab.
2. Dispense 1.4 ml of each type of culture into a microfuge tube.
3. Centrifuge for 30 seconds in the microfuge at room temperature.
4. Decant all of the medium supernatant.
5. Add 5 μl of autoclaved H_2O to the bacterial cell precipitant.
6. Resuspend the cells by mixing to obtain a homogeneous solution.
7. Add 0.3 ml of TENS buffer and quickly mix by inverting the tubes. At this point the solutions should be clear and viscous.
8. Add 0.15 ml of cold 3 M sodium acetate and mix by inversion to neutralize the bacterial lysates.
9. Allow sample tubes to sit for 10 minutes on ice.
10. Centrifuge for 5 minutes at 12,000 rpm, then transfer the supernatants to fresh microfuge tubes.
11. Add 0.9 ml of cold (-20°C) 95% ethanol and mix well to precipitate the DNA. Store at -20°C for 10 minutes.
12. Spin the tubes at 4°C for 10 minutes then carefully remove and discard the supernatants.
13. Wash the pellets in 70% ethanol, spin briefly, then drain the tubes.
14. Dry the pellets for 2-3 minutes in a lyophilizer.
15. Suspend the pellets into 35 μl TE buffer.
16. Add 1 μl of 1 mg/ml heat-treated (deoxyribonuclease destroyed) ribonuclease (RNase). One microliter should yield sufficient DNA to see on an ethidium-stained gel.

NOTE: Plasmid preparations may be pooled to produce "preparative amounts" of materials, for use in subsequent steps.

2. *Hin*dIII Restriction Endonuclease Cleavage of Plasmids

NOTE: All procedures use sterile H_2O. Do not use your pipet tips twice.

1. Combine the following in the listed order in a 0.6-ml microfuge tube:

7 μl	sterile H_2O
2 μl	10x buffer (enzyme-specific)
10 μl	plasmid DNA from the "minipreps"
+ 1 μl	*Hin*dIII
20 μl	V_{total}

2. Preparative protocol:

pUR288	*p2D and pW*	
5 μl	27μl	sterile H_2O
3 μl	10 μl	10x buffer (enzyme-specific)
20 μl	60 μl	plasmid DNA
+ 2 μl	+ 3 μl	*Hin*dIII
30 μl = V_{total}	100 μl = V_{total}	

3. Incubate for 2.5 to 3 hours at 37°C. The TA will remove the reaction tubes from the water bath and store them at -20°C.
4. If desired, the *Hin*dIII-digested pUR288 may be dephosphorylated as follows:
   a. Add 1 U calf-intestinal alkaline phosphatase (CIP) directly to the tube containing the *Hin*dIII digest of pUR288 (it is usually not necessary to exchange buffers at this point; CIP will work in a variety of different buffers).
   b. Incubate at 37°C for 1 hour.

c. Add 0.5 M EDTA to a final concentration of 5 mM in order to stop the reaction.
d. Heat to 75°C for 10 minutes. [CIP is inactivated by heat, bacterial alkaline phosphatase (BAP) is not.]

5. The dephosphorylated vector may then be purified by extracting once with phenol:chloroform and once with chloroform. Next, precipitate the DNA by adding a 1/10 volume 10 M ammonium acetate and one volume of isopropanol. Wash the DNA with 70% ethanol and dry under vacuum. Dissolve your DNA samples in appropriate volumes of TE or water. An alternate method is to simply precipitate the DNA and dissolve in a small volume of TE or water. Gel-purify the DNA using the powdered glass method as described in Labs 1.2 and 2.1.

NOTE: Ammonium inhibit kinase, do not use it if the DNA is to be phosphorylated at a later time.

Notes:

1. Restriction enzyme digestion:
   a. Most Type II restriction endonucleases require Mg^{2+}.
   b. EDTA chelates Mg^{2+}, which limits the amount of DNA cleavage.
   c. This is not true for RNA, which is readily degraded by Mg^{2+}.
   d. Treating DNA with alkali and Mg^{2+} produces a useless precipitate. Be certain EDTA is present before adding alkali.
2. Quantitating DNA and protein concentrations:
   a. "OD_{280}" refers to the optical density at 280 nm, an older nomenclature. This term is sometimes used for bacterial cultures because light is both absorbed and scattered.
   b. 1 A_{260} (absorbance unit), approximately 50 µg/ml for double-stranded (ds) DNA with 50% G+C content.
   c. RNA, proteins, detergents, and organic solvents affect the absorbance at 260 nm because they absorb there.
   d. A_{260} is the absorption maximum for DNA; A_{280} is the aromatic absorption maximum for tyrosine(s), phenylalanine(s), and tryptophan(s) in proteins.
   e. The ratio A_{260}/A_{280} is approximately 1.8 for pure DNA.

References:

1. Zhou *et al.* (1990) Biotechniques 8: 172-173.

PROCESS RATIONALE
pUR278 AND P2D RESTRICTION MAPS

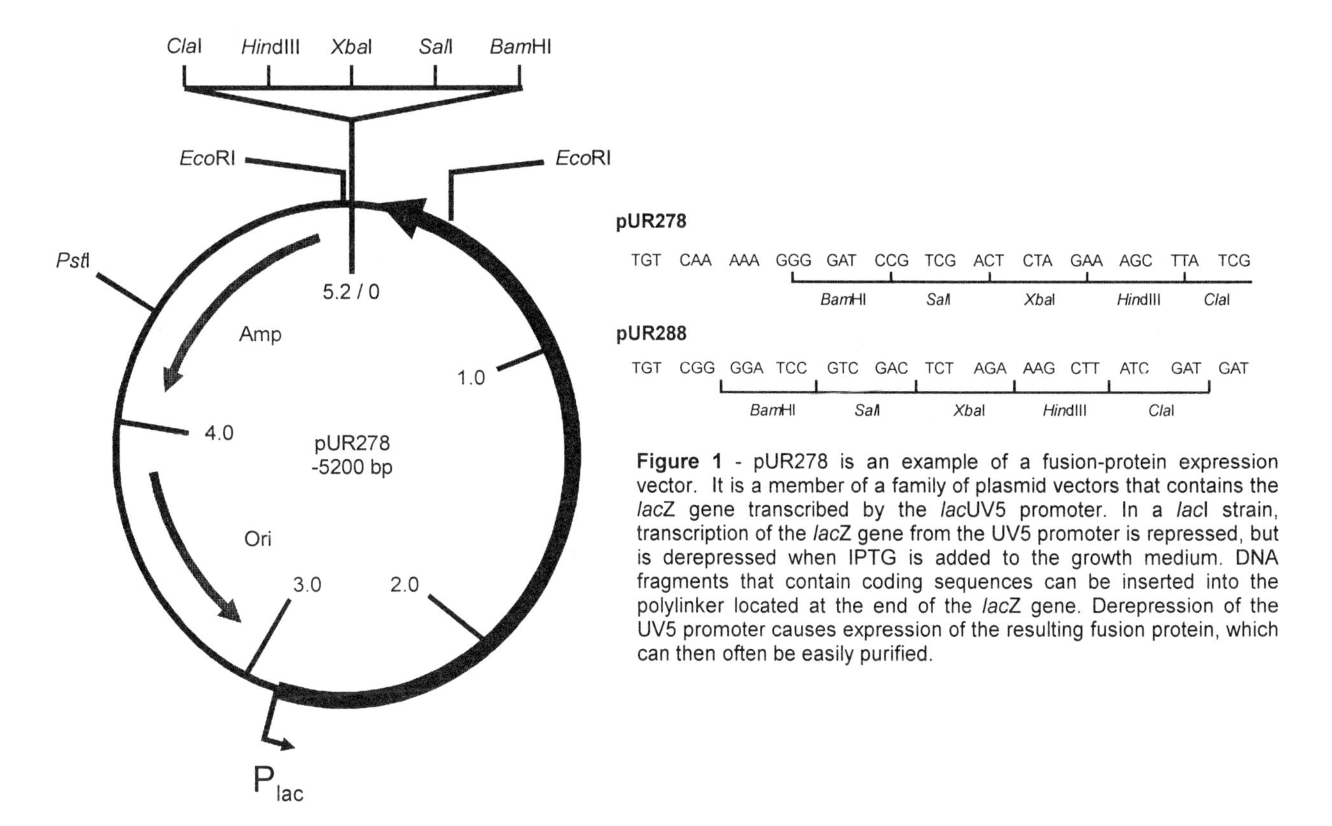

Figure 1 - pUR278 is an example of a fusion-protein expression vector. It is a member of a family of plasmid vectors that contains the *lac*Z gene transcribed by the *lac*UV5 promoter. In a *lac*I strain, transcription of the *lac*Z gene from the UV5 promoter is repressed, but is derepressed when IPTG is added to the growth medium. DNA fragments that contain coding sequences can be inserted into the polylinker located at the end of the *lac*Z gene. Derepression of the UV5 promoter causes expression of the resulting fusion protein, which can then often be easily purified.

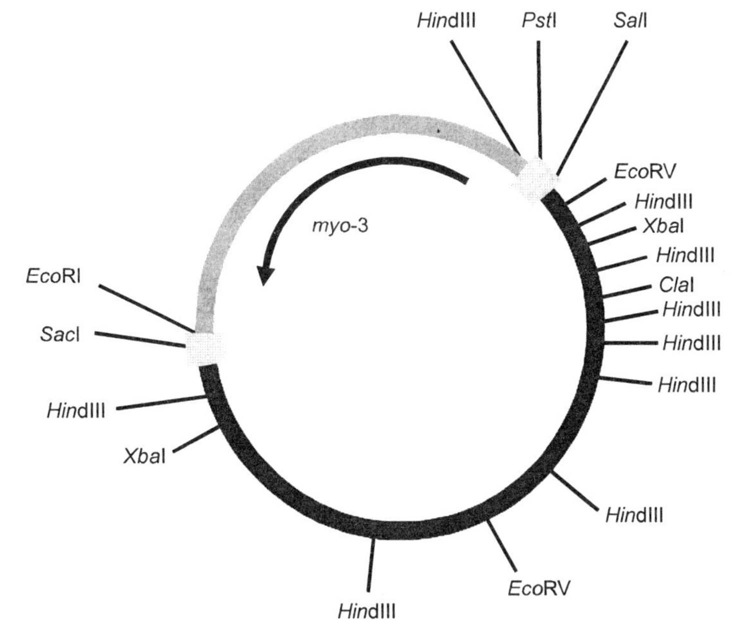

Figure 2 - The plasmid p2D was constructed from pUC12 vector (2.7 kb) and a 3.8-kb fragment of genomic *C. elegans* DNA containing the *myo*-3 gene.

Process Rationale

Cloning the *myo*-3 Gene from *C. elegans* and Construction of an Expression Vector

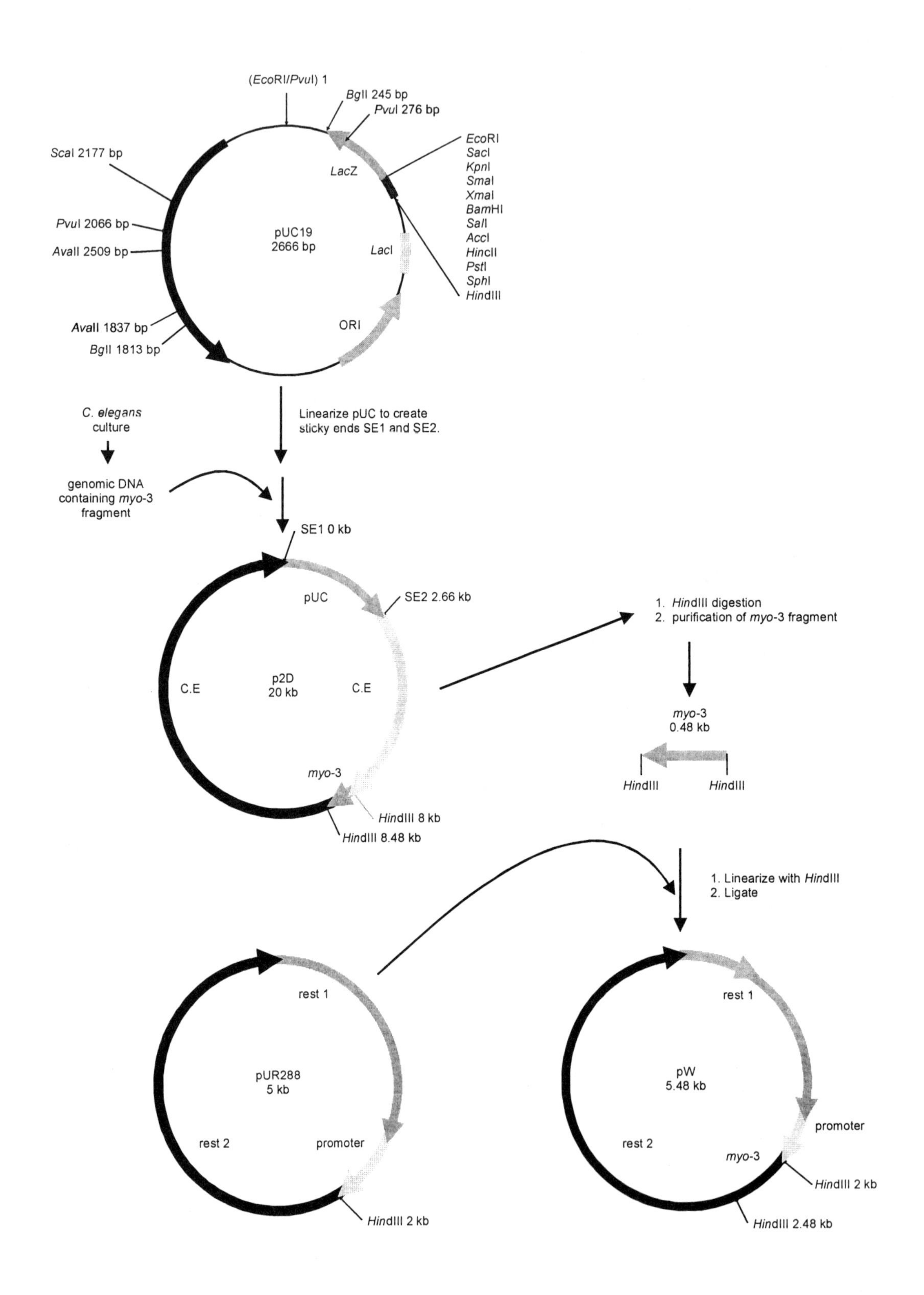

C. ELEGANS MYO-3 GENE IN pUR288

The 480-bp *Hin*dIII fragment is marked by bold letters. Notice that the ends of the fragment correspond to the *Hin*dIII restriction enzyme recognition sequence.

```
6201    GGCAAGACTA CCAGACTCAA GAAGAGGCTG CCGAGGCTGC TAAAGCCGGA
6251    CAGACTGCAG GAGGAAAGCG TGGAAAATCA TCTTCATTCG CTACCGTCTC
6301    TATGATCTAC AGAGAGTCTC TCAACAACTT GATGAACATG CTTTATCAAA
6351    CTCACCCACA TTTCATCCGT TGTATCATTC CAAACGAGAA GAAGGCTTCT
6401    GGAGTTATTG ACTCTGCGCT TGTTCTTAAC CAGCTTACAT GCAACGGAGT
6451    GTTGGAAGGA ATCAGAATTT GCCGTAAGGG ATTCCCCAAC AGAATGCTCT
6501    ATCCAGATTT CAAGCATCGT TACGCTATCC TTGCCGCTGA TGCCGCTAAG
6551    GAATCTGATC CAAAGAAGGC TTCCGTCGGA ATCCTTGACA AGATCTCTGT
6601    TGATGGAAAC TTGACCGATG AAGAGTTCAA GGTTGGAGAG ACCAAGATCT
6651    TCTTCAAGGC CGGAGTTCTT GCTAAGCTTG AGGACTTGCG TGACGAGATT
6701    CTTTCTAGAA TCGTTACCAT GTTCCAATCC CGCATTCGTT CTTACCTTGC
6751    CAAGGCCGAA GTTCGTCGTC GTTACGAACA ACAAACTGGA TTGCTTGTTG
6801    TTCAACGCAA TGTTCGTGCT TGGTGCACAC TCAGAACCTG GGAATGGTTC
```

```
6851  5'AAGCTTTTCG GAAAGGTCAA GCCAATGCTC AAGGCCGGTA AGGAACAAGA
6901    AGCAATGGGA GAGCTTGCCG TAAAGATCCA AAAGCTCGAA GAAGCAGTTC
6951    AACGCGGAGA AATTGCTCGT TCCCAATTGG AATCCCAAGT TGCCGATTTG
7001    GTTGAAGAGA AGAACGCATT GTTCTTGAGC CTTGAAACTG AAAAGGCTAA
7051    TCTTGCCGAT GCTGAAGAAA GAAACGAGAA GCTCAACCAA CTCAAGGCTA
7101    CCCTTGAGAG CAAGCTGTCC GATATCACTG GACAACTTGA AGACATGCAA
7151    GAACGTAACG AGGATCTTGC TCGCCAAAAG AAGAAGACTG ATCAAGAACT
7201    TTCTGATACC AAGAAGCATG TTCAAGATTT GGAGCTTTCT TTGAGAAAGG
7251    CTGAGCAGGA GAAGCAGAGC CGTGATCATA ACATCCGCTC CCTTCAAGAT
7301    GAGATGGCTA ATCAAGATGA AGCTGTCGCA AAGCTT3'AACA AGGAGAAGAA
```

```
7351    GCATCAAGAA GAATCCAACC GCAAATTGAA CGAAGATCTT CAATCTGAAG
7401    AAGSCGGGGT TAACCATCTT GAGAAGATCC GCAACAAGCT CGAGCAACAA
7451    ATGGATGAGC TCGAGGAAAA TATTGACCGC GAGAAGAGAT CTCGTGGTGA
7501    TATTGAAAAG GCAAAGAGAA AGGTCGAGGG AGACTTGAAG GTTGCGCAAG
7551    AGAACATCGA TGAGATTACC AAGCAGAAGC ATGACGTTGA AACTACTTTG
7601    AAGAGAAAGG AAGAAGATCT TCACCACACC AATGCTAAGC TCGCTGAGAA
7651    CAACTCTATC ATTGCTAAGC TTCAACGGCT TATCAAGGAG CTCACTGCTA
7701    GAAATGCCGA ACTCGAAGAA GAACTTGAAG CCGAGAGAAA CTCTCGCCAG
7501    AAATCTGATA GATCCCGTAG CGAGGCTGAA CGCGAGTTGG AAGAACTCAC
```

VENDOR LITERATURE

RESTRICTION ENZYMES *HIND*III AND *BSTE*II; λ DNA DIGESTS

New England Biolabs, Beverly, MA, ph. (508) 927-5054; Promega Corporation; Madison WI, ph. (800) 356-9526. (Adapted)

1. *Bst*EII Enzyme

G$^{\vee}$GTNACC
CCANTG$_{\wedge}$G

Size	Concentration	Catalog #	Price
2,000	8-12 U/μl	R6641	$50
10,000	8-12 U/μl	R6642	$200

1.1 Source. Bacillus stearothermophilus.

1.2 Incubation Cond. Buffer D, 60°C.

1.3 Test Substrate. DNA.

1.4 Percent Activity in 4-CORE® Buffer System.

*1.5 Isoschizomers. Bst*PI.

1.6 Storage Buffer. 10 mM Tris-HCl, pH 7.4; 50 mM NaCl; 0.1 mM EDTA; 1 mM DTT; 0.5 mg/ml BSA; 50% glycerol store at -20°C.

*1.7 Comments. Bst*EII exhibits 25-50% activity at 37°C.

2. Lambda DNA *Bst*EII Digest

#301-4S 150 μg $50
#301-4L 750 μg $200

Fragment	Base Pair	MW (Daltons x10^6)
1	8,454	5.49
2	7,242	4.71
3	6,369	4.14
4	5,686	3.70
5	4,822	3.13
6	4,924	2.81
7	3,675	2.39
8	2,323	1.51
9	1.929	1.25
10	1.371	0.69
11	1,264	0.82
12	702	0.46
13	224	0.15
14	117	0.08

2.1 Description. The *Bst*EII digest of lambda DNA (c1857ind λ Sam 7) yields 14 fragments suitable for use as molecular weight standards for agarose gel electrophoresis. The cohesive ends of fragments 1 and 4 may be separated by heating to 60°C for 3 minutes. The biotinylated markers are suitable for biotin based chemiluminescent DNA detection methods including the NEBlot Phototoped chemiluminescent Southern/Northern blotting kit.

2.2 Preparation. The double-stranded DNA is digested to completion with *Bst*EII, phenol extracted and dialyzed against 10 mM Tris-HCI (pH 8.0) and 1 mM EDTA.

The biotinylated markers are prepared by visible light irradiation of digested DNA in the presence of a photoactivatable biotin analog, followed by extraction with n-butanol to remove the unreacted analog. Unmodified markers are then added as carrier DNA, which is necessary for consistent banding patterns. This method introduces approximately 1 biotin per 100 nucleotides, which, based on comparison with the unmodified markers, increases the apparent molecular weight of each fragment by less than 2%.

2.3 Concentration and Shipping. 500 μg/ml (unbiotinylated) or 100 μg/ml (biotinylated, including carrier DNA). Supplied in 10 mM Tris-HCl (pH 8.0) and 1 mM EDTA. Store at -20°C.

NOTE: Dilute in TE or other buffer of minimal ionic strength. DNA may denature if diluted in distilled H_2O.

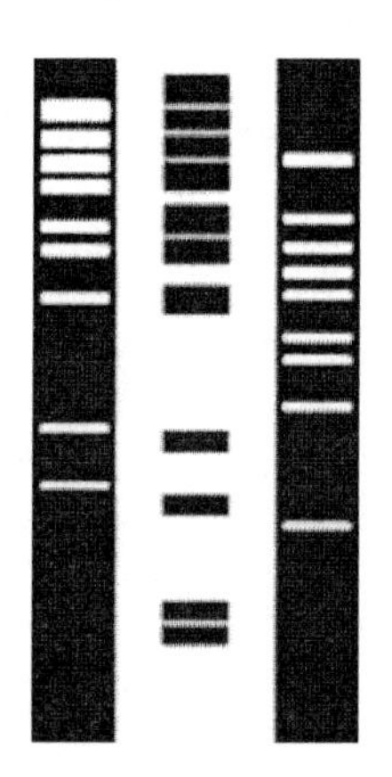

DNA-*Bst*EII digest of λ phage visualized by ethidium bromide staining (left) and chemi-luminescent detection of biotinylated form (center), 1% agarose gel. Lambda DNA, 0.7% agarose (*Bst*EII) (right). (idealized patterns shown)

3. *Hin*dIII Enzyme

A$^{\vee}$AGCTT
TTCGA$_{\wedge}$A

Size	Concentration	Catalog #	Price
3,000U	8-12 U/μl	R6041	$15
15,000U	8-12 U/μl	R6042	$60
15,000U	8-12 U/μl	R6045	$60
15,000U	40-80 U/μl	R6044	$60

3.1 Source. Haemophilus influenzae fd.
Blue/White Cloning Qualified

3.2 Incubation Conditions. Buffer D, 60°C.

3.3 Test Substrate. DNA

3.4 Isoschizomers. None.

3.5 Compatible Ends. None.

3.6 Storage Buffer. 10 mM Tris-HCl, pH 7.4, 50 mM NaCl, 0.1 mM EDTA, 1 mM DTT, 0.5 mg/ml BSA, 50% glycerol. Store at -20°C

3.7 Comments.

1. Consult site specific methylation chart.
2. Star activity may be observed in the presence of Mn^{2+}

4. Lambda DNA *Hin*dIII Digest

#301-2S 150 μg $50
#301-2L 750 μg $200

Fragment	Base Pair	MW (Daltons x10^6)
1	23,130	15.00
2	9,614	6.12
3	6,557	4.26
4	4,361	2.83
5	2,322	1.51
6	2,027	1.32
7	564	0.37
8	125	0.08

Biotinylated Form:
#301-2BTS 50 lanes $50
#301-2BTL 250 lanes $200

4.1 Description. The *Hin*dIII digest of lambda DNA (c1857ind λ Sam 7) yields 8 fragments suitable for use as molecular weight standards for agarose gel electrophoresis. The cohesive ends of fragments 1 and 4 may be separated by heating to 60°C for 3 minutes.

The biotinylated markers are suitable for biotin based chemiluminescent DNA detection methods including the NEBlot® Phototope® chemiluminescent Southern Northern blotting kit.

4.2 Preparation. The double-stranded DNA is digested to completion with *Hin*dIII, phenol extracted and dialyzed against 10 mM Tris-HCl (pH 8) and 1 mM EDTA. The biotinylated markers are prepared by visible light irradiation of digested DNA in the presence of a photoactivatable biotin analog (2), followed by extraction with n-butanol to remove the unreacted analog. Unmodified markers are then added as carrier DNA, which is necessary for consistent banding patterns. This method introduces approximately 1 biotin per 100 nucleotides, which, based on comparison with the unmodified markers, increases the apparent molecular weight of each fragment by less than 2%.

4.3 Concentration and Shipping. 500 μg/ml (unbiotinylated) or 100 μg/ml (biotinylated, including carrier DNA). Supplied in 1 mM Tris-HCl (pH 8) and 1 mM EDTA. Store at -20°C.

NOTE: Dilute in TE or other buffer of minimal ionic strength. DNA may denature if diluted in dH_2O.

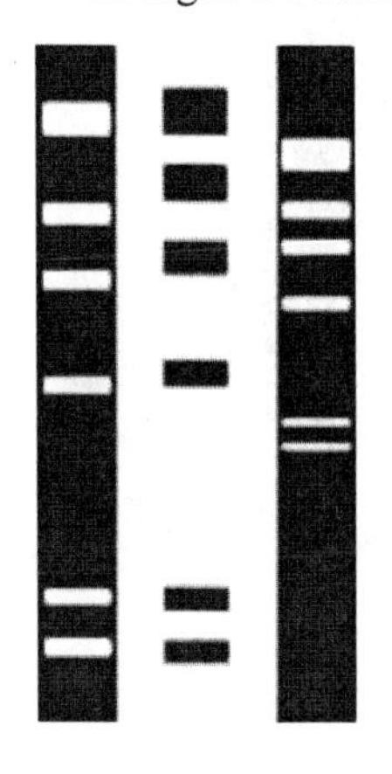

Lambda DNA-*Hin*dIII Digest visualized by ethidium bromide staining (left) and chemiluminescent detection of biotinylated form (center), 1% agarose gel. Lambda DNA 0.7% agarose (right). (idealized patterns shown)

References:

1. Daniels, D. L. *et al.* (1983) in *Lambda-II*, Eds., Hendrix, R. W., Roberts, I. W., Stahl, F. W. and Weisberg, R. A., Cold Spring Harbor Laboratory, N.Y.
2. Forster, A. C. *et al.* (1985) Nucleic Acids Res. 13, 745-761.

PROCESS RATIONALE

PHAGE λ *BST*EII DIGEST

The following show (idealized) duplicate agarose gel electrophoresis results illustrating variations in banding patterns between electrophoretic runs of identical restriction endonuclease-digested DNA samples. Variations in electrophoresis banding patterns can result from inconsistent sample loading, incomplete or inaccurate digestion, differing conditions in gel composition or electrophoretic parameters, or variations in the amount of ethidium bromide stain used.

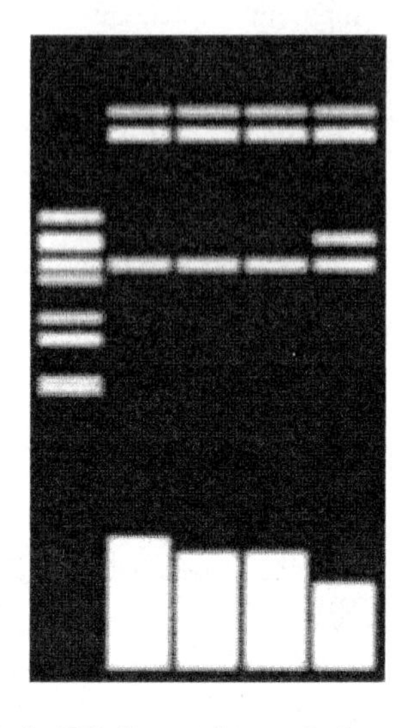

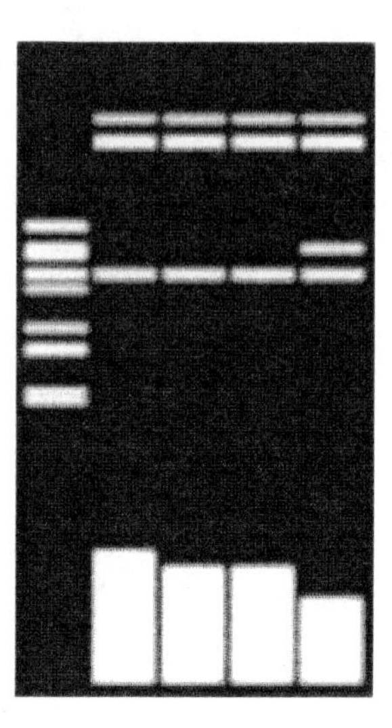

Lane 1: l, *Bst*EII digest. Lanes 2, 3: *myo*-3 insert (*Hin*dIII restriction endonuclease digest). Lanes 4, 5: pUR288 (idealized patterns shown)

LAB 1.2:
AGAROSE GEL ELECTROPHORESIS; PHOTOGRAPHY OF *HINDIII* PLASMID DIGESTS VISUALIZED BY FLUORESCENCE OF INTERCALATED ETHIDIUM

Overview:

In the first part of the lab, you will concentrate your digested *myo*-3 DNA. This step is necessary to adjust the dilute DNA to a concentration that can be visualized following agarose gel electrophoresis. Next, an agarose minigel will be run in the presence of the fluorescent DNA-binding chromophore ethidium in order to analyze the products of the *Hin*dIII restriction endonuclease reactions. The DNA fragments from the restriction digest will separate by size during the agarose gel electrophoresis. An agarose gel is used because its large pore size is sufficient to separate the large DNA digestion fragments. Since smaller fragments run faster and further than larger fragments in the gel, the size of each fragment can be determined by comparing its migration distance to standard DNA fragments of known size. You will then learn how to photograph the gel while visualizing the DNA on a UV "gel box" to get a record of your experimental results. Finally, you will isolate your DNA by excising the visualized DNA band in preparation for the Elu-Quik clean-up and DNA ligase reactions in the next lab.

Experimental Outline:

1. Concentration determination of digested plasmid DNA.
2. Analysis of *Hin*dIII reaction products on an agarose minigel.
3. Photography of the ethidium-stained gel pattern.
4. Extraction of stained DNA bands from the agarose gel.

Required Materials:

isopropanol; 10 M NH_4OAc; vacuum concentrator; solid agarose (not the low-temperature melting type); TAE buffer: 40 mM Tris-acetate (pH 8), 1 mM EDTA (per liter of) 50x stock: 242 g Tris base, 57.1 ml glacial acetic acid, 100 ml 0.5 M EDTA (pH 8); ethidium bromide stock: 10 mg/ml; DNA "loading dye": 0.25% bromophenol blue, 0.25% xylene cyanol, 40% (w/v) sucrose; ϕX174 *Hae*III DNA marker; TE buffer (recipe in Lab 1.1); camera film; gel slices containing DNA bands (from Lab 1.1); UV gel illuminator box; ethanol-rinsed scalpel; centrifuge; microfuge tubes

Procedures:

1. Concentrating Digested Plasmid DNA

1. Mix the following in a microfuge tube

digested DNA	100 µl
10 M NH_4OAc	25 µl
+ isopropanol	250 µl
V_{tot} =	275 µl

NOTE: Add the digested DNA and ammonium acetate first and mix. Then add the isopropanol.

2. Centrifuge for 15 minutes at 12,000 rpm.
3. Wash the pellet with 70% ethanol.
4. Centrifuge for 2 minutes at 12,000 rpm.
5. Concentrate the DNA in a vacuum for 10 minutes.
6. Resuspend the pellet in 25 µl of distilled H_2O.

2. Making a 1% Agarose Gel

1. Combine 50 ml 1x TAE buffer and 0.5 g agarose in a 250-ml Erlenmeyer flask.
2. Microwave for 2 minutes making sure that it doesn't boil over. Afterward, check to be sure that all of the agarose dissolved, swirl the flask if not (plus a 5-10 second burst in the microwave if necessary).
3. Allow the agarose solution to cool for 5 minutes.
4. While the agarose is cooling, set up the gel apparatus. Using masking tape, tape the ends of the gel tray so that a 1-cm deep rectangular mold is formed to pour the gel into; the ends should be tightly sealed. Next place the tray on the "table" in the electrophoresis apparatus. Make sure the apparatus is level.
5. After the agarose cools to a temperature at which the flask can be handled safely, add 5 µl of the stock ethidium bromide (10 mg/ml) to the gel solution and swirl the flask to distribute the dye evenly.
6. Pour the agarose solution into the gel tray. No leakage should occur.

7. Place the gel comb into the gel perpendicular to the top edge of the gel.
8. Allow the agarose gel to solidify.

3. Running the Gel

1. While the gel is solidifying, prepare the digested DNA samples by adding 2 μl of the DNA "loading dye" to each of the *Hin*dIII plasmid digestion products.
2. Prepare a MW standard of the commercial ϕX174 *Hae*III restriction "digest" as follows:

TE buffer	7.5 μl	
DNA running dye	2.0 μl	
+ 0.5 mg/ml ϕX174 *Hae*III	1.5 μl	(0.75 μg DNA total)
V_{tot} =	11 μl	

3. Load the samples into the wells in the gel using a pipet. Do not expel any air bubbles into the wells or allow the sample to diffuse over into the adjacent wells. Be sure to record the order and orientations of the samples as loaded (suggestion below).

Lane 1:	ϕX174 *Hae*III DNA marker
Lane 2:	uncut pUR288
Lane 3:	cut pUR288
Lane 4:	uncut pW
Lanes 5 and 6:	cut pW
Lanes 7 and 8:	cut p2D

4. Attach the leads from the electrophoresis apparatus to the power supply (red to +, black to -).
5. Run the gel at 60 V for 1.5 to 2 hours.
6. Turn off the power supply when the dye has migrated to half-way across the gel.
7. Detach the leads before removing the gel from the apparatus for photography.

4. Photographing the Gel

1. Load the film into the camera as instructed. Use an orange filter for the photography.
2. Place the gel directly on the light box along with a ruler.
3. Turn on the UV lights on the light box (wear a face safety shield) and turn off the overhead lights.
4. Photograph the gel, then turn the UV light off, and turn on the overhead lights.
5. Allow the film to develop.

5. Excising the *Hin*dIII DNA Fragments from the Gel

1. Place the gel slab in the 'cradle' on top of the light box. Slide the gel off of the cradle onto a clean sheet of plastic wrap placed on the surface of the UV light box. (The plastic wrap prevents damage to the glass surface of the transilluminator.)
2. Photograph the gel.

NOTE: Since the gels were run in the presence of ethidium, the resolved DNA fragments will fluoresce pale orange when observed on the UV light box. **For the SAFETY of you and your sample**, before using the light box, put on the UV face shield. Do the manipulations described below as quickly as possible because **UV light will mutate DNA, including those in your face and hands**.

3. Rinse the scalpel or razor blade with ethanol.
4. Turn on the UV lamp and observe the bands.
5. Quickly draw out the relevant band pattern and turn off the UV lamp or quickly cut the band from the agarose matrix. Be careful to minimize the amount of excess contaminating agarose.
6. Freeze these slices in labeled microfuge tubes for use in the next lab.

Notes:

1. Ethidium bromide is a MUTAGEN and SUSPECTED CARCINOGEN. Wear gloves when handling this material!
2. The *myo*-3 DNA fragment will electrophorese toward the bottom of the gel just above the relatively diffuse bromophenol blue dye band. Since less dye will be bound to the fragment relative to the larger fragments, it will appear only as a light and diffuse band.

3. Ethidium binds between every other set of base pairs ("**neighbor exclusion binding**"), so the number of binding sites on a shorter duplex fragment is smaller. We will use this fact and the known length of marker DNA bands to **quantitate the amount of DNA** to be used in the DNA ligase reactions in the next set of experiments.
4. TAE buffer was used in these gels because it is more soluble in sodium perchlorate than the more commonly used buffer Tris-Borate (TBE). Sodium perchlorate treatment will be necessary during the gel digestion steps in the first part of the next lab.

Additional Report Requirements:

1. Photograph of agarose gel

RESTRICTION MAPPING EXERCISES

Restriction Mapping Exercise #1

1. A plasmid was isolated and treated with several restriction enzymes. The fragment lengths were determined for each digest. From the data below, determine the restriction map for this plasmid. Include the restriction sites and fragment lengths in your diagram.

*Eco*RI	*Bam*HI	*Eco*RI + *Bam*HI	*Eco*RI + *Sma*III	*Bam*HI + *Sma*III
1.7	1.3	1.2	1.4	1.2
0.8	1.2	0.7	0.8	0.9
		0.5	0.3	0.4
		0.1		

2.

*Bam*HI	*Eco*RI	*Bam*HI + *Sma*III	*Eco*RI + *Sma*III	*Bam*HI + *Eco*RI
1.7	1.8	1.0	1.6	1.2
0.9	0.8	0.9	0.8	0.6
		0.7	0.2	0.5
				0.3

3. A linear DNA fragment was digested with 2 restriction enzymes; fragment lengths were determined and are given below. Draw the restriction map for this linear DNA sequence. Include the restriction sites and fragment lengths.

*Rsa*I	*Bam*HI	*Rsa*I + *Bam*HI
0.9	0.6	0.5
0.2	0.5	0.4
		0.2

4. Construct a restriction fragment map for a plasmid based on these data. Indicate the location of the restriction sites for each enzyme and the fragment sizes between adjacent cuts.

*Rsa*I	*Bam*HI	*Rsa*I + *Bam*HI	*Rsa*I + *Sma*III	*Bam*HI + *Sma*III
2.6	2.8	1.8	1.5	2.5
0.9	0.7	0.9	1.1	0.7
		0.7	0.9	0.3
		0.1		

Restriction Mapping Exercise #2

1. Digestion of bacteriophage lambda (λ) with *Hin*dIII:

Migration (mm)	Size (kb)
7	23.1
19	9.4
28	6.6
68	2.3
74	2.0
90	1.353
98.5	1.078
105.5	0.872
117.5	0.560
134.5	0.310

Digestion of plasmid pUC12 with various restriction enzymes:

Migration distance (in mm) and molecular size (in kb).

*Eco*RI		*Xma*III and *Eco*RI		*Rsa*I		*Eco*RI and *Rsa*I		*Xma*III and *Rsa*I	
mm	kb	mm	kb	mm	kb	mm	kb	mm	kb
42.5	3.25	52	2.8	71	1.98	71	1.98	83	1.55
		103.5	0.94	83	1.55	83	1.55	88	1.35
				112	0.75	120	0.55	107	0.84
						141	0.20	112	0.75

Draw the restriction map of plasmid pUC12:

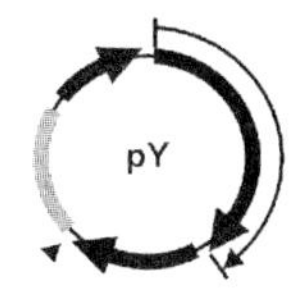

UNIT 2: CONSTRUCTION OF RECOMBINANT PLASMIDS

1. Isolating DNA Samples from Electrophoretic Gels

It is often necessary to recover biologically active DNA fragments from gels. Several good methods have been developed in recent years.

1.1 Electroelution from Agarose Gels. The gel is stained and the desired DNA fragment is excised from the gel. The gel slice is then put into a dialysis bag and placed into an electrophoresis chamber. The DNA is electroeluted out of the gel and into the surrounding buffer within the dialysis bag. Once out of the gel, the DNA remains in the dialysis bag free of contaminating gel. Alternatively, the DNA could be electrophoresed onto a DEAE cellulose membrane and later eluted from the membrane. Another variation is to electrophorese the DNA fragment into a trough that is cut into the gel just ahead of the band and recover the eluate.

1.2 Recovery from Low Melting Point Agarose. Many types of agarose have been developed for a wide spectrum of applications. One type is low melting point agarose, which melts at about 65°C; well below the melting point of regular agarose and below the melting point of most duplex DNAs. Once the DNA fragments have been separated on the gel, they can be excised with a blade. The gel slices can be melted, then extracted with phenol and chloroform to remove impurities. Once the DNA is precipitated and redissolved it is ready for use in ligations, restriction endonuclease digestions, etc.

1.3 DNA Binding to Powdered Glass Beads. Under high salt conditions, DNA binds to microscopic glass beads. In this technique, a solution of a chaotropic salt such as NaI or $NaClO_4$ is used to dissolve the agarose. After electrophoresis and excision of the gel slice containing the DNA, it is dissolved in the chaotrope solution. Next, some of a powdered glass suspension is added and the mixture is incubated while gently shaken. Eventually the DNA binds quantitatively to the powdered glass. The glass beads can then be recovered by centrifugation and washed to remove impurities, under conditions which still favor DNA binding to the beads. DNA is then eluted with H_2O or a low salt buffer.

1.4 Elution of DNA Fragments from Polyacrylamide Gels. This method is particularly useful for isolating very small DNA fragments since small fragments are much better resolved in polyacrylamide gels than in agarose. DNA fragments are separated on a nondenaturing polyacrylamide gel then stained with ethidium bromide. The DNA fragments of interest are excised from the gel, placed in small tubes, then crushed followed by addition of about 2 volumes of elution buffer. The tubes are then incubated with shaking at 37°C for several hours to overnight. The buffer containing the eluted DNA is collected and the DNA is precipitated with ethanol.

1.5 Other Methods for Recovering DNA from Gels Are Also Available. One example, an enzymatic gel digestion method, is called GELase (by Epicentre).

2. Quantitating DNA Concentration

2.1 A_{260}/A_{280} ratio. Spectrophotometric reading of DNA samples at 260 and 280 nm gives an estimation of the quantity (from A_{260} readings) and quality of the sample (from A_{260}/A_{280} ratios). An absorbance of 1.0 at 260 nm corresponds to an approximate concentration of 50 μg/ml. The A_{260}/A_{280} ratio for pure DNA is ca. 1.8; the value for pure RNA is about 2.0. This method of quantitating DNA is reasonably accurate, but is not practical when working with small amounts.

2.2 Quantitation by Estimation of Ethidium Bromide Fluorescence. The most common method of visualizing DNA in gels is by *ethidium bromide* staining. Ethidium bromide is a planar aromatic fluorescent dye that *intercalates* between the *stacked* base pairs of the DNA. The intercalation of ethidium bromide and other intercalating compounds follows the "nearest-neighbor exclusion principle." Because of the geometric distortion of the DNA molecule caused by intercalation, ethidium bromide molecules can intercalate only at every second available intercalation site. Ethidium bromide is a fluorescent dye that absorbs ultraviolet radiation at 302 nm and 366 nm. The energy emitted by the dye has a maximum intensity at 590 nm, in the red region of the visible light spectrum. The fluorescence yield of bound ethidium bromide is much greater than for unbound ethidium bromide (the fluorescence is enhanced by moving into a less polar environment), thus ethidium bromide-DNA complexes fluoresce much more brightly than unbound ethidium bromide in the surrounding gel. Ethidium bromide is a potent carcinogen and must be handled with care.

DNA concentration may be estimated using ethidium bromide in several ways. Since ethidium bromide intercalates between every other base pair the intensity of fluorescence is directly related to the mass of DNA present in each band on a gel. The best way to estimate DNA concentration is by resolving the unknown DNA samples on an agarose gel along with a parallel lane containing a series of DNA standards of known concentrations. One then identifies a band in one of the standards that is about equal in intensity to the unknown DNA fragment. This allows one to calculate the amount of DNA in the standard band, and from this value estimate the amount of DNA in the unknown. It is best to pick a standard band of about the same size as the unknown. The structure of ethidium bromide is shown below.

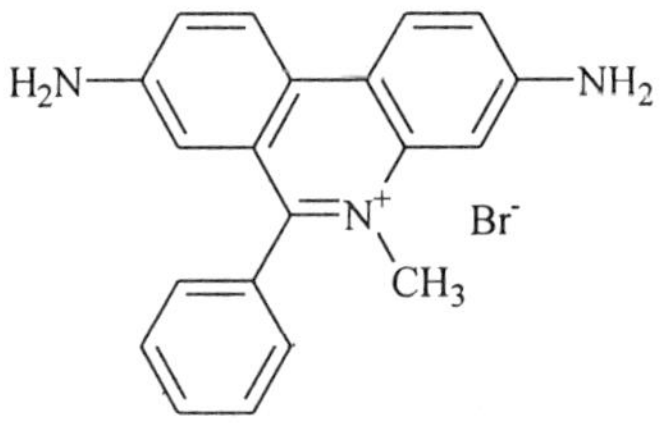

H_2N NH_2 N^+ Br^- CH_3

3. DNA Ligase Reactions

DNA ligase is an enzyme that links 5'- and 3'-terminal nucleotides of two different DNA strands by forming a phosphodiester bond (Figure 1 and 2). The discovery of this enzyme led to the first successful recombinant DNA experiments in the early 1970s. When DNA fragments having complimentary cohesive ends (i.e., DNA fragments produced with the same restriction enzyme) are allowed to hybridize, DNA ligase can join the backbone of the two strands by forming a phosphodiester bond. DNA ligase requires the presence of ATP and a Mg^{2+} cofactor to provide the energy necessary for phosphodiester bond formation. DNA fragments with blunt ends may also be joined by DNA ligase, however, higher concentrations of all the reaction components are necessary since the reaction is far less efficient than ligation of DNA fragments with complementary tails. Minimal *RNA ligase* substrates are a 5'-terminal trinucleotide and a 3'-terminal pNp. The lower case p to the right of the nucleotide (N) represents a 5'-terminal phosphate group. The lower case p to the left of the nucleotide (N) represents a 3'-terminal phosphate. This enzyme is used with ^{32}P-labeled pCp to radioactively tag RNAs (Figure 1).

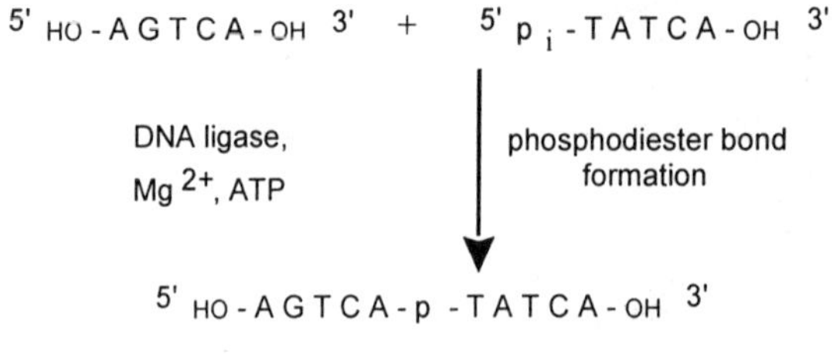

5' HO - A G T C A - OH 3' + 5' p_i - T A T C A - OH 3'

DNA ligase, Mg^{2+}, ATP — phosphodiester bond formation

5' HO - A G T C A - p - T A T C A - OH 3'

Figure 1 - DNA ligase reactions.

The concentration and ratio of linearized vector to insert DNA can be critical to the success of a ligation reaction. If vector and insert DNAs have identical cohesive termini, it is possible that the vector molecule will recircularize by ligating back to itself. This is especially likely if the total DNA concentration is low (because it is difficult to find a partner and easier to find its own other end), or if vector DNA is in excess. A maximum yield of chimeric (insert-bearing) molecules can be obtained by using a molar ratio (vector termini to insert termini) of about 1:2. This problem may also be circumvented by *alkaline phosphatase* treatment of the vector DNA or by *directional cloning* with two restriction endonucleases (Figure 2).

In vitro ligation reactions require that each DNA strand contains a 5' phosphate group and a 3' hydroxyl group, respectively. Removing the 5'

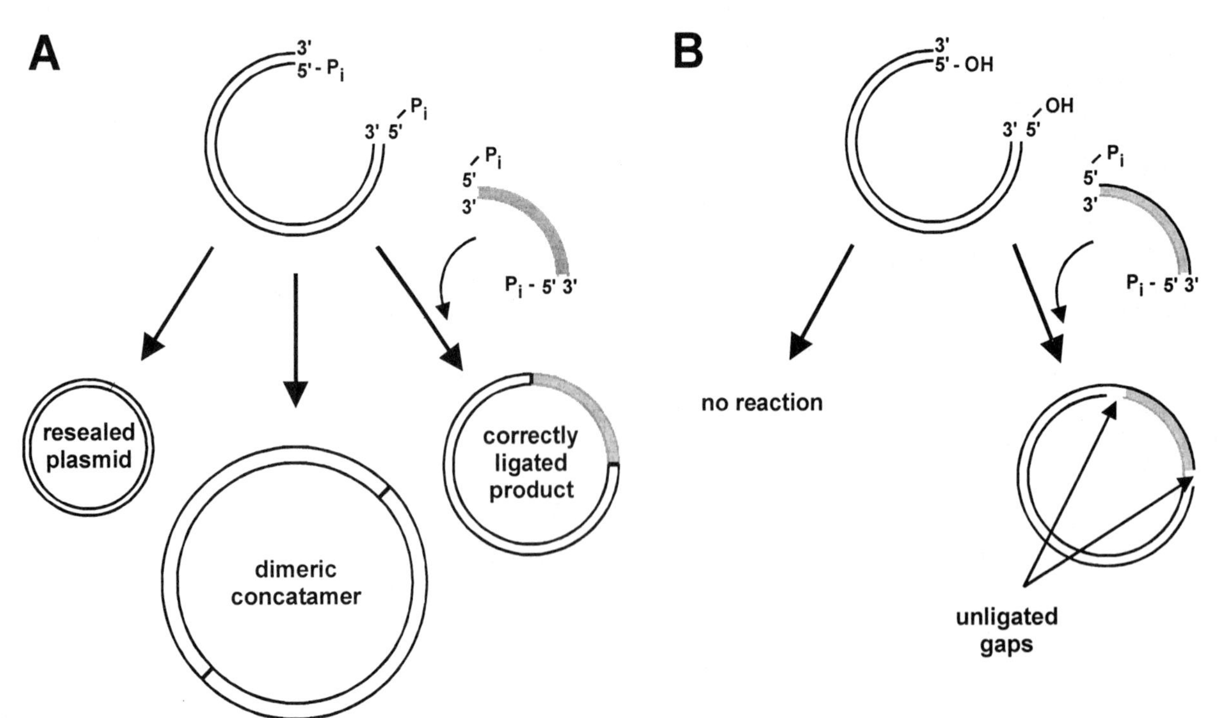

Figure 2 - Phosphate termini required for catalysis by DNA ligase.

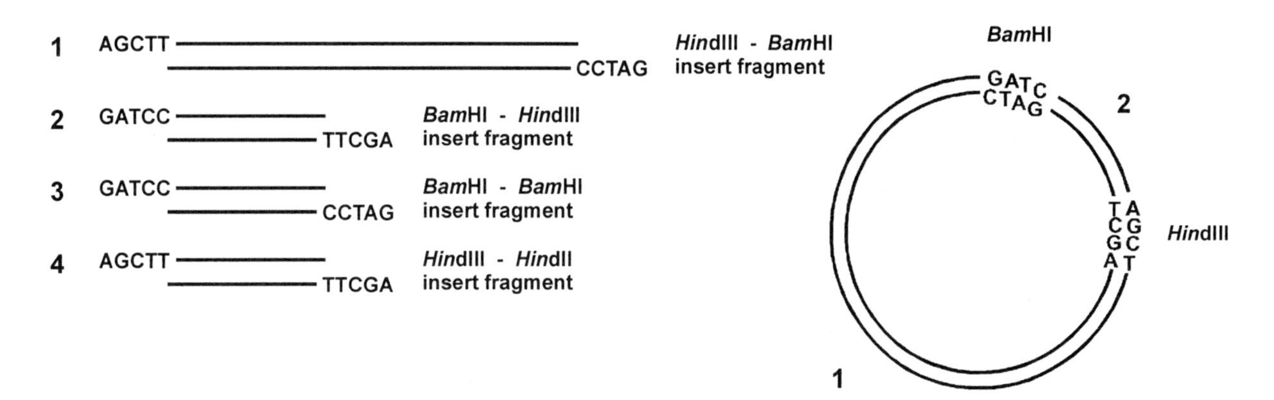

Figure 3 - Vector (1) can ligate only to fragment 2, producing the recombinant plasmid shown to the right. The vector cannot self-ligate (it is possible, however, for two vector molecules to ligate together).

phosphate from the linearized vector DNA strands with alkaline phosphatase renders the vector unable to self-ligate. Since the insert DNA still contains 5' phosphate groups, open-circular chimeric molecules (containing 2 single-stranded nicks) may still be formed between the dephosphorylated vector DNA and foreign DNA that has a 5' phosphate.

Directional cloning is accomplished by digesting the vector and insert DNA with two different enzymes. The cohesive termini created by the two enzymes are incompatible. For example, a *Bam*HI - *Bam*HI fragment won't fit into a *Bam*HI - *Hin*dIII site. Therefore, intramolecular ligation is eliminated. Also, insertion can occur only in one of the two orientations.

INNOVATION/INSIGHT

PROTECTING AND MANIPULATING LARGE DNA SUBSTRATES

1. Introduction

The advent of yeast artificial chromosomes (YACs) forced the development of two new technologies, *in situ* enzymology (in gels) and pulsed gel electrophoresis, designed to protect large labile DNA constructs. The basic difficulty goes back to the origins of molecular biology. Early workers thought DNAs were limited to sizes in the ca. 1 to 5 kilobase range. They later realized that they were producing an experimental artifact by shearing the larger pieces apart in the process of pipeting. In fact, chromosomes could be huge, microscopically observable aggregates of nucleic acids and proteins. Further, in macroscopic cells, the chromosomes are connected together via their telomeres under some circumstances, the cells means of orchestrating their movements *en masse*. *Shear forces* develop between the walls of the tip and the solution, but only if the pieces have enough mass to be broken off.

2. *In situ* Enzymology

To prevent this breakage, reactions involving very large constructs such as YACs and natural chromosome isolates are often carried out *in situ* within low melting point agarose. This provides a support matrix that helps to protect the DNA while the reaction is occurring. Many of the enzymes are compatible with the agarose matrix. In fact, transcription has been accomplished using nuclei that have been lightly permeabilized within agarose. This important advance paves the way toward the important goal of determining and reconstructing the three-dimensional structural organization of the nucleus, which will aid in our understanding of how morphological localization and spatial separation of factors and substrates affect the course of genetic phenomena.

3. Pulsed-Field Gel Electrophoresis

Very large chromosomal constructs present a difficult problem in separation science. Since these structures are of very high molecular weights, they essentially do not move in normal gels at typically used voltages. To solve this problem, the method of pulsed-field gel electrophoresis was developed. The basis of the method resides in the concept of *centripetal force*. Recall the feeling of acceleration you get when you go around a curve in the road in a car. One can obtain an analogous effect in an electrophoretic protocol if one alternates the field across a gel between the two perpendicular directions, back and forth between the two proscribed directions. The AC duty cycle is used to alternate the field between them. How often do we want to alternate the direction of the field to maximize the separation? As often as possible because otherwise the DNA is traveling linearly and does not accelerate.

References:

1. Wittig, B., Dorbic, T. and Rich, A. (1991) Transcription is associated with Z-DNA formation in metabolically active permeabilized mammalian cell nuclei. Proc. Natl. Acad. Sci. USA 88: 2259-2263.
2. Nyberg, S. L., Hardin, J. A., Matos, L. E., Rivera, D. J., Misra, S. P. and Gores, G. J. (2000) Cytoprotective influence of (the anti-poptotic agent) ZVAD-fmk and Glycine in a bioartificial liver. Surgery 127: 447-455.
3. (a) Bio-Rad (1997) Life Science Research Products (catalog), pp. 175, 185, Clamped Homogeneous Pulsed Electric Field Gel Electrophoresis DNA Plug Kits; (b) Zhang, T. Y., Smith, C. L. and Cantor, C. R. (1991) Nucleic Acids Research, 1291.

INNOVATION/INSIGHT

YEAST OF BURDEN - YOKING THE YAC

Roberts, S. S. (1990) Journal of NIH Research 2: 77. (Adapted)

1. Introduction

In only a few short years, yeast artificial chromosome (YAC) technology has matured from a stripling method for studying chromosomes into a powerful cloning technique. YACs will lighten the load for researchers mapping the human genome by allowing them to clone it in large pieces. YAC technology recently pulled its weight when researchers harnessed it in tandem with other techniques to locate the type-1 neurofibromatosis gene.

But the technique has not been completely tamed. The complexity of making and manipulating YACs limits the number of laboratories able to work with them and can make research with YACs tedious and slow.

Despite the technique's intricacies, the principles behind YACs are simple. At their most basic, YACs begin as circular pieces of bacterial DNA called *plasmids*, with *telomere* sequences (chromosome ends), *centromere* sequences (the central region of the chromosome) and one *autonomous replication sequence* (ARS) inserted. Although telomere sequences can come from the chromosomes of yeast (the eventual host for the YAC), complete or even partial telomeres from other organisms can also work.

The plasmid is then *cut apart* to make it *linear* and the new hybrid chromosome is inserted into the cells of baker and brewer yeast (*Saccharomyces cerevisiae*). If the telomere sequences on the YAC are similar enough to yeast sequences for the yeast cell to recognize them, it alters the YAC telomere sequences to make them more like natural yeast telomeres. The YAC then behaves much like any other chromosome and replicates during cell division. Researchers take advantage of this replicability to make copies (clones) of DNA sequences by inserting them into YACs and letting the yeast grow. [Also, see a series of reviews scheduled to be published in *Genetic Analysis: Techniques and Applications* 7(5) (August, 1990)].

The basic conceptual simplicity of the YAC led researchers working independently at the Dana-Farber Cancer Institute and Harvard Medical School in Boston and at the Fred Hutchinson Cancer Research Center in Seattle to invent YACs in 1983. The breakthrough, says Virginia Zakian, who led the Seattle group, was the demonstration in 1982 that the easy-to-get telomeres of ciliates (a class of protozoans) could function as the ends of the chromosomes of yeast as well. Building on this discovery she and postdoctoral associate Ginger Martin Dani constructed some of the first YACs to investigate whether plasmid stability changed when a plasmid was made linear or when centromeres or telomeres were added.

At the same time, Jack Szostak and Andrew Murray in Boston also invented the YAC. "The original reason that we developed the technology [was] to use it as a way of probing chromosome structure and function in yeast, particularly segregation, the process by which each daughter cell receives only one chromosome of each pair during cell division," says Murray now at the University of California at San Francisco.

Today, YACs are still valuable to the study of chromosomes because researchers can rearrange their DNA, delete it, and alter it – procedures that would not be possible with a real chromosome because the cell would die. "We use the YAC because it's completely dispensable" says Zakian. "When the cell loses it, it doesn't care."

For example, she says, "we have a "mutant hunt" to identify genes that are important for telomere biology." Zakian and postdoctoral associate Kurt Runge induce mutations in yeast cells that have YACs and then isolate strains that lose the YAC at high rates. They then insert a new YAC or a circular YAC into the mutant yeast strains. Mutations affecting the centromere or the ARS (where DNA replication starts) destabilize both circular and linear YACs, but mutations affecting telomeres destabilize only linear molecules, allowing the researchers to tell the two types of mutations apart.

Rodney Rothstein of Columbia University's College of Physicians and Surgeons in New York uses YACs to study the genetic control of recombination. "We may be able to study recombination at hot spots of recombination in yeast by cloning them into YACs," he says. The greater the similarity between two sequences, the greater the probability that they will recombine. Thus, the repeated sequences of DNA that Rothstein's lab is examining would be expected to recombine at high rates. However, when his group puts so-called *Alu repeats (repetitive sequences* about 300 base pairs long that occur in the human genome every 5 to 20 kilobases, kb) into YACs, they do not recombine very often. This hints that "there's enough diversity among Alu repeats to keep the recombination in check," says Rothstein.

Researchers also can clone genes with YACs. YACs can incorporate pieces of foreign DNA 100 to 1,000 kb long; large enough to combine whole genes and far longer than the 20 to 40 kb possible with cloning methods based on bacteria. Researchers may also be able to clone DNA sequences with YACs that are difficult to clone by other methods; for example, *palindromic sequences* are hard to clone in bacteria because the bacteria frequently delete them.

At a May 16 seminar at NIH sponsored by the National Center for Human Genome Research, David Schlessinger (Washington University School of Medicine in St. Louis) said that YACs are crucial for efforts to sequence the human genome. Because YACs contain an average of 250 to 300 kb, he said, five copies of the 3 million kb human genome can be contained in only 60,000 clones (a tiny number compared with the nearly 900,000 clones needed with 17 kb phages).

Researchers can map genes at three levels of precision with YACs, says Schlessinger. First, at a gross level, YAC sequences can be localized to a particular band on a chromosome by *in situ* hybridization. Second, researchers can put YACs, and thus the sequences they contain, in order along the chromosome by techniques such as DNA fingerprinting and YAC cross-hybridization. Third, researchers can start from a known sequence in a YAC and then sequence nearby areas with chromosome walking.

As soon as YACs were invented, researchers recognized that it might be possible to clone genes with them. Murray David Burke (Princeton University) concurs: "It was pretty obvious to a lot of yeast geneticists at that point that there was the potential for [making clones]. But it just became a matter of putting together the right technical pieces to get it to work at a reasonable proficiency." In 1987, the technical hurdles were first overcome by a Washington University team consisting of Maynard Olson, Burke, and Georges Carle, now at the Université de Nice.

Today, cloning with YACs remains time consuming and labor intensive, and YAC cloning and mapping efforts are concentrated at a very few universities and national laboratories, among them Washington University. Its Center for Genetics in Medicine, which Olson and Schlessinger direct, maintains and screens a library of the total human genome in YACs [see B.H. Brownstein *et al.*, Science 244, 1348 (1989)].

In addition, Schlessinger's group – in conjunction with Michele D'Urso's group at the Istituto Internazionale di Genetica e Biofisica in Naples – is mapping the region between Xq24 and Xq28 – the tip of the long arm of the X chromosome, which contains the genes aberrant in Lesch-Nyhan syndrome and some forms of hemophilia and colorblindness. Schlessinger's YAC data confirm and extend maps made previously by other methods, bolstering the working assumption that sequences on YACs are usually not rearranged or altered. He anticipates that 200 YACs may suffice to cover the tip of the X chromosome with clones that overlap in sequence.

Olson's lab has recently isolated the 250 kb cystic fibrosis gene discovered by Francis Collins of the University of Michigan at Ann Arbor and Lap-Chee Tsui of the Hospital for Sick Children in Toronto. "Their limitation was that using the conventional [bacteria-based] technology available, [they could not] isolate an entire gene of that size in one piece," says Eric Green, a postdoctoral researcher in Olson's lab.

For this study, forthcoming in *Science*, the Washington University researchers inserted into yeast cells two YACs, each containing portions of the cystic fibrosis gene, including an overlapping sequence. "Yeast are naturally good at recombining across regions of common DNA," says Green. When the yeast recombined the two YACs, the researchers got one gigantic YAC with a 790-kb stretch of human DNA containing the entire cystic fibrosis gene.

Another large YAC research effort is under way at Johns Hopkins School of Medicine in Baltimore. One part of the program is a multilab study of a mouse model of Down syndrome. According to Philip Hieter, one of the researchers, the goal of the research is to use YACs to clone a region of DNA found both in humans on chromosome 21 and in mice on chromosome 16, map the region, determine the gene order, and then put some of the genes back into mice. Project director John Gearhart says that by creating transgenic mice with extra copies of various genes from the shared region, they may discover which genes or chromosome segments are responsible for each physical manifestation of Down syndrome.

YACs are valuable to this project because of their size. "The YACs that we're putting into mouse cells are bigger than the pieces of DNA that anyone was able to put into mouse cells before," says Roger Reeves, another member of the team.

The future for YACs seems bright. YACs have established themselves as a research tool valuable to growing research areas such as the human genome project and genetic engineering. If researchers can make cloning with YACs easier, many more laboratories will want to harness the YAC.

2. Cloning Around with YACs

Yeast artificial chromosomes [YACs] are an important advance in the cloning techniques by which researchers make multiple copies of sequences of DNA. As its name implies, a YAC is an artificially constructed DNA molecule disguised as a yeast chromosome. Sometimes, the yeast cell is fooled by the masquerade and accepts the YAC as part of its genome.

The deception is possible because only some elements of the chromosome are necessary for it to be recognized as such by a cell. Just as a stranger can be identified as a priest only on the basis of a Roman collar, so is a yeast chromosome identified by having a centromere (a central region), two telomeres (the "caps" on the ends that keep them from unraveling), and an autonomously replicating sequence, or ARS (where DNA replication begins).

Because YACs contain these three elements, the cell treats them as real yeast chromosomes. However, in many ways they are quite alien. A YAC starts its existence as a circular piece of bacterial DNA called a plasmid. Researchers cut this circle at various places with restriction enzymes to insert not only the three disguising elements – a centromere, an ARS, and telomeres – but also other sequences, such as cutting sites for other restriction enzymes and yeast genes to act as markers. The resulting vector is a hybrid circular molecule ready to be turned into a YAC. Today, such vectors are widely available, and few researchers bother to make their own.

To construct a YAC, a researcher cuts the vector apart. The first cut makes the molecule linear; the second yields two segments, called the left and right arms, each with a telomere and at least one marker gene. The researcher also cuts up the DNA that is to be cloned and mixes these pieces with the arms. Complete chromosomes may form, consisting of a left arm, one or more pieces of source DNA in the middle, and a right arm.

The researcher then strips the cell walls from yeast cells, leaving spheroplasts, and inserts the new chromosomes into them. The delicate spheroplasts

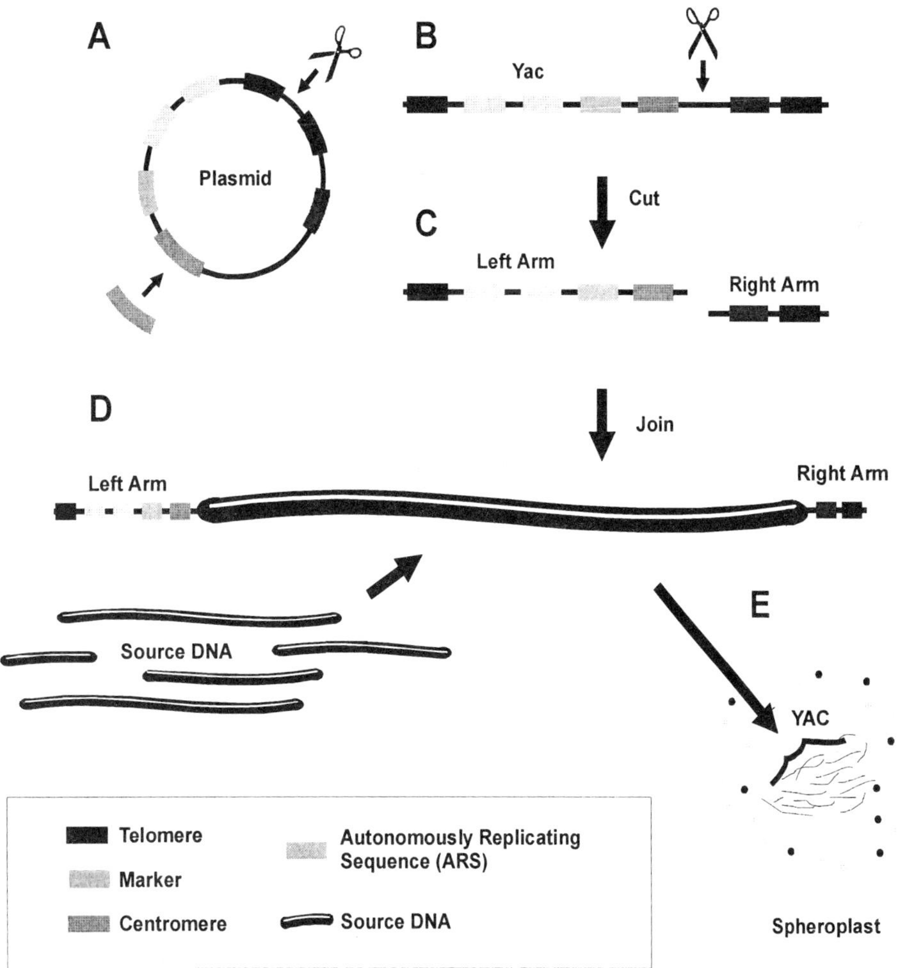

A. To make a YAC, the researcher starts with a circular bacterial plasmid into which have been inserted foreign elements, including markers, telomeres, a centromere, and an autonomously replicating sequence.
B. The hybrid plasmid is cut to make it linear.
C. The plasmid is again cut into two separate arms.
D. The researcher cuts the source DNA that is to be cloned into pieces and inserts them between the left arm and right arm pieces. The researcher inserts the artificial chromosomes into yeast cells that have had their cell walls removed. When the yeast cells divide and multiply, the artificial chromosomes replicate along with the real ones.

must be embedded in agar for support; this allows the cells to regenerate their walls and begin reproducing.

To be able to winnow out the cells lacking YACs, the researcher selects as the YAC host a yeast strain that does not produce an essential nutrient. By growing the yeast in a medium also lacking the nutrient, the researcher ensures that cells with YACs – whose marker genes help to produce the nutrient – are the only ones to survive and reproduce. Each arm of the YAC needs a different marker, because "there's a tendency for these DNA fragments to go into yeast and circularize," says Roger Reeves of Johns Hopkins University School of Medicine in Baltimore. If only one arm contained a marker, the researcher would be unable to weed out cells that had a circularized arm with that marker instead of a complete YAC.

The source DNA multiplies along with the rest of the YAC, but recovering the DNA is a time-consuming chore. For example, one problem is the agar in which the YACs need to grow, "You have to go in there and individually dig out every colony," says Reeves, "and that takes a while when you're talking about fifty or sixty thousand colonies." Once removed from the agar, the colonies are generally stored in separate wells in microtiter plates.

Additional Reading:

1. D. T. Burke, G. F. Carle, and M. V. Olson, Cloning of large segments of exogenous DNA into yeast by means of artificial chromosome vectors, Science 236, 806 (1987).

LAB 2.1:
EXTRACTION AND CLEANUP OF DNA BANDS CUT FROM AGAROSE GELS, QUANTITATION OF YIELDS, AND LIGATION OF *MYO*-3 *HIN*DIII DNA INSERT FRAGMENT INTO LINEARIZED β-GAL PLASMID DNA

Overview:

In the first part of this lab, you will use the Elu-Quik™ procedure (Schleicher & Schuell™) to digest the remaining agarose from the gel slice prepared in the previous lab. DNA fragment cleanup is necessary for a future ligation procedure. The Elu-Quik™ system provides a convenient method for simple and rapid recovery of DNA from gels. The purity of the recovered material is adequate for subsequent biochemical manipulation, such as ligation. This kit uses glass particles to isolate single- or double-stranded DNA of sizes from 500 bp to 200 kb. It uses a combination of heating and high salt concentration in the binding buffer to melt the gel. When the glass is added, it binds to the DNA. Several washing/centrifugation steps remove the agarose and other contaminants. The DNA is then eluted with buffer or water.

Next, you will run an agarose minigel to quantitate the amounts of DNA by comparing the intensities of the ethidium-stained bands with known amounts of ϕX174 *Hae*III DNA standards. Finally, you will incubate appropriate amounts of the *myo*-3 insert and linearized pUR288 plasmid DNAs into tubes and add DNA ligase to initiate the ligation reactions. DNA ligase catalyzes the linkage of 5'- and 3'-termini at each end of the *myo*-3 fragment with the plasmid ends to produce a recombinant DNA molecule that can be translated efficiently. pUR288 is a fusion protein expression vector that contains the *lac*Z gene that is transcribed by the *lac* UV5 promoter. In the lacI strain the promoter is repressed, but derepressed with addition of IPTG, a β-galactosidase analog. Insertion of the *myo*-3 gene into a protein fusion expression vector will be necessary for future protein purification experiments. The stoichiometric ratio of insert to plasmid plays a vital role in the ligation reactions.

Experimental Outline:

1. Elu-quik™ cleanup of plasmid digests; bands cut from agarose gel.
2. Quantitation of yields.
3. Ligation of *myo*-3 *Hin*dIII DNA insert fragment into dephosphorylated and phosphorylated linearized β-gal plasmid DNA.

Required Materials:

Elu-Quik™ (Schleicher & Schuell) kit components: 1 ml "glass milk" particles in 2 ml screw-cap vials (5 in kit), 30 ml amber bottle with 20 ml guanidinium thiocyanate in "cell lysis buffer," 125 ml amber bottle with 95 ml $NaClO_4$ "DNA binding buffer," 125 ml bottle with 125 ml Tris/NaCl/EDTA "wash buffer" – 2x concentrate to be diluted in ethanol, 125 ml alcohol-based desalting buffer, pipet tips – designed to minimize DNA shearing during the wash steps; DNase-free ddH_2O (distilled, deionized); ethanol; TE buffer (recipe in Lab 1.1), low-speed centrifuge; 4300 x *g* fixed-angle rotor/centrifuge; microfuge; rocking platform/shaker; 50°C water bath; 1.5-, 15-, and 50-ml plastic capped centrifuge tubes); ϕX174 *Hae*III DNA; *Hin*dIII-linearized pUR288 DNA; *Hin*dIII digested *myo*-3 fragment; agarose minigel apparatus; T4 DNA ligase

Procedures:

1. Elu-Quik™ Cleanup of DNA Gel Slice Bands

1. Centrifuge the DNA/agarose material at 7000 x *g* for 1 minute to draw the fragments from the sides of the tube. Determine the approximate volume of gel in the tube by comparison with known volumes of pipeted H_2O in a second tube.
2. If isolating DNA from TAE gels, add 2.8 volumes (e.g., 280 μl for 100 μl of gel) of "binding buffer." If isolating DNA from TBE gels, use 3.8 volumes (see the NOTES at the end of the previous lab).
3. Fragment the gel by agitation using a plastic pipet tip.
4. Place the tubes in a floating microfuge tube holder.
5. Incubate at 50°C for 5-10 minutes, mixing periodically. If fragments remain, add another 100 μl of binding buffer and repeat the 5 minutes/50°C incubation.
6. Resuspend the "glass milk" provided by the instructor by stirring with a pipet tip.

7. Add 20 µl of the "glass milk" to the gel slice/binding buffer sample, then tap the side of the tube to mix the contents.
8. Let the solution stand for 10 minutes at room temperature, inverting the tube at one minute intervals. A rocking platform or shaker can be used for scaled-up volumes.

NOTE: If the amount of DNA is greater than 2.5 µg/100 µl, increase the amount of "glass milk" added proportionally. The ratio of binding buffer to sample solution should be kept at 3 to 1.

9. Centrifuge at 7000 x *g* for 30 seconds in a microfuge.
10. Discard as much supernate as possible using a pipet.
11. Add 500 µl of "wash buffer." Using a non-shear pipet tip (supplied with the kit), tease the glass pellet from the wall of the microfuge tube then gently pipet the glass pellet up and down until it becomes flocculant (loosened). Invert the tube. Obtaining a flocculant pellet is important for proper buffer exchange.
12. Centrifuge at 7000 x *g* for 30 seconds. Discard the supernatant.
13. Repeat steps 6 and 7.
14. Add 500 µl of "reduction buffer." Gently resuspend the glass pellet until flocculant. Invert the tube and centrifuge at 7000 x *g* for 2 minutes. Discard the supernate. Reducing the salt improves the efficiency of DNA elution from the silica beads.
15. Centrifuge the glass pellet at 7000 x *g* for 30 seconds. Carefully withdraw as much of the residual supernate as possible, discard it.
16. Dry the pellet under vacuum for 15 to 30 minutes at room temperature. This removes traces of ethanol that may interfere with some enzymes (e.g., *Bam*HI). Drying the DNA does not impair elution of fragments less than 50 kb from the matrix.
17. Add 10 µl of TE buffer. Resuspend the glass beads by flicking the tube repeatedly.
18. Incubate the mixture at 50°C for 5 minutes.
19. Centrifuge at 7000 x *g* for 30 seconds and withdraw the supernate to a separate tube for analysis or storage (stable for 2 months at -18°C).
20. Repeat step 12 with 5 µl of TE buffer. Try to avoid sucking up the beads.

NOTE: Some glass may be left in the tube. It should not interfere with subsequent reactions. If too much glass remains, centrifuge the DNA with a quick burst in the microfuge before use.

2. Quantitation of Fragment Yield

Rough quantitative amounts of vector and insert DNAs are used to calculate and adjust the ratio of the two types of DNA in the DNA ligase reactions. The *plasmid to insert ratio* is adjusted to approximately 1 to 3 in order to drive the bimolecular plasmid-insert linkage reactions to product. Using a relatively *small molar excess* of insert (300%) avoids nonproductive A-A or B-B encounters, leading to plasmids that do not contain inserts. Unfortunately, excessive insert can increase the amount of linear multi-insert products (concatamers) that link into the vector. It will be interesting to see if we can produce a fusion protein with two or more linked *myo*-3 protein sequence domains.

1. Make a 1.5% agarose gel as described in Experiment 1.2. Use the comb with larger teeth to form larger wells (8 wells; 4-5 mm wide).
2. Pipet known quantities of the molecular weight marker solution into two of your gel lanes (use e.g., 0.5 µl in one lane, 1 µl in another). Place 1 µl of the Elu-Quik™ cleaned insert fragment in a third lane and the *Hin*dIII-linearized pUR288 vector DNA in a fourth lane. Prepare as follows:

Insert	*Vector*	*Sample(M_r std)*
__ µl sample	__ µl sample	0.5 or 1 µl or (0.5 ng/µl φx174 *Hae*III or λ *Hin*dIII digest)
2 µl dye	2 µl dye	2 µl dye
+8 µl TE	+8 µl TE	+8 µl TE
__ = V_{tot}	__ = V_{tot}	__ = V_{tot}

3. Electrophorese for 15-20 minutes. If it runs longer the DNA will become diffuse and it will be difficult to quantitate the amount of DNA accurately.
4. Stop the gel.
5. Estimate the amount of DNA recovered from the Elu-Quik™ procedure and the amount of plasmid DNA by comparing the intensities of the respective fluorescent bands with those of the marker DNA bands.

a. Record the source DNA and restriction enzyme used to make the marker digest:
   DNA: ϕx174 and λ Source: ___________

   Enzyme: *Hae*III and *Hind*III   Source: ___________

6. Calculate the concentrations of plasmid and insert DNAs and their molar ratios.
   a. Find a band that has approximately the same visual intensity as the DNA of interest. The sizes of the DNA bands produced in a *Hae*III digestion of ϕx174 DNA and *Hind*III digested λ phage are reproduced below:

Fragment	# of Base pairs ϕx174 *Hae*III	# of Base Pairs λ *Hind*III
1	1,353	23,130
2	1,078	9,416
3	872	6,557
4	603	4,361
5	310	2,322
6	281	2,027
7	271	564
8	234	125
9	194	total = 48.5 kb
10	118	
11	72	
	total = 5.386 kb	

7. Use the following example as a template to calculate the amount of plasmid (5200 bp) and insert (488 bp) DNA recovered from the Elu-Quick procedure.

NOTE: Let's assume that the total mount of ϕx174 *Hae*III *fragment DNA* (**5.386 kb** total length) was determined spectroscopically. Use the approximate equivalence 50 μg = 1 A_{260} to be **0.5 μg/μl**, and that you used 1 μl in your sample for electrophoresis.

NOTE: Generally, the molecular weight marker protein literature will provide the marker concentration.

In this example, we'll assume that the **intensity** of your *myo*-3 488 bp insert band **matches** that of the 310 bp phage ϕx174 *Hae*III fragment and the **intensity** of vector 5200 bp vector band **matches** that of the 610 bp phage ϕx174 *Hae*III fragment. We will also assume that 2 μl of both insert and plasmid DNA were electrophoresed.

Calculate the amounts and concentrations of your insert and plasmid DNA as shown below:

a. **Insert**: (assume same intensity as 310 bp ϕx174 *Hae*III fragment DNA)
   example: *amount of DNA in our insert and its corresponding band:*
   = 1 μl x 0.5 μg/μl x (310 bp/5.386 kb) = 0.02878 μg = **28.78 ng**
   concentration of DNA in our insert:
   = 28.78 ng/2 μl DNA loaded = **14.39 ng/μl**

b. **Plasmid**: (assume same intensity as 603 bp ϕx174 *Hae*III fragment DNA)
   example: *amount of DNA in our insert and its corresponding band:*
   = 1 μl x 0.5 μg/μl x (603 bp/5.386 kb) = 0.05598 μg = **55.98 ng**
   concentration of DNA in our insert:
   = 55.98 ng/2 μl DNA loaded = **27.99 ng/μl**

NOTE: For longer DNAs, one sees more intense bands because more ethidium is bound to a strand (or mole of strands) relative to shorter DNAs. Since the amount that binds is directly related to the length of the duplex (1 ethidium per 2 bp), one has to account for DNA length when using fluorescence intensity to characterize the molar amount of DNA.

8. To determine the *amount of DNA total in our sample,* convert this calculated *μg of DNA nucleotides* value to units of *moles of DNA bps*. Since this is done on a *per bp basis*, the length does *not* affect the calculation. This is why μg units are often used for DNAs ranging from short oligonucleotides to polynucleotide

duplexes. The average molecular weight of a bp is 660 g/mole. First, calculate the total amount of your insert and plasmid DNA. (This amount is determined from the volume of DNA extracted from the Elu-Quick procedure remaining after gel electrophoresis to quantitate fragment yield.)

NOTE: In these examples the volumes of insert and plasmid DNA extracted from the Elu-Quick procedure remaining after gel electrophoresis to quantitate fragment yield are assumed to be **9 μl**. To calculate the amount of your insert and plasmid DNA, use the volume remaining from your Elu-Quick extraction after gel electrophoresis.

a. Calculate the values for your **insert** DNA (from 7a):
example: **14.39 ng/μl** x 9 μl = **129.51 ng**

Calculate the molecular weight of the DNA in your band:
example: 488 bp *myo*-3 insert x (660 g/mol bp) = **3.22 x 10^5 g/mole fragment**

Calculate the number of moles of fragment (from 8a):
example: (129.51 x 10^{-9} g/9 μl)/(3.22 x 10^5 g/mol) = **4.469 x 10^{-8} mol of insert DNA in 9 μl**

b. Calculate the total amount of **plasmid** DNA (from 7b):
example: **27.99 ng/μl** x 9 μl = **251.91 ng**

Calculate the molecular weight of the DNA in your band:
example: 5,200 bp pUR288 plasmid x (660 g/mol bp) = **3.43 x 10^6 g/mole fragment**

Calculate the number of moles of fragment (from 8b):
example: (251.91 x 10^{-9} g/9 μl)/(3.43 x 10^6 g/mol) = **8.147 x 10^{-9} mole**

Follow the example calculations above to quantitate the amount of your insert and plasmid DNA. Use these values in the next section to calculate the volumes of plasmid and insert DNA to add to the ligation mixture.

3. Ligation of the *myo*-3 *Hin*dIII DNA Product into pUR288

1. Pipet the following components into tubes 1 and 2. Calculate the volumes of DNA needed to obtain a 1 to 3 molar ratio of vector (in moles of strands) to insert DNA. Use the maximum volume of insert and vector DNA in the ligation reaction (maintain the 1 to 3 molar ratio) and adjust the reaction size to the desired volume with H_2O. The first tube contains the normal reaction. The second tube is a *control reaction* in which you will *leave out the myo-3 fragment*, to determine what the gel pattern will look like when ligation reactions are run in the absence of insert. The two patterns will be compared with those of the molecular weight standards to determine how well the reaction worked.

Tube 1:
____ μl pUR288/*Hin*dIII
____ μl *myo*-3 insert
____ μl H_2O
1 μl 10x ligase buffer
+ 1 μl T4 DNA ligase
10 μl

Tube 2:
____ μl pUR288/*Hin*dIII
____ μl H_2O
1 μl 10x ligase buffer
+ 1 μl T4 DNA ligase
10 μl

Incubate the ligation reactions overnight at 12°C. The reactions will be stored and used for the transformations in the next lab period.

Enzymology Notes:

1. One *unit* of *restriction endonuclease* is typically designated as the amount required to digest 1 μg of substrate DNA to completion in 1 hour at 37°C.
2. The enzyme is stored in *glycerol* as a preservative. The concentration of glycerol should be < 5% since it inhibits some enzymes. Enzymes are often heat inactivated to stop the reaction by placing the tube containing the reaction mixture in a 65°C water bath or heating block. Gel loading buffer also stops the reaction. Not all enzymes have optimal activity at 37°C.
3. There are three types of restriction endonuclease cuts: 3' overhang, 5' overhang, and blunt end.
4. *"Star" activity* refers to altered site recognition and cleavage, usually due to suboptimal reaction conditions (*e.g.*, produced by low ionic strength, high glycerol concentration).
5. If you cut with two different enzymes, the fragment can only be inserted in one orientation.
6. Double restriction digestion eliminates nonproductive vector religation. However, two or more vectors can ligate to form *linear concatamers* (long sequence repeats).
7. *Linkers* usually hexanucleotides that encode a particular type of restriction site. Used to place one or a set of restriction sites at the linker location (often at a blunt-end). The linker is then treated with restriction enzyme to make the "sticky end." *Polylinkers* contain several types of sites and allow more flexibility in cloning after only a single construction step. If contaminating nuclease activity is present and you cut for too long, you'll ruin the product.
8. *DNA ligase* seals a covalent bond between 3' OH and 5' PO_4^{2-}. The *phosphatase* activity prevents religation of the vector to itself. The two ends are attached to each other through the linearized chain of the vectors so they are more likely to find each other than if they came from two independent, unconnected ends. *T4 bacteriophage DNA ligase* seals *blunt ends*. Bacterial ligases do not.
9. The following factors affect the efficiency of ligation:
   a. Temperature
   b. DNA concentrations
   c. (Vector:Insert) stoichiometry
10. *T4 bacteriophage DNA ligase*:
    a. Reaction: phosphodiester bond formation accompanied by dehydration.
    b. Increased concentration of DNA increases the reaction rate.
    c. ATP increases the efficiency
    d. Blunted-end ligations are more difficult, and therefore less efficient, than staggered-end reactions
11. *ATP* should be fresh because it's unstable. It is usually stored frozen in aliquots so you don't have to take too much out at once, helping to preserve your stock. Tricine buffer is added to stabilize the ATP.
12. *Chloroform extraction*: including, isoamyl alcohol, helps to separate the phases and decreases protein foaming.
13. *Ethanol precipitation* of DNA/RNA is used to concentrate the sample for resuspension in another solvent. 95% ethanol is better than 100% because the latter has benzene in it to eliminate water. Salts used in ethanol precipitation usually include potassium (KOAc) or sodium acetate (NaOAc); the latter is more soluble in aqueous ethanol. However, it must be carefully removed because it interferes with ligation. Drying DNA from ethanol can cause formation of odd structures that do not revert easily to usable molecules.
14. *Phenol* is immiscible with water. You get two phases, an aqueous/DNA phase and the phenol/protein phase.
15. Ammonium acetate and alkylated derivatives are used in a lot of biological applications (e.g., HPLC) because they are volatile and lyophilize away fairly easily.
16. You see the bound ethidium molecules on transilluminated gels, not the DNA. Ethidium binds to double-stranded DNAs without much preference for sequence. It binds by "*intercalating*" between every other set of base pairs and leaves the alternate sites empty (called "*neighbor exclusion binding*"). You only "see" the bound ethidium because fluorescence is stimulated by intercalation and the bound form dominates the fluorescence yield. This is caused by electronic interactions between the ethidium, the stacked base pairs and neighboring bound ethidiums. Relaxation from the electronically excited state(s) is reduced when the intercalator becomes shielded from the H_2O relaxation "sink" external to the DNA helix.
17. *Bacterial alkaline phosphatase* (BAP). Alkaline phosphatase removes 5' PO_4^{2-} groups from DNA and RNA molecules; preparations are sometimes not particularly pure. Phosphatase activity is eliminated by phenol extraction (phenol denatures the protein). Zinc is a required cofactor, so the enzyme can also be inactivated with a Zn^{2+} chelator (e.g., *nitrilotriacetic acid*, NTA).

18. *Agarose gel electrophoresis* separates molecules by size differences. The gel matrix is a solid support; the pore size depends on the agarose concentration.
19. *Ethidium bromide* (the Br^- is not attached) alters the mobility of DNA when it becomes intercalated within it, sometimes in unpredictable ways depending upon the specific structure.
20. Transformation efficiency = colony forming units (cfu) /μg DNA.
21. Definitions and Conversions
    a. Spectrophotometry approximations:
        1 A_{260} unit of double stranded DNA = 50 μg/ml
        1 A_{260} unit of single stranded RNA = 40 μg/ml
        1 A_{260} unit of single stranded DNA = 33 μg/ml
    b. Molar end calculations: phosphatase treatments, ligation and primer applications:
        i. Picomoles ends/μg of linear DNA = 2×10^6/(660 g/mole x number of bases); 660 g/mole is the average of molecular weight of a base pair.
        ii. 1 μg of 1,000 bp linear DNA = 1.52 pmole (3.03 pmole ends)
    c. Molecular weight markers:
        i. DNA: λ *Hind*III digest, MW of fragments in kilobase pairs (kb): 23, 9.0, 6.0, 4.3, 2.3, 2.0, 0.56
        ii. Proteins: the average molecular weight of an amino acid is 114 Da.

Additional Report Requirements:

1. Photograph of agarose gel
2. Fragment yield calculations
3. Ligation calculations

References:

1. Elu-Quik™ DNA Purification Kit™, Schleicher & Schuell, Keene, NH.

VENDOR LITERATURE

GIBCO BRL™ T4 DNA LIGASE

Gibco BRL Life Technologies, Gaithersburg, MD, ph. (800) 828-6686. (Adapted)

1. Introduction

T4 DNA ligase catalyzes the ATP-dependent formation of a phosphodiester bond between the 3' hydroxyl end of a double-stranded DNA fragment and the 5' phosphate end of the same or another DNA fragment (Figure 1). It is widely used for generating recombinant DNA molecules in which DNA fragments of interest are covalently inserted into vector molecules. The enzyme also can be used to add linkers to the ends of DNA fragments in order to introduce the appropriate cohesive ends prior to cloning into a vector with cohesive ends (Figure 2). This bulletin describes conditions for using T4 DNA ligase for direct cloning into plasmids or other circular DNA vectors such as M13, for cloning into DNA vectors, and for adding linkers to DNA fragments. See Table 1 for a summary of the reaction parameters in the protocols. When transforming bacteria with plasmid ligation products, the goal is to maximize the number of colonies containing recombinant plasmids with single inserts. This requires ligation reaction conditions that maximize the number of monomeric, and circular molecules, since concatamers and linear molecules do not transform as efficiently. A number of ligation reaction parameters are important, including DNA concentration, the molar ratio of insert to vector, temperature, buffer composition, and enzyme concentration. Different protocols are recommended for cloning into circular vectors with cohesive ("sticky") ends and those with blunt ends. Because ligating blunt ends is less efficient, the process requires higher concentrations of enzyme and DNA, a longer incubation time, and a lower reaction temperature than used for cohesive ends.[1,2,3]

After ligation of inserts into a λ vector, the reaction produces are packaged *in vitro* into bacteriophage, a procedure that works most efficiently with linear concatamers of DNA. Therefore, the protocol for cloning into λ-DNA vectors with cohesive ends uses a relatively high vector DNA concentration that favors concatamer formation.[4]

When ligation reaction products are used to transform bacteria, a *high background* of colonies containing nonrecombinant plasmids may result from the *religation of the ends of the vector molecules.* This background can be reduced by treating the linear vector DNA with *bacterial alkaline phosphatase* (BAP) prior to insert ligation.[5] BAP removes the 5' phosphatase groups from each strand of the vector molecule, preventing T4 DNA ligase from forming phosphodiester bonds between the two ends of the vector. BAP treatment of the restriction endonucleases generates incompatible ends. The use of BAP may not be necessary if a method is employed to distinguish colonies containing vectors with inserts from those containing vectors without inserts (e.g., "blue/white" screening by *α*-

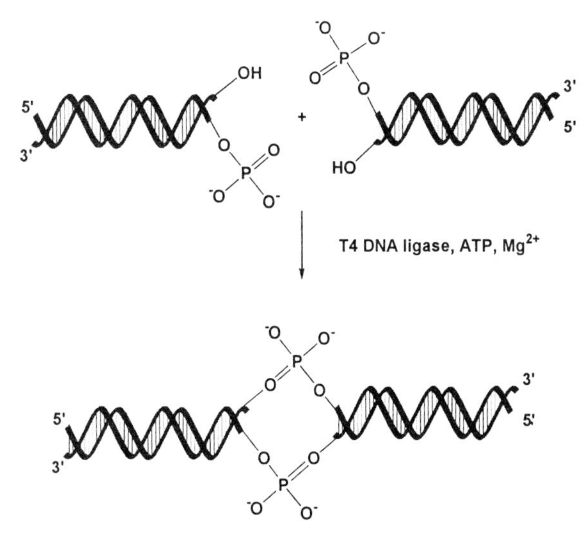

Figure 1 - T4 Ligase catalyzed ATP-dependent phosphodiester bond formation.

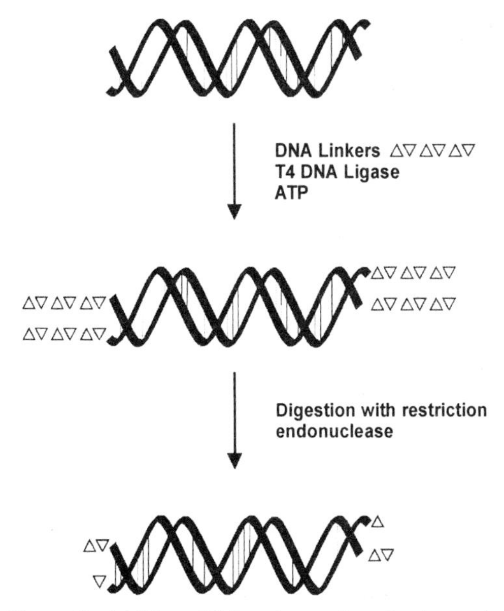

Figure 2 - Addition of linkers to create cohesive ends.

complementation), but BAP treatment will help maximize the number of recombinants. BAP treatment is also used to reduce background in λ vector ligations.

In making recombinant DNA molecules, having the flexibility to change the restriction endonuclease sites at the ends of fragments is often helpful. Linkers are used for this purpose (Figure 2). *Linkers* are small double-stranded oligodeoxyribonucleotides that include the recognition sequence for a restriction endonuclease. Following ligation of the blunt-ended linkers to blunt-ended DNA fragments, digestion with the appropriate restriction endonuclease generates cohesive ends. The insert with its *new cohesive ends* then can be ligated into a vector with compatible cohesive ends. Linker ligation is done with a large molar excess of linkers. This increases the likelihood that every insert will have at least one linker ligated to each end and therefore generates DNA fragments with multiple linkers on each end. Restriction endonuclease digestion followed by gel-exclusion chromatography, ion-exchange chromatography, or gel electrophoresis removes the extra linkers from the insert. To protect the DNA inserts, internal restriction endonuclease site inserts that contain the same restriction endonuclease site as the linker are treated with the corresponding methylase before the linkers are added. Commercially available linkers often lack the 5' phosphate groups required for ligation and must be phosphorylated using T4 polynucleotide kinase before ligation.

T4 DNA ligase is the *product of gene 30 of Escherichia coli bacteriophage T4*. It is a monomer with a molecular weight of 68 ± 6.8 kDa.[6] BRL's T4 DNA Ligase is purified from *E. coli* L lysogen NM9S89[7] and is supplied in 10 mM Tris-HCl (pH 7.5), 50 mM KCl, 1 mM dithiothreitol, 50% (v/v) glycerol at a concentration of 0.5 to 2 units/μl.

Unit Definition: One Weiss unit of T4 ligase catalyzes the *exchange of 1 nmol* ^{32}*P-labeled pyrophosphate into ATP* in 20 minutes at 37°C.[8] BRL™ T4 DNA ligase is measured in Weiss units. Other suppliers may use *cohesive-end ligation units*. One Weiss unit is equal to ~300 cohesive-end units.

Table 1 - DNA Ligation Tips

Possible Causes	Suggested solution
Inhibitors of DNA ligase are present in the insert or vector (See Analysis of Ligations for Plasmid Cloning, for more information).	Extract the DNA with buffer saturated phenol,[14] extract with chloroform:isoamyl alcohol, then precipitate with ammonium acetate and ethanol.[15] Be sure the DNA is free of phenol and that the phosphate concentration is <25 mM and the NaCl concentration is <50 mM. Be sure the DNA is free of contaminating DNA that might compete for ligase in the insert or vector (e.g., linker fragments, DNA fragments from which the insert was incompletely purified).
DNA ligase is inactive (see procedures to check ligase activity).	ATP in the reaction buffer has degraded. Use fresh ligase. Use 5x Ligase Reaction Buffer that is <6 months old. Store the buffer at -20°C.
Insert and vector DNA (or linkers) have incompatible ends.	Confirm that the vector and insert have been digested with the same restriction endonuclease or with different restriction endonucleases that generate compatible ends. If the vector has been digested with two restriction endonucleases, be sure that both digestions are complete.[16] If ligating blunt-ended molecules, treat the vector and/or insert with T4 DNA polymerase to be sure that the ends are blunt.[17] (Do not treat linkers.)
DNA is degraded by nonspecific endodeoxyribonucleases contaminating reaction mixture.	Use fresh buffer, fresh ligase and distilled autoclaved water. (Test this by performing a mock ligation reaction with the supercoiled vector; contamination will result in lower transformation efficiency compared to supercoiled vector that has not been exposed to the ligation reaction components.)
Restriction endonucleases are present causing digestion of ligased treated products.	After digestion of the vector and insert DNA, remove restriction endonucleases by extraction with buffer-saturated phenol,[14] extraction with chloroform:isoamyl alcohol, and ethanol precipitation.[15] Concentrations of DNAs are not known; determine by gel electrophoresis.

2. Experimental Materials

In addition to T4 DNA ligase and the insert and vector DNAs, the following reagents and equipment are required for the protocols described below:

1. *5x ligase reaction buffer*: 250 mM Tris-HCl (pH 7.6), 50 mM $MgCl_2$, 5 mM ATP, 5 mM dithiothreitol, 25% (w/v) polyethylene glycol 8000. Store at -20°C. This buffer is included

with BRL's T4 DNA Ligase. Autoclaved 1.5 ml microcentrifuge tubes, autoclaved distilled water, microcentrifuge (15,000 x g), 70°C water bath.

2. *Ligations for plasmid cloning*: 14°C water bath (required for ligation of blunt-ended fragments only), 0.5 M EDTA.
3. *Ligation of DNA linkers to blunt-ended DNA*: DNA linkers, T4 polynucleotide kinase, 5x T4 polynucleotide *kinase forward reaction buffer* [300 mM Tris-HCl (pH 7.8), 75 mM 2-mercaptoethanol, 50 mM $MgCl_2$, 3 mM ATP], 16°C water bath, 37°C water bath.
4. *Ligation of DNA linkers to blunt-ended DNA and λ cloning of DNA fragments with cohesive ends*: *Buffer-saturated phenol*, *chloroform-isoamyl alcohol* [24:1 (v/v)], 3 M sodium acetate, absolute ethanol, 70% (v/v) ethanol.

3. Troubleshooting

Because *ligation is assayed by transformation of bacteria*, it is important to ensure that the transformation procedure is working. The controls described in *Analysis of Ligation for Plasmid Cloning* will help to distinguish between problems with the *ligation reaction* and problems with the *competent cells*, the *selection medium*, the restriction endonuclease *digestion of the vector*, and the *phosphatase* treatment of the vector.

Because transformation of some competent cells is inhibited by components of the ligation reaction, the restriction should be diluted five-fold before being used for transformation.[13] Some possible causes of unsuccessful ligation are listed below along with suggested solutions.

3.1 Calculation of Molar Ratios. One microgram of a 1000 bp DNA fragment is equivalent to 3000 fmol (10^{-15} mol) double-stranded DNA ends (factor 1). The amount of double-stranded DNA corresponding to a given number of ends is calculated as follows. The units are in parentheses following each quantity. Factor 1 and 2 are conversion factors which are used to express the DNA concentration in particular units.

$$[\mathrm{DNA}]\left(\frac{\mu\mathrm{g}}{\text{pair of ends}}\right) = [\mathrm{DNA}](\mathrm{fmol}) \times \text{factor1} \times \text{factor2}$$

$$[\mathrm{DNA}](\mu\mathrm{g}) = [\mathrm{DNA}](\mathrm{fmol}) \times \frac{1\,\mu\mathrm{g}}{3000\,\mathrm{fmol}} \times \frac{\text{size DNA bp}}{1000\,\mathrm{bp}}$$

$$= 0.016\,\mu\mathrm{g} = \frac{16\,\mathrm{ng}}{\text{pair of ends}}$$

Example. A 1600 bp insert with cohesive ends is to be cloned into a 2700 bp vector. Ten fmol of vector will be used. An insert:vector molar ratio of 3:1 is desired. The number of micrograms of DNA corresponding to 10 fmol of ends of vector is:

$$[\text{vector}](\mu\mathrm{g}) = 10\ \mathrm{fmol} \times \frac{1\,\mu\mathrm{g}}{3000\ \mathrm{fmol}} \times \frac{2700\ \mathrm{bp}}{1000\ \mathrm{bp}} = 0.009\,\mu\mathrm{g}$$

The number of femtomoles of insert ends required is 3 times the number of moles of vector ends, or 30 fmol. The amount of double-stranded insert DNA required is:

$$[\text{insert}](\mu\mathrm{g}) = 30\mathrm{fmol} \times \frac{1\,\mu\mathrm{g}}{3000\ \mathrm{fmol}} \times \frac{1600\ \mathrm{bp}}{1000\ \mathrm{bp}} = 0.016\,\mu\mathrm{g}$$

3.2 Analysis of Ligations for Plasmid Cloning. Ligation reactions for transformation are best analyzed by actual transformation of bacteria. Ligation reagents can be tested by gel analysis of ligation reaction products.

4. Transformation

For most applications, transformation of competent cells is the ultimate test of a ligation reaction because not all of the high molecular weight forms created in a reaction will transform cells efficiently. Several ligation controls necessary to identify the source of any ligation problems are described below.

1. *Supercoiled vector*: control DNA provided with BRL's competent cells. Perform a transformation reaction and plate the number of cells expected to generate 50 to 100 colonies per plate (based on the anticipated transformation efficiency of the competent cells). The expected number of colonies should be seen, indicating that the competent cells are transforming with high efficiency. The control DNA provided with BRL's competent cells is supercoiled monomer. Vector DNA preparations that contain other forms will not transform as efficiently. Transformation efficiencies will be ~10-fold lower to ligate inserts to vectors versus to intact control DNA.
2. *Restriction endonuclease-digested vector*. Perform a transformation with an amount of vector DNA equivalent to that contained in the fraction of the ligation reaction used to test the reaction initially. Few or no colonies should be

seen, indicating *complete restriction endonuclease digestion* of the vector. The presence of colonies indicates incomplete digestion of the vector. This population of *background* colonies contains nonrecombinant vector in the experimental reactions.

3. *Restriction endonuclease-digested religated vector.* Set up a ligation reaction using the same amount of vector DNA used in tests and use it to transform competent cells. Religation of vectors with cohesive ends should result in = 50% of the number of colonies obtained with supercoiled vector DNA, indicating that the components of the ligation reaction are working. Religation of vectors with blunt ends should yield 10% to 20% of the number of colonies obtained with supercoiled vector DNA. This is an appropriate control only with vectors that have been digested with a single restriction endonuclease. Double-digested vectors may not religate because the ends are incompatible and the small DNA fragment that is released from between the two sites is sometimes lost during ethanol precipitation of the DNA.
4. *Restriction endonuclease-digested, dephosphorylated religated vector.* Set up a ligation reaction using the same amount of vector that is used in the experimental ligations and use it to transform competent cells. Few or no colonies should be observed, indicating *complete dephosphorylation* of the vector. The dephosphorylated vector is not ligated by T4 DNA ligase.
5. *No vector.* Perform a mock transformation of competent cells to which no DNA is added. No colonies should be seen, indicating that the selecting antibiotic on the agar plates is potent and that the competent cells are pure.

5. Gel Analysis

The ligation reactions can be tested by performing a ligation reaction with a molecular size marker such as the λ DNA *Hin*dIII fragment ladder.[9,10] Compare the ligation reaction products to unligated DNA on an agarose gel. The ligation reaction should contain a high molecular weight smear and a few low molecular weight bands. *To test for the presence of ligation inhibitors*, perform a ligation reaction in which some of the insert or vector DNA is included along with some marker DNA. If ligation of the DNA marker fragments occurs alone but is not observed when other DNA is added, an inhibitor is present in the vector or insert DNA.

Gel analysis is not a good method of evaluating ligation reactions for transformation. Although gel electrophoresis of a ligation reaction should reveal the disappearance of the vector and insert fragments and the appearance of higher molecular weight forms, not all high molecular weight forms will transform efficiently.

6. Ligation Protocol for λ Cloning of DNA Fragments with Cohesive Ends

The following reaction conditions are for ligation of DNA inserts to dephosphorylated λ DNA "arms" that have been previously ligated at their *cos* (host integration/recombination) sites.

Component	Amount
5x Ligase Reaction Buffer	2 μl
λ vector DNA	1 μl
Insert DNA	50 ng
Autoclaved distilled water	to 8 μl

1. In an autoclaved 1.5 ml microcentrifuge tube, dilute an aliquot of T4 DNA ligase to 0.1 units/μl in 1x Ligase Reaction Buffer.

NOTE: The ligase must be used immediately after dilution.

2. To an autoclaved 1.5 ml microcentrifuge tube, add the following:
3. Add 2 μl (1 unit) of the diluted T4 DNA ligase. Mix gently. Centrifuge briefly to bring the contents to the bottom of the tube.
4. Incubate at 4°C for at least 16 hours.
5. Heat the reaction at 70°C for 10 minutes to inactivate the T4 DNA ligase.
6. Package the reaction products into bacteriophage.

NOTE: Packaging may be inhibited by polyethylene glycol concentrations >1%. Because the ligation reaction contains 5% polyethylene glycol, it is necessary to dilute the ligation reaction if it contributes more than 20% of the final volume of the packaging reaction.

7. Inactivation

T4 DNA ligase can be inactivated by heating to 70°C For 10 minutes or by adding EDTA to 25 mM to chelate the Mg^{2+} ions in the reaction mixture.

8. Other Activities of T4 DNA Ligase

T4 DNA ligase will ligate DNA/RNA hybrid molecules with extremely low efficiency, joining either DNA to DNA or DNA to RNA. It will ligate double-stranded RNA molecules with extremely low efficiency.[6] The enzyme can be used to repair nicks in double-stranded DNA.[6,8] Single-stranded nucleic acids are not substrates for T4 DNA ligase.

9. Analysis of Linker Ligations

The ligation of linkers to an insert and the subsequent restriction endonuclease digestion of the products can be evaluated by electrophoresis of reaction products containing ~125 ng of linkers on a 6% (w/v) polyacrylamide. Stain the gel with 0.5 µg/ml ethidium bromide in water for 15 minutes. This method does not directly assess ligation of linkers to the insert DNA because the insert is frequently too large to enter the gel, but a "ladder" of ligated linkers indicates that the linkers are phosphorylated and that the ligation reagents are working. The "ladder" disappears after restriction endonuclease digestion.

Linker phosphorylation and ligation can be monitored by agarose gel electrophoresis and autoradiography if [γ-^{32}P]ATP is included in the linker phosphorylation reaction. Successful phosphorylation and ligation will result in the addition of ^{32}P -labeled linkers to the ends of the insert molecule. Insert DNA that has been restriction endonuclease digested and electrophoresed should appear as a distinct band on an autoradiogram.

10. Reaction Conditions

Temperature affects the rate and extent of ligation (Table 2). Blunt-end ligation at 26°C produces 90% as many transformants in 4 hours as are obtained in 23 hours, and 25-fold more transformants than are obtained at 4°C in 4 hours incubation at room temperature.[1] Overnight incubation at 14°C produces 4-fold more transformants than a 4 hour incubation at room temperature.[2]

T4 DNA ligase *requires* Mg^{2+}. Using the unit assay conditions, the optimal concentration is 10 mM. At 3 mM Mg^{2+}, 35% of maximal activity observed; 80% of maximal activity is observed at 30 mM Mg^{2+}, while Mn^{2+} can substitute for Mg^{2+}. At the optimal Mn^{2+} concentration of 10 mM, the activity is 25% of that observed with 10 mM Mg^{2+}.[5] The *optimal pH range* for T4 DNA ligase is 7.2 to 7.8. At pH 6.9 the enzyme is 46% as active as it is at pH 7.6; at pH 8, it is 65% as active as at pH 7.6.[6]

High molecular weight polymers, such as polyethylene glycol (PEG), Ficoll bovine serum albumin and glycogen, have been shown to enhance both blunt-end and cohesive-end ligation by T4 DNA ligase.[18] When ligation is assayed by the number of transformants obtained, a sharp optimum is seen at a PEG 8000 concentration of 5%.[1]

Hexamine cobalt chloride has been shown to enhance ligation of blunt-ended fragments when ligation is assayed by gel electrophoresis,[19] but it does not increase the number of transformants obtained when ligations are performed according to the protocols described above.[1]

Although T4 RNA ligase was reported to enhance the ligation of blunt-ended DNA fragments by T4 DNA ligase,[20] subsequent studies were unable to confirm this finding .[18]

dATP is a *competitive inhibitor* of T4 DNA ligase resulting in a 60% decrease in activity at a concentration of 66 µM. The following compounds do not inhibit ligase at a concentration of 66 µM: GTP, CTP, UTP, dGTP, dCTP, dTTP, ADP, ATP, NAD^+.[8] Transfer RNA does not inhibit DNA ligase.[21]

The following ions inhibit T4 DNA ligase almost completely at concentrations of 0.2 M: K^+, Cs^+, Li^+, NH_4^+; however, NH_4^+ is not inhibitory at a concentration of 10 mM. [6,22] NaCl at a concentration of 50 mM causes negligible inhibition of blunt-end ligation and 15% inhibition of cohesive-end ligation; at a concentration of 100 mM, NaCl inhibits blunt-end ligation by 50% and cohesive-end ligation by 40%. Potassium phosphate at a concentration of 50 mM inhibits blunt-end ligation by 50% and cohesive-end ligation by 5%.[23] EDTA inhibits T4 DNA ligase by chelating Mg^{2+} ions.

Molar reaction parameters are located in Table 2.

Table 2 - Reaction Parameters for Ligations.

Parameter	Plasmid Cohesive Ends	Linker Ends	Blunt Ends	λ
T4 DNA ligase (units)	0.1 (1*)	1	2	1
amount of DNA				
Vector (fmol)	3-30 (30*)	15-60	-	-
Insert (fmol)	9-90 (90*)	45-180	1400	-
Linkers (pmol)			150	-
Vector (ng)	-	-		1000
Insert (ng)	-	-		
molar ratio				
Insert:Vector	3:1	3:1	-	-
Linker:Insert	-	-	100:1	-
Temperature (°C)	23-26	14	14	4
Time (hours)	∃1	24	16	∃16
Reaction volume (µl)	20	20	50	10

11. Storage and Stability

T4 DNA ligase is stable for one year at -20°C. The 5x Ligase Reaction Buffer is stable for 6 months at -20°C.

NOTE: The ligase must be used immediately after dilution. Do not store diluted ligase.

NOTE: Phenol extraction, chloroform:isoamyl alcohol extraction and ethanol precipitation of the DNA are recommended before restriction endonuclease digestion to ensure complete digestion. If you omit these steps, it will be necessary to inactivate the ligase by heating to 70°C for 10 minutes and to decrease the polyethylene glycol concentration by diluting the ligation reaction at least 2.5-fold before performing the restriction endonuclease digestion.

References:

1. King, P. W. and Blakesley, R. W. (1986) Focus 8:1, 1.
2. Focus (1986) 8:3, 13.
3. Focus (1986) 8:4, 12.
4. Dugaiczyk, A., Boyer, H. W. and Goodman, H. M. (1975) J. Mol. Biol. 96, 171.
5. Focus on Applications, Bacterial Alkaline Phosphatase, Technical Bulletin 8011-1 (1989) Life Technologies Incorporated.
6. Engler, M. J. and Richardson, C. C. (1982) in *The Enzymes*, Vol. XV, p. 3, Academic Press, New York.
7. Murray, N. E., Bruce, S. A. and Murray, K. (1979) J. Mol. Biol. 132, 493.
8. Weiss, B., Jacquemin-Sablon, A., Live, T. R., Fareed, G. C. and Richardson, C. C. (1968) J. Biol. Chem. 243,
9. Focus (1985) 7:2, 14.
10. Focus (1989) 11, 28.
11. Focus (1984) 6:2, 10.
12. Zeugin, J. and Hartley, J. (1985) Focus 7:4, 1.
13. Jessee, J. (1984) Focus 6:4, 5.
14. Karger, B. D. (1989) Focus 11, 14.
15. Crouse, J. and Amorese, D. (1987) Focus 9:2, 3.
16. Crouse, J. and Amorese, D. (1986) Focus 8:3, 9.
17. Focus on Applications, T4 DNA Polymerase, Technical Bulletin 8005-1 (1990) Life Technologies Incorporated.
18. Pheiffer, B. H. and S. B. Zimmerman (1983) Nucl. Acids Res. 11, 7853.
19. Rusche, J. R. and Howard-Flanders, P. (1985) Nucl. Acids Res. 13, 1997.
20. Sugino, A., Goodman, H. M., Heyneker, H. L., Shine, J., Boyer, H. W. and Cozzarelli, N. R. (1977) J. Biol. Chem. 252, 3987.
21. Ausubel, F. M., Brent, R., Kingston, R. E., Moore, D. D., Seidman, I. G., Smith, J. A. and Struhl, K. (1989) *Current Protocols in Molecular Biology*, p. 3.14.2, John Wiley and Sons, New York.
22. Lehman, I. R. (1974) Science l86, 790.
23. Focus (1983) 5:1, 12.

VENDOR LITERATURE

DNA PURIFICATION KIT (NAI/GLASS BEAD METHOD)

Schleicher and Schuell, Keene, NH, ph. (800) 245-4024. (Adapted)

1. Introduction

The Elu-quik™ DNA Purification Kit is recommended both for the isolation of genomic DNA from whole cell suspensions and for the isolation of single- and double-stranded DNA molecules less than 50 kb in size from a variety of sources. By exploiting the *affinity of DNA for glass in sodium perchlorate* genomic DNA fragments as large as 200 kb can be purified using procedures designed to minimize shearing of the DNA (Figure 1). The effect of DNA binding to glass particles is well established and has been demonstrated for DNA extraction from agarose,[1,2] plasmid purification from bacterial cell lysates[3] and preparation of single-stranded M13 sequencing templates.[4] Thompson *et al.* have reported the use of glass in sodium perchlorate buffer to extract cellular DNA from crude cell lysates.[5] Many commercially available *silica based* DNA isolation systems bind the DNA sample in a *sodium iodide* binding buffer. Recent investigations indicate irreversible binding of DNA to glass particles in the presence of NaI.[6] This is a function of both the *concentration of NaI* and the *duration of exposure* and can result in *decreased sample recoveries.*

Traditional labor-intensive methods generally yielding no more than 400 µg of DNA per 10^8 cells

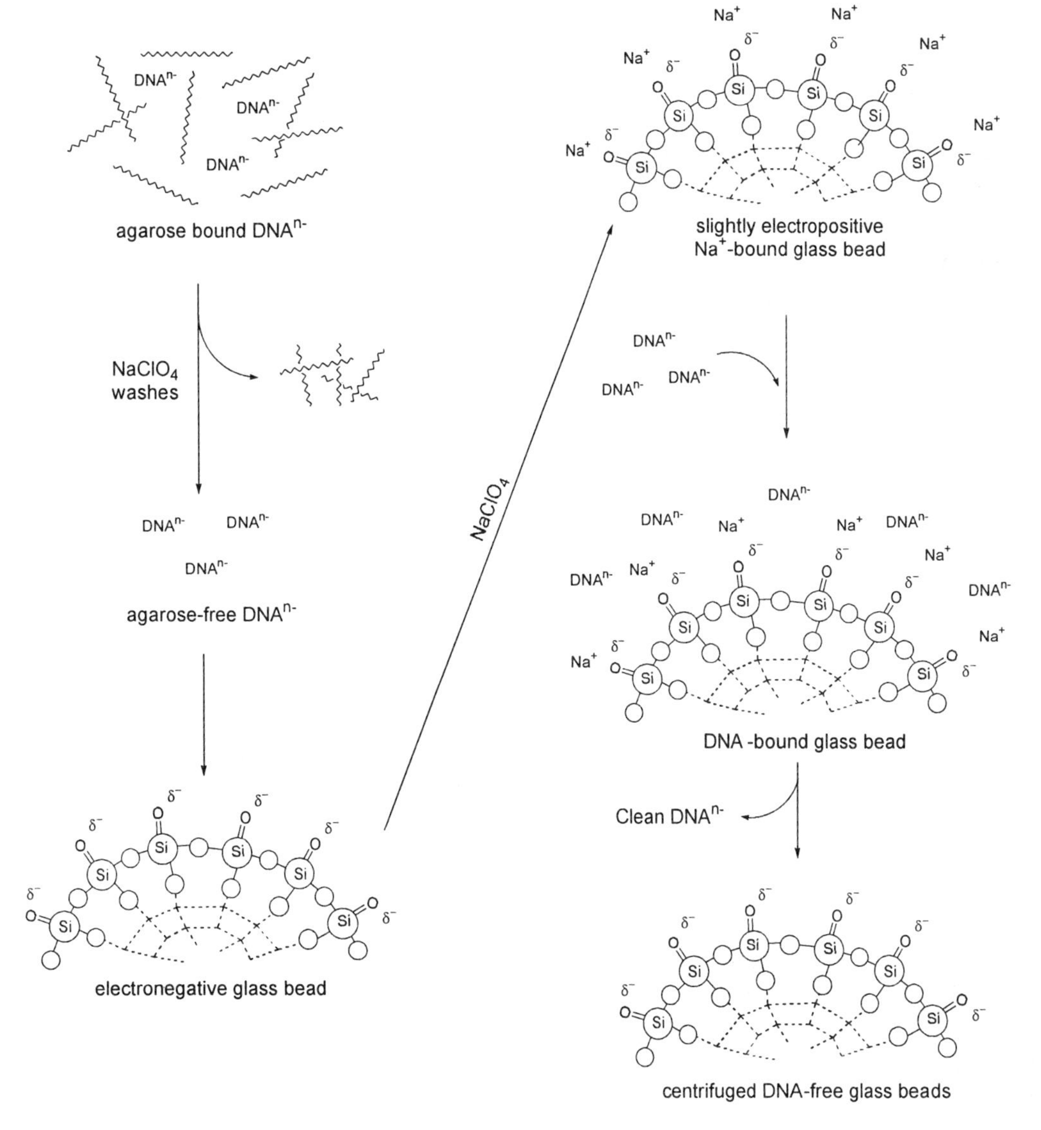

Figure 1 - DNA purification using glass beads in NaClO$_4$. The "n-" superscript on the DNA refers to their net ionic charge.

involve phenol/chloroform extractions recovery of DNA using ethanol and may take up to 24 hours to complete.[7,8] Using the Elu-Quik™ Kit, recovery of more than 650 μg of genomic DNA from 10^8 cells can be accomplished within 90 minutes without ethanol precipitation. *Use of toxic materials such as phenol and chloroform is avoided.* No cell preparation is necessary; complete materials are included in the Elu-Quik™ Kit for lysing the cells.

The following procedures for isolating DNA using the Elu-Quik™ Kit can be subdivided into three areas:

1. The *first* involves a *group of protocols* for isolation of DNA of less than 50 kb, including isolations from agarose, from alkaline or boiling *minipreps*, and from *labeling*, *nick-translation* and *enzymatic amplification* (PCR) reactions.
2. The *second protocol* is for *purification* of smaller quantities (*ca.* 5-10 μg) genomic DNA from 2 x 10^6 cells for use in *library screening*, *restriction fragment length polymorphism studies* and other research procedures.
3. The *third set* of instructions details the isolation of large quantities (>650 μg) of genomic DNA from 10^8 cells for *Southern transfers* or cloning in λ bacteriophage vectors. Yields will vary according to cell type.

2. Equipment (not supplied with kit)

Low-speed centrifuge;
High-speed centrifuge with a capability of at least 4300 x *g* (fixed angle rotor preferred);
Microcentrifuge;
Rocking platform/shaker;
50°C water bath

3. Supplies (not supplied with kit)

15 or 50 ml conical-bottom plastic centrifuge tubes;
50 ml round-bottom plastic centrifuge tubes;
0.50 ml plastic microcentrifuge tubes;
Clean 250 ml bottle with cap

4. Reagents (not supplied with kit)

1. Distilled deionized water (DNase free)
2. Ethanol (95% non-denatured or 100% distilled over benzene preferred)
3. 1x TE buffer (10 mM Tris-HCl containing 1 mM EDTA, pH 8)

5. Kit Components

Kit Components are listed in Table 1.

6. Setting Up the Elu-Quik™ DNA Purification Kit

1. Rinse a clean 250 ml bottle with distilled, deionized water.
2. Add the entire contents of the bottle marked "Wash buffer concentrate."
3. Add 125 ml of ethanol.
4. Cap tightly and invert three times to mix contents.
5. Affix the label "Elu-Quik™ Wash Buffer" included in your Kit and store at -18°C.
6. Remove the bottle labeled "Elu-Quik™ Salt Reduction Buffer" to storage at -18°C.

NOTE: All other components of the kit should be stored in a dark place (i.e., a drawer or cabinet) at room temperature. This storage requirement is important because the vials of glass concentrate and binding buffer both contain light-sensitive ingredients.

6.1 Isolation of DNA Less Than 50 Kilobases in Length: Purification Procedure.

1. Add 280 μl of binding buffer to 100 μl of DNA sample to be purified in a microcentrifuge tube.
2. After resuspending glass using a plastic transfer pipette, add 20 μl of Elu-Quik™ glass concentrate. Tap tube briefly to mix the contents.

Table 1 – Kit Components

Glass concentrate	(5) 2 ml screw-cap vials with 1 ml glass particles suspended in binding buffer.
Lysis buffer	(1) Amber 30 ml bottle with 20 ml of a guanidine thiocyanate-containing cell lysis buffer.
Binding buffer	(1) Amber 125 ml bottle with 95 ml of sodium perchlorate binding buffer.
Wash buffer concentrate	(1) 125 ml bottle with 125 ml of ml Tris/NaCl/EDTA wash buffer (1x concentrate) to be diluted in ethanol (see below).
Salt reduction buffer	(1) 125 ml bottle with 125 ml of an alcohol-based desalting buffer.
Nonshear pipette tips	(20) Disposable pipette tips designed to minimize DNA shearing during the washing step.
Floating tube rack	(1) Floating rack for holding up to ten 1.5 ml microcentrifuge tubes for heating in a water bath.

NOTE: Tests with the Elu-Quik™ Kit show that a *3 to 1 ratio of binding buffer plus glass concentrate* (both of which contain *sodium perchlorate*) *to sample solution is optimal* to purify DNA fragments *under 50 kb in size.* The concentration of glass particles will also influence the efficiency of binding. If the concentration of DNA is greater than 2.5 μg per 100 μl sample, increase the amount of glass proportionately. The ratio of binding buffer plus glass concentrate to sample solution should be kept at 3 to l.

3. Let stand at room temperature for 10 minutes. Invert the tube at one minute intervals. (Alternately, a rocking platform or shaker can be used for scaled-up volumes.)

NOTE: For mixture volumes of less than 400 μl, it is sufficient to incubate at room temperature for 10 minutes without rocking while occasionally flicking the tube.

4. Centrifuge at 7000 x *g* for 30 seconds. Discard as much supernatant as possible using a transfer pipette.
5. Add 500 μl of wash buffer. Using a non-shear pipette tip (supplied with the kit), tease the glass pellet from the wall of the microcentrifuge tube. Gently pipette glass pellet up and down until pellet becomes flocculent. Invert the tube once.

NOTE: Achieving a flocculent pellet is important for buffer exchange.

6. Centrifuge at 7000 x *g* for 30 seconds. Discard supernatant.
7. Repeat steps 5 and 6.
8. Add 500 μl salt reduction buffer. Resuspend glass pellet until flocculant with the non-shear pipette tip as described above. Invert the tube once and centrifuge at 7000 x *g* for 2 minutes. Discard supernatant.

NOTE: The salt reduction buffer is an important feature of this kit. Reducing the salt concentration improves elution efficiency with glass matrices. Tests with the Elu-Quik™ DNA Purification Kit confirm a 25% improvement in yield when substituting this buffer for the final wash buffer recommended with other silica-based procedures.

9. Centrifuge glass pellet at 7000 x *g* for 30 seconds to condense pellet. Using a 200 μl pipette tip, carefully withdraw as much residual supernatant as possible and discard.
10. Dry the glass pellet under vacuum for 15 to 30 minutes at room temperature.

NOTE: It is possible to skip this step and proceed directly to elution (step 11); however, the drying step is recommended to remove trace amounts of ethanol which may interfere with some enzymes (e.g., *Bam*HI). Drying the glass pellet does not impair the elution of DNA fragments 50 kb or less from the matrix.

11. Add 10-50 μl distilled, deionized water, or 1x TE buffer to glass pellet and resuspend by flicking the tube repeatedly. Incubate mixture at 50°C for 5 minutes, occasionally flicking the tube. Centrifuge at 7000 x *g* for 30 seconds in a microcentrifuge and withdraw supernatant containing the isolated DNA to a fresh tube.
12. Withdraw 5-10 μl for immediate analysis, or store at -18°C until ready for use.

NOTE: Sample can be eluted in water if it is to be used immediately, otherwise elutions should be done using 1x TE. Eluted material is stable for at least 2 months at -18°C in 1 x TE. A small amount of residual glass may be apparent in the eluted material. We have not observed this material to interfere with subsequent restriction enzyme digestion. Prior to withdrawing an aliquot for analysis, the residual glass may be pelleted by centrifugation for 30 seconds at 7000 x *g*.

1.1.1 Isolation of DNA from Agarose Gel Slice.

1. Rinse a scalpel with ethanol. Stain gel with ethidium bromide. Cut out the band of interest under UV light and transfer it to a clean microcentrifuge tube.
2. Centrifuge at 7000 x *g* for 1 minute to draw agarose fragments down to a discrete level. Determine the approx-imate volume of gel in the tube.
3. If isolating DNA from TAE gels, add 2.8 volumes binding buffer (e.g., 280 μl per 100 μl gel slice). If isolating DNA from TBE gel slices, add 3.8 volumes of binding buffer.
4. Fragment the gel using a plastic transfer pipette. Place the tube in the floating tube rack supplied with the kit. Insert each tube until the cap is flush with the surface of the floating tube rack. Incubate in a 50°C water bath for 5 minutes.

NOTE: Since TBE gels are somewhat less soluble in sodium perchlorate, more binding buffer is necessary when working with TBE gel slices. When working with either TAE or TBE gel slices, if agarose fragments remain after incubation for 5 minutes at 50°C, add another 100 μl of binding buffer and repeat the above incubation. Additional gentle pipeting with the transfer pipette may aid in dissolution of the gel slice in these cases.

5. Proceed with step 2 of The Purification Procedure.

6.1.2 Isolation of Plasmid DNA from Boiling Miniprep.

1. Inoculate 5 ml of medium containing the appropriate antibiotic with a single bacterial colony. Incubate at 37°C overnight with vigorous shaking.
2. Pour 1.5 ml of the culture into a microcentrifuge tube. Centrifuge for 1 minute. Store the remainder of the culture overnight at 4°C.
3. Remove the medium by aspiration, leaving the bacterial pellet as dry as possible.
4. Resuspend the cell pellet in 350 µl of *STET buffer*:
   0.1 M NaCl
   10 mM Tris-HCl, pH 8
   1 mM EDTA
   5% (w/v) Triton X-100
5. Add 25 µl of freshly prepared lysozyme (10 mg/ml in 10 mM Tris-HCl, pH 8). Mix by vortexing for 3 seconds.
6. Place the tube in a boiling water bath for 40 seconds.
7. Centrifuge immediately at 12,000 x *g* for 10 minutes at room temperature in a microcentrifuge.
8. Proceed with step 1 of The Purification Procedure using 100 µl of the supernatant isolated in the previous step.

NOTE: Up to 300 µl of supernatant may be used by scaling up the purification procedure (*e.g.*, for 300 µl of supernatant, add 840 µl of binding buffer and 60 µl of glass concentrate). If the amount of glass concentrate is scaled up, wash buffer salt reduction buffer and elution volume should also be increased (*e.g.*, elute DNA with 30 µl of 1x TE if using 60 µl of glass concentrate).

6.1.3 Isolation of Plasmid DNA from Alkaline Miniprep.

1. Inoculate 5 ml of medium containing the appropriate antibiotic with a single bacterial colony. Incubate at 37°C overnight with vigorous shaking.
2. Pour 1.5 ml of the culture into a microcentrifuge tube. Centrifuge for 1 minute. Store the remainder of the culture overnight at 4°C.
3. Remove the medium by aspiration, leaving the bacterial pellet as dry as possible.
4. Resuspend the pellet by vortexing in 100 µl of an ice-cold solution of:
   50 mM glucose
   10 mM EDTA
   25 mM Tris-HCl, pH 8
5. Add 200 µl of a freshly prepared, ice-cold solution of:
   0.2 N NaOH
   1% SDS
   Close the top of the tube and mix the contents by inverting the tube rapidly two or three times. Do not vortex. Store the tubes on ice for 5 minutes.
6. Add 150 µl of an ice-cold solution of *potassium acetate* (approximately pH 4.8) made up as follows: to 60 ml of 5 M potassium acetate, add 11.5 ml glacial acetic acid and 28.5 ml distilled, deionized water. The resulting solution is 3 M with respect to potassium and 5 M with respect to acetate. Close the cap of the tube and vortex it gently in an inverted position for 10 seconds. Store on ice for 5 minutes.
7. Centrifuge at 12,000 x *g* for 5 minutes at 4 °C in a microcentrifuge.
8. Proceed with step 1 of The Purification Procedure using 100 µl of the supernatant isolated in the previous step.

NOTE: Up to 300 µl of supernatant may be used by *scaling up* the purification procedure (e.g., for 300 µl of supernatant add 840 µl of binding buffer and 60 µl of glass concentrate). If the amount of glass concentrate is scaled up, the wash buffer, salt reduction buffer, and elution volume should also be increased (e.g., elute DNA with 30 µl of 1x TE if using 60 µl glass concentrate).

6.1.4 Probe Purification from Nick-Translation Reaction. Proceed with step 1 of The Purification Procedure using 10-100 µl of the nick-translation reaction.

6.1.5 Purification of Reaction Product Following PCR. Proceed with step 1 of The Purification Procedure using 10-100 µl of the reaction mixture from the enzymatic amplification.

6.1.6 Preparation of Labeled DNA Insert from Unreacted Labeled Nucleotides. Proceed with step 1 of The Purification Procedure using 10-100 µl of the labeling mixture.

6.2 Isolation of Genomic DNA in Small Quantities for Screening Properties.

1. Collect 2 x 10^6 cells (i.e., from tissue culture, or isolated from *in vivo* source; actual number of cells may be varied depending on cell type) in a

15 ml conical tube by centrifugation at 300 x *g* for 10 minutes in a low-speed centrifuge at room temperature.

2. Carefully pour off and discard the supernatant. Using a 1 ml pipette tip remove as much residual supernatant as possible without disturbing the cell pellet. Resuspend in remaining supernatant cells by vortexing.
3. Add 40 µl lysis buffer immediately after resuspending cells. Flick the tube until the suspension turns clear and the mixture is uniform.
4. Add 80 µl binding buffer. Flick tube briefly to obtain mixing of binding buffer and the lysed material. Using a non-shear pipette tip (supplied with kit), carefully transfer mixture to microcentrifuge tube.
5. Add 25 µl of glass concentrate (alter resuspending glass with a plastic transfer pipette). Mix viscous solution by rapid inversion. Let this solution stand at room temperature with occasional inversion (e.g., once a minute) for 10 minutes.
6. Centrifuge at 7000 x *g* for 30 seconds in a microcentrifuge. Discard as much supernatant as possible by inverting tube and gently tapping. Leave any viscous material at the interface behind.

NOTE: Presence of a viscous material at the interface may interfere with complete removal of the supernatant. This viscous material indicates a good yield of isolated DNA. Any residual binding buffer and cell components resulting from lysis, which remain with the interface, will be removed in subsequent washes.

7. Add 500 µl of wash buffer. Using a non-shear pipette tip (supplied with kit), gently tease the glass pellet from the wall of the microcentrifuge tube. Resuspend pellet until flocculent by gently pipeting it up and down. Centrifuge at 7000 x *g* for 30 seconds in a microcentrifuge and discard the supernatant. Repeat wash one time.

NOTE: Achieving a flocculent pellet is important for buffer exchange.

8. Add 500 µl of salt reduction buffer. Resuspend pellet until flocculent as above. Centrifuge at 7000 x *g* for 2 minutes in a microcentrifuge and discard supernatant.

NOTE: The salt reduction buffer is an important feature of this kit. Reducing the salt concentration improves elution efficiency with glass matrices. Tests with the Elu-Quik™ DNA Purification Kit confirm a 25% improvement in yield when substituting this buffer for the final wash buffer recommended with competitive silica-based procedures.

9. Centrifuge again at 7000 x *g* for 30 seconds in a microcentrifuge. Withdraw residual salt reduction buffer with a 200 µl pipette tip.

NOTE: This step removes any excess ethanol which may interfere with subsequent enzyme reactions. *DO NOT DRY the DNA/glass pellet when isolating genomic DNA.*

10. Add 10 µl of distilled, deionized water, or 1x TE buffer to glass pellet and resuspend by flicking the tube repeatedly. Use the floating tube rack (supplied with kit). Insert each tube until the cap is flush with the surface of the floating tube rack. Incubate mixture at 50°C for 15 minutes. Centrifuge at 7000 x *g* for 30 seconds in a microcentrifuge and withdraw supernatant containing the isolated DNA to a fresh tube.

NOTE: Sample can be eluted in water if it is to be used immediately, otherwise elutions should be done using lx TE Eluted material is stable for or at least 2 months at -18°C in 1x TE.

11. Repeat elution in step 10, and combine supernatants. Withdraw 10-20 µl for immediate analysis, or store at -18°C until ready for use.

Yield: 5-10 µg from 2 x 10^6 cells (may vary depending on cell type)

Purity: $A_{260}/A_{280} > 2.0$

References:

1. Vogelstein, B. and D. Gillespie. 1979. Proc. Natl. Acad. Sci. USA. 76: 615.
2. Robert C., A. Yang, J. Lis and R. Wu. 1979. In *Methods in Enzymology*, R. Wu (Ed.).Academic Press, New York. 68: 176-182.
3. Marko, M. A., R. Chipperfield and H. C. Birnboim. 1982. Anal. Biochem. 121: 382-387.
4. Kristensen, T. H. Voss and W. Ansorge. 1987. Nucl. Acids Res. 15: 5507-5516.
5. Thompson, J. D., K. K. Cuddy, D. S. Haines and D. Gillespie. 1990. Nucl. Acids Res. 18: 1074.
6. Gillespie, D. 1990. Personal communication.
7. Maniatus, T., E. F. Fritsch and J. Sambrook. 1989. *Molecular Cloning II: A Laboratory Manual*. Cold Spring Harbor Laboratory, Cold Spring Harbor, New York.
8. Wilson, K. 1987. *Current Protocols in Molecular Biology*, F. Ausubel *et al.* (Eds.). John Wiley & Sons, New York, pp. 2.4.1-2.4.5.

ALTERNATIVE APPROACH

THE USE OF β-AGARASE TO RECOVER DNA FROM GEL SLICES

Krall, J. A. United States Biochemical Corporation, Cleveland, OH, ph. (216) 765-5000. (Adapted)

It is frustrating to see gray streaks and undetectable ladders on a sequencing film due to dirty template DNA. This is especially true when the template is a PCR product. One thinks "I don't know what happened! The template DNA looked really good in the agarose gel. There was just one sharp band before I started."

The importance of a homogeneous and clean template for DNA sequencing, especially when the template is a PCR product, cannot be stressed enough. The homogeneity of the template DNA is determined by its source and in some cases its preparation (e.g., PCR products). The purity of the template is determined by the method of preparation and purification.

This article outlines a method for recovering clean DNA from agarose gel slices. The enzyme β-agarase I can be used to efficiently and effectively clean and recover DNA samples from low-melt agarose gel slices. *β-Agarase I digests agarose by cleaving the agarose subunit to neoagaro-oligosaccharides.* The recovered DNA is ready to use for sequencing, ligation or other applications requiring gel purified DNA. The recommended protocol involves the following:

1. Run gels in a minimum of 0.8% low-melt agarose.
2. Cut out the band of interest.
3. Equilibrate gel slice in at least 2 volumes of 50 mM Bis-Tris (pH 6), and 10 mM EDTA.
4. Spin briefly and remove supernatant. Incubate at 70°C for 10 minutes to melt the gel slice. Cool to 37°C and maintain temperature at 37 °C.
5. Add an equal volume of Bis-Tris buffer (as described in 3).
6. Add 1 unit of β-Agarase I for every 50 μl of agarose gel (do not include the volume of buffer for this determination). Incubate 12-18 hours at 37°C.

NOTE: One may increase the number of units by 1.5-fold and incubate for 2-3 hours.

7. Extract 1 time with phenol. Extract 1 time with phenol/chloroform. Extract 2 times with chloroform.
8. Add 2 volumes of TE, pH 7.6, to prevent precipitation of oligosaccharides. Precipitate with 0.05 volumes of 5 M NaCl and 2 volumes of 95% EtOH. Incubate at -20°C for 30 minutes, then spin in a microcentrifuge. Wash the pellet 2 times with 70% EtOH. Resuspend the pellet in an appropriate volume of buffer.

This protocol for using β-agarase can be modified in several ways. Our customers have substituted 10 mM Tris, 5 mM EDTA, and 100 mM NaCl for Bis-Tris in the equilibration and reaction mixture. In addition, customers have also omitted the phenol/chloroform extractions from the protocol and have directly ligated the agarase-treated DNA for subcloning. For DNA sequencing, however, it is recommended that the DNA recovered from the gel slice be extracted with the phenol/chloroform. If the recovered DNA template is a double-stranded PCR product, the DNA can be treated with *T7 gene 6 exonuclease* to form single-stranded templates.[1] The DNA can then be sequenced using the standard *Sequenase*™ *protocol*. Yields of DNA recovered using β-agarase I vary according to the protocol used. Expect a 10-40% loss upon phenol/chloroform extractions and precipitation.

References:

1. Ruan, C.C., and Fuller, C.W., 1991, Comments 18, No. 1, pg. 1, United States Biochemical Corporation, Cleveland, OH, 800-321-9322.

ALTERNATIVE APPROACH

GELASE™

Epicentre™ Technologies, 1202 Ann St., Madison, WI 53713, ph. (800) 284-8474. (Adapted)

What is GELase? A novel enzyme preparation that digests the carbohydrate backbone of agarose into small soluble oligosaccharides, yielding a clear liquid that will not become viscous or gel even on cooling in an ice bath. It permits simple and quantitative recovery of intact DNA or RNA from low melting point (LMP) agarose gels. GELase™ contains no contaminating DNase, RNase or phosphatase.

Here's why GELase™ may replace NaI/glass bead kits for purifying DNA from low melting point agarose gels:

1. Recovery of DNA is about 100% using GELase™. Glass bead/NaI kits give about 50% recovery for 2-15 kb DNA and much less outside of that size range.

   Recoveries of 1.2 to 8.8 kb DNA fragments excised from a 1% LMP-agarose gel were consistently much greater using GELase™ than using a commercial kit containing sodium iodide and silica matrix.
2. High molecular weight DNA, even megabase DNA, is not damaged using GELase™. DNA larger than 15 kb is sheared using NaI/glass bead kits.
3. GELase™ is easy to use. Just melt the gel slice with GELase™ buffer, add GELase™ and incubate at 45°C to digest. To concentrate the DNA, add ethanol. The gel digestion products are soluble and will not precipitate with the DNA.
4. GELase™ is inexpensive. One unit of GELase™ digests 600 mg of a 1% LMP-agarose gel in 1 hour in GELase™ buffer. With a 10-hour incubation instead of 1 hour, the 200-unit size of GELase™ is enough to digest more than a kilogram of a 1% gel.
5. DNA purified using GELase™ is ready to use and biologically active. Some companies recommend two rounds of purification with a NaI/glass bead kit to obtain DNA for cloning. That is not necessary with GELase™. DNA recovered using GELase™ is ready for use in restriction mapping, cloning, labeling, sequencing or other molecular biological experiments.
6. GELase™ is active in electrophoresis buffers. It digests gels in TAE, TBE, MOPS and phosphate buffers. Special NaI/glass bead kits are needed for gels in TBE buffer.
7. Protocols for using GELase™ are the same for RNA as for DNA. GELase™ is RNase-free and active in MOPS or phosphate buffers that are used for RNA gels. In contrast, a special version of NaI/glass bead kit is needed for purification of RNA

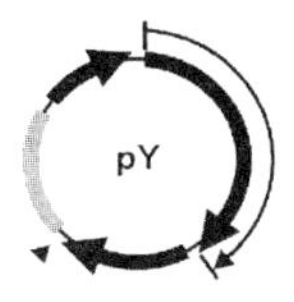

UNIT 3: THE POLYMERASE CHAIN REACTION

Timmer, W. C. and Villalobos, J. M. (1993) *Journal of Chemical Education*, 70(4): 273-280.

1. Introduction

The polymerase chain reaction (PCR) is an enzymatic process which is employed to amplify specific sequences of DNA or RNA. PCR consists of three stepwise operations which are performed sequentially:

1. DNA denaturation
2. primer annealing, and
3. primer extension/amplification.

These operations involve varying the temperature and are repeated a number of times to yield amplified products. The DNA products are then separated by gel electrophoresis and visualized by either ethidium bromide fluorescence or autoradiography.

Although the theoretical basis of PCR was first enumerated by Kleppe *et al.*,[1] the technique did not arouse general interest until Mullis and co-workers developed PCR into a technique that could be used to generate large amounts of single copy genes from genomic DNA.[2] PCR requires two synthetic oligonucleotides, or primers, of about 20 base pairs (bp) each; a DNA polymerase; the four nucleotide triphosphates: dNTPs: dATP, dCTP, dGTP, and dTTP; magnesium ion; and the template DNA that is to be amplified. The primers are complementary to sequences that flank the target sequence on opposite strands of the template DNA. They are oriented with their 3' ends directed toward each other, so synthesis by DNA polymerase "fills in" the DNA segment between them, producing new strands.

The first reaction step involves denaturation of the template DNA at 95°C. This is performed with a molar excess of the primers, dNTPs and Mg^{2+} in the presence of the template DNA. The temperature is then lowered to between 37°C and 55°C. The concentration difference between the primers and denatured DNA favors formation of a primer-template complex over reannealing of the two DNA strands. The primers, now annealed to their target sequences, are then extended with DNA polymerase at 72°C. The cycle of denaturation, annealing and DNA synthesis is then repeated (Figure 1). The *products of one round of amplification serve as templates for the next round*; thus each successive cycle essentially doubles the amount of desired material, resulting in nearly exponential accumulation of target DNA.

2. *Taq* DNA Polymerase

DNA polymerase synthesizes new DNA strands from a primed template strand. The original protocols for PCR used the Klenow fragment of DNA polymerase I from *E. coli* to catalyze the extension of the annealed primers. (Klenow is the large fragment of *E. coli* DNA polymerase I.) A major drawback of using the Klenow fragment was that it is inactivated at temperatures necessary to denature DNA. Thus, each reaction cycle required adding a fresh aliquot of

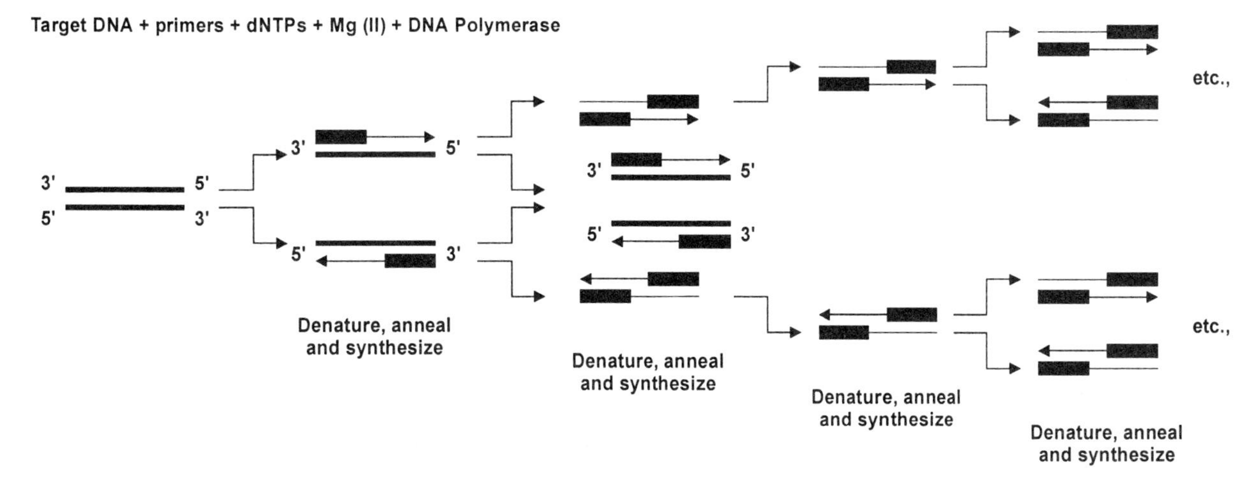

Figure 1 - PCR is initiated by heating DNA duplexes at 95°C to denature them, producing single unpaired strands. Primers are annealed to the target DNA strands in the presence of DNA polymerase, Mg^{2+} and excess dNTPs at a temperature between 37 and 55°C, depending on the details of the PCR incubation time protocol. Amplification/extension is then carried out by raising the temperature to *ca.* 72°C. The first cycle generates a product of indeterminate length; the second cycle generates the short product which accumulates exponentially with each round of amplification. PCR primers (▬), reactant DNAs (———) and newly synthesized DNAs (——).

enzyme. The use of a *thermostable DNA polymerase* (*Taq*), originally discovered in a hot spring in Yellowstone National Park and purified from the thermophilic bacterium *Thermus aquaticus*, eliminated this problem.

Taq DNA polymerase can survive extended incubation at 95°C and its activity is not substantially reduced during the repeated denaturation steps. Mispriming (incorrectly annealed primers and mis-extension of incorrect nucleotides) is also greatly reduced since the annealing and extension steps are carried out at higher temperatures. This results in substantial improvement in specificity and yield of the amplified product. In theory, the amplification yield is exponential: 2^n, where n is the number of cycles. A 30-cycle amplification would be expected to produce 10^8 copies of new DNA. In practice, however, the overall amplification is limited by the amplification efficiency of each cycle. After a certain number of cycles, highly dependent upon individual reaction conditions, the amplified product stops accumulating exponentially and enters a linear or stationary phase. This is known as the reaction plateau. A variety of experimental parameters, for example, the *Taq* concentration, which becomes rate limiting, cause reaction plateau. The efficiency of amplification has been estimated to be between 60 and 85%. This results in a yield which is calculated as follows:

$$\text{yield} = (1+\text{efficiency})^n$$

translating, for a 30-cycle amplification, into 10^6-10^8 copies of new DNA. But the true power of PCR can be discerned by examining the initial quantity of template DNA necessary for PCR. Typically, between 0.05-1 μg of single copy genomic targets, as long as 2-10 kb in length, can be obtained from a 30-35 cycle amplification from 1 pg of starting template DNA.

A *limitation* of *Taq* is its *error rate*, or rate of misincorporation of nts, which results from the enzyme not having 3'-5' exonuclease (proofreading) activity. Recently, two new DNA polymerases have become available, both of which possess 3'-5' exonuclease activity. *Vent polymerase* and *Pfu polymerase*, isolated from *Thermococcus litoralis* and *Pyrococcus furiosus*, respectively, were discovered in geothermal vents on the ocean floor. These enzymes have error rates several-fold less than *Taq* polymerase. Regardless, individual error rates vary according to the DNA sequence to be amplified and variations in reaction conditions. For most applications, the error rate will not be a problem.

3. Variations of PCR

3.1 Amplification of RNA. RNA may be obtained from cells via standard methods such as guanidinium isothiocyanate/CsCl gradient centrifugation.[3] The enzyme *reverse transcriptase* (RT) is then used to prepare complementary DNA (cDNA) prior to conventional PCR (Figure 2). Two types of RT are commercially available: a form isolated from avian myeloblastosis virus (AMV) and a recombinant form from the RT gene of the Moloney Murine Leukemia Virus (MMLV). Both forms lack 3'-5' exonuclease activity. This is quite important because *RT is prone to misincorporation errors*; in the presence of high concentrations of dNTPs and Mg^{2+}, the misincorporation rate is about 1 in every 500 bases.

Regardless of which method is used to isolate RNA, a finite but nominal concentration of DNA will be present. Therefore, prior to the RT step, it is advisable to treat the RNA with the enzyme DNase I. DNase I is an endonuclease that hydrolyzes DNA at sites adjacent to pyrimidine nucleotides. In the presence of Mg^{2+}, DNase I attacks each strand of DNA independently; the cleavage sites are distributed in a statistically random fashion. In order to check on the effectiveness of the DNase treatment, a no RT control should be included and subsequently amplified to check for the presence of amplified DNA.

3.2 Anchored PCR. One disadvantage of *ordinary* (*symmetric*) *PCR* is that the sequence at the *ends of the DNA fragments must be known in order to construct the appropriate primers*. This is often a limitation since the sequence of a specific gene may not be known. In this case, a novel technique known

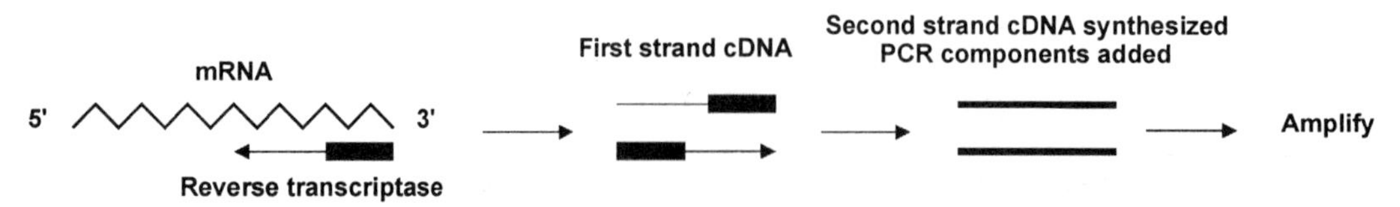

Figure 2 - Amplification of RNA. The RNA is copied into a cDNA by the enzyme reverse transcriptase, ordinary (symmetric) PCR is then performed on the double-strand cDNA. The strand shown beneath the "zigzag" mRNA strand in the first panel is shown in its completed form above the newly synthesized strand in the second panel. The completed PCR product DNA is shown in the third panel. Strand termini result from the completed first round of synthesis whose ends are defined by the primer sequence.

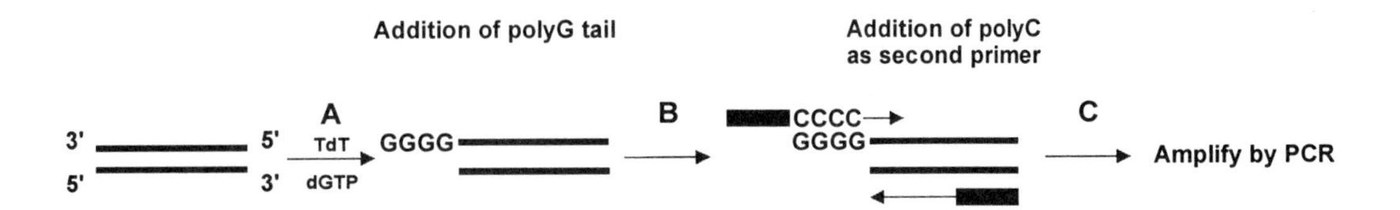

Figure 3 - Anchored PCR. (A) A poly(dG) tail is added enzymatically by TdT to one strand of DNA containing unknown sequences. (B) The anchor, consisting of one primer bonded to poly(dC), is then added. (C) PCR is then performed in the normal manner.

as anchored PCR may be of use. Anchored PCR involves *adding an artificial primer*, such as a poly(dG) tail, to one DNA strand with the enzyme *terminal deoxynucleotidyl transferase* (TdT) (Figure 3). TdT is a primer-dependent DNA polymerase that catalyzes the repetitive addition of deoxynucleotides to the 3' terminus of DNA. The template is then amplified with a specific primer and a second primer consisting of, in this case, poly(dC), the *anchor*.[4]

3.3 Inverse PCR. Inverse PCR is used to amplify DNA outside a known sequence. The method involves digesting the DNA with restriction enzymes and linking the ends of each fragment to form a circle (Figure 4). The circle then contains the known sequence flanking it. Primers are then bound to the known sequence. However, they are designed so that DNA synthesis proceeds outward; that is, DNA synthesis takes the long way around the circle to the other primer, rather than inward, as in conventional PCR. The first round of amplification yields a linear product in which the sequences are inverted; the unknown sequences are bracketed by primer sites and ordinary (symmetric) PCR can begin.[5]

3.4 Asymmetric PCR. Asymmetric PCR is used to amplify a specific strand of DNA by using an unequal, or asymmetric, concentration of primers. During the initial thermal cycles, exponentially derived double-stranded (ds) DNA accumulates. As the low concentration primer becomes depleted, further cycles generate an excess of one of the two strands, depending on the limiting primer. This single-stranded (ss)-DNA accumulates linearly and is complementary to the strands that were amplified previously using the limiting primer.

Asymmetric PCR is less efficient than standard PCR; hence, more cycles (30-40) are required to obtain a maximum yield of ss-DNA. Primer ratios lie in the 50-100:1 range. The concentration of target DNA is important; yields will be low if too little is included, while background can be excessive if too much is included. Low ss-DNA yields can be increased by carefully increasing the *Taq* concentration; lower dNTP concentrations, which minimize mispriming at non-target sites, are used to limit background.

The major problem with asymmetric PCR has been that the amount and quality of template differ among preparations. This depends primarily on the

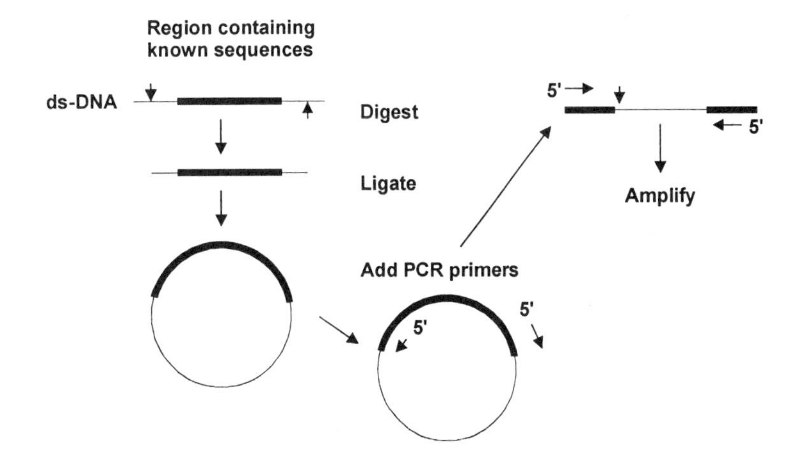

Figure 4 - Inverse PCR. Double-stranded DNA is digested with a restriction enzyme and the ends are linked together to form a circle. Primers, designed so that DNA synthesis proceeds outward from each primer, are then annealed to the known sequences. A second restriction enzyme cleavage within the known sequences results in a linear template with primers located at the termini. Ordinary (symmetric) PCR is then performed. Restriction enzyme sites are indicated by ↓ and ↑.

primers and target DNA. Many empirical studies may be necessary in order to produce good ss-templates. In addition, asymmetric PCR does not always result in reproducibly high yields of ss product. This can be overcome by performing symmetric PCR on ds-DNA, isolating and purifying the products, followed by reamplification with a single primer. This asymmetric reamplification would supply a sufficient quantity of ss-template for dideoxy sequencing.

3.5 Multiplex PCR. Multiplex PCR refers to the simultaneous amplification of different genomic regions during a single PCR. Chamberlain *et al.* pioneered multiplex PCR for the diagnosis of Duchenne Muscular Dystrophy (DMD).[6] DMD is a neuromuscular disorder that results from mutations at the human dystrophin gene; it is among the most common human genetic disorders, affecting approximately 1 in 3500 male births. One-third of all cases arise via a new mutation; of this amount, about 60% are associated with DNA deletions. Deletions at the dystrophin gene are commonly detected by restriction enzyme digestion followed by Southern blotting. This method results in more than 65 DNA fragments requiring hybridization (hydrogen bonding of probe sequences to target sequences) with 7-9 cDNA probes, which is clearly unsuitable for routine use in clinical laboratories. Multiplex PCR is capable of detecting 80-90% of all dystrophin gene deletions. The technique relies on the observation that dystrophin gene deletions are concentrated in two specific regions of the 2 megabase DMD gene. Consequently, only a small percentage of the gene need be analyzed. Nine primer sets are used in the reaction. The PCR products are analyzed by gel electrophoresis. Deletions correlated with the disease are easily identified by the absence of expected DNA fragments. Analysis of 150 patients by multiplex PCR was confirmed by Southern analysis; only two cases required further analysis. A recent multicenter clinical study determined that multiplex PCR detected 82% of the deletions detected by Southern blotting, with a false positive rate of 0.013%.[7]

4. Quantitation

In order to understand gene expression, it is necessary to measure messenger RNA (mRNA) concentrations accurately. Differences in mRNA expression can provide information for the diagnosis of infectious disease or cancer. Conventional methods (Northern analysis) are not sensitive enough to detect low copy number mRNA. Hence, PCR should be ideal for quantitation. However, quantitation is difficult due to the exponential increase of PCR product per amplification cycle. Small differences in any one experimental variable greatly affect PCR product yield.

Several experimental approaches can be employed to overcome these limitations. The most obvious approach is to use an internal control in the same tube prior to amplification. This results in the normalization of amplification variables such as Mg^{2+} and dNTP concentrations, cycle number, temperature, and length of cycle steps. The internal control also verifies whether a result is a true false negative when no signal is obtained from the sequences of interest. Of course, the internal control PCR product should be different enough in size to be resolved from the product of interest.

Wang *et al.* constructed a synthetic gene to use as an internal control.[8] They cloned (inserted) it into a control plasmid containing the 5' primers of 12 different target mRNAs, followed by a *polylinker* (an artificially constructed sequence of restriction enzyme sites), then followed by the 12 complementary 3' primers. Placing polylinker between the 5' upstream and 3' downstream primer, allowed cloning using all 12 primer pairs. Prior to PCR, the plasmid was linearized at a restriction enzyme site and then transcribed with T7 polymerase. This resulted in a cRNA, which was reverse-transcribed and amplified with the target mRNA in the same tube. Thus, the plasmid ultimately served two functions: as internal mRNA control for the reverse transcription reaction and to generate a standard curve for quantitating the target mRNAs. Differences in primer efficiency were minimized since the same primer set was used for both templates. In addition, tube-to-tube variations were eliminated since the quantitation step was performed in the same tube. The control plasmid, pAW109, is commercially available.

The above method was applied to quantitate several *lymphokines* (soluble protein factors secreted by lymphocytes) after stimulating human *macrophages* with bacterial endotoxin. PCR was performed on total cellular RNA with radiolabeled primers. After electrophoresis of the products, the amounts of radioactivity in the excised bands were plotted against either template concentration or cycle number. Relative to the internal control, it was determined that 1 ng of RNA contained 10^4 molecules of interleukin la (IL-1a). When plotted against cycle number, the rate of amplification was found to be exponential between 14 and 22 cycles; further amplification resulted in a decreased rate. The efficiency of amplification, calculated from the slope, was found to be 88% for both the template and standard.

Several other mRNA species were analyzed before and after stimulation. The concentration of IL-la and IL-1α and IL-1β mRNA increased 50-fold after stimulation. The levels of platelet-derived growth factor A (PDGF-A), macrophage colony stimulating factor (M-CSF), and tumor necrosis factor increased 5-10-fold, while PDGF-B mRNA remained constant. Quantitative PCR performed by this method thus provided a detailed picture of the *transcriptional activity* in both resting and induced states in human macrophages.

One of the problems with the above method is that it depends on the assumption that both the unknown template and reporter cDNA have the same efficiency of amplification. In this case, however, the two amplification reactions differ in the dissociation temperatures of the amplified template DNA and primer/template duplexes, as well as in lengths and concentrations of template. As such, amplification efficiencies of the two unrelated cDNAs may be different due to different initial reaction rates, different decreases in reaction rates, and different amplification efficiencies per cycle.

Gilliland *et al.* developed an alternate technique to overcome these limitations.[9] It involves co-amplifying a competitive DNA template and target DNA template with identical primers. The competitive template was identical to the cDNA of interest except for a mutated restriction site. Thus, the ratio of product to competitor remained constant throughout the amplification, regardless of any changes in cycle number or reaction variables. The rate should thus reflect the initial concentration of the unknown cDNA versus that of the competitor cDNA. One limitation of this method is that the efficiency of reverse transcription is assumed to be 100%. If it is not, the amount of actual mRNA present will be underestimated.

As an example, *granulocyte-macrophage colony stimulating factor* (GM-CSF) mRNA was quantified from a stimulated cell line using genomic GM-CSF (gDNA) as the competitive template. Both the cDNA and gDNA should be amplified with the same efficiencies since the primers are identical. In effect, the gDNA and cDNA templates compete for PCR substrate. Quantitation was achieved by titrating an unknown amount of cDNA against serial dilutions of known gDNA concentrations. Northern analysis could not detect any GM-CSF mRNA from less than 10^6 cells. Competitive PCR detected GM-CSF mRNA at concentrations ranging from 0.47-0.55 attomol, in as few as 200 cells. This value corresponds to each cell having about 1500 GM-CSF molecules.

5. Sequencing PCR Products

Dideoxy chain termination ("the Sanger method") is one of the most commonly used techniques for *DNA sequencing*. Either single- or double-stranded templates may be used; however, the best results are obtained from single-stranded templates, which require the use of an *M13 sequencing vector*. In order to use M13 DNA, which is a single-stranded *E. coli* bacteriophage (a bacterial virus), the portion of DNA to be sequenced must be cloned into the vector. A sequencing primer is then annealed to the vector and the sequencing protocol initiated. While PCR can be used to amplify specific fragments, each fragment must be cloned into M13, which is time-consuming. A better technique would be to sequence PCR products directly. In addition to simplicity, direct sequencing would greatly reduce the potential for errors due to imperfect PCR fidelity. An ideal method to generate single-stranded template would be with asymmetric PCR. The single-stranded template could then be sequenced with either the limiting primer or a third internal primer. Of course, two separate asymmetric PCR reactions would be necessary in order to sequence both DNA standards.

Gyllenstein and Erlich used *asymmetric PCR* to sequence a portion of the *major histocompatibility complex* (MHC).[10] The MHC consists of a family of surface cell glycoproteins; they are also known as HLA (human leukocyte antigens) as they were first discovered on leukocytes (white blood cells). These highly *polymorphic proteins* are involved in specific immune responses. Knowledge of the specific alleles (alternate forms of a gene) within this region is useful in tissue typing for transplantation to minimize graft rejection by selecting HLA-matched donors and recipients. HLA typing is also useful in *forensic science* and *paternity determinations*.

In order to study polymorphisms, a particular gene must be examined in a population of individuals. In this case, asymmetric PCR was employed to examine the sequence diversity of the HLA genes. Following amplification and product clean-up, the amplified DNA was sequenced with the limiting primer. Of the 25 individuals examined, eight sequences were identified as *allelic variation* within the HLA genetic locus. This sequence information can then be used to design *allele-specific oligonucleotide* (ASO) *probes*; that is, probes that recognize single base-pair mutations. Once the allelic sequences are known, their frequencies in patient and control populations can be determined in PCR amplified DNA by *dot blot hybridization* with the ASO probes. Dot blot hybridization involves the transfer of spots or dots of amplified DNA to a nitrocellulose filter. The spotted filter is then hybridized with a labeled probe.

Dideoxy DNA sequencing from double-stranded templates (i.e., plasmid DNA) has become increasingly popular. The advantages of this method are that either DNA strand may be sequenced and the M13 cloning step is eliminated. However, prior to sequencing, the double-stranded template must be denatured. The trick is to maintain sufficient dissociation of the rapidly reannealing strands during the annealing and extension of the sequencing primer. Strand reassociation can be slowed by quickly cooling the denatured template on ice. *Double-stranded sequencing* is particularly useful to screen for specific plasmid constructs from DNA minipreparations.

Recent efforts have been directed toward the development of direct PCR double-stranded sequencing protocols. Several problems, none of which is insurmountable, persist with double-stranded sequencing. After the PCR is performed, the amplification primers, buffer, and dNTPs must be removed from the PCR products before sequencing. In particular, this eliminates any competition between the amplification and sequencing primer(s). Removal can be effected either by anion exchange columns or excision of the appropriate bands from an agarose gel. A second problem is that small, linear, PCR-generated fragments tend to have regions of secondary structure that impede the polymerase as it progresses along the template. This can be overcome either by rapid cooling after denaturation or by sequencing with *Taq* polymerase. In this latter case, reaction temperatures between 55 and 75°C will eliminate most of the secondary structure of the templates.

A recent advance in this area involves multiple cycles of *linear PCR* (*LPCR*).[11] After symmetric PCR is performed, the products are isolated and subjected to a sequencing protocol that uses *Taq* polymerase. The reactants are cycled through a 95°C denaturation step, a 50°C annealing step and a 72°C extension step for about 20 cycles. This results in a linear accumulation of a population of sequenced templates. As a result, *femtomole* (10^{-15} mol) *quantities of template* can be *used to generate up to 500 bases of readable sequence*. Additional advantages are that the 95°C denaturation step limits template reannealing to only correctly paired duplexes and the formation of trapped, incorrectly folded secondary structures (if you're lucky). *Taq* DNA polymerase sequencing kits are available from several vendors.

6. Application to AIDS

Human immunodeficiency virus (HIV) is the *etiological agent* of acquired immune deficiency syndrome (AIDS). HIV is a retrovirus: genetic information is contained in RNA rather than DNA. The HIV genome carries the enzyme *reverse transcriptase* (RT), which catalyzes the synthesis of viral RNA into cDNA. This process is carried out in the cytoplasm of the infected cell. The cDNA, or *proviral DNA*, may exist either as an *unintegrated* form in the *cytoplasm*, or it may be *translocated* to the *nucleus* where it is *integrated* into the genome of the infected cell.

HIV infection can be confirmed by several immunological assays, including a test that detects binding of antibodies, made against viral proteins to those proteins (or related species) in blood sera (or other fluids), called the *Western blot* assay. However, very few peripheral blood cells contain proviral DNA early in HIV infection and there is a variable period of time when an individual is infected and anti-HIV antibodies can be detected (usually about 6 months). The potential of PCR to identify proviral DNA was quickly realized and has since been widely employed in basic research and clinical studies.

The most important criterion for HIV analysis by PCR is identification of regions of the viral genome for amplification. *HIV has a high degree of genetic variation*; only invariant regions of the genome that lack significant secondary structure can be usefully targeted for amplification. Highly conserved sequences have been identified; primers and probes for these regions have been designed.[12] With this information, PCR has been used to confirm the presence of proviral DNA in seropositive individuals (with anti-HIV antibodies in sera) in all stages of disease progression.[13] It has also been used to detect virus in high-risk seronegative individuals prior to seroconversion[14] and in brain tissue, which has been hypothesized to correlate with the *pathogenesis of HIV encephalitis*.[15] Conversely, PCR can detect HIV sequences in seropositive asymptomatic individuals who are negative by other detection assays.[16]

An important application of PCR has been in the identification of HIV-infected infants. Infants born to HIV-infected mothers have maternal antibodies acquired transplacentally and thus are seropositive. Maternal antibodies may circulate up to 15 months in newborns, making it difficult to differentiate between maternal and infant antibodies. Recent PCR-based studies have shown that approximately 20-60% of children born from seropositive mothers are infected. About 6000 HIV-infected women give birth in the United States each year, and thus a minimum of 1200 children become infected per year. Rogers *et al.* detected proviral DNA in five of seven neonates (1-16 days old) who later developed AIDS[18] while HIV DNA was detected in 20 of 50 infants born to HIV-positive mothers.[19] These results are important since AIDS develops more rapidly in infants and children, and early identification allows for corrective therapies.

Serological assays are used to screen blood and blood products for the presence of anti-HIV antibodies. Samples that are repeatedly reactive by ELISA determination are retested with the Western blot assay (WB). Although sensitive, WB assays can produce indeterminate results. Individuals within the determinant WBs are either HIV infected and in the early stage of seroconversion or truly HIV negative. As expected, indeterminate WBs result in the rejection of blood from prospective donors. The use of PCR for direct detection of HIV is therefore more desirable. In one study, 20 out of a total of 100 blood donors had indeterminate WBs. PCR analysis demonstrated that all 20 were HIV negative.[20]

Schnittman *et al.* developed an amplification scheme capable of detecting HIV-specific mRNA by RNA PCR.[21] In this way it was possible to study the *in vivo* state of HIV expression in HIV-infected individuals. After RT of HIV-specific mRNAs, the cDNA was amplified with primers specific for the mRNAs of individual genes. Approximately 84% of the patients had ongoing HIV transcription as evidenced by the detection of at least one HIV-specific mRNA. This frequency was observed whether the patients were asymptomatic or had already developed full-blown AIDS.

PCR has also been used to quantitate proviral DNA to determine the number of infected cells in HIV-positive individuals. Quantitation was achieved by serial dilution against a standard amount of infected cells prior to amplification. In seropositive individuals, approximately 1 in 10^4 T-cells which express the *CD4+ surface molecule* (the HIV receptor) contain proviral DNA. In AIDS patients, the ratio was 1 in 10^2 CD4+ cells.[22] These results are important since infection with HIV results in the selective depletion of the CD4+ T-cell subpopulation. Aoki *et al.* studied the distribution of proviral DNA in peripheral blood mononuclear cells (PBMC) after *fractionation into B- and T-cells.*[23] PCR analysis showed that 94% of HIV proviral DNA resided in the T-cell fraction and 4% resided in the B-cell fraction. There was a significant decrease in HIV proviral DNA after treatment with the recently FDA approved anti-retroviral drug *2',3'-dideoxyinosine.* For example, one patient's proviral DNA decreased about 75% over a 15-week treatment.

7. DNA Fingerprinting

DNA fingerprinting refers to the characterization of a portion of an individual's genome. Every individual has a characteristic phenotype because every individual possesses a unique hereditary composition (except identical twins who have identical DNA sequences). To put this into perspective, the human genome contains 3 x 10^9 nt, which encode information for about 100,000 genes. Given this amount of information, it is not surprising that considerable genetic variation is found from person to person. This uniqueness is defined by *genetic markers* inherited from one's parents. These genetic markers can be analyzed at the level of either protein variation or DNA sequence variation.

Two experimental techniques are employed in DNA fingerprinting: *restriction fragment length polymorphism (RFLP)* and PCR-based typing methods. RFLPs are different DNA fragment lengths generated by the action of a restriction enzyme in different individuals. The fragments are separated electrophoretically, transferred to a membrane (Southern blotting), hybridized to a radioactive cDNA probe, and visualized via autoradiography. The original set of fingerprinting probes was developed by Jefferies *et al.*[24] The human genome contains a number of tandem repetitive minisatellite regions that have extensive allelic variation. These regions also share a common core repeat sequence. Hybridization of DNA to a cDNA probe, consisting of multiple copies of the core sequence, produces a unique DNA fingerprint for each individual.

PCR-based methods involve the use of PCR amplified DNA with *ASO probes.* ASO probes will only hybridize to sequences that match it perfectly; a single bp mismatch is sufficient to prevent hybridization. Of course, the DNA sequence of all allelic variants must be known to design the appropriate ASO probe. Once this has been established, the presence of specific alleles can be examined by *dot blot hybridization.* Saiki *et al.* reversed the dot blot hybridization procedure; that is, they attached ASO probes to the nitrocellulose support and hybridized the amplified DNA to the individual probes.[25] Using this procedure, an entire series of sequences could be analyzed by a single hybridization reaction. An additional advantage was that the ASO probes were sufficiently sensitive to use nonradioactive detection methods.

The development of *PCR-based DNA typing* represents one of the more recent advances in forensic science. PCR-based methods have several advantages over RFLP methods. First, they are easier to perform, do not necessarily require the use of radioacivity, and are amenable to automation, which can reduce operator error and increase sample throughput. A second advantage is sensitivity; DNA can be extracted from *single human hairs* as well as *small amounts of blood, semen* and *saliva stains.* In addition, sample analysis can be repeated easily since PCR-based methods consume so little sample.

At present, only a few *DNA marker systems* based on PCR are sufficiently well characterized for use in forensic applications. The first and most well-developed system is the HLA-DQa (a subset of the HLA region) system. The specific sequences of this region, previously determined by Gyllenstein and Erlich,[10] were used to design PCR primers and ASO probes. A 242-bp segment of the HLA-DQJ region was amplified. Reverse dot blot hybridization with ASO probes identified six alleles defining 21 genotypes with population frequencies ranging from 0.005 to 0.15. The *discriminating power* (DP) of this typing system, that is, the probability of distinguishing between two individual chosen at random, is 0.93. This compares favorably with the discriminating power of traditional markers such as ABO blood groups with a DP of 0.60. The HLA-DQa-typing system is currently used for forensic applications and is available commercially as a kit.[26]

8. The Human Genome

The Human Genome project is an international research effort to catalogue and analyze the estimated 100,000 human genes that lie within human DNA. The ultimate goal is to determine the complete sequence of the entire human genome. Continuously evolving technologies, not necessarily in existence today, will be used to obtain this information. Knowledge of the human genome and how it functions will broadly impact all aspects of biological and medical research.

The human genome project will probably be performed in two steps. First, detailed maps will be constructed covering the entire genome. Three *types of maps* will be constructed: *cytogenetic*, *genetic* and *physical*. Cytogenetic maps are based on the distinctive banding pattern of stained chromosomes when examined microscopically. They are low-resolution maps that delineate specific genes or specific chromosomal regions. Genetic linkage maps reflect the relative locations of genetic markers, such as a physical trait or a particular medical syndrome, by their patterns of inheritance. Physical maps, measured in base pairs, consist of ordered landmarks at known distances. For example, a low resolution physical map would be a chromosomal banding pattern obtained with a defined set of restriction endonucleases; a high resolution map would be the complete nucleotide sequences of the entire set of chromosomes. This information will provide us with the sequences of individual segments corresponding to all known map positions, a necessary prerequisite for being able to generate the expected experimental restriction fragment pattern(s) for any genetically characterized disease.

PCR techniques will be involved in nearly every aspect of the project. An important PCR topic involves the use of *sequence tagged sites* (STS) as landmarks for the construction of genetic and physical maps.[27] The basic premise of this concept is that all DNA fragments used to construct these maps would be identified by a 200-500 bp site which is unique to that fragment. Each site would be sequenced and primers would be designed to enable specific PCR amplification at that site. Thus, PCR could be used to regenerate a DNA fragment from any genomic region. The final product from the STS proposal would be a STS map of the human genome. The map would be a composite of both genetic and physical maps, with genetic markers and physical landmarks translated into STSs and positioned relative to each other.

Individual laboratories could use whatever mapping techniques they chose; results would always be reported in terms of the STS markers. A database would be generated containing the location, sequence ,and primers necessary to amplify each STS. The database thus would be laboratory-independent. The STS approach would eliminate the need to physically store clones and/or distribute them to laboratories worldwide. Finally, it would provide a common language and landmarks for mapping.[28]

The high resolution physical map may require as many as 10,000 polymorphic DNA markers. One approach to generate these markers involves primers directed to *interspersed repeated sequences* (*IRS PCR*). IRS PCR involves using primers directed to *Alu sequences* (*Alu PCR*) to amplify human DNA falling between Alu repeats in the genome. The Alu sequence is a tetranucleotide that can be cleaved by the restriction enzyme *Alu*I. This family is the most common *short interspersed repeat sequence* (*SINE*) in the human genome, with one 300 bp sequence occurring approximately every 4-10 kb. Alu PCR was originally devised to identify human DNA in human-rodent hybrid cell lines. However, it has been demonstrated to be an efficient tool for generating probes from specific chromosomal regions.[29]

The IRS concept can be extended to other sequences. This is actually quite important due to the uneven distribution of other repeated sequences throughout the human genome. For example, Kpn, a *long, interspersed repeat sequence* (*LINE*) recognized by the restriction enzyme *Kpn*I, is the second most common repeated sequence in the human genome. The Alu and Kpn sequences represent more than 90% of *middle repetitive elements* within the human genome.[30] At present, optimizing IRS-PCR to develop probes from any targeted chromosomal region is limited by knowledge of repeat sequence distribution within the human genome. Thus, a variety of primers and methods involving PCR will need to be developed to obtain uniform coverage of the entire genome with DNA markers derived by IRS-PCR.

Other applications of PCR to mapping and sequencing include "*chromosomal walking*" and expansion of yeast artificial chromosome (YAC) clones,[31] mapping and analysis of expressed sequences from particular genomic regions[32] and direct PCR-coupled DNA sequencing.[33]

9. Review Articles

Several recent review articles are worth noting. Erlich *et al.* describe recent advances in PCR,[34] while Bej *et al.*,[35] Lizardi *et al.*,[36] and Gibbs[37] provide general overviews; a summary of nucleic acid amplification strategies has also been presented.[38] Bloch has presented an in-depth discussion of PCR reaction principles.[39] Wetmur has summarized the use of cDNA probes for hybridizations.[40] PCR sequencing methods have also been reviewed[41]; optimum conditions for ds-sequencing have been reviewed.[42,43] Methods for the quantitative determination of mRNA concentrations have been summarized.[44] Kumar presented a detailed experimental protocol for PCR.[45] Muul[12] and Poiesz *et al.*[46] discuss PCR for the detection of AIDS; both include experimental protocols. Reynolds and Sensabaugh discuss PCR-based forensic analysis.[47] Finally, Linz discuss several experimental strategies for optimizing PCR.[48]

10. Summary

PCR has revolutionized recombinant DNA techniques by permitting the generation of almost any nucleic acid sequence *in vitro*. Manipulation of these amplified sequences has broadly impacted basic and clinical sciences. Other amplification strategies, aside from those discussed, are possible. Future refinements in PCR technology will involve easier quantitation, automated systems for better reproducibility and overall improvements in methodology in order to achieve routine analyses.

Editorial note: This field has progressed tremendously since this article was written. Current reviews, available on the Web, should update the reader regarding the latest status of "human genomics." A large portion of the initial sequence determination task has been completed and the emphasis is rapidly shifting toward structural reconstitution and physical characterization of subcomponents, assemblages, and entire genetic systems. Nevertheless, the present article is enlightening both technically and, dare we say, "historically."

References:

1. Kleppe, K; Ohtsuka, E.; Kleppe, R; Molineux; Khorana, H. G., J. Mol. Biol. 1971, 56, 341.
2. Mullis, K. B.; Faloona, F.; Scharf, S. J.; Saiki, R. K.; Horn, G. T.; Erlich, H. A., Cold Spring Harbor Symp. Quant. Biol. 1986, 51, 236.
3. Chirgwin, J. J.; Przbyla, A. E.; MacDonald, R. J.; Rutter, W. J., Biochemistry 1979, 18, 5294.
4. Loh, E. Y.; Elliott, J. F.; Cwirla, S.; Lanier, L. L.; Davis, M. M., Science 1989, 243, 217.
5. Triglia, T.; Peterson, M. G.; Kemp, D. J., Nucleic Acids Res. 1988, 16, 8186.
6. Chamberlain, J. S.; Gibbs, R. A.; Ranier, J. E.; Nguyen, P. N.; Caskey, C. T., Nucleic Acids Res. 1988, 16, 11141.
7. Chamberlain, J. S., *et al.*, JAMA 1992, 267(19), 2609.
8. Wang, A. M.; Doyle, M. V.; Mark, D. F., Proc. Natl. Acad. Sci. USA 1989, 86, 9717.
9. Gilliland, G.; Perrin, S.; Blanchard, K., Proc. Natl. Acad. Sci. USA 1990, 87, 2725.
10. Gyllenstein, U.; Erlich, H., Proc. Natl. Acad. Sci. USA 1988, 85, 7652.
11. Murray, V., Nucleic Acids Res. 1989, 17, 8889.
12. Muul, L., AIDS Updates 1990, 3(4), 1.
13. Jackson, J. B.; Kwok, S.; Sininsky, J. J.; Hopsicker; Sannerud, K. J.; Rhame, F. S.; Henry, K.; Simpson, M.; Balfour, H. J., J. Clin. Microbiol. 1990, 28, 16.
14. Hewlett, I. K.; Gregg, R. A.; Ou, C. Y.; Hawthorne, C. A.; Mayner, R. E.; Schumacher, R. T.; Schochetman, G.; Epstein, J. E., J. Clin. Immunoassay 1988, 11, 161.
15. Pang, S.; Koyanagi, Y.; Miles, S.; Wiley, C.; Vinters, H. V.; Chen, L. S. Y., Nature 1990, 343, 85.
16. Ou, C. Y.; Kwok, S.; Mitchell, S. W.; Mack, D. H.; Sninsky, J. J.; Krebs, J. W.; Feorino, P.;

Warfield, D.; Schochetman, G., Science 1988, 238, 295.
17. Laure, F.; Courgnaud, V.; Rouzioux, C.; Blanche, S.; Veber, F.; Burgard, M.; Jacomet, C.; Griscelli, C.; Brecholt, C., Lancet 1988, II, 538.
18. Rogers, M. F.; Ou, C. Y.; Rayfield, M.; Thomas, P. A., *et al.*, N. Engl. J. Med. 1989, 320, 1649.
19. Loche, M.; Mach, B., Lancet 1988, II, 418.
20. Genesca, J.; Shih, J. W. K.; Jett, B. W.; Hewlett, L. K.; Epstein, J. S.; Alter, H. J., Lancet 1989, II, 1023.
21. Schnittman, S. M.; Greenhouse, J.; Lane, H. C.; Pierce, P. F.; Fauci, A. S., AIDS Res. Hum. Retroviruses 1991, 7, 361.
22. Spector, S. A.; Hsia, K.; Denaro, F.; Spector, D. H., Clin. Chem. 1989, 35, 1581.
23. Aoki, S.; Yarchoan, R.; Thomas, R. V.; Pluda, J.; Marczyk, K.; Broder, S.; Mitsuya, H., AIDS Res. Hum. Retroviruses 1990, 6, 1331.
24. Jefferies, A. J.; Wilson, V.; Thein, S. L., Nature 1985, 314(6006), 67.
25. Saiki, R. K; Walsh, P. S.; Levenson, C. H.; Erlich, H. A., Proc. Natl. Acad. Sci. 1989, *86*, 6230.
26. Erlich, H. A., Ed. PCR Technology, Stockton Press: NewYork, 1990.
27. Olsen, M.; Hood, L.; Cantor, C.; Botstein, D., Science 1989, 245, 1434.
28. Rose, E. A., FASEB J. 1991, 5, 46.
29. Nelson, D. L.; Ledbetter, S. A.; Corbo, L.; Victoria, M.F.; Ramirez-Solis, R.; Webster, T. D.; Ledbetter, D. H.; Caskey, C. T., Proc. Natl. Acad. Sci. USA 1989, 86, 6686.
30 Moysiz, R. K.; Torney, D. C.; Meyne, J. M.; Buckingham, J. M.; Wu, J. R.; Burks, C.; Sirotkin, K. M.; Goad, W. B., Genomics 1989, 4, 273.
31. Green, E. D.; Olsen, M. V., Proc. Natl. Acad. Sci. USA 1990, 87, 1213.
32. Corbo, L.; Maley, J. A.; Nelson, D. L.; Caskey, C. T., Science 1990, 249, 652.
33. Innis, M.; Myambo, K. B.; Gelfand, D. H.; Brow, M. A., Proc. Natl. Acad. Sci. USA 1988, 85, 7652.
34. Erlich, H. A.; Gelfand, D.; Sninsky, J. J., Science 1991, 252, 1643.
35 Bej, A. K.; Mahbubani, M. H.; Atlas, R. M., Critical Rev. Biochem. Mol. Biol. 1991, 26, 301.
36. Lizardi, P. M., Kramer, F. R., Trends Biol. Sci. 1991, 9, 53.
37. Gibbs, R. A., Anal. Chem. 1990, 62, 1202.
38. Anonymous., J. NIH Res. 1991, 3, 81.
39. Bloch, W., Biochemistry 1991, 30(11), 2735.
40. Wetmur, J. G., Crit. Rev. Biochem. Mol. Biol. 1991, 26, 277.
41. Gyllenstein, U., Biotechniques 1989, *7*, 700.
42. Casanova, J.; Pannetier, C.; Janlin, C.; Kourilsky P., Nucleic Acids Res. 1990, 18, 4028.
43. Kusukawa, N.; Uemori, T.; Asada, K.; Kato, I.; Shuzo, T., Biotechniques 1990, 9(1) 66.
44. Becker-Andre, M., *Meth. Mol. Cell Biol.* 1991, 2, 189.
45. Kumar, R., Technique 1989, 1, 133.
46. Poiesz, B. J.; Erlich, G. D.; Byrne, B. C.; Wells, K; Kwok, S.; Sninsky, J. Med. Virol. 1990, 9, 47.
47. Reynolds, R.; Sensabaugh, G., Anal. Chem. 1991, 63, 2.
48. Linz, U., Methods Mol. Cell Biol. 1991, 2, 98.

INNOVATION/INSIGHT

POLYMERASE CHAIN REACTION USED FOR ANTIGEN DETECTION

Chemical and Engineering News Brief, p. 56. October 5, 1992. (Adapted)

A technique that harnesses the polymerase chain reaction (PCR) to detect vanishingly small amounts of an antigen has been developed by Charles R. Cantor, Cassandra L. Smith, and Takeshi Sano, of the University of California, Berkeley, and Lawrence Berkeley Laboratory (Science 258, 120, 1992). In the technique, dubbed "immuno-PCR," antigen is immobilized and exposed to an antibody specific for it. Then a molecule that binds both DNA and antibodies is used to attach a DNA marker to the antigen-antibody complex, resulting in the formation of a specific antigen-antibody-DNA conjugate. The marker DNA is then amplified by PCR. The presence of PCR products demonstrates that marker DNA molecules have attached specifically to antigen-antibody complexes. The absence of PCR products shows that no antibody was present for binding to antibody. Using agarose gel electrophoresis to detect PCR products, the researchers were able to detect the presence of *as few as 580 immobilized antigen molecules*, a level of sensitivity about 100,000 times greater than achievable with conventional enzyme-linked immunosorbant assays. The researchers suggest that relatively straight forward improvement in the technique could, in principle, allow the *detection of a single antigen molecule.*

IMMUNO-PCR: VERY SENSITIVE ANTIGEN DETECTION BY MEANS OF SPECIFIC ANTIBODY-DNA CONJUGATES

Takeshi, S., C.L. Smith and C.R. Cantor (1992) Science 258: 120. (Adapted)

1. Abstract

An antigen detection system, termed immuno-polymerase chain reaction (immuno-PCR), was developed in which a specific DNA molecule is used as the marker. A *streptavidin-protein A chimera* that possesses tight and specific binding affinity both for *biotin* and *immunoglobulin G* was used to attach a biotinylated DNA specifically to antigen-monoclonal antibody complexes that had been immobilized on microtiter plate wells. Then, a segment of the attached DNA was amplified by PCR. Analysis of the PCR products by agarose gel electrophoresis after staining with ethidium bromide allowed as few as 580 antigen molecules (9.6 x 10^{-22} mol) to be readily and reproducibly detected. Direct comparison with enzyme-linked immunosorbent assay with the use of a *chimera-alkaline phosphatase conjugate* demonstrates that enhancement (approximately x 10^5) in detection sensitivity was obtained with the use of immuno-PCR. Given the enormous amplification capability and specificity of PCR, this immuno-PCR technology has a sensitivity greater than any existing antigen detection system and, in principle, could be applied to the detection of single antigen molecules.

Antibody-based detection systems for specific antigens are a versatile and powerful tool for various molecular and cellular analyses and clinical diagnostics. The power of such systems originates from the considerable specificity of antibodies for their particular epitopes. A number of recent antibody technologies, including *genetic engineering of antibody molecules*[1] and the production of *catalytic antibodies*[2] and bispecific antibodies,[3] are allowing a rapid expansion in the applications of antibodies. We were interested in further enhancing the sensitivity of antigen detection systems. This should facilitate the specific detection of rare antigens, which are present only in very small numbers, and thus could expand the application of antibodies to a wider variety of biological and nonbiological systems.

Polymerase chain reaction (PCR) technology[4] has become a powerful tool in molecular biology and genetic engineering.[5] The *efficacy* of PCR is based on its ability to amplify a specific DNA sequence flanked by a set of primers. The enormous amplification capability of PCR allows the production of large amounts of specific DNA products, which can be detected by various methods. The extremely high specificity of PCR for a target sequence defined by a set of primers should avoid the generation of false signals from other nucleic acid molecules present in samples. We reasoned that the capability of antigen detection systems could be considerably enhanced and potentially broadened by coupling to PCR. Following these ideas, we have developed an *antigen detection system*, termed *immuno-PCR*, in which a specific antibody-DNA conjugate is used to detect antigens.

In immuno-PCR, a linker molecule with bispecific binding affinity for DNA and antibodies is used to attach a DNA molecule (marker) specifically to an antigen-antibody complex, resulting in the formation of a specific antigen-antibody-DNA conjugate. The attached marker DNA can be amplified by PCR with the appropriate primers. The presence of specific PCR products demonstrates that marker DNA molecules are attached specifically to antigen-antibody complexes, which indicates the presence of antigen. A *streptavidin-protein A chimera* that we recently designed[6] was used as the linker. The chimera has two independent specific binding abilities; one is to biotin, derived from the streptavidin moiety, and the other is to the F_c portion of an immunoglobulin G (IgG) molecule, derived from the protein A moiety. This bifunctional specificity both for biotin and antibody allows the specific conjugation of any biotinylated DNA molecule to antigen-antibody complexes.

To test the feasibility of this concept, we immobilized various amounts of an antigen on the surface of microtiter plate wells and detected them by immuno-PCR. Bovine serum albumin (BSA) was used as the antigen because of the availability of pure protein and monoclonal antibodies against it. The *detection procedure* used is similar to conventional enzyme-linked immunosorbent assay (ELISA). Instead of an enzyme-conjugated secondary antibody directed against the primary antibody, as in typical ELISA, a *biotinylated linear plasmid DNA* (pUC19) *conjugated to the streptavidin-protein A chimera* was targeted to the antigen-antibody complexes. A segment of the attached marker DNA was amplified by PCR with appropriate primers, and the resulting PCR products were analyzed by agarose gel electrophoresis after staining with ethidium bromide.

A specific 260 bp PCR product was observed in all the lanes that contained immobilized BSA, which indicates that the *biotinylated pUC19* was *specifically attached to the antigen-monoclonal antibody complexes* by the chimera. (Note: Original figures are not reproduced in this adapted version.) In contrast, almost no 260-bp fragment was observed in lanes 10 through 12, which came from wells without immobilized antigen. Quantitation of the 260-bp PCR product demonstrates that background PCR signals generated by nonspecific binding of the antibody or the chimera-pUC19 conjugate were sufficiently small to allow clear discrimination of positive signals from background. This also indicates that the specificity of PCR amplification is high enough to avoid the generation of false signals from other DNA molecules present in the wells. Because the sequences of a marker DNA and its amplified segment are purely arbitrary they can be changed frequently, if needed, to prevent deterioration of signal-to-noise ratios caused by contamination.

The result demonstrates the specific detection of immobilized antigen by immuno-PCR. The 260-bp fragment was clearly observed even with only 580 antigen molecules (9.6 x 10^{22} mol). Direct comparison with ELISA with the use of a chimera-alkaline phosphatase conjugate demonstrated that enhancement (approximately x 10^{5}) in detection sensitivity was obtained with the use of immuno-PCR instead of ELISA. A consideration of the detection limits of typical *radioimmunoassays*, in which *sensitivity is primarily determined by the specific radioactivity of antigens or antibodies* used,[7] indicates that immuno-PCR is likely to be several orders of magnitude more sensitive than radioimmunoassays.

This extremely high sensitivity of immuno-PCR was achieved just with the use of agarose gel electrophoresis to detect PCR products. The sensitivity and versatility could be enhanced considerably with the use of better detection methods for PCR products. For example, direct incorporation of a label, such as *radioisotopes*, *fluorochromes*, and *enzymes*, into PCR products with the use of label-conjugated primers or nucleotides allows simple analytical formats. Alternatively, gel electrophoresis could be used to detect many different antigen molecules simultaneously, each of which is labeled with a differently sized marker DNA.

The amount of the 260-bp fragment decreased with decreasing amounts of immobilized antigen from lanes 6 to 9, which demonstrates that the PCR amplification was not saturated below 0.96 attomol of BSA. For wells that contained more antigen, the PCR amplification was saturated. In principle, *quantitation of PCR products below saturation should provide an estimate of the number of antigens* (epitopes) *after appropriate calibration* with known numbers of antigen molecules. When more dilute chimera-pUCl9 conjugates were used, saturation of PCR amplification occurred with larger amounts of the immobilized antigen. Thus, one can *control the sensitivity* of the system by *varying the concentration of the conjugate*. Other key factors, such as the concentration of antibody, the number of PCR amplification cycles, and the detection method for PCR products, can also be used to control the overall sensitivity of the system.

In principle, the extremely high sensitivity of immuno-PCR should enable this technology to be applied to the detection of single antigen molecules; no method is currently available for this. The sensitivity of current antigen detection systems can be enhanced by at least a few orders of magnitude simply by the introduction of PCR. The controllable sensitivity and the simple procedure of immuno-PCR should allow the development of fully automated assay systems without loss in sensitivity, with a great potential promise for applications in clinical diagnostics.

2. Methods

1. Various amounts [6.4 ng to 6.4 ag (attograms); 9.6×10^{-14} to 9.6×10^{-22} mol] of BSA in 45 μl of 150 mM NaCl, 20 mM Tris-Cl (pH 9.5), and 0.02% NaN_3, prepared by ten-fold serial dilutions, were placed in wells of a microtiter plate (Falcon 3911; Becton Dickinson). The microtiter plate was incubated at 4°C overnight (*ca.* 15 hours) to immobilize BSA molecules on the surface of the wells. The same solution without BSA was used as the control. The wells were washed three times, each for 5 minutes with 250 μl of Tris-buffered saline [*TBS*; 150 mM NaCl, 20 mM Tris-Cl (pH 7.5), and 0.02% NaN_3]. Then, 200 μl of *ETBS* (TBS plus 0.1 mM EDTA) that contained 4.5% *nonfat dried milk* and *denatured salmon sperm DNA* (1 mg/ml) was added to each well. The microtiter plate was incubated at 37°C for 80 minutes to block reactable sites on the surface of the wells to avoid nonspecific binding in subsequent steps, and then the wells were washed seven times, each with 150 μl of *TETBS* (TBS plus 0.1 mM EDTA and 0.1% Tween 20) for 5 minutes. To each well, 50 μl of TETBS containing 0.45% nonfat dried milk and denatured salmon sperm DNA (0.1 mg/ml), and diluted (8000-fold) monoclonal anti-body against BSA (mouse *ascites fluid*, IgG2a, clone BSA-33; Sigma) was added. The microtiter plate was incubated at room temperature for 45 minutes to allow the antibody to bind to immobilized BSA molecules. The wells were washed 15 times, each with 250 μl of TETBS for 10 minutes, to remove unbound antibody molecules, and 50 μl of TETBS containing 0.45%, nonfat dried milk, denatured salmon sperm DNA (0.1 mg/ml), and 1.4×10^{-16} g/mol of biotinylated pUC19 conjugated to the streptavidin-protein. A chimera was added to each well. The microtiter PCR plate was incubated at room temperature for 50 minutes to allow the chimera-pUCl9 conjugates to bind to the antigen-antibody complexes, and then the wells were washed 15 times, each with 250 μl of TETBS for 10 minutes, to remove unbound conjugates. The wells were washed three times with TBS without NaN_3, and the microtiter plate was subjected to PCR amplification; each reaction mixture was analyzed by agarose gel electrophoresis.
2. The biotinylated pUC19 used was a linear 2.67-kb *Hin*dIII-*Acc*I fragment, in which one biotin molecule had been incorporated at its *Hin*dIII terminus by a *filling in reaction* with Sequenase version 2.0 (U.S. Biochemical) in the presence of a biotinylated nucleotide (biotin-14-deoxy-adenosine phosphate; BRL). By gel retard-ation, dependent on streptavidin binding, almost 100% of the 2.67-kb fragment contained biotin.
3. A chimera was expressed in *Escherichia coli* by means of the expression vector pTSAPA-2 and purified to homogeneity. The purified chimera was stored frozen at -70°C in 150 mM NaCl, 20 mM Tris-Cl (pH 7.5), 0.02% NaN_3 and 6% sucrose. No appreciable changes in the biotin- and the IgG-binding ability were observed upon frozen storage.
4. We prepared the chimera-pUC19 conjugate by mixing the purified chimera and the biotinylated pUC19 at a molar ratio of biotin : binding site of 1. The resulting conjugates contain four biotinylated pUC19 per chimera, which possesses four biotin binding sites.[6]
5. PCR was carried out under the following conditions: 50 mM KCl, 10 mM Tris-Cl (pH 8.3 at 20°C), 1.5 mM $MgCl_2$, gelatin (10 μg/ml), 0.8 mM deoxyribonucleoside triphosphates (dNTPs) (0.2 mM each), 2 μM primers (bla-1 and bla-2, 1 μM each), and *Taq* DNA polymerase (50 unit/ml) (Boehringer Mannheim). *Pre-PCR mixtures sterilized by ultraviolet* (UV) *irradiation at 254 nm* were added to the wells of a microtiter plate (40 μl per well), and *mineral oil* sterilized by UV irradiation was layered (20 μl per well) on the reaction mixture. PCR was performed with an *automated thermal cycler* (PTC-100-96 Thermal Cycler, MJ Research, Inc.), with the use of the following temperature profile: initial denaturation. 94°C, 5 minutes; 30 cycles of denaturation (94°C, 1 minutes), annealing (58°C, 1 minute), and extension (72°C, 1 minute); and final extension, 72°C, 5 minutes. The 30-mer primers, bla-1 and bla-2, hybridize to a segment of the *bla* gene and should generate a 261-bp fragment upon PCR amplification.
6. These numbers indicate the amounts of antigen added to each well. However, it is unlikely that all of the added antigen molecules were immobilized on the wells. Furthermore, some molecules that were initially immobilized may have been released in subsequent steps. Therefore, the actual number of antigens in each well is very likely to be lower than indicated.
7. Various amounts of BSA (640 μg to 64 fg; 9.6×10^{-9} to 9.6×10^{-19} mol) in 50 μl of 150 mM NaCl, 20 mM Tris-Cl (pH 9.5), and 0.02% NaN_3,

prepared by ten-fold serial dilutions, were placed in wells of a microtiter plate. The plate was incubated at 4°C overnight (15 hours) to immobilize BSA molecules on the surface of the wells. The same solution without BSA was used as the control. The wells were washed three times, each with 100 µl of TBS for 5 minutes, and then 200 µl of ETBS containing 4.5% nonfat dried milk was added to each well. The microtiter plate was incubated at room temperature (22°C) for 60 minutes to block reactable sites on the surface of the wells, and the wells were washed three times each with 200 µl of TETBS (TBS plus 0.1 mM EDTA and 0.04% Tween 20) for 5 minutes. To each well, 75 µl of TETBS containing 0.45% nonfat dried milk and diluted (500-fold) monoclonal antibody against BSA[7] was added. The microtiter plate was incubated at room temperature for 60 minutes to allow the antibody to bind to immobilized BSA molecules. The wells were washed six times, each with 200 µl of TETBS for 5 to 10 minutes, to remove unbound antibody and 75 µl of TETBS containing 0.45% nonfat dried milk and 6 pmol of biotinylated alkaline phosphatase (Boehringer Mannheim) conjugated to the streptavidin-protein A chimera[9] was added to each well. The microtiter plate was incubated at room temperature for 60 minutes to allow the chimera-alkaline phosphatase conjugate to bind to the antigen-antibody complexes, and then the wells were washed six times, each with 200 µl of TETBS for 5 to 10 minutes. The wells were washed once with 200 µl of TBS without NaN_3 and then with 200 µl of 1 M diethanolamine (pH 9.8) and 0.5 mM $MgCl_2$. To each well, 200 µl of 1 M diethanolamine (pH 9.8) and 0.5 mM $MgCl_2$ containing 10 mM *p*-nitrophenylphosphate were added, and a color development reaction was performed at 37°C for 60 minutes.

References:

1. S. L. Morrison *et al.*, Clin. Chem. 34, 1668 (1988); S. L Morrison and V. T. Oi, Adv. Immunol. 44, 65 (1989); J. D. Rodwell, Nature 342, 99 (1989); A. Pluckthun, *ibid.* 347, 497 (1990); G. Winter and C. Milstein, *ibid.* 349, 293 (1991); A. Pluckthun, Bio/Technology 9, 545 (1991); R. Wetzel, Protein Eng. 4, 971 (1991); M. J. Geisow, Trends Biotechnol. 10, 75 (1992); O. J. Chiswell and J. McCaffery, *ibid.* p. 85.
2. R. A. Lerner and A. Tramanto, Trends Biochem. Sci. 12, 27 (1987); K. M. Shokat and P. G. Schultz, Annu. Rev. Immunol. 8, 935 (1990); P. G. Schultz., Science 240, 426 (1988); S. J. Benkovic, J. A. Adams., C. L. Borders, Jr., K. D. Janda, R. A. Lerner, *ibid.* 250, 1135 (1990); R. A. Lerner, S. J. Benkovic, and P. G. Schultz. *ibid.* 252, 659 (1991).
3. O. Nolan and R. O'Kennedy, Biochem. Biophys. Acta 1040, 1 (1990); M. F. Wagner and P. M. Guyre, Trends Biotechnol. 9, 375 (1991); R. L. H. Bolhuis, E. Stuhn, J. W. Gratama, E. Braakman, J. Cell. Biochem. 47, 306 (1991).
4. R. K. Saiki *et al.*, Science 230, 1350 (1985).
5. T. J. White, N. Arnheim, and H. A. Erlich, Trends Genet. 5, 185 (1989); R. A. Gibbs, Anal. Chem. 62, 1202 (1990); W. Bloch, Biochemistry 30, 2735 (1991); R. A. Gibbs, Curr. Opin. Biotechnol. 2, 69 (1991); H. A. Erlich, D. Gelfand, and J. J. Sninsky, Science 252, 1643 (1991).
6. T. Sano and C. R. Cantor, Bio/Technology 9, 1378 (1991).
7. H. Van Vunakis, Methods Enzymol. 70, 201 (1980); C. N. Hales and J. S. Woodhead, *ibid.* 334; J. P. Feber, Adv. Clin. Chem. 20, 129 (1978); Editorial, Lancet ii, 406 (1976).

Lab 3.1:

Polymerase Chain Reaction Test for *myo*-3 Gene Insert Orientation

Overview:

In this lab, the ligation of the *C. elegans myo*-3 gene into linearized plasmid DNA will be verified using the polymerase chain reaction (PCR). We will also obtain the orientation of the insert in the circularized insert-plasmid ligation product. PCR allows one to synthesize a large number of copies of DNA, provided that two oligonucleotide primers are available that hybridize to the sequences on the complementary DNA strands. This reaction requires target DNA, the two primers, all four deoxyribonucleotide triphosphates, and a thermostable DNA polymerase, such as *Taq* DNA polymerase. A PCR cycle typically consists of three steps: denaturation, primer annealing, and elongation. This cycle is repeated for a set number of times depending on the desired degree of amplification. Ligation accuracy will be confirmed with primers designed to generate DNA from plasmids with correctly ligated inserts.

Next, agarose gel electrophoresis will be run. PCR amplification of the *myo*-3 gene DNA will be detected by the presence of intercalated (bound) ethidium. An image of the DNA band pattern in the gel will be obtained by "transilluminating" it on a UV "gel box." PCR reactions showing an amplified DNA band indicate correct ligation. To preserve a record of our result, we will photograph the gels.

Experimental Outline:

1. PCR of the recombinant plasmid.
2. Agarose gel electrophoresis.
3. Photography of the agarose gel.

Required Materials:

bacteria containing recombinant plasmid; aerosol-resistant pipet tips; PCR assay buffer (composition from the *Taq* DNA polymerase product literature; see NOTES); PCR lysis buffer (recipe below); thermocycler; PCR amplification buffer (recipe below); centrifuge; mineral oil; PCR primers (sequence below); agarose gel electrophoresis apparatus; ethidium bromide stock: 10 mg/ml; 'DNA loading dye': 0.25% bromophenol blue, 0.25% xylene cyanol, 40% (w/v) sucrose; *Hae*III digested ϕX174 DNA marker fragments; TE buffer (recipe in Lab 1.1); camera; film; face shield; UV gel illuminator box

Prelab Preparations:

1. Prepare 'PCR lysis buffer' as follows:

PCR assay buffer	1x
NP-40	0.45%
Tween-20	0.45%
Proteinase K	0.06 μg/ml

2. Acquire the following primers (e.g., from a DNA synthesis facility or company):
   Primer 1: AAGCTTTGCGACAGCTTC
   Primer 2: GGATCCGTCGACTCTAGA
3. Prepare 'Amplification reaction mixture' as follows:

primers	125 ng/ml
dNTPs	0.25 mM each
Taq polymerase	100 U/ml

Procedures:

1. PCR Amplification of Cloned *myo*-3 Gene

1. Touch the top of a bacterial colony that contains the recombinant plasmid with an aerosol-resistant pipet tip.
2. Resuspend the bacteria in 25 μl of PCR lysis buffer.
3. Heat the tube at 55°C for one hour.

4. Heat the tube at 95°C for 10 minutes.
5. Incubate the sample on ice (within the tube.)
6. Add 25 μl of PCR amplification reaction mixture and mix well.
7. Centrifuge the tube and contents for 3 seconds.
8. Layer 50 μl of mineral oil over the reaction mixture.
9. Run PCR according to the following script of reaction cycles:

    Step 1: 95°C for 1 minute,
    Step 2: 50°C for 1 minute,
    Step 3: 72°C for 1 minute,
    Run the process for 30 cycles;
    Final cycle: 72°C for 10 minutes.

2. Making a 1% Agarose Gel

1. Combine 50 ml of 1x TAE buffer and 0.5 g of agarose in a 250-ml Erlenmeyer flask.
2. Microwave for 2 minutes (be sure that the solution doesn't boil over). Afterward, check to be certain that all of the agarose dissolved; if not, swirl the flask (plus a 5-10 second burst in the microwave if necessary).
3. Allow the gel solution to cool for 5 minutes.
4. While the agarose is cooling, set up the gel apparatus. Using masking tape, tape the ends of the gel tray so that a ca. 1 cm deep rectangular mold is formed; the ends should be tightly sealed. Place the tray on the platform in the electrophoresis apparatus. The apparatus must be level; check it by placing the small round bubble-level on the platform and centering the bubble.
5. After the agarose cools to a temperature that the flask can be handled safely, add 5 μl of the stock ethidium bromide to the gel solution and swirl the flask to distribute the dye evenly.
6. Pour this solution into the gel tray; no leakage should occur.
7. Place the gel comb into the gel perpendicular to the running direction (parallel to the tape) about 1 cm from the top of the gel.
8. Allow the gel to solidify.

3. Running the Gel

1. While the gel is solidifying, prepare the samples. Add 2 μl of the DNA "loading dye" to 10 μl of each PCR product. They are now ready to be loaded.
2. Prepare the MW standard, a commercial ϕX174 *Hae*III restriction endonuclease "digest" of bacteriophage λ DNA by placing the following materials into a 0.6 ml microfuge tube:

	TE buffer	7.5 μl	
	DNA running dye	2.0 μl	
+	0.5 mg/ml *Hae*III digested ϕX174	1.5 μl	(0.75 μg DNA total)
		V_{tot} = 11 μl	

3. Mix the contents well then carefully load the samples into the wells in the gel using a pipet. Do not expel any air bubbles into the wells or allow the sample to diffuse over into the adjacent wells. Be sure to record the order and orientations of the samples as loaded (i.e., listed from left to right with the wells at the top of the gel.)

    Lane 1: *Hae*III digested ϕX174 marker DNA fragments
    Lane 2:
    Lane 3:
    Lane 4:
    Lane 5:
    Lane 6:
    Lane 7:
    Lane 8:

4. Attach the leads from the electrophoresis apparatus to the power supply (red to "+," black to "-").
5. Run the gel at 60 V for 1.5 to 2 hours.

4. Photographing the Gel

1. Load the film into the camera as instructed. Use an orange filter for the photography.
2. Place the gel directly on the light box "transilluminator" along with the ruler.
3. While wearing the face safety shield, turn on the UV lights on the light box. Turn off the overhead lights.
4. Photograph the gel; turn off the UV lightstand and turn on the overhead lights.
5. Allow the film to develop.
6. Label the lanes above or below the photograph with information about each sample or control reaction.
7. Store the photographs by taping them into your annotated notebook.

NOTES:

1. Ethidium bromide is a MUTAGEN and SUSPECTED CARCINOGEN. Wear gloves when handling this material!
2. Ethidium binds between every other set of base pairs ("*neighbor exclusion binding*"), so the number of binding sites on a shorter duplex fragment is smaller.
3. If one does not desire to include ethidium bromide directly in the gel, it may be stained after electrophoresis with a 5 μg/μl solution.
4. The primers were designed with sequences complementary to the 5' end of the antisense strand of the *myo*-3 insert and to the 3' end of the sense strand of the plasmid. Nesting one primer in the insert and one primer in the plasmid will produce successful amplification of a product if the *myo*-3 fragment was ligated into the plasmid in the correct orientation.
5. The PCR assay buffer is supplied by the manufacturer to accompany the *Taq* polymerase. This buffer is designed to optimize *Taq* polymerase activity. The recommended assay buffer accompanying FisherBiotech brand 10x *Taq* DNA polymerase contains: 100 mM Tris-HCl, pH 8.0 (at room temperature), 100 mM KCl, 0.1 mM EDTA, 1 mM DTT, 50% glycerol, 0.5% Tween 20, 0.5% Nonidet P40.

ADDITIONAL REPORT REQUIREMENTS:

1. Photograph of agarose gel

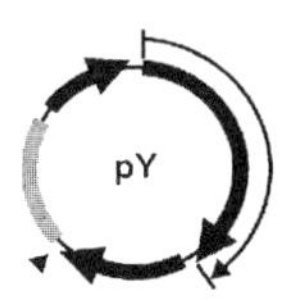

UNIT 4: TRANSCRIPTION OF GENOMIC DNA AND ANALYSIS OF THE RESULTING mRNAS

1. Gene Expression

The genomes of prokaryotes and eukaryotes are composed of millions of nucleotides that encode the thousands of proteins in each individual organism. Transcription converts the encoded DNA into usable products, RNAs first by transcription, then proteins by translation. For example, the genome of *C. elegans* contains 97 million nucleotides and the transcription apparatus produces mRNAs that encode 19,000 proteins. An entire issue of *Science* was recently devoted to a discussion of the *C. elegans* genome, the encoded proteins, and the importance of this organism as a paradigm for our understanding of how an organism develops, starting from a defined genetic background.[1]

2. Gene Transcription

The basal transcription machinery involves catalysis by "core proteins" of RNA polymerase II, which produce copies of DNA sequences by adding one nucleotide at a time to the 3' terminus using the complementary DNA strand as a template. RNA and DNA are always polymerized in the "5' to 3' direction" by adding nucleotides to the 3' terminus of the growing chain.[2] As diagrammed in Figure 1, many factors are required for efficient and accurate synthesis of mRNA. An entire issue of *Trends in Biochemical Sciences* (TIBS) was devoted to the topic of eukaryotic transcription.[3] Briefly, the DNA "TATA" sequence located 30 nucleotides upstream from the genomic transcription start site binds the "TATA binding protein" (TBP). This important positioning step is followed rapidly by binding of "TBP-associated factors" (TAFs) to TBP.

TAFs are composed of a conglomerate of specialized proteins (probably more than 12) that communicate with transcriptional inducers or repressors. TAFs have names such as TFIID, which indicates that they are *T*ranscription *F*actors for RNA polymerase *II*. The letter "D" merely means they were eluted in fraction D from a protein-separation column. Other transcription factors have different functions. TFIIA helps stabilize the DNA/TBP/TAF complex and helps recruit TFIIB, which then recruits RNA polymerase II and TFIIF to the growing complex. Once RNA polymerase II is bound, the helicase activity of TFIIE unwinds the DNA strands and associates the polymerase with the proper nucleotides to begin RNA synthesis. Finally, TFIIH causes further local melting of the DNA and phosphorylates the carboxyl terminal tail of RNA polymerase, which triggers RNA synthesis. These factors are placed schematically in Figure 1.

DNA sequences called enhancers and silencers, which bind transcriptional trans-activating factors (inducers and repressors), are also shown. Inducers or repressors bind these DNA sequences in the presence of the assembled transcriptional apparatus, interacting with it to increase or decrease the rate of transcription, respectively. A relatively new class of molecules, termed co-activators, are also shown. They are thought to function as intermediaries between transcriptional trans-activating factors (bound to enhancer regions) and TAFs. These co-activators have histone acetyltransferase activity and function to render DNA available for transcription.[4]

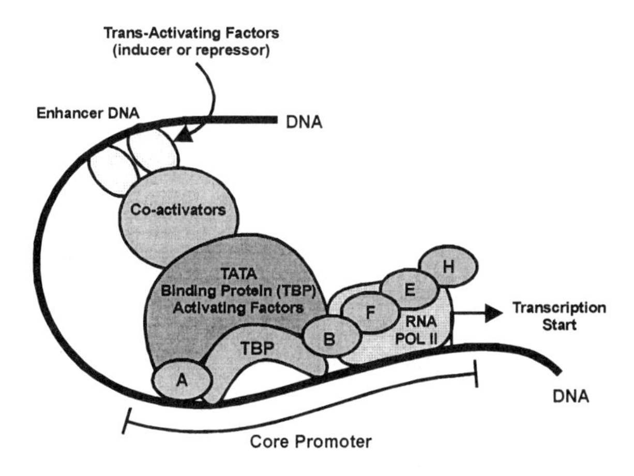

Figure 1 - Schematic of the transcription initiation complex.

3. Analysis of Messenger RNA (mRNA)

Every tissue expresses its own unique set of mRNAs. Many of these mRNAs encode enzymes needed to maintain cell structure or produce energy, processes that are required by all cells to survive (housekeeping mRNAs). Many mRNAs are expressed uniquely in tissues; examples include the rhodopsin mRNA, which is required to sense light in the eye; another set encodes steroid-producing enzymes found mostly in gonadal tissues. Many cells have dramatic variations in gene expression as a result of hormonal and/or electrical stimulation, often producing sizable increases or declines in the amounts of specific mRNAs. Understanding tissue regulation usually requires analysis of the cell-dependent occurrence and variations in the amounts of specific mRNAs.

3.1 Northern Blot Analysis. Northern blot analysis is a popular method for quantitating mRNAs. Although this text does not describe methods to isolate total RNA or mRNA, many techniques are available.[5] Approximately 2% to 3% of total RNA is mRNA; 95% is ribosomal RNA and 3% is tRNA. All RNA isolation procedures include strong denaturing agents (such as phenol) to destroy the actions of ubiquitous ribonucleases (RNases). Once isolated, mRNAs can be fractionated by polyacrylamide electrophoresis and then transferred to nylon membranes, using a process called Northern blotting, which produces a mirror image of the electrophoretic pattern. The membranes bind mRNAs tightly, permitting hybridization of the bound mRNA to labeled complementary cDNAs or cRNA probes. The positions and intensities of the hybridized probes on the membranes indicate the sizes and relative amounts of specific mRNAs in the sample. Typically, specifically labeled DNA or RNA probes are hybridized to the mRNA of interest.

As an internal control of probe hybridization, another mRNA encoding a housekeeping gene is included on the same blot. While the amount of mRNA of interest might change with treatment, the amount of mRNA produced by the housekeeping gene does not change and acts as an internal quantitation standard. Quantitation of Northern blots is most efficiently performed using a PhosphorImager, which provides linear quantitation of radioactive bands over 5 orders of magnitude. Autoradiography of radioactivity or chemiluminescense can also be used to visualize signals on a Northern blot, but regardless of the quantitation method, variations are limited to only 1.5 orders of magnitude.

3.2 RNase Protection Analysis. RNase protection Analysis (RPA) of mRNA is more sensitive than Northern blotting. Northern blot analysis detects 1 molecule of mRNA per cell, but RNase protection analysis is 10- to 100-fold more sensitive. For an RPA, labeled RNA is hybridized to its complementary mRNA in solution (not on a nylon membrane). Accurate hybridization in solution yields a highly ordered RNA:RNA duplex that is impervious to degradation by RNases. When the solution is treated with an RNase, unhybridized or poorly hybridized RNA is digested leaving only the true hybrids undigested. These fully protected RNAs are then quantitated by either collection on a filter (followed by scintillation counting) or by fractionation by polyacrylamide gel electrophoresis (followed by PhosphorImaging). Electrophoretic fractionation permits analysis of the specific size of RNA of interest and quantitation of a second "internal standard" like the housekeeping mRNA described above.

3.3 DNA Chip Analysis. DNA chip analysis of mRNA is on the horizon. This technique expands single RNA analysis (Northern blot or RPAs) into the realm of quantitating >100,000 different mRNAs simultaneously. Currently, the best chip arrays contain hundreds of thousands of oligonucleotides that allow the quantitation of 40,000 mRNAs at one time.[6] Multiple internal standard mRNAs can be quantitated, allowing one to determine a very reliable level of basal expression. One can assess whether the levels of specific mRNAs increase or decrease relative to those of the housekeeping mRNAs. Such information can be obtained for thousands of mRNAs using one DNA chip. DNA chip technology is very powerful but currently very expensive. These techniques are expected to provide valuable tools for use in the rapidly growing field of genomics. A recent article in TIBS focused on DNA chips and their analytical power.[5]

References:

1. Struhl, K. (1998) Enzymatic Manipulation of DNA and RNA, in *Current Protocols in Molecular Biology* (Eds.: Ausubel, F. M., Brent, R., Kingston, R. E., Moore, D. D., Smith, J. A., Seidman, J. G., and Struhl, K.), John Wiley & Sons, New York, Section 3.
2. Kornberg, R. (1996) RNA polymerase II transcription control. Trends Biochem. Sci. *21*: 325-356.
3. Imhof, A., Yang, X. J., Ogryzko, V. V., Nakatani, Y., Wolffe, A. P., Ge, H. (1997) Acetylation of general transcription factors by histone acetyltransferases. Curr. Biol. 7: 689-692.
4. Kingston, R. Preparation and analysis of RNA, in *Current Protocols in Molecular Biology* (Eds.: Ausubel, F. M., Brent, R., Kingston, R. E., Moore, D. D., Smith, J. A., Seidman, J. G., and Struhl, K.) John Wiley & Sons, New York, Section 3.
5. Gerhold, D., Rushmore, T., Caskey, C. T. (1999) DNA chips: Promising toys have become powerful tools. *Trends Biochem. Sci.* 24: 168-173.
6. Genome Sequence of the Nematode *C. elegans*: A Platform for Investigating Biology. The *C. elegans* (1998) Science 282: 1945-2140. [For a list of authors see http://genome.wustl.edu/gsc/C_elegans/#Authors].

ALTERNATIVE APPROACH

ISOLATION OF TOTAL RNA FROM *E. COLI* CELLS

Molecular Research Center, Inc., Cincinnati, OH. (Adapted)

1. Introduction

TRI REAGENT™ (patent pending) is a complete and ready to use reagent for the simultaneous isolation of RNA, DNA and proteins from samples of *human, animal, plant, yeast, bacterial,* and *viral origin.* Simultaneous isolation of RNA, DNA and proteins from the same sample by TRI REAGENT is accomplished by the *single step liquid-phase separation.*[1] The TRI REAGENT method is the improved version of the popular single-step method of isolation of total RNA.[2] This highly reliable technique *performs well with small and large quantities of tissues or cultured cells* and *allows simultaneous processing of a large number of samples.*

The composition of TRI REAGENT includes phenol and guanidine thiocyanate in a monophase solution. A biological *sample is homogenized or lysed in TRI REAGENT.* Then, *after the addition of chloroform and centrifugation, the homogenate separates into three phases*: *aqueous*, *interphase* and *organic*. RNA remains exclusively in the *aqueous* phase, DNA in the *interphase*, and proteins remain in the *organic* phase. RNA *is precipitated* from the aqueous phase *by addition of isopropanol*, washed with ethanol and solubilized. DNA and proteins are *sequentially precipitated* from the interphase, and organic phase *with ethanol and isopropanol*, washed with ethanol and solubilized.

TRI REAGENT retains all of its properties when stored at 2-8°C for at least 9 months from the date of purchase. If left at room temperature, the reagent is stable for up to 3 months. Protect from long exposure (hours) to light.

CAUTION: TRI REAGENT contains poison (phenol) and irritant (guanidine thiocyanate). **IT CAN BE FATAL!** When working with TRI REAGENT *use gloves and eye protection* (shield, safety goggles). Do not get on skin or clothing. Avoid breathing vapor. Read the warning note on the bottle. In case of contact, immediately flush eyes or skin with a large amount of water for at least 15 minutes and seek immediate medical attention.

2. RNA Isolation

Reagents required but not supplied: chloroform, isopropanol, and ethanol. We recommend the use of disposable polypropylene tubes provided by Molecular Research Center, Inc. Tubes from other suppliers should be tested to ensure integrity during centrifugation at 12,000 x *g* with TRI REAGENT.

The protocol includes the following steps (*2.1-2.5*): Unless stated otherwise the procedure is carried out at room temperature.

2.1 Homogenization.

a. *Tissues.* Homogenize tissue samples in TRI REAGENT (1 ml/50-100 mg tissue) using a glass-Teflon or Polytron homogenizer. The sample volume should not exceed 10% of the volume of TRI REAGENT used for homogenization.
b. *Cells.* Cells grown in monolayer should be lysed directly in a culture dish. Add TRI REAGENT and pass cell lysate several times through a pipette. Use 1 ml of the reagent per 10 cm^2 area of a culture dish. Cells grown in suspension should be sedimented first and then lysed in TRI REAGENT by repetitive pipeting. Use 1 ml of the reagent per 5 to 10 x 10^6 of animal, plant, or yeast cells, or per 10^7 bacterial cells. Washing cells before addition of TRI REAGENT should be avoided as this increases the possibility of mRNA degradation. Disruption of some yeast and bacterial cells may require the use of a homogenizer.
c. *Liquid Samples.* These (whole blood, serum, etc.) should be processed using TRI REAGENT (cat. no. TS-120).

2.2 Phase Separation. Store the homogenized samples for 5 minutes at room temperature to permit the complete dissociation of nucleoprotein complexes. Next, add 0.2 ml of chloroform per 1 ml of TRI REAGENT, cover the samples tightly, shake vigorously for 15 seconds, and store them at room temperature for 2-15 minutes. Centrifuge the resulting mixture at 12,000 x *g* (max) for 15 minutes at 4°C. Following centrifugation, the mixture separates into a lower red, phenol-chloroform phase, interphase, and the upper colorless aqueous phase. RNA remains exclusively in the aqueous phase whereas DNA and proteins are in the interphase and organic phase. The volume of the aqueous phase is about 60% of the volume of the TRI REAGENT used for homogenization.

NOTE: Chloroform used for phase separation should not contain isoamyl alcohol or any other additive. For isolation of the poly$(A)^+$ fraction from the aqueous phase, see below.

2.3 RNA Precipitation. Transfer the aqueous phase to a fresh tube and save the interphase and organic phase at 4°C for subsequent isolation of DNA and proteins. Precipitate RNA from the aqueous phase by mixing with isopropanol using 0.5 ml of isopropanol per 1 ml of TRI REAGENT (used for the initial homogenization). Store samples at room temperature for 5-10 minutes and centrifuge at 12,000 x *g* (max) for 10 minutes at 4°C. RNA precipitates (often invisible before centrifugation) and forms a gel-like pellet on the side and bottom of the tube.

2.4 RNA Wash. Remove supernatant and wash the RNA pellet once with 75% ethanol by vortexing and subsequent centrifugation at 7,500 x *g* (max) for 5 minutes at 4°C. Add at least 1 ml of 75% ethanol per 1 ml of TRI REAGENT (used for the initial homogenization).

NOTE: If the RNA pellet accumulates on the side of a tube and has a tendency to float, perform the ethanol wash at 12,000 x *g*.

2.5 RNA Solubilization. At the end of the procedure, briefly dry the RNA pellet by air-drying or under vacuum (5-10 min.). It is *important* not to let the RNA pellet dry completely as this will greatly decrease its solubility. Do not dry RNA by centrifugation under vacuum. Dissolve RNA in stabilized formamide (FORMAzol™, cat. no. FM-121), water or in 0.5% SDS solution by passing the solution a few times through a pipette tip, and incubating for 10-15 minutes at 55-60°C. Water or the SDS solution used for RNA solubilization should be made Rnase free by DEPC treatment.

TRI REAGENT isolates a whole spectrum of RNA molecules, rarely observed in RNA preparations isolated by other methods. The ethidium bromide staining of the RNA separated in agarose gel (or methylene blue staining of a hybridization membrane after the RNA transfer) visualizes two predominant bands of small (~ 2 kb) and large (~ 5 kb) ribosomal RNA, low molecular weight RNA (0.1-0.3 kb), and discrete bands of high molecular weight RNA (7-15 kb).

The final preparation of total RNA is free of DNA and proteins; the A_{260}/A_{280} ratio is 1.6-1.8. Expected yields are:

a. *tissues* (RNA/mg tissue): liver, spleen, 6-10 μg; skeletal muscles, brain: 3-4 μg; placenta, 1-4 μg;
b. *cultured cells* (RNA/10^6 cells): epithelial cells, 8-15 μg; fibroblasts, 5-7 μg.

NOTES and COMMENTS:

1. To facilitate isolation of RNA from small samples (<10^6 cells or <10 mg tissue), perform homogenization (or lysis) of samples in 0.8 ml of TRI REAGENT supplemented with 5-8 μl of Microcarrier™ Gel-TR (cat. no. MR-123). Following homogenization, add chloroform and proceed with the phase separation and other steps of isolation as described above.
2. After homogenization (before addition of chloroform), samples can be stored at -70°C for at least one month. The RNA precipitate (step 2.4, RNA Wash) can be stored in 75% ethanol at 4°C for at least one year at -20°C.
3. For cells grown in monolayer, use the amount of TRI REAGENT based on the area of a culture dish and not on cell number. The use of an insufficient amount of TRI REAGENT may result in contamination of the isolated RNA with DNA. As a more economical alternative, TRI REAGENT can be recommended for isolation of RNA from cells grown in monolayer in large flasks or dishes. Perform cell lysis by the addition 0.3-0.4 ml of TRI REAGENT™LS per 10 cm^2 of the flask, and follow the isolation protocol described in the brochure.
4. *Hands* and *dust* may be the major source of the RNase contamination. Use gloves and keep tubes closed throughout the procedure.
5. An additional isolation step may be required for samples with high content of proteins, fat, polysaccharides, or extracellular material – such as muscles, fat tissue, and tuberous parts of plants. Following homogenization, remove insoluble material from the homogenate by centrifugation at 12,000 x *g* for 10 minutes at 4°C. The resulting pellet contains extracellular membranes, polysaccharides and high molecular weight DNA, while the supernatant contains RNA. In samples from fat tissue, an excess of fat collects as a top layer; it should be removed. Transfer the clear supernatant to a fresh tube and proceed with the phase separation and other steps of RNA isolation as described above. High molecular weight DNA can be recovered from the pellet by following steps 2 and 3 of the DNA isolation protocol.

3. Troubleshooting Guide

3.1 RNA Isolation.

1. Low yield.
   a. incomplete homogenization or lysis of samples,
   b. incomplete solubilization of the final RNA pellet.
2. The A_{260}/A_{280} ratio is <1.65.
   a. volume of reagent used for sample homogenization was too small,

b. following homogenization, samples were not stored for 5 minutes at room temperature,
c. contamination of the aqueous phase with phenol phase,
d. incomplete solubilization of the final RNA pellet.

3. RNA degradation.
   a. tissues were not immediately processed or frozen after removing from the animal,
   b. samples used for isolation, or the isolated RNA preparations, were stored at -20°C rather than -70°C,
   c. cells were dispersed by trypsin digestion,
   d. aqueous solutions or tubes used for solubilization of RNA were not Rnase free,
   e. formaldehyde used for the agarose-gel electrophoresis had a pH < 3.5.
4. DNA contamination.
   a. volume of reagent used for sample homogenization was too small,
   b. samples used for the isolation contained organic solvents (ethanol, DMSO), strong buffers, or alkaline solution.

3.2 Quantitation of DNA. Take an aliquot of the DNA preparation solubilized in 8 mM NaOH, mix it with water, and measure the A_{260} of the resulting solution.

1. Calculate the *DNA content* using the A_{260} value for the double-stranded DNA: one A_{260} unit equals 50 μg of double-stranded DNA/ml.
2. For calculation of a *cell number* in analyzed samples, assume that the amount of DNA per ml of diploid cells of human, rat, and mouse origin equals: 7.1 μg, 6.5 μg, and 5.8 μg, respectively.[3]

A typical preparation of DNA isolated from tissues is composed of 60-100 kb DNA (70%) and ca. 20 kb DNA (30%). The DNA isolated from cultured cells contains >80% of 60-100 kb DNA and <10% 20 kb DNA. The preparation of DNA is free of RNA and proteins and has an A_{260}/A_{280} ratio of > 1.7.

3. Expected yield:
   a. *tissues* (DNA/mg tissue): liver, kidney, 3-4 μg; skeletal muscles, brain, placenta, 2-3 μg.
   b. *cultures*: human, rat and mouse cells 5-7 μg DNA/10^6 cells.

3.3 DNA Isolation.

1. Low yield.
   a. incomplete homogenization or lysis of samples,
   b. incomplete solubilization of the final DNA pellet.
2. A_{260}/A_{280} ratio < 1.70.

Phenol was not sufficiently removed from the DNA preparation. Add one more wash of the DNA pellet with the 10% ethanol – 0.1 M sodium citrate solution.

3. DNA degradation.
   a. tissues were not immediately processed or frozen after removing from the animal,
   b. samples used for isolation were stored at -20°C instead of at -70°C,
   c. samples were homogenized with a Polytron or other high speed homogenizer.
4. RNA contamination.
   a. too large volume of aqueous phase remained with the interphase and organic phase,
   b. DNA pellet was not sufficiently washed with 10% ethanol-0.1 M sodium citrate solution.

References:

1. Chomczynski, P. (1993) A reagent for the single-step simultaneous isolation of RNA, DNA and proteins from cell and tissue samples. BioTechniques 15, 532-537.
2. Chomczynski, P. and Sacchi, N. (1987) Single step method of RNA isolation by acid guanidinium thiocyanate-phenol-chloroform extraction. Anal. Biochem 162, 156-159.
3. Ausubel, F. M., Brent, R., Kingston, R. E., Moore, D. D., Seidman, J. G., Smith, J. A. and Struhl, K. (1990) Appendix 1, in *Current Protocols in Molecular Biology, Vol. 2*, p. A.1.5, Greene Publishing Assoc. & Wiley-Interscience, New York.

ALTERNATIVE APPROACH

PROMEGA™ POLYATRACT™ SYSTEM 1000

Promega Corporation, Madison, WI, ph. (608) 274-4330
Other product trademarks: Corex™ from Corning, Inc., Tissuemizer™ from Tekmar, Inc. (Adapted)

1. Description

The PolyATract System 1000 isolates messenger RNA directly from crude cell or tissue lysates and eliminates the need for total RNA isolations. The System utilizes Promega's MagneSphere™ Technology for the purification of poly(A)$^+$ RNA, eliminating the need for oligo(dT) cellulose columns.

The PolyATract System 1000 does not have lengthy ethanol precipitation steps, phenol: chloroform extractions, overnight ultracentrifugation through cesium chloride gradients, and lithium chloride (LiCl) precipitations. Increased mRNA yields are obtained because losses that occur during the organic extractions and precipitations are eliminated. The isolated mRNA is suitable for all molecular biology applications, including *in vitro* translation, cDNA synthesis, PCR analysis, and Northern blots. The increased yield of mRNA using this system allows the detection of low copy number RNAs in relatively small amounts of material using Northern blot analysis.

The benefits obtained by eliminating the intermediate purification of total RNA include:

1. Time savings – 30 minutes to mRNA vs. 4.5 hours with previous systems.
2. Increased yields of poly(A)$^+$ RNA – up to two-fold greater than other methods.

The PolyATract System 1000 protocol can be adjusted to the amount of starting material available,

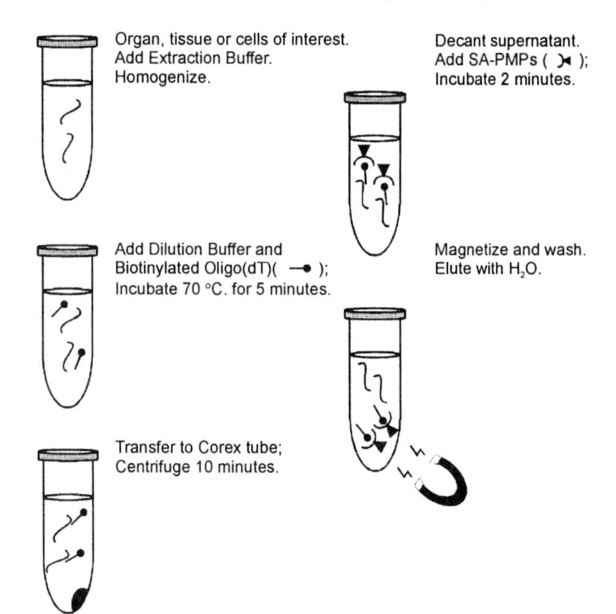
Organ, tissue or cells of interest.
Add Extraction Buffer.
Homogenize.
Add Dilution Buffer and
Biotinylated Oligo(dT)( );
Incubate 70 °C. for 5 minutes.
Transfer to Corex tube;
Centrifuge 10 minutes.
Decant supernatant.
Add SA-PMPs ( );
Incubate 2 minutes.
Magnetize and wash.
Elute with H₂O.

Figure 1 - PolyATract purification.

thus maximizing the number of purifications obtainable with this system. Sufficient reagents are supplied to isolate mRNA from up to 2 g of starting material in any combination of reaction sizes up to 1 g per isolation. All system components are guaranteed to be free of contaminating ribonucleases and are thoroughly tested to ensure optimal performance.

2. General Considerations

2.1 Direct Purification of mRNA from Tissue. The successful isolation of intact RNA by any procedure requires that four important steps be performed: (1) effective disruption of cells or tissue, (2) denaturation of nucleoprotein complexes, (3) inactivation of endogenous ribonuclease (RNase) activity, and (4) purification of RNA away from contaminating DNA and protein. The most important of these is the immediate inactivation of endogenous RNase activity – which is released from membrane-bound organelles upon cell disruption.

The PolyATract System 1000 combines the disruptive and protective properties of guanidine thiocyanate and β-mercaptoethanol[1] to inactivate the RNases present in cell extracts. Guanidine thiocyanate (GTC), in association with SDS, acts to disrupt nucleoprotein complexes, allowing RNA to be released into solution and isolated free of protein. The final GTC concentration allows for hybridization between the poly(A)$^+$ sequence of most mature eukaryotic mRNA species and a synthetic biotinylated oligo(dT) probe, yet maintains complete inhibition of the cellular RNases. Hybridization occurs during the brief centrifugation step that removes cellular debris and precipitated proteins. After centrifugation, the *biotinylated oligo(dT): mRNA hybrids* are captured with *Streptavidin Paramagnetic Particles* (SA-PMPs). The particles are washed at high stringency and purified mRNA is eluted by the simple addition of nuclease-free deionized water. This procedure yields an essentially pure fraction of mature mRNA after only a single round of magnetic separation without organic extractions or precipitations. A summary of the procedure is provided in Figure 1.

The PolyATract System 1000 protocol was developed and optimized for mRNA isolation from tissues with a broad spectrum of mRNA expression levels. Because rigorous performance specifications require quantitative capture from tissues high in mRNA, the protocol has been designed with an excess mRNA binding capacity to ensure recovery of all available mRNA. In order to maintain sufficient molar concentrations of the biotinylated oligo(dT) and the mRNA for hybridization, it is necessary to keep the volumes of the extraction and dilution buffers to a minimum. Depending on the amount of starting material, either of two protocols should be followed. For sample sizes less than 0.125 g of tissue, perform the small-scale protocol. For sample sizes between 0.125 g and 1 g, perform the large-scale protocol. Do not use more than one gram of tissue because the increased lysate viscosity will lead to irreversible clumping of the particles during magnetic capture, resulting in lower RNA yields. In addition, mRNA isolations can be performed on 10^7 to 10^8 cultured cells.

If the desired yield of mRNA is known in advance and starting material is not limiting, the PolyATract System 1000 can be tailored to give the desired yield. In this case, consult the top of the Probe and Particle Usage charts and purify the mRNA from the corresponding amount of starting material indicated on the chart. Note that these charts were generated using mouse liver tissue (ICR strain), that has relatively high levels of mRNA per gram weight of tissue, and the corresponding yields are higher than expected from other sources. Additional tissue can be processed as necessary to supply the desired amount of mRNA.

2.2 Storage Conditions. The PolyATract System 1000 components are stable for 6 months from the date of purchase if stored at 4 to 25°C. Do not freeze the Streptavidin MagneSphere Particles, as this will result in failure of the system.

3. Magnetic Particle Separation of Macromolecules

Paramagnetic particles incorporate iron oxide into submicron-sized particles that have no magnetic field but form a magnetic dipole when exposed to a magnetic field. The use of paramagnetic particles eliminates the need for traditional column chromatography, centrifugation, or any other special equipment. These particles have been successfully used in the development of *immunoassays*,[2] *probe diagnostic* assays,[3] and for *measuring RNA* in cell lysates *using dT-tailed capture probes.*[4]

Unlike procedures that use direct coupling of probes to paramagnetic particles,[4] the PolyATtract System 1000 uses a *biotinylated oligonucleotide probe* to hybridize in solution to the targeted nucleic acid. The hybrids are then captured using *covalently coupled* SA-PMPs. This approach combines the speed and efficiency of solution hybridization with the convenience and speed (< 1 minute) of magnetic separation.

Promega uses its own highly purified streptavidin for the production of particles. These Streptavidin Paramagnetic Particles exhibit a high binding capacity for biotinylated oligonucleotides and low nonspecific binding of nucleic acids. The binding capacity of the particles varies with the specific oligonucleotide probe used. For biotinylated oligo-(dT), the calculated binding capacity is roughly 1 nmol of free probe captured per milligram of SA-PMPs.

4. Creating a Ribonuclease-Free Environment

Ribonuclease is difficult to inactivate. Great care should be taken to avoid inadvertently introducing RNase activity into your RNA during or after the isolation procedure. This is especially important if the starting material has been difficult to obtain or is irreplaceable. The following notes may help you to prevent accidental contamination of your sample.

1. Two of the most common sources of RNase contamination are the user's hands and bacteria or molds that may be present on airborne dust particles. To prevent contamination from these sources, sterile technique should be observed when handling the reagents supplied with the kit. Gloves should be worn at all times.
2. Whenever possible, sterile disposable plasticware should be used for handling RNA. These materials are generally Rnase free and thus do not require pretreatment to inactivate RNase.
3. Nondisposable glassware and plasticware should be treated before use to ensure that it is Rnase free. Glassware should be baked at 200°C overnight and plasticware should be thoroughly rinsed before use with 0.1 N NaOH, 1 mM EDTA, followed by RNase-free water.

NOTE: Corex tubes should be *rendered Rnase free by treatment with diethylpyrocarbonate* (DEPC) and not by baking. This will reduce the failure rate of this type of tube during centrifugation.

4. Solutions supplied by the user should be treated with *0.5% DEPC* overnight at room temperature and then autoclaved for 30 minutes to remove trace amounts.
5. Tris buffers cannot be treated with DEPC.

NOTE: DEPC degrades to produce TOXIC *carbonyl radical* when heated.

6. We have found that *many good sources of distilled H_2O are free of contaminating RNase activity.* You may want to test your water source for the presence of RNase activity.

5. Protocols

5.1 Materials to be Supplied by the User.

- Tissuemizer™ or equivalent homogenizer
- sterile screw cap conical tubes

Table 1 - Volumes of PolyATtract® System 1000 Reagents Required for Various Starting Amounts of Tissue (Small Scale)

	Tissue Amount Prepared (mg)[1]				
Component	5	10	25	50	100
GTC Extraction Buffer	40 μl	80 μl	200 μl	400 μl	800 μl
Dilution Buffer	80 μl	160 μl	400 μl	800 μl	1.6 ml
Oligo(dT) Probe[2]	5 pmol	10 pmol	25 pmol	50 pmol	100 pmol
SA-PMP	60 μl	120 μl	300 μl	600 μl	1.2 ml
SSC 0.5x Solution[3]	500 μl	500 μl	1 ml	1 ml	1 ml
Nuclease-Free Water	40 μl	80 μl	200 μl	400 μl	800 μl

- sterile Corex™ centrifuge tubes
- 70°C water bath
- Beckman model J2-21 centrifuge or equivalent
- 1x PBS (for isolation from cell cultures).

NOTES:
1. Do not exceed the 1 g tissue capacity; the increased lysate viscosity leads to irreversible clumping of the Particles during magnetic capture and decreased system performance.
2. The Paramagnetic Particles cannot be reused because biotinylated oligo(dT) remains attached, decreasing the capacity of the particles. In addition, trace amounts of nucleic acid contaminants may be present.
3. Centrifuges and rotors should be equilibriated at room temperature.

5.2 Protocol for mRNA Isolation from Tissue Samples.

1. Remove the GTC Extraction Buffer, Biotinylated Oligo(dT) Probe, Nuclease-Free Water, and SSC 0.5x Solution from the refrigerator and warm to room temperature. Preheat the Dilution Buffer to 70°C.
2. In an appropriately sized tube (see Table), add 41 μl of β-mercaptoethanol ("BME," 48.7%) to 1 ml of the Extraction Buffer (Extraction/BME Buffer). The final concentration of BME is 2%. Use RNase-free pipets and wear gloves to reduce the chance of RNase contamin-ation. Weigh the tube containing the buffer and record the weight.

CAUTION: BME is highly toxic. Dispense in a fume hood and wear appropriate personal protective equipment.

Table 2 - Volumes of PolyATtract® System 1000 Reagents Required for Various Starting Amounts of Tissue (Large Scale)

	Tissue Amount Prepared (mg)[1]				
Component	125	250	500	750	1000
GTC Extraction Buffer	1 ml	1 ml	2 ml	3 ml	4 ml
Dilution Buffer	2 ml	2 ml	4 ml	6 ml	8 ml
Oligo(dT) Probe[2]	125 pmol	250 pmol	500 pmol	750 pmol	1000 pmol
SA-PMP	1.5 ml	3 ml	6 ml	9 ml	12 ml
SSC 0.5x Solution[3]	2 ml	2 ml	2 ml	2 ml	2 ml
Nuclease-Free Water	1 ml	2 ml	3 ml	-	800 μl

3. Working as quickly as possible, extract the tissue of interest and place into the tube containing the Extraction/BME Buffer. Homogenize the tissue at high speed using a small homogenizer (such as a Tekmar Tissuemizer™) until no visible tissue fragments remain.
   If a small homogenizer is unavailable, quickly cut the tissue into small pieces with a sterile razor blade, freeze in liquid nitrogen, and grind using a mortar and pestle under liquid nitrogen. Transfer the liquid nitrogen and the ground tissue to a sterile 50 ml screw cap conical tube, allow the liquid nitrogen to evaporate and then immediately add the Extraction/ BME Buffer. Mix thoroughly by inversion.
4. Weigh the tube containing the tissue in Extraction/BME Buffer. Calculate the tissue mass by subtracting the weight obtained in Step 2 from this new weight.
5. Determine the amount of Biotinylated Oligo(dT) Probe and SA-PMPs necess-ary for the tissue mass calculated in Step 4. When adding <50 pmol Biotinylated Oligo(dT) Probe per sample, dilute the Probe 1:10 in Nuclease-Free Water. The diluted Probe is stable when stored at 4 to 25°C.
6. Aliquot the preheated Dilution Buffer to a sterile tube and add 20.5 μl of BME (48.7%) per milliliter of Dilution Buffer. The final concentration of BME is 1%. Add this to the homogenate and mix thoroughly by inversion. Add the amount of probe determined in Step 5 and mix well by shaking. Incubate this mixture at 70°C for 5 minutes.
7. Transfer the lysate to a clean, sterile 15 ml Corex tube® or other appropriately sized tube. Centrifuge at 12,000 x *g* for 10 minutes at room temperature to clear the homogenate of cell debris and precipitated proteins. During the

centrifugation, completely resuspend the SA-PMPs by gently rocking the bottle.

NOTE: The particles should appear as a homogeneous mixture and be fully suspended in the liquid.[5]

Transfer the amount of particles determined in Step 5 to a sterile 50 ml or other appropriately sized, conical screw cap tube. Place the tube with the SA-PMPs on the Magnetic Stand. Slowly move the Stand toward the horizontal position until the particles are collected at the tube side. Carefully pour off the storage buffer by tilting the tube so that the solution runs over the captured particles.

8. Resuspend the SA-PMPs in an equal volume of SSC 0.5x Solution and capture using the magnetic stand. Pour off the SSC Solution as described in Step 7. Repeat this wash step twice more for a total of three times. Resuspend to the original volume with SSC 0.5x Solution.
9. When centrifugation of the homogenate is complete, carefully remove the supernatant with a sterile 10 ml pipet, avoiding the pellet. The homogenate will be translucent and brownish in color. Add this cleared homogenate to the tube, containing the washed particles in SSC 0.5x Solution, away from the magnetic stand to ensure proper mixing of the homogenate and particles. Mix by inversion.
10. Incubate the homogenate/SA-PMP mixture at room temperature for 2 minutes. Capture the SA-PMPs using the magnetic stand in the horizontal position until the homogenate clears and then carefully pour off the supernatant as in Step 7. Save the supernatant in a sterile tube on ice until you are certain that satisfactory binding and elution of the mRNA have occurred.
11. Resuspend the particles in SSC 0.5x Solution to the volume indicated in Table 1 or Table 2 away from the magnetic stand. Mix gently by flicking the tube. Transfer the particle mixture to one of the 2 ml mRNA User Tubes provided. Capture the particles by placing the tube in the magnetic stand. Carefully pipet off the SSC solution and discard. Repeat this wash step twice. After the final wash, pipet off as much of the SSC solution as possible without disturbing the SA-PMP particle cake.
12. To elute the mRNA, add the amount of Nuclease-Free Water indicated in Table 1 or Table 2 to the SA-PMP cake. Resuspend the particles by flicking the tube gently.
13. Magnetically capture the SA-PMPs, placing the stand in a horizontal position. Transfer the liquid containing the eluted mRNA to a sterile microcentrifuge tube. Resuspend the particles in Nuclease-Free Water and save on ice until the yield and purity of the mRNA have been determined. Proceed to the ethanol precipitation step.

NOTE: If SA-PMP particle carryover occurs, which is often the case, centrifuge the supernatants for 1 minute at 12,000 x *g* and transfer the supernatant to a fresh tube. Residual particles do not inhibit many of the common enzymatic reactions used to process RNA; they actually help in locating the pellets after precipitations.

5.2 Precipitation and Concentration of mRNA.

1. *For cDNA cloning*: Add 0.1 volume of 3 M sodium acetate and 1 volume of isopropanol to the eluate and incubate at -20°C overnight.
   For translation in vitro: Add 0.1 volume of 3 M potassium or ammonium acetate and 1 volume of isopropanol to the eluate and incubate at -20°C overnight.
2. Centrifuge at >12,000 x *g* for 10 minutes. Resuspend the RNA pellet in 1 ml of 70% ethanol and centrifuge again.
3. *For short-term storage*: Dry the pellet in a vacuum desiccator for about 15 minutes, resuspend in RNase-free, deionized water at 0.5-1 μg/μl and store at -70°C.
   For long-term storage: Store the RNA pellet in 70% ethanol at -70°C.

5.3 Determination of mRNA Concentration and Purity. The concentration and purity of the eluted mRNA can be determined approximately by spectrophotometry. Determine the absorbance readings at 290, 280, 260, and 230 nm (A_{290}, A_{280}, A_{260}, and A_{230}). Absorbance readings should be greater than 0.05 to ensure significance. When the mRNA has been isolated from small amounts of tissue, the expected yields are such that absorbance must be read directly, without dilution of the sample. Be certain that cuvettes are RNase free so that samples can be recovered after spectrophotometry. This can be done by washing the cuvettes briefly with 0.1 N NaOH, 1 mM EDTA, followed by a brief rinsing with RNase-free water. Pure mRNA will have an A_{260}/A_{280} absorbance ratio of 2. To estimate the mRNA concentration, assume that a 40 μg/ml mRNA solution will have an absorbance ratio of 1 at 260 nm. Also determine the A_{260}/A_{230} ratios, which will provide information on the purity of the sample. An A_{260}/A_{230} ratio of less than 2 indicates that GTC or BME from the Extraction Buffer is still present. If this is the case, precipitate the RNA again.

The quality of the isolated mRNA may also be checked by denaturing agarose gel electrophoresis.[5] The mRNA should appear as a smear extending from approximately 8 kb to approximately 0.5 kb (depending on the tissue). The bulk of the mRNAs should be clustered around 2 kb. Expect to see very little ribosomal RNA using the PolyATract System 1000; however, the appearance of some ribosomal bands does not indicate poor performance of the system. Figure 4 shows a Northern blot of mRNA which contains visible amounts of both 28S and 18S ribosomal.

The blot was probed with an alkaline phosphatase oligonucleotide conjugate made using Promega's LIGHTSMITH™ I System. Little or no hybridization is seen in the total RNA lane, whereas the mRNA prepared from the PolyATract System 1000 shows a significant enrichment, despite the presence of minor amounts of ribosomal RNAs. A small amount of ribosomal contamination should not affect the functionality of the mRNA and is suitable for most applications.

6. Solutions

6.1 GTC Extraction Buffer.

4 M guanidine thiocyanate
25 mM sodium citrate, pH 7.1

6.2 Dilution Buffer.

6x SSC
10 mM Tris-HCl, pH 7.4
1 mM EDTA
0.25% SDS

6.3 20x SSC (500 ml).

87.7 g NaCl
44.1 g sodium citrate

Dissolve in 400 ml of distilled H_2O. Adjust pH to 7.2 with NaOH and bring the volume to 500 ml. Dispense into aliquots. Sterilize by autoclaving.

6.4 10x PBS.

11.5 g Na_2HPO_4
2 g KH_2PO_4
80 g NaCl
2 g KCl

Dissolve in 1 liter of distilled water. The pH of the 1x PBS will be 7.4.

7. Troubleshooting

(See Table 3.)

References:

1. Chirgwin, J. M., *et al.* (1979) Biochemistry 18, 5294.
2. Birkmeyer, R. C., *et al.* (1987) Clin. Chem. 33, 1543.
3. Morrissey, D. V., *et al.* (1989) Anal. Biochem. 181, 345.
4. Thompson, J., *et al.* (1989) Anal. Biochem. 181, 371.
5. Sambrook, J., *et al.* (1989) *Molecular Cloning: A Laboratory Manual*, Cold Spring Harbor Laboratory, Cold Spring Harbor, NY.

Table 3 - Troubleshooting Chart

Problem	Possible Causes	Comments
No mRNA eluted	Salt not eliminated during elution	Wash last SA-PMP pellet again with deionized H_2O and check A_{280} of this eluate.
	RNase contaminated during mRNA isolation	Repeat entire procedure. Read "Crèating an RNase-Free Environment"
Low A_{260}/A_{280}	Protein contamination	Several methods may be used for further removal of contaminating protein from RNA. The most expedient method is to perform an additional phenol:chloroform extraction on the purified RNA. Add an equal volume of phenol:chloroform to the final resuspended RNA pellet. This procedure should yield higher A_{260}/A_{280} ratios. However, some loss of RNA (up to 40% may be expected).
Low A_{260}/A_{280} ratios	Contamination with guanidine thiocyanate or β-mercaptoethanol	Precipitate the mRNA again as described.

ALTERNATIVE APPROACH
ELECTROPHORESIS AND NORTHERN BLOTTING OF RNA

1. Preparation of a 2% Agarose-Formaldehyde Gel

NOTE: It is recommended that all steps in the following procedure be conducted in a fume hood due to the caustic nature of formaldehyde fumes. Besides, IT STINKS!

Dissolve agarose in 18.5 ml H_2O by heating in a microwave. When completely dissolved, cool at room temperature until the solution reaches 60 °C. Add 5.9 ml 5x MOPS electrophoresis buffer and 5.4 ml 37% formaldehyde. Mix and pour the gel immediately.

2. Quantitating RNA

1. Dilute a 5 µl aliquot of isolated RNA to 1 ml with dH_2O.
2. Determine the A_{260}/A_{280} ratio of each sample.
3. Determine how many mg of RNA you have per ml (assume 40 µg RNA per A_{260} unit). Make sure the sample contains 10 µg of RNA in 3 µl. If the sample is concentrated enough, dilute it in water to give a final concentration of 10 µg RNA/3 µl. Treat this RNA as follows:
4. Place a 5 µl sample in the palm of your hand. Mix it around but keep it in a discrete drop. Finally, transfer it to a 0.5 ml tube and allow it to incubate for 15 minutes at room temperature. This procedure is used to see how much RNase you have on your hand!
   NOTE: CHANGE YOUR PIPET TIP!!!
5. Take a second sample of the RNA and dilute it to 10 µg/3 µl. Prepare the RNA samples by mixing the following:
   a. Incubate the sample for 15 minutes at 60°C, cool on ice.
   b. Add 2 µl of loading buffer and 1 µl of ethidium bromide (1 µg/ml). Molecular weight standard (*Bst*EII digested λ DNA) will be prepared for you.
6. Prepare your electrophoretic gel apparatus while your RNA is denaturing. Add 1x MOPS buffer to the apparatus so that it just covers the gel.
7. Load the samples in the gel wells after the RNA is denatured.
8. Electrophorese at 100 V for 45 minutes to 1 hour. During this time prepare Genius™ membranes and filter papers as explained in the next step.

3. Transfer of Formaldehyde-Denatured RNA to the Membrane

1. Cut the Genius™ membrane to the exact size of the gel. Wear gloves!
2. Just prior to the transfer, wet the membrane first with water and then with 10x SSC for 15 minutes. Be careful not to trap air bubbles beneath the membrane or it will saturate unevenly.
3. Cut 3 mm blotting paper, 9-mm blotting paper and a 5-8 cm stack of absorbent paper towels to the exact size as the gel.
4. Fill a tray with 10x SSC and allow a filter paper wick to absorb the buffer.
5. Place the gel on top of the filter paper wick; remove any air bubble(s) with a pipette.
6. Place the buffer-saturated Genius™ membrane on the gel and remove any air bubbles.
7. Place the buffer saturated 3-mm and 9-mm blotting paper on top of the membrane and remove any air bubbles.
8. Place the stack of absorbent paper towels on top of the blotting paper.
9. Place a glass plate on top of the paper towels.
10. Place a 500-g weight on top of the glass plate.
11. Allow the RNA transfer to proceed overnight (at least 12 hours).

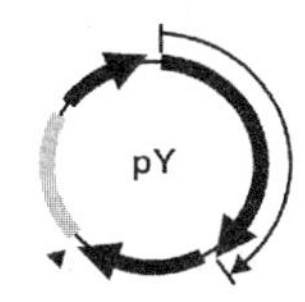

UNIT 5: TRANSFORMATION AND GENE EXPRESSION

1. Competence and Transformation of *E. coli* Cells

The breakthrough discovery of genetic transformation by Griffith in 1928 led to the discovery of DNA as the transforming genetic material and laid the groundwork for the development of molecular biology. Transformation is the process by which cells take up and incorporate DNA from the environment. Cells that can undergo transformation are said to be competent. Many groups of bacteria can undergo natural transformation. Usually this depends on the phase of growth. Many types of bacteria do not become competent naturally, including the one most often used for molecular biology work – *E. coli.*

Fortunately, *E. coli* can be made competent artificially by treatment with chemicals such as high concentrations of cold $CaCl_2$. *E. coli* and/or other cells that are more difficult to transform often work, but efficiencies can vary from acceptable levels to zero, depending on the cell type, growth conditions, disease state, presence of virus, mutagens, and other factors.

- Treatment with other cations, e.g., LiCl, Li glutamate, etc. can induce much higher transformation efficiencies (up to 1000x).
- Application of short bursts of very high voltage electricity (electroporation) can induce transformation. The highest reported *E. coli* transformation efficiencies are obtained using this method.
- Projectile methods using small particles coated with the transforming DNA have also been successful. These particles are shot into the recipient cells using a "gene gun." This technique is not used for *E. coli* and is typically reserved for cells that are difficult to transform. Efficiencies are generally very low.

For most routine biotechnology applications, competent *E. coli* cells prepared by the $CaCl_2$ method are sufficient; however, some applications require highly competent cells. Methods described above may be used or competent *E. coli* cells may be purchased from commercial suppliers. Commercial competent cells are generally extremely efficient, but they are also very expensive. If *E. coli* transformations are being performed routinely, it is much cheaper to render the cells competent yourself.

Competent cells are prepared by growing the appropriate *E. coli* strain to mid-exponential phase, then harvesting and suspending them in a $CaCl_2$ solution for a short time. Plasmid DNA is added and allowed to attach to the competent cells. Transformation is completed by heat shock at 42°C, a short incubation on ice and a recovery period, involving incubation in growth medium, to allow cell repair and expression of enzymes that confer antibiotic resistance. The exact mechanism of competence and transformation is not fully understood, but is believed to involve "freezing" of outer membranes and separation of stacked lipid hydrocarbon tails, which is presumed to produce "cracks" that allow DNA molecules to pass through.

2. Plasmids as Protein Expression Vectors

As discussed in Unit 1, plasmids that are used to carry foreign genes into bacterial hosts are referred to as cloning vectors. If the foreign genes are to be expressed in *E. coli* – used to transcribe mRNA that is translated into protein products – one must use a bacterial expression vector. If the cloned gene is from a eukaryotic organism such as *C. elegans*, one usually uses a bacterial expression vector to clone the genes; bacterial RNA polymerases do not recognize *C. elegans* RNA polymerase binding and initiation sites. If the cloned gene is from another bacterial species, it can sometimes be expressed without using an expression vector, however, small product yields are typical. Bacterial expression vectors contain necessary elements for bacterial transcription and translation, including a strong bacterial promoter containing appropriate recognition sequences (-10 and -35 consensus sequences) for RNA polymerase. A suitable ribosome-binding site (Shine-Dalgarno sequence) is required for efficient translation initiation. These sequences are placed in the plasmid an appropriate distance downstream from the inserted DNA fragment.

It is important that a gene be inserted into the plasmid in the proper reading frame to ensure that the correct protein will be expressed. This problem is usually solved by using a family of three cloning vectors that have been altered by one base pair each. In other words, if the same DNA fragment is inserted into the same unique restriction enzyme site in each of the three different vectors, it will be in the correct reading frame at one of the three (given the 1 of 3 odds resulting from the triplet nature of the genetic code).

An efficient expression vector system can convert a bacterial cell into a protein-manufacturing machine. Yields as high as 50% of the total cellular protein can, with careful attention to nutrient levels, be produced from the introduced foreign gene. Since foreign proteins are frequently very toxic to host cells, the genes must be under very strict control. These vectors generally contain inducible promoters that can be controlled by the presence or absence of an inducing agent. For example, the *E. coli lac* operon can be induced by the addition of lactose analogs such as isopropyl-thio-β-d-galactoside (IPTG) (Figure 1). A very low, basal level of transcription occurs in the absence of such inducers. Growth in the absence of protein production allows one to make plasmid DNA and perform other manipulations without toxic side effects.

A fusion protein expression vector contains sequences that encode a particular predesigned protein and are situated on the vector so that they will be adjacent to the cloned gene. When the cloned gene is transcribed, the resulting mRNA contains sequences from both the cloned gene and the fusion gene. When the mRNA is translated, a fusion protein is produced – which contains protein domains corresponding to both the cloned and vector genes. The vector typically encodes a protein, or fragment thereof, that is derived from an enzyme that can be assayed, e.g., β-galactosidase. The advantages of using a fusion protein expression vector are:

1. Attaching an enzyme to your protein allows one to follow the activity and fate of the protein produced by gene expression easily. β-Galactosidase assays are fast and relatively simple to perform.
2. Attaching a polypeptide segment to the cloned protein frequently makes the fusion protein much more stable than cloned protein alone.
3. Protein encoded by the vector can often be purified easily by affinity chromatography using galactose-derivative matrices.

3. Colony Immunoblotting

When potential clones have been obtained, it is necessary to screen them for the presence of your gene of interest. If antibodies that bind the protein gene product are available and a good expression vector is employed, colony immunoblotting may be used as a screening method. Immunoblotting is based on the ELISA (enzyme-linked immunosorbent assay), procedure (see Unit 9).

Clones are replica-plated onto two sets of plates containing the appropriate antibiotic. Nitrocellulose filters are soaked in an inducing substance then placed on culture-containing agar plates. IPTG may be used if the expressed genes are under the control of a lactose operon promoter. A set of plates that does not contain the filters is grown overnight and used as "master plates" from which positive clones may be recovered later. The plates containing the filters are incubated for several hours, long enough for the cloned genes to be expressed. The clone-containing cells are then lysed and the expressed proteins are allowed to bind to filters. The filters may then be probed with antibodies that have been elicited specifically to recognize the protein of interest (a primary antibody). After washing away unbound antibodies, the filters are incubated with a second antibody that is specific for the F_c portion of the primary antibody (e.g., goat anti-mouse antibody, when the primary antibody is produced in mice). An enzyme is bound covalently to this secondary antibody, usually alkaline phosphatase or horseradish peroxidase. The filters are washed again to remove unbound secondary antibody, then incubated with a chromogenic substrate that can be processed by the

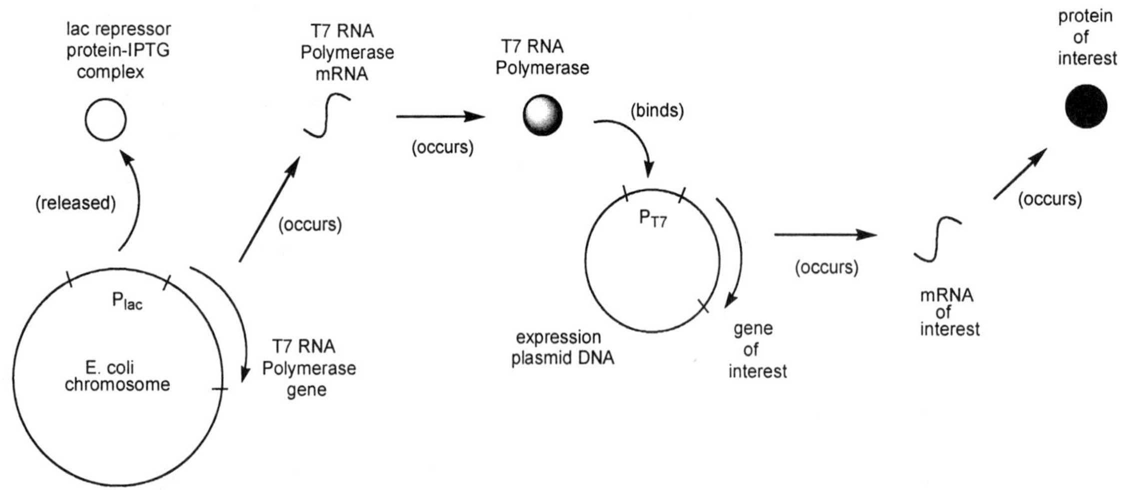

Figure 1 - IPTG-induced protein expression by coupled transcription-translation. Plasmid mRNA induction and protein synthesis. 1. The added IPTG binds lac repressor protein, causing it to come off the P_{lac} lac promoter site on the *E. coli* chromosome (only if the appropriately engineered *E. coli* cell is used). 2. As a result, the T7 RNA Pol gene is activated and *E. coli* RNA Pol can be transcribe it to produce T7 RNA Pol mRNA. 3. T7 RNA Pol mRNA is translated to produce the T7 RNA Pol protein, which binds the P_{T7} T7 promoter and transcribes the plasmid gene of interest (*myo*-3-β-lac mRNA is made). This mRNA is translated to produce the *myo*-3-β-lac fusion protein. Therefore, the addition of IPTG ultimately results in production of the plasmid-encoded protein through a process that involves the *E. coli* chromosome, the lac repressor, the plasmid, and two different RNA polymerases with different promoter specificities.

enzyme, if it and the antibody have adhered to the primary antibody, which has bound the translated fusion protein. The product of this enzyme reaction is a colored precipitate that remains bound to the filter. In this way, clones are specifically identified based on the expressed product of the gene of interest.

Innovation/Insight

How Cells Respond to Stress

Welch, W. J. (May, 1993) Scientific American 44-56. (Adapted)

During emergencies, cells produce stress proteins that repair damage. Inquiry into how they work offers promise for coping with infection, autoimmune disease, and repair. Immediately after a sudden increase in temperature, all cells – from the simplest bacterium to the most highly differentiated neuron – increase production of a certain class of molecules that buffer them from harm. When biologists first observed this 30 years ago, they called it the heat-shock response. Subsequent studies revealed that the same response takes place when cells are subjected to a wide variety of other environmental assaults, including toxic metals, alcohols, and many metabolic poisons. It occurs in traumatized cells growing in culture, in the tissues of feverish children, and in the organs of heart-attack victims and cancer patients receiving chemotherapy. Because so many different stimuli elicit the same cellular defense mechanism, researchers now commonly refer to it as the *stress response* and to the expressed molecules as *stress* proteins (Figure 1).

In their pursuit of the *structure and function* of the stress proteins, biologists have learned that they are far more than just defensive molecules. Throughout the life of a cell, many of these proteins participate in essential metabolic processes, including the pathways by which all other cellular proteins are synthesized and assembled. Some stress proteins appear to orchestrate the activities of molecules that regulate cell growth and differentiation.

The understanding of stress proteins is still incomplete. Nevertheless, investigators are already beginning to find new ways to put the stress response to good use. It already shows great potential for pollution monitoring and better toxicological testing. The promise of medical applications for fighting infection, cancer, and immunological disorders is perhaps more distant, but it is clearly on the horizon.

Such uses were far from the minds of the investigators who first discovered the stress response; as happens so often in science, it was serendipitous. In the early 1960s, biologists studying the genetic basis of animal development were focusing much of their attention on the *fruit fly Drosophila melanogaster. Drosophila* is a convenient organism to study the maturation of an embryo into an adult, in part because it has an unusual genetic feature. Cells in its salivary glands carry four chromosomes in which the normal amount of DNA has been *duplicated thousands of times*; all the copies align beside one another. These so-called *polytene chromosomes* are so large that they can be seen through a light microscope. During each stage of the developmental process, distinct regions along the polytene chromosomes *puff* out, or enlarge. Each puff is the result of a specific change in gene expression.

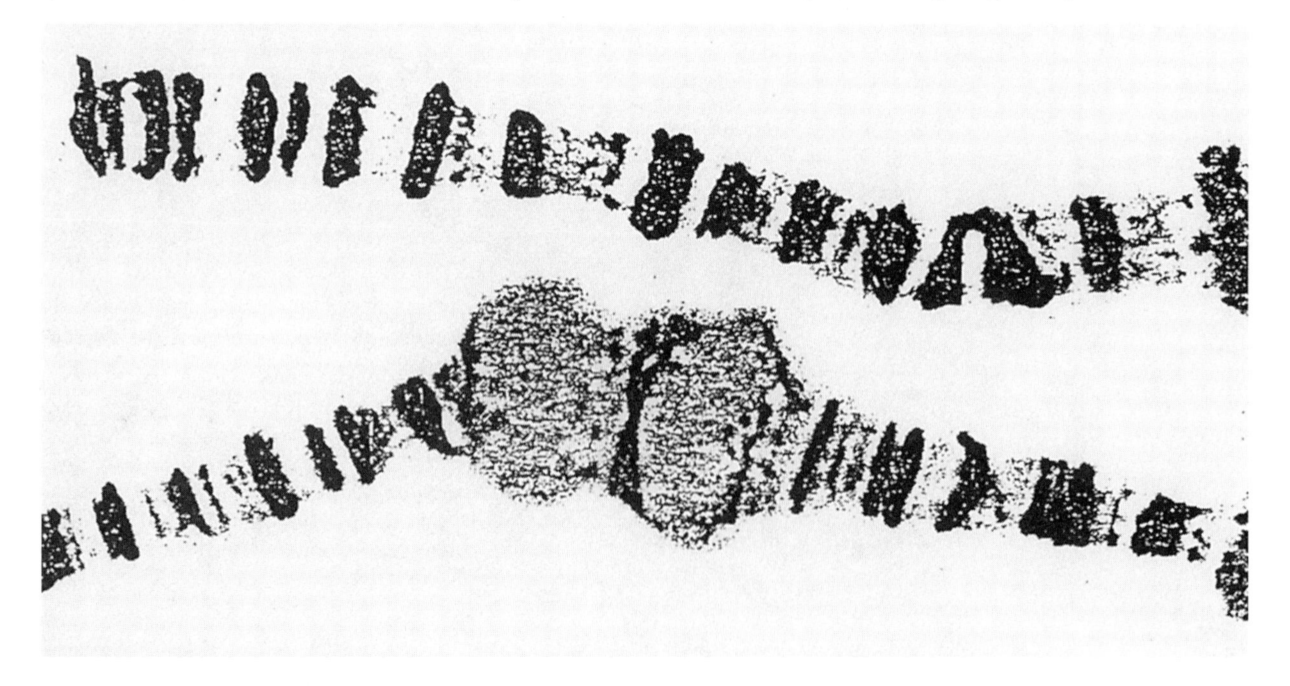

Figure 1 - Normal control (top), and heat-shocked chromosome (bottom).

During the course of his studies, F. M. Ritossa (of the International Laboratory of Genetics and Biophysics in Naples) saw that a new pattern of chromosomal puffing followed exposure of isolated salivary glands to temperatures slightly above those optimum for the fly's normal growth and development. The puffing pattern appeared within a minute or two after the temperature rise and puffs continued to increase in size for as long as 30 to 40 minutes. Over the next decade, other investigators built on Ritossa's findings.

In 1974, Alfred Tissieres, a visiting scientist (from the University of Geneva), and Herschel K. Mitchell (of the California Institute of Technology) demonstrated that heat-induced chromosomal puffing was accompanied by high-level expression of a unique set of "heat-shock" proteins. Those new chromosomal puffs represented sites in the DNA where specific messenger RNA molecules were made; these messenger RNAs carried the genetic information for synthesizing the individual heat-shock proteins.

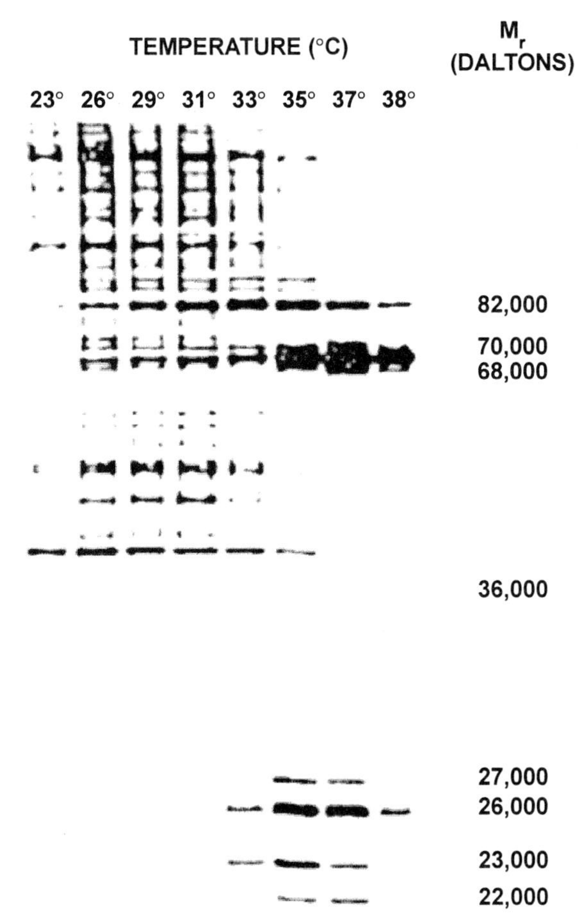

Figure 2 - Heat shock proteins separated on a denaturing gel.

Heatshock protein levels rise in cells as the temperature increases. In these electrophoretic gels, each horizontal band is a protein found in the cells of *Drosophila*. As the temperature rises, the cells stop making most proteins and produce far more of the heat-shock proteins. The most prevalent of these belong to the hsp70 family, which has molecular weights around 70 kDa (Figure 2).

By the end of the 1970s, evidence was accumulating that the heat-shock response was a general property of all cells. Following a sudden increase in temperature, bacteria, yeast, plants, and animal cells grown in culture all increased their expression of proteins that were similar in size to the *Drosophila* heat-shock proteins. Moreover, investigators were finding that cells produced one or more heat-shock proteins whenever they were exposed to heavy metals, alcohols, and various other metabolic poisons.

Because so many different toxic stimuli brought on similar changes in gene expression, researchers started referring to the heat-shock response more generally as the stress response and to the accompanying products as stress proteins. They began to suspect that this universal response to adverse changes in the environment represented a basic cellular defense mechanism. The stress proteins, which seemed to be expressed only in times of trouble, were presumably part of that response.

Mounting evidence during the next few years confirmed that stress proteins did play an active role in cellular defense. Researchers were able to identify and isolate the genes that encoded individual stress proteins. Mutations in those genes produced interesting cellular abnormalities. For example, bacteria carrying mutations in the genes encoding several of the stress proteins exhibited defects in DNA and RNA synthesis, lost their ability to undergo normal cell division, and appeared unable to degrade proteins properly. Such mutants were also incapable of growth at high temperatures.

Cell biologists soon discovered that, as in bacteria, the stress response played an important role in the ability of animal cells to withstand brief exposures to high temperatures. Animal cells given a mild heat shock – one sufficient to increase the levels of the stress proteins – were better protected against a second heat treatment that would otherwise have been lethal. Moreover, those thermotolerant cells were also less susceptible to other toxic agents. Investigators became convinced that the stress response somehow protected cells against varied environmental insults.

As scientists continued to isolate and characterize the genes encoding the stress proteins from different organisms, two unexpected results emerged. First, many of the genes that encoded the stress proteins were remarkably similar in all organisms. Elizabeth A. Craig and her colleagues (at the University of Wisconsin) reported-that the genes for heat-shock protein (*hsp*l) 70, the most highly induced stress protein were more than 50 percent identical in bacteria, yeast, and *Drosophila*. Apparently, the stress proteins had been conserved throughout evolution and likely served a similar and important function in all organisms.

The second unexpected finding was that many stress proteins were expressed in normal and unstressed cells, not only in traumatized ones. Consequently, researchers subdivided the stress proteins into two groups: those constitutively expressed under normal growth conditions and those induced only in cells experiencing stress. Investigators were still perplexed as to how so many seemingly different toxic stimuli always led to the increased expression of the same group of proteins.

In 1980, Lawrence E. Hightower (working at the University of Connecticut) provided a possible answer. He noticed that many of the agents that induced the stress response were protein denaturants – that is, they caused proteins to lose their shapes. A protein consists of long chains of amino acids folded into a precise conformation. Any disturbance of the folded conformation can lead to the loss of protein function.

Hightower suggested that the accumulation of denatured or abnormally folded proteins in a cell initiated a stress response. The stress proteins, he reasoned, might somehow facilitate the identification and removal of denatured proteins from the traumatized cell. Within a few years, Richard Voellmy (University of Miami) and Alfred L. Goldberg (Harvard University) tested and confirmed Hightower's proposal. In a landmark study, they showed that injecting denatured proteins into living cells was sufficient to induce a stress response. Thereafter, several laboratories set out to purify and characterize the biochemical properties of the stress proteins.

The most highly inducible *heat-shock protein*, hsp70, was the focus of much of this work. Using molecular probes, researchers learned that after a heat shock, much hsp70 accumulated inside a nuclear structure called the nucleolus. The *nucleolus manufactures ribosomes*, the organelles that synthesize proteins. That location for hsp70 was intriguing: previous work had demonstrated that after heat shock cells stopped making ribosomes. Indeed, the nucleolus became awash in denatured ribosomal particles. Hugh R. B. Pelham (MRC, Laboratory of Molecular Biology, Cambridge, England) therefore suggested that hsp70 protein might somehow recognize denatured intracellular proteins and restore them to their correctly folded, biologically active shape.

In 1986, Pelham and his colleague Sean Munro succeeded in isolating several genes, all of which encoded proteins related to hsp70. They noticed that one form of hsp70 was identical to *immunoglobulin binding protein (BiP)*. Other researchers had shown that BiP was involved in the preparation of immunoglobulins, or antibodies, as well as other proteins for secretion. BiP bound to newly synthesized proteins as they were being folded or assembled into their mature form. If the *proteins failed to fold or assemble properly,* they *remained bound to BiP* and were eventually *degraded.* In addition, under conditions in which *abnormally folded proteins accumulated, the cell synthesized more BiP*.

Taken together, those observations indicated that BiP helped to orchestrate the early events associated with protein secretion. *BiP* seemed to act as a *molecular overseer of quality control* (a selective "chaperonin"), allowing properly folded proteins to enter the secretory pathway, but holding back those unable to fold correctly.

As more genes encoding proteins similar to hsp70 and BiP came to light, it became evident that there was an entire family of hsp70-related proteins. All of them shared certain properties, including an *avid affinity for adenosine triphosphate (ATP)*, the molecule that serves as the universal, intracellular fuel. With only one exception, all these related proteins were present in cells growing under normal conditions (they were *constitutive*), yet in cells experiencing metabolic success they were synthesized at much higher levels. Moreover, all of them *mediated* the *maturation* of other cellular proteins, much as BiP did. For example, the cytoplasmic forms of hsp70 interacted with many other proteins that were being synthesized by ribosomes.

In healthy or unstressed cells, the interaction of the hsp70 family member with immature proteins was transient and ATP-dependent. Under conditions of metabolic stress, however, in which newly synthesized proteins experienced problems maturing normally, the proteins remained stably bound to an hsp70 escort.

The idea that members of the hsp70 family participated in the early steps of protein maturation paralleled the results emerging from studies of a different family of stress proteins. Pioneering work by Costa Georgopoulos (University of Utah) and others showed that mutations in the genes for two related stress proteins, *groEL* and *groES*, render bacteria unable to support the growth of small viruses that depend on the cellular machinery provided by their hosts. In the absence of functional groEL or groES, many *viral proteins fail to assemble properly*.

Proteins similar to the bacterial groEL and groES stress proteins were eventually found in plant, yeast and animal cells. Those proteins, which are known as hsp10 and hsp60, have been seen only in mitochondria and chloroplasts. Recent evidence suggests that more forms probably appear in other intracellular compartments.

Biochemical studies have provided compelling evidence that hsp10 and hsp60 are essential to protein folding and assembly. The *hsp60 molecule consists of two seven-membered rings stacked one atop the other.* This large structure appears to serve as a "*work-bench*" into which *unfolded proteins bind and acquire their final three-dimensional structure.* According to current thought, the *folding process is extremely dynamic and involves a series of binding and release events.* Each event requires *energy*, which *is provided by the enzymatic splitting of ATP*, and the participation of the small hsp10 molecules. Through *multiple rounds of binding and release*, the *protein* undergoes *conformational changes leading to a stable, properly folded state.* Investigators suspect that both the hsp60 and the hsp70 families work together to facilitate protein maturation. As a new polypeptide emerges from a ribosome, it is likely to become bound to a form of hsp70 in the cytoplasm or inside an organelle. Such an interaction may prevent the growing polypeptide chain from folding prematurely. Once its synthesis is complete, the new polypeptide, still bound to its hsp70 escort, would be transferred to a form of hsp60, on which folding of the protein and its assembly with other protein components would commence.

Protein folding occurs spontaneously because of thermodynamic constraints imposed by the protein's sequence of hydrophilic and hydrophobic amino acids (Figure 3). Although proteins can fold themselves into biologically functional configurations (self-assembly), errors in folding can occasionally occur. Stress proteins seem to help ensure that cellular proteins fold themselves rapidly and with high fidelity.

These new observations regarding the properties of hsp70 and hsp60 have forced scientists to reconsider previous models of protein folding. Work done in the 1950s and 1960s established that a denatured protein could spontaneously refold after the denaturing agent was removed. This work led to the concept of *protein self-assembly*, for which Christian B. Anfinsen received a Nobel Prize in Chemistry in 1972. According to that model, the process of folding was dictated solely by the sequence of amino acids in the polypeptide. *Hydrophobic* amino acids (those that are not water soluble) would position themselves inside the coiling molecule while *hydrophilic* amino acids (those that are water soluble) would move to the surface of the protein to ensure their exposure to the aqueous cellular environment. Folding is driven entirely by thermodynamic constraints.

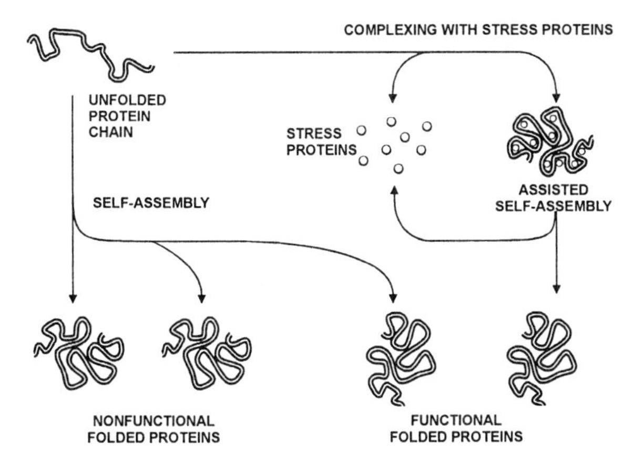

Figure 3 - Folding and function of proteins mediated by stress proteins.

The principle of self-assembly is still regarded as the primary force that drives proteins into their final conformation. Many investigators now suspect that protein folding requires the activity of other cellular components, including the members of the hsp60 and hsp70 families of stress proteins.

Accordingly, R. John Ellis of the University of Warwick and others, scientists have begun to refer to hsp60, hsp70, and other stress proteins as "molecular *chaperones*." Although the molecules do not convey information for the folding or assembly of proteins, they do ensure that those processes occur quickly and with high fidelity. They expediate self-assembly by reducing the possibility that a maturing protein will head down an inappropriate folding pathway.

Having established a role for some stress proteins as molecular chaperones in healthy and unstressed cells, investigators have turned their attention to determining why those proteins are expressed at higher levels in times of stress. One clue is the conditions that increase the expression of the stress proteins. Temperatures that are sufficient to activate the stress response may eventually denature some proteins inside cells. *Heat-denatured proteins*, like newly synthesized and unfolded proteins, would therefore represent *targets to which hsp70 and hsp60 can bind.* Over time, as more *thermally denatured proteins become bound to hsp60 and hsp70*, the *levels of available molecular chaperones drop* and begin to limit the ability of the cell to produce new proteins. The cell somehow senses this reduction and responds by *increasing the synthesis of* new *stress proteins* that serve as *molecular chaperones*.

Researchers suspect that a rise in the expression of stress proteins may also be a requirement for the *ability of cells to recover from a metabolic insult.* If heat or other metabolic insults irreversibly denature many cellular proteins, the cell will have to replace them. Raising the levels of those stress proteins that act as molecular chaperones will help facilitate the synthesis and assembly of new proteins. In addition, higher levels of stress proteins may prevent thermal denaturation of other cellular proteins.

The *repair and synthesis of proteins* are vital jobs in themselves. Nevertheless, stress proteins also serve a pivotal role in the *regulation of other systems of proteins and cellular responses*. Another family of stress proteins, epitomized by one called *hsp90*, is particularly noteworthy in this regard.

Initial interest in hsp90 was fueled by reports of its association with some *cancer-causing viruses*. In the late 1970s and early 1980s, cancer biologists focused attention on the mechanism by which certain *viruses infect cells* and *cause them to become malignant*. In the case of Rous sarcoma virus, investigators had pinpointed a viral gene that was responsible for the development of malignant properties. The *enzyme* it produced, *pp60^{src}, acted on other proteins that* probably *regulated cellular growth*. Three laboratories independently reported that after its synthesis in the cytoplasm, pp60^{src} rapidly associates with two proteins, p50 and hsp90.

When pp60^{src} is in the cytoplasm and *is linked to its two escorts, it is enzymatically inactive*. As the trio of molecules moves to the plasma membrane, the *hsp90 and the p50 fall away and allow the pp60^{src} to deposit itself in the membrane and become active*. Similar interactions between hsp90, p50, and cancer-causing enzymes encoded by several other tumor viruses have been discovered. *When bound to hsp90 and p50, these viral enzymes seem incapable of acting on the cellular targets necessary for the development of the malignant state*.

Some studies have also linked hsp90 to another important class of molecules in mammalian cells, the steroid hormone receptors (Figure 4). Steroid hormones mediate several vital biological processes in animals. For example, the glucocorticoid steroids help to *suppress inflammation*. Other steroid hormones play important roles in *sexual differentiation and development*. When a *steroid receptor binds to its specific hormone*, the receptor becomes capable of interacting with DNA and either *activating or repressing the expression of* certain *genes*.

A crucial question concerned how steroid receptors were kept inactive inside a cell. The answer became clear following the characterization of both the active and inactive forms of the progesterone receptor. In the absence of hormone, the receptor associates with several cellular proteins, among them hsp90, which maintain it in an inactive state. After binding to progesterone, the receptor is released from the hsp90 and experiences a series of events that allows it to bind with DNA. As with the viral enzymes, hsp90 seems to regulate the biological activity of steroid hormone receptors.

Scientists are beginning to realize practical applications for the stress response. Medicine is one area that stands to benefit. When an individual suffers a heart attack or stroke, the *delivery of blood to the heart or brain is temporarily compromised*, a condition referred to as *ischemia*. While deprived of oxygen, the affected organ cannot maintain its normal levels of ATP, which causes essential metabolic processes to falter. When blood flow is restored, the ischemic organ is rapidly reoxygenated – yet that too can be harmful. Often the rapid reexposure to oxygen generates *highly reactive molecular species*, known as *free radicals*, that can do other damage.

In animal studies, researchers have observed the induction of stress responses in both the heart and brain after brief episodes of ischemia and reperfusion. The magnitude of the resulting success response appears to correlate directly with the relative severity of the damage. Clinicians are therefore beginning to examine the utility of using changes in stress protein levels as markers for tissue and organ injury. Cells that produce high levels of stress proteins appear better able to survive the ischemic damage than cells that do not. Consequently, raising the levels of stress proteins, perhaps by pharmacological means, may provide additional protection to injured tissues and organs. Such a therapeutic approach might reduce the tissue damage from ischemia incurred during surgery or help to safeguard isolated organs used for transplantation, which often suffer from ischemia and reperfusion injury.

One exciting development concerns the role of the stress response in immunology and infectious diseases. Tuberculosis, malaria, leprosy, schistosomiasis, and other diseases that affect millions of people every year are a consequence of infection by bacteria or parasitic microorganisms. Immunologists have found that the stress proteins made by these organisms are often the *major antigens*, or protein *targets, that the immune system uses to recognize and destroy the invaders*. The human immune system may be constantly on the lookout for alien forms of stress proteins. The stress proteins of various pathogens, when produced in the laboratory by recombinant-DNA techniques, may therefore have potential as vaccines for preventing microbial infections. In addition, because they are so immunogenic, microbial stress proteins are being considered as *adjuvants*. Linked to viral proteins, they could *enhance immune responses* against viral infections.

Immunologists have also discovered a possible connection between stress proteins and autoimmune diseases. Most *autoimmune diseases* arise when the immune system turns against antigens in healthy tissues. In some of these diseases, including rheumatoid arthritis, ankylosing spondylitis, and systemic lupus erythematosus, antibodies against the patient's own stress proteins are sometimes observed. If those observations are confirmed on a large number of patients, they may prove helpful in the diagnosis and perhaps the treatment of autoimmune disorders.

Because microbial stress proteins are so similar in structure to human stress proteins, the immune system may constantly be obliged to discern minor differences between the stress proteins of the body and those of invading microorganisms. The possibility that the stress proteins are uniquely positioned at the interface between *tolerance* to an infectious organism and autoimmunity is an intriguing idea that continues to spark debate among researchers.

The presence of antibodies against microbial stress proteins may prove useful in diagnostics. For example, the bacterium *Chlamydia trachomatis* causes a number of diseases, including trachoma, probably the world's leading cause of preventable blindness, and pelvic inflammatory disease, a major cause of infertility in women. Infection with chlamydia generally triggers the production of antibodies against chlamydial antigens, some of which are stress proteins. The resulting immune response is often effective and eventually eliminates the pathogen. Yet in some individuals, particularly *those who have had repeated or chronic chlamydial infections, the immune response is overly aggressive and causes injury and scarring in the surrounding tissues.*

Richard S. Stephens and his colleagues at the University of California at San Francisco have observed that more than 30 percent of women with *pelvic inflammatory disease* and more than 80 percent of women who have had *ectopic pregnancies* possess abnormally high levels of antibodies against the chlamydial groEL stress protein. Measurement of antibodies against chlamydial stress proteins may prove useful for identifying women at high risk for *ectopic pregnancies* or *infertility*.

The link between stress proteins, the immune response, and autoimmune diseases becomes even more intriguing in light of other recent discoveries. Some members of the hsp70 family of *stress proteins are remarkably similar in structure and function to the histocompatibility antigens* (Figure 5). The latter proteins participate in the very early stages of immune responses by presenting foreign antigens to cells of the immune system.

Researchers have wondered how any one histocompatibility protein could bind to a diverse array of different antigenic peptides. Recently, Don C. Wiley and his colleagues (Harvard University) helped to resolve that issue by determining the three-dimensional structure of the class I histocompatibility proteins. A pocket (groove) on the class I molecule is able to bind to different antigenic peptides. Simultaneously, James E. Rothman (then at Princeton University) reported that members of the hsp70 family of stress proteins were also capable of binding short peptides. This property of hsp70 is consistent with its role in binding parts of unfolded or newly made polypeptide chains.

Computer models revealed that hsp70 probably has a peptide-binding site analogous to that of the *class I histocompatibility protein.* The apparent resemblance between the two classes of proteins appears even more intriguing because several of the genes that encode hsp70 are located very near genes for histocompatibility proteins. Taken together, the observations continue to support the idea that *stress proteins are integral components of the immune system.*

The ability to manipulate the stress response may also prove important in developing new *approaches to treating cancer.* Tumors are often more thermally sensitive than normal tissues. Elevating the temperature of tissues to eradicate tumors is one idea that is still at the experimental stage. Nevertheless, in early tests, the use of site-directed hyperthermia, alone or in conjunction with radiation or other conventional therapies, has brought about the regression of certain types of tumors.

The stress response is not necessarily the physician's ally in the treatment of cancer – it may also be an obstacle. Because stress proteins protect cells, *anticancer therapies that induce a stress response may make a tumor more resistant to subsequent treatments.* Still researchers may yet discover ways to inhibit the ability of a tumor to

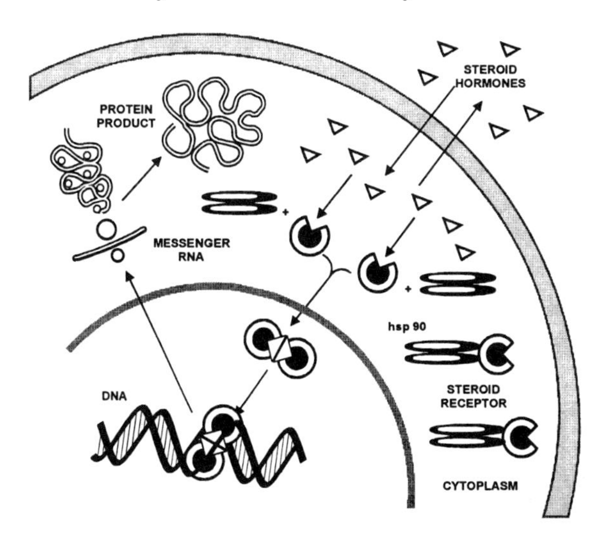

Figure 4 - Responses to steroid hormones. Controlled in part by hsp90, a stress protein, which helps to maintain steroid receptors in their inactive form. When hormones are present, they bind to the receptor and hsp90 is released. The activated receptor complex can then interact with DNA and initiate the expression of genes for certain proteins.

mount a stress response and thereby render it defenseless against a particular therapy.

Scientists are also exploring the potential use of the stress response in toxicology. Changes in the levels of the stress proteins, particularly those produced only in traumatized cells, may prove useful for assessing the toxicity of drugs, cosmetics, food additives, and other products. Such work is only at a preliminary stage of development, but several application strategies are already showing signs of success.

Employing recombinant DNA technologies, researchers have constructed cultured lines of "stress reporter" cells that ought to be used to screen for biological hazards. In such cells, the *DNA sequences that control the activity of the stress protein genes* are *linked* to a *reporter gene* that encodes an enzyme, such as β-galactosidase. When these cells experience *metabolic stress*, they produce stress protein and linked reporter enzyme, which can be *detected easily* by various assays. The amount of β-galactosidase expressed in a cell can be measured by adding a chemical substrate. If the reporter enzyme is present, the cell turns blue, and the intensity of the color is directly proportional to the concentration of the enzyme in the cell.

Using such reporter cells, investigators can easily determine the extent of the stress response induced by chemical agents or treatments. If such assays prove reliable, they could ultimately reduce or possibly replace the use of animals in toxicology testing.

An extension of the technique could also be used to monitor the dangers of environmental pollutants, many of which evoke stress responses. Toward that end, scientists have begun developing transgenic stress reporter organisms. Eve G. Stringharn and E. Peter M. Candido (University of British Columbia) and Stressgen Biotechnologies (Victoria, B.C.) have created transgenic worms in which a reporter gene for β-galactosidase is under the control of the promoter for a heat-shock protein. When these transgenic worms are exposed to environmental pollutants, they express the reporter enzyme and turn blue. Candido's laboratory is currently determining whether stress reporter worms might be useful for monitoring a wide variety of pollutants.

Voellmy and Nicole Bournias-Vardiabasis (City of Hope Medical Center, Duarte, CA) have used a similar approach to create a line of stress reporter fruit flies. The fruit flies turn blue when exposed to *teratogens*, agents that cause abnormal fetal development. Significantly, that bioassay is responsive to many of the teratogens that are known to cause birth defects in humans. The door appeared open for the development of other stress reporter organisms that could prove useful in toxicologicalal and environmental testing.

More than 30 years ago, heat-shock and stress responses seemed like mere molecular curiosities in fruit flies. Today they are at the heart of an active and vital area of research. Studies of the structure and function of stress proteins have brought new insights into essential cellular processes, including the pathways of protein maturation. Scientists are also learning how to apply their understanding of the stress response to solve problems in the medical and environmental sciences. I suspect that we have only begun to realize all the implications of this age-old response by which cells cope with stress.

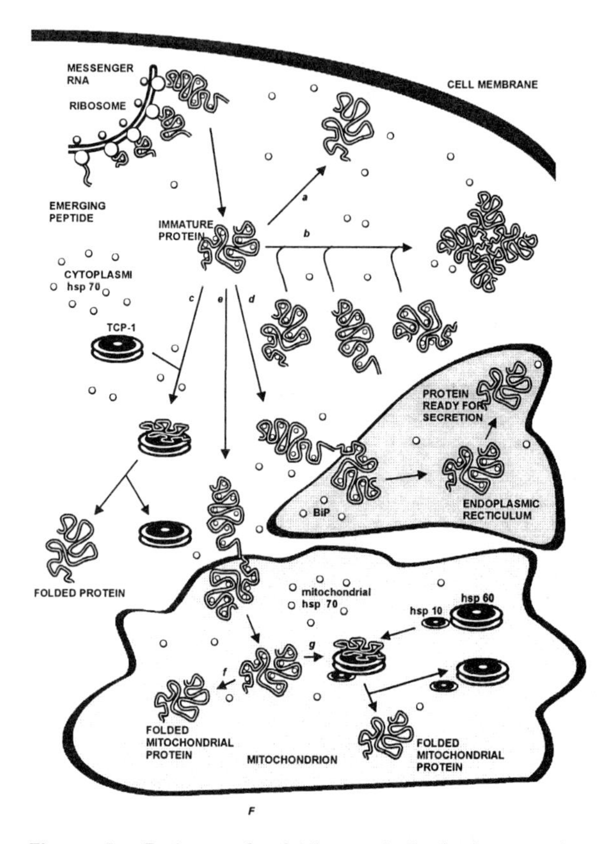

Figure 5 - Pathways for folding and distribution proteins inside the cell managed by stress proteins. In many cases, different stress proteins seem to work in tandem. The cytoplasmic form of hsp70 binds to proteins being produced by the ribosomes to prevent premature folding. The hsp70 may dissociate from the protein and allow it to fold into its functional shape (*a*) or associate with other proteins and thereby form larger multimeric complexes (*b*). In some cases, proteins are passed from hsp70 to another stress protein, TCP-1, before final folding and assembly occur (*c*). If the protein is destined for secretion, it may be carried to the endoplasmic reticulum and given to BiP or another related stress protein that directs its final folding (*d*). Other proteins are transferred to mitochondria or other organelles (*e*). Inside the mitochondrion, another specialized form of hsp70 sometimes assists the protein in its final folding (*f*), but in many cases the protein is passed onto a complex of hsp60 and hsp10 (*g*). The hsp60 molecules seem to serve as a "workbench" on which mitochondrial proteins fold.

References:

1. The Induction of Gene Activity in *Drosophila* by Heat Shock, M. Ashburner and J. J. Bonner (1979) Cell 17, 241-254.
2. *Stress Proteins in Biology and Medicine* (1990) R. I. Morimoto, A. Tissieres and C. Georgopoulos, Cold Spring Harbor Laboratory Press.
3. Molecular Chaperones (1991) R. J. Ellis and S. M. Van der Vies, Annu. Rev. Biochem. 60, 321-347.
4. Successive Action of DNAK, DNAJ and GroEL Along the Pathway of Chaperone-Mediated Protein Folding (1992) T. Langer, C. Lu, H. Echols, J. Flanagan, M. K. Hayer and F. U. Hartl, Nature 356, 683-689.
5. Mammalian Stress Response: Cell Physiology, Structure/Function of Stress Proteins, and Implications for Medicine and Disease (1992) W. J. Welch, Physiol. Rev. 72, 1063-1081.

LAB 5.1:
PREPARATION OF FRESH TRANSFORMATION-COMPETENT CELLS

Overview:

In this lab, *E. coli* cells will be treated with calcium chloride to make them competent for the uptake of plasmid DNA. The uptake of DNA by bacterial cells is called transformation. The recombinant plasmids from Lab 2.1 will be transformed into competent *E. coli* cells. Once the bacterial cells accept the DNA, they are said to have been transfected by the recombinant plasmid. Bacterial cells can be purchased from biochemical companies in a competent state; however, the practice of making them from scratch is appropriate for understanding the theory behind their production. Again, only cells containing the recombinant plasmid will be viable on antibiotic selection plates.

Experimental Outline:

1. Bacterial cell growth.
2. Prepare fresh transformation-competent cells.
3. Bacterial cell transformation.

Required Materials:

SOB media (salt optimizing buffer media); 250 mM solution of KCl; 5 N NaOH (~0.2 ml); pH meter; 2 M $MgCl_2$; SOC media; 0.22-µm filter; *E. coli* cells; ice cold 0.1 M $CaCl_2$; centrifuge; ice/water bath; incubator; LB-ampicillin plates

Prelab Preparations:

1. Cell Growth

Be quick and keep the samples cold in this procedure. The procedure is done most safely (for the sample) in the cold room.

1. Prepare Solutions.
   a. *SOB media*: (1 liter)
      i. To deionized H_2O add:

bactotryptone	20 g
bactoyeast extract	5 g
NaCl	0.5 g

      ii. Shake until the solutes have dissolved.
      iii. Add 10 ml of a 250 mM solution of KCl. (This solution is made by dissolving 1.86 g of KCl into 100 ml of deionized H_2O.)
      iv. Adjust the pH to 7 with 5 N NaOH (~0.2 ml).
      v. Adjust the volume of the solution to 1 liter with deionized H_2O.
      vi. Sterilize the media by autoclaving for 20 minutes at 15 lb/in^2 on liquid cycle.
      vii. Just before use, add 5 ml of a sterile solution of 2 M $MgCl_2$. This solution is made by dissolving 19 g of $MgCl_2$ to 90 ml of deionized H_2O. Adjust the volume of the solution to 100 ml with deionized H_2O and sterilize by autoclaving for 20 minutes at 15 lb/in^2 on liquid cycle.
   b. *SOC media*: SOC medium is identical to SOB media, except that it contains 20 mM glucose.
      i. After the SOB medium has been autoclaved, allow it to cool to at least 60°C.
      ii. Add 20 ml of a sterile 1 M solution of glucose. [This solution is made by dissolving 18 g of glucose in 90 ml of deionized H_2O. After the sugar has dissolved, adjust the volume of the solution to 100 ml with deionized H_2O and render it (essentially) aseptic by filtration through a 0.22-µm micron filter.]
2. Inoculate a culture (in SOB) with a 1/500 volume of a stationary phase culture of *E. coli* cells.
3. Grow the cells at 37°C until A_{600} = 1 (about 10^8 cells/ml; 2.5 to 3.5 hours).

Procedures:

1. Preparation of Fresh Transformation-Competent Cells

1. Chill the bacterial culture on ice for 10 minutes. Swirl the flask periodically to speed cooling.
2. Harvest the cells by centrifugation; 8,000 x g for 10 minutes.
3. Pour off the supernatant and drain the remaining drops with a Pasteur pipet.

4. Resuspend the cells in a 1/10 volume of ice cold 0.1 M $CaCl_2$. Resuspend gently by pipetting up and down.
5. Incubate the cells at 4 °C in an ice/water bath for 20 minutes.
6. Harvest the cells by centrifugation at 4,000 x *g* for 10 minutes.
7. Pour off the supernate and resuspend in a 1/50 volume (relative to starting volume of culture) of ice cold 0.1 M $CaCl_2$. Be gentle here.
8. Pipet 200 µl aliquots into cold eppendorf tubes. Store on ice.

2. Transformations

1. Add the recombinant plasmid DNA (50 ng in 10 µl) to 200 µl of competent *E. coli* cells.
2. Mix by inversion and place undisturbed on ice for 30 minutes to allow the plasmid DNA to bind to the cellular membrane.
2. Heat shock at 42°C for 90 seconds.
3. Add 800 µl of SOC medium and incubate/shake at 37°C for 45 minutes.
4. Concentrate the cells by centrifugation and bring up to the desired volume for plating.
5. Plate transformed cells in three different volumes (50 µl, 100 µl, 200 µl) on LB plates containing ampicillin.
6. Inoculate an overnight culture at 37°C.

Notes:

1. Transformation efficiencies are consistently 1 to 4 x 10^7 cfu/µg pBR322. The transformation efficiency is measured by determining the number of cfu/µg DNA. Supercoiled DNA from pUC19 or pBR322 is often used as a standard.
2. If the recovery of transformants is low, the cells can be concentrated by centrifugation and plated in a sterile manner.
3. The ions provided by KCl in the SOB media are included to offset the transport of sodium ions from the interior of the cell. Disrupting the movement of sodium ions out of the cell allows the plasmid DNA to enter the cell with greater ease.
4. The ions provided by $MgCl_2$ in the SOB media are included to neutralize the negative charge on the DNA. Once the charge is neutralized, the DNA binds the cellular membrane for eventual uptake by the cell.
5. $CaCl_2$ is used to transform bacterial cells by disrupting the phopholipids in the cell membrane. These disruptions allow plasmids to enter the cell through the cell membrane.
6. Only colonies containing pBR322 plasmid (or other plasmid containing the ampicillin resistance gene) will grow on LB/ampicillin plates.

ALTERNATIVE APPROACH

ULTRACOMP ™ TRANSFORMATION KIT

Ultracomp™ Manual, Invitrogen Corporation, Carlsbad, CA; ph. (800) 955-6288. Portions copyright 1997 Invitrogen Corp. (Adapted)

1. Introduction

E. coli cells have been rendered competent for transformation or transfection and can be stored for extended periods at or below -70°C. The cells have been prepared by a modified procedure based on published protocols (Hanahan, D., 1988, J. Mol. Biol. 168: 557-580). This method produces a high proportion of cells competent for transformation as well as very high transformation efficiencies.

2. Components

Competent cells, 5 vials, 0.3 ml each
100 ng lyophilized supercoiled pUC18
SOC transformation medium, 5 vials, 2 ml each

3. Strain Specification

Genotype: *Hsd*520 (ra^-, ma^-), *sup*E44, *ara*14, *gal*K2, *lac*Y1, *pro*A2, *rsp*L20(Smr), *xy*$^-$ 5, *deo*R^+, *mtl*$^-$ 1, *rec*A13, *mcr*A^+, *mcr*B^-, λ^-.

4. Quality Control Data

The competent *E. coli* cells and all reagents provided with the kit have been tested and lot qualified for optimum performance. The transformation efficiencies have been evaluated in duplicate.

$$\text{Transformation Efficiency}: \frac{1.09\times 10^{8}(\text{amp}^{r})}{\mu\text{g pUC18}}$$

$$\text{Transformation Efficiency}: \frac{1.18\times 10^{8}(\text{amp}^{r})}{\mu\text{g pUC18}}$$

Test transformations were performed as follows: 0.1 ml aliquots of cells were transformed with 10 pg of supercoiled test plasmid. The transformation procedure is outlined in the instructions provided. Samples of the transformed cultures (0.1 ml) are spread on LB plates containing 50 mg/liter ampicillin. Transformation efficiencies are calculated according to the number of colonies per plate after overnight growth at 37°C.

5. Storage Conditions

Competent *E. coli* cells are stored at -70 °C and are stable for more than 4 months. Avoid repeated thawing and freezing. We do not recommend using the cells for high-efficiency transformation after the expiration date. The pUC18 plasmid preparation should be stored at -20°C after reconstitution in water.

6. Transformation of Ultracomp™ *E. coli*

The following protocol is for transformation of Invitrogen Ultracomp™ strains: MC1061/P3, INVα, INVαF, HB101, NM522, JM101, JM109, TOP10F.

1. Remove tube(s) containing competent cells from -70°C and thaw on ice. When the cell suspension is liquid, mix gently by hand and return to ice immediately. Keep cells cold at all times! Changes in temperature will result in decreased transformation efficiencies.
2. Place a 100 ml cell aliquot into a pre-chilled 15 ml polypropylene tube (Falcon 2059). For more than one transformation, use additional 100 µl aliquots in separate tubes. If you wish to store the excess bacteria, place the vial(s) at -70°C immediately. Be aware that each freeze-thaw cycle will lower the transformation efficiency.
3. Do not perform this step with HB101 cells! Make a fresh dilution of 500 mM β-mercaptoethanol in high quality sterile water. Add 5 µl to each 100 µl aliquot of cells and swirl gently. Allow to stand on ice for 10 minutes, swirling gently every 2 minutes.
4. Add the DNA solution in a volume of 1 to 5 µl per tube and swirl the tube(s) to mix the DNA evenly with the cells. Incubate on ice for 30 minutes.
5. Heat shock the cells by placing the tube(s) in a 42°C water bath for 75 seconds, then return the tubes to ice to chill for 2 minutes.
6. Add 900 µl of SOC (see below) medium (provided with the kit) and incubate at 37°C with moderate agitation (225 rpm) for 60 minutes.
7. Cells may be plated directly using 100 µl per plate. If you desire a higher cell density, cells can be concentrated by centrifugation at 4000 x *g* (maximum) for 5 minutes and resuspended in a small volume of SOC. Plate on LB plates containing the appropriate antibiotic(s).

7. Alternative Heat-Shock Recommendations

Volume of Cells	Tube INVαF	MC1061/P3M
100 μl	Flacon 2059	75 seconds
1 ml	50 ml conical	120 seconds
2 ml	50 ml conical	N.A.
4 ml	50 ml conical	150 seconds

8. Media Preparation

8.1 SOC medium.

- Bactotryptone – 2%
- Bactoyeast Extract – 0.5%
- Glucose – 20 mM
- NaCl – 10 mM
- KCl – 2.5 mM
- $MgCl_2$ – 10 mM
- $MgSO_4$ – 10 mM

Combine tryptone yeast extract NaCl and KCl in (Nanopure) water and autoclave for 30-40 minutes. Make a 2 M stock of Mg^{2+} consisting of 1 M $MgCl_2$ and 1 M $MgSO_4$. Sterilize by filtration through a 0.22 μl membrane. Prepare a 2 M stock of glucose, filter and store at -20°C. Just prior to use, combine the medium with Mg^{2+} and sterilize by filtration.

Reference:

1. *DNA Cloning*, 1, D. M. Glover, Ed., p. 127, IRL Press.

LAB 5.2:
COLONY IMMUNOBLOTTING TO SCREEN FOR TRANSFORMANTS

Overview:

In this lab, we will perform colony immunoblotting. Procedures will include cell growth, protein expression, immunoblotting with a monoclonal antibody specific for the fusion protein, and color production. Prior to the lab, *E. coli* cells containing the recombinant plasmid will be grown on ampicillin plates overnight. Isolated colonies will be transferred to both a new plate and a plate with an IPTG-saturated nitrocellulose membrane. The IPTG will induce protein synthesis in bacterial colonies harboring the recombinant plasmid. The membrane and the plate on which the colonies have been grown will be marked in order to orient them for later identification of the positive colonies.

In this immunobinding assay, the nitrocellulose-bound proteins are first incubated with a primary antibody. Later, a labeled secondary antibody is adhered to the primary antibody to locate colonies that "screen positively" for immunological detection of recombinant proteins. The antibody is an immunoconjugate that produces a color change enzymatically when subjected to an appropriate substrate. Therefore, color production is an indication that a certain bacterial colony harbors transformed plasmids containing the *myo*-3 insert. Colony immunoblotting is used to verify transformed colonies by confirming that they can translate mRNA generated from the plasmid DNA template, and can produce authentic *myo*-3. This, in turn, verifies that the plasmid entered the cells during transformation, produced mRNA and supports translation, yielding authentic *myo*-3 protein.

Experimental Outline:

1. Grow cells on nitrocellulose soaked with IPTG to induce fusion protein production.
2. Immunoblotting.

Required Materials:

mouse monoclonal antibody to *myo*-3 (provided by the instructors); nitrocellulose filter; IPTG; LB-ampicillin plates; toothpicks; incubator/shaker; chloroform; lysis buffer (BSA blocking solution with 40 μg/ml lysozyme and 1 μg/ml pancreatic DNase); 10x Tris buffered saline (TBS); BSA blocking solution; TTBS (100 ml 10x TBS + 899 ml H_2O + 1 ml Tween 20); goat anti-mouse peroxidase (GAMP) conjugate (1:500 dilution of GAMP antibody conjugate in BSA blocking solution); Stain Solution (60 mg 4-chloro-1-napthol + 20 ml ice cold methanol). Just before use, add to 100 ml of TBS along with 60 μl of 30% hydrogen peroxide.

Procedures:

1. Fusion Protein Induction

1. Prepare a sterile filter to be probed with the *mouse monoclonal antibody to myo*-3, mAb 5-6, referred to as *mAb* in this protocol.

NOTE: Never touch nitrocellulose or nylon membranes with your fingers. Oils will inhibit transfer of solutions to the membranes and proteins from your fingers will bind irreversibly. Use gloves and blunt ended forceps to manipulate the filters.

2. Saturate a nitrocellulose filter with 750 μl of 20 mg/ml IPTG solution in an empty petri dish.
3. Place the filter on an LB/amp plate. To prevent air bubbles, use two forceps and bend the filter so the middle part touches the plate first. Lightly roll the rest of the membrane onto the plate.
4. Using toothpicks, make a *master plate* by "picking" single bacterial colonies containing plasmids that produce fusion protein mRNA, then scraping them onto a fresh LB/amp plate and onto the plate containing the IPTG-soaked nitrocellulose. Both plates should be inoculated in the same orientation from a single colony. You'll use the pattern to identify clones, so transcribe the pattern into your notebook and onto the filter correctly.

NOTE:

a. Using a different toothpick for each colony, use the blunt end to "pick" samples of cells from your plates of transformants.
b. Place the two LB/amp plates on a sheet containing a grid. This aids in the "picking" and transcription (of the information into your notebook) processes.

c. Lightly drag the toothpick in one direction and then repeat the process in an identical position and motion on the second plate.
d. Pick about 40 recombinant clones, or as many as you have.
e. Use scissors to make a small notch that will be used to orient your filter with the master plate.
f. Orient the filter so the notch is at the base of the positive clone.
g. Be certain to pick your stocks for positive and negative controls.

6. Place the plate containing the IPTG-soaked nitrocellulose filter and LB/amp plate in a 37°C incubator for 3-5 hours.
7. During this waiting period, you'll make up the following solutions: Tris and NaCl Stock Solutions, BSA Blocking Solution, Lysis Solution, and 10x Tris buffered saline (TBS).
   a. *1 M Tris-HCl buffer* (pH 7.4): Tris base; 121.1 g/liter (adjust the pH with HCl)
   b. *10x TBS*:
      1 M Tris HCl (pH 7.4) 100 ml
      NaCl 43.8 g
      Add H_2O to 500 ml. Store at room temperature.
   c. *1x TBS*: Mix 200 ml of 10x TBS with 1800 ml of H_2O, yielding 2 liter of 1x buffer. Store at room temperature.
   d. *BSA blocking solution*: 1x TBS containing 3% BSA.
   e. *TTBS*: Mix 100 ml 10x TBS and 899 ml H_2O. While stirring, slowly add 1 ml of Tween 20. Store at room temperature.
8. Lyse the bacterial cells by exposing them to chloroform vapors for 30 minutes in the fume hood. Do not remove the filter from the LB plate at this point. It will remain attached to the agar during the time required for lysis.

NOTE: Chloroform will disintegrate plastic petri dishes. Be certain that the chloroform is in a glass container. It is also a CARCINOGEN.

9. Wash the filters overnight in a petri dish in 7.5 ml of Lysis Buffer to remove bacterial debris.
10. Cover the plates with parafilm to preserve them overnight at room temperature. Only 30 minutes is required for this step; however, an overnight incubation can be substituted if convenient.

2. Immunoprobing with MAbs

1. Incubate the mAb filter in 10 ml of BSA blocking solution (without lysozyme or DNase) at room temperature for 10 minutes.
2. Rinse in 10 ml of TTBS plus 0.1% BSA for 10 minutes at room temperature.
3. Rinse in TBS with 0.1% BSA for 15 minutes.
4. Incubate in 7.5 ml of BSA blocking solution with 75 μl mAb, diluted 1:100 in Blocking solution and agitate for 1 hour on the shaker.
5. Place the filter in a new petri dish.
6. Wash the filters for 10 minutes in each of the following buffers.
   a. TTBS containing 0.1% BSA
   b. TBS containing 0.1% BSA
7. Discard the TBS in the sink, invert the petri dish, and drain the remaining TBS on a paper towel.
8. Add 7.5 ml of GAMP conjugate, diluted 1:500 in BSA blocking solution (15 μl GAMP in 7.485 ml of BSA blocking solution).
9. Incubate for 30 minutes with shaking.
10. Place the filter in a new petri dish.
11. Wash as in step 5, for 10 minutes with each of the following solutions (change petri plates between washes):
    a. TTBS containing 0.1% BSA
    b. TBS containing 0.1% BSA
12. Incubate the filters in 7.5 ml of Peroxide Stain Solution for 10 minutes at room temperature. An intense purple color should develop for positive clone colonies.
13. Wash the filters in 2 changes of distilled water.
14. Dry the filters on a paper towel. Keep in the dark to maintain the color.

Notes:

1. Tris buffered saline (TBS) is an aqueous solution of NaCl salt and Tris-HCl buffer.
2. TTBS is Tris buffered saline containing the detergent Tween 20.
3. TBS and TTBS are used to wash unbound antibody and contaminants from the immunoblot.

Additional Report Requirements:

1. Diagram of plates showing correct transformants.
2. Statistical percentage calculations and figures indicating which clones were transformed.

Alternative Approach

THE QIA*EXPRESSIONIST*, QIAGEN™

QIAGEN, Valencia, CA, ph. (800) 426-8157. (Adapted)

1. Introduction

The QIA*express* system provides materials for expression, purification, detection, and assay of 6xHis-tagged proteins. The QIA*expressionist* covers expression and purification products (pQE vectors, host strains, and Ni-NTA chromatographic matrices), which allow the fast and efficient production and purification of the heterologously expressed 6xHis-tagged proteins.

1.1 QIAexpress PQE Vectors. High-level expression of 6xHis-tagged proteins in E. *coli* using the QIA*express* pQE vectors is based on the T5 promoter transcription-translation system pQE plasmids belong to the pDS family of plasmids[3] and were derived from plasmids pDS56/RBS11 and pDS781/RBS11-DHFRS.[15] These low-copy plasmids (Figure 1) have the following features:

- Optimized promoter-operator element consisting of phage T5 promoter (recognized by the *E. coli* RNA polymerase) and two *lac* operator sequences that increase lac repressor binding and ensure efficient repression of the powerful T5 promoter.
- Synthetic ribosomal binding site, RBS11, for high translation rates.
- 6xHis-tag coding sequence located either 5'- or 2'- terminal to the cloning region.
- Multiple cloning site and translation stop codons in all reading frames for convenient preparation of expression constructs
- Two strong transcriptional terminators: t_0 from phage lambda,[14] and T1 from the operon in *E. coli*, to prevent read through transcription and ensure stability of the expression construct
- β-Lactamase gene (*bla*) confers resistance to ampicillin[18] at 100 μg/ml [the chloramphenicol acetyltransferase gene (CAT) is not expressed].
- *Col*E1 origin of replication.[18]

1.2 Regulation of Expression – pREP4 Plasmid. The extremely high transcription rate initiated at the T5 promoter can be efficiently regulated and repressed only by the presence of high levels of the lac repressor protein. *E. coli* host strains used in the QIA*express* system contain the low-copy pREP4 plasmid that confers kanamycin resistance and constitutively expresses the lac repressor protein encoded by the *lac*I^q gene.[5] The pREP4 plasmid is derived from pACYC and contains the p15A replicon. Multiple copies of pREP4 are present in the host cells, ensuring the production of high levels of the lac repressor protein, which binds operator sequences and tightly regulates recombinant protein expression. The pREP4 plasmid is compatible with all plasmids carrying the *Col*E1 origin of replication, and is maintained in *E. coli* in the presence of kanamycin at a concentration of 25 μg/ml.

Expression of recombinant proteins encoded by pQE vectors is rapidly induced by the addition of isopropyl-β-d-thiogalactoside (IPTG), which binds to the *lac* repressor protein and inactivates it. The promoter/operator region is extremely efficient and can be prevented only by the presence of high levels of *lac* repressor. The transcripts produced are then

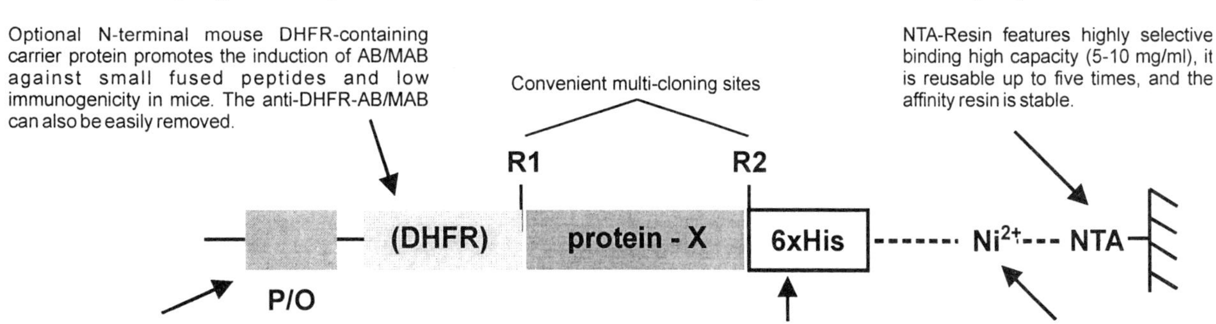

Optional N-terminal mouse DHFR-containing carrier protein promotes the induction of AB/MAB against small fused peptides and low immunogenicity in mice. The anti-DHFR-AB/MAB can also be easily removed.
Convenient multi-cloning sites
NTA-Resin features highly selective binding high capacity (5-10 mg/ml), it is reusable up to five times, and the affinity resin is stable.
R1
R2
(DHFR)
protein - X
6xHis
Ni2+--- NTA
P/O
The E. coli phage T5 regulated promoter contains two lac operator sequences, a strong promoter and tight expression control led by the lac I. Repressor encoded on separate plasmid with chemical induction by IPTG
The 6xHis tag, N- or C-terminal features a short length, poor immunogenicity and high-affinity binding to Ni-NTA. Proteins can be purified under native or denaturing conditions and can be renatured on the column. High specificity allows efficient removal of background with no need to remove the 6xHis tag. Requires mild elution conditions.
Tightly complexed Ni2+ permits high affinity protein binding and no leaching of metal ions.

Figure 1 - Features of the QIA*express* system. Abbreviations: Ab - antibody, DHFR - dihydrofolate reductase, IPTG – isopropyl-β-d-thiogalactoside, mAb - monoclonal antibody, and NTA - nitrilotriacetic acid.

translated into recombinant protein. This special "double operator" system in the pQE expression vectors, in combination with the high levels of the lac repressor protein generated by pREP4, ensure tight control at the transcriptional level of expression. The pREP4 plasmid is already present in the QIA*express* *E. coli* strains M15 (pREP4) and SG13009 (pREP4).

1.3 E. coli Host Strains. Any *E. coli* host strains containing both the expression (pQE) and the repressor (pREP4) plasmids can be used to produce recombinant proteins. The QIA*express* system uses *E. coli* strain M15 (pREP4), permitting high-level expression and easy handling. Strain SG13000(pREP4)[6] is also supplied, but some proteins may be produced at levels that are so high that they are toxic to the cells. SG13009 may be useful for the production of proteins that are poorly expressed in M15(pREP4). Both the M15 and SG13009 strains derived from *E. coli* K12 and have the phenotype Nal^s, Str^s, rif^s, lac^-, ara^-, gal^-, mtl^-, F^-, $recA^+$, uvr^+.

E. coli strains that harbor the *lac*I^q mutation, such as XL1 Blue, JM109, and TG1, produce enough lac repressor to efficiently block transcription, and are ideal for storing and propagating pQE plasmids. These strains can also be used as expression hosts for expressing nontoxic proteins, but they may be less efficient than the M15 and SG13009, which do not harbor a chromosomal copy of the *lac*I^q mutation. As a result, pREP4 must be maintained by selection for kanamycin resistance.

1.4 The 6xHis tag. The 6xHis affinity tag facilitates binding to Ni-NTA. Using pQE vectors, it can be placed at the C- or N-terminus of the protein of interest. It is poorly immunogenic, and at pH 8 the tag is small, uncharged, and therefore does not generally affect secretion, compartmentalization, or folding of the fusion protein within the cell. In most cases, the 6xHis tag does not interfere with the structure or function of the purified protein, as demonstrated for a wide variety of proteins, including enzymes, transcription factors, and vaccines. A further advantage of the 6xHis tag is that it allows the immobilization of the protein on metal-chelating surfaces, such as Ni-NTA HisSorb Strips or Plates, and, therefore, simplifies many types of protein interaction studies. In addition, anti-His antibodies can be used for detection.

1.5 Ni-NTA Technology. Immobilized-metal affinity chromatography (IMAC) was first used to purify proteins in 1975[11] using the chelating ligand iminodiacetic acid (IDA). IDA was charged with metal ions – such as Zn^{2+}, Cu^{2+}, or Ni^{2+} – and then used to purify a variety of different proteins and peptides.[17] IDA has only 3 metal-chelating sites and cannot bind metal ions tightly. Weak binding leads to ion leaching upon loading with strongly chelating proteins and peptides or during wash steps. This results in low yields, impure products, and metal-ion contamination of isolated proteins.

Nitrilotriacetic acid (NTA), exclusively available from QIAGEN, is a tetradentate chelating adsorbent developed at Hoffmann-La Roche that overcomes these problems. NTA occupies four of the six ligand binding sites in the coordination sphere of the nickel ion, leaving two sites free to interact with the 6xHis tag. NTA binds metal ions far more stably than other available chelating resins[7] and retains the ions under a wide variety of conditions, especially under stringent wash conditions. The unique, patented NTA matrices can, therefore, bind 6xHis-tagged proteins more tightly than IDA matrices, allowing the purification of proteins from less than 1% of the total protein preparation to more than 95% homogeneity in just one.[10]

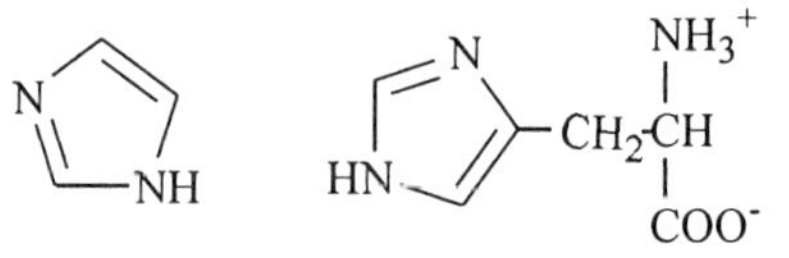

Figure 2 - Chemical structures of imidazole and histidine.

2. The Purification System

The QIA*express* purification system employs a novel metal chelate adsorbent – Ni-NTA resin – which provides an elegant yet simple one-step method for rapid protein purification.[7] The power of the QIA*express* system is based on the remarkable affinity of Ni-NTA resin for proteins and peptides that contain six consecutive histidine residues – the 6xHis affinity tag – at either their N- or C-terminus.[8]

2.1 Protein Binding. Proteins containing one or more 6xHis affinity tags, located at either the amino or carboxyl terminus of the protein, bind to the Ni-NTA resin with an affinity ($K_d = 10^{-13}$ M^{-1}, pH 8) far greater than the affinity between most antibodies and antigens, or enzymes and substrates. This means that any host proteins that bind nonspecifically to the NTA resin can be easily washed away under relatively stringent conditions, without affecting the binding of 6xHis-tagged proteins. Zinc finger proteins, for example, will bind to the Ni-NTA resin, but can be easily discriminated from the proteins carrying the 6xHis tag, which bind more tightly. The high binding constant also allows proteins in very dilute solutions, such as those expressed at low levels or secreted into the media, to be efficiently bound to the resin.

The binding of tagged proteins to the resin does not require any functional protein structure and is thus unaffected by strong denaturants such as 6 M guanidine hydrochloride or 8 M urea. This means that unlike purification systems that rely on antigen/antibody or enzyme/substrate reactions, Ni-NTA can be used to purify almost any protein – even those that are insoluble under nondenaturing conditions.

The stability of the 6xHis/Ni-NTA interaction in the presence of low levels of β-mercaptoethanol (1-10 mM) prevents the copurification of host proteins that may have formed disulfide bonds with the protein of interest during cell lysis. The presence of detergents such as Triton™ X-100 and Tween™-20 (0.1-1%), or high salt concentrations (0.1-1 M), also has no effect on protein binding. This facilitates the complete removal of proteins that would normally co-purify due to various nonspecific interactions. Nucleic acids that might associate with certain DNA and RNA binding proteins can also be efficiently removed without affecting recovery of the tagged protein.

2.2 Protein Elution. Elution of the tagged proteins from the column can be achieved by several methods. Reducing the pH will cause the histidine residues to become protonated, and to dissociate from the Ni-NTA. Monomers are generally eluted at approximately pH 5.9, while aggregates and proteins that contain more than one 6xHis-tag elute around pH 4.5.

Elution can also be achieved by competition with imidazole, which binds to the Ni-NTA and displaces the tagged protein. Low levels of imidazole can also be used to selectively elute contaminants that bind less strongly to the resin.[10]

2.3 Binding Capacity. The Ni-NTA resin is composed of a high surface concentration of NTA ligand attached to Sepharose CL-6B, sufficient for the binding of approximately 5-10 mg of 6xHis-tagged protein per ml of resin. The Ni-NTA resin can be reused 3-5 times for purification of the same protein and is very stable and easy to handle. It retains full activity even after prolonged storage.

2.4 The 6xHis Affinity Tag. The affinity tag that binds to the Ni-NTA resin consists of just six

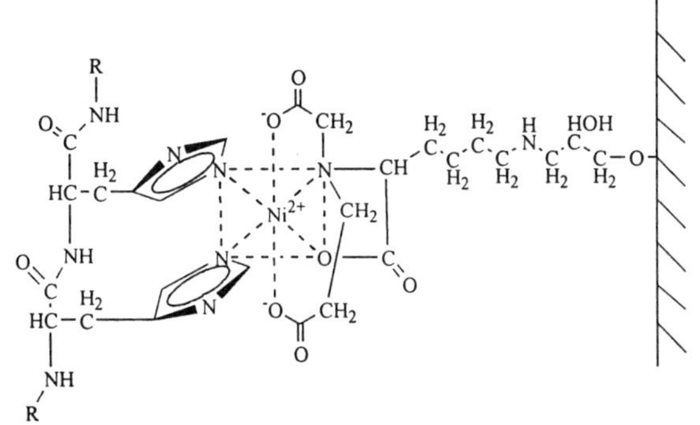

Figure 3 - Interaction between neighboring residues in the 6xHis tag and Ni-NTA matrix.

consecutive histidine residues. Its small size means that there is minimal addition of extra amino acids to the recombinant protein. It is nonimmunogenic – or at most, very poorly immunogenic – in all species except some monkeys. It is uncharged at physiological pH and generally does not affect the secretion, compartmentalization, or folding of the protein to which it is attached. In over 150 proteins purified using this system at Hoffmann-La Roche, they have never found the 6xHis tag to interfere with the structure or function of the purified protein. This has been examined for a wide variety of proteins, including enzymes,[4] transcription factors[10] antigens,[16,19] and membrane proteins (D. Meyer, personal communication), to name just a few.

For all of the above reasons, there is rarely any need to remove the 6xHis affinity tag from the recombinant protein after purification. Proteolytic removal of affinity tags is time-consuming, often inefficient, and rarely necessary with the QIA*express* system. If it is desirable for some reason to remove the tag (from very small proteins, for example, where the tag represents a large protein sequence) a protease cleavage site can be inserted easily into the construct.

The small size of the 6xHis tag makes it ideally suited for inclusion in a variety of other expression systems, whether prokaryotic, mammalian, yeast, baculovirus, or other hosts are used. A short sequence coding for the 6xHis tag, $(CATCAC)_3$, can be introduced into the expression construct via PCR, *in vitro* mutagenesis, or fragment insertion.

3. Choosing a QIA*express* Construct

Recombinant QIA*express* constructs based on the pQE vectors can be produced by placing the 6xHis tag at the N-terminus of the protein of interest (Type IV), at the C-terminus (Type III), at the C-terminus with the protein beginning with its natural ATG start codon (Type ATG), or at the N-terminus combined with the sequence for mouse DHFR (Type II).

3.1 Intended Use of Recombinant Proteins and pQE Vector Choice. The intended use of the recombinant protein to be purified will determine the choice of the pQE vector and the design of the fusion protein. Type IV constructs, with the 6xHis tag on the N-terminus of the protein, are the most commonly used and are generally the easiest to prepare. Only the 5' end of the open reading frame must be ligated in frame. Since pQE vectors have termination codons in all three reading frames, the entire coding sequence at the 3' end of the insert must not be accurately determined. When using C-terminus tags (for example Type III constructs), the insert must be cloned in frame with both the ATG start codon and the 6xHis coding sequence. Only full proteins generated by Type III constructs are purified. An undetected frame shift or translational stop codon would lead to premature termination, resulting in peptides lacking the 6xHis tag.

Type ATG constructs the ribosome binding consensus (Shine-Dalgarno) sequence is provided by the expression vector. These vectors allow the translation of the coding fragment to initiate at the original start codon without fusing additional amino acids to the N-terminus. The optimized translation initiation region in pQE expression vectors enables the expression of proteins in Type IV constructs 2-4 times more efficiently than proteins with a C-terminal affinity tag.

For many applications, the number of amino acids added to the protein along with the 6xHis tag should be kept to a minimum. This can be achieved with a Type ATG construct, or by cloning the insert at the 5' end of the Type IV multiple cloning site (MCS). If the number of additional amino acids is not important, the most convenient restriction sites in the Type IV MCS may not be used, but many of the advantages conferred by the small size of the 6xHis tag will be lost. If it is desirable to detect the protein or peptide being expressed by using the RGS·His™ Antibody (QIA*express* Detection System), the RGS·His epitope must be present. Vectors that encode the RGS·His epitope include pQE-9-11, pQE-30-32, and pQE-40-42. Penta·His™ and Tetra·His™ Antibody can detect any 6xHis-tagged protein expressed with pQE vectors.

3.2 Protein Size. Very small proteins and peptides are sometimes difficult to express stably in *E. coli* – because they cannot fold correctly and are often subject to proteolytic degradation. These proteins can be stabilized by expressing them fused to a large protein such as the mouse DHFR protein encoded in a Type II construct. The DHFR protein is poorly immunogenic in mice and rats and protects the attached peptides from proteolysis after immunization. DHFR enhances the general antigenicity of the peptides to which it is attached – by allowing them to fold properly.

Very long recombinant proteins may be subject to premature termination (see below). Placing the 6xHis tag at the C-terminus (Type III or ATG construct) will select only for full-length proteins during purification.

3.3 Codon Usage. Some codons are rarely used in *E. coli*. For example, the arginine codons AGG and AGA are the least frequently used, and the tRNAs that recognize them are among the least abundant. Consecutive AGG or AGA codons can lead to a high level of frame shifting.[12] If the DNA sequence encoding the recombinant protein contains several such codons, a variety of truncated protein products, particularly from large recombinant proteins, are likely to be synthesized. By placing the 6xHis affinity tag at the C-terminus, only the full-length proteins will bind to the Ni-NTA resin during the purification procedure; the truncated forms will not bind and will be removed in the flowthrough and wash.

3.4 Internal Start Sites. Initiation of translation at internal start sites can occur when a Shine-Dalgarno sequence is present 5' adjacent to an internal ATG or GTG codon. Placement of the tag at the N-terminus of the protein will prevent the copurification of shorter proteins, whose synthesis is initiated at these start sites, but internal "starts" should be eliminated by deleting any Shine-Delgarno sequence in the coding region.

3.5 Inefficient Translation. Some DNA sequences contain regions that interfere with the interaction between the *E. coli* ribosomes and the ribosome binding sites provided by the expression vector. This may be the result of stable stem-loop structures formed in the presence of inverted repeats. In most cases this interference can be minimized by modifying the 5' end of the insert by increasing the A-T content, or by constructing a fusion protein in which the sequence is inserted at the 3' end of a fusion partner such as the DHFR protein (Type II construct). Placing the 6xHis-tag sequence at the 5' end of the gene often increases expression levels.

3.6 Secretion. Secretion of proteins in *E. coli* is mediated by an N-terminal signal sequence that is cleaved after protein translocation. Expression of certain gene products as secreted proteins may in some cases be necessary to promote proper folding and disulfide bond formation or to direct toxic proteins out of the cell. The 6xHis tag must be placed at the C-terminus to prevent it from interfering with the signal sequence and to prevent its loss during N-terminus processing. The location of the 6xHis tag at the C-terminus has no effect on secretion. One major drawback of secretion into the periplasm is the lower yield that is generally obtained due to the hydrophobic nature of the signal peptide. Expression may also be complicated by the formation of inclusion bodies in the periplasmic space.[2]

3.7 Removal of the 6xHis Affinity Tag. In most

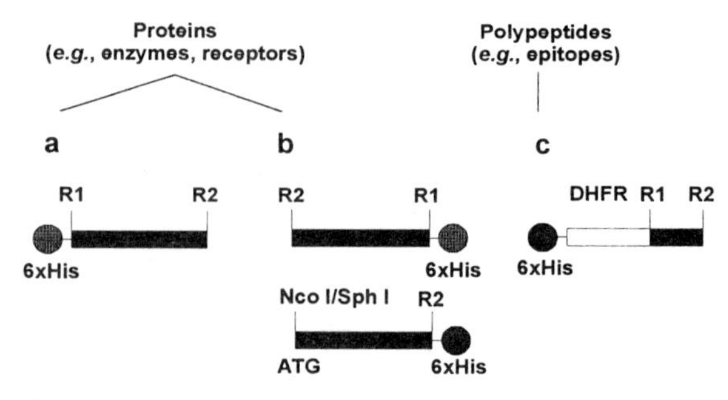

Figure 4 - QIA*express* constructs. (a) Type IV construct, 6xHis-protein. (b) Type III construct, 6xHis protein. (c) Type II construct, 6xHis-DHFR fusion.

cases, there is no need to remove the 6xHis affinity tag from the purified proteins. If it must be removed, a protease cleavage site should be inserted between the 6xHis sequence and the N-terminus of the protein. *Enterokinase* recognizes the sequence Asp_4-Lys (DDDDK), and cleaves after the lysine. This sequence can be inserted into the construct by site-directed mutagenesis or PCR, after checking that it is not already present in the protein sequence. Other protein cleavage sites can also be used.

Carboxypeptidase A can be used to remove C-terminal 6xHis tags.[8] The enzyme efficiently removes aromatic C-terminal residues until it encounters a basic residue, at which point removal is terminated. Care should be taken to ensure that one or more basic residues (such as arginine) is/(are) at the C-terminus of the recombinant protein.

4. Cloning Procedures

4.1 Preparation of pQE Expression Constructs. The construction of expression vectors is generally straightforward. The gene or cDNA sequence encoding the protein sequence to be expressed is cloned into the appropriate pQE vector in the same reading frame as the 6xHis affinity tag. The choice of the vector depends on where the 6xHis tag is to be placed (N-terminal or C-terminal), whether the protein is to be expressed as a fusion protein with DHFR, and how the DNA fragment will be inserted in the correct reading frame. For every type of construct there are vectors for cloning in each of the three reading frames.

The pQE expression construct is created by ligation and then transformed into the M15 or other host strain carrying the pREP4 repressor plasmids. Transformants are selected on plates containing both ampicillin and kanamycin. It is recommended that those unfamiliar with these procedures consult a practical manual, such as *Molecular Cloning: A Laboratory Manual*,[13] or *Current Protocols in Molecular Biology*.[1]

4.2 Vector Preparation. The vector from the QIA*express* Kit is prepared by first dissolving it in TE buffer. The addition of 10 μl TE to 5 μg of plasmid DNA is generally convenient. A 2 μl aliquot (i.e., 1 μg) can then be linearized using the appropriate restriction enzyme according to the enzyme manufacturer's recommended buffer and incubation conditions. Many combinations of enzymes are compatible when used together in the same buffer. However, different enzymes cut with different efficiencies, especially when the two sites are close together. If restriction enzymes are sufficiently active in a given buffer and their sites are more than 10 bp apart, they can be used in the same reaction. If this is not the case, the digestion should be carried out separately with a clean-up step in between. QIAquick™ Kits provide a very fast and efficient method to purify DNA.

Expression vectors such as the pQE plasmids do not allow "*color selection*" of clones that contain plasmids with inserts. Care should therefore be taken to ensure that vectors are digested to completion before subcloning. If the insert is to be cloned into a single restriction site, it is especially important to dephosphorylate the vector ends after digestion. In vectors cut with two enzymes, when the sites are close together, or if one of the enzymes cuts inefficiently, dephosphorylation decreases the nonrecombinant background caused by incomplete digestion with one of the enzymes. Following digestion, it is usually worthwhile to gel-purify the vector prior to insert ligation in order to remove residual nicked and supercoiled plasmid. The latter will transform *E. coli* much more efficiently than ligated plasmids. Please note that positive expression clones can be detected directly on colony blots using one of the anti-His antibodies or Ni-NTA conjugates.

4.3 Insert Preparation. In general, it is only necessary to prepare the insert by restriction digestion and gel purification. The fragment can be ligated to the vector directly. If there is no appropriate restriction site, if it is desirable to minimize the number of extra codons. If the construct must be optimized in some other way, more complicated

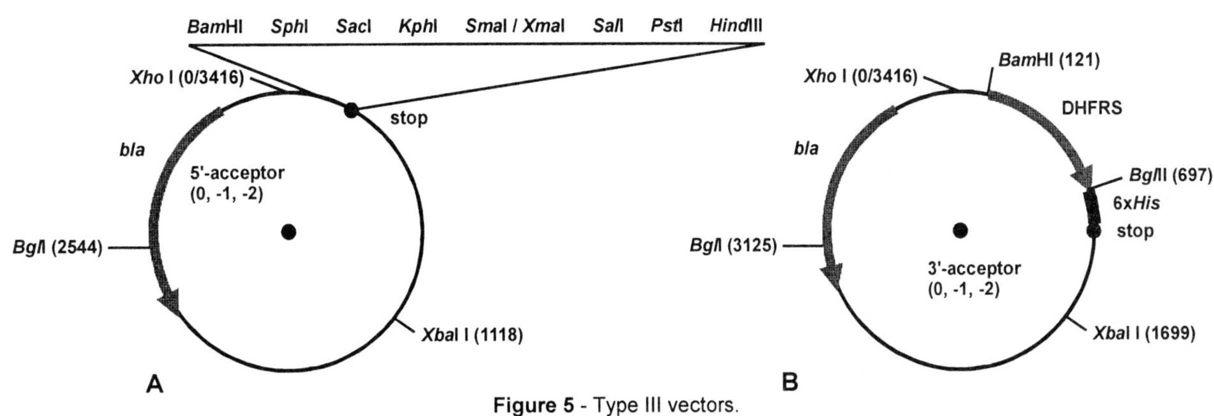

Figure 5 - Type III vectors.

manipulation may become necessary. (See specific construct sections below.) The ends of coding fragments can be modified by PCR, *in vitro* mutagenesis or addition of linkers.[1,13]

4.4 Type II and Type IV Constructs (N-Terminal 6xHis Tag). The relevant pQE vectors for these constructs contain the 6xHis tag (plus DHFR, in the Type II constructs) 5'-adjacent to the polylinker. The appropriate vector must only be digested with the necessary restriction enzyme(s), purified to obtain the linear form, and ligated with the insert containing the desired coding region. If only one restriction site is used, we recommend that the ends of the vector be dephosphorylated to prevent religation.

The Shine-Delgarno sequence and ATG initiation codon should be removed from the fragment that is to be inserted into these vectors. "Internal starts," from control sequences provided by the inserted fragment itself, will result in the expression of proteins that lack the 6xHis tag, and thus cannot be purified. The endogenous stop codons can be retained in the inserted fragment, but are not required since pQE vectors provide translational stop codons in all three reading frames.

4.5 Type III Constructs (C-Terminal 6xHis Tag). Type III constructs place the 6xHis affinity tag at the C-terminus of the protein. The insert must be "in frame" with both the 5' start codon and the 3'-adjacent sequences coding for the 6xHis tag. This is achieved by preparing 5' and 3' "acceptor arms" in the appropriate reading frames, and then recreating the expression construct with a three-arm ligation between these acceptors and the coding fragment. The vectors pQE-50, -51, and -52 provide the 5' acceptor arms, the vectors pQE-16, -17, and -18, provide the 3' acceptor arms.

4.6 Type ATG Constructs (C-Terminal 6xHis Tag). Type ATG constructs allow the expressed protein to initiate with the authentic ATG, but using the optimized Shine-Dalgarno region of the pQE vector. They are prepared in the same way as standard Type III constructs, with all the same considerations, except that the sequence around the authentic ATG must be modified to create either an

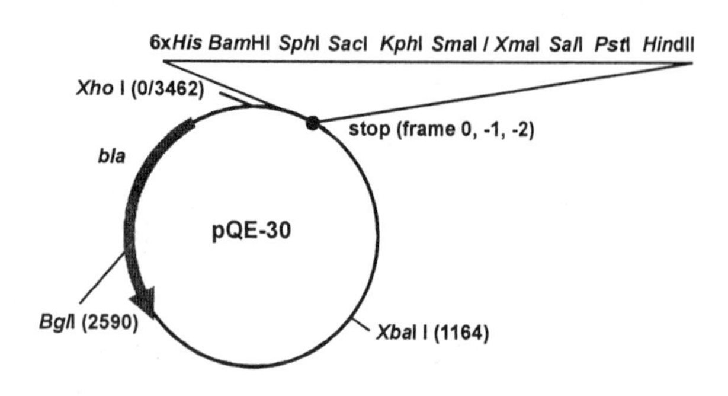

Figure 6 - Type II vector.

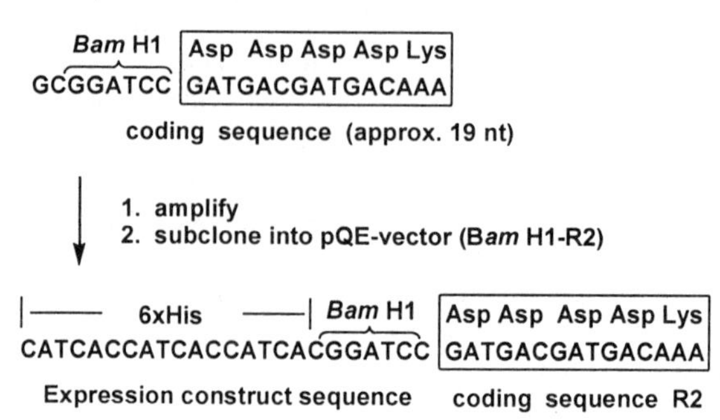

Figure 7 - Modification of inserts by PCR.

*Nco*I site (pQE-60) or an *Sph*I site (pQE-70). Cleavage of both the vector and the coding fragment with the appropriate enzyme allows the coding fragment to be ligated into the vector such that the authentic ATG codon replaces the vector ATG codon. If the 6xHis tag cannot be cloned in the correct reading frame, a 3' acceptor arm can be prepared from vectors pQE-16, -17, or -18 as described above, and the expression construct created by a three-arm ligation (three pieces at once).

4.7 Ligation, Transformation, and Screening. The ligation of the insert with the prepared vector is usually carried out using T4 DNA ligase under standard conditions.[13]

A procedure for the transformation of M15 (pREP4) cells is presented later in the QIA*expressionist* manual. The efficiency of the subcloning procedure should always be monitored by the transformation of nonligated and self-ligated vector constructs. A control transformation without DNA should be performed to ensure that the antibiotic ampicillin is working effectively, and a control of transformation efficiency with a known amount of intact plasmid DNA should also be included (e.g., 1 or 10 ng per transformation).

Transformants may be screened for correct insertion of the coding fragment by restriction analysis of the pQE plasmid DNA, by sequencing the cloning junctions, by directly screening bacterial colonies for the expressed protein (colony blotting procedure), or by preparing small-scale expression cultures. We prefer and recommend screening the transformants directly for expression using the colony-blotting procedure – because it allows simultaneous screening of transformants with the correct coding fragment, expression levels, and in-frame translation of the 6xHis tag. Clones with religated vectors, that do not express a fusion protein, will not generate a false positive signal.

5. PCR Mutagenesis to Insert a 6xHis Tag

If the expression vector or the existing construct being used does not encode a 6xHis tag, the tag can be easily engineered into the insert sequence itself by employing PCR techniques. One PCR primer includes the 5' restriction site [R1], the CAT CAC CAT CAC CAT CAC sequence for the 6xHis tag, and a priming sequence that is homologous to the 5' end of the insert DNA corresponding to the N-terminus of the protein. The second primer contains another restriction site [R2] that is located within the insert. The PCR product and the original construct are then both cut with the enzymes R1 and R2 and ligated together. The recombinant plasmid encoding the 6xHis tag can be inserted into a construct following a similar approach and using an appropriate set of primers.

If you are unfamiliar with PCR techniques, consult the technical guidelines for "Constructing Recombinant DNA Molecules by the Polymerase Chain Reaction"[1] or refer to the *Taq* PCR Handbook available at http://www.QIAGEN.com/literature/handbooks.

6. Propagation of pQE Plasmids and Constructs

QIA*express* pQE vectors are supplied lyophilized with sucrose and bromophenol blue for visualization, and should be resuspended in a convenient volume of TE (e.g., 10 μl) and stored at –20°C. Sucrose and bromophenol blue do not interfere with restriction reactions.

QIA*express* pQE vectors and constructs can be maintained in any *E. coli* strain that is ampicillin sensitive and carries the pREP4 repressor plasmid, or harbors the *lacI* gene or the F-factor (episome). M15 carries *lacI* on pREP4, but XL1-Blue or the JM series contain an episomal copy of *lacI*^{q}. *LacI*^{q} is a mutation of *lacI* that produces very high levels of the lac repressor. Initial cloning and propagation using XL1-Blue are recommended because plasmid preparations derived from QIA*express* host strains will also contain pREP4 DNA, which could make clone analysis more difficult.

For the most stable propagation of expression constructs, especially when toxic proteins are expressed, we recommend M15(pREP4) because of the higher repressor levels. In this case, both ampicillin and kanamycin must be applied to maintain the expression and repressor constructs.

7. Protein Purification

7.1 Batch or Column Purification. Proteins may be purified on Ni-NTA resins in either batch or column procedures. The batch procedure entails binding the protein to the Ni-NTA resin in solution and then packing the protein-resin complex into a column for washing and elution steps. This strategy promotes efficient binding of the 6xHis-tagged protein, especially when the 6xHis tag is not fully accessible or when the protein in the lysate is present at very low concentration.

In the column procedure, the Ni-NTA resin is first packed into the column and equilibrated with the lysis buffer, then cell lysate is slowly applied to the column. Washing and elution steps are identical in the batch and column procedure.

7.2 Protein Binding. Proteins containing one or more 6xHis affinity tags located at either the amino and/or carboxyl terminus of the protein can bind to the Ni-NTA groups on the matrix with an affinity greater than that of antibody-antigen or enzyme-substrate interactions. Binding of the 6xHis tag does not depend on the three-dimensional structure of the protein. Even when the tag is not completely accessible, it will bind as long as more than two histidine residues are available to interact with the nickel ion. In general, the smaller the number of accessible histidine residues, the weaker the binding will be. Untagged proteins that have histidine residues in close proximity on their surface will also bind to Ni-NTA, but in most cases this interaction will be much weaker than the binding of the 6xHis tag. Any host proteins that bind nonspecifically to the NTA resin itself can easily be washed away under relatively stringent conditions that do not affect the binding of 6xHis-tagged proteins.

Binding can be carried out in a batch or column mode (see "Batch or Column Purification"). If the concentration of 6xHis-tagged proteins is low, or if they are expressed at low levels, or secreted into the media, the proteins should be bound to Ni-NTA in a batch procedure, and under conditions in which background proteins do not compete for the binding sites – i.e., at a slightly reduced pH or in the presence of low imidazole concentrations (10-20 mM). At low expression levels under native conditions, binding can be optimized for every protein by adjusting the imidazole concentration and/or pH of the lysis buffer. If high background levels are still present, equilibrating the Ni-NTA matrix with lysis buffer containing 10-20 mM imidazole prior to binding is recommended. The matrix is thus "shielded," and nonspecific binding of proteins that weakly interact is significantly reduced.

7.3 Wash. Endogenous proteins with histidine residues that interact with the Ni-NTA groups can be washed out of the matrix with stringent conditions achieved by lowering the pH to 6.3 or by adding imidazole at a 10-50 mM concentration. In bacterial expression systems, the recombinant proteins are usually expressed at high levels, and the level of copurifying contaminant proteins is relatively low. Therefore, it generally is not necessary to wash the bound 6xHis-tagged protein under very stringent conditions. In lysates derived from eukaryotic systems, the relative abundance of proteins that may contain neighboring histidines is higher. The resulting background problem becomes more critical, especially when nondenaturing procedures are employed. In these instances, it becomes necessary to increase the stringency of the wash steps considerably. This can be performed most effectively by gradually decreasing the pH of the wash buffer or by slowly increasing the concentration of imidazole in defined steps. "Step gradients" are preferable because they are much more effective than linear gradients when metal affinity chromatography methods are employed. The optimal pH and/or imidazole concentrations for the washes will vary slightly for each protein and must be determined empirically.

7.4 Protein Elution. The histidine residues in the 6xHis tag have a pK_a of approximately 6.0 and will become protonated if the pH is reduced (to pH 4.5-5.3). Under these conditions, the 6xHis-tagged protein can no longer bind the nickel ions and will dissociate from the Ni-NTA resin. Similarly, if the imidazole concentration is increased to 100-250 mM, the 6xHis-tagged proteins will also dissociate because they can no longer compete for binding sites on the Ni-NTA resin.

Elution conditions are highly reproducible, but must be determined for each 6xHis-tagged protein that is being purified. Monomers generally elute at approximately pH 5.9, whereas aggregates and proteins that contain more than one 6xHis tag elute at approximately pH 4.5.

Reagents such as EDTA or EGTA chelate the nickel ions and remove them from the NTA groups. This causes the 6xHis-tagged protein to elute as a protein-metal complex. NTA resins that have lost their nickel ions become white in color and must be recharged if they are to be reused.

Whereas all elution methods (imidazole, pH, and EDTA) are equally effective, imidazole is mildest and is recommended under native conditions, when the protein would be damaged by a reduction in pH, or when the presence of metal ions in the eluate may have an adverse effect on the purified protein.

8. Purification on Ni-NTA-Resin

While many proteins remain soluble during expression and can be purified in their native form under nondenaturing conditions on Ni-NTA resin, many others form insoluble precipitates. Since almost all of these proteins are soluble in 6 M guanidine hydrochloride, Ni-NTA chromatography and the 6xHis tag provide a universal system for the purification of recombinant proteins.

8.1 Protein Solubility and Cellular Location. Before deciding on a purification strategy, it is important to determine whether the protein is soluble in the cytoplasm, located in cytoplasmic inclusion bodies, or secreted into the periplasmic space. Many proteins form inclusion bodies when they are expressed at high levels in bacteria, while others are well tolerated by the cell and remain in the cytoplasm in their native configuration.

8.2 Batch or Column Purification. Proteins have been purified on Ni-NTA resin in either a batch or a column procedure. Both procedures are equally efficient; the choice is up to the user. Batch procedures entail binding the protein to the Ni-NTA resin in solution, and then packing the protein/resin complex into a column for the washing and elution steps. This may promote more efficient binding of the tagged protein, and reduce the amount of debris loaded onto the column. Batch binding for an extended time is recommended when purifying very dilute proteins.

Proteins that contain neighboring histidines are not common in bacteria, but are quite abundant in mammalian cells. These proteins bind to the resin much more weakly than proteins with a 6xHis tag and can be easily washed away – even when they are much more abundant than the tagged protein.[10] The addition of 10 mM β-ME to the loading buffer will reduce background due to cross-linked proteins. This is important when purifying proteins, which contain cysteine residues. Do not use DTT or DTE – they reduce the Ni^{2+} ions.

8.3 Protein Refolding. Each denaturing protein must be refolded according to a specially optimized protocol. Refolding is generally carried out by gradual dilution of the denaturing agents, together with careful reformation of the disulfide bridges. While there are few hard and fast rules about protein refolding, the following guidelines may prove helpful.

As a general rule, protein refolding should take place slowly and in dilute solution to avoid the formation of insoluble aggregates. It should be carried out at a redox potential, which is close to the equilibrium of Cys-SH and Cys-S-S-Cys. Disulfide bridges will more likely become stabilized if they are trapped in a strong and correct tertiary structure. The addition of detergents, salt, and alcohol to the refolding buffer may aid in keeping the refolded proteins soluble.

Refolding is generally carried out by step-wise dilution of denaturants in a dialysis procedure. One folding buffer system used successfully in zinc finger protein refolding is 1 M urea, 0.05 M Tris-HCl, 0.0005% Tween™-80, 2 mM reduced glutathione, 0.02 mM oxidized glutathione, pH 8; stir at 4°C for 30 hours.[9] Many proteins that are insoluble when refolding is attempted in solution can be successfully refolded while immobilized on the Ni-NTA column. It may be that immobilizing one end of the protein during renaturation prevents the formation of misfolded aggregates. The recommended renaturation conditions are as follows: use a linear 6 M to 1 M urea or guanidinium chloride gradient in 500 mM NaCl, 20% glycerol, Tris-HCl, pH 7.4, containing protease inhibitors.

References:

1. Ausubel, F. M., Brent, R., Kingston, R. E., Moore, D. D., Sedman, J. G., Smith, J. A., and Struhl, K. eds. (1995) "*Current Protocols in Molecular Biology*." John Wiley and Sons, New York.
2. Bowden, G. A., and Georgiou, G. (1990) Folding and aggregation of β-lactamase in the periplasmic space of *Escherichia coli*. J. Biol. Chem. 265, 16760-16766.
3. Bujard, H., Gentz, R., Lanzer, M., Stüber, D., Müller, M., Ibrahimi, I., Häuptle, M. T., and Dobberstein, B. (1987). A T5 promoter based transcription-translation system for the analysis of proteins *in vivo* and *in vitro*. Methods Enzymol. 155, 416-433
4. Döbeli, H., Trecziak, A., Gillessen, D., Matile, H., Srivastava, I. K., Perrin, L. H., Jakob, P. E. and Certa, U. (1990). Expression, purification, biochemical characterization and inhibition of recombinant *Plasmodium falciparum* aldolase. Mol. Biochem. Parasitol. 41, 259-268.
5. Farabaugh, P. J. (1978). Sequence of the lacl gene. Nature 274, 765-769.
6. Gottesmann, S., Halpern, E., and Trisler, P. (1981). J. Bacteriol. 148, 265-273.
7. Hochuli, E. (1989). Aufarbeitung von Bioproteinen: Elegant und wirtschaftlich. Chem. Ind. 12, 69-70.
8. Hochuli, E., Bannwarth, W., Dobeli, H., Gentz, R., and Stüber, D. (1988) Genetic approach to facilitate purification of recombinant proteins with a novel metal chelate adsorbent. Bio/Technology 6, 1321-1325.
9. Jaenicke, R. ànd Rudolph, R. (1990). Folding proteins. In "*Protein Structure - A Practical Approach*" (T. E. Creighton, ed.). IRL Press, New York.
10. Janknect, R., de Martynoff, G. Lou, J., Hipskind, R. A., Nordheim, A., and Stunnenberg, H. G. (1991) Rapid and efficient purification of native histidine-tagged protein expressed by recombinant vaccinia virus. Proc. Natl. Acad. Sci. USA 88, 8972-8976.
11. Porath, J., Carlsson, J. Olsson, I., and Belfrage, G. (1975) Metal chelate affinity chromatography, a new approach to protein fractionation. Nature 258, 598-599.
12. Rosenberg, A. H., Goldman, E., Dunn, J. J., Studier, F. W., and Zubay, G. (1993) Effects of consecutive AGG codons on translation in *Escherichia coli* demonstrated with a versatile codon test system. J. Bacteriol. 175, 716-722.
13. Sambrook, J., Fritsch, E. F., and Maniatis, T. (1989). "*Molecular Cloning: A Laboratory Manual*" (2nd Edition). Cold Spring Harbor Laboratory Press.
14. Schwortz, E., Scherrer, G., Hobom, G., and Kössel, H. (1987). Nucleotide sequence of cro, cII and part of the 0 gene in phage lambda DNA. Nature 272, 410-414.
15. Stüber, D., Bannwarth, W., Pink, J. R. L., Meloen, R. H., and Matile H. (1990). New B cell epitopes in the *Plasmodium falciparum* malaria circumsporozoite protein. Eur. J. Immunol. 20, 819-824.
16. Stüber, D., Matile, H., and Garotta, G. (1990). System for high-level production in *Escherichia coli* and rapid purification of recombinant proteins: Application to epitope mapping, preparation of antibodies, and structure-function analysis. In "*Immunological Methods*" (I. Lefkovits and B. Pernis, eds.), Vol. IV, 121-152.
17. Sulkowski, E. (1985) Purification of proteins by IMAC. Trends Biotechnol. 3, 1-7.
18. Sutcliffe, J. G. (1979). Complete nucleotide sequence of the *E. coli* plasmid pBR322. Cold Spring Harbor Symp. Quant. Biol. 43, 77-90.
19. Takacs, B. J., and Girard, M. F. (1991). Preparation of clinical grade proteins produced by recombinant DNA technologies. J. Immunol. Methods 143, 231-240.

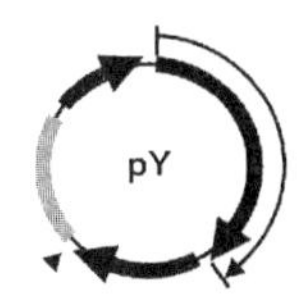

UNIT 6: ANALYSIS OF DNA OR RNA BY DUPLEX HYBRIDIZATION: DNA ISOLATION, LABELING, AND PROBING

1. Introduction

The specific, high affinity binding that one DNA strand has for its complementary strand of DNA (or RNA) is a key requirement for maintaining life on earth. These interactions convert the genetic code, preserved as DNA, into active molecules such as RNAs and proteins, which are used to operate cellular machinery. Another key function of DNA is to preserve the genetic information, ensuring that it continues to pass from one generation to the next as a result of replication and mitosis or meiosis. Binding of one DNA strand to another is called *duplex pairing* or *"hybridization,"* and the high fidelity and tremendous stability of these processes have provided one of the most sensitive and accurate types of *in vitro* assays ever devised.

An assay that utilizes DNA:DNA or DNA:RNA duplex hybridization can measure "attomoles" (10^{-18} mole) of DNA or RNA. Attomole sensitivity is more than one million-fold better than the most sensitive protein assays (20 x 10^{-12} mole[1]) and a thousand-fold more sensitive than the most sensitive immunoassays (10^{-15} mole[2]). Given that 1 attomole contains about 600,000 molecules, it is possible to detect one molecule per cell if just 600,000 cells are analyzed, corresponding to about ≈ 1 mg wet weight of animal tissue. When an amplification step is used prior to DNA analysis, such as polymerase chain reaction (PCR) or reverse transcriptase PCR (RT-PCR) for mRNA, to produce more material, the assay sensitivity can be increased another billion-fold.

2. Isolation of Genomic DNA from *C. elegans* and Preparation for Southern Blotting

The nuclei of all eukaryotic cells contain a highly condensed form of chromatin. To free the genomic DNA from its surrounding protein, strong denaturants and reducing reagents are typically required, such as sodium dodecyl sulfate (SDS), β-mercaptoethanol (β-ME), or dithiothreitol (DTT). β-ME and DTT reduce disulfide bonds, preventing them from stabilizing protein structure. SDS is a detergent that solubilizes lipid and nuclear membranes and dissociates most chromatin proteins from genomic DNA strands. It does so by denaturing the hydrophobic interiors of these structures, which impairs proper recognition and assembly. Because cells also contain extremely potent and stable DNases and RNases the β-ME/SDS or DTT/SDS solution is fortified with proteinase K, a protease that is very stable in 1% SDS and degrades DNases and RNases that have been rendered inactive by SDS and β-ME or DTT.

Following the initial extraction of DNA from cells, phenol is usually used to further remove proteins from DNA. Phenol is an excellent protein denaturant and either solubilizes the proteins or confines them to a white precipitate at the phenol/water interface. Afterward the phenol and water phases are separated by centrifugation. DNA remains in the aqueous phase of the extraction mixture. Chloroform is added to enhance the denaturation and aid in separating the phenol from the aqueous phase. After the aqueous portion of the extract (top, lighter portion) is separated from the phenol/chloroform phase, two volumes of cold 95% ethanol are added. The sample is then placed at -80°C for 5 minutes or 11°C overnight. The precipitates (> 99% of the DNA) are pelleted by centrifugation (10,000 x *g*) and redissolved in buffer with minimal salt. DNA can often remain double helical in H_2O at room temperature. Extraction and precipitation are affected by salt concentration as well as pH. Depending on salt or pH conditions, the DNA, RNA, or protein can move back and forth between aqueous and phenol phases. Optimum salt and pH conditions have been determined experimentally by many investigators and differ depending on the overall procedure being used, so care should be taken to follow the instructions exactly.

DNA solution concentrations can be quantitated with reasonable accuracy using absorbance spectrophotometry. Based on the average molar extinction coefficient for DNA, 50 µg/ml of DNA should have an absorbance of approximately 1 at 260 nm. One test for DNA purity relies on the spectral characteristics of pure DNA; DNA has an absorbance at 260 nm which is about twice that at 280 nm, so pure DNA has an A_{260}/A_{280} ratio of 2. Pure protein has spectral characteristics which give an A_{260}/A_{280} ratio of 0.5. Therefore, if DNA has an A_{260}/A_{280} ratio that is less than 2, it suggests that protein is contaminating the preparation. In practice, a ratio of 1.7-2 is a sign that the DNA is highly pure. It should be noted that phenol absorbs at 260 nm, so if phenol is not completely removed from the extraction mixture the A_{260}/A_{280} ratio could be found to be above 2, even for the less than pure samples.

3. Labeling DNA

DNA can be 3' or 5' end labeled, but the highest specific activities are obtained using DNA-dependent polymerase (Klenow fragment) to copy a strand of DNA in the presence of one (or more) labeled nucleotides. This method, called "body labeling," permits an average incorporation of one nucleotide for every four nucleotides polymerized. Thus, a 1-kb strand of DNA would contain about 250 labeled nucleotides.

The Klenow fragment of DNA-dependent DNA polymerase requires a primer whose 3' end is extended until it reaches the terminus of the complementary DNA or encounters a previously annealed or synthesized DNA strand in its path. Klenow DNA polymerase has no 5' to 3' nuclease activity, but does contain 3' to 5' nuclease ("proof-reading") activities. When sufficient nucleotide triphosphates are present, Klenow fragment rapidly produces complementary DNA from the 3' end of a DNA primer.[3]

The most efficient way to label cDNA is the "random priming" method. This technique relies on random hybridization of a mixture of short DNA oligomers (6 to 9 nucleotides long) to parent DNA strands; these oligomers are then used as distributed templates for Klenow DNA polymerase, which extends their 3' ends by filling the gaps with new "body-labeled" strands. The general form of the products is shown in Figure 1.

4. Analyzing DNA by Southern Hybridization ("Blotting") and RNA by Northern Blotting

Labeled complementary nucleic acids ("probes") can be used in several ways to analyze DNA or RNA sequences for occurrence and quantitative Watson-Crick binding. A method that provides reasonably accurate information about both *size* and *amount* of a specific DNA, developed by Southern,[4] is called Southern blot hybridization analysis. An analogous technique used for RNA analysis is called "Northern blotting" to distinguish it from the Southern technique that produces DNA:DNA* products (where DNA* is the labeled probe). Northern blots involve RNA:DNA* complexes. For typical Southern analyses, samples of restriction-digested plasmid or genomic DNA are fractionated on an agarose gel. The agarose gel is then placed against a nylon or nitrocellulose membrane and buffer is induced to flow (by blotting liquid through the gel onto the membrane where the DNA is deposited). The membrane retains a replica of the agarose gel pattern. The advantage is that the membrane retains the fractionated (but denatured) nucleic acid fragments, which can be treated easily with DNA probes to analyze for specific binding. The original agarose gel is too fragile for such manipulations.

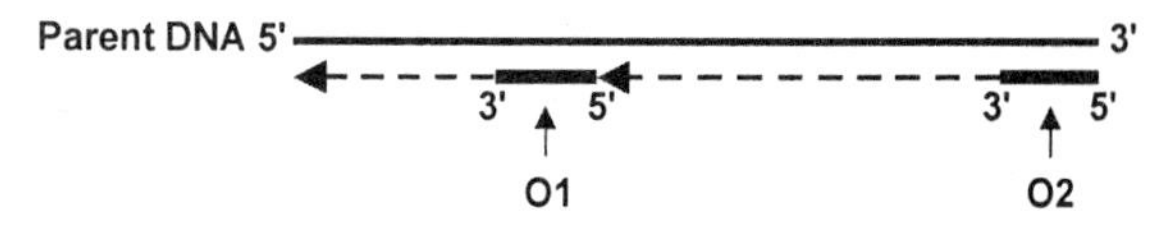

Figure 1 - The parent (template) DNA is bound by two complementary oligonucleotides (O1 and O2). The 3' termini of the oligonucleotides are then extended by Klenow DNA polymerase to produce labeled stands (←----).

Membranes used to bind DNA are typically positively charged, binding the negatively charged DNA or RNA backbone electrostatically. Once the DNA (or RNA) binds the membrane, it is treated with saturating amounts of nonspecific DNA (such as fragmented salmon sperm DNA) and protein (such as casein) so that all unoccupied sites on the membrane are coated ("blocked") from binding other DNAs (e.g., DNA probes) or proteins (e.g., antibodies) non-specifically. Finally, the labeled probes are hybridized to the DNA (or RNA) at very specific salt, pH, temperature, and nucleic acid concentrations. Formamide is often used in the hybridization solution to destabilize the complexes and thereby lower the hybridization temperatures. The melting temperature (T_m) of a long duplex in the absence of formamide is often between 68 and 75°C. Formamide lowers DNA:DNA or DNA:RNA T_m values to ca. 42°C. Another reagent often found in hybridization solutions is polyethylene glycol (PEG), which decreases the concentration of "free" water in the hybridization solution, effectively increasing the concentration of labeled probe and promoting faster hybridization of the probe to the membrane-bound DNA or RNA. After hybridization, the membrane is washed several times and analyzed for retained labeled probe.

5. Membrane-Bound DNA (or RNA) Analysis Using Colorimetry and Chemiluminescence

Labeled DNA probes retained by "Southern" or "Northern" membranes can be detected using colorimetry, chemiluminescence, or radioactivity. *Colorimetric methods* are least sensitive, quite rapid (minutes to hours), do not depend on film or sophisticated detection equipment, and yield a non-hazardous record that can be copied or retained permanently in a notebook. Colorimetry analysis is often used when only semiquantitative results are needed rapidly.[5]

Chemiluminescense[6] and *radioactivity*-based techniques are sensitive, allowing one to detect as few as picomoles of probe. Unfortunately, they can take hours to days to complete with small quantities (including film/plate exposure time), and often require computer analysis. The final results will have characterized precision levels, allowing quantitative statistical comparisons among several samples (e.g., at different stages in a purification scheme). Both techniques have similar sensitivities. Chemiluminescence is measured on film or using reusable image deposition screens (storage plates). This limits the range of concentrations compatible with linear responses (to ca. 15-fold). Technical aspects of the densitometer also limit linear responses (ca. 2.5 to 3 orders of magnitude). Phosphorimagers are used to detect radioactivity, which can be quantitated accurately over 5 orders of magnitude. Thus, ^{32}P end-labeling is the preferred method used to perform and analyze Southern or Northern blots in most laboratories.

A nonisotopic technique used in Lab 5.2 involves labeling the probe DNA with digoxigenin. The student is introduced to a sensitive and versatile set of methods to study DNA and RNA probes by hybridization analyses. The structure of the DNA body-labeling functional group, Digoxigenin-dUTP, is shown in the Genius® kit literature.

Colorimetric detection is useful to estimate the amount of labeled DNA probe made. Digoxigenin-UTP is incorporated into a complementary DNA probe (via random priming). The labeled probe is next spotted onto a positively charged membrane and exposed to antidigoxigenin antibodies that are conjugated to alkaline phosphatase. Thus, alkaline phosphatase becomes specifically associated with labeled probe via the antibody attraction and a colored spot is created by adding BCIP (5-bromo-4-chloro-3-indoyl phosphate). When BCIP is dephosphorylated, it is oxidized by a nitroblue tetrazolium salt that produces a colored precipitate whose intensity is proportional to the amount of incorporated DNA label.[6]

Chemiluminescent detection is the method of choice for quantitating genomic DNA fragments on a Southern blot (remember this is product literature). Following hybridization of the digoxigenin-labeled probe to blot-bound DNA, the blot is exposed to anti-digoxigenin antibodies as above. It is then exposed to a substrate (Lumi-phos 530) that produces light by phosphorescence, which is recorded by exposure to film or imaging plates.

References:

1. Walker, J. M. (1994) The bicinchoninic acid (BCA) assay for protein quantitation. *Methods* Mol. Biol. 32: 5-8.
2. Joyce, C. M. and Grindley, N. D. F. (1983) Construction of a plasmid that overproduces the large proteolytic fragment (Klenow fragment) of DNA polymerase I of *Escherichia coli*. Proc. Natl. Acad. Sci. U.S.A. 80: 1830-1834.
3. Feinberg, A. P. and Vogelstein, B. (1983) A technique for radiolabeling DNA restriction endonuclease framents to high specific activity. Anal. Biochem. 132: 6-13.
4. Southern E. M. (1975) Detection of specific sequences among DNA fragments separated by gel electrophoresis. J. Mol. Biol. 98: 503-517.
5. McGadley U. (1970) A tetrazolium method for non-specific alkaline phosphatase. Histochemie 23: 180-184.
6. Bronstein I., Edwards B. and Voyta J. C. (1989) 1,2-Dioxetanees; novel chemiluminescent substrates: Applications to immunoassays. *J.* Biolumin. Chemilumin. 4: 99-111.

INNOVATION/INSIGHT

REDUCTION OF BACKGROUND PROBLEMS IN NONRADIOACTIVE NORTHERN AND SOUTHERN BLOT ANALYSES ENABLES HIGHER SENSITIVITY THAN ^{32}P-BASED HYBRIDIZATIONS

Engler-Blum, G., Meier, M., Frank, J. and Muller, G. A. (1993) Analytical Biochemistry, 210, p 235-244. (Adapted)

1. Introduction

An improved chemiluminescence-based RNA or DNA detection procedure, offering a widely applicable alternative to the conventional ^{32}P labeling employed in molecular biology, is described. Even highly sensitive applications such as Northern blot analysis of low-copy RNAs are shown to be feasible now without radioactive labeling. Improved quality of nonradioactive detection was obtained by the use of digoxigenin-labeled nucleotide in combination with dioxetane substrates, which are decomposed by the hydrolysis of alkaline phosphatase. Previously existing problems involving unacceptably high background signals in nonradioactive labeling procedures were eliminated by the application of a modified RNA/DNA transfer, hybridization, and detection protocol. The data presented here delineate a system consistently superior to radioactivity and should considerably increase the usefulness of nonradioactively-labeled probes detected by chemiluminescence. [*Note*: Portions of this article have been deleted in adaptation.]

Currently, nucleic acid hybridizations are most frequently performed by radioactive detection methods. Particularly, the ^{32}P labeling of nucleic acid probes remains the preferred technique for the detection and quantification of specific RNA or DNA sequences at low concentrations. Different nonradioactive methods have been introduced as possible alternatives to radioactive hybridizations assays, using probes labeled with fluorescent agents, enzymes, fluorescein, biotin or digoxigenin (DIG).[a,1-9] Besides their potentially greater sensitivity and stability, nonradioactive systems lack the health, hazards, cost, and disposal problems of radioactive detection, implying that radiolabeled probes could be replaced by nonradiolabeled counterparts as soon as practical protocols are available.

Maximum sensitivity in RNA/DNA detection is frequently achieved with digoxigenin- or biotin-labeled probes in combination with secondary enzyme-labeled antibody or avidin complexes and chemiluminescent substrates.[10] Enzymatic conversion of chromogenic or chemiluminescent substrates is already widely used in detection systems for nonradioactive hybridization assays because (by their turnover) enzymes provide an inherent amplification mechanism. Most enzyme-based DNA detection assays utilize *alkaline phosphatase* as the preferred label, due to its unusual resistance to environmental conditions and *thermal stability*.[11-14] Combined with ultrasensitive luminescent detection methods, sensitivity can be further improved. Recently described new chemiluminescent substrates, such as the dioxetane-based AMPPD or CSPD, are a further development toward a sensitive and fast nonradioactive alternative to isotopic labeling.[15,16]

Nevertheless, sensitivity is determined by the ability to detect signals above background. Nonradioactive labeling and detection systems available today are limited in their sensitivity by the nonspecific background, mainly caused by the indirect detection steps involved. For this reason, nonradioactive methods have not been readily applicable in all fields of molecular biology, especially in Northern blot analysis of low-copy RNA, where *high sensitivity* and *longer exposure times* are of critical importance.

Here we describe conditions for nonradioactive RNA/DNA analysis without the problem of limited sensitivity due to unacceptably high background signals. The method detailed below includes digoxigenin-labeling techniques combined with indirect detection of labeled probes via alkaline phosphatase conjugates and ultrasensitive chemiluminescent substrates together with an appropriate transfer, hybridization, and detection protocol. Variations of different parameters with respect to their influence on sensitivity and quality of the background are described. Sensitive applications such as the frequently used Northern blot analysis of rare mRNAs and detection of single-copy genes in Southern blot analysis are demonstrated.

2. Materials and Methods

2.1 Reagents. The DIG-oligonucleotide 3' end labeling kit, the DIG-oligonucleotide tailing kit, the DIG-DNA labeling and detection kit, blocking reagent (No. 1096176), anti-DIG-AP:F_{ab} fragments, and AMPPD were obtained from Boehringer-Mannheim. The *new chemiluminescent substrate CSPD* was obtained from Serva/Tropix. [α-^{32}P]dCTP (3000 Ci/mmol) and the Multiprime labeling system for standard labeling of cDNA probes as well as Hyperfilm MP were purchased from Amersham Buchler (Braunschweig, Germany). *Nylon membranes* were from Amersham Buchler (Hybond N$^+$

and Hybond N), Boehringer-Mannheim (positively charged nylon membrane) Serva/Tropix (Heidelberg, Germany; Tropilon), GIBCO Karlsruhe (Photo Gene), Stratagene (Heidelberg, Germany; Duralon-UV), Pall (Biodyne A), Schleicher & Schull, Dassel (N413, BA83), Satorius (Gottingen, Germany; Sartolon) and Millipore (Bedford, MA; Immobilon).

A specific 30mer GAPDH oligonucleotide was synthesized by phosphoramidite chemistry using a DNA synthesizer (Gene Assembler from Pharmacia, Freiburg, Germany). Following deprotection, the crude oligomer was desalted by Sephadex G50 chromatography (NAP-10 columns from Pharmacia) and assayed by UV.

The sequence for a specific 30mer GAPDH oligonucleotide probe was derived from a 5' coding region, using recently published sequence data.[17] A coding region was chosen which showed no obvious homology with other mRNAs recorded in the European Molecular Biology Laboratory sequence database.

2.2 Probe Labeling. Either digoxigenin-11-dUTP from Boehringer-Mannheim or [α-^{32}P]dCTP from Amersham Buchler was used for the random primed labeling reactions of cDNA probes. The labeling reactions were carried out according to the manufacturer's instructions. The 3' end labeling of 30mer oligonucleotides was done by the DIG-oligonucleotide 3'-OH end labeling kit or the DIG-oligonucleotide tailing kit from Boehringer-Mannheim.

2.3 Preparation of Northern and Southern Blots. Total RNA for Northern blots was isolated from cultured normal epithelial cells derived from human kidneys. Cells were lysed in guanidine isothiocyanate as described earlier.[19] Normally, 20 μg of total RNA was loaded on each slot and separated by electrophoresis through 1% agarose formaldehyde gels. RNA was blotted after a serial wash of the gel in distilled water and 20x SSC. The transfer of RNA was carried out either by alkali transfer or by capillary blotting with 20x SSC for 15-17 hours. RNA was then fixed to membranes by different fixation methods as described under Results.

*Bgl*II-digested genomic DNA isolated from transfected mouse fibroblasts was separated on a 1% agarose gel and transferred to membrane by capillary blotting as described earlier.[19] The DNA was covalently crosslinked to the membrane in the same way as RNA.

2.4 Hybridization of DIG-Labeled cDNA Probes and Oligonucleotides. The transferred or dotted nucleic acids were uv crosslinked and/or fixed by baking at 80°C. Prehybridization was performed in hybridization solution with 20 ml/100 cm^2 of membrane for 1 hour without the labeled probe. Prehybridization and hybridization temperature was 42°C when formamide was added to the solutions at 68°C without formamide. The compositions of the various prehybridization and hybridization solutions tested so far are described in Table 1. The concentration of labeled probes in hybridization solutions was 2.5 ng/ ml for cDNAs (or 10 pmol/ml of 30mer oligonucleotides). For a membrane size of 100 cm^2 a volume of 10 ml hybridization solution was applied. Hybridization was carried out for 1 hour at 60°C with oligonucleotides and for 16 hours at 68°C with cDNA probes. After hybridization, the membranes were washed 3 times at 65/58°C with 50 ml of washing buffer. In association with SSC hybridizations, washing buffer A (0.1x SSC, 0.1% SDS) was used, whereas in combination with Church and Gilbert hybridizations, washing buffer B (20 mM Na_2HPO_4, 1 mM EDTA, 1% SDS) was applied. Subsequently, the membranes were used directly for the immunological detection.

Table 1 - Modifications of the SCC Hybridization Protocol Proposed by Boehringer-Mannheim.

Modification	1	2	3	4	5
SSC	5x	5x	5x	5x	5x
N-Laurylsarcosine	0.1	0.1	0.1	0.1	0.1
Blocking reagent (%)	5	1	1	1	-
Denhardt's SDS (%)	7	7	7	0.02	0.5
Formamide (%)	50	50	-	-	50
Hybridization temperature (°C)	2	42	68	68	42

2.5 Immunological Detection of DIG-Labeled Nucleotides. Because the detection procedures proposed by the manufacturers were found to be unsatisfactory, variations of the luminescent detection protocol (Boehringer-Mannheim) were performed as described under Results. During the immunological detection, salt concentrations and pH values were increased. The salt concentration of the maleic acid-based washing buffer 1, blocking buffer 2, and conjugate buffer 3 was increased from 0.15 to 3 M NaCl. The pH of these buffers was also increased to pH 8 and the concentration of blocking reagent (from Boehringer-Mannheim) was decreased from 1 to 0.5%. Anti-DIG:AP conjugates were diluted 1:15,000 in blocking buffer 2. The same alterations were performed for the PBS-based detection system proposed by Tropix. Substrate buffers and concentrations (1:100) were not altered. The documentation of chemiluminescent signals was carried out by sealing the wet membrane in a plastic bag and placing it in close contact with the Amersham Hyperfilm MP for various exposure times. For *reprobing the blots*, the probes were removed from membranes according to the following protocol: after a washing step for 5 minutes in distilled water at room temperature, the membrane was equilibrated in 5x SSC for 20 minutes followed by a 2 minute stripping of the probe by an incubation in 0.1% SDS at 95/75°C.

2.6 Radioactive Hybridization. The cDNA probe for GHPDH was radiolabeled with [^{32}P]dCTP by primer extension using random hexanucleotides. Hybridizations were performed as described previously.[18] Filters were preincubated for 2 hours at 68°C in a solution containing 0.25 M Na_2HPO_4, 1% BSA, 1 mM EDTA, and 7% SDS. The hybridization was carried out for 16-20 hours at 68°C in the same solution containing 2.5 ng/ml of labeled cDNA. After hybridization the *filters were washed at the final stringency* of 20 mM phosphate buffer, 1 mM EDTA, and 1% SDS at 68°C. Filters were exposed to an Amersham Hyperfilm MP at -70°C for up to 5 days depending on the level of expression.

3. Results

The aim of these studies was the development of an improved protocol for nonradioactive Northern hybridization in combination with chemiluminescent detection. Northern hybridizations often require longer exposure periods which lead to nonspecific background and ultimately limits the sensitivity of chemiluminescent assays. For this reason, almost every step in this procedure was optimized to achieve maximum sensitivity and a significantly lower background.

3.1 Optimization of Northern Blotting and Hybridization. Before capillary elution of RNA from gels to nylon membranes, the formaldehyde-containing gels were rinsed several times in sterile redistilled water and 20x SSC to remove the *formaldehyde*, which may cause an increase in unspecific background. In addition, alkali blotting without any further fixation was compared to RNA transfer with neutral transfer buffer (20x SSC). Capillary blotting in a *neutral transfer buffer*, in combination with an appropriate fixation method, resulted in enhanced sensitivity, whereas background was not affected by these two different transfer buffers. The fixation methods selected for this study were (1) UV crosslinking with a Biometra Fluo-Linker and/or (2) baking at 80°C. Various UV doses were tested to obtain optimal sensitivity. Maximal sensitivity was achieved with an *irradiation* of 0.12 J/cm^2 on positively charged or neutral nylon membranes. *Baking* at 80°C for 30 minutes showed a slightly reduced sensitivity compared with optimized uv crosslinking. The combination of UV crosslinking and baking revealed high sensitivity but resulted in no further enhancement of the signal (data not shown).

For testing the influence of different membranes on sensitivity and background, unlabeled target DNA or RNA was applied in a dot-blot format to charged and uncharged nylon membranes. Best results were obtained by the use of the positively charged nylon membranes from Boehringer-Mannheim and Amersham (Hybond N$^+$). *No signal* could be detected on uncharged nylon membrane Hybond N for either RNA or DNA, presumably due to a bad lot, whereas other neutral membranes such as Tropilon were also compatible with the protocol (data not shown). However, uncharged membranes in general are not recommended because the quality, especially for detection of RNA, varied from lot to lot.

To achieve a further improvement of the signal-to-noise ratio, variations of different hybridization protocols were performed as given in Tables 1 and 2. The combination and concentrations of the components were checked precisely; *optimum results* were obtained when a *modified hybridization protocol of Church and Gilbert* was used (Table 2). The blocking reagent in this buffer as well as the SDS concentration were important factors in reducing nonspecific background. The concentration of the blocking reagent, as well as its composition, seemed to be critical, since the use of blocking reagent (e.g., from Boehringer) resulted in reduced background compared to the commonly used BSA. For this reason, blocking and SDS concentrations were checked carefully (Table 2).

3.1.1 Solutions:

Prehybridization solution: 0.25 M Na_2HPO_4, (pH 7.2), 1 mM EDTA, 20% SDS (for cDNA probes) or 10% SDS (for oligonucleotides) 0.5% blocking reagent.

Hybridization solution: prehybridization

Table 2 - Modifications of the Hybridization Protocol Described by Church and Gilbert[18]

Modification	1	2	3	4	5	7	8	9	10
Na_2HPO_4, (pH 7.2) (M)	0.25	0.25	0.25	0.25	0.25	0.25	0.25	0.25	0.25
EDTA (mM)	1	1	1	1	1	7	20	20	20
SDS (%)	7	7	7	1	10	7	20	20	20
Blocking reagent	1% BSA[a]	1% BBR[a]	5% BBR	5% BBR	1% BBR	0.5% BBR	0.5% BBR	0.25 % BBR	0.5% BBR
Hybridization temperature (°C)	68	68	68	68	68	68	68	68	72

[a]BSA, bovine serum albumin; BBR, Boehringer blocking reagent.

solution + 2.5 ng cDNA probe/ml (or 10 pmol DIG-tailed oligo probe/ml).

Washing buffer: 20 mM Na_2HPO_4, 1 mM EDTA, 1% SDS.

After an appropriate transfer and fixation of RNA (DNA) on positively charged nylon membranes, hybridization is performed according to modification 8 described in Table 2.

1. Incubate membrane for 60 minutes in prewarmed prehybridization solution at 68°C with 20 ml/100 cm^2 of membrane.
2. Replace the solution with 10 ml prewarmed hybridization solution containing 25 ng DIG-labeled cDNA probe.
3. Incubate the filters for at least 6 hours at 68/60°C and redistribute the solutions occasionally.
4. Wash the filters 3 x 20 minutes at 65/58°C with at least 50 ml of prewarmed wash buffer. Filters can now be used directly for detection. Do not allow the filters to dry out during the entire procedure.

3.1.2 Detection:

Buffers:

Washing buffer: 0.1 M maleic acid, 3 M NaCl, 0.3% Tween 20 (pH 8).

Blocking buffer 2: washing buffer + 0.5% blocking reagent.

Conjugate buffer 3: blocking buffer + anti-DIG-AP conjugate (1:15,000).

Substrate buffer 4: 0.1 M Tris-HCl, 0.1 M NaCl, 50 mM $MgCl_2$ (pH 9.5).

Substrate solution 5: substrate buffer 4 + 0.24 mM AMPPD or CSPD.

1. Wash membranes briefly (5 minutes) in buffer 1.
2. Incubate for 60 minutes in blocking buffer 2.
3. Dilute anti-DIG-AP conjugate 1:15,000 in blocking buffer 2.
4. Incubate membranes for 30 minutes with about 10 ml of diluted antibody-conjugate solution.
5. Remove unbound antibody-conjugate by washing 4 x 10 minutes with 50 ml of washing buffer 1.
6. Equilibrate membranes for 5 minutes in substrate buffer 4.
7. Dilute AMPPD (or CSPD) 1:100 in substrate buffer 4.
8. Incubate filter for 5 minutes in approximately 10 ml substrate solution 5.
9. Let excess liquid drip off the membrane and seal the membrane in a hybridization bag.
10. Expose the membrane for various times to X-ray film, depending on the signal intensity.
11. For reprobing, the membranes must be kept wet.

Best results were obtained with *0.5% blocking reagent* and *20% SDS*. Furthermore, the use of Na_2HPO_4 and EDTA in hybridization solutions, as described by Church and Gilbert for hybridization,[18] was superior to the SSC hybridization (Table 1). The often-suggested *formamide* led to a further deterioration of the background.

3.2 Optimization of Immunological Detection. The indirect immunological detection of digoxigenin-labeled probes by the F_{ab} fragments of a highly specific polyclonal sheep antibody conjugated with alkaline phosphatase[20] can cause additional background signals by nonspecific binding of the F_{ab}-AP conjugates to the membrane. To achieve a further suppression of background signals, immunological detection of DIG-labeled probes was optimized.

Detection was first carried out as described by the manufacturers. The maleic acid-based detection buffers (according to the Boehringer protocol) were compared to the phosphate buffer system suggested by Tropix for the use of the dioxetane substrates AMPPD and CSPD. Although the results were dependent on the membrane type used, the maleic acid based protocol was superior to most membranes tested.

To obtain a further reduction of nonspecific background signals, variations of the immunological detection method were performed. Antibody-antigen interactions are equilibrium reactions, which are affected by temperature, pH, and solvent. An essential decrease in nonspecific binding of the Fab-AP conjugates was obtained by a strong shift of salt concentration and pH values. Best results were obtained with an additional supplementation of the blocking, conjugate, and wash buffers with 3 M NaCl and a shift of the pH value to 8-10. The variation of the salt concentration and pH values also resulted in better RNA detection by DNA probe and oligonucleotides. A shift to pH 10 still gave a better background, but it seemed that the specific signal was negatively affected (data not shown).

Temperature shifts to 40°C and 50°C during the immunological detection and the following washing steps had no influence on background or specific signal detection.

Nonradioactive detection via biotin-labeled cDNA probes resulted in a lower sensitivity caused by strong background signals. Even in combination with the improved hybridization and detection procedure, the nonspecific noise could not be reduced as strikingly as in combination with DIG-labeled probes.

In a comparison of the two dioxetane substrates AMPPD and CSPD, over an extended time interval CSPD was not found to be superior despite the delayed light emission of AMPPD. *Enforced stringency during the detection step* by the use of 3 M NaCl, and a *shift of the pH value to 8 during the immunological detection* step, resulted in an improved signal-to-noise ratio. Even after a prolonged exposure period, there was a clear background on membrane A compared to membrane B, that was detected according to the manufacturer's instructions. Interestingly, RNA could only be visualized after an alteration of the protocol toward higher salt concentrations and pH values. The chemiluminescent substrate was AMPPD and the film was exposed after an overnight incubation of the membranes. During the first 4 hours of the enzymatic reaction, CSPD did not reveal a stronger signal than AMPPD. Even after overnight incubations of the membranes, there was no advantage of CSPD compared to AMPPD. Film exposures of membranes detected with AMPPD revealed a higher sensitivity.

3.3 Sensitive Applications. To verify the applicability of this protocol, sensitive Northern hybridizations were performed with probes specific for rare mRNA, such as the cytokines Il-6 and PDGF, which are weakly expressed by epithelial cells. Total RNA of unstimulated renal epithelial cells was separated, blotted, and hybridized as described above. Even for these hybridizations, a good signal-to-noise ratio was obtained within short exposure times.

A further application that requires maximum sensitivity was tested, by using differently labeled oligonucleotides for Northern blot hybridizations instead of random-primed labeled cDNA probes. *By normal end labeling reactions, only one digoxigenin molecule can be introduced into one oligonucleotide.* However, a more effective labeling of oligonucleotides is attainable by an *enzymatic tailing of the 3'-OH end.*[20] To achieve a direct control comparable to the random-primed labeled GAPDH cDNA probe, a specific 30mer oligonucleotide was chosen from the GAPDH sequence[17] and labeled either by the 3' tailing reaction or by 3' end labeling. Excellent results could be obtained even after prolonged exposure times. As expected, the tailing reaction was the more effective labeling procedure.

The applicability of the improved protocol to Southern analysis was demonstrated by the detection of a single-copy gene. On genomic Southern blots of HLA class 1-positive and -negative cell lines, the gene could be detected in 10 μg of total genomic DNA by a 1400-bp cDNA probe specific for HLA class 1 antigens.

3.4 Sensitivity of Nonradioactive Labeling and Radioactive Labeling. The quality of the optimized protocol of nonradioactive RNA detection was compared to the traditionally used ^{32}P Northern hybridization as described by Church and Gilbert.[18] As a control, a 1.2-kb cDNA probe for GAPDH was used. GAPDH is a key enzyme in the control of glycolysis, which is normally applied to the quantification of other mRNA species. The intensity of the nonradioactive and radioactive labeled bands in exposures A and B are comparable, indicating a multiple increase in the speed of detection, although the nonradioactive detection has been performed during the less sensitive starting kinetics of light emission. A faster detection can be performed after overnight incubations.

To compare maximal sensitivity of the elaborated nonradioactive Northern hybridization with our standard radioactive technique (described under Materials and Methods), various dilution series of total RNA were separated by gel electrophoresis and subsequently subjected to Northern hybridization. Application of DIG labeling and chemiluminescence enabled detection of mRNA specific for GAPDH down to *500 ng of total RNA* within 10-20 minutes when the steady state of light emission was reached. Within 5-16 hours the mRNA of GAPDH was detectable down to 50 ng of total RNA. The signals of a 5-day autoradiography of a ^{32}P labeling from the identical blot were weaker than short-term exposures (<20 minutes) of the nonradioactive detections and allowed recording of mRNA specific for GAPDH only down to 500 ng. With the dot-blot format, mRNA of GAPDH could be detected down to *17 ng of total RNA* with DIG-labeled cDNA probes, whereas homologous DNA was visible with dot blots down to 3 fg. When fresh hybridization and substrate solutions were used, a detection down to 1.5 fg of homologous DNA was possible.

4. Discussion

In this paper, we present the development of an optimized chemiluminescent procedure with increased sensitivity and its adaptation to frequently used molecular biological techniques, which require excellent signal-to-noise ratios. Based on a new totally revised hybridization and detection protocol, a simple, rapid, and sensitive chemiluminescent detection system was developed that is now also applicable to frequently used Northern hybridizations as well as to Southern analysis of single-copy genes. High sensitivity of the assays was obtained irrespectively of whether low abundant RNA or DNA species were detected by cloned DNA probes with a length of 300-2000 bp or by short synthetic oligonucleotide probes (30mer). In comparison to radioactive labeling techniques, chemiluminescent detection in combination with digoxigenin-labeled probes allowed a much more rapid detection of nucleic acids. Besides an accelerated speed of detection, a greater sensitivity is the major criterion for the acceptance of this improved method.

Chemiluminescence has been developed as an alternative detection method to radioisotopes in the differentiation of nucleic acids and proteins.[15, 16, 21-23] In combination with nonradioactively labeled

DNA/RNA probes or antibodies and indirect detection via alkaline phosphatase conjugates, chemiluminescence provided efficient and versatile nonradioactive test systems. Its major limitation, however, is a loss of sensitivity caused by nonspecific background in the course of longer exposure periods.

Nonradioactive detection of proteins or nucleic acids via digoxigenin or biotin-labeled probes are the alternatives to radioactive detection assays most often employed. In comparison to radioactive labeling, a major drawback of the nonradioactive techniques is still the additional detection steps, which lead to low signal-to-noise ratios. The biotin:streptavidin system, which depends on the naturally occurring high avidity of streptavidin to biotin,[1,2] not only causes nonspecific reactions with endogenous biotin, but also shows a marked adherence of streptavidin even on blocked membranes. A second often described nonradioactive detection system is based on the labeling of nucleic acids with DIG, a hapten that is detected by a highly specific antibody.[20] Although lacking problems of endogenous cross-reactions, this system is still limited in its applicability on Northern blots due to unacceptable background signals, especially after longer exposure times.

Starting with total cellular RNA on dot blots and on Northern blots, together with a DIG-labeled cDNA probe specific for GAPDH, a human single-copy gene, parameters of hybridization and detection were improved.

Reduction of background signals was achieved with a modified hybridization protocol, first described by Church and Gilbert.[18] For this analysis, the use of 0.5% blocking reagent instead of 1% BSA and an increase in SDS concentration to 20%, as well as a hybridization temperature of 68°C, can be recommended as the most efficient modifications for suppression of background signals and increasing sensitivity. Moreover, the effectiveness was further improved by the development of an appropriate detection procedure. Highest stringency during hybridization, but also during the immunological detection of the DIG-labeled probes, was applied to obtain the best sensitivity. Optimum results were obtained after changes in salt concentrations and pH values toward unphysiological conditions and the addition of blocking reagent at a concentration of 0.5%.

The increase of salt concentrations seems to be the most critical point because *RNA-DNA or DNA-DNA hybrids are probably stabilized by monovalent cations during the detection.* The *labeled probes are* therefore *protected from a washout* during the washing steps of detection, resulting in higher sensitivity and an appearance of RNA signals that could not be detected by previous protocols. A second synergistic effect may be the depression of nonspecific anti-DIG-AP, binding also forced by higher salt concentrations and a shift to higher pH values, with the result of a clearer background, enabling longer exposure periods to film.

Signal intensity, moreover, was dependent on the membranes used as well as on transfer and fixation techniques. In contrast to recently published data,[10] our results show superiority of charged nylon membranes over uncharged membranes because of their higher binding capacity for nucleic acids. Thus, interfering background signals no longer limited use of charged nylon membranes. Best results were obtained by SSC transfer combined with UV crosslinking. However, a further increase in sensitivity may be achieved by utilizing other transfer buffers such as 1 M NH_4Ac.[24]

Compared with radiolabeling, exposure times could be greatly reduced from 5 days down to a few minutes using the chemiluminescent substrates AMPPD or CSPD, without appearance of interfering background. In addition to the acceleration of detection and a more than 10-fold increase of sensitivity, the applicability for quantification of mRNA is demonstrated. The signal from CSPD as well as from AMPPD is extremely well localized, with a resolution comparable to or better than that from ^{32}P. Hybridization of homologous DNA with a 1.2 kb cDNA probe enabled the detection of 0.04 to 0.004 amol of complementary DNA, indicating a sensitivity of about 2300 DNA copies.

By this optimized nonradioactive procedure, single-copy genes and rare mRNA species, such as cytokines in total cellular RNA extracts of kidney epithelial cells, have been detected very efficiently. In contrast to previously published instructions, which strongly recommend RNA detection probes because of their higher binding capacity, it is now also possible to apply commonly used cDNA probes as well as 3' end labeled oligonucleotides. According to a recent publication, a novel, more sensitive DIG labeling of oligonucleotides could be performed, with an incorporation of DIG-modified dUTP molecules into a tail by the enzyme-terminal transferase.[20] The tailing of oligonucleotides resulted in a stronger signal compared to normal 3'-OH end labeling reactions and therefore presents an efficient method to detect mRNAs on Northern blots by the use of oligonucleotides which can be easily synthesized.

In the continued development of chemiluminescent substrates, CSPD has been recently introduced as another dioxetane-based enzyme substrate with an improved sensitivity.[15,16] Although CSPD could be applied very efficiently with our buffer system in the detection of nucleic acids, this new substrate was not superior to AMPPD in our applications.

Thus, a nonradioactive protocol for the detection of membrane-bound nucleic acids was established in this study which meets most requirements of very high sensitivity and rapid performance and therefore offers new means for replacement of radioactivity in molecular biological test systems.

[a]To whom correspondence should be addressed.

[b]Abbreviations used: AMPPD, disodium 3-(4-methoxyspiro-[1,2-dioxetane-3-2'-tricyclo [3.3.1.1^{3,7}]-decane]-4-yl) phenyl phosphate; CSPD, disodium 3-(4-methoxyspiro-[1,2-dioxetane-3-2'-(5'-chloro) tricyclo [3.3.1.1^{3,7}] decane]-4-yl) phenyl phosphate; AP, alkaline phosphatase; BSA, bovine serum albumin; BBR, blocking reagent from Boehringer; DIG, digoxigenin; EDTA, ethylenediaminetetraacetic acid; GAPDH; glyceraldehyde 3-phosphate dehydrogenase; IL-6, interleukin 6; PDGF, platelet-derived growth factor; SSC, standard saline citrate; SDS, sodium dodecyl sulfate; PBS, phosphate-buffered saline.

References:

1. Studencki, A. B., and Wallace, R. B. (1984) DNA 3, 7-15.
2. Green, N. M. (1975) Adv. Protein Chem. 29, 85-133.
3. Matthews, J. A., and Kricka, L. J. (1988) Anal. Biochem. 169, 1-25.
4. Wilchek, M., and Bayer, E. A. (1988) Anal. Biochem. 171, 1-32.
5. Leary J. J., Brigate, D. J., and Wa, D. C. (1983) Proc. Natl. Acad. Sci. USA 80, 4045-4049.
6. Seibl, R., Hoeltke, J., Burg, J., Muhlegger, K., Mattes, R., and Kessler, C. (1988) Fresenius' Z. Anal. Chem. 330, 305.
7. Dooley, S., Radtke, J., Blin, N., and Unteregger, G. (1988) Nucleic Acids Res. 16, 11839.
8. Lanzillo, J. J. (1990) BioTechniques 8, 621-622.
9. Kesster, C., Seibl, R., Ruger, R., and Sagner, G. (1989) J. Cell. Biochem. 13E, 292.
10. Lanzillo, J. J. (1991) Anal. Biochem. 194, 45-53.
11. Garen, A., and Levinthal, C. (1960) Biochem. Biophys. Acta 38, 470-483.
12. Schlesinger, M. J. (1965) J. Biol. Chem. 240, 4293-4298.
13. Schlesinger, M. J., and Barret, K. (1965) J. Biol. Chem. 240, 4284-4292.
14. Yeh, M. F., and Trela, J. M. (1976) J. Biol. Chem. 251, 3134-3139.
15. Martin, C., Bresnick, L., Juo, R.R., Voyta, J. C., and Bronstein, I. (1991) BioTechniques 11(1), 110-113.
16. Tizard, R., Cste, R. L. Ramachandran, U. L., Wysk, M., Voyta, J. C., Murphy, O. J., and Bronstein, I. (1990) Proc. Natl Acad. Sci. USA 87, 4514-4518.
17. Tokunada, K., Nakamura, Y., Sakata, K., Fujimori, K., Ohkubo, M., Sawada, K., and Sakiyama, S. (1987) Cancer Res. 47, 5616-5619.
18. Church, G. M., and Gilbert, W. (1984) Proc. Natl. Acad. Sci. USA 81, 1991-1995.
19. Sambrook, J., Fritsch, E. F., and Maniatis, T. (1989) *Molecular Cloning–A Laboratory Manual*, Cold Spring Harbor Laboratory Press, Cold Spring Harbor, NY.
20. Schmitz, G. G., Walter, T., Seibl, R., and Kessler, C. (1991) Anal. Biochem. 192, 222-231.
21. Bronstein, I., Edwards, B., and Voyta, J. C. (1988) J. Biolumin. Chemilumin. 2, 186.
22. Schaap, A. P., Sandison, M. D., and Handley, R. S. (1987) Tetrahedron Lett. 28, 1159-1162.
23. Edwards, B., Sparks, A., Voyt, J. C., and Bronstein, I. (1989) J. Biolumin. Chemilumin. 5, 1-4.
24. Allefs, J. J. H. M., Salentijn, E. M. J., Krens, F. A., and Rouwendal, G. J. A. (1990) Nucleic Acids Res. 18(10), 3099-3100.

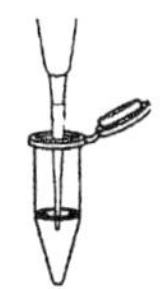

LAB 6.1: LABELING OF DNA AND PROBE CONSTRUCTION FROM CLONED *C. ELEGANS MYO*-3 GENE; QUANTITATION OF DNA CONCENTRATION

Overview:

In this lab, you will construct a digoxygenin-labeled DNA probe from our cloned *C. elegans myo*-3 DNA gene. You will then quantitate the concentration of the synthesized probe by "dot blot." This procedure involves immobilizing both your labeled probe and substrates of known concentrations on a membrane and then detecting them with the appropriate antibodies and colored substrate. The color intensities are then used to estimate the approximate probe concentration by comparison with standards.

In lab 6.3, this probe will be hybridized to the *C. elegans* genome to study the number and distribution of binding sites. The rate of formation of double stranded hybrids depends on the concentration of the two single-stranded species. The concentration and number of binding sites will be measured by hybridization with this labeled DNA probe (i.e., a DNA fragment that is complementary to the nucleic acid being assayed). This probe must be labeled in order to detect the formation of hybrids between it and the target nucleic acid. Many times, labeling involves incorporating a radioisotope; however, nonradioactive chemical labels are often preferred. In this lab, you will label our DNA probe with digoxygenin (DIG), a steroid derivative, by using DIG-labeled nucleotides during synthesis. Hybrids containing the DIG-labeled probe can then be detected with anti-DIG antibodies linked to moieties that produce color upon addition of color forming substrate. After excess unbound probe is washed away, the amount of labeled probe indicates the concentration of target nucleic acid in the sample.

Experimental Outline:

1. *Myo*-3 probe construction.
2. Quantitation of *myo*-3 probe using dot-blot hybridization and control reactions as quantitative standards.

Required Materials:

DNA template; hexanucleotide mixture (10x); dNTP labeling mixture (containing DIG-labeled nucleotide); Klenow enzyme, labeling grade; water bath; dry ice or ethanol; labeled control DNA; anti-DIG-alkaline phosphatase; NBT solution; X-phosphatase solution; DNA dilution buffer; blocking reagent; washing buffer; blocking solution; detection buffer (all reagents available through Genius™ nucleic acid labeling system); TE buffer (recipe in Lab 1.1)

Procedures:

1. *Myo*-3 Probe Construction (Genius™)

1. Add the reagents to a sterile microfuge tube (on ice) in the following order:

DNA template	3 µl
Hexanucleotide mixture (10x)	2 µl
dNTP labeling mixture (10x)	2 µl
H_2O	12 µl
Klenow enzyme, labeling grade (100 units/ml)	+ 1 µl
Total volume	20 µl

NOTE: The DNA template should be heat-denatured in a boiling water bath for 10 minutes and quickly chilled on dry ice/ethanol for 30 seconds before use. We have found that denaturation using a heating block is less effective and may result in lowered labeling efficiency.

2. Incubate the reaction at 37°C for at least 60 minutes. Longer incubations (up to 20 hours) will increase the yield of DIG-labeled DNA.
   a. The amount of synthesized labeled DNA depends on the amount of template DNA in the labeling reaction and on the length of the incubation time at 37°C. See the following table.

Amount of template DNA per labeling reaction (ng)	10	30	100	300	1,000	3,000
Amount of synthesized (after 1 hour)	15	30	60	120	260	530
DIG-labeled DNA (ng) (after 20 hours)	50	20	260	500	780	890

3. Add 2 μl of EDTA to terminate the reaction.

2. Quantitation of *myo*-3 Probe

1. Mix 45 μl of NBT solution and 35 μl of X-phosphate solution in 10 ml of Detection Buffer. This freshly prepared Color Substrate Solution will be used in step 13.
2. Make serial dilutions of the labeled control DNA in DNA Dilution Buffer. See table below. Diluted DNAs can be stored at -20°C for at least 1 year. (Use of siliconized tubes helps prevent adsorption of DNA to walls of the tubes.)

Starting Concentration	Stepwise Dilution	Final Concentration (Dilution Name)	Total Dilution
5 ng/μl	2 μl DNA/8 μl TE buffer	1 ng/μl (A)	1:5
1 ng/μl (dilution A)	2 μl DNA/18 μl TE buffer	100 pg/μl (B)	1:50
100 pg/μl (dilution B)	2 μl DNA/8 μl TE buffer	10 pg/μl (C)	1:500
10 pg/μl (dilution C)	2 μl DNA/8 μl TE buffer	1 pg/μl (D)	1:5,000
1 pg/μl (dilution D)	2 μl DNA/8 μl TE buffer	0.1 pg/μl (E)	1:50,000

3. Spot 1 μl of each diluted DNAs onto a positively charged nylon membrane, marking the membrane lightly with a pencil to identify each dilution.
4. Make serial dilutions of the newly labeled experimental DNA probe (of unknown starting concentration) in DNA Dilution Buffer according to step 2 until a 1:50,000 dilution is made.
5. Spot 1 μl of each of the 1:50 to 1:50,000 dilutions made in step 4 onto the same nylon membrane, marking the membrane lightly with a pencil to identify each dilution.
6. Fix the DNAs to the membrane by cross-linking with UV light.
7. Wet the membrane with a small amount of Washing Buffer.
8. Incubate the membrane in Blocking Solution for 5 minutes at room temperature.
9. Dilute the anti-DIG-alkaline phosphatase 1:5,000 in Blocking Solution.
10. Incubate the membrane in the diluted anti-DIG-alkaline phosphatase antibody for 15 minutes at room temperature. The diluted antibody solution must cover the entire membrane.
11. Wash the membrane twice, 15 minutes per wash, in Washing Buffer at room temperature.
12. Incubate the membrane in Detection Buffer for 2 minutes. Application of this buffer "activates" the alkaline phosphatase that is conjugated to the antibody.
13. Pour off the Detection Buffer and add the Color Substrate Solution. Allow color development to occur in the dark for 30-60 minutes, monitoring occasionally. The 1:5,000 dilution (1 pg) of the DIG-labeled control DNA should be visible 30 minutes after adding the Color Substrate Solution. Since the incubations times are shortened in the interest of getting a quick assay result, sensitivity of this assay may not be great enough to visualize the 0.1 pg control spot.
14. Compare the spot intensities of the control and experimental dilutions to estimate the concentration of the experimental probe.

Notes:

1. The Color Substrate Solution contains Nitroblue tetrazolium salt (NBT) and dimethylformamide as its stabilizing solvent.

Additional Report Requirements:

1. Dot blot representation.
2. Probe concentration calculations.

References:

1. Genius™ Nucleic Acid Labeling System, Boehringer-Mannheim, Indianapolis, IN.

VENDOR LITERATURE

DIGOXIGENIN LABELING OF DNA: GENIUS™ NUCLEIC ACID LABELING SYSTEM

Boehringer-Mannheim, Indianapolis, IN, ph. (800) 262-1640. (Adapted)

1. Introduction

Scientists reluctant to incorporate nonradioactive labeling and detection into their research have held that nonradioactive products lacked sufficient sensitivity. However, recent technological advancements with chemiluminescent detection have allowed select products to meet or exceed the sensitivity levels achievable with ^{32}P-labeling and detection.

The Genius™ System with Lumi-Phos™ 530 from Boehringer Mannheim guarantees detection of 0.1 pg of target nucleic acid in less than one hour. In addition, it allows detection of single copy genes in human DNA in less than 30 minutes when used in Southern blotting techniques.

LIGHT

Unstable intermediate

The Genius products rely on the very same procedures already used for radioactive labeling and provide results days earlier. Results can be recorded on X-ray film and, because the chemiluminescence lasts more than 48 hours, multiple exposures are possible. Those films can be easily scanned and quantified by densitometry.

1.1 Southern Blotting. The accepted wait for accurate single-copy gene detection using ^{32}P labeled probes has always been one to three days. Identical results can now be achieved in 30 minutes with the Genius System and its chemiluminescent substrate. In addition, the probes are stable, allowing researchers to reuse them multiple times without relabeling. Furthermore, Southern blots developed with the Genius System can be repeatedly stripped and reprobed, using the same procedures as the older radioactive method.

1.2 Library Screening. Using probes labeled with the Genius System, researchers are detecting positive colonies and plaques within one hour versus the overnight exposures required with radioactive techniques.

Because the Genius System can use DNA, RNA or oligonucleotide probes, many research needs can be met with a single system. Probe labeling utilizes the same techniques currently employed to prepare radioactive probes, and multiple filters can be processed simultaneously.

1.3 Dot Blots. With exposure times reduced from overnight to 30-60 minutes, critical answers are obtained quickly.

Digoxigenin-labeled probes are stable for at least one year and can be reused multiple times, offering savings in terms of the time and money normally spent relabeling decayed radioactive probes. By using the same probe, experimental variables are reduced and reproducibility between experiments is provided.

The Genius System uses digoxigenin, a steroid hapten, to label RNA and DNA for subsequent detection.[1] DNA probes are *labeled* with DIG-dUTP via random primed labeling, nick translation, cDNA synthesis, or PCR (A). Oligonucleotide probes are 3'-end labeled with DIG-ddUTP, tailed with DIG-dUTP by terminal transferase, or labeled with digoxigenin-NHS ester. Alternatively, RNA is labeled with DIG-UTP by SP6, T3 or T7 RNA polymerase in an *in vitro* transcription reaction. The DIG-labeled probes are *hybridized* to a membrane-bound nucleic acid (A, B) through *Northern*, *Southern*, *dot* or *slot blotting*, or *colony plaque lift hybridization* procedures. These hybridized probes are immuno-detected with an *alkaline phosphatase-conjugated anti-digoxigenin antibody* (C) and then visualized with the *chemiluminescent substrate* Lumi-Phos 530 or with the *colorimetric substrates* NBT and X-Phosphate (D).

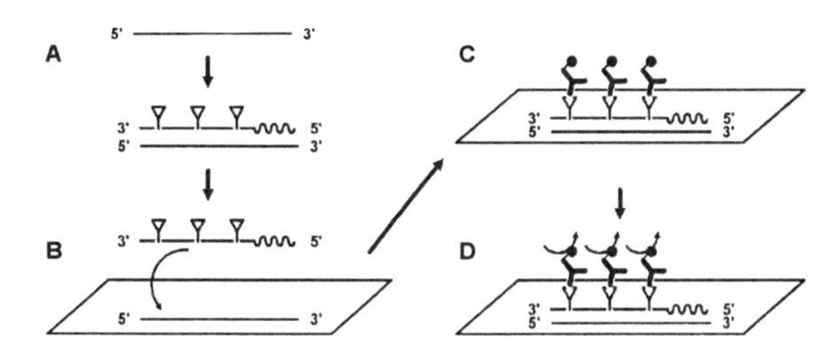

2. General Considerations for Labeling

2.1 Template Purity. In general, the higher the purity of the DNA template, the better the labeling efficiency. We routinely phenol/chloroform extract our DNA templates prior to the labeling reaction. In addition, for the random primed DNA labeling method, it is critical that you linearize and heat-denature the template prior to the labeling reaction. Oligonucleotides should be gel purified or HPLC purified prior to 3'-end labeling or 3'-tailing.

2.2 Labeling Procedures. When using the Genius System, DNA probes can be labeled by a number of methods. RNA probes are labeled by *in vitro* transcription. Oligonucleotides are labeled by 3'-end labeling, 3'-tailing, or labeling with DIG-NHS ester. Table 1 provides a guide to the different probe-labeling options, the sensitivity of the probes produced, and their major applications. Each of the labeling procedures previewed in Table 1 is described in detail in this User's Guide.

3. Kit Contents

1. Unlabeled Control λ DNA: One vial with 20 μl pBR328 DNA, 100 μg/ml. pBR328 is digested separately with *Bam*HI, *Bgl*I and *Hin*dI. The separate digests are combined in a ratio of 2:3:3. Sizes of the 16 pBR328 fragments: 4907, 2176, 1766, 1230, 1033, 653, 517, 453, 394, 298 (2x), 234 (2x), 220 and 154 (2x) bp.
2. Unlabeled Control DNA 2: One vial with 20 μl pBR328, 200 μg/ml, linearized with *Bam*HI.
3. DNA Dilution Buffer: Two vials with 1 ml herring sperm DNA, 50 μg/ml, in Tris-HCI, 10 mmol/l; EDTA, 1 mmol/l; pH 8 (20°C), each.
4. Labeled Control DNA: One vial with 50 μl linearized pBR328 DNA, labeled with digoxigenin according to the standard protocol containing 1 μg template DNA and approx. 260 ng synthesized labeled DNA.
5. Hexanucleotide Mixture: One vial with 50 μl of 10x concentrated hexanucleotide reaction mixture.
6. dNTP Labeling Mixture: One vial with 50 μl of 10x concentrated dNTP labeling mixture containing dATP 1 mM; dCTP, 1 mM; dGTP, 1 mM; dTTP 0.65 mM; DIG-dUTP, 0.95 mM; pH 7.5 (20°C).
7. Klenow Enzyme, labeling grade: One vial with 25 μl Klenow enzyme, labeling grade, 2 units/μl.
8. Anti-digoxigenin-AP conjugate: One vial with 200 μl polyclonal sheep anti-digoxigenin Fab-fragments, conjugated to alkaline phosphatase, 750 units/ml.
9. NBT: Two vials with 1.25 ml nitroblue tetrazolium salt, 75 mg/ml in dimethyl-formamide, 70% (v/v), each.
10. X-Phosphate: Two vials with 0.9 ml 5-bromo-4-chloro-3-indolyl phosphate, toluidinium salt, 50 mg/ml, in dimethylformamide, each.

Table 1 - Methods for Labeling Probes with Digoxigenin

Labeling Method	Probe Sensitivity	Application
Random primed DNA labeling	0.1 pg	• Northern blots Southern blots Library screening Dot/slot blots
Nick translation with DIG-dUTP	0.1-0.5 pg	• Northern blots Southern blots Library screening Dot/slot blots
3'-End labeling oligonucleotides with DIG-ddUTP	>10 pg	• *In situ* hybridization Northern blots Southern blots Library screening
3'-Tailing oligonucleotides with DIG-dUTP/dATP	>1 pg	• Northern blots Southern blots Library screening
5'-End labeling oligonucleotides with digoxigenin-NHS ester	>10 pg	• Library screening Dot/slot blots *In situ* hybridization Primer extension
Labeling RNA by *in vitro* transcription	0.1 pg	• Northern blots Southern blots *In situ* hybridization
Preparing cDNA with DIG-dUTP	0.1 pg	• Library screening Subtraction hybridization Northern blots Southern blots
Labeling DNA by PCR	0.2 pg	• Dot/slot blots Northern blots Southern blots Library screening
Labeling Nucleic Acids with Photodigoxigenin	>1 pg	• Labeling DNA molecular weight markers Labeling RNA molecular weight markers

11. Blocking Reagent*: Two bottles with 50 g powder each.

4. Stability

The components of the kit are stable at -20°C. Once opened, the antibody conjugate (vial 8) should be stored at 4°C. The blocking reagent (vial 11) can also be stored dry at 4°C or at room temperature. DIG-labeled DNA is stable for at least one year when stored in hybridization solution or TE-buffer (see below) at -20°C.

5. Procedure

Using the following protocol 0.1 pg of homologous DNA can be detected in a dot blot. We strongly recommend performing one reaction with the control reagents included in the kit for familiarization with this system.

5.1 Required Reagents.

- Nylon Membranes; positively charged, from Boehringer Mannheim (Cat. No. 1209272, 1209299, 1417240)
- Nitrocellulose (e.g., Schieicher & Schuell, BA85) or Nylon filter (e.g., Schleicher & Schuell, Nytran 13)
- Other: Ethanol, LiCl, NaCl, Tris, EDTA, $MgCl_2$, Na-citrate, SDS, N-Lauroylsarcosine (Na^+ salt), sterile redist. H_2O.

5.2 Required solutions.

- EDTA, 0.2 M; pH 8 (20°C)
- LiCl, 4 M
- Ethanol, 70% (v/v)
- *N*-Lauroylsarcosine, 10% (w/v)
- SDS, 10% (w/v)
- 20x SSC: NaCl, 3 M; Na-citrate, 0.3 M; pH 7
- TE buffer: Tris-HCl, 10 mM; EDTA, 1 M; pH 8 (20°C)
- Buffer 1: maleic acid, 0.1 M; NaCl, 0.15 M; pH 7.5 (20°C), adjusted with solid or concentrated NaOH, autoclaved
- Blocking stock solution: blocking reagent (vial 6) is dissolved in buffer 1 to a final concentration of 10% (w/v) with shaking and heating either on a heating block or in a microwave oven. This stock solution is autoclaved and stored at 4°C or -20°C
- 2x SSC; SDS: 0.1% (w/v); sterile.
- 0.1x SSC; SDS: 0.1% (w/v); sterile.
- Washing buffers:
  Buffer 1 + Tween-20™, 0.3% (w/v)
  Buffer 2: blocking stock solution, diluted 1:10 in buffer 1 (final concentration = 1% blocking reagent)
  Buffer 3: Tris-HCl, 100 mM; NaCl, 100 mM; $MgC1_2$, 50 mM; pH 9.5 (20°C).
  Buffer 4: Tris-HCl, 10 mM; EDTA, 1 mM; pH 8 (20°C).
- Color-substrate solution (freshly prepared): 45 µl NBT solution (vial 4) and 35 µl X-phosphate solution (vial 5) are added to 10 ml of buffer 3.

6. DNA Labeling

10 ng to 3 µg of linear DNA can be labeled per standard reaction. Larger amounts can be labeled by scaling up all components and volumes. Linear DNA is labeled more efficiently than circular and supercoiled DNA.

6.1 Control and Standard Assay.

1. We recommend purifying the linearized DNA by phenol:chloroform extraction and ethanol precipitation.
2. The DNA must be denatured by heating in a boiling water bath (100°C) for 10 minutes and chilling quickly on ice/NaCl. Complete denaturation is essential for efficient labeling.
3. Add the following to a microfuge tube on ice:
- 1 µg freshly denatured DNA, corres-ponding to 5 µl of control DNA (vial 2),
- 2 µl hexanucleotide mixture (vial 5),
- 2 µl dNTP labeling mixture (vial 6),
- make up to 19 µl with H_2O (the control reaction requires 10 µl H_2O), and add 1 µl Klenow enzyme (vial 7).
4. Centrifuge briefly and incubate for at least 60 minutes at 37°C. Longer incubation (up to 20 hours) can increase the amount of labeled DNA.
5. Stop the reaction by adding 2 µl EDTA solution, 0.2 M; pH 8.
6. Precipitate the labeled DNA with 2.5 µl LiCl, 4 M, and 75 µl prechilled (-20°C) ethanol. Mix well.
7. Leave for at least 30 minutes at -70°C or 2 hours at -20°C.
8. Centrifuge (at 12000 x g), wash the pellet with cold ethanol, 70% (v/v), dry under vacuum, and dissolve in 50 µl TE buffer. If the DNA has been labeled in the presence of low melting point agarose the ethanol precipitation must be replaced by gel filtration.

9. The amount of newly synthesized labeled DNA depends on the amount and purity of the template DNA. In the standard reaction with 1 µg DNA per assay, ca. 10% of the nucleotides are incorporated into about 250 ng of newly synthesized labeled DNA within 1 hour and approx. 30% of the nucleotides into about 750 ng after 20 hours. Reactions with smaller amounts of template DNA (e.g., 30 ng) results in a 1:1 ratio of labeled to unlabeled DNA (see Table 2).
10. The amount of labeled DNA generated can be approximated by comparison with the labeled control DNA (vial 4) in hybridization or direct detection. It is recommended to routinely check the labeling efficiency by direct detection.
11. If desired, the kinetics of the labeling reaction can be followed and the amount of newly synthesized labeled DNA determined by adding radioactively labeled tracer dNTP and trichloroacetic acid, or ethanol precipitation.

6.2 Hybridization. Hybridization buffer: 5x SSC; blocking reagent, 1% (w/v) (added from 10% sterile blocking stock solution); N-lauroylsarcosine, 0.1% (w/v); SDS, 0.02% (w/v). The hybridization buffer can be stored frozen at -20°C. One can also add formamide to 50% (v/v) to the hybridization buffer. In this case, the concentration of the blocking reagent must be increased to 2% (w/v). Hybridize with formamide at 42°C.

6.3 Control.

1. Prepare nitrocellulose membranes by presoaking in water and then 20x SSC. Nitrocellulose membranes must be dried before loading with DNA. Nylon membranes can be used without pretreatment.
2. Make a dilution series of the unlabeled control-DNA (vial 1) in dilution buffer (vial 3) from 10^{-4} (10 pg/µl) to 10^{-7} (10 fg/µl).

3a. Dot blot: Denature DNA by heating in a boiling water bath and chilling quickly on ice. Spot 1 µl of the diluted control or heterologous DNA (vial 3) onto the dry membrane.

3b. Southern blot: Load the diluted control DNA (e.g., 100 pg - 1 pg per lane) onto an agarose gel, separate the fragments and subsequently perform a Southern transfer[3] to membrane.

4. Bind the DNA to the nitrocellulose by baking for 2 hours in a vacuum at 80°C, and to nylon membranes either by baking for 1 hour (vacuum not required) or by UV crosslinking with a transillumination device for 3 minutes.[4]

NOTE: We recommend using Boehringer Mannheim Nylon Membranes, positively charged, to obtain optimal results. In this case, use the following protocol:

a. Fixation of nucleic acids to the membrane:
   1. Bake for 15-30 minutes at 120°C instead of 80°C, or
   2. Crosslink for 3 minutes with UV transilluminator. (It is also possible to use the Stratalinker from Stratagene according to manu-facturers instructions). UV-cross-linking is not recommended after alkaline transfer (see below).

b. Alkaline Transfer:

   Transfer DNA using NaOH, 0.4 M as transfer buffer from agarose-gels directly after electrophoresis without prior denaturation or neutralization steps. Fix DNA by baking at 120°C. After alkaline transfer, UV-crosslinking is less effective.

5. Prehybridize filters in a sealed plastic bag or box with at least 20 ml hybridization buffer per 100 cm^2 of filter at 68°C for at least one hour. Distribute the solution from time to time. Do not allow the filters to dry out between prehybridization and hybridization.
6. Replace the solution with about 2.5 ml per 100 cm^2 filter of hybridization buffer containing 5 µl (26 ng) of freshly denatured labeled control DNA (vial 4) per ml. The amount of labeled DNA in the hybridization solution (12.5 µl = 65 ng) is approx. 25% of the total labeled DNA obtained in our standard labeling assay. Very small filters may require slightly more than 2.5 ml per 100 cm^2 filter of hybridization solution.
7. Incubate the filters for at least 6 hours at 68°C, redistribute the solution occasionally. Higher DNA concentrations in the hybridization can be used to minimize the hybridization times down to approximately 2 hours.
8. Wash filters 2 times for 5 minutes at room temperature with at least 50 ml of 2x SSC; SDS, 0.1% (w/v), per 100 cm^2 filter and 2 x 15

Table 2 - The amount of Synthesized DIG Labeled DNA Depends on the Amount of Template DNA in the Labeling Reaction and on the Length of the Incubation Time at 37°C

Amount of Template	10 ng	30 ng	100 ng	300 ng	1000 ng	3000 ng
Amount of synthesized DIG-labeled DNA						
after 1 hour	15 ng	90 ng	60 ng	120 ng	260 ng	530 ng
after 20 hour	50 ng	120 ng	260 ng	500 ng	780 ng	890 ng

Table 3 - Sensitivity Depends on Both the Concentration of Labeled DNA and Incubation Time

	Concentration of labeled DNA (ng/ml).						
Sensitivity (pg) at:	0.5	2	5	10	20	30	50
1 hour	-	10	5	2	1	1	0.5
3 hours	10	2	1	0.5	0.5	0.5	0.2
16 hours	2	0.5	0.2	0.1	0.1	0.05	0.05

minutes at 68°C with 0.1x SSC; SDS, 0.1% (w/v).

9. Filters can then be used directly for detection of hybridized DNA or stored air-dried for later detection.

6.4 Standard Assay.

1. If necessary, prepare dilutions of the DNA to be probed and filters as described for the control reaction. The specificity of the reaction can be controlled, if necessary, by making similar dilutions of a heterologous DNA (vial 3).
2. Transfer the DNA (or RNA) to be probed to a nitrocellulose or nylon membrane by dot blot, plaque lift, colony hybridization, Southern or Northern transfer.
3. Label the DNA probe according to the standard assay procedure.
4. Prehybridize filters as described for the control reaction. Do not allow the filters to dry out between prehybridization and hybridization.
5. Replace the prehybridization solution with about 2.5 ml per 100 cm^2 filter of hybridization solution containing freshly denatured probe DNA. The optimal concentration of labeled DNA in the hybridization mixture depends on the amount of DNA to be detected on the filter. Usually 10-100 ng of labeled DNA (see Table 2, row 2 and Table 3) per ml hybridization solution is used. About 2.5 ml of hybridization solution per 100 cm^2 of filter; for very small filters, slightly more is required. (For hybridization to membrane-bound sample RNA, we recommend using hybridization solution containing form-amide.)
6. Hybridize and wash the filters as described for the control reaction.
7. The hybridization solution containing labeled DNA may be stored at -20°C and reused several times. Immediately before use, redenature the probe by heating the hybridization solution at 90°C for 10 minutes This step also redissolves any precipitate, which may have formed during storage.
8. The filters can then be used directly for detection of hybridized DNA or stored air-dried.

7. Immunological Detection

7.1 Control and Standard Assay. All of the following incubations are performed at room temperature and, except for the color reaction, with shaking or mixing. The volumes of the solutions are calculated for a 100 cm^2 filter-size and should be adjusted for other filter sizes.

1. Wash filters briefly (1 minute) in buffer 1.
2. Incubate for 30 minutes with about 100 ml of buffer 2.
3. Dilute antibody conjugate (vial 8) to 150 mU/ml (1:5000) in buffer 2. (Diluted antibody-conjugate solutions are stable for only about 12 hours at 4°C.)
4. Incubate filters for 30 minutes with about 20 ml of diluted antibody-conjugate solution.
5. Remove unbound antibody conjugate by washing 2 times for 15 minutes with 100 ml of buffer 1.
6. Equilibrate membrane for 2 minutes with 20 ml of buffer 3.
7. Incubate filter with ca. 10 ml of color solution, sealed in a plastic bag or in a suitable box, in the dark. The colored precipitate starts to form within a few minutes and the reaction is usually complete after 16 hours. Do not shake or mix while color is developing.
8. When the desired spots or bands are detected,

Table 4 - Products Required

Reagent name	Description	Available as
Hexanucleotide mixture (10x)	62.5 A_{260} units/ml random hexanucleotides, 500 mM Tris-HCl, 100 mM $MgCl_2$, 1 mM dithioerythritol (DTE)	1. Vial 5, Genius 1 2. Vial 5, Genius2 3. Hexanucleotide
2 mg/ml BSA, pH 7.2	Mixture	1. Cat. No. 1277 081
dNTP labeling mixture (10x)	1 mM dATP, 1 mM dCTP, 1 mM dGTP, 0.65 mM dTTP, 0.35 mM DIG-dUTP, pH 6.5	1. Vial 6. Genius 1 2. Vial 6, Genius 2 3. DIG DNA labeling 4. Cat. No. 1277 065
Klenow enzyme, labeling grade	2 units/μl DNA polymerase I (Klenow enzyme, large fragment), labeling grade from *E. coli*	1. Vial 7, Genius 1 2. Vial 7, Genius 2 3. Cat. No. 1008 404, 1008 412
glycogen solution	20 mg/ml glycogen in redistilled water	1. Cat. No. 901 393

Table 5 - Products Required

Name in Procedure	Description	Available as
Labeled control DNA	Digoxigenin-labeled pBR328 DNA that has been random primed labeled according to the standard labeling procedure; the total DNA concentration in the vial is 20 μg/ml, but only 5 μg/ml of it is DIG-labeled DNA	Vial 4, Genius 1 Vial 4, Genius 2 Vial 1, Genius 3
Anti-DIG-alkaline phosphatase	Antidigoxigenin (F_{ab}) conjugated to alkaline phosphatase	Vial 8, Genius 1 Vial 3, Genius 3 Cat. No. 1093 274
NBT solution	75 mg/ml nitroblue tetrazolium salt in 70% (v/v) dimethylformamide	Vial 9, Genius 1 Vial 4, Genius 3 Cat. No. 1383 221
X-Phosphate solution	50 mg/ml 5-bromo-4-chloro-3-indolyl phosphate (X-phosphate), toluidinium salt in 100% dimethylformamide	Vial 10, Genius 1 Vial 5, Genius 3 Cat. No. 1383 221
DNA dilution buffer	50 μg/ml herring sperm DNA, in 10 mM Tris-HCl, 1 mM EDTA; pH 8.0 (25°C)	Vial 3, Genius 1 Vial 3, Genius 2 Vial 2, Genius 3
Blocking reagent	Blocking reagent for nucleic acid hybridization white powder	Vial 11, Genius 1 Vial 6, Genius 3 Cat. No. 1096 176

stop the reaction by washing the membrane for 5 minutes with 50 ml of buffer 4.

9. The results can be documented by photocopying the wet filter or by photography. Photocopying onto overhead transparencies allows for densitometric scanning. (For this purpose, the color reaction can be interrupted for a short time and continued afterwards.)
10. The filter may then be dried at room temperature or by baking at 80°C and stored. The colors fade upon drying. Membranes must not be allowed to dry if they are to be re-probed.
11. The color can be revitalized by wetting the membrane with buffer 4. The filters can also be stored in sealed plastic bags containing buffer 4. The color then remains unchanged.

8. Troubleshooting

If your DNA is not labeled efficiently, i.e., if it does not give the same sensitivity as the included labeled control DNA supplied with the kit (vial 4), check the following points:

- Repurify the DNA by phenol:chloroform extraction and/or ethanol precipitation.
- Incubate the labeling reaction for a longer period (up to 20 hours instead of one hour).
- Perhaps the DNA was not denatured completely; this may be especially important for long DNA fragments. Heat to 100°C for 10 minutes and chill immediately in an ice/salt or ice/alcohol bath.

If the expected sensitivity is not attained in your assay, check the following points:

- Efficiency of DNA-labeling (see above).
- The concentration of labeled DNA or the incubation time in the hybridization may need to be increased.
- The time for color development may need to be increased up to 3 days.
- If you observe excessive background coloring:
- Decrease the amount of labeled DNA in the hybridization.
- Increase the volume of the prehybridization solution to allow the filter to float freely.
- There may be difficulties with certain types of nylon membranes. We recommend using Boehringer Mannheim Nylon Membranes, positively charged.
- Add heterologous nucleic acid to prehybridization and hybridization (i.e., herring sperm DNA or yeast total RNA).

8.1 Sensitivity and Specificity. Using the unlabeled control-DNA 2 (vial 2), labeled as described in the protocol, or the labeled control DNA (vial 4), it is possible to detect 0.1 pg of homologous DNA (vial 1), diluted with 50 ng of heterologous DNA (vial 3) in a dot blot or Southern blot. Under identical conditions, 100 ng heterologous DNA are not detected. A single copy human gene (tissue plasminogen activator gene) is detected in a Southern blot of 1 μg digested placental DNA.

8.2 Quality control. Using the unlabeled control DNA 2 (vial 2), isolated as described in the protocol, 0.1 pg of homologous control DNA (vial 1) is detected in a dot blot after a 16 hour color development. (One pg of homologous DNA can be detected after 1 hour color development.)

9. DNA Labeling

9.1 Random Primed DNA Labeling. DNA can be labeled with digoxigenin-11-dUTP using the random primed method (Table 4). Procedures for standard and scaled-up random primed labeling reactions are given below. The scaled-up labeling reaction requires the same amount of DNA template as the standard reaction but produces a larger amount of DIG-labeled DNA.

For optimal results with either method, the template DNA should be linearized and purified by at least one phenol/chloroform extraction and ethanol precipitation prior to labeling. Also, templates of 100-10,000 bp label efficiently and produce probes with maximal sensitivity; therefore, templates >10 kb should be restriction-digested prior to labeling.

Products required (Tables 5 and 6): The standard and scaled-up random primed reactions require the same reagents, most of which are available separately, in the Genius 1 DNA Labeling and Detection Kit (Cat. No. 1093 657) or in the Genius 2 DNA Labeling Kit (Cat. No. 1175 033).

9.1.1 Standard Random Primed DNA Labeling Reaction.

The aim of the standard random primed DNA labeling reaction is to produce a sufficient amount of a digoxigenin-labeled probe in the shortest amount of time (Table 7); this 20 μl reaction will yield a minimum of 260 ng of digoxigenin-labeled probe from 1 μg of DNA template (Table 5). In this standard reaction, one digoxigenin molecule is incorporated in every 25-35 nucleotides.

9.1.2 Procedure.

1. Add reagents to a sterile microfuge tube (on ice) in the following order:

The DNA template should be heat-denatured in a boiling water bath for 100 minutes and quickly chilled on dry ice/ethanol for 30 seconds before use. We have found that denaturation using a heating block is less effective and may result in lowered labeling efficiency.

2. Incubate the reaction at 37°C for at least 60 minutes. Longer incubations (up to 20 hours) will increase the yield of DIG-labeled DNA.
3. Add 2 μl EDTA to terminate the reaction.
4. Add 1 μl glycogen solution to the reaction tube.
5. Precipitate the labeled DNA with 0.1 volume of LiCl and 2.5-3 volumes of chilled ethanol. Mix well and incubate at -70°C for 30 minutes.

Table 6 - Required Solutions

Required Solution	Description
Washing buffer	100 mM Tris-HCl, 150 mM NaCl; pH 7.5 (20°C)
Blocking solution	2% (w/v) Blocking reagent for nucleic acid hybridization dissolved in Washing buffer
Detection buffer	100 mM Tris-HCl, 100 mM NaCl, 50 mM $MgCl_2$, pH 9.5 (20°C)

6. Remove from the -70°C incubation and thaw briefly at room temperature. Centrifuge the reaction at 13,000 x *g* for 15 minutes in a microcentrifuge.
7. Decant the ethanol and wash the pellet with 100 μl of 70% ethanol. Centrifuge at 13,000 x *g* for 5 minutes in the microcentrifuge, then remove the 70% ethanol.
8. Dry the pellet and resuspend in 50 μl of TE/SDS buffer. If not used immediately, store the labeled probe at -20°C.

NOTE: The SDS may be left out if siliconized tubes are used during precipitation.

NOTE: Resuspension of the digoxigenin-labeled pellet may require heating to 37°C for 10 minutes with frequent vortexing.

NOTE: Failure to adhere to these guidelines will result in the loss of approximately 20% of the digoxigenin-labeled DNA.

NOTE: See table 6 and 7 for required products and solutions.

9.1.3 Labeled Control DNA Procedure.

1. Mix 45 μl NBT solution and 35 μl X-phosphate solution in 10 ml of Detection buffer. This freshly prepared Color Substrate Solution will be used in step 13.
2. Make serial dilutions of the Labeled Control DNA in DNA dilution buffer (Table 8). Use of siliconized tubes helps prevent adsorption of DNA to the walls of the tubes.

Table 7 - The Amount of Synthesized Labeled DNA Depends on the Amount of Template DNA in the Labeling Reaction and on the Time at 37°C.

Amount of Template DNA per Labeling Reaction	10 ng	30 ng	100 ng	300 ng	1000 ng	3000 ng
Amount of synthesized DIG-labeled DNA:						
after 1 hour	15 ng	30 ng	60 ng	120 ng	260 ng	530 ng
after 20 hours	50 ng	20 ng	260 ng	500 ng	780 ng	890 ng

Table 8 - Labeled Control DNA Dilutions

Labeled Control DNA Starting Concentration	Stepwise Dilution	Final Concentration (Dilution Name)	Total Dilution
5 ng/μl	2 μl/8 μl DNA dilution buffer	1 ng/μl (A)	1:5
1 ng/μl (dilution A)	2 μl/18 μl DNA dilution buffer	100 pg/μl (B)	1:50
100 pg/μl (dilution B)	2 μl/18 μl DNA dilution buffer	10 pg/μl (C)	1:500
10 pg/μl (dilution C)	2 μl/18 μl DNA dilution buffer	1 pg/μl (D)	1:5,000
1 pg/μl (dilution D)	2 μl/18 μl DNA dilution buffer	0.1 pg/μl (E)	1:50,000

3. Spot 1 μl each of diluted DNAs above onto a positively charged nylon membrane, marking the membrane lightly with a pencil to identify each dilution.
4. Make serial dilutions of the newly labeled experimental DNA probe (of unknown starting concentration) in DNA dilution buffer according to step 2 until a 1:50,000 dilution is made.
5. Spot 1 μl of each of the 1:50 to 1:50,000 dilutions made in step 4 onto the same nylon membrane, marking the membrane lightly with a pencil to identify each dilution.
6. Fix the DNAs to the membrane by cross-linking with UV light or by baking for 30 minutes at 80°C.
7. Wet the membrane with a small amount of Washing buffer.
8. Incubate the membrane in Blocking solution for 15 minutes at room temperature.
9. Dilute anti-DIG-alkaline phosphatase 1:5,000 in Blocking solution.
10. Incubate the membrane in the diluted antibody for 15 minutes at room temperature. The diluted antibody solution must cover the entire membrane.
11. Wash the membrane twice, 15 minutes per wash, in Washing buffer at room temperature.
12. Incubate the membrane in Detection buffer for 2 minutes. Application of this buffer "activates" the alkaline phosphatase that is conjugated to the antibody.
13. Pour off the Detection buffer and add the Color Substrate Solution. Allow color development to occur in the dark for 30-60 minutes, monitoring occasionally. The 1:5,000 dilution (1 pg) of the DIG-labeled control DNA should be visible 30 minutes after adding the Color Substrate Solution.
    Since the incubations times are shortened in the interest of getting a quick assay result, sensitivity of this assay may not be great enough to visualize the 0.1 pg control spot.
14. Compare spot intensities of the control and experimental dilutions to estimate the concentration of the experimental probe.

9.2 Oligonucleotide Labeling.

9.2.1 A Comparison of Oligonucleotide-labeling Methods. Synthetic oligonucleotide probes are widely used in library-screening procedures, Southern and

Table 9 - Types of Labeling Reactions.

Method	Amount of Oligo Produced	Probe Sensitivity	Probe Specificity	Major Characteristics	Application
3'-End labeling	100 pmol per reaction	>10 pg	+++	Addition of a single DIG residue	Library screening, dot/slot blotting, *in situ* hybridization
3' tailing	100 pmol per reaction	>1 pg	++	Addition of multiple DIG residues	Southern blotting, Northern blotting, *in situ* hybridization
5"-End tailing with DIG-NHS	100 nmol per reaction	>10 pg	+++	Oligo must be synthesized with ester; good for large-scale labeling chemical reaction	Library screening AminoLink

Table 10 - Required Materials for Northern Blotting.

Name in Procedure	Product Description	Available as
Anti-DIG-alkaline phosphatase	Anti-digoxigenin [Fab] conjugated to alkaline phosphatase	1. Vial 8, Genius 1 2. Vial 3, Genius 3 3. Cat. No. 1093 274
NBT solution	75 mg/ml nitroblue tetrazolium salt in 70% (v/v) dimethyl-formamide	1. Vial 9, Genius 1 2. Vial 4, Genius 3 3. Cat. No. 1383 213
X-Phosphate solution	50 mg/ml 5-bromo-4-chloro-3-indolyl phosphate (X-phosphate), tolui-dinium salt in 100% dimethyl-formamide	1. Vial 10, Genius 1 2. Vial 5, Genius 3 3. Cat. No. 1383 211
Lumi-Phos 530	0.33 mM Lumigen PPD [4-methoxy-4-(3-phosphate-phenyl)-spiro(1,2-dioxetane-3, 2"-adamantane) disodium salt]; 750 mM store at 4°C, protected from direct light, warm to room temperature before using	Cat. Nos. 1413 155, 1275 470, 1413 163 2-amino-2-methyl-1-pro-panol buffer (pH 9.6); 0.88 mM $MgCl_2$; 1.13 mM cetyltrimethyl ammonium bromide; 0.035 mM fluorescein surfactant

Northern blots, dot blots, and *in situ* hybridization experiments. To provide researchers with maximum flexibility, Boehringer Mannheim has developed three methods for labeling oligonucleotides with digoxigenin. This section briefly outlines the three oligonucleotide-labeling methods, each of which produces probes that are optimized for specific applications (Table 9).

9.2.2 Oligonucleotide 3'-End Labeling (Genius 5 Kit). The Genius 5 Oligonucleotide 3'-End Labeling Kit (Cat. No. 1362 372) is designed for the addition of digoxigenin-11-ddUTP (DIG-ddUTP) to the 3' end of a synthetic oligonucleotide 14-100 nucleotides in length. Using the enzyme terminal transferase, only one digoxigenin residue is added per oligonucleotide because chain elongation cannot proceed past the dideoxy nucleotide. Probes labeled with this method retain a high degree of specificity. In addition, this method enables nonradioactive DIG-labeling of conventionally synthesized oligonucleotides; therefore, the nonradioactive label DIG-ddUTP can be linked to the oligonucleotide without using any special reagents for oligonucleotide synthesis.

9.2.3 Oligonucleotide 3'-Tailing. (Genius 6 Kit). The Genius 6 Oligonucleotide Tailing Kit (Cat. No. 1417 231) is designed for the addition of a tail of residues of unlabeled ranging from 5-50 bases in length. In the 3'-tailing reaction, terminal transferase adds a mixture of unlabeled nucleotides and digoxigenin-11-dUTP, producing a tail containing multiple digoxigenin residues. The resulting probes are about ten times more sensitive than 3'-end-labeled probes produced with the Genius 5 Kit. Although tailed oligonucleotide probes are more sensitive than 3'-end-labeled probes, they can also produce nonspecific background due to the presence of the longer tail. For example, if the unlabeled nucleotide used in the tailing reaction is dATP, the probe may anneal to T-rich regions in complex nucleic acid mixtures.

9.2.4 5'-End Labeling Oligonucleotides with Digoxigenin-NHS Ester. (Digoxigenin-3-O-methyl-carbonyl-ε-aminocaproic acid-*N*-hydroxysuccinimide ester). Oligonucleotides can be chemically tagged with digoxigenin at the 5' end using a two-step procedure.

In the first step, the oligonucleotide is synthesized with an Aminolink residue on its 5' end.

After the synthetic oligonucleotide is purified, the second step involves the covalent linkage of digoxigenin-NHS ester to the free 5'-amino residue. The resulting probes can be produced in large quantities (100 nmol per reaction), are specific, and have a sensitivity comparable to that of 3'-end-labeled probes. Another useful feature of 5'-end-labeled oligonucleotides is that the 3' end is free to act as a primer for DNA synthesis reactions. Thus, extension reactions such as PCR can be conducted with labeled primers, allowing the nonradioactive tagging of the reaction products. Subsequently, the labeled extension products can be detected or purified by affinity chromatography using anti-digoxigenin antibodies.

10. Northern Blotting

Because of the unique conditions surrounding Northern blots, the hybridization and detection sections of this User's Guide are combined for this application. General recommendations on labeling a probe and on preparing the blot are presented first. These are followed by detailed procedures for hybridization chemiluminescent detection with Lumi-Phos 530, and colorimetric detection with NBT and X-phosphate.

Products and solutions required for Northern blotting are listed in Table 10 and 11. Depending on the detection method, either Lumi-Phos 530 or both NBT solution and X-phosphate solution should be purchased. Lumi-Phos 530 is only available as a single reagent. The reagents required for colorimetric detection are available separately in the Genius 1 DNA Labeling and Detection Kit (Cat. No. 1093 657) or in the Genius 3 Nucleic Acid Detection Kit (Cat. No. 1175 041). The anti-DIG-alkaline phosphatase is required by both detection methods.

10.1 Labeling a Probe. Because oligonucleotide probes are not as sensitive as longer probes, they should only be used to detect fairly abundant mRNA.

10.2 Preparing the Blot. Standard techniques and conditions may be used during electrophoresis and transfer. For best results, Boehringer Mannheim's Nylon membranes (Cat. Nos. 1209 299, 1209 272, 1417 240) should be used for the transfer. This membrane has an optimal charge density, allowing it to bind the RNA tightly without producing high backgrounds. Our nylon membrane is also specifically tested with the Genius System to ensure optimal background characteristics. Other membranes that work well are Amersham's Hybond-N and M.S.I.'s Magnagraph nylon membranes, but these are not tested with the Genius System and may produce backgrounds due to lot-to-lot variability. After the transfer, the RNA must be fixed to the membrane by either baking or UV crosslinking.

10.3 Hybridization Procedure.

1. Place the blot in a hybridization bag containing 20 ml Northern prehybridization solution per 100 cm^2 of membrane surface area. Seal the bag and prehybridize at the anticipated hybridization temperature for 2 hours. Longer prehybridization times are acceptable.
   The optimal hybridization temperature for a specific probe will depend on the length of the probe and the extent of sequence homology with the target sequence. Therefore, the optimal hybridization temperature must be determined empirically. See Table 12 for acceptable temperatures for different types of probes.
2. When using double-stranded DNA probes, heat in a boiling water bath for 10 minutes to denature the DNA. Single-stranded RNA probes and oligonucleotide probes do not require denaturation prior to dilution unless extensive secondary structure is predicted from the sequence.
3. Dilute the probe in Northern hybridization solution. See Table 12 for optimal probe concentrations. See Table 13 for the required amount of Northern hybridization solution.
4. Discard the Northern prehybridization solution from the bag. Add the Northern hybridization solution containing the DIG-labeled probe. Allow the probe to hybridize. See Table 13 for optimal hybridization conditions.
5. At the end of the hybridization, pour the Northern hybridization solution from the bag into a capped tube that can withstand freezing and boiling (e.g., 50 ml polypropylene). This Northern hybrid-ization solution contains unannealed DIG-labeled probe. The entire solution can be reused in future hybridization experiments. Label and date the tube and store at -20°C. DIG-labeled probes stored in this manner are stable for at least 1 year. For reuse, thaw and denature by heating to 68°C for 10 minutes.

Table 11 - Optimal Hybridization Conditions for Different Probe Types.

Required solution	Description
Northern prehybridization solution	5x SSC, 50% formamide, 0.02% sodium dodecyl sulfate (SDS), 0.1% *N*-lauroylsarcosine, 2% (w/v) solution Blocking reagent for nucleic acid hybridization, 20 mM sodium maleate, pH 7.5 (added from the Blocking Reagent Stock Solution)
Blocking reagent stock solution	100 mM sodium maleate pH 7.5 containing, 10% (w/v) Blocking reagent for nucleic acid hybridization. The sodium maleate allows treatment of the Blocking Reagent Stock Solution with DEPC. Any RNase activity associated with the Blocking reagent preparation is destroyed by this treatment
Northern hybridization solution	DIG-labeled probe diluted in Northern prehybridization solution
Maleate buffer	100 mM maleic acid, pH 7.5; 150 mM NaCl
Northern blocking solution	2% (w/v) Blocking reagent for nucleic acid hybridization dissolved in 100 mM maleic acid, pH 7.5; 150 mM NaCl
Detection buffer	100 mM Tris-HCl, pH 9.5; 100 mM NaCl; 50 mM $MgCl_2$ Note: If the blot will be reprobed in future experiments, the $MgCl_2$ should be deleted from this buffer
2x wash solution	2x SSC containing 0.1% SDS
0.5x wash solution	0.5x SSC containing 0.1% SDS

Table 12 – Optimal Proble Concentration and Hybridization Conditions.

Probe Type	Probe Concentration	Hybridization Conditions
RNA	10-50 ng/ml	Hybridize overnight at 68°C
DNA	5-20 ng/ml	Hybridize overnight at 42°C
Oligonucl eotide	0.5-2 pmol/ml	Hybridize for 1-6 hours; hybridization temperature varies considerably and can be approximated by considering probe length and G plus C content

Table 13 - Solutions Required for Colorimetric Detection with NBT and X-Phosphatase

Number of Blots	Size	Required Amount of Northern Hybridization Solution
1	10 x 10 cm	Minimum of 10 ml
1	>20 x 20 cm	Minimum of 25 ml
∃2	>10 x 10 cm	Enough to allow each blot to move freely in the solution

6. Wash the membrane twice, 5 minutes per wash, in 2x Wash solution at room temperature. These washes (steps 6 and 7) remove unbound probe, which will lead to high backgrounds if not removed.
7. Wash the membrane twice, 15 minutes per wash, in 0.5x Wash solution. Long probes (>100 bp) should be washed at 65°C. For shorter probes, the wash temperature must be determined empirically.

11. Chemiluminescent Detection with Lumi-Phos 530

1. Place the membrane in a clean plastic tray or hybridization bag containing 50 ml of maleate buffer and wash for 1 minute at room temperature.
2. Incubate the membrane in 50 ml of Northern blocking solution for 60 minutes at room temperature.
3. Just prior to completing step 2, dilute anti-DIG-alkaline phosphatase 1:5,000 in Northern Blocking solution at room temperature.
4. Replace the Northern Blocking solution with the diluted antibody and incubate the membrane for 30 minutes at room temperature with gentle agitation. This step can be done in a plastic bag or tray. Make certain that the membrane is completely covered with the solution.
5. Pour off the diluted antibody and transfer to a new dish or hybridization bag. Wash twice, 15 minutes per wash, with 200 ml maleate buffer at room temperature. These washes remove unbound antibody.
6. Equilibrate the membrane for 5 minutes in 50 ml Detection buffer. Pick up the membrane with forceps and let the buffer drain off for 5 to 10 seconds. If the blot will be reprobed in future experiments, the $MgCl_2$ should be deleted from this buffer. High $MgCl_2$ concentration causes the RNA on the blot to degrade, preventing any future experiments. The alkaline phosphatase will still function in the absence of any free Mg^{2+}, although detection times will be longer.
7. Place the membrane between two sheets of acetate (plastic page protectors). Gently lift the top sheet of plastic and, with a sterile pipet, add approximately 0.5 ml (per 100 cm^2) of Lumi-Phos 530 to the top surface of the membrane. Then, scatter the drops of Lumi-Phos 530 over the surface. Rock the membrane gently to distribute the reagent over the surface. After 1 minute drain the excess Lumi-Phos from the membrane. Lower the top sheet of plastic and gently smooth out any air bubbles.
8. Expose the membrane to x-ray film for 60 minutes initially. If this exposure is not ideal, shorten or lengthen the next exposure time accordingly. If $MgCl_2$ has been deleted from the Detection buffer, the exposure may take several hours.

12. Colorimetric Detection with NBT and X-Phosphate

Perform all incubations at room temperature.

NOTE: See Table 15 for required solutions.

1. Mix 45 µl of NBT solution and 35 µl of X-phosphate solution in 10 ml of Detection buffer. This freshly prepared Color Substrate Solution will be used in step 9.
2. After hybridization and post-hybridization washes, equilibrate the membrane in filtered Washing buffer for 1 minute.
3. Using a freshly washed dish or bag, block the membrane by gently agitating it in Blocking solution for 30 minutes. Longer blocking times are also acceptable. For sufficient blocking, there must be ample room in the bag or dish to allow for unrestricted shaking of the membrane. If you are using more than one membrane, add enough solution to cover all membranes.
   Near the end of the thirty-minute blocking period, prepare the antibody solution.
4. Dilute the anti-DIG-alkaline phosphatase conjugate 1:5,000 in Blocking solution for a working concentration of 150 mU/ml. Mix gently by inversion. For example, add 6 µl anti-DIG-alkaline phosphatase conjugate to 30 ml Blocking solution and mix. This working antibody solution is stable for about 12 hours at 4°C.
5. Pour off the Blocking solution and incubate the membrane for 30 minutes in the antibody solution prepared in step 4.
6. Transfer the membrane to a new bag or a dish that has never come into contact with anti-DIG-alkaline phosphatase conjugate.
7. Wash twice, 15 minutes per wash, in 100 ml of Washing buffer. These washes remove unbound antibody.

Table 14 - Troubleshooting Guide

Symptom	Possible Cause	Remedy
Low sensitivity	Probe was not labeled properly	Assay the labeling reaction with the DIG-labeled controls provided in the Genius kits
	DNA probe was not denatured before hybridization	Boil the probe for 10 minutes before diluting in hybridization buffer
	Antibody conjugate has lost activity	Check antibody control date; store at 4°C
	Probe and target have unusual hybridization characteristics	Empirically determine optimal hybridization conditions
	Incompatible membrane	Use a different type of membrane (e.g., nylon) or a membrane from a different source
High background	Exposure time was too long	Decrease exposure time; chemiluminescent detection typically requires only 60 minutes at room temperature
	Probe concentration was too high	Use 10-50 ng/ml for RNA probes; 5-20 ng/ml for DNA probes; 0.5-2 pmol/ml for oligo probes
	Insufficient blocking was observed with Blocking solution or Northern blocking solution	Prepare fresh blocking solution and/or increase blocking times
	Incompatible membrane was used	Use a different type of membrane (e.g., nylon), or a membrane from a different source
	Insufficient solution volume was used in prehybridization, hybridization, or blocking steps	Increase the volume to allow the membranes to move freely in solution
	An incorrect antibody dilution was used	Check the antibody dilution factor; 1:5,000 is optimal

8. Equilibrate the membrane in 20 ml of Detection buffer for 2 minutes.
9. Add approximately 10 ml of Color Substrate Solution to the membrane. Incubate the membrane in a sealed plastic bag or box in the dark. Do not shake the container while the color is developing. The membrane can be exposed to light for short periods to monitor the color development. The color precipitate starts to form within a few minutes and the reaction is usually complete after 12 hours.
10. Once the desired spots or bands are detected wash the membrane with 50 ml of Washing buffer for 5 minutes to stop the reaction.

13. Troubleshooting Guide

See Table 14.

Table 15 - Colorimetric Detection Solutions

Required solution	Description
Washing buffer	100 mM Tris-HCl, 150 mM NaCl; pH 7.5 (20°C) Filter this buffer through a 0.45 μm membrane filter before use.
Blocking solution	2%, (w/v) Blocking reagent for nucleic acid hybridization dissolved in Washing buffer. Blocking solution is a cloudy solution and should not be filtered. It is stable for at least two weeks when stored at 4°C, but must be brought to room temperature before use.
Detection buffer	100 mM Tris-HCl, 100 mM NaCl, 50 mM $MgCl_2$; pH 9.5 (20°C)
Genius buffer 4	10 mM Tris-HCl, 1 mM EDTA, pH 8.0 (20°C)

References:

1. Martin, R., Hoover, C., Grimme, S., Grogan, C., Holtke, J. & Kessler, C. (1990) BioTechniques 9(6): 762-768.
2. Feinberg, A. F. & Vogelstein, B. (1983) Anal. Biochem. 132, 6.
3. Gebeyehu, G. *et al.* (1987) Nucl. Acids Res. 15, 4513.
4. Southern, E. M. (1975) J. Mol. Biol. 98, 503.
5. Khandjian, E. W. (1967) BioTechnology 5, 165.

LAB 6.2:
ISOLATION OF *C. ELEGANS* GENOMIC DNA, QUANTITATION OF DNA CONCENTRATION, AND DIGESTION TO EXTRACT THE *MYO*-3 GENE

Overview:

In this lab you will isolate the genomes from two strains of the nematode *C. elegans*: N2 (wild type) and CB1407 (a mutant containing a duplicate copy of the *C. elegans myo*-3 gene). Genomic DNA is isolated by disrupting the bacterial cell wall and separation from other cellular components by a series of extraction steps specific for its properties. Once the genomes are isolated, the concentration of genomic DNA will be calculated.

Quantitation of genomic DNA concentration is accomplished using the intrinsic property that double-stranded DNA absorbs light at a wavelength of 260 nm. Each unit of absorbance equals approximately 50 μg/ml double stranded DNA. Restriction enzyme digests will then be performed to extract the *myo*-3 gene from genomic DNA. In the next lab, the genomic *myo*-3 gene will be probed to verify which strains contain the duplication.

Experimental Outline:

1. Isolation of genome from two *C. elegans* strains (N2 and CB1407).
2. Quantitation of *C. elegans* genomic DNA concentration by absorbance spectroscopy at 260 nm.
3. *Hin*dIII digestion of *myo*-3 gene from *C. elegans*.

Required Materials:

DNA extraction buffer; incubator; 70% ethanol; 95% ethanol; vacuum; TE buffer (recipe in Lab 1.1); spectrophotometer; cuvette; *Hin*dIII restriction endonuclease; 10x buffer (enzyme-specific)

Procedures:

1. Prepare DNA Extraction Buffer

100 mM NaCl
100 mM Tris-HCl (pH 8.5)
50 mM EDTA (pH 7.4)
1% SDS
1% β-mercaptoethanol
100 μg/ml proteinase K
Distilled H_2O to 0.5 ml

2. Isolation of *C. elegans* Genomic DNAs from N2 and CB1407 Strains

1. Thaw a microfuge tube containing 1.5 μl of frozen *C. elegans*.
2. Immediately add 0.5 ml of DNA extraction buffer.
3. Incubate with occasional agitation for 30 minutes at 65°C.
4. Add 0.5 ml of phenol to the microfuge tube.
5. Vortex the microfuge tube and centrifuge at 12,000 x *g* for 5 minutes. You should see an interface between the two layers.
6. Remove the aqueous (viscous) top layer containing the genomic DNA with a cut pipet tip and put in a new microfuge tube.
7. Add 0.5 ml of phenol/chloroform to the transferred aqueous phase.
8. Mix well by inversion for 2 minutes (do not vortex!) and centrifuge at 12,000 x *g* for 5 minutes.
9. Remove the aqueous (viscous) top layer containing the genomic DNA with a cut pipet tip and place it in a new microfuge tube.
10. Add 0.5 ml of chloroform to the transferred aqueous phase.
11. Mix well by inversion for 2 minutes and centrifuge at 12,000 x *g* for 5 minutes.
12. Remove the aqueous (viscous) top layer (genomic DNA) and transfer to a new microfuge tube.
13. Add 0.5 ml of isopropanol to precipitate the DNA.
14. Mix well by inversion for 2 minutes (do not vortex!) and centrifuge at 12,000 x *g* for a short burst. Allow the centrifuge to reach full speed and then immediately turn off the centrifuge. Centrifuge just long enough to pellet the DNA.
15. Remove and discard the supernate without disturbing the pellet.

16. Wash the pellet with 70% ethanol, repeat.
17. Wash the pellet with 95% ethanol.
18. Dry the pellet in a lyophilyzer.
19. Resuspend the pellet in 50 μl of TE buffer.

3. Quantitation of *C. elegans* Genome by Absorbance at 260 nm

1. Zero the spectrophotometer with H_2O.
2. Measure the absorbance of each genomic DNA sample at 260 nm.
3. Calculate the DNA concentration from each *C. elegans* strain (1 A_{260} unit of DNA = 50 μg/ml).

4. *Hind*III Digestion of the *myo*-3 Gene from *C. elegans*

NOTE: All procedures are done in sterile H_2O. Do not use pipet tips twice.

1. Combine the following contents in the listed order in a microfuge tube:
   75 μl sterile H_2O
   10 μl 10x buffer (enzyme-specific)
   10 μl plasmid genomic DNA from the "minipreps"
   \+ 5 μl *Hind*III
   100 μl = V_{total}

2. Incubate for 2.5 to 3 hours at 37°C. The TA will remove the reactions from the water bath and store them at -20°C.

Notes:

1. Phenol and chloroform extract the cellular components and proteins by precipitation.
2. Greater than 70% ethanol precipitates DNAs from aqueous samples by excluding water from nucleic acids.

LAB 6.3:
SOUTHERN BLOTTING

Overview:

In this lab, you will electrophorese the digested *C. elegans myo*-3 gene on an agarose gel. Electrophoresis is used to separate and determine the size of the DNA molecules. The digested DNAs, still in the gel, will then be denatured and neutralized as a preliminary step for Southern blotting. The DNA will then be transferred and fixed to a blotting membrane that binds the DNA fragments, thereby retaining the band pattern from the gel following transfer. The *myo*-3 digoxigenin (DIG) probe you made in Lab 6.2 will then be hybridized to the *C. elegans* DNA contained on the blot.

Next, hybridization of the DIG-labeled probe to the genomic DNA will be detected by chemiluminescence. After hybridization, the probe will contain only bound DNA fragments; complementary sequences and excess probe will be washed away. Chemiluminescent detection is a three-step process. In the first step, membranes are treated with Blocking Reagent to prevent nonspecific adhesion of antibody to the membrane. Next, membranes are incubated with anti-DIG F_{ab} fragments, conjugated (attached) to alkaline phosphatase (a bifunctional immunoenzyme conjugate). In the third step, the membranes carrying the hybridized probe and bound antibody conjugate are reacted with a chemiluminescent substrate and exposed to X-ray film to record the chemiluminescent signals.

Experimental Outline:

1. Gel electrophoresis of *C. elegans* DNA.
2. Denaturation and neutralization of *C. elegans* DNA in an agarose gel.
3. Transfer and fixation of DNA to the membrane.
4. Prehybridization and hybridization of *myo*-3 probe (to the *C. elegans* DNA).
5. Detection of DIG-labeled nucleic acids.

Required Materials:

Electrophoresis apparatus; 1x TAE buffer; Denaturing Solution (GENIUS™); Neutralization Solution (GENIUS™); SSC buffer; GENIUS™ membrane; blotting paper; paper towels; glass plate; weight for blot plate, Prehybridization Solution (GENIUS™); Hybridization Solution (GENIUS™); 2x Wash Buffer; 0.5x Wash Buffer; anti-DIG-alkaline phosphatase; Lumi-Phos 530 (GENIUS™); Blocking Solution (GENIUS™); Detection Buffer (GENIUS™); TE buffer (recipe in Lab 1.1); DEPC-treated H_2O; clean trays or bags; autoclaved blunt-ended forceps; X-ray film

Procedures:

1. Gel Electrophoresis of *myo*-3 DNA

1. Prepare a electrophoretic gel with the appropriate agarose composition.
2. Add 1x TAE buffer to the apparatus so it just covers the gel.
3. Load the samples. The instructor or TA will tell you which sample in each lane and how much to load. The amount to be loaded depends on the DNA concentration. Record the detailed information regarding the contents of the electrophoresis gel lanes in the space below.
   Lane 1
   Lane 2
   Lane 3
   Lane 4
   Lane 5
   Lane 6
   Lane 7
4. Connect the apparatus to a power supply and run the gel at 100 V for 45 minutes to 1 hour. If desired, ethidium bromide staining can be done to visualize the DNA fragments. Staining will confirm Southern transfer to the membrane.

2. Denaturation and Neutralization of *myo*-3 DNA

1. Submerge and shake the agarose gel in Denaturing Solution for 30-60 minutes at room temperature to denature the DNA prior to transfer.

2. Neutralize the gel by submerging in Neutralization Solution for 30-60 minutes.

3. Transfer of DNA to the Membrane

1. Cut the Genius™ membrane to the exact size of the gel. Wear gloves!
2. Just prior to the transfer, wet the membrane first with water and then with 10x SSC for 15 minutes. Be careful not to trap air bubbles beneath the membrane or it will saturate unevenly.
3. Cut 3-mm blotting paper, 9-mm blotting paper and a 5-8 cm stack of absorbent paper towels to the exact size as the gel.
4. Fill a tray with 10x SSC and allow a filter paper wick to absorb the buffer.
5. Place the gel on top of the filter paper wick; remove any air bubble(s) with a pipet.
6. Place the buffer saturated Genius™ membrane on the gel and remove any air bubbles.
7. Place the buffer-saturated 3-mm and 9-mm blotting paper on top of the membrane and remove any air bubbles.
8. Place the stack of absorbent paper towels on top of the blotting paper.
9. Place a glass plate on top of the paper towels.
10. Place a 500 g weight on top of the glass plate.
11. Allow the DNA transfer to proceed overnight (at least 12 hours).

4. DNA Fixation to Membrane

1. Rinse the membrane in 5x SSC buffer for 1 minute at room temperature to wash the gel debris and particulate contaminants.
2. Carefully place the damp membrane on a piece of blotting paper.
3. Bake the membrane and blotting paper in an oven at 80°C for 1 hour or 120°C for 15-30 minutes or fix by crosslinking.

5. Prehybridization and Hybridization

1. Place the blot in a hybridization bag or tray containing 20 ml of Prehybridization Solution per 100 cm^2 of membrane surface area. Prehybridize at 50-68°C for 2 hours.
2. Heat double-stranded DNA probes in a boiling water bath for 10 minutes to denature the DNA. Chill directly on ice.
3. Dilute the probe in Hybridization Solution to a concentration of 5-25 ng/ml.
4. Discard the Prehybridization Solution from the bag or tray.
5. Add the standard buffer Hybridization Solution containing the DIG-labeled probe.
6. Hybridize overnight at 50-68°C. The required amount of Hybridization Solution for a 100-cm^2 blot is at least 20 ml.
7. After hybridization, wash the membrane twice, 5 minutes per wash, in 2x Wash Buffer at room temperature to remove unbound probe and reduce background.
8. Wash the membrane twice, 15 minutes per wash, in 0.5x Wash Buffer to remove unbound probe and reduce background.

6. Detection of DIG-Labeled Nucleic Acids

1. In a tray or bag, equilibrate the membrane in Wash Buffer for 1 minute after hybridization and post-hybridization washes.
2. Allow the Lumi-Phos 530 chemiluminescent substrate to come to room temperature.
3. In a fresh tray, block the membrane by gently agitating it in Blocking Solution for 30-60 minutes.
4. Prepare a 1:10,000 dilution of anti-DIG-alkaline phosphatase in Blocking Solution.
5. Mix gently by inversion. This antibody solution is stable for about 12 hours at 4°C.
6. Pour off the Blocking Solution and incubate the membrane for 30 minutes in the antibody solution prepared in step 4.
7. Discard the antibody solution.
8. Gently wash the membrane twice, 15 minutes per wash, in Wash Buffer.
9. Pour off Wash Buffer and equilibrate the membrane in Detection Buffer for two minutes. Be sure that the filter is kept wet before the chemiluminescent substrate is applied.
10. Pipet 5-10 ml of the appropriate chemiluminescent substrate into the center of a sterile dish. Place the membrane in the dish and make sure the entire membrane is saturated.
11. Remove the membrane from the substrate and allow any excess liquid to drip off. (Do not let the membrane dry out.)

12. Cover the damp membrane by placing it between two page protectors or clear acetate sheets.
13. Wipe the top sheet with a damp lab tissue to remove any bubbles present between the sheet and the membrane.
14. For the briefest exposure to X-ray film, the alkaline phosphatase chemiluminescent reaction must be at a steady state. At room temperature, a 7 to 8 hour incubation is required to reach a steady-state reaction. Once a steady state is reached, single-copy gene detection on a human genomic blot can be obtained with an approximate exposure time of 15 minutes. If the membrane is exposed before the steady state is reached, approximately 60 minutes of exposure is required for single-copy gene detection on a human genomic blot. Therefore, to shorten exposure times, it is recommended that the membrane be incubated for 15 minutes at 37°C directly after addition of the chemiluminescent substrate and before exposure to X-ray film.
15. Expose the membrane to standard X-ray film to detect the chemiluminescent signal.
16. Visualize the Southern blot for the probed nematode *myo*-3 gene.

Notes:

1. The highest probe concentration that gives an acceptable background should be used for hybridization.
2. Perform all incubations at room temperature.
3. Incubations can be performed in a sealed hybridization bag or clean plastic tray. If a tray is used, agitate the tray gently to ensure that the membrane is always covered.
4. To minimize background staining, avoid touching the membrane with your fingers. Use washed and autoclaved blunt-ended forceps to pick up membranes at the membrane edge.
5. SSC buffer refers to saline sodium citrate. This buffer increases the salt concentration of the Genius™ membrane, which stabilizes DNA charges during the blotting procedure.
6. Blotting paper comes in a variety of sizes. The membranes used in this blotting protocol are 9 mm and 3 mm thick. The blotting paper is used to absorb and transfer buffer through the gel and onto the membrane during blotting.

Additional Report Requirements:

1. Picture of Southern Blot

Reference:

1. *Genius® System User's Guide for Membrane Hybridization.* Version 2.0, Indianapolis, Indiana, 1992.

PART 2

Protein Purification

INTRODUCTION

This set of experiments uses the expression plasmid produced in Part 1 to generate the β-gal-*myo*-3 fusion protein. Lab 7.1 begins with a computer simulation exercise that introduces parameters to be considered when setting out to purify a protein of interest. The program simulates some typical decision-making pathways encountered when a researcher attempts to use any of a range of different chromatography procedures, applied in any chosen order by the student.

These lessons are brought to life in the following set of linked experimental exercises. Three linked pathways all begin with Lab 7.2, induction of β-gal *myo*-3 fusion protein synthesis using the β-galactosidase substrate to mimic isopropyl-β-d-thiogalactoside (IPTG). Next, one of three methods is used to attempt the purification of our expressed *myo-3-β-gal* fusion protein, gel filtration chromatography (GFC, Lab 7.3), ion exchange chromatography (IEC, Lab 7.5), and substrate affinity chromatography (AC, Lab 7.6). The goal is to separate the fusion protein from impurities while simultaneously preserving its catalytic activity. The chromatography procedures resolve the input material into a set of separated "fractions" that are each tested ("assayed") to determine if they contain β-gal-*myo*-3 fusion protein or not. The presence of protein activity is accomplished on a semi-quantitative basis using the "dot blot" assay in Lab 7.4. The goal requires maximizing the amount of activity per amount of protein (the "specific activity"). Since some fractions have a much higher ratio, it is necessary to pool together only the "peak" samples. Lab 7.7 involves measuring quantitative activities and protein concentrations to obtain the specific activities and thereby determine which of the three chromatography techniques purified our fusion protein most successfully. Lab 8.1 involves the use of gel electrophoresis to assess the purities of these fusion protein samples. In Lab 9.1, the student transfers the proteins from the gel to a Western immunoblot membrane, retaining the separated protein pattern. These proteins are then probed to detect which bands contain epitopes that bind a specific antibody for the *myo*-3 portion of the fusion protein. This allows the student researcher to determine the relative amounts of purified and contaminating proteins.

In summary, students make the fusion protein and then try to use each of three different chromatography methods to purify it from the unfractionated cellular extract. Next, they determine their success using each method quantitatively. They student then determine the purity of each sample by separating the proteins in the product samples using gel electrophoresis. Finally, an antibody that recognizes the *myo*-3 protein fragment of the fusion protein is used to characterize its presence by Western blotting.

Flowchart Part 2

Unit 7: Lab 7.1

1. Chromatography practice: simulations using the computer program "Protein Purifier."

Unit 7: Lab 7.2

1. Induce production of recombinant protein.

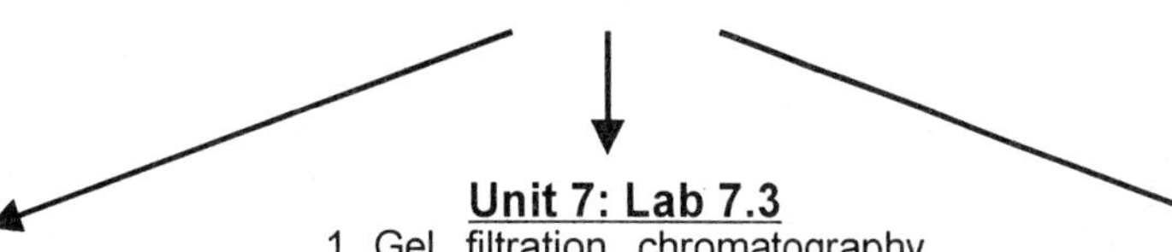

Unit 7: Lab 7.5

1. Ion exchange chromatography purification attempt with β-gal-*myo*-3 recombinant protein.
2. B-Gal catalytic activity dot-blot assay to determine fractions containing recombinant protein.

Unit 7: Lab 7.3

1. Gel filtration chromatography purification attempt with β-gal-*myo*-3 recombinant protein.
   a. Column calibration by comparing elution volume of uncharacterized protein with those of commerical proteins (M_r standards).
   b. Column chromatography with recombinant protein (uncharacterized MW_r).

Unit 7: Lab 7.6

1. *p*-Aminobenzyl-1-thio-β-galactopyranoside-agarose affinity chromatography purification attempt.
2. β-Gal catalytic activity dot-blot assay to determine fractions containing recombinant protein.

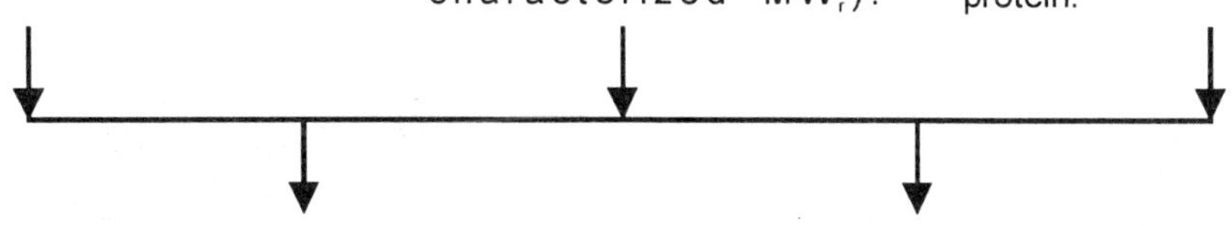

Unit 7: Lab 7.4

1. B-Gal catalytic activity dot-blot assay to determine fractions containing recombinant protein.

Unit 7: Lab 7.7

1. Quantitative recombinant β-gal-*myo*-3 protein concentration and activity assays.
2. Construct table that recounts stepwise enzyme-specific activity increases (purification).

Unit 8: Lab 8.1

1. SDS discontinuous gel electrophoresis of SDS denatured β-gal-*myo*-3 protein to characterize presence/absence of impurities. Proteins resolve across a range of hydrodynamic radii.

Unit 9: Lab 9.1

1. Western blot characterization: Immunological recognition of epitopes exposed blots containing transferred protein from SDS gels.

UNIT 7: PROTEIN PURIFICATION

1. Protein Purification: Developing Strategies Using the Literature

This unit introduces the student to the basic steps in a protein expression, extraction, and purification scheme. Developing a protein purification scheme involves optimizing the methods to best maintain and improve the functional activity of the macromolecule. This requires minimizing the number of consecutive methods employed, possibly varying their order, and testing different elution conditions and reagents. It is common to systematically vary the gradient conditions by adjusting lower and upper salt concentrations. Specific cation concentrations (e.g., Mg^{2+}) and pH are also manipulated. Teaching students how to approach the purification of a previously uninvestigated protein system involves describing each of the different types of substeps and then explaining how each variable can be manipulated to obtain pure protein. Revealing "hidden" inhibitors and/or problems caused by interactions between two or more components can also lead to success. Another important goal is sample preservation. Reviewing the points made in the Introductory Unit will refresh some important caveats to keep in mind when trying to optimize the procedures. These ideas are addressed in more detail in appendices in this Unit that discuss protein manipulations in preparation for crystallographic and protein folding studies.

One can present a reasonably useful model for a general approach to characterize the purity of a material and some important quantitative aspects to consider when doing so. Students should realize that this manual merely introduces the subject. The broad range of assays used to characterize the diverse activities of proteins is only summarized. Filling this gap for a specific application is one of the primary reasons to use "biochemical literature." This literature includes journals, proceedings, reviews, product information – the sources of protocols, results, and interpretations of studies that are available to the prospective researcher (see appendix). Many research groups also maintain their own Web sites to describe their projects and sometimes to distribute protocols, example data, etc.

An experimental report must explain the purpose(s) of the study, provide tested and verified protocols, and present the data. The latter includes all primary results, calculated functional parameters ("activities") and statistical analyses of errors. Any conclusions drawn from the results about the purpose of the study should be stated clearly. Finally, one must present a thorough analysis of the breadth of situations in which these conclusions can be expected to apply.

When one wants to investigate a problem, the first step is to consult the literature, generally by photocopying the key precedents or printing the file from the Web site, and reading them thoroughly. Next, plan the approach with the new system based on what"s been learned and accomplished with analogous systems. Obviously, these precedent systems are chosen by selecting properties that approximate those of the target material as closely as possible. Many activities and purification techniques are described in the literature. A wide range of very clever methods has been developed to achieve, measure, and even regain purity. In some cases, two or more alternative approaches are available. As a result, when consulting the literature, it is not necessarily sufficient to find a method or two before starting to plan the experiment. Also consult studies published later by the same authors or by other investigators with the same or analogous systems. You might also be able to extract new leads by studying a competitor's approach.

Caution should be exercised when pursuing the "Web search" approach to finding literature based on keywords or names. Many journals are on-line, but older volumes have been deemed "too old" to include in the electronic version; some useful journals are not on-line. A lot of protocols are summarized in their most thorough form in *Methods in Enzymology* chapters, books published following meetings, and theses that are not typically accessed by the Web "search engines." *Biological* and *Chemical Abstracts*, *Medline*, etc. can be very useful. Finding "hidden gem" protocols and "buried" information requires an intuition that is learned both from practice and by observing how more experienced practitioners approach their searches. As you work on problems, approaches become more generalized. Lessons learned in researching previous problems can be helpful. One sometimes has to read "between the lines" of reports, develop hypotheses, then test them in "pilot experiments" (small "scouting/fishing trip" attempts to solve a problem). Editors and referees of manuscripts submitted for publication are generally charged with ensuring that only the most critical information to prove the point(s) of the study be included. As a result, key information about subtle aspects of the preparation, storage, and maintenance of activities is relegated to theses or lab notebooks and, therefore, difficult or impossible to obtain.

2. Chromatography

Chromatography is typically accomplished by passing a dissolved biomolecule and its accompanying impurities through a column. A "column" is composed of chromatography matrix packed in a solid tube, terminated on one end by a porous "frit," and exit port where flow can be controlled by a valve. Methods are classified on the basis of the separation technique and summarized below.

Column tubes used to support the matrix and establish a flowing system vary according to the type(s) of stresses imposed by the separation conditions. Low-pressure chromatography is done using columns constructed of quartz (glass), plastic, or mixed construction polymers. Glass columns should be coated with a salinization reagent to prevent samples from sticking. Also consider whether chemicals, whose concentrations are changed during the separation processes, might adhere to the walls or matrix. Medium and higher pressure chromatographies (FPLC, HPLC) require use of columns composed of materials that resist breakage at separation pumping pressures ranging between ca. 200 and 500 psi in the former methods and 1000 and 5000 psi in the latter.

2.1 Gel Filtration Chromatography. We begin our summary with the most commonly used technique, gel filtration (or "size exclusion") chromatography. Gel filtration involves preferential retention of smaller molecules with a gradient in separation capability that depends on the size of molecules that are excluded from or included within the matrix material to varying extents. In an important application, a set of standard proteins that spans a range of sizes (molecular masses) is used to calibrate the column, permitting one to determine the apparent mass of a newly purified or otherwise uncharacterized protein of interest.

2.2 Ion-exchange Chromatography. Ion-exchange chromatography involves adhesion of the biomolecule to the matrix by ionic interactions and elution by applying a gradient of salt solution that overcomes matrix-biomolecule interaction by shielding their respective charges. Separation depends on differences in interaction (binding) constants for different biomolecules, which results in different characteristic elution positions in the salt gradient. An example of a typical ion-exchange matrix is diethylaminoethyl (DEAE) cellulose. This matrix binds a variety of proteins and nucleic acids at a lower salt concentration and then releases them at a characteristic salt concentration, typically in the 100 mM to 1 M range, which depends on the negative charge of the biomolecule. Since most contaminants have charge properties different from the biomolecule of interest, this technique can lead to substantial purification.

In some cases, molecules bind irreversibly to the column. Sometimes the chromatography resin can inactivate the biomolecule yet release it to the unsuspecting researcher. Such situations can be trouble-shot by running small-scale "batch" partitioning tests in Eppendorf tubes, as described in the Pharmacia literature.

2.3 Reverse-Phase Chromatography. Reverse-phase chromatography separates molecules based on the combined effects of ion-exchange ligands that contain a nonpolar moiety ("amphiphilic" functional groups). A typical resin is benzoylated DEAE (bz-DEAE) cellulose. Proteins and nucleic acids (e.g., transfer RNAs) are separated using this technique. For example, one can separate and purify many of the natural tRNAs using elution gradients involving increased [NaCl] and then adding ethanol to decrease the solvent polarity, even though the ionic strength has been increased. The combined effect liberates more hydrophobic species, which bind relatively irreversibly to DEAE cellulose, thus illustrating the distinction between the resins.

2.4 Substrate Affinity Chromatography. Substrate affinity chromatography involves separation by using a small- to intermediate-sized ligand (e.g., nucleotide, peptide fragment, enzymatic substrate, metal complex, nucleic acid, etc.), which is either attached to the column matrix or to the biomolecule that is to be purified. This column-bound ligand is recognized by the protein and bound while contaminants pass through the column. An example of one such ligand is ATP, which binds ATP utilizing proteins. Many variations on the basic theme have been developed. An example of the latter technique has garnered a lot of attention as a means to purify cloned proteins. A very common application involves attaching a hexa-histidine tail ("his-tag") to the N- or C-terminus of the expressed protein. This is accomplished by encoding the corresponding DNA in the correct translational reading frame with the cloned gene. The protein binds ligand-bound nickel on the column while his-tag-deficient proteins pass through. The purified protein is then eluted with imidazole, which displaces the his-tag from the nickel resin.

2.5 Hydrocarbon Interaction Chromatography. Hydrocarbon interaction chromatography (HIC) describes methods that differ from reverse-phase chromatography in that the column matrices contain attached hydrocarbon chains rather than amphipathic ligands such as bz-DEAE. Less polar and virtually nonpolar molecules are separated based on different affinities for the hydrocarbons. Different chain lengths produce different separation properties. Both solvent polarity and ionic strength may be manipulated, but one generally focuses on low polarity solvents with different chemical characteristics. These ligands resemble those used in chromatography to separate organic chemicals by partitioning them in the gas phase. This technique is used to separate intrinsic membrane proteins based on their lower polarities, low polarity nucleic acid components and coenzymes, and analogous applications.

2.6 Ion-Pairing Chromatography. Ion-pairing chromatography (IPC) involves complex formation between the material to be separated and IPC reagent molecules via salt bridging. Complex formation produces a unique set of physical properties for the coated biomolecule, which can be exploited by chromatography, usually over a reverse-phase resin. This is common practice in the separation of nucleic acids and their fragments using HPLC.

2.7 Thin-Layer Chromatography (TLC). Many applications have been developed to use impregnated TLC surfaces to separate protein and nucleic acid fragments and components, coenzymes, lipids and derivatives, carbohydrates, etc. The ion-exchange, reverse-phase, and HIC chromatography methods all work in certain situations. One "spots" the sample onto the starting positions of the TLC plates and then places it in the solvent of interest. One can increase the spot "concentrations" by spotting and drying samples repetitively. Capillary flow forces separate the molecules of interest from contaminants as they are carried upward with the absorbed liquid through the thin-layer materials from the bottom of the chamber.

Much of this work has now been relegated to "old literature" (little used and deemed relatively passe´). *Methods in Enzymology* articles from the 1950-1980 period provide a wealth of informative ideas and some useful perspectives that will never become intellectually "passe´." The techniques often required using volatile organic-based solvents, sometimes presenting valid safety and disposal concerns. TLC approaches were used more for analytical chemistry applications (e.g., determining modified nucleosides in tRNAs and protein modifications) than for preparative protocols. Shorter oligonucleotides can often be recovered and used for further analytical, spectroscopic, or biological experiments.

3. Proteins Produced in *E. coli*

Escherichia coli has been the workhorse for protein production since the early days of biochemical analysis. The rudimentary details of molecular genetics were worked out by researchers using materials typically prepared from *E. coli* extracts. In the process, they discovered and learned how to manipulate DNA polymerases, nucleic acid substrates and products and catalogued a large number of required or stimulatory activator proteins and inhibitors (e.g., repressors), often referred to as "factors."

Important advantages of bacteria relative to eukaryotic systems include rapid growth in fully or relatively well-defined media, and the ability to use soluble inhibitors to control growth or expression patterns and to verify retention of plasmids via the expressed functional activities of their proteins. Materials can be produced with and without infection by viruses, e.g., the "T-even and -odd phages," ϕx174, fd, M13, and Qβ. The early research enterprise was enlightening because it showed how virus activities are induced by subjecting the host cells to stressful conditions. They also demonstrated how viruses use key functional components of the host to accomplish their cohabitation with, or dominance over, the host cell.

Obtaining proteins from these relatively well understood organisms involves growing smaller pilot cultures from stored materials (on "plates" or "slants") then inoculating larger cultures and allowing the cells to grow until the appropriate time in the growth curve (typically "mid-" or "late log phase"). At that time, cells are harvested, concentrated by centrifugation or by using a filtration device, and "wet weight" yields are determined. Problems encountered in subsequent steps are summarized below.

4. Cell Lysis and Protein Extraction

Cells are typically lysed using either chaotropic, amphipathic or nonpolar chemicals, a lysozyme treatment, physical disruption by sonication (using a metal probe), or a combination thereof. A chaotrope (e.g., urea, guanidium) "induces chaos" in the structure of H_2O surrounding the membranes and intrinsic core components and thereby disperses the lipids and unfolds the proteins. An amphipathic material (e.g., sodium dodecyl sulfate, "SDS") contains both polar and nonpolar substructures and essentially opens up the hydrophobic interiors of structures (by binding to them via their nonpolar moiety) and then coats them (by extending the polar moiety into the surrounding solution).

Typically, the cellular extract is fractionated into soluble and insoluble subfractions by one or a set of "salting in" or "salting out" procedures using ammonium sulfate followed by centrifugation to separate the liquid and solid phases. In some cases, the former is retained, in others the latter. The solid can then be redissolved, dialyzed, and concentrated for subsequent steps, assays, electrophoresis, and other procedures.

To "concentrate samples" (actually to "increase their concentrations") one typically uses centrifugation through a filter (ultrafiltration, e.g., Centricon filtration) or controlled lyophilization. Lyophilization involves placing the sample in an opened Eppendorf or culture tube, placing parafilm over the opening, and poking small holes in it, creating a surface that helps retain materials if they "spatter" during lyophilization. (This "spattering" is called "bumping" in chemists slang and means to bubble out, burst, and possibly spread the material around or out of the sample holding device as a result of "boiling" during errant sublimation.) Next, the lower portion of the sample tube is either quick-frozen, by immersing it in an ethanol/dry ice bath, or frozen conventionally. Samples are placed in a rack and then into a bottle that is attached to the cooling/vacuum pump apparatus. Concentrated (but not "freeze-dried") materials are obtained by terminating the process prior to removing all of the liquid. Cautiously releasing the lyophilized or concentrated materials is essential to avoid losing the material to "bumping."

The Speed-Vac centrifugation-lyophilizer is mentioned as an example of how to look for "hidden causes" in preparative protocols. The solid-liquid is centrifuged and sublimes at decreased pressures by spinning the material in an Eppendorf tube, continually impacting the sample into the bottom of the tube by centripetal force. In contrast, a conventional lyophilizer sublimes the sample but does not compact it. The difference in these methods is of practical interest because Speed-Vac samples can be much more difficult to dissolve or reconstitute from concentrates because the impaction affects the properties of the protein. The Centricon "concentration-dialysis" device removes small salts and molecules as centrifugation decreases the volume; the Speed-Vac does not. Consider dialyzing samples in the Centricon or a microdialysis apparatus prior to the lyophilization step. In fact, to produce spectroscopic samples (e.g., CD, Raman, NMR) protocols are developed for such procedures, sometimes involving several rounds of concentration, dialysis, desalting, and solvent-exchange steps (e.g., H_2O to D_2O).

An important consideration for some proteins involves the necessity of isolating them from "inclusion bodies" and then refolding them. Guanidinium chloride denatures protein secondary and tertiary structures. One follows the denaturation step with stepwise dialyses performed at progressively lower guanidinium concentrations to gently rid the protein of denaturant and promote faithful refolding.

5. β-Galactosidase Activity Assay

Assessing protein purity involves assaying its specific activity. Activities span a range of processes but generally involve either catalysis of some chemical transformation from a supplied reactant to a product or binding the protein to a functionally relevant substrate. *Specific activity* characterization involves determining the number of transformations catalyzed per unit time per amount of catalytic protein. Thus, one must also determine the molar quantity of the protein. This in turn requires knowing its *molecular mass* and its *percentage* relative to the *total amount of material.* The goal of the purification is to improve the specific activity of the biomolecule of interest, in a preparation at a given stage. This is achieved by decreasing contaminant concentrations, while retaining as much of the original activity and pure material as possible from previous stages.

The protein β-D-galactosidase catalyzes the cleavage of the disaccharide lactose and produces the monosaccharides galactose and glucose. The protein can also convert a number of synthetic colorimetric or fluorescent substrates into products. Their presence demonstrates that the β-galactosidase gene was transcribed correctly, the mRNA was translated properly, and that the protein folded into the correct domain structure to support enzymatic catalysis. These are three very powerful pieces of information. Unfortunately, if no color is produced, one must dig deeper to figure out whether the problem occurred at the transcription, translation, or folding level. A fourth possibility is that some component in the cytoplasm, cell extract, or purification reagents inhibits the assay and the protein functions properly if it's removed. One can also infer, at least in principle, that your attached protein of interest, your "β-galactosidase fusion partner," has also adopted a functionally viable domain structure. Obtaining β-galactosidase activity *only implies* that the functional activity of the connected fusion protein will also work properly. This assumption can only be confirmed or denied based on further functional studies with your activity.

Note that the fusion partner enzyme β-galactosidase should not be confused with the much discussed "*lac repressor*." The latter protein is bound by compounds such as isopropylthiogalactoside (IPTG), an "inducer," which results in the disassociation of the lac repressor from the operator DNA sequence and thereby induces transcription. IPTG is built to resist breakdown by β-galactosidase so it remains present in the cell or extract.

5.1 *Enzymatic Assays. Enzymatic assays* characterize the protein"s ability to catalyze reactions in kinetic terms. One learns how fast product is made per mole of substrate(s) at fixed enzyme concentration. The Michaelis-Menten kinetics analysis technique is treated in detail in nearly all biochemistry textbooks. Many students find the mathematical frameworks imposing and assume that they are impenetrable; as a result, they lose sight of the goal.

The *Michaelis-Menten constant* (K_m) and the *maximum velocity* (V_{max}, the maximum "initial rate") are the key parameters of interest. The parameters characterize the velocity and the catalytic rate, and incorporate information about the substrate concentration. To determine them, one holds the enzyme concentration constant and measures the reaction rates (products made per unit time). The early linear portion of the time course is used to determine the initial rate ("velocity"). Little substrate (S) is depleted during this time, so this velocity is the maximum value obtained at this particular [S]. From this experiment one data pair is acquired, $[S]_i$ and the initial velocity $V_{o,i}$. The same procedure is then carried out with a range of [S], producing a range of data pairs, which are used to construct "secondary plots." The plot $V_{o,i}$ versus $[S]_i$ for the data pairs (i = 1, 2 ...), with $V_{o,i}$ as the y-axis and $[S]_i$ as the x-axis, is essentially linear at low $[S]_i$ then the slope decreases in experiments that contain sufficient [S] to approach V_{max}. The fixed amount of enzyme can only produce so much product, despite adding more and more reactant. K_m is the substrate concentration [S] that gives half of V_{max}, and is determined using a double-reciprocal plot (see most textbook treatments). This treatment assumes that the [protein], temperature, and solution conditions remain constant.

We refer students to the biochemical literature to verify that they can actually understand how Michaelis-Menten analyses are used to characterize specific activities. Typical university libraries have a range of books that discuss the details. Basic characterization techniques are typically described, then a series of models, designed to capture the details of the specific mechanism, are usually developed. Understanding the actions of medicines requires that these concepts be thoroughly understood since pharmaceuticals usually interact with enzymatic active sites.

6. Gel Filtration Analysis of Molecular Mass (Using Standards)

Articles in this Unit, especially Lab 7.2 demonstrate how to measure an apparent molecular mass using gel filtration. Gel filtration chromatography (GFC) and polyacrylamide gel electrophoresis (PAGE) behaviors of macromolecules are dependent upon their ability to diffuse through the respective resolving matrices. This capability is subject to restrictions imposed by the matrix and the Stokes radius (r_{Stokes}) of the particle of interest and those of the standard materials used to construct the calibration plot, and the conditions of the experiment. The r_{Stokes} parameter characterizes the tumbling behavior of a particle as a function of its apparent size as a weighted average of its three cartesian coordinates. When all three coordinates are the same length, the particle is spherical and tumbles "isotropically." When different radii are present the tumbling is called "anisotropic" and occurs more predominantly about the longer axis. This accounts for the effect of the nonspherical shape of the particle (oblate, prolate ellipsoids).

Proteins diffuse/tumble through the gel filtration "resin" or polyacrylamide gel driven by either the flow force of the elution buffer (GFC) or attraction to the electropositive pole of the electrophoretic charge gradient (PAGE). Mobilities in GFC and PAGE are affected in opposite ways by the size of the particle. In GFC, the smaller particles diffuse into the interior space within the gel bead particles, which is composed of an intermeshed webwork of carbohydrate chains. In order to travel down the length of the chromatography column, they must follow a more torturous path than larger particles, which are less "included" as they flow. Sufficiently large particles are "excluded" from the bead interior. As a result, smaller proteins elute later than larger ones in GFC profiles and are resolved across a gradient of sizes of included particles; larger molecules elute at small elution volume, smaller ones at larger elution volumes. In contrast, in gel electrophoresis, smaller proteins migrate more rapidly than larger ones, thus reaching the bottom of the gel earlier and developing a molecular size gradient from larger molecules (top of the gel) to smaller (further down the gel "lane").

In each technique, the measured "molecular mass" is merely a relative value obtained with reference to a series of precalibrated proteins (or carbohydrates such as "blue dextran"). This relative value is referred to as molecular weight and is symbolized variously as MW, M_r, Mol. Wt., or m (for mass), but it must be kept in mind that it is not an absolute value, as would be obtained by, for example, mass spectrometry. Additional details are described in the lab and included Pharmacia literature.

7. Enzyme Purification Table

When presenting a new purification protocol in a thesis, professional report, or publication, it is essential that the progress of the purity be documented at each stage. The series of steps are documented as vertical entries down a set of columns containing the correlated information. A generic example is presented below: The first column presents the title of the technique used in the step and the second recounts additional considerations. For example, a method might involve running a column with the same resin and simply changing the pH of the solvent to produce different separation characteristics in the step. The third and fourth columns present experimental activity and protein concentration, respectively. The fifth presents the calculated specific activity. By following the progressive increase in specific activity and the net yield, one can assess the extent of purification. This is reported as net yield of protein and specific activity of each step.

8. Pooling Chromatography Fractions

Optimize the steps by continually improving the parameters for each consecutive step. Removing unwanted protein decreases the net amount but improves the amount of activity per unit of protein obtained. One tries to separate and discard inactive material without losing any more of the desired active protein than necessary. "Pooling" fractions means to choose and combine the tubes from the chromatography that retain the activity. This is accomplished by measuring the amount of protein (often only approximately using the A_{280}) and the activity of selected samples distributed across the output fractions. Plot both sets of data (as vertical axes) on the same graph versus the fraction number (horizontal axis). Typically, the activity peaks while the protein concentration remains similar or even decreases. Pooling as much activity as possible and minimizing the amount of inactive protein involves "cutting" fractions from the wings of the activity peak. Keep in mind that obtaining more activity when pooling material (retaining the wings, "being greedy") may decrease the extent of purification and/or yield. Leaving some of the material behind often results in a higher specific activity; purer, but at a net loss. Fewer impurities means you have less chance of inhibition or other uncontrolled effects when the material is reconstituted into a reaction mixture, expressed *in vivo*, etc. Curiously, removing other proteins, even unrelated to that of interest, can sometimes produce a loss of activity. This lost activity is recovered in some cases by adding a known protein, typically BSA, to the reaction. The added protein increases the extent of molecular crowding, decreasing the net volume and thereby increasing the concentrations of enzyme and substrates. Ultimately, this increases the reaction rate and product concentration. Another effect of BSA is its ability to keep purified proteins from binding non-specifically to tube walls.

Purification of ______________________________ **batch no.**________ **Date**_____________
Notebook(s) no. (ppg.):
Notes:

Step no. and method	**Additional info.**	**Activity** *Units*:	**Net amount of protein obtained:** *Units*:	**Specific activity** *Units*:
1. Cell extract	Includes ammonium sulfate salting out, centrifugation, dialysis, Centricon			
2. Gel filtration chromatography				
3. DEAE cellulose chromatography	Gradient: 50 mM to 1.2 M NaCl in 20 mM Tris-HCl, pH 7			
4. (same matrix)	Same procedure except at pH 8.7			

THEORY/PRINCIPLES

PREPARATION AND HANDLING OF BIOLOGICAL MACROMOLECULES FOR CRYSTALLIZATION

Lorber, B. and Giege, R. (1992) *Crystallization of Nucleic Acids and Proteins: A Practical Approach* (A. Ducruix and R. Giege, Eds.), pp. 19-45, IRL Press. (Adapted. Referred to as "CNAP" in text.)

1. Introduction

In the crystallization of biological macromolecules, the quality and quantity of the required material are important. Although experimenters may have no choice, difficulties in crystal growth may be linked to the nature or source of the biological material. Purification, stabilization, storage, and handling of macromolecules are therefore essential steps prior to crystallization. As a general rule purity and homogeneity are regarded as conditions *sine qua non*; however, quality of the macromolecules will depend upon the way they are prepared. Although some structures were solved with less than 1 mg of material, a few milligrams should be available for the first crystallization trials. Once crystals can be produced which are suitable for X-ray analysis, additional material will be needed to improve their quality and size and to prepare heavy-atom derivatives. For these reasons it is essential that isolation procedures are able to supply enough fresh macromolecules of reproducible quality.

The aim of this chapter is to give a brief overview of biochemical methods used to prepare and characterize biological macromolecules. Practical aspects concerning manipulation and qualitative analysis of macromolecules intended for crystallization experiments will be emphasized. The cases of nucleic acids and membrane proteins are described in, more detail in Chapters 7 and 8 (of the book, CNAP). Finally, methods for the rapid characterization of the macromolecular content of crystals will be presented as well.

2. The Biological Material

2.1 Sources of Macromolecules. Many specific biological functions are sustained by classes of proteins and nucleic acids universally present in living organisms. Therefore, the source of the macromolecules may seem unimportant. In fact, better crystallization conditions or molecules that form better diffracting crystals are frequently found by switching from one organism to another. This is because variability in sequences between heterologous species may lead to different conformations, and consequently to different crystallization behaviors. Also, because of the high solvent content (50-80%) and the existence of relatively few contacts in macromolecular crystal lattices, differences in crystal quality or habitat may result from addition or suppression of intermolecular contacts. In practice, proteins isolated from eukaryotes are frequently more difficult to crystallize than their prokaryotic counterparts. Often their degree of structural complexity is higher. They can possess additional domains that may contribute to less compact and/or more flexible structures.

Posttranslational modifications are often responsible for *structural or conformational microheterogeneity*. Proteins isolated from *thermophilic microorganisms* are more stable at higher temperatures than those from other cells. As a result, they may be more amenable to crystallization. Better crystallizations of aminoacyl-tRNA synthetases and ribosomes isolated from extreme thermophiles illustrate this well. Proteins from halophilic microorganisms are alternative candidates because they have their optimal stability in the presence of high salt concentrations close to those used to reach supersaturation. Finally, "freshness" of the starting material and physiological state of cells may be important. Indeed, certain proteins from unicellular organisms are isolated in their native state only when cells are in exponential or prestationary growth phase. Also, since catabolic processes are predominant in tissues of dead organisms, such materials should not be stored before use, unless they have been frozen immediately post mortem.

2.2 Macromolecules Produced in Host Cells and in Vitro. Macromolecules that are most stable, easiest to isolate, and most abundant are usually studied and crystallized first. Researchers typically deal with biological molecules that are present only in trace amounts. Thus, preparing quantities sufficient for crystallization assays is a limiting step even when using micromethods. In a number of cases, this problem can be circumvented owing to the advancement of genetic engineering methods which make it possible to clone and overexpress genes in bacterial or eukaryotic cells. To use these methods, proteins must be isolated and a small part of their sequence determined. In engineered cells, overproduced proteins can accumulate to levels ranging up to a quarter of the total proteins in the cell (typically about a hundred milligrams of protein can be isolated from less than a hundred grams of bacterial cells). Unfortunately, high intracellular concentrations of certain proteins may lead to the formation of inclusion bodies of denatured, aggregated, or pseudo-crystallized material that require the use of adapted isolation procedures. Therefore, overproduction levels should be optimized to obtain functionally active proteins in their natural conformation. The separation of foreign macromolecules from endogenous ones can be difficult when no specific biological assay is available. Finally, the presence of ligands or chaperone proteins may be indispensable to maintain the native conformation of certain proteins. All of the above factors have to be kept in mind when planning a purification strategy.

One of the advantages of recombinant DNA technology is the possibility of expressing the same macromolecule in different cell lines. Maturation enzymes responsible for co- or posttranslational modifications in host cells may not work with the same efficiency on recombinant macromolecules and different end products may be obtained. For example, glycosylation occurs to various extents in eukaryotic cells but it is not known to occur in prokaryotes. Consequently, different modification patterns will lead to structural variants and this may be amplified when modification enzymes are limiting in the presence of overproduced proteins. In the future, continuous cell-free translation systems may represent an alternative tool to produce natural or fictitious proteins for crystallization.

A similar situation is encountered when genes for nucleic acids are transcribed in host cells or *in vitro*. For tRNAs produced *in vivo*, the modification of bases may be partial. The tRNAs transcribed *in vitro* by polymerases (from bacteriophages SP6 or T7) do not possess modified bases since enzymes responsible for posttranscriptional maturation are absent. T*hese transcripts can also be heterogeneous in length because individual polymerases do not terminate transcription at a unique position.*

2.3 Design of Modified Macromolecules. Using site-directed mutagenesis, unlimited changes can be introduced in the sequence of genes coding for polypeptide chains or RNAs. Such alterations can have various consequences, including conformational changes, decreases in stability, loss of activity, incomplete folding, changes in solubility, variations in the extent of postsynthetic modifications, or unforeseen degradation. Design of protein mutants may include:

a. substitution of amino acid residues to increase stability or solubility,
b. deletion to remove parts of the polypeptide chain forming flexible domains,
c. fusion with sequences recognizable by immobilized ligands in affinity chromatography.

The production of mutants more suitable for the preparation of heavy-atom derivatives is another potential application of genetic engineering. This was first shown in the case of phage T4 lysozyme, but also in that of selenomethionine-containing proteins for anomalous scattering studies.

3. Isolation and Storage of Pure Macromolecules

The topic of macromolecule purification has been reviewed extensively. Here, only a brief overview will be given. Readers are also encouraged to consult manufacturers, or suppliers, catalogues and application notes containing updated technical information.

3.1 Preparative Isolation Methods.

3.1.1 Generals. Purification procedures differ widely because individual macromolecules have peculiar properties. For that reason it is difficult to give a general scheme to facilitate the purification of an unknown macromolecule, each having to be considered as unique. Development, improvement, and optimization of purification protocols are mainly achieved by trial-and-error approaches.

In crude extracts, macromolecules may be protected by interaction with ligands or other molecules, which will probably be eliminated during purification. Consequently, the sequence of events in purification protocols is important. In any case, macromolecules must be separated as quickly as possible from harmful compounds that can damage them (e.g., hydrolytic enzymes). Under no circumstances should these harmful compounds be enriched or cofractionated. Major methods used in purification processes are listed in Table 1 together with the appropriate equipment.

Two major stages common to most purification protocols are the *preparation of a cellular extract* (except for secreted macromolecules) and the *subsequent fractionation of its components*. Intracellular macromolecules have to be released from cells or tissues using physical, chemical, or enzymatic disruption methods. Hydrophobic macromolecules must be solubilized and extracts have to be clarified by centrifugation or ultrafiltration. Extracellular compounds, i.e., those secreted by the cells in the culture medium, and macromolecules synthesized *in vitro*, may be recovered either by ultrafiltration, centrifugation, flocculation, or liquid-liquid partitioning.

3.1.2 Proteins. Isolation procedures for proteins usually involve a gross fractionation done by one or several precipitations induced either by addition of salts (e.g., ammonium sulfate), organic solvents (e.g., acetone), organic polymers (e.g., PEG), or by physical treatments such as pH or temperature changes, in order to decrease solubility or denature unwanted macromolecules. Fractionation between two liquid phases and selective precipitation (e.g., of nucleic acids by protamine) are additional methods. The next steps involve more resolutive separation methods, generally a combination of column chromatographies. These are based on *separation by charge* (absorption, anion or cation exchange chromatography, chromatofocusing), *hydrophobicity* (hydrophobic interaction or reverse-phase chromatography), *size* (exclusion chromatography or ultracentrifugation), *peculiar structural features* (e.g., affinity chromatography on matrices substituted by heparin, antibodies, metal ions, or sulfhydryl-containing compounds), or *activity* (affinity chromatography based on catalytic site or receptor/ligand recognition, biomimetic chromatography). HPLC yields higher resolution than the standard technique because of the monodispersity and small size of spherical matrix particles. Free-flow electrophoresis and preparative isoelectric focusing (IEF) [in gels or in a Rotofor cell commercialized by BioRad] are alternative separation methods. Monitoring specific activities during the purification procedure allows detection of unsatisfactory steps in which macromolecules are lost or inactivated.

Table 1 - Methods and Equipment Useful for the Purification of Biological Macromolecules

Cell culture	Fermentors, culture plates, thermostated cabinets, high capacity centrifuges or filtration devices for cell recovery
Cell disruption	Mechanical disruption devices (grinders, glass bead mills, French press). Chemical or biochemical treatments (e.g., permeation of cells by enzymes, phenol treatment for recovery of small RNAs) Others (e.g., sonication, freezing/thawing)
Centrifugation	Centrifuges (low speed to eliminate cell debris or recover precipitates and high speed to fractionate subcellular components)
Dialysis and ultrafiltration	Dialysis tubing, hollow fibers or membranes (various porosities and size) Concentrators (various capacities), high flow rate membranes with low macromolecule binding and various cut-offs)
Chromatography (use metal-free systems)	Low-pressure chromatography or HPLC columns and matrices Pumps, programmer, on-line absorbance detector, fraction collector, recorder
Preparative electrophoresis or isoelectric focusing	Electrophoresis apparatus for large tube or slab gels Preparative liquid IEF apparatus (column or horizontal rotating cell)
Detection, characterization, and quantification	Spectrophotometer, fluorimeter, pH meter, conductimeter, refractometer (for monitoring solutions and chromatographic elutions) Liquid scintillation counter (for radioactivity detection) Analytical electrophoresis and IEF equipment.

Guidelines for effective protein purification:

1. work in the cold room (i.e., at 4°C) with chilled equipment and solutions if the protein is unstable at higher temperatures;
2. use precipitation steps to speed up fractionation;
3. limit the number of chromatographic steps to not more than three or four, if possible;
4. avoid the use of time-consuming assays for the characterization of macro-molecules;
5. use quick intermediate treatments (e.g., dialysis and concentration) that are non-denaturing and efficient;
6. prevent inactivation, denaturation, and degradation by adding protectors, stabilizing agents, and/or protease inhibitors.

3.1.3 Nucleic Acids. The purification of nucleic acids requires appropriate methods. For tRNAs, the first step is separation from proteins by phenol extraction, usually followed by a gross fractionation by counter-current distribution, or chromatography on benzoyl-DEAE-cellulose or BD-Sepharose. Further purification is based on anion-exchange, adsorption, reverse-phase, hydrophobic interaction, or affinity chromatographies. Intermediary treatments include precipitation by ethanol, dialysis, and concentration by evaporating under vacuum. Improvements have been made by using HPLC on various matrices (e.g., ion exchange, reverse-phase, adsorption, hydrophobic interaction, or mixed-made matrices). Recently, chromatography on a matrix made of silica substituted by short aliphatic chains has proven useful. An alternative approach is the benzoylation of DEAE groups bound to resin or silica-based matrices. Oligo-nucleotides used for crystallization are obtained by automated chemical synthesis on solid-phase sup-ports. Since final products are often contaminated by abortive sequences, a purification step (usually by reverse-phase HPLC) is needed. Ribosomal 5S RNA, which was used in crystallization attempts, was separated from other ribosomal components by molecular sieving.

3.2 Stabilization and Storage of Pure Macromolecules. Biological macromolecules are frequently fragile, sensitive to variations of their environment, and can easily lose their native structure or activity. Once they have been removed from their natural medium, they must therefore be placed in solvents having properties close to those of the cellular medium, to maintain their native conformation intact and active.

Improper storage conditions may spoil the precious macromolecules obtained after long and hard work. Buffers whose pK_a are only weakly affected by temperature must be chosen so that samples stored at -20°C and assayed at room or higher temperatures are not subjected to important pH variations. Ionic strength should also be controlled since macromolecules may require a minimal salt concentration to stay soluble, although all ions may not be compatible with their native structure or activity. In most cases, denaturation can be minimized by avoiding denaturing treatments, pH or temperature extremes, as well as contact with organic solvents, chaotropic agents, or oxidants. Proteins containing cysteine residues usually require a *reducing agent* (e.g., DTE, DTT, 2-mercaptoethanol, or glutathione). Finally, proteins should not be stored as diluted solutions because they may adsorb onto the walls of glass or plastic containers.

Glycerol at high concentration (e.g., 50-60% v/v) is a good *stabilizing agent* for storage of proteins. It has the advantage of staying liquid at -20°C so that denaturation by ice formation can be minimized (its high viscosity also reduces diffusion by about two orders of magnitude). Storage as a suspension in ammonium sulfate solution is also recommendable but sometimes less convenient. Ligands can help to increase protein stability. Most proteins to be crystallized should not be dried or lyophilized; the latter process removes bound water molecules that belong to the macromolecule solvation shell. A bactericidal agent (e.g., sodium azide, ethyl-mercuri-thiosalicylate – Highly Toxic!) may be added for storage. Sodium azide may also be an additive in crystallization attempts, as well as the antifungal and volatile thymol, which may be added in reservoirs during crystallizations by vapor diffusion methods.

Nucleic acids may be stored either as alcoholic precipitates or dry, but should never stay in the presence of phenol – which leads to alkaline-type hydrolysis. Transfer RNA molecules should be stored at rather acidic pH (4.5-6) in the presence of Mg^{2+}.

4. Analytical Biochemical Methods

4.1 Gel Electrophoresis. Electrophoresis allows estimation of the apparent size of proteins and nucleic acids. IEF in polyacrylamide or agarose gels separates proteins according to their charge and gives a good estimate of their isoelectric point. In free or immobilized pH gradients, it is a very resolutive method that can reveal differences in isoelectric points smaller that 0.01 pH unit.

Electrophoretic titration of proteins in a pH gradient may be used to confirm their purity. Capillary electrophoresis is an emerging technique well adapted for rapid purity and homogeneity analysis. HPLC can be useful to separate small amounts of macromolecules for further analysis.

Sequence analysis is a good control, more sensitive than analysis of amino acid composition. The choice of appropriate methods to quantitate macromolecules is dictated by the amount and concentration of samples and the required degree of specificity. Spectrophotometry is direct and non-destructive and can be very accurate when extinction coefficients are known. The theoretical molar extinction coefficient of pure proteins (at 280 nm, ε_{280}; units: liter mole^{-1} cm^{-1}) can be calculated from tryptophan and tyrosine content using

$$\varepsilon_{280} = 5690\ n_x + 1280\ n_y$$

where 5690 and 1280 are the molar absorption coefficients of tryptophan and tyrosine at 280 nm, and n_X and n_y are the numbers of tryptophan and tyrosine residues in polypeptide chains, respectively. Thus, protein concentrations are obtained from:

$$c(\text{mg/ml}) = c(\text{mole/l})M_r(\text{g/mole}) = A_{280}M_r/(\varepsilon_{280}b)$$

The coefficients can be determined from ponderal, spectrophotometric, or refractive index measurements, as well as from colorimetric or dye binding assays.

In mixtures, protein concentrations are estimated using empirical formulas based on absorption differences at two wavelengths in order to eliminate the contribution of nucleic acids. For cells of l cm optical pathways or for a better estimate:

$$c(\text{mg/ml}) = (1.55A_{280} - 0.76A_{280})/I$$

Approximate concentrations of RNA are obtained by assuming that 1 A_{260} unit (for 1 cm pathlength) corresponds to 0.040 mg/ml. Values are more accurate when extinction coefficients are known. They are obtained from absorbance and phosphorus content measurements.

Proteins can also be quantitated by colorimetric methods such as the Lowry or Bradford assays. Results are skewed when macromolecular compositions deviate from that of the standard protein or when contaminants give interference. Activity assays are often more specific. For enzymes, active-site titration monitors the activity of individual molecules, but this method cannot detect micro-heterogeneous molecules whose activity is not affected. Immunological properties may be used to assess the ability of antibodies to recognize conformational states or parts of molecules.

4.2 Handling of Macromolecules. Pure macromolecules require special care to ensure they are not damaged or lost before or during the crystallization assays. Diluted solutions are concentrated by ultrafiltration in devices using pressure or centrifugal force (in cylindrical cells or parallel flow plate systems) or by dialysis against hygroscopic compounds (e.g., dry high M_r PEG or gel filtration matrices). In stirred pressure cells, loss of material by adsorption can be prevented by choosing appropriate membranes with low binding capacity. Adsorption onto membrane surfaces and damage by shearing are prevented by optimizing the stir-rate. Foaming and air bubbles should be avoided because they create an oxidizing environment, where disulfide bonds in proteins can break. Aggregates forming as a consequence of a decrease in pH, oxidation, or an increase in salt or protein concentration must be removed by centrifugation.

Concentration is also carried out by precipitation, e.g., by adding ammonium sulfate, followed by redissolution of the precipitate in a small volume. Adsorption on a chromatographic matrix followed by elution at high ionic strength is an alternative approach. Unfortunately, some techniques may also concentrate contaminants, including hydrolytic enzymes. Other techniques may be used as additional purification steps; among them concentration over membranes to eliminate small M_r compounds (provided appropriate cut-offs are chosen) and exchange of buffers in microscale set-ups. Finally, it must be mentioned that lyophilization leads to denaturation of many macromolecules.

Mild nonionic detergents (e.g., octyl glucoside) help to solubilize proteins. High concentrations of denaturing agents like guanidinium chloride, urea, or chaotropic detergents should be avoided because they inactivate or unfold macromolecules. For biological macromolecules, the term "purity" means absence of contaminants, whereas "homogeneity" implies absolute identity of all macromolecules in a sample.

Never repeatedly freeze and thaw macromolecules, to avoid denaturation, and do experimentation with separately packaged aliquots to limit repeated handling of single stock solutions. Remove undesired molecules that might hinder crystallization, by dialysis (e.g., glycerol), ultrafiltration, or size-exclusion chromatography. In practice, it may also be important to prepare macromolecules with or without their ligands (e.g., coenzyme, metal ions) or try additives (e.g., ions, reducing agents, chelators) because either form may crystallize more readily.

Contaminants may compete for sites on the growing crystals and generate lattice errors leading to internal disorder, dislocations, irregular faces and secondary nucleation, twinning, poor diffraction, or early cessation of growth. The *large number of molecules in a single crystal (about 10^{20} in a cubic crystal measuring 1 mm on the edge)* means that ppm amounts of contaminant (large numbers of molecules) are present to interfere with crystal growth.

4.2.1 *Small size contaminants.* Small molecules like peptides, oligonucleotides, amino acids, carbohydrates, or nucleotides, as well as uncontrolled ions, should be considered as contaminants. Buffer molecules remaining from a former purification step can be responsible for irreproducibility of crystallization assays (e.g., phosphate ions are relatively difficult to remove and may crystallize in the presence of other salts). Ions act as counterions and play a critical role in the packing of macromolecules. Often macromolecules do not crystallize or yield the same crystal environments in the presence of various buffers adjusted at the same pH. For these reasons, "purity" also means that all reagents used with pure macromolecules (e.g., precipitants, buffers, or detergents) should be of the highest grade. Commercial reagents contaminated by trace impurities should be purified. This is especially true for precipitants since they are used at relatively high concentrations. For instance, a contamination by Fe ions at a level of 0.001% w/w in a 2 M ammonium sulfate solution corresponds to 0.1 mM impurities, equal to a protein concentration of 0.1 mM (e.g., 5 mg/ml for a protein of M_r of 50,000). Purification techniques for common precipitants are listed in Table 2. Salts must be recrystallized. Several techniques have been developed for the purification of PEG, but also for detergents. MPD is purified either by distillation or column chromato-graphy.[63] High-grade products (e.g., HPLC-grade ammonium sulfate) in which contaminants do not exceed a few ppm are commercially available, but the label "ultrapure" is sometimes exaggerated or misleading.

Table 2 - Techniques for the Purification of Major Crystallization Agents.

Chemicals	Major Contaminants in Commercial Batches	Purification Techniques and REeferences
Ammonium Sulfate	Ca^{2+}, Fe ions, Mg^{2+}, PbSO4,[a] CN^-, NO_3^-	Recrystallization
PEG	Cl^-, F^-, NO_3^-, PO_4^{2-}, SO_4^{2-}, peroxides, aldehydes	Column chromatography[b,59,60]
MPD	Cl^-, K^+, Na^+, SO_4^{2-}	Distillation[d] column chromatography[c, 63]

[a] Nonsoluble species. [b] See Protocol 1 in CNAP Chapter 4. [c]Technique for the purification of PEG-6000: Dissolve 30 g PEG-6000 in 240 ml acetone (ACS grade) at 40°C in the water bath. Chill on ice to obtain a white precipitate. Add 240 ml diethylether and dry in, then under vacuum. Store dry. [d]MPD is distilled under vacuum.

Molecules released from chromatographic matrices are another type of contaminant. Molecules originating from noninert matrices susceptible to enzymatic digestion (e.g., Sephadex, celluloses) or from desorption of organic compounds (e.g., organic phases bound to silica matrices).

4.3 Microheterogeneity of Macromolecular Samples. Heterogeneity in pure macromolecules is called "microheterogeneity" because generally it is *only revealed by high-resolution methods* (see Section 4.4). As Table 3 shows, its causes are multiple, the most common being *uncontrolled fragmentation* and *postsynthetic modifications*.

4.3.1 Fragmentation. Proteolysis is often the major difficulty to overcome in protein isolation. In cells, proteolysis is a controlled mechanism involved in major physiological processes (e.g., maturation, regulation of enzymatic activity, and catabolism). For general reviews, see refs. 65-68. Proteases (also termed peptidases, proteinases, and peptide hydrolases) are enzymes with M_r values in the range

Table 3 - Possible Sources of Microheterogeneity in Proteins and Ribonucleic Acids

- Variation in primary structure (genetic variations, degradations)
- Variation in secondary structure (errors in folding or partial unfolding)
- Variation in tertiary structure (conformers)
- Variation in quaternary structure (oligomerization)
- Molecular dynamics (flexible domains)
- Variation in post-transcriptional or posttranslational modifications
- Partial binding of ligands or foreign molecules
- Various aggregation states
- Partial oxidation (e.g., sulfhydryl groups in proteins)
- Aging (e.g., deamidation in proteins)

20,000-800,000; they are formed both localized in various cellular compartments and secreted into the extracellular medium. *Four classes* have been characterized according to the structure of their catalytic site: the *serine-*, *aspartic acid-*, *cysteine-*, and *metallo-proteases*. They can be inhibited by a number of commercially available compounds (see Table 4). Some proteases cannot be inhibited by the usual inhibitors. During isolation of intracellular proteins, control over proteolysis is lost after all disruption due to the mixing of the contents of cellular compartments and to contact with extracellular proteases. Proteolysis can decrease a protein' size, make it less stable, and modify its charge, hydrophobicity, activity as a catalyst, or immunological properties.

Hydrolysis by endo- or exonucleases is a frequent source of heterogeneity in nucleic acids. Contaminant proteases or nucleases may not be detectable even when overloading electrophoresis gels – but can cause damage during concentration or storage of samples. Chemical hydrolysis is another problem encountered when working with RNAs. It occurs mainly in the presence of *metal ions or at alkaline* pH.

Table 4 - Some Commercially Available Protease or Nuclease Inhibitors

Proteases or Nucleases	Inhibitors [a]
All protease classes	Possibly α_2-macroglobulin or DEPC
Serine proteases	DIFP, PMSF, Pefabloc SC[b] (from Pentapharm Ltd), aminobenzamidine, 3,4-dichloroisocoumarin, antipain, chymostatin, elastinal, leupeptin, boronic acids, cyclic peptides, trypsin inhibitors (e.g., aprotinin, peptidyl-chloromethyl ketone)
Aspartic acid proteases	Pepstatins and statin-derived inhibitors
Cysteine proteases	Peptidyldiazomethanes, epoxysuccinyl peptides (e.g., E-64), cystatins, all thiol-binding reagents, peptidyl chloromethanes
Metallo-proteases	Chelators (e.g., EDTA, EGTA), phosphoramido and phosphorus-containing inhibitors, bestatin, amastatin, and structurally related inhibitors, thiol derivatives hydroxamic acid
Ribonucleases	RNasin (Promega), ribonucleoside-vanadyl complexes, DEPC
Deoxyribo nucleases	DEPC, chelatants (e.g., EDTA, EGTA)

[a] These compounds are dangerous for human health and must be manipulated with caution. See also ref. 32.
[b]According to the manufacturer, Pefabloc SC [4-(2-aminoethyl)-benzene-sulfonyl fluoride] is a nontoxic alternative to PMSF and DIFP.

4.3.2 Postsynthetic Modifications. Modifications occurring co- or posttranslationally are other common sources of microheterogeneity in proteins. Either all modification sites are occupied but the added groups are heterogeneous (e.g., glycosylation-introducing oligosaccharide chains or various sequences), or the modifications are complete but unevenly distributed over the polypeptide chains (e.g., when all sites are not substituted). Over a hundred modifications are known – some of which are listed in Table 5. Most of them require special methods for their analysis. Some modifications are reversible (e.g., phosphorylation) whereas others are not (e.g., glycosylation, methylation). Heterogeneity in carbohydrate chains, either N-linked at asparagine or O-linked at serine or threonine residues, is frequent in eukaryotic proteins. Partial posttranslational modifications may be minimized by removing added heterogeneous groups, as was done for carbohydrates using mild hydrolysis by oligosaccharide-cleaving enzymes. Microheterogeneities may appear during storage, e.g., by deamidation of asparagine or glutamine residues.

Numerous nucleotides are modified or hypermodified post-transcriptionally in tRNAs. Some of them modulate physicochemical properties, in particular, charge and hydrophobicity, of individual tRNA species. Also, their content varies with the physiological state of cells, so that it is understandable that microheterogeneities in crude tRNA batches affect purification and crystallization of individual species.

4.3.3 Conformational Heterogeneity. Conformational heterogeneity may have several origins: *binding of ligands*, *intrinsic flexibility* of *macromolecular backbones*, *oxidation of cysteine residues*, or *partial denaturation.* Macromolecules should be prepared in both forms, one deprived of its ligands and the other saturated with them. Controlled fragmentation may be helpful. Oxidation of a single cysteine residue can lead to complex mixtures of molecular species, diminishing the chances of growing good quality crystals. Redox effects like these can often be reversed by reducing agents.

4.4 Probing Purity and Homogeneity. A combination of several independent analytical methods is always required to assert the absence of contaminants and microheterogeneities in a macromolecular sample.

Spectrophotometry and fluorimetry are helpful to obtain information about the quality of macromolecules if they (or their contaminants) have special characteristics in their absorbance or emission spectra. Chromatographic techniques are generally not resolutive enough to be used as the only reference. Occurrence of a sharp peak in HPLC is not sufficient to ensure the quality of the product. Analysis of a protein sample by SDS-PAGE gives an idea about the presence and the size of polypeptidic contaminants, if they are detectable by staining, autoradiography, or immunodetection. Unfortunately, nonproteic contaminants are usually not as easy to detect. Gel IEF gives an estimate of the isoelectric point of protein components of a mixture; electrophoretic titration shows the mobility of proteins as a function of pH. Electrophoretic titration can also suggest the type of chromatography (i.e., anion or cation exchange) or chromatofocusing suitable for further purification. They serve to guide one' experiments toward chromatographies based on adsorption, size exclusion, hydrophobic interaction, or affinity. Capillary electrophoresis is also well adapted for sensitive control of purity; when known, the amino acid composition can be used as a control. If the primary structure of the polypeptide is known, sequencing of its N- and C-termini is used to control its integrity.

4.5 Improving Purity and Homogeneity. Some practical advice. Start with fresh material, change the sequence of events (by inverting chromatographic steps), or change the steps themselves (by using other chromatographic matrices). Avoid crosscontamination; never mix batches of pure macromolecules even when they look identical. A small shift in the elution from a chromatography column, or a preparation done on the same column but at another scale or temperature, may introduce other contaminants into the fractions. Such variability can sometimes be detected by IEF. All solutions in contact with pure macromolecules should be cleaned and sterilized by filtration (e.g., over 0.22 mm

Table 5 - Some Co- or Posttranslational Modifications of Proteins

Amino Acid Residues or Chemical Groups	Modifications or Chemical Groups Added
Amino terminal $-NH^{3+}$	formyl-,acetyl-, glycosyl-, amino acyl-, cyclization of Gln
Carboxy terminal-COOH	amide, amino acyl-
Arg	ADP ribosyl-, methyl-, ornithine
Asp	carboxyl-, methyl-
Asn, Gln	glycosyl-, deamination
Cys	seleno-, heme, flavin-
His	flavin-, phospho-, methyl-
Lys	glycosyl-, pyridoxyl-, biotinyl-, phospho-, lipoyl-, acetyl-, methyl-
Met	seleno-
Phe	hydroxyl-
Pro	hydroxyl-
Ser	phospho-, glycosyl-, ADP ribosyl-
Thr	phospho-, glycosyl-, methyl-
Tyr	iodo-, hydroxy-, bromo-, chloro-

porosity filters). To avoid release of molecules from chromatography columns, the best remedy is to use chemically inert and autoclavable matrices (e.g., Trisacryl supports from IBF Biotechnics or TSK gels from Merck).

A priori, macromolecules can be rendered more homogeneous in various ways. Although there is no guaranteed method to prevent unwanted proteolysis occurring *in vivo* or *in vitro*, a convenient approach is to add protease inhibitors. A variety of such inhibitors are commercially available (Table 4). Their efficacy can be checked using assays based on the solubilization of clotted protein or employing labeled proteins.[32,81] A cocktail of inhibitors should contain at least one compound specific for each class of proteases. Useful methods for detecting conformational heterogeneity are shown in Table 6.

Removal of undesirable parts of multidomain macromolecules by controlled fragmentation may help to obtain more compact and homogeneous structures. Indeed, some proteins crystallize or yield better crystals when proteolysed prior to crystallization (In controlled situations, unsuspected contaminations by proteases may lead to a similar result.) After controlled treatment, the macromolecular core must be purified in order to remove proteases and small fragments. To be reproducible, an enzymatic tool must be free of contaminant proteases that could generate undesirable cleavages. Unfortunately, preferential attack of accessible domains often introduces microheterogeneities that are detrimental to crystallization. Thus, purification of single species may be essential to grow large monocrystals; improved purifications by HPLC or IEF often yield better crystals.

On a small scale, chromatography over a column of immobilized protease inhibitors (e.g., α-macroglobulin) or substrate analogues (like arginine or benzamidine) may be designed to trap proteases. The major drawback of inhibitors lies in the possibility of their binding to, or inactivating, the proteins they should protect. An alternative solution to proteolysis or degradation is to express proteins in strains deprived of harmful protease activities.

Table 6 - Selected Techniques to Detect Macromolecular Conformational Heterogeneity.

Technique	Expected Information
Activity assay	Biological activity
Active site titration	Ligand binding, affinity
Gel electrophoresis	Mobility, size, charge, shape
Gel filtration	Size, shape
IEF, titration curve	Charge
Immunological titration	Antigenic determinants
Scattering methods (light, X-rays, neutrons)	Size, shape
Spectrophotometry, fluorimetry	Absorption or emission properties
Ultracentrifugation	Size

5. Preparation of Buffer Solutions Containing Protease Inhibitors and Stabilizing Agents

1. Prepare buffer solutions containing 10% glycerol and 10^{-3} M EDTA.
2. Prepare a stock solution of DIFP[a] by diluting a 1 ml commercial sample (~1 g) in 50 ml cold and anhydrous isopropanol to obtain a 0.1 M solution. This solution should be stored at -20°C. The peptidic inhibitors[b] (pepstatin, bestatin, and E-64) are prepared as 10^{-3} M stock solutions in ethanol/water (50/50).
3. Add DIFP, peptidic inhibitors and 2-mercaptoethanol (also DTE or DTT)[c] in buffers just prior to use (final concentrations 5 x 10^{-4} M, 5 x 10^{-6} M, and 5 x 10^{-3} M, respectively).
4. Add inhibitors fresh at each step of the isolation procedure.

[a]IMPORTANT: DIFP has to be handled with extreme caution since it is a powerful inhibitor of human acetylcholine esterase (POISON!).

[b]Experimenters should be aware of possible limitations: low ligand solubilities, the toxicity of the reagents, limited stabilities, excessively high binding affinities (and irreversibility), and matrix breakdown.

[c] Final concentration 10^{-4} M.

6. Homogeneity of Nucleic Acids

The purity of fractionated transfer RNA species is checked by electrophoresis under denaturing conditions in urea gels. Radioactive end-labeling produces low levels of cleavage in ribose-phosphate chains, which could hamper crystallization. Purity can be evaluated by specific activity measurements (usually expressed in units/mg), allowing one to estimate the efficiency of the individual steps in a purification procedure. Recall, the action of nucleases can often be minimized by adding inhibitors (Table 4). Ribonucleoside-vanadyl complexes trap the transition states of the SN2 displacement mechanism of many RNases. inhibitors. Diethylpyrocarbonate reacts with histidine residues at their surface and may inactivate these enzymes by eliminating a critical acid-base catalyst group catalytic site. *RNasin* (Promega), an M_r 51,000 protein isolated from human placenta, inactivates ribonucleases by stoichiometric noncovalent and noncompetitive binding. It can be added in cell-free transcription media to protect *in vitro*-synthesized RNAs but has to be removed prior to crystallization. An alternative way to prevent hydrolysis is to remove nucleases by affinity chromatography over *blue dextran-agarose* or *over 5''-(4-aminophenyl)-uridine-(2'',3'')-phosphate-agarose*. *Metal ions* involved in chemical hydrolysis of RNAs are removed from solutions by *ion exchange* or *chelation* (e.g., with *EDTA*, *EGTA*, or chelators immobilized on agarose beads). Magnesium at neutral pH is an exception to this rule and should be added to stabilize the structure of tRNAs. But beware, some phosphodiester bonds are particularly susceptible to hydrolysis when heated in Mg^{2+}-containing solutions. The pH of the medium should never be very alkaline, i.e., above 7.5.

THEORY/PRINCIPLES

SOLUTION STRUCTURE OF BIOMACROMOLECULES IN IONIC SOLUTIONS

Von Hippel, P. H. and Schleich, T. (1969) Accounts of Chemical Research 2: 257-265. (Adapted)

1. Introduction

Biological macromolecules, particularly the proteins and nucleic, consist basically of one or more linear copolymer chains. These chains contain about twenty (proteins) or about four (nucleic acids) chemically distinct types of monomer units, linked together by peptide or sugar phosphate bonds in a genetically determined sequence, and range from 10^2 to 10^6 residue units in length. Thus, within the limits of monomer sequence and backbone stereochemistry, any particular chain or set of interacting chains is potentially capable of assuming an enormous number of three-dimensional conformations.

Yet in aqueous solution at moderate temperature and pH, a given set of polypeptide or polynucleotide chains is generally found to be disposed in space in one or a small number of closely related structures. Furthermore, these are almost certainly equilibrium structures, since it has been shown with several systems that macromolecules can be reduced to random coil polymers and then "reannealed" into the original conformation by judicious manipulation of the solvent environment. It is thus generally believed that the particular "native" structure of a protein or nucleic acid represents a state of minimum free energy, and furthermore that the free energy of this state is the algebraic sum of the free energies of the individual residue-residue and residue-solvent contacts involved in the equilibrium conformation.

In marked contrast to the structural properties of the usual small molecule, the native conformation of most proteins and nucleic acids is only marginally stable, and often a decrease of only a few hundred calories per mole in one of the stabilizing interactions involved in maintaining this form is enough to trigger a cooperative transition to a quite different structure. This borderline structural stability can easily be rationalized since the existence of specific conformations requires a delicate balance of intermolecular interactions. Thus, some preferential residue-residue interactions are required to give the macromolecule an "inside" and a specific three-dimensional structure (fully preferential residue-solvent interaction leads, of course, to a random coil conformation [or a "molten globule intermediate"]). In addition, some favorable residue-solvent contacts are required to maintain a stable "outside" of the macromolecule and to prevent uncontrolled intermolecular interaction resulting in precipitation or phase separation.

This marginal conformational stability, together with the cooperative character of macromolecular transconformation reactions, is also important in functional terms since it permits the easy interconversions from one form to another; this may accompany the activity of these molecules as enzymes, genetic templates, and so forth. However, it represents a major headache to those physical biochemists who attempt to predict macromolecular conformations by summing up the free energies of the intra- and interchain interactions involved in various structures, or who attempt to evaluate the contributions of individual types of residue-residue or residue-solvent interactions to overall structural stability.

Nevertheless, in this account we will describe some aspects of the structural stability of biological macromolecules and consider how this stability might be altered by intersection of various components of the macromolecule with a variety of structural perturbants (particularly electrolytes), which can be added to the solvent environment.

2. Classification of Perturbants

Most work on the effects of perturbants on macromolecular structure has been done with proteins or protein models, and the first widely recognized perturbants used to destabilize ("denature") native protein structures were agents with obvious hydrogen-bonding potential, such as urea and guanidinium chloride. Since it had long been thought that proteins are largely held in their specific secondary-tertiary-quaternary structure by hydrogen bonds, and since several molar solutions of urea and guanidinium chloride had been found to bring about the conversion of compact globular proteins to disordered random coil forms, it was assumed that these agents worked by competing more successfully than water for the donors and acceptors involved in intramolecular hydrogen bonding in the native protein.[1] This view, while probably not entirely incorrect, must be considerably modified by the finding that these same agents increase appreciably the solubility of alkanes and nonpolar amino acids,[2] which clearly shows that competitive hydrogen bonding is not the whole story. Also, the development of the concept of hydrophobic bonding as a major determinant of protein stability[3] has tended to undermine the older simplistic view. Finally, and most obviously in nucleic acids, charge-charge interactions can also be shown to play a major role in determining the stability of macromolecular conformations.

Thus, to a first approximation, we can attempt to classify perturbants as potential hydrogen-bond, hydrophobic-bond,[4] and electrostatic-interaction affectors. However, it is immediately apparent that very few structural perturbants can be classified as "pure" affectors of one of the above types of interactions, and it is further apparent that ionic species can potentially affect all three of these types of interactions.

Specifically, then, we shall adopt the structural and operational concept of a biological macromolecule implied above (which follows directly from the basic ideas developed by Kauzmann[3] and Tanford[5] and their collaborators, and many others). In this view, that a macromolecule consists of residues that are predisposed to forming an "inside," consisting of residues that are not in contact with solvent, and an "outside," consisting of residues that are in contact with solvent. The major functional expression of perturbants, which increase or decrease the stability with respect to conversion to a fully solvated random coil, is to increase or decrease the free energy of transfer of residues from the inside of the protein into the solvent environment.

This thermodynamic effect can obviously have its origin in either of two general types of mechanisms (or a combination of both). (1) Small molecule perturbants can bind directly to the various interior groups in the folded macromolecule (which is in dynamic equilibrium with various unfolded forms so that its interior residues are continuously, though transiently, exposed to the solvent), thus increasing or decreasing the solubility of these structural elements in the external surroundings. (2) Perturbants can modify the organization solvent around potentially exposed groups, thus increasing or decreasing the free energy of transfer of these groups to the exterior through effects on local solvent structure. Since most biological macromolecules seem to attain the biologically active native conformation only in water or largely aqueous solutions, we infer that the properties of water and its potential for interaction with the various functional groups of macromolecules must play a decisive role in any final explication of conformational stability.

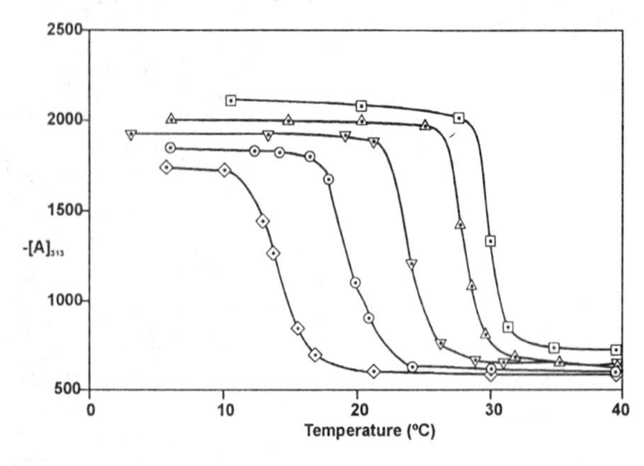

Figure 1 - Melting curves for icthyocol collagen in various concentrations of $CaCl_2$, pH 7, ~1 mg of collagen/ml and 0.12 M $CaCl_2$ (⊡); 0.26 M $CaCl_2$ (△); 0.5 M $CaCl_2$ (▽); 0.75 M $CaCl_2$ (⊙); or 1 M $CaCl_2$ (◇).

3. Macromolecule Manifestations

Often the simplest way to induce a reversible folding/unfolding or order/disorder transformation in a protein or nucleic acid is to increase the temperature of the solution by 10 to 50°C. If this is done gradually and the effect monitored by some physicochemical criterion sensitive to conformation, such as viscosity, ultraviolet hypo- or hyperchromism, optical activity, etc., it is seen that unfolding occurs quite abruptly over a rather narrow temperature range. By analogy with the not dissimilar behavior of macroscopic phase transitions, the temperature of the midpoint of this unfolding range is called the melting temperature (T_m). Thus, an experimentally simple way to measure the net effect of adding a small molecule perturbant to the solvent environment is to determine the change in T_m with changing additive concentration.

This type of effect is illustrated in Figure 1 for the thermally induced conversion of collagen from the triple-helical ordered form to the disordered random coil conformation in the presence of various concentrations of $CaCl_2$. As this figure implies, neither the structure of the ordered nor of the disordered form of collagen is changed appreciably by the addition of $CaCl_2$, but T_m (the temperature at which the helical and the coil forms are equally stable, i.e., at which $\Delta G = 0$ and therefore ($T_m = \Delta H/\Delta S$ for a two-state transition) is moved to progressively lower temperatures with increasing concentrations of $CaCl_2$. Furthermore, the shift in T_m is approximately linear with molarity of $CaCl_2$ (T_m decreases ~16°/mol of $CaCl_2$ added), and the effect is nonsaturating, at least at concentrations up to 1 M.

Such experiments are generally carried out at macromolecule concentrations of ~1 mg/ml. Thus, for proteins, the residue concentration is generally less than 10^{-2} M. Therefore, the reaction scheme for unfolding of the macromolecule in the presence of a particular perturbant is eqs. 1 and 2, where eq. 1 represents the transfer of an interior residue into the solvent environment and eq. 2 represents the binding of a single molecule of perturbant (C) to each residue.

$$r_{\text{interior}} \rightleftharpoons r_{\text{exterior}} \qquad (1)$$

$$r_{\text{exterior}} + C \rightleftharpoons r_{\text{exterior}} + C \qquad (2)$$

If the observed effects are due entirely to preferential binding to the newly exposed residues, then the association constant for eq. 2 must be of the order of 0.01 M^{-1} (see below).

Plots similar to that shown in Figure 1 could have been presented to document the similar destabilizing effect of $CaCl_2$ on the temperature-induced folding/unfolding transitions of macromolecules as diverse as globular proteins (ribonuclease), α-helical fibrous proteins (myosin), and nucleic acids (DNA), suggesting that the effects of this salt on the stability of the ordered form transcend chemical and conformational details of macromolecular structure.[7]

The effects produced on the thermal transition temperature of ribonuclease by a number of neutral salts at various concentrations are shown in Figure 2.[8] This figure shows that while a given salt may have a similar effect on the stability of a variety of macromolecules, the magnitude and the direction of the effects differ widely from one salt to another. Thus, while some salts (KSCN, $CaCl_2$) are potent destabilizers of the ordered structure, others [(KH_2PO_4, $(NH_4)_2SO_4$)] are potent stabilizers of the ordered form, and still others (KCl, NaCl) are quite inert as either stabilizers or destabilizers.

In general, these plots of T_m vs. C_s (salt concentration) are relatively linear, especially at higher salt concentrations, so the effect of any particular salt on macromolecular stability can be described by an approximate empirical equation of the form

$$T_m - T_m^o = K_m C_s \qquad (3)$$

where $T_m°$ and T_m are the melting temperatures of the macromolecule in pure water and in the aqueous perturbant solution of concentration C_s, and K_m is the slope of a plot of T_m vs. C_s. Thus, K_m may be thought of as a measure of the molar effectiveness of a given salt as a perturbant of macromolecular stability, where a negative value characterizes a destabilizing salt, and a positive value a stabilizing salt.[9]

This generality of the effects of ions on the stability of macromolecular structures transcends intramolecular folding/unfolding reactions and applies to all systems in which there is a net transfer of residues from an unsolvated to a solvated environment.[10] Thus, as Figure 3 shows, ions that effectively stabilize macromolecular conformations, i.e., which increase the ΔG for transfer of residues from the inside of the macromolecule to the outside, also are effective salting-out agents for macromolecules. They increase the chemical potential of the exterior groups of the macromolecule, thus favoring the formation of a separate insoluble phase. Similar effects can be seen on protein subunit association/dissociation equilibria, on macroscopic phase transitions (melting of polymers, shortening of tendons), on enzyme activities (the maintenance of the active site is, of course, dependent on the overall native geometry of the macromolecule), and on the kinetics of macromolecular folding/unfolding reactions.[10]

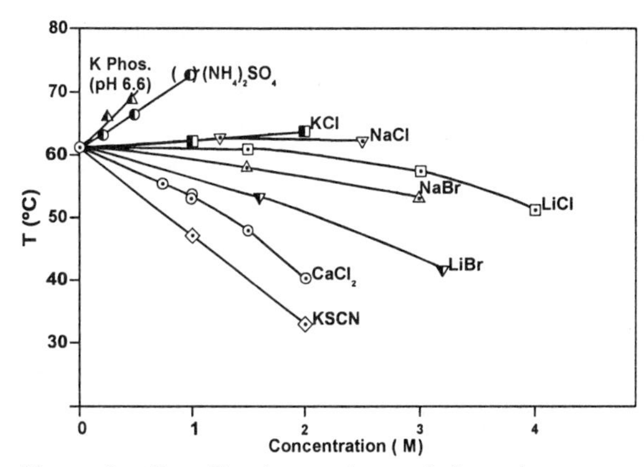

Figure 2 - Transition temperatures of ribonuclease as a function of concentration of various added salts. All of the solutions were adjusted to pH 7 and contained 0.15 M KCl and 0.013 M sodium cacodylate; ribonuclease concentration, ~5 mg/ml.

Figures 2 and 3, by implication, also bring out the following additional facts about the effects of the various ions on macromolecular structure (for details and documentation, see ref. 10).

(1) Cations and anions are additively effective in altering T_m (and thus the free energy of transfer; see below). In fact, this additivity has been shown to extend well beyond the simple neutral salt system. For example, the destabilizing effects of $CaCl_2$ and ethylene glycol on the T_m of the ribonuclease transition sum quantitatively. Also, the destabilizing effects of the guanidinium salts are greatly increased if the guanidinium cation (Gu^+) is coupled with an anion such SCN^- or ClO_4^-, rather than with the conventional by relatively inert Cl^-. On the other hand, Cu_2SO_4, is a net stabilizing agent, the T_m-depressing effect of the Cu^+ ion being more than offset by the stabilizing activity of the SO_4^- moiety.[8]

(2) In order of effectiveness as salting-out agent (Figure 3) the anions and cations both follow closely the classical series first demonstrated by Hofmeister for the salting-out of euglobulins. This series has also been shown to apply to the salting-out of a number of simple nonpolar nonelectrolytes from solution (for an extensive compilation of such data, see Long and McDevit).[11]

(3) The tetraalkylammonium ions increase destabilizing effectiveness with increasing length alkyl side chain. It has been shown that the effectiveness of these cations, per side chain, is directly comparable to that of the analogous aliphatic alcohol or acid, which is one methylene unit shorter.[8]

(4) Certain polymers (e.g., polyvinylmethyloxazolidinone or poly-l-proline) form a separate phase when heated to a critical temperature (T_c).

← Helix Native, Salting-out; Cell Denatured, Salting-in →

Collagen-Gelatin:

$SO_4^{2-} > CH_3COO^- > Cl^- > Br^- > NO_3^- > ClO_4^- > I^- > CN^-$
$(CH_3)_4N^+ < NH_4^+ < Rb^+, K^+, Na^+, Cs^+ < Li^+ < Mg^{2+} < Ca^{2+} < Ba^{2+}$
$(CH_3)_4N^+ < (C_2H_5)_4N^+ < (C_5H_7)_4N^+$

Ribonuclease:
$SO_4^{2-} < CH_3COO^- < Cl^- < Br^- < ClO_4^- < CH_5^-$
$(CH_5)_4H^+, NH_4^+, K^+, Na^+, Li^+ < Ca^{2+}$
$(CH_3)_4N^+ < (C_2H_5)_4N^+ < (C_3H_7)_4N^+ < (C_4H_9)_4N^+$

DNA:
$Cl^-, Br^- < CH_3COO^- < I^- < ClO_4^- < SCN^-$
$(CH_3)_4N^+ < K^+ < Na^+ < Li^+$

Polyvinylmethyloxazolidine:
$SO_4^{2-} < CO_3^{2-} < F^- < Cl^- < Br^- < ClO_4^- < SCN^-$
$Na^+, K^+ < NH_4^+ < Li^+$

Figure 3 - Relative effectiveness of various ions in stabilizing or destabilizing the "native" form of collagen, ribonuclease, DNA and polyvinylmethyloxazolidinone in aqueous solution.

Salting-out agents tend to lower T_c (stabilize the insoluble phase) while salting-in or macromolecular destabilizing agents raise T_c in accord with the rankings shown in Figure 3.[12]

It is clear that the various ions have specific and predictable effects on the stability of a wide variety of macromolecular structures and complexes. What then are the mechanisms by which perturbants bring about these effects?

4. Free Energy of Transfer of Model Compounds

In several places above we have stated that the effects of the various structural perturbants on the stability of macromolecules in aqueous solution can be viewed as effects on the free energy of transfer of the interior groups of the macromolecule that are exposed to the solvent as a consequence of the unfolding transition. One way to subject this idea to quantitative test is to measure the free energy of transfer of appropriate model compounds from a nonpolar to an aqueous phase containing different concentrations of the various structural perturbants under consideration. Charged groups are not generally found inside macromolecules, presumably because the "thermodynamic cost" of burying such groups in the nonpolar interior is too great. The individual residues that comprise a polypeptide chain are held together by peptide bond. Thus, burying a highly nonpolar sequence of amino acid residues in the macromolecular interior also buries these polar moieties. Likewise, the heterocyclic purine and pyrimidine bases that "stack' upon one another in the native DNA structure, and as a consequence are partially removed from contact with the solvent, also contain polar -NHCO- groups as well as nonpolar and presumably hydrophobic methylene groups. It is this partially hydrophobic and partially hydrophilic interior of biological macromolecules that gives them a specific structure, and simultaneously makes the native form only marginally stable when compared, for example, to a *micelle of sodium dodecyl sulfate*, where the *interior is completely nonpolar* and like a "liquid drop"; i.e., it has no specific internal structure. Thus, good models for such free energy of transfer studies include compounds containing both peptide groups and nonpolar components or purine and pyrimidine rings. Such studies have been carried out by a number of workers," and Figures 4 and 5 show some representative results, obtained with acetyltetraglycyl ethyl ester [(Ac(Gly)$_4$Et)] and adenine, respectively.

These studies differ slightly from the transfer experiments described above in that, for technical reasons, these workers have chosen to study the transfer of the model compound from a *solid phase to the aqueous solution*. Thus, what is actually being measured is the change in solubility of the model compound in aqueous solutions containing different concentrations of various neutral salts. This change in solubility has been defined in terms of the *activity coefficient* of the model compound in the aqueous phase. The activity coefficient (γ) is defined by eq. 4

$$\gamma = \frac{S_o}{S} \tag{4}$$

where S_o is the solubility of the model compound in pure water and S is the solubility in aqueous solution containing perturbant. (γ_o, the activity coefficient for the model compound in pure water, is defined as unity in this formulation.) It is well known that the *salting-out of nonelectrolytes* can often be represented by the empirical Setschenow equation,[11] which we may write

$$\log \gamma = \log\left(\frac{S_o}{S}\right) = K_s C_s \tag{5}$$

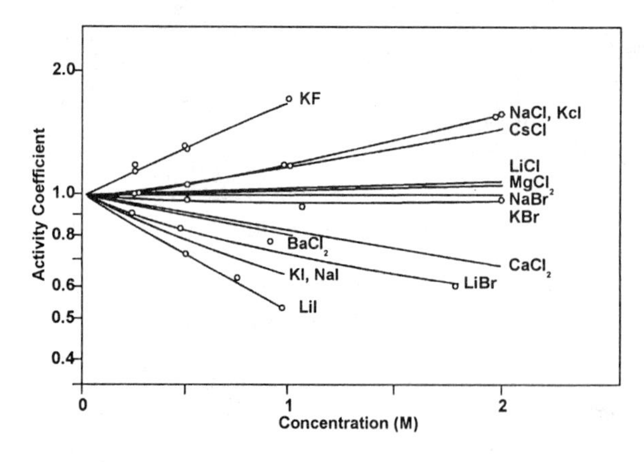

Figure 4 - Activity coefficients of acetyltetraglycyl ethyl ester Ac(Gly)$_4$Et in solutions of various salts at 25°C.

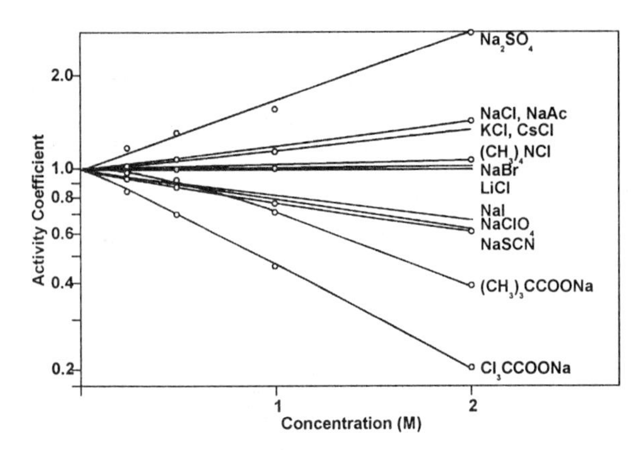

Figure 5 - Activity coefficients of adenine in aqueous salt solutions at 25°C.

where K_s is the salting-out coefficient for a given salt and C_s is salt molarity. By analogy with the empirical equation, which applies to macromolecular stability (eq. 3), effective salting-out agents are characterized by positive values of K_s and salting-in agents by negative K_s values.

Figures 4 and 5 show that the salting-out (or -in) of Ac(Gly)$_4$Et and adenine follow the Setschenow equation quite well. Furthermore, with some exceptions, that the relative effectiveness of various ions as salting-in or -out agents is very comparable to the Hofmeister series, which applies to macromolecules (Figure 3). However, there are significant quantitative differences; thus salts like NaCl, which had essentially *no effect* on macromolecular stability ($K_m \cong 0$), are significantly effective salting-out agents for Ac(Gly)$_4$Et and adenine ($K_s > 0$), while agents like LiCl and NaBr, which have appreciable *destabilizing potency* (K_m < Q), are essentially *ineffective as perturbers* of the solubility of the model compounds ($K_S \cong 0$). Some clue as to the significance of these differences may be had by studying the salting-out behavior of benzene (Figure 6). Here we see that all the salts tested, including NaI and NaClO$_4$, act as salting-out agents for this totally nonpolar compound, though the relative effectiveness of the various salts still follows the *Hofmeister series*.

The differences between these various effects have been put on a quantitative basis by Schrier and Schrier,[13] who carried out a comparative study of the salting-out behavior of *N*-methylacetamide (NMA) and *N*-methylpropionamide (NMP). They found that all the neutral salts tested were more effective salting-out agents for the slightly more nonpolar *N*-methylpropionamide than for *N*-methylacetamide. They used the data obtained (plus a few reasonable assumptions) to break up the model compounds into constituent groups, each characterized by a group-specific salting-out constant, as given in eqs. 6 and 7.

$$K_{s,\,NMA} = 2k_{s,\,CH3} + k_{s,\,A} \tag{6}$$

$$K_{s,\,NMA} = 2k_{s,\,CH2} + k_{s,\,CH2} + k_{s,\,A} \tag{7}$$

They then used the salting-out coefficients calculated for the methylene (k_{s,CH_2}) and ethyl acetate (k_{s,CH_2}) groups for each salt to calculate an amide group salting-out coefficient ($k_{s,Ac(Gly)_4Et}$) from the Ac(Gly)$_4$ Et data of Robinson and Jencks, as in eq. 8.

$$K_{s,\,Ac(Gly)4\,Et} = k_{s,\,EtOAc} + 4k_{s,\,CH4} + 4k_{s,\,A} \tag{8}$$

Some results are summarized in Table 1. We may note that all salts are effective salting-out agents for the nonpolar methylene groups ($k > 0$), but that the relative effectiveness of the various salts as salting-out agents varies as predicted by the Hofmeister series. On the other hand, k seems to be negative for all the salts and is reasonably constant (with some exceptions) for all salts of a particular charge type, such as the monovalent salts. Furthermore, the values of k obtained have been shown to be in reasonable accord with what one might calculate for the salting-out of a polar nonelectrolyte using a modified *Debye-McAuley equation.*[13] Schrier and Schrier[13] have taken the analogy between the macroscopic K and the group salting-out *coefficient* k still further. By combining a relation between the *melting temperature of crystalline polymers as a function of diluent concentration*[14] with the *Setschenow equation*, they obtained eq. 9,

$$T_m - T_m{}^{\circ} = \left[\frac{R(T_m{}^{\circ})^2}{\Delta H_{res}}\right] C_s \sum n_i k_i \tag{9}$$

where $T_m{}^{\circ}$ and T_m are the melting temperatures of the macromolecule in pure water (or standard dilute salt) and in the presence of a particular salt at concentration C_s. The enthalpy of melting per residue of the polymer under consideration (ΔH_{res}) is assumed to be independent of temperature, N_i is the mole fraction of polar or nonpolar groups exposed in the transition, and k_i is the relevant salting-out coefficient for the groups in the particular salts involved. Making some reasonable assumptions about the type and number of groups exposed in the transition and using the values of k_i obtained from the free energy of transfer of model compound studies, these workers were able to obtain rather good agreement between calculated values of T_m and experimental melting data on collagen and ribonuclease. These studies suggest the following broad picture:

(1) The *effects* of salts on the stability of macromolecules can be quantitatively correlated with the effects of these agents on the free energy of

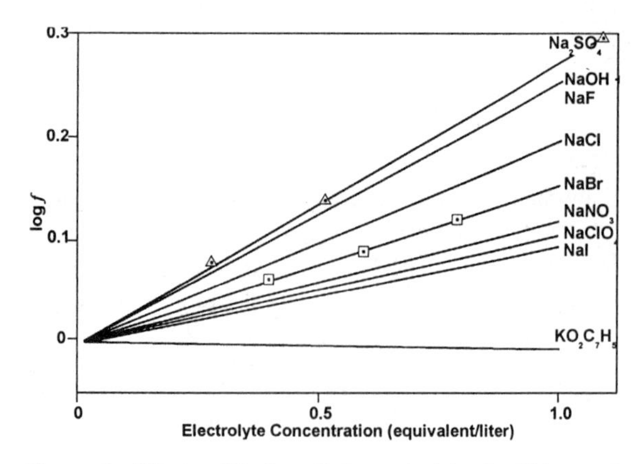

Figure 6 - Effects of Sodium Salts on Activity Coefficients of Benzene, 25°C.

transfer of appropriate model compounds from a nonaqueous environment into the aqueous milieu.

(2) The specific effect of the various ions arise primarily through their effect on the *chemical potential* of the nonpolar portions of groups exposed in a macromolecular unfolding or depolymerization reaction, while interaction with the newly exposed amide (or polar) groups provides a constant salting-in bias.[16]

(3) Whether a particular salt will serve as a stabilizer or destabilizer of the "native" state (or as a salting-in or salting-out agent for a particular model compound) depends on the ratio of polar to nonpolar groups in the portion of the macromolecule exposed by the unfolding or depolymerization reaction (or by the polar/nonpolar ratio of the model compound under study). Apparently the observation that the same set of salts is characterized by $K_m \cong 0$ for most macromolecules tested may be attributed to the fact that the polar/nonpolar ratio of the groups exposed in thermally induced unfolding or depolymerization reactions at moderate temperatures is approximately the same in these macromolecules.

References:

1. A. E. Mirsky and L. Pauling, Proc. Natl. Acad. Sci. U.S.A. 22, 439 (1936).
2. Y. Nozaki and C. Tanford, J. Biol. Chem. 238, 4074 (1963); D. B. Wetlaufer, S. K. Malik, L. Stollen, and R. L. Coffin, J. Am. Chem. Soc., 86, 508 (1964).
3. W. Kauzmann, Advan. Protein Chem., 14, 1 (1959).
4. For present purposes we shall operationally define a hydrophobic bond affector as one that increases the solubility (decreases the activity coefficient) of a hydrocarbon-like model compound relative to the solubility of that compound in pure water.
5. C. Tanford, J. Am. Chem. Soc., 84, 4240 (1962).
6. P. H. von Hippel and K.Y. Wong, Biochemistry, 2, 1387 (1963).
7. P. H. von Hippel and K.Y. Wong, Science, 145, 577 (1964).
8. P. H. von Hippel and K.Y. Wong, J. Biol. Chem., 240, 3909 (1965).
9. These are obviously not nonspecific electrostatic effects, since both the magnitude and the direction of the effects depend on the specific ions involved. However, at low salt concentrations, macromolecules with large net charge do show a dependence of T_m on salt concentration, which deviates markedly from the simple behavior predicted by eq. 3 and may be superimposed upon it. For example, at *low salt concentrations*, the T_m of DNA shows a logarithmic dependence on the concentration of monovalent cation, T_m increasing about 18°C per ten-fold increase in salt concentration. Theoretical calculations have shown that this effect of salt can be almost completely attributed to suppression of the interchain electrostatic repulsion between negatively charged phosphates in native DNA. Comparable effects (though smaller in mag-nitude) should also be expected in proteins with appreciable net charge.
10. P. H. von Hippel and T. Schleich, in *Biological Macromolecules*, Vol. II, 5. Timasheff and G. Fasman, Ed., Marcel Dekker, New York, N. Y., 1969, p. 417.
11. F. A. Long and W. F. McDevit, Chem. Rev., 51, 119 (1962).
12. I. M. Klotz, Fed. Proc., 24, S-24 (1965); A. Ciferri and T. A. Orofino, J. Phys. Chem., 70, 3277 (1988).
13. (a) D. R. Robinson and W. P. Jencks, J. Am. Chem., 87, 2470 (1965); (b) D. R. Robinson and M. E. Grant, J. Biol. Chem., 241, 4030 (1966); (c) E. E. Schrier and E. B. Schrier, J. Phys. Chem., 71, 1851 (1967).
14. P. J. Flory, J. Cell. Comp. Physiol., 49, Suppl. 1, 175 (1957).
15. P. H. von Hippel and K. Y. Wong, Biochemistry, 1, 664 (1962).
16. It should be noted, however, that the magnitude of this bias will often depend on a number of factors, such as the dipole moment of the polar portion of the exposed group. Furthermore, specific salt effects in addition to those involving the nonpolar portions may arise if the exposed groups are appreciably acidic or basic.[11]

THEORY/PRINCIPLES

SOLUBILITY AS A FUNCTION OF PROTEIN STRUCTURE AND SOLVENT COMPONENTS

Schein, C. H. (1990) Nature Biotechnology 8: 308. (Adapted)

1. Introduction

This review deals with ways of stabilizing proteins against aggregation and with methods to determine, predict, and increase solubility. Solvent additives (osmolytes) that stabilize proteins are listed with a description of their effects on proteins and on the solvation properties of water. Special attention is given to areas where solubility limitations pose major problems, as in the preparation of highly concentrated solutions of recombinant proteins for structural determination with NMR and X-ray crystallography, refolding of inclusion body proteins, studies of membrane protein dynamics, and in the formulation of proteins for pharmaceutical use. Structural factors relating to solubility and possibilities for protein engineering are analyzed.

It is generally known that proteins must be stored in an appropriate temperature and pH range to retain activity and prevent aggregation.

Proteins are often most soluble in solution conditions mimicking their natural environment. Serum proteins are soluble in a pH and salt range where mature insulin, which is stored in acidic granules in the cell, precipitates. Bacterial proteins may prefer buffers containing glutamate or betaine, compounds that accumulate in response to high concentrations of Cl^- in the medium.[2] Caseins and other Ca^{2+}-associated proteins may require small amounts of the ion to maintain their native structure during purification.[4,5] The stability of lactase (β-galactosidase) is greatly increased in the presence of milk proteins.[6] But for most proteins, experimental determination of the solution properties can help in solvent design.

Low solubility in aqueous solvents is often regarded as an indication that a protein is "hydrophobic" as aggregation of integral membrane proteins after transfer to a hydrophilic environment is a well-described phenomenon.[7] But all proteins are to some extent hydrophobic, with tightly packed cores that exclude water.[8,9] As native, properly folded structures aggregate less than unfolded, denatured ones, there is an intimate relationship between solubility and stability. The free energy of stabilization of proteins in aqueous solution is very low (ca. 12 kcal/mole at 30°C);[10] consequently, proteins are on the verge of denaturation.[10,11] Protein stability can be increased by solvent additives or by alteration of the protein structure itself.

2. The Properties of Proteins in Solution

2.1 Defining Solubility. The chemist"s definition of solubility, parts purified substance per 100 parts of pure water, is not useful in a biological frame, as proteins in nature are never found in pure water. Blood and eukaryotic cytoplasm contain on the order of 0.15 M salt, with large quantities of trace metals, lipids, and other proteins. The cytoplasm of bacteria is more variable, with salt concentration ranging from 0.3 to 0.6 M.[2] The solubilizing effects of small molecules and even other proteins means that protein solubility does not correlate with purity.[12]

Operationally, solubility is the maximum amount of protein in the presence of specified cosolutes that is not sedimented by 30,000 x *g* centrifugation for 30 minutes.[13] An even stricter criterion, function retained after centrifuging for 1 hours at 105,000 x *g*, has been suggested for membrane proteins.[14] If one has a pure, lyophilized protein or a salt precipitate, one can determine solubility by adding increasing amounts of weighed solid, centrifuging, and measuring the protein content of the supernatant. Dissolved protein should reach a maximum (solubility) and level off. However, in the food industry, solubility is defined by sediment (in ml)

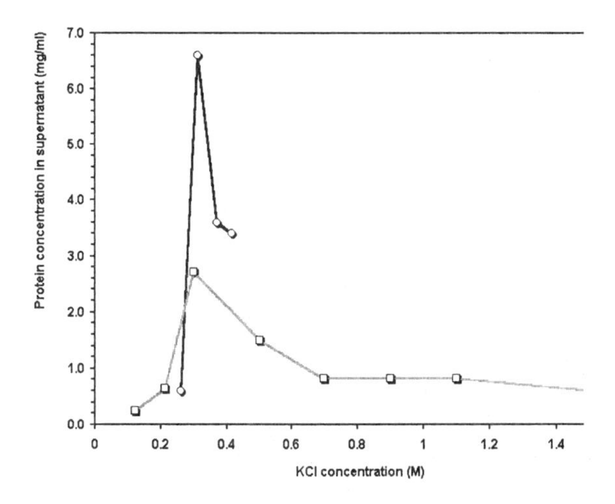

Figure 1 - Solubility of T7 RNA polymerase as a function of salt concentration in 10 mM cacodylate buffer, 1 mM DTT, and 0.02 mM PMSF. The polymerase solution (*ca.* 1 mg/ml in 0.1 M KCl, 20 mM Tris-HCl, pH 7.9, 5% glycerol) was diluted in 1:10 with the indicated buffers and each sample was individually concentrated in 30-kDa MW cutoff "Centricons" (Amicon). The protein concentrating in the supernatant (measured by the Coomassie blue assay) after concentration is indicated (Protein A, squares). The top curve (Protein B, circles) is from a second measurement using a finer salt gradient and more protein per sample.

remaining after centrifuging; the solubility index is thus inverse to the actual solubility.[13]

The method described in Figure 1 allows definition of the solubility range of a protein in solution. A protein solution is diluted into a buffer series and the samples are centrifuged in microconcentrators. As one can conveniently concentrate about 50-fold, a relatively small amount of protein is sufficient for the estimation.

2.2 Measuring Stability. Methods for determining the thermodynamic stability of proteins use pH and temperature extremes, or high concentrations of denaturants.[10] Although useful for discerning changes in the structural stability of mutant proteins that are not clear from activity data, they are not directly correlated with the half-life of proteins in solution. Since aggregation occurs at temperatures well below the T_d for proteins; additives that stabilize proteins against aggregation may not necessarily affect the T_d.[15]

The major problem with using thermodynamic measurements is their failure to account for the kinetic effects that lead to aggregation. Both the enthalpy (ΔH) and entropy (ΔS) of hydration vary greatly with temperature, but they cancel to give a relatively small measured free energy (ΔG) of hydration that seems to vary little with temperature. Most of the temperature-dependent kinetic contribution, which is the more important in explaining hydrophobic effects, dissipates in alterations of the solvent structure around the protein and reversible deformation of the protein structure itself.[10,16] Accurate discrimination of hydration shells can be done only from crystal structures. Clearly, other methods of determining protein stability are needed.

Proteins with shorter half-lives generally have larger subunit molecular weights, lower isoelectric points (p*I*), higher affinity for hydrophobic surfaces, and greater susceptibility to proteases. Both of the latter characteristics can be used as the basis for determining enzyme stability in less extreme environments as well as the effect of additives on stability. As less stable proteins have a higher tendency to adsorb to surfaces,[17] resistance to mechanical shaking may be a useful indicator of solution half-life.[18] Trypsin digestion has been used to define the salt stabilization of hyalin.[5]

2.3 Determining Surface Charge. Isoelectric focusing gives the p*I*, the pH at which the protein shows no net charge in isoionic conditions. However, due to the binding of salt, one cannot assume that a protein in solution will be negatively charged at pHs above its p*I* (e.g., acidic caseins bind Ca^{2+} and appear positively charged at pH 7.4). At pH 7.5 and 50 mM salt, most proteins will bind to DEAE-coupled resins if they are negatively charged and to phospho- and other negatively charged resins if they are positively charged. The charge strength can be estimated from the salt concentration required for elution. Gel methods for following the changes in surface charge during protein folding and aggregation have also been developed.[19]

Generally, charged proteins can be "salted in" by counterions. Binding of salts to proteins decreases bound water as well as the net charge at the surface. The solubility of lysozyme, a positively charged protein, was shown to vary more with the anion added than the cation; the anion dependence followed the Hofmeister series.[20]

The solubility of caseins with p*I* between pH 3 and 5 varies with the cation: sodium, potassium, and ammonium caseinates are all more soluble than those prepared with calcium or aluminum.[4,15]

2.4 Determining Hydrophobicity. Binding to resins coupled with hydrophobic groups, like Phenylsepharose (Pharmacia), indicates the presence of hydrophobic residues at the protein surface. Proteins are applied in high salt (0.7-1 M ammonium sulfate), which furthers hydrophobic interactions, and then eluted with a decreasing salt gradient. Most proteins elute between 0.5 and 0.1 M salt; very hydrophobic proteins will not elute into low salt buffer unless the polarity is decreased by adding ethylene glycol. If a protein does not bind to phenylsepharose, it either has a very hydrophilic surface (e.g., RNase A) or it is aggregated.

One can determine the hydrophobicity of a purified protein or follow changes in exposure of hydrophobic groups during folding by measuring interaction with a hydrophobic dye or radioactive tracer (e.g., 1-anilino-8-naphthalenesulfonate[21] or ^{125}I-TID, 3-(trifluoromethyl)-3-(*m*-[^{125}*I*]iodophenyl) diaz-irine[3]).

2.5 Aggregation and Precipitation. Precipitation via any agent can be:

2.5.1 Reversible, as after precipitation with salts or large organic molecules like polyethylene glycol (PEG). Because PEG molecules are excluded from the surface of the protein, a two phase system develops and the protein is concentrated into a smaller volume, where its chances of interacting with another protein molecule to form an aggregate are increased ("excluded volume" model).[22,23] When the precipitant is removed, the water layer around the original molecule can reform and the protein molecules separate into soluble monomers.

The protein structure does not significantly change during reversible aggregation. A plot of protein in solution versus the concentration of the precipitant should look the same whether it is made with increasing precipitant (to precipitation) or decreasing precipitant (to solubility). Reversibility is assumed for most mathematical models of salting out[12] as well as some recent models of low salt aggregation phenomena.[25,26]

Table 1 - Protein Cosolutes

Compounds	Mode of Action	Amount Used
A. Osmotic stabilizers	Generally have little direct interactions with proteins; they affect the bulk solution properties of water.	
i. Polyols and sugars • glycerol, erythritol, arabitol, sorbitol, mannitol, xylitol, mannisidomannitol (Man-Man), glucosylglycerol, glucose, fructose, sucrose, trehalose, isofluoroside	These stabilize the lattice structure of water, thus increasing surface tension and viscosity. They stabilize hydration shells and protect against aggregation by increasing the molecular density of the solution without changing the dielectric constant.	10 – 40%
ii. Polymers • dextrans, levans, polyethylene glycol	Polymers increase the molecular density and solvent viscosity thus lowering protein aggregation in a single phase system. At high polymer concentration, a two phase system develops and the protein aggregates in the phase where its concentration is the highest.	1 – 15%
iii. Amino acids and derivatives • glycine, alanine, proline, taurine, betaine, octopine, glutamate, sarcosine, α-aminobutyric acid, trimethylamine N-oxide (TMAO)	Small amino acids with no net charge, like Gly and Ala, have weak electrostatic interactions with proteins. Octopine is a derivative of Arg that is less denaturing to proteins. TMAO stabilizes proteins, even in the presence of denaturants like urea. Most of these compounds increase the surface tension of water.	20 – 500 mM
B. Ionic compounds	These affect enzyme reactions and their stabilizing effects on protein occur in a much narrower concentration range than the above compounds.	
i. Stabilizing • citrate, sulfates, acetate, phosphates, quaternary amines	Larger anions shield charges and can stabilize proteins at low concentrations. They lead to precipitation by competing for water molecules.	20 – 400 mM
ii. Destabilizing • chlorides, nitrates, thiocyanates	These are generally less stabilizing than large ions but are also useful in charge shielding at lower concentrations.	20 – 400 mM[1]
C. Denaturing chaotrophs • urea, guanidinium salts, trichloroacetates, cetylmethylammonium salts, organic solvents	Denatures; either stabilize the unfolded state of proteins (urea) or perturb protein structure by interfering with hydrogen bonding or disturbing the hydration shell.	0.2 – 8 M
D. Other common additives (mostly nonphysiological)	Either interact directly with proteins or specifically affect impurities in the buffer but do not change the bulk solvent physical properties.	
• 2 – mercaptoethanol, dithiothreitol (DTT)	Reductants, protect free sulfhydryls from oxidation and prevent intermolecular glutathione sulfhydryl cross-linking.	1 – 5 mM 0.1 – 1 mM[2] 1 – 4 mM
• phenylmethylsulfonyl fluoride (PMSF), benzamideine	Inhibit serine proteases by reacting with the active site serine hydroxyl group.	0.02 – 0.05 mM[3] < 1 mM
• leupeptin, peptides	Protect from protease attack by serving as alternate substrates.	
• ethylenediaminetetraacetic acid (EDTA), ethylene-bis(oxyethylenenitrilo)tetra-acetic acid (EGTA)	Chelate divalent metal ions which may react with proteins; inhibit metalloproteases; EDTA at 5-20 mM aids in the lysis of G-bacteria by lysozyme.	0.01 – 0.1 mM for buffers
• catechols, phenolics, NaN_3	Bacteriocides	< 0.05%

[1]Higher concentrations may be used when working with enzymes from halophilic organisms.
[2]DTT is a potential denaturant of proteins at higher temperatures and has limited solubility in high salt. The concentrations indicated should not be exceeded.
[3]Dissolve PMSF to 20 mM in isopropanol. The indicated concentrations represent the maximum solubility in aqueous buffers.

2.5.2 Partially reversible, a behavior frequently seen in pH-induced precipitation. Proteins precipitate around their pI and resolubilize as the pH is adjusted upward or downward. But during the pH adjustment, residues may change orientation. When the pH is readjusted, they may not be able to regain their former position and a mixture of structures (isomers) results. Even primary structure can change if a protein is held at acid pH for long periods of time, as for example the deamidation of as asparagine 21 of insulin.[1] A plot of protein in solution as a function of pH will depend on whether the protein has already precipitated. Kinetic modeling of pH-dependent aggregation has been attempted by linear regression.[27] Models could also use *hysteresis* in the folding-unfolding equations (thermodynamic cycles do not exactly connect.)

2.5.3 Irreversible, which is usually initiated by extreme changes in the solvent leading to protein denaturation. But some proteins (Figure 1) also precipitate irreversibly when concentrated above their maximum solubility in a given buffer. Inactive flakes of protein form and remain insoluble even on redilution of the sample or transfer to a buffer of the correct salt concentration. The nature of this tight intermolecular binding is not easy to study, as the aggregates arise from *many-body interactions* potentially involving all parts of the protein. The initiation could be direct interaction of surface hydrophobic residues, or, as aggregation shows cooperativity, partial disturbance of the hydration sheath, or unfolding of the protein structure allowing interaction between normally "buried" residues. Irreversible protein denaturation is not easily modeled. Thermal denaturation curves are calculated at very low protein concentration to avoid aggregation terms in the equations.[10]

3. Buffer Design for Maximizing Solubility

3.1 The Properties of Water as a Solvent. Water"s high dielectric constant and its tendency to solvate ions make it an active copartner in enzyme reactions. When NaCl crystals are added to water, the atoms attract each other with about 1/80th of the force in the dry state and the crystal dissolves. Analogously, dissolved proteins are coated with a "hydration shell" around charged and polar groups that prevents self-binding. This bound water does not freeze (some proteins are even efficient antifreezes)[28] and has different properties than in the surrounding solvent molecules.[29] Bulk water molecules and the protein are in continual fluctuation, which leads to instability in the system.[11,30,31]

3.2 *Protein Stability in the Solid State vs. Solution.* On the other hand, a protein completely stripped of its hydration shell is difficult to redissolve, as intermolecular hydrophobic forces must be broken. Lyophilization and other drying methods should thus be used with caution and osmotic stabilizers added where necessary to ensure that the protein can be rehydrated. The water content of dry milk powder is a compromise between shelf life, which decreases with increasing content of water, and solubility, which increases with hydration index.[13]

Proteins in the solid state have different levels of reactivity depending on the water content. Dried protein with a water content below 22-25%, the minimum required for conformational flexibility and activity, is thermostable. *Glycerol*, which stabilizes proteins in solution, acts as a humectant on the powder and causes decomposition (as indicated by the Maillard browning reaction) at much lower water content (5-15%). Conversely, *sorbitol* competes for the hydration water of the protein and does not enhance denaturation.[31]

3.3 Solvent Additives. There are many potential stabilizing cosolutes for proteins (Table 1). Buffers are described in several excellent reviews[32,33] and will not be covered here. Table 1 is separated into groups of compounds that have varying effects on the solvation properties of water: (1) *dielectric constant*, (2) *chemical potential*, (3) *viscosity*, and (4) the clathric tendency (*surface tension*). The first two qualities are related to protein *polarity*; the last two relate to the *diffusion* of the protein, its *partial molar volume*, and to *hydrophobic hydration*.

3.4 Osmolytic Stabilizers. The first group of compounds is osmolytes, which are not strongly charged and have little effect on enzyme activity up to at least 1 M concentration.[34] Their major effects are on the viscosity and surface tension of water, and hence on solvent ordering. Many of these compounds are used *in vivo* to control the osmotic pressure of eukaryotic and bacterial cells.

Osmolytes can be polyols, sugars, polysaccharides, neutral polymers, amino acids and their derivatives, and large dipolar molecules like TMAO. Glycerol is the most commonly used osmolyte, as it is easily removed by dialysis and does not interfere with ion-exchange chromatography. It does not alter the dielectric constant of the medium significantly and its stabilization effect on proteins seems to be due to its ability to enter into anc strengthen the water lattice structure. High concentrations of glycerol decrease the diffusivity and the partial molar volume of proteins,[15] thus lowering the rate of aggregate-producing solute interactions.

Glycerol has major drawbacks, however, especially for large-scale work, as it is an excellent substrate for bacteria. Xylitol, a potential substitute, is not degraded well by bacteria and can be recycled from buffers by alcohol precipitation. PEG can be added to *in vitro* systems for nucleic acid and protein synthesis, where sufficient molecular density but low ionic strength is needed.

3.5 Ionic Stabilizers. Ionic compounds and salts can stabilize protein structure by shielding surface charges. Salts are also considered as osmolytes and are used to some extent as such *in vivo*. *E. coli* transiently accumulated K^+ and glutamate after osmotic shock, but within 30 minutes switches to carbohydrates as osmoprotectants.[2] Most ionic compounds will affect the dielectric constant and the chemical potential of the solvent and the protein at concentrations well below where they affect the other bulk properties of the solution. Normal bacterial and mammalian enzymes function at a rather low salt concentration and are inhibited by high salt. *Halophilic* organisms, which can accumulate as much as 7 molal K^+ intracellularly, have adapted their enzymes to function in very high salt concentrations.[34]

There is no general rule on *salting in* of proteins; models that work for one protein are not necessarily applicable to another.[12] The salt concentration for maximum solubility frequently falls within a very narrow range. As shown in Figure 1, a 50 mM change in salt concentration gave as much as a 20-fold increase in dissolved T7 RNA polymerase. The solubilizing effect of ions is dependent on the size and charge distribution, but because polar groups on proteins are so diverse, it is hard to say a priori which ion will be best. Large ions are generally better at stabilizing proteins than small ones; in general, the more electronegative the ion, the more it interacts with and destabilizes protein structure.

The finest experimental work on the effects of salts on protein solubility (usually during salting out) has been done by crystallographers.[20,35,36] The assumed mechanism for salting out by small molecules is that they compete for water molecules until the concentration is too low to maintain the hydration sheath around the protein.[35]

3.6 Divalent Cations. These components have extremely pronounced solubility effects at very low concentrations. Even 1 mM Ca^{2+} induces a conformational change characterized by insensitivity to trypsin in sea urchin hyalin, and Ca^{2+} in the range of 1-20 mM encourage self-association.[5] Zn^{2+} aids in insulin solubilization as well as crystallization.[1] As even tiny amounts of Cu, Zn, and Mn (among others) can also induce aggregation, chelators are often added to buffers.

3.7 Denaturants, Chaotrophs, Cryoprotectants and Other Additives. One can solubilize almost any protein (usually at the expense of its activity) by chemical denaturation with perturbing ions. Urea stabilizes the unfolded states of proteins because essentially all protein parts, from the backbone to the tryptophan side chains, are more soluble in 6 M urea than water as evidenced by the free energy of transfer into this solvent.[37] Another class of denaturants, "chaotrophs," like guanidinium, cetyltrimethylammonium salts, trichloroacetate, and thiocyanate ions, disrupt hydrogen bond formation and disturb the hydration shell around proteins.[15] Detergents, amphiphilic compounds that lower the surface tension of water, bind to hydrophobic areas of proteins.

Another class of denaturants, organic solvents, lowers the dielectric constant of water. The denaturing activity of hydrophobic solvents is due to a limited detergent effect and a competing interaction for the intramolecular hydrophobic interactions responsible for a stable tertiary structure. Some proteins are remarkably resistant to the denaturing effects of protic, hydrophilic organic solvents. The original method for isolation of insulin and human interferon-α from tissue and bacteria used extraction with acidic ethanol;[38] crambin can be crystallized from 60% ethanol.[39]

Two organic solvents frequently used as cryoprotectants, dimethyl sulfoxide (DMSO) and ethylene glycol, can also denature proteins. DMSO encourages unfolding by favoring peptide N-H • O=S solvent bonds over peptide N-H • O=C peptide bonds.[40] Ethylene glycol, by reducing solvent polarity, weakens structural hydrophobic interactions.

Unless a protein is to be used *in vivo*, it is general practice to include protease inhibitors, sulfhydryl reductants, bacteriocides, and chelating agents in small amounts to all protein solvents.

3.8 Concentrating Proteins. Limits on the maximum protein concentration one can achieve are due to the structure of the protein, the buffer components, and the purity of the protein preparation. Overloading the preparation on SDS acrylamide gels may not detect proteases that cause damage during concentration or storage. To minimize this contamination, during purification from bacterial extracts, the protein should completely change buffer at least three times. Suitable *transfer methods* are salt precipitation and dissolving in fresh buffer, binding to an affinity resin or HPLC column and elution, or gel filtration. Dialysis, flow-through affinity steps, and redissolving lyophilized samples do not count as buffer transfers. All purification buffers should be made with ultrapure water and HPLC grade chemicals where possible, and sterilized to avoid the reintroduction of bacterial contaminants.

The most commonly used *methods for concentration* are salt precipitation, affinity chromatography, ultrafiltration, and occasionally, chromatofocusing, electrofocusing, and freeze condensation (for cryoresistant proteins). Very stable proteins and peptides can be lyophilized or spray-dried and redissolved. One should get the preparation to as high a concentration as possible by judicious elution during the last affinity step.

The easiest method for concentrating proteins that cannot be lyophilized is ultrafiltration. Micro-concentrators are useful for volumes up to 10 ml. Stirred pressure cells (Amicon, Millipore, or equivalent) are available for volumes between 10 and 500 ml, and membrane type can be selected according to the size and hydrophobicity of the protein. Pressure cells did not work for Mx protein or T7 RNA polymerase, however, as aggregation at the membrane surface was too high. The stir rate should be kept to a minimum as concentrated protein solutions are shear sensitive. For T7 RNA polymerase, losses were lowest with the Sartorius vacuum dialysis system, where concentrations up to 40-50 mg/ml were obtained in 0.2 M ammonium sulfate (pH 7).

Hollow fibers or parallel plate continuous flow systems can be scaled up to any size. The Minitan system from Millipore is a good intermediate size for lab use. Protein loss on the membranes is significantly higher than the maximum predicted by the manufacturers.

4. Situations Where Protein Solubility Becomes Limiting

4.1 Refolding Inclusion Body (IB) Proteins. IBs behave like protein that has been irreversibly precipitated. To obtain active protein, high concentrations of chaotrophic agents in the presence of sulfhydryl reducing agents are used to unfold the chains, which must then be refolded during removal of the denaturants. The primary refolding problem is aggregation of partially unfolded protein. In one study, the maximum protein concentration for efficient refolding was only 20 μg/ml;[41] for interleukin-2 the maximum was only 1 μg/ml.[42] Concentration by ultrafiltration after refolding is possible, but losses due to proteolysis, aggregation of isomers, and membrane binding are frequently very high. For tissue plasminogen activator (t-PA), the folding to intermediate states is rapid but the proper disulfide bonds formed much more slowly. As the close to folded forms are relatively soluble, timed addition of more unfolded protein concentrate (a sort of "feedback') can allow much higher final concentration of the extract.[41a] Residual denaturant can also stabilize the native state of the protein; its optimal concentration in the final extract should also be determined. TMAO may be a useful osmolyte when refolding proteins from urea solution.[34]

Every protein contaminant present during refolding increases the total dilution necessary to avoid aggregation. In addition, partially unfolded proteins are excellent protease substrates. Thus, one of the major advances in IB protein refolding has been the development of purification steps that can be used in the presence of the denaturant. These include gel filtration, certain types of affinity chromato-graphy,[43] and a new method based on the interaction between a poly-histidine peptide fused to the protein of interest and a nickel chelate column.[44]

Alternate methods for refolding, such as binding denatured protein to thiol-Sepharose columns or other affinity matrices and eluting with denaturant-free buffers,[45] are also being explored. It is possible that activated thiol-Sepharose mimics the structure of protein disulfide isomerase.[84] Serine proteinases[46] and interleukin-4[47] refolding yields were greatly improved by pretreatment with glutathione. Interleukin-2 was renatured by dilution and autooxidation in the presence of Cu^{2+}.[42]

Appropriate choice of buffer during the refolding step can also improve yields at higher concentrations of protein.[11] As optimal refolding conditions vary with the protein, one should either dilute the denatured sample into or dialyze it against many different buffers, and measure active or soluble protein after centrifugation.

4.2 Solubilization and Reconstitution of Membrane Enzyme Systems. Difficulties in solubilizing proteins from membranes have greatly limited structure and function studies.[48] Membrane proteins function in an amphiphilic environment and fold differently from cytoplasmic proteins: they turn their hydrophobic sites outward rather than inward. This probably accounts for why computer programs developed from soluble proteins predict the opposite of the known X-ray structures for membrane proteins.[49] This structural difference also accounts for the failure of detergents to solubilize IB proteins.

The only way to isolate most *integral membrane proteins* is to extract them from their lipid environment with bulky detergents (typically Triton X-100 or Emulphogen BC-720). The protein is integrated into a detergent micelle with detergent replacing phospholipids or proteins that were previously in contact with the hydrophobic surfaces.[14] Even if the protein is not inactivated by this treatment, *low critical micelle concentration (CMC) detergents* interfere with protein concentration (by giving a gel), functional assays, and further purification steps (as the detergent"s properties dominate the protein"s).

Thus, proteins are transferred after the initial extraction to less harsh detergents forming smaller micelle[48] via gel filtration. For detergents with CMCs too low to allow for efficient dilution into monomers, one may need to use highly polar micelle dispersing agents like ethanediol or bile salts.[50] NMR structural studies of small membrane proteins in micelles[51] are possible.

A major advance in membrane protein crystallization is the use of "small amphiphiles" to replace detergents binding to the face of the protein. One can thus prevent some of the problems caused by phase separation at higher salt and protein precipitation as the detergent in the micelles becomes too concentrated.

Osmolytic stabilizers (20% glycerol), or high salt (0.3-0.4 M KCl) added before the detergent, may stabilize the tertiary structure of the protein during extraction and dilution into proteoliposomes.[52] Glycerol or PEG is needed for efficient elution of membrane proteins from chromatofocusing columns. *In vitro* assays of transport systems from bacteria,[24,52] signal peptidase from yeast,[54] and the tamoxifen binding protein from a breast cancer cell line[55] were only possible by judicious control of the salt concentration during detergent extraction of the membrane.

As there is some evidence that high salt concentrations can stabilize secondary structural elements even during tertiary structure disruption,[21] the need for osmolytic stabilization may indicate that membrane proteins can undergo a transition phase *"molten globule" state* during solubilization. This state is defined for soluble proteins as an intermediate during reversible unfolding which retains comparable structure and a CD spectrum similar to the native state, but shows other evidence (e.g., increased binding of a hydrophobic dye) of a nonnative tertiary structure.

4.3 Very Concentrated Protein Solutions and NMR Work. As growth factors and enzymes are so active, one generally works with solutions containing less than 1 μg/ml. But much more concentrated solutions are required for microinjection into cells, for clinical trials of drugs, and for analytical studies of protein structure. There are many references on preparing proteins for X-ray studies.[1,35] As it has only recently been shown to be a general method for protein structure determination,[56] less has been written on preparing proteins for NMR. The major requirement for good spectra is absolutely pure protein at high concentration (1-20 mM).

Most structural determination by ^1H NMR use solutions in D_2O and H_2O at acid pH. Acid conditions encourage aggregation and protein unfolding, which shortens sample life. Solvent protons can significantly obscure regions of interest in the protein spectrum (C_α protons), so buffers are usually phosphate or deuterated Tris. Some groups prefer to work without ionic stabilizers, as they can blur peak profiles and cause excessive heat-up of the sample during measurements. These stringent requirements obvious-ly limit the proteins that can be studied by the technique to small, stable ones.

Assuming the solubility requirements are met, structure analysis for up to 80 amino acid proteins is almost routine.[57] The recent descriptions of well resolved (but very complex) 2-D NOESY and COSY spectra for urokinase (54 kDa; solution was 1.5 mM in D_2O at pH 4.5),[58] as well as the interaction of pepsin (35 kDa) with its ^{15}N-labeled inhibitor[59] show that investigation of even larger proteins is possible. Isotope-edited NMR spectroscopy, which selectively detects protons bound to labeled (^{15}N,^{13}C) nuclei, allows larger proteins to be analyzed and widens the choice of noninterfering buffers.[33,55-61]

The solution should be stable during the measurement, which for target proteins means addition of some salt. Staphylococcal nuclease was solubilized in 0.3 M NaCl at pH 7.6, ovomucoid domains (55 amino acids) were soluble to 12-15 mM in 0.2 M KCl at pH 8,[62] and yeast phosphoglycerate kinase substrate binding was studied in 0.1 M Na ^2H-acetate buffer at pH 7.1 (unspecified enzyme concentration).[63] Narrowest line widths were obtained for a solution of thrombin (35 kDa) concentrated to 0.5 mM in 0.2 M KCl at neutral pH. Significant line broadening [population heterogeneity] was seen if the protein concentration was increased or at lower salt concentrations at the same pH (Gerhard Wagner, personal communication).

5. Protein Engineering to Increase Solubility

5.1 Amino Acid Solubility and Water Affinity. Individual amino acids vary greatly in solubility and affinity for water (Table 2). Protein solubility is based on the ability of soluble, polar residues to interact with water in such a way that the rest of the protein can maintain an active structure. According to the "hydrophobic collapse" model of protein folding, the driving force for folding is hydrophobic amino acid clustering to avoid water, with the eventual secondary and tertiary structure further stabilized by hydrogen bonding and electrostatic interactions.[8,9] The distri-bution of polarity toward the surface is so typical that it has been used as a criterion for protein design.[64]

The data in Table 2 show that the tendency of residues to be "buried" in a protein (definitions range from less than 5% of the residue surface exposed to solvent[65] to up to 30%)[9] agrees with these generalizations. Most positively charged and amide side chain residues (His, Lys, Arg, Gln, Asn) were on the surfaces of the proteins studied, and the interiors were primarily composed of the aliphatics Gly, Ala, Ile, Leu, Val and the aromatic Phe. But only 23% of the Trp residues and 13% of the Tyr in the structures were nonaccessible to solvent, similar to that of the negative polar residues Glu (20%) and Asp (14.5%). One could argue that the large volume that Trp and Tyr residues occupy makes them difficult to completely bury in a small protein, but more likely the aqueous affinity of the tryptophan imidazole ring and the hydroxyl group of tyrosine were underestimated by early hydrophobicity measure-ments.

5.2 Peptide Solubility. There is also a great difference in solubility in secondary structural elements, as illustrated by peptides designed to adopt one conformation or another. For peptides of more than 8 amino acids, sequences favoring α-helix/random coil structures are more soluble in polar solvents than those forming β-sheet structures. The sum of the *Chou-Fasman coil index* (Table 2) for individual amino acids correlated with the solubility of a series of peptides. The tendency of peptides to form β-sheets could be significantly reduced by the strategic positioning of tertiary peptide bonds (protected residues or prolines) at intervals in the sequence.[40]

The small membrane-interacting protein melittin has a positively charged end, which makes it soluble in water, but the protein spontaneously forms a tetramer through interactions at the hydrophobic end.[26] For other peptides, insertion of arg-NO_2 residues, or replacement of hydrophobic residues, improved solubility and lowered aggregation tendencies.[66]

5.3 Primary Structure Alterations. Small changes in protein primary structure can have drastic effects on stability and solubility. Replacement of the hydrophobic (-EGNFFGKIIDYIKLMFHHWFG) car-boxy-terminal amino acids of *E. coli* penicillin-binding protein 5 with a shorter hydrophilic sequence (-IRRPAAKLE) made the protein water soluble and allowed crystallization.[67] A 13 residue deletion (EVLNENLLRFFVA) in α-casein makes the molecule more soluble.[4] Note that both of the deleted sequences contained FF. *Phenylalanine residues are likely to self-interact and are frequently found at subunit interfaces.*[68]

A series of point mutations altered the stability and solubility of insulin without significantly affecting the biological activity.[1] In particular, it was possible to replace the *asparagine* at position 21, which *deamidates in acid solution and leads to dimer formation*, with Gly, Ser, Thr, Asp, His, and Arg. Similarly, the tendency of yeast cytochrome *c* to autoreduction and dimerization was eliminated by substituting a Thr for Cys-107.[69] A hybrid interferon-α protein precipitated at low salt, unlike either of the parent molecules.[70]

The fragility of protein structure is the major limiting factor on the industrial use of enzymes. Thus the question of what makes proteins stable at high temperatures and in organic solvents and whether the two correlate is not purely academic. Specific sequence changes in proteins from *thermophilic organisms* show a tendency to r*eplace lysine and glutamic acid with arginine and aspartic acid*, and a preference for the hydrophobic amino acids Phe, Val, and Ile over Leu, Ala, and Met.[71] *Most* of these *changes occur in α-helical regions and increase the net hydrophobicity of the residue.*[72] Crambin, a plant toxin that is extremely stable in polar organic solvents, contains no Met and has a higher content of Phe, Val, and Ile residues than the hydrophilic plant toxins to which it is related.[93]

5.4 Post-Isolation Alterations. One can alter the solubility of isolated proteins *in vitro* by coupling to polyethylene glycol. Such modifications have been shown to significantly increase the activity, aqueous solubility, and *in vivo* half-life of interleukin-2.[76] Native lipase M from *Candida rugosa*, which acts on nonpolar substrates, is soluble up to 12% in water but insoluble in benzene. PEG 5000-lipase dissolved rapidly and was active in benzene, toluene, chloroform, and trichloroethane.[77]

5.5 Designer Proteins. As site-directed mutagenesis is relatively straightforward for recombinant proteins, one might simply replace surface hydrophobic amino acids with acidic residues when aggregation problems arise. But which residues are at the surface and what will the changes do to the tertiary structure and the enzymatic activity? Obviously, the problem of designing soluble proteins is greatly dependent on the ability to predict protein structure.

The *Chou-Fasman rules*, like most programs used to predict secondary structure from primary sequence data, are based on the study of known structures and the pattern of amino acid usage discerned from them.[78] The learning capabilities of *neural networks* may be the basis for the next generation of predictive programs.[79] Although the determined "code" can predict where α-helices are likely to occur, β-sheets and turns are less easy to locate. Faster computing techniques have allowed the development of local energy minimization of conformations to predict stable structures,[80] but the problem of dealing with solvent energies remains. Further, many intermediate secondary structures disappear before the native state is reached,[81] and no program in use today correctly predicts tertiary structures. Thus, [directed] mutation is still [somewhat] guesswork.

Mutation of proteins like T4 lysozyme[69] or RNase A (unpublished) may aid in structure-based stability design. For example, conversion of a single Thr residue near the carboxy terminus of T4 lysozyme to Ile, Gln, Ser, Arg, or His lowered the stability of the molecule compared to the wild type.[69] Such mutations are rather easy to produce but time-consuming to characterize; even selective mutagenesis may deplete the graduate student supply long before all the possibilities are exhausted. It may be easier to design a soluble protein from scratch, and make "designer proteins" from designer genes.[82] The potential usefulness of this approach was recently demonstrated by the production in *E. coli* of an α-helical protein designed from "first principles." The tetramer was soluble in the bacteria and seems to be both α-helical (by its CD spectrum) and very stable

(-22 kcal/mol). Betabellin, a predominantly β-sheet engineered protein, which is being made synthetically, may also be coming into solution.

These proteins show that although the folding language is not understood, a primitive but internally consistent translation is available. If this subcode really works, the next molecules should be stable, soluble, and active.

References:

1. Markussen, J., Diers, I., Hougaard, P., Langkjaer, L., Norris, K., Snel, L., Sorensen, A. R., Sorensen, E., and Voigt, H. O. 1988. Soluble, prolonged-acting insulin derivatives. 111. Degree of protraction, crystallizability, and chemical stability of insulins substituted in positions A2l, B13, B23, B27 and B30. Protein Eng. 2:157-166.
2. Dinnbier, U., Limpinsel, E., Schmid, R., and Bakker, E. P. 1988. Transient accumulation of potassium glutamate and its replacement by trehalose during adaptation of growing cells of *Escherichia coli* K-12 to elevated sodium chloride concentrations. Arch. Microbiol. 150:348-357.
3. Mitchell, R. D., Simmerman, H. K. B., and Jones, L. R. 1988. Ca^{2+}-binding effects on protein conformation and protein interactions of

Table 2 - The Aqueous Solubility and Affinity of the Amino Acids, Their Relative Tendency to Exist in a Coil Conformation (P_c), and Accessibility to Solvent in Protein Crystal Structures (Percent Buried). The Calculations in the Last Two Columns are Based on Table II of Ref. 65. The Amino Acid Names are Followed by the One Letter Codes in Parentheses.

Amino Acid	Solubility[a]	F[b]	W	P_c[c]	% Buried[d]	VR(A_3)[e]
i. Aliphatics						
glycine (G)	25.0	0	2.39	1.5	37% (10)	66
alanine (A)	16.7	0.91	1.94	0.7	38% (12)	92
isoleucine (I)	4.1	1.8	2.15	0.66	65% (12)	169
leucine (L)	2.4	1.7	2.28	0.68	41% (10)	168
valine (V)	8.9	1.22	1.99	0.62	56% (15)	142
ii. Aromatics						
phenylalanine (F)	2.97	1.79	-.076	0.71	48% (5)	203
tryptophan (W)	1.14	2.25	-5.88	0.75	23% (1.5)	240
tyrosine (Y)	0.045	0.96	-6.11	1.06	13% (2.2)	203
iii. Hydroxy/sulfur						
serine (S)	5.0	-0.004	-5.06	1.82	24% (8)	99
threonine (T)	s	0.26	-4.88	1.07	25% (5.5)	122
methionine (M)	3.4	1.23	-1.48	0.58	50% (2)	171
cystine	0.01					
cysteine (C)	s	1.54	-1.24	1.18	47% (8)	106
iv. Proline						
proline (P)	(160)	0.72	NA	1.59	24% (3)	129
hydroxy-l-proline	36.1					
v. Charged/Amides						
aspartic acid (D)	0.5	-0.77	-10.95	1.2	14.5% (5)	125
glutamic acid (E)	0.86	-0.64	-10.20	0.83	20% (2)	155
asparagine (N)	3.1	-0.6	-9.68	1.35	10% (2)	135
glutamine (Q)	3.6	-0.22	-9.98	0.86	6.3% (2.2)	161
histidine (H)	4.2	0.13	-10.27	1.06	19% (1.2)	167
lysine (K)	s	-0.99	-9.52	0.98	4.2% (0.1)	171
arginine (R)	15	-1.01	-19.92	1.04	0	225

[a]Solubility of the amino acids in g/100 g water at 25°C. Source: *CRC Handbook of Chemistry and Physics*, 68th Edition (1987-88) and *Lang"s Handbook of Chemistry* (12th edition). Sigma l-proline was not soluble at more than 1 g/ml, even at 40°C; s = freely soluble.

[b]Two different scales are shown. **F** is the hydrophobicity scale of Fauchere *et al.,*[73] which is based on the partition coefficient of the Na-acetyl-amino acid amides in octanol/water relative to glycine. **W** is the hydration potential (water affinity) of the amino acid side chain as calculated from the free energy of transfer of the side chain (e.g., methane for A) from the vapor phase to water (see ref. 74 for details). Note that both of these scales differ from the frequently used Nozaki and Tanford[75] scale which assigns values only to residues considered hydrophobic (A:0.5; I:1.8; L:1.8; V:1.22; M:1.3; C:0.5; F:2.5; Y:2.3; W:3.4; all other amino acids: 0).

[c]Coil conformation parameter based on Chou-Fasman data.[40] The parameter is based on the frequency with which a residue is present in a coil relative to its overall occurrence in the 29 proteins studied.

[d]This column represents the tendency of an amino acid to be buried (less than 5% of residues available to solvent) in the interior of a protein, and is based on the structures of 9 proteins [total of about 2000 individual residues studied, with 587 of these (29%) buried]. The first number indicates how often each amino acid was found buried, relative to the number of residues of this amino acid found in the proteins. The number in parentheses indicates the number of buried residues of this amino acid found relative to all buried residues. For other calculation methods with similar results, refs. 9 and 74a.

[e]Average volume of buried residues, calculated from the surface area of the side chain (refs. 29, 64).

canine cardiac calsequestrin. J. Biol. Chem. 263:11376-11381.
4. Farrell, H. M., Kumosinski, T. F., Pulaski, P., and Thompson, M. P. 1988. Calcium-induced associations of the caseins: a thermodynamic linkage approach to precipitation and resolubilization. Arch. Biochem. Biophys. 265:146-158.
5. Robinson, J. J. 1988. Roles for Ca^{2+}, Mg^{2+} and NaCl in modulating the self-association reaction of hyalin, a major protein component of the sea-urchin extraembryonic hyaline layer. Biochem. J. 256:225-228.
6. Mahoney, R., Wilder, T., and Chan, B. S. 1988. Substrate-induced thermal stabilization of lactase (*E. coli*) in milk. Ann. N.Y. Acad. Sci. 542:274-278.
7. McCloskey, M. and Poo, M. 1984. Protein diffusion in cell membranes: some biological implications. Int. Rev. Cytol. 87: 19-81.
8. Wright, P. E., Dyson, H. J., and Lerner, R. A. 1988. Conformation of peptide fragments of proteins in aqueous solution: implications for initiation of protein folding. Biochemistry 27:7167-7175.
9. Rose, C. D., Geselowitz, A. R., Lesser, G. J., Lee, R. H., and Zehfus, M. H. 1985. Hydrophobicity of amino acid residues in globular protein. Science 229:834-838.
10. Privalov, P. L. 1979. Stability of proteins, small globular proteins. Adv. Protein Chem. 33:167-241.
11. Jaenicke, R. 1988. Stability and self organization of proteins. Naturwissenschaften 75:604-610.
12. Arakawa, T. and Timasheff, S. N. 1985. Theory of protein solubility. Methods Enzymol. 114:49-77.
13. Kinsella, J. E 1984. Milk proteins: Physicochemical and functional properties. CRC Crit. Rev. Food Sci. Nut. 21:197-262.
14. Hjelmeland, L. M. and Chrambach, A. 1984. Solubilization of functional membrane proteins. Methods Enzymol. 104:305-318.
15. Gekko, K. and Timasheff, S. 1981. Thermodynamic and kinetic examination of protein stabilization by glycerol. Biochemistry 20:4677-4686.
16. Huot, J. Y. and Jolicoeur, C. 1985. Hydrophobic effects in ionic hydration and interactions, P. 417-471. In: *The Chemical Physics of Solvation.* (Dogonadze, R. R. et al., Eds.). Elsevier Science Publications, Amsterdam and New York.
17. Horbett, T. A. and Brash, J. L. 1987. Proteins at interfaces: Current issues and future prospects. In: *Proteins at Interfaces: Physiochemical and Biochemical Studies.* (Brash, J. L. and Horbett, T. A., Eds.) Am. Chem. Soc., Washington, D.C.
18. Mann, D. F. and Shah, K., Stein, D., and Snead, G. A. 1984. Protein hydrophobicity and stability support the thermodynamic theory of protein degradation. Biochim. Biophys. Acta 788:17-22.
19. van den Oetelaar, P. J. M., de Man, B. M., and Hoenders, H. J. 1989. Protein folding and aggregation studies by isoelectric focusing across a urea gradient and isoelectric focusing in two dimensions. Biochim. Biophys. Acta 995:82-90.
20. Ries-Kautt, M. M. and Ducruix, A. F. 1989. Relative effectiveness of various ions on the solubility and crystal growth of lysozyme. J. Biol. Chem. 264:745-748.
21. Goto, Y. and Fink, A. L. 1989. Conformational states of β-lactamase: molten globule states at acidic and alkaline pH with high salt. Biochemistry 38:945-952.
22. Zimmerman, S. B. and Trach, S. O. 1988. Effects of macromolecular crowding on the association of *E. coli* ribosomal particles. Nucleic Acids Res. 16:6309-6326.
23. Ingham, K. C. 1984. Protein precipitation with polyethylene glycol. Meth. Enzymol. 104:351-356.
24. Hanada, K., Yamato, I., and Anraku, Y. 1988. Solubilization and reconstitution of proline carrier in *Eschericia coli*; quantitative analysis and optimal conditions. Biochim. Biophys. Acta 939:282-288.
25. Brenner, S. L. Zlotnick, A., and Griffith, J. D. 1988. RecA protein self-assembly. Multiple discrete aggregation states. J. Mol. Biol. 204:959-972.
26. Schwarz, G. and Beschiaschvili, G. 1988. Kinetics of melittin self-association in aqueous solution. Biochemistry 27:7826-7831.
27. Zimmerle, C. T. and Frieden, C. 1988. Effect of pH on the mechanism of actin polymerization. Biochemistry 27:7766-7772.
28. Yang, D. S. C., Sax M., Chakrabartty, A., and Hew, C. L. 1988. Crystal structure of an antifreeze polypeptide and its mechanistic implications. Nature 333:232-237.
29. Richards, F. M. 1977. Areas, volumes, packing, and protein structure. Annu. Rev. Biophys. Bioeng. 6:151-176.
30. Eisenberg, D., Wilcox, W., and McLachlan, A. D. 1986. Hydrophobicity and amphiphilicity in protein structure. J. Cell. Biochem. 31:11-17.
31. Hageman, M. J. 1988. The role of moisture in protein stability. Drug Dev. and Ind. Pharm. 14:2047-2070.
32. Good, N. E. and Izawa, S. 1972. Hydrogen ion buffers. Methods Enzmol. 24:53-68.
33. Blanchard, J. S. 1984. Buffers for enzymes. Methods Enzymol. 104:404-414.
34. Yancey, P. H., Clark, M. E., Hand, S. C., Bowlus, R. D., and Somero, G. N. 1982. Living with water stress; evolution of osmolyte systems. Science 217:1214-1222.

35. McPherson, A. 1982. *Preparation and Analysis of Protein Crystals*. John Wiley and Sons., N.Y.
36. Feher, G. and Kam, Z. 1985. Nucleation and growth of protein crystals; general principles and assays. Methods Enzymol. 114:77-111.
37. Kamoun, P. P. 1988. Denaturation of globular proteins by urea: breakdown of hydrogen or hydrophobic bonds? TIBS 15:424-425.
38. Schellekens, H., de Reus, A., Boluis, R., Fountoulakis, M. Schein, C., Easodi, J., Nagata, S., and Weissmann, C. 1981. Comparative antiviral efficiency of leukocyte and bacterially produced human α-interferon in rhesus monkeys. Nature 292:775-776.
39. Arnold, F. H. 1988. Protein design for nonaqueous solvents. Protein Eng. 2:21-25.
40. Narita, M., Ishikawa, K., Chen, J. Y., and Kim, Y. 1984. Prediction and improvement of protected peptide solubility in organic solvents. Int. J. Peptide Protein Res. 24:580-587.
41. Janicke, R. and Rudolph, R. 1989. Folding proteins, P. 191-223. In: *Protein Structure: a Practical Approach.* (T. E. Creighton, Ed.). Oxford University Press, U.K.
41a. Rudolph, R. and Fisher, S. 1987. Verfahrung zur Renaturierung van Proteinen. Eur. Patent Appl. 0241-022.
42. Weir, M. P. and Sparks, J. 1987. Purification and renaturation of recombinant human interleukin-2. Biochem. J. 245:85-91.
43. Marston, F. A. O. 1986. The purification of eukaryotic polypeptides synthesized in *Escherichia coli.* Biochem. J. 240:1-12.
44. Hochuli, E. Bannarth, W., Dobeli, H., Gentz, R., and Stuber, D. 1988. Genetic approach to facilitate purification of recombinant proteins with a novel metal chelate adsorbent. Bio/Technology 6:1321-1325.
45. Smith, D. C. and Hider, R. C. 1988. Thiol exchange catalyzed refolding of small proteins utilizing solid-phase supports. Biophys. Chem. 31:21-28.
46. Light, A., Duda, C. T., Odorzynski, T. W., and Moore, W. G. I. 1986. Refolding of serine proteinases. J. Cell. Biochem. 31:19-26.
47. van Kimmenade, A., Bond, M. W., Schumacher, J. H., Laquoi, C., and Kastelein, R. A. 1988. Expression, renaturation and purification of interleukin-4 from *Escherichia coli.* Eur. J. Biochem. 173:109-117.
48. Kuhlbrandt, W. 1988. Three-dimensional crystallization of membrane proteins. Quart. Rev. Biophys. 21:429-477.
49. Wallace, B. A., Cascio, M., and Mielke, D. L. 1986. Evaluation of methods for the prediction of membrane protein secondary structures. Proc. Natl. Acad. Sci. U.S.A. 81:9423-9427.
50. Furth, A. J., Bolton, H., Potter, J., and Priddle, J. D. 1984; Separating detergent from proteins. Methods Enzymol. 104:318-328.
51. Lee, K. H., Fitton, J. E., and Wuthrich, K. 1987. Nuclear magnetic resonance investigation of the conformation of δ-haemolysin bound to dodecylphosphocholine micelles. Biochim. Biophys. Acta 911:144-153.
52. Maloney, P. C. and Ambudkar, S. V. 1989. Functional reconstitution of prokaryotic and eukaryotic membrane proteins. Arch. Biochem. Biophys. 269:1-10.
53. Welte, W. and Wacker T. 1989. Protein-detergent micellar solutions for the crystallization of a membrane protein. Some general approaches and experiences with the crystallization of pigment-protein complexes from purple bacteria. In: *Membrane Protein Crystallization.* (Michel, H., Ed.). CRC Press, Inc., Boca Raton, FL.
54. Yadeau, J. T. and Blobel, G. 1989. Solubilization and characterization of yeast signal peptidase. J. Biol. Chem. 264:2928-2934.
55. Fargin, A., Faye, J. C., le Maire, M., Bayard, F., Potier, M., and Beauregard, C. 1988. Solubilization of a tamoxifen-binding protein. Biochem. J. 256:229-236.
56. Kline, A. D., Braun, W., and Wuthrich, K. 1988. Determination of the complete three-dimensional structure of the α-amylase inhibitor tendamistat in aqueous solution by nuclear magnetic resonance and distance geometry. J. Mol. Biol. 204:675-724.
57. Montelione, G. T., Wuthrich, K., Nice, E. C., Burgess, A. W., and Scheraga, H. A. 1987. Solution structure of murine epidermal growth factor determination of the polypeptide backbone chain-fold by nuclear magnetic resonance and distance geometry. Proc. Natl. Acad. Sci. U.S.A. 84:5226-5230.
58. Oswald, R. E., Bogusky, M. J., Bamberger, M., Smith, R. A. G., and Dobson, C. M. 1989. Dynamics of the multidomain fibrinolytic protein urokinase from two-dimensional NMR. Nature 337:579-582.
59. Fesik, S. W. 1988. Isotope-edited NMR spectroscopy. Nature 332:865-866.
60. Oh, B. H., Westler, W. M., Darba, P., and Markley, J. L. 1988. Protein carbon-13 spin systems by a single two-dimension nuclear magnetic resonance experiment. Science 240:908-911.
61. Senn, H., Eugster, A., Otting G., Suter, F. and Wuthrich, K. 1987. ^{15}N-labeled P22 c2 repressor for nuclear magnetic resonance studies of protein-DNA interactions. Eur. J. Biophy. 14:301-306.

62. Markley, J. L. 1987. One- and two-dimensional NMR spectroscopic investigation of the consequences of amino acid replacements in proteins, P. 15-33. In: *Protein Engineering* (Oxender, D. L. and Fox, C. F., Eds.), Alan R. Liss, Inc., New York.
63. Fairbrother, W. J., Hall, L., Littlechild, J. A., Walker, P. A., Watson, H. C., and Williams, R. J. P. 1988. Probing the 3-phosphoglycerate-binding site of yeast phosphoglycerate kinase using site-specific mutants and ^1H nuclear magnetic resonance spectroscopy. Biochem. Soc. Proc. 16:724-725.
64. Baumann; G., Frommel, C., and Sander, C. 1989. Polarity as a criterion in protein design. Protein Eng. 2:239-334.
65. Richards. F. M. 1986. Protein design: are we ready. UCLA Symp. Mol. Cell. Biol. 39:171-196.
66. Toniolo, C., Bonora, G. M., Moretto, V., and Bodanszky, M. 1985. Self-association and solubility of peptides. Int. J. Peptide Protein Res. 25:425-430.
67. Ferreira, L. C. S., Schwarz, U., Keck, W., Charlier, P., Dideberg, O., and Ghuysen, J. M. 1988. Properties and crystallization of a genetically engineered, water-soluble derivative of penicillin-binding protein 5 of *Escherichia coli* K12. Eur. J. Biochem. 171:11-16.
68. Argos, P. 1988. An investigation of protein subunit and domain interfaces. Protein Eng. 2:101-113.
69. Shaw, W. V. 1987. Protein engineering. The design, synthesis and characterization of factitious proteins. Biochem. J. 246:1-17.
70. Le, H. V., Syto, R., Schwartz, J., Nagabhushan, T. L, and Trotta, P. P. 1988. Purification and properties of a novel recombinant human hybrid interferon, $\delta 4$ $\alpha 2$ / $\alpha 1$. Biochim. Biophys. Acta 957:143-151.
71. Zuber, H. 1988. Temperature adaptation of lactate dehydrogenase. Structural, functional and genetic aspects. Biophys. Chem. 29:171-179.
72. Menendez-Arias, L. and Argos, P. 1989. Engineering protein thermal stability. Sequence statistics point to residue substitutions in α-helices. J. Mol. Biol. 206:397-406.
73. Fauchere, J., Charton, M., Kier, L. B., Verloop, A., and Pliska, V. 1988. Amino acid side chain parameters for correlation studies in biology and pharmacology. Int. J. Peptide Protein Res. 32:269-278.
74. Wolfenden, R., Andersson, L., Cullis, P. M., and Southgate, C. C. 198l. Affinities of amino acid side chains for solvent water. Biochemistry 20:849-855.
74a. Janin, J. 1979. Surface and inside volumes in globular proteins. Nature 277:491-492.
75. Nozaki, Y. and Tanford, C. 1971. The solubility of amino acids and two glycine peptides in aqueous ethanol and dioxane solutions. Establishment of a hydrophobicity scale. J. Biol. Chem. 246:2221-2227.
76. Knauf, M. J., Bell, D. P., Hirtzer, P., Luo, Z. P., Young, J. D., and Katre, N. V. 1988. Relationship of effective molecular size to systematic clearance in rats of recombinant interleukin-2 chemically modified with water soluble polymers. J. Biol. Chem. 263:15064-15070.
77. Baillargeon, M. W. and Sonnet, P. E 1988. Lipase modified for solubility in organic solvents. Ann. N.Y. Acad. Sci. 542:244-49.
78. Kikuchi, T., Nemethy, G., and Sheraga, H. A. 1988. Prediction of probable pathways of folding in globular proteins. J. Prot. Chem. 7:491-507.
79. Holley, L. H. and Karplus, M. 1989. Protein secondary structure prediction with a neural network. Proc. Natl. Acad. Sci. U.S.A. 86:152-156.
80. Karplus, M. 1987. The prediction and analysis of mutant structures, P. 55-44. In: *Protein Engineering* (Oxender, D. L. and Fox, C. F., Eds.), Alan R. Liss, Inc. New York.
81. Creighton, T. E. 1988. On the relevance of non-random polypeptide conformations for protein folding. Biophys. Chem. 31:155-162.
82. DeGrado, W. F., Wasserman, Z. R., and Lear, J. D. 1989. Protein design, a minimalist approach. Science 243:622-628.
83. Kim, P. S. 1988. Passing the first milestone in protein design. Protein Eng. 2:249-250.
84. Schein, C. 1989. Production of soluble recombinant proteins in bacteria. Bio/Technology 7:1141-1149.

THEORY/PRINCIPLES

DOMINANT FORCES IN PROTEIN FOLDING

Dill, K. A. (1990) Biochemistry 29 (31): 7133-7155.

1. Introduction

The purpose of this review is to assess the nature and magnitudes of the dominant forces in protein folding. Since proteins are only marginally stable at room temperature, no type of molecular interaction is unimportant, and even small interactions can contribute significantly (positively or negatively) to stability.[3,4,189,190] However, the present review aims to identify only the largest forces that lead to the structural features of globular proteins: their extraordinary compactness, their core of nonpolar residues, and their considerable amounts of internal architecture.

This review explores contributions to the free energy of folding arising from electrostatics (classical charge repulsions and ion pairing), hydrogen-bonding and van der Waals interactions, intrinsic propensities, and hydrophobic interactions. An earlier review by Kauzmann[149] introduced the importance of hydrophobic interactions. His insights were particularly remarkable considering that he did not have the benefit of known *protein structures*, *model studies*, *high-resolution calorimetry*, *mutational methods*, or *force-field* or *statistical mechanical results*. The present review aims to provide a reassessment of the factors important for folding in light of current knowledge. Also considered here are the opposing forces, *conformational entropy* and *electrostatics*.

The process of protein folding has been known for about 60 years. In 1902, Emil Fischer and Franz Hofmeister independently concluded that proteins were chains of covalently linked amino acids[126] but deeper understanding of protein structure and conformational change was hindered because of the difficulty in finding conditions for solubilization. Chick and Martin[52] were the first to discover the process of denaturation and to distinguish it from the process of aggregation. By 1925, the *denaturation process* was considered to be either *hydrolysis* of the peptide bond[322,14] or *dehydration* of the protein.[260] The view that protein denaturation was an *unfolding* process was first put forward by Wu.[320,321] He proposed that native proteins involve regular repeated patterns of folding of the *chain* into a three-dimensional network somewhat resembling a *crystal*, held together by *noncovalent linkages*. "Denaturation is the breaking up of these labile linkages. Instead of being compact, the protein now becomes a diffuse structure. The *surface* is altered and the *interior* of the molecule is exposed."[320] "Denaturation is disorganization of the natural protein molecule, the change from the regular arrangement of a rigid structure to the irregular, diffuse arrangement of the flexible open chain."[321]

Before discussing forces, we ask: Is the native structure *thermodynamically stable* (the "thermodynamic hypothesis")[12] or *metastable*, determined, for example, as the protein leaves the ribosome? To prove thermodynamic stability, it is sufficient to demonstrate that the native structure is only a *function of state* and does not depend on the process or initial conditions leading to that state. By definition, such a state would be at the *global minimum of free energy* relative to all other states accessible on that time scale. Experiments of Anson and Mirsky[15] and Anson[13] showed that hemoglobin folding is *reversible* as evidenced by similarities in the following properties of native and renatured protein: solubility, crystallizability, characteristic spectrum, binding to O_2 and CO_2, and inaccessibility to trypsin digestion. The folding of serum albumin and other proteins was shown to be similarly reversible by these coarse measures of native structure.[211,13,178] It was then demonstrated that denaturation is also *thermodynamically* reversible for some proteins[87,40] and involved large conformational changes.[125,271] Recent high-resolution calorimetry experiments show thermodynamic reversibility for many small single-domain globular proteins[244,245,269,34,220] and also for some multidomain and coiled-coil proteins.[248] Reversibility was tested much more specifically by the experiments of Anfinsen.[12] in which the *disulfide bonds* of bovine pancreatic ribonuclease were "scrambled" to random distributions of the 105 possible binding patterns and reacquisition of native structure and activity was observed upon *renaturation*.[12] The advantage of monitoring disulfide bonds is that they are uniquely *trappable* and identifiable. Similarly, two *circularly permuted proteins* refold to their original native states.[177,115] Therefore, despite often extreme difficulties in the achievement of reversibility, the thermodynamic hypothesis has now been widely established. These experiments do not necessarily imply reversibility is completely general for other conditions, for other proteins, or even for other parts of a given protein than those monitored by the given experiment. It is clear that the folding of some proteins can be catalyzed by other assisting proteins, such as polypeptide binding or *chaperone proteins*.[266,219,88,12] Nevertheless, the existence of chaperones bears only on the *rate* that a protein is folded (provided the chaperone is a true *catalyst*) and has no bearing on the thermodynamic hypothesis, on the nature of the native state (if the native structure is otherwise reversible on the experimental time scale), or on the driving forces that cause it. The present discussion addresses only those proteins and conditions for which reversibility holds.

In this discussion of the nature of forces, it is useful to distinguish *long-ranged* and *short-ranged forces*, on the one hand, from *local* and *nonlocal* forces, on the other. The distance dependence defines the range: energies that depend on distance r as r^{-p} are long-ranged if $p < 3$ (*ion-ion* and *ion-dipole interactions*, for example) or short-ranged if $p > 3$ (*Lennard-Jones attractions* and *repulsions*, for example). This inverse third-power dependence is the natural division because for simple pure media the integral that gives the total energy of a system diverges, according to this definition, for long-ranged forces and converges for short-ranged forces.[133] (This relates to the three degrees of freedom resulting from three spatial dimensions in Cartesian space.) For polymer chains such as proteins, segment position in the chain is also important, in addition to the range of force. "Local" interactions are those among chain segments that are "connected" neighbors (i, $i+1$), or near neighbors, in the sequence (see Figure 1). "Nonlocal" refers to interactions among residues that are significantly apart in the sequence. Local interactions can arise from either long- or short-ranged forces, as can nonlocal interactions.

2. Long-Ranged Interactions: Electrostatics

Because acids and bases were among the earliest known denaturants of proteins, the folding forces were first assumed to be electrostatic in nature. The signature of electrostatically driven processes is a dependence on pH and/or ionic strength. Whereas the pH determines the total charge on the protein, the salt determines the extent of interaction among those charges – since salts shield charges. The first *quantitative model of electrostatic interactions* in native proteins was proposed by Linderstrom-Lang[174] (when he was 27 years old!). This work appeared less than 1 year after the *Debye-Huckel* theory on which it was based. By treating a *native protein* as a multivalent impenetrable spherical particle with its net charge uniformly distributed at the surface, Linderstrom-Lang predicted the number of protons bound and the net charge as functions of the hydrogen ion concentration, i.e., the pH titration curve. The view that protein electrostatics can be represented in terms of charges on a sphere of low dielectric constant in a higher dielectric medium has remained useful. Recent improvements have included (*i*) the consideration of discrete charges located at specific positions on the spherical native protein,[308,309,188,187] (*ii*) modeling native structural deviations from spherical shape,[112,113] and (*iii*) the development of electrostatic theory for the unfolded state and therefore for free energy contributions to

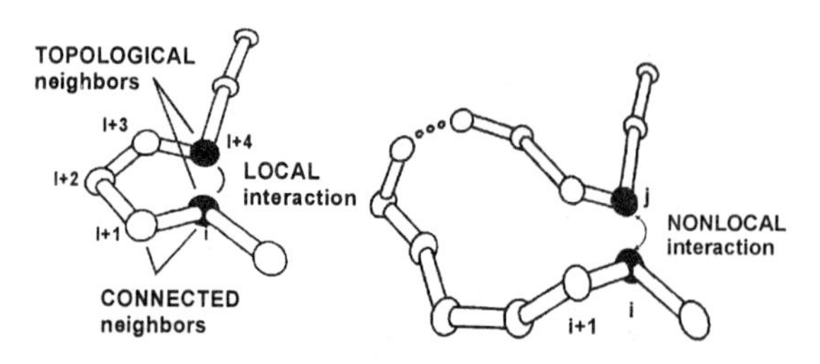

Figure 1 - Spatially neighboring residues (i, j) are defined as *connected* neighbors if they share a backbone bond, $j = i + 1$; otherwise, they are *topological* neighbors. Interactions are *local* or *nonlocal* depending on the separation along the chain of the interacting residues.

stability.[293,294] There are two different ways in which electrostatic interactions can affect protein stability.

2.1 Classical Electrostatic Effects. Classical electrostatic effects are the nonspecific repulsions that arise when a protein is highly charged, for example, at extremes of pH. The traditional view[302,148,174] of these effects has been that the electrostatic free energy depends on the square of the net charge. Hence, no electrostatic contribution to protein stability is expected near the isoelectric point. As the net charge on the native protein is increased by increasing acidity or basicity of the solution, the increasing charge repulsion will destabilize the folded protein because the charge density on the folded molecule is greater than on the unfolded molecule. Thus, the process of unfolding leads to a state of lower electrostatic free energy. Hence, acids and bases destabilize native proteins (see Figure 2).

2.2 Specific Charge Interactions. Specific charge interactions can also affect stability. For example, *ion pairing* (*salt bridging*) occurs when oppositely charged amino acid side chains are in close spatial proximity. Whereas the classical mechanism predicts that increasing the charge could only destabilize the folded state, ion pairing could stabilize it. It has traditionally been held that classical and ion-pairing effects could be distinguished by experiments on the effects of salt concentration (below about 0.1-1.0 M) or the dielectric constant of the solution. Only at low concentrations does salt predominantly affect electrostatic shielding; at higher concentrations the electrostatic shielding is saturated, so that then the dominant effects of salt, like any other additives, are on the solvent properties of the solution.[202] In the traditional view, it is assumed that salts and the dielectric constant of the medium do not affect the net charge on the molecule and that they affect the native state more than the denatured state. It has often been assumed therefore that either adding salt or increasing the dielectric constant of the solution should stabilize proteins if classical effects are dominant or destabilize them if ion pairing is important. It is now clear, however, that effects of neither salt nor dielectric constant can be interpreted

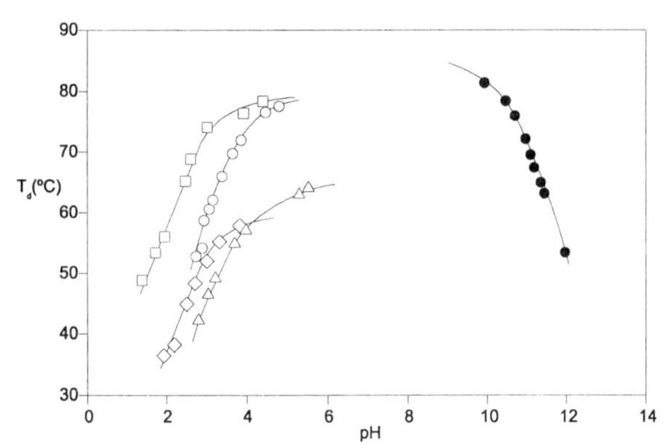

Figure 2 - Denaturation temperature vs. pH for (Δ) metmyoglobin, (•) ribonuclease A, (○) cytochrome c, (◇) α-chymotrypsin and (□) lysozyme. Increased charge on the protein at extremes of pH (low or high) favors unfolding. Reproduced from Privalov, P. L., & Kechinashvili, N. N. (1974) *J. Mol. Biol. 86*, 665. Copyright (1974) Academic Press Inc. (London) Ltd.

so simply, for the following reasons. First, salts strongly affect the unfolded state (see below). Second, *ion-pair bonds* are generally much shorter than the *Debye lengths* in salt solutions, so salt should have little effect on ion-pairing stability. [One exception is a Glu-2$^-$···Arg-10$^+$ salt bridge in the C-peptide helix,[283] which is screened by 1 M salt, but this may be a solvent-separated ion pair[24].] Third, although a decreased *dielectric constant* will lead to increased charge interactions, it will also decrease the total *ionization* since charging is energetically more costly in a low-dielectric medium. Moreover, the dielectric constant is also correlated with other solvent properties such as *hydrophobicity* and is not a simple diagnostic for charge effects alone. Therefore, discriminating between classical and ion-pairing electrostatics contributions to stability has been difficult.

During the 1930s, ion pairing was considered to be the dominant contributor to protein stability.[59,198,91] Mirsky and Pauling suggested that folding was driven by the ion pairing of carboxyl and amino groups on the side chains of the charged amino acids.

If ion pairing is important for protein stability, then such stability must arise from charged pairs at protein surfaces rather than from charged pairs buried in the protein core. The first evidence that few ion pairs are buried was due to Jacobsen and Linderstrom-Lang[139] on model compounds. An important signature of electrostatic effects in solution is a change in volume: the *local volume* of water decreases around a molecule of increasing charge. The electrostatic field of the charged molecule orients and orders neighboring water dipoles (*electrostriction*), decreasing the entropy and volume of the local water molecules. At low pH where only the carboxyl groups are titratable, Jacobsen and Linderstrom-Lang noted that the volume increase upon protonation of COO^- to COOH, of about 10 ml/mol in proteins, is the same as in model carboxyl compounds in water, suggesting that ion pairs in proteins must be exposed. More recent studies of known protein structures show that indeed few ion pairs are buried (on average, only about one ion pair per 150-residue protein is buried).[25] This follows from the very high *Born energy* required to transfer a charged ion from aqueous solution to the low-dielectric interior of the protein, ranging from 19 kcal/mol for full burial to 4 kcal/mol for a half-exposed ion at the surface, 7 kcal/mol for complete burial of an ion pair.[137,136] Thus, unless other specific interactions are involved, only surface ion pairs could generally stabilize native states.

It is clear that ion pairing can contribute to protein stability. Studies of X-ray crystal structures of known proteins[315,25] show that ion pairing is common on the surfaces of proteins. Also, variations in sequence that affect ion pairing can change stability by about 1-3 kcal/mol of ion pairs;[94,235] Asp-70–His-31 in T4 lysozyme has recently been found to stabilize by -3 to -5 kcal/mol.[11] Similarly, ion binding sites designed into proteins can affect stability.[223] However, it is clear that ion pairing is not the dominant force of protein folding. The first evidence emerged from the pivotal paper of Jacobsen and Linderstrom-Lang.[139] They interpreted the models of Eyring and Stearn[91] and Mirsky and Pauling[198] as shown in Figure 3. The hydrogen protonates the carboxyl group in the unfolded state, so both carboxyl and amino groups are uncharged, whereas in the folded state the hydrogen protonates the amino group, so that the carboxyl and amino groups form an ion pair. Jacobsen and Linderstrom-Lang presumed that the charges remained solvated upon folding. Model compounds show that the *protonation* of NH_2 to NH_3^+ leads to an *electrostriction* of about -4

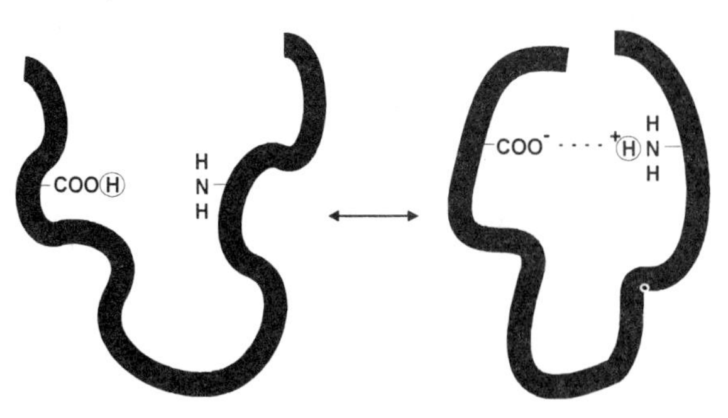

Figure 3 - Early model in which protein folding was proposed to be driven by ion-paired hydrogen bonding among side chains,[198,91] shown by Jacobsen and Linderstrom-Lang[139] to be inconsistent with partial molar volumes.

ml/mol and *deprotonation* of COOH to COO^- leads to -10 ml/mol, as noted above. Folding should then result in a volume change of -14 ml/mol per ion pair. In contrast to this model, experiments show that folding leads to an *increase* in volume.[139,329,41,82]

Also inconsistent with ion pairing as the dominant force of folding is the observation that the stabilities of proteins show little dependence on pH or salt (at low salt concentrations) near the *isoelectric point*.[304,,129,1,313] [For some proteins, the pH of maximum stability does not coincide with the isoelectric pH, but this can be accounted for within the classical model by the burial in the hydrocarbon core of some of the titratable groups (often histidines).[293]] As further evidence that charge generally contributes only weakly to protein stability, Hollecker and Creighton[134] found little effect of changing the charges on several different amino groups in three different proteins.

Third, perhaps the most persuasive evidence that ion pairing is not the dominant force of folding comes from the structural studies of Barlow and Thornton.[25] They have observed that ion pairs are not highly *conserved in evolution*. More importantly, the number of ion pairs in proteins is small. They observe about five ion pairs per 150 residues of protein (about one of which is buried, noted above). It is unlikely that any interaction involving only 10 residues, less than 10% of the molecule, could be the dominant folding force. Using the estimate of -1 to -3 kcal/mol[94,235] for the stabilization per ion pair leads to a value of -5 to -15 kcal/mol stabilization. *Even though ion pairing would thus contribute a free energy equal to that of the net stability of the protein, this is still 5- to 10-fold smaller than the hydrophobic interaction discussed below.*

A similar estimate, of about 10 kcal/mol stabilization due to ion pairing, has been made by Friend and Gurd.[104] They observed decreased stability of sperm whale ferrimyoglobin with increased salt and interpreted this as evidence for ion pairing. They assumed *salt* predominantly affects the native state, on the basis of the difference in *titration behavior* of native and denatured states. However, these results do not necessarily imply the electrostatic stabilization comes from ion pairing. Salt can affect the relative free energies differently than it affects the titration behavior. A recent *polyelectrolyte model* of proteins[293,294] shows instead that increasing salt, by classical effects alone, will reduce the electrostatic free energy of the unfolded state of myoglobin more than the folded state. Increased salt shields the charge repulsions in the unfolded molecule more effectively than in the folded molecule at low pH, probably because of better penetration of the salt solution into the unfolded molecule. The model is consistent with an additional experimental observation that is otherwise difficult to explain on the basis of ion pairing. For β-lactamase, similar to the myoglobin experiments of Friend and Gurd, Goto and Fink[117] observe that salt *destabilizes* the native state when the molecule is highly charged at low pH, but they also find that salt *stabilizes* the native structure when the molecule is charged at high pH. It is interesting that a significant fraction of the electrostatic free energy is predicted to arise from the *entropy of proton release*,[292,293] rather than simply from the charge energetics.

3. Hydrogen Bonding and van der Waals Interactions

van der Waals attractions arise from interactions among *fixed* or *induced dipoles*. A *hydrogen bond* occurs when a hydrogen atom is shared between generally two electronegative atoms. Hydrogen-bond strength, which depends on the electronegativity and orientation of the bonding atoms, is in the range of -2 to -10 kcal/mol.[227] For example, the water-water hydrogen bond in the vapor phase is -6.4 kcal/mol.[316] A hydrogen bond is primarily a linear arrangement of *donor*, *hydrogen*, and *acceptor* and is comprised of electrostatic, dispersion, charge-transfer, and steric repulsion interactions.[312] The dominant component of a hydrogen bond is electrostatic.[227,64,312] In this section we ask: do hydrogen bonds and van der Waals interactions contribute *differently* to folded and unfolded states of proteins, and therefore to stability? While these two types of force are microscopically quite different, there are few simple *macroscopic diagnostics* that can distinguish between them; hence, in this section we consider them together. The evidence cited below suggests that they may play an important role in protein folding, but the magnitudes, among all the types of force contributing to protein folding, are currently perhaps the most difficult to assess.

Mirsky and Pauling[198] were the first to suggest that hydrogen bonding was the dominant force of protein folding. Although their focus appears to have been the electrostatic hydrogen bonds arising from ion-paired side chains (see preceding section), they also suggested that hydrogen bonding could occur between the carbonyl C=O and amide NH groups of the peptide backbone. It is the peptide hydrogen bonds we consider here. Their proposal led to the X-ray crystallography studies of amino acid crystals by Pauling *et al.* begun in 1937, culminating in the discovery of the α-helix and parallel and antiparallel sheets in 1951.[232,228-231] These were first called "*secondary structures*" by Linderstrom-Lang.[175]

During the 1950s, Doty and his colleagues found a model system, poly(γ-benzyl-L-glutamate), for studying the *driving forces* in the formation of polypeptide helices in *nonaqueous solution.*[75-78] Soon thereafter a theoretical framework emerged for understanding the balance of forces driving the *helix-coil transition.* The first theoretical model was due to Schellman.[274] Many other elegant treatments followed, principally based on the one-dimensional *Ising model.*[233,108,328,327,98,240] In these models, the *intrachain hydrogen bond* is considered energetically favorable relative to the hydrogen bond with the solvent. However, to form the first such bond requires overcoming *configurational entropy* to arrange the immediately adjacent bonds into a helical configuration. At low temperatures with simple solvents, the *enthalpic contribution* dominates, and the molecule forms a helix; at high temperatures, the entropy dominates and the molecule is configured as a *random coil.*[282,182,180,238] A *sharp transition* between these states results from this subtle balance between the large forces. The *entropy is local* insofar as it involves the configurations of only immediately neighboring bonds along the chain and thus is assumed to be independent of aspects of the chain configurations more distant. Theoretical helix-coil transition models successfully predict (*i*) this temperature dependence and (*ii*) that helices become more stable and that transitions sharpen with increasing chain length.[240] The models also predict the influence of pH on the helix-coil transition: greater charge on the molecule destabilizes the helix since the coil has lower charge density and thus lower electrostatic free energy.[233,327] The helix-coil transition has inverted temperature dependence in some mixed solvents.[328,180] Solvents that bind to the peptide bond will favor the coil; one example is formic acid, which protonates the bond.[176] Consistent with the view that hydrogen bonding is the principal driving force of the helix-coil transition, solvents that form strong hydrogen bonds compete more effectively with the peptide and destabilize the helix relative to the coil. For example, chloroform, dimethylformamide, 2-chloroethanol, trifluoroethanol, and other alcohols favor the helix, relative to formic acid, dichloroacetic acid, or trifluoroacetic acid.[75-78,180,61,210,207] A similar theory has been developed for *β-sheet* formation.[192]

For three reasons, it is natural to assume that hydrogen bonding and van der Waals interactions will be important for the conformational changes of proteins. First, the amino acids that comprise proteins are dipolar and are capable of hydrogen bonding. Second, helices are common features of globular proteins, and the studies cited above show that the helix-coil transition is largely driven by hydrogen bonding. Similarly, intramolecular sheets are also formed by hydrogen bonding.[17] Third, the conformational forces for *nonelectrolyte* polymers in nonelectrolyte solvents are short ranged, arising from differences in *monomer-monomer attractions* of the chain relative to *monomer-solvent attractions.*[97,68] If monomer and solvent interactions are short ranged, then classical polymer theories would predict that chains should usually be relatively self-attractive, with radius changes characterized by a temperature-independent enthalpy.[97] Such a temperature-independent enthalpy has been inferred to contribute to protein folding on the basis of model assumptions about the contribution of the hydrophobic interaction.[24,244,74] This *residual folding force* becomes more favorable as the *number of polar groups* increases.[246] For hen lysozyme, the magnitude of this enthalpy is 0.43 kcal/mol of residues.[24]

Although these short-ranged forces are therefore undoubtedly important, Kauzmann[148,149] concluded that they are probably not the dominant forces that fold proteins in water. *A fundamental criterion for a dominant driving force is that it must explain why the folded state is advantageous relative to the unfolded state.* He argued that hydrogen bonding would not satisfy this criterion, because there was no basis for believing that the intrachain hydrogen bonds in the folded state would have lower free energy than those of the unfolded chain to water. In support of this view, the distribution of *hydrogen-bond angles* in proteins is observed to be about the same as in small-molecule compounds.[23] It follows however that folded proteins must contain many hydrogen bonds; for otherwise, the protein would denature.

Kauzmann"s hypothesis led to model studies on analogues to determine the free energy of peptide hydrogen bonds in water. The several models of the peptide hydrogen bond include urea,[273] valerolactam,[298] *N*-methylacetamide (NMA),[157,160] and cyclic dipeptides, the diketopiperazines.[109] For reasons described below, however, these model studies have not yet yielded definitive estimates for the contribution of hydrogen bonds to protein stability. The *dimerization binding constants* and their temperature dependences have been measured for these molecules in water, in order to obtain free energies and enthalpies of dimerization. Because the *concentration dependences* in these experiments are linear at low concentrations, the bound species is presumed to be predominantly in the form of dimers, rather than higher multimers. At 25°C, dimerization in water is disfavored; the equilibrium ratio of dimers to monomers is only 4.1 x 10^{-2} for urea and 5.0 x 10^{-2} for diketopiperazine. Thus, the *free energy* of dimerization is positive ($\Delta G_{dimerization}$ = +1.89 kcal/mol for urea). However, the enthalpy of dimer formation is negative (-2.1 kcal/mol for diketopiperazine, -2.1 kcal/mol for urea, -2.8 kcal/mol for valerolactam), except for *N*-methylacetamide for which it is approximately zero.[157] On the assumption that the hydrogen bond is the only attraction driving dimerization, it has been generally concluded that the hydrogen bond in water is enthalpically favored relative to the monomer-water bond. For *N*-methylacetamide, Jorgensen[142] has shown by *Monte Carlo simulation* that this assumption is probably not valid: in water, the amides *stack* rather than form hydrogen bonds. In contrast, in chloroform, the amides form good hydrogen bonds.[142] Susi and Ard[299] have suggested that ε-caprolactam dimerization may also be driven by some mechanism other than hydrogen bonding. Thus, in addition to hydrogen bonding, van der Waals and other interactions may also contribute significantly, but their importance for the other model compounds is not yet clear.

An additional problem prevents unequivocal determination of the free energy of hydrogen-bond formation from these model studies. In all the model compounds, there are two ways a dimer can form: singly bonded, wherein one partner in the dimer will have considerable rotational freedom relative to the other, or doubly bonded, with one partner considerably *restricted in its rotation* relative to the other. The experiments find only the ratio of "complexed" molecules (of singly-bonded plus doubly-bonded dimers) to unbound monomers. To obtain the free energy of hydrogen-bond formation from these data requires additional knowledge of the relative numbers of singly-bonded and doubly-bonded dimers, not currently available for these model compounds. The dilemma is illustrated by the following comparison. Suppose, on the one hand, that the only species in solution was known to be the singly-bonded dimer; then the measured positive free energy implies hydrogen bonding is *disfavored* in water. Suppose alternatively that the only species in solution was known to be the doubly-bonded dimer. Then the binding free energy will include contributions from the two hydrogen bonds and an unfavorable entropy of rotational restriction. If this rotational restriction is sufficiently unfavorable, contributing a large enough positive free energy to the overall dimerization free energy, then the intrinsic free energy of hydrogen-bond formation will be inferred to be negative, implying that hydrogen bonding is *favored* in water. Schellman[273] estimated this entropic restriction to be 3-6 eu and concluded that the free energy of formation of a hydrogen bond is negative but probably small. Thus the inference as to whether hydrogen-bond formation is favored or disfavored in water depends on (*i*) which compound is chosen as a model, (*ii*) the importance of interactions other than hydrogen bonding for the association processes of those model compounds in water, and (*iii*) estimation of the magnitude of a *rotational restriction entropy*, presently unknown but probably of about the same magnitude as the free energy of hydrogen bonding itself.

Moreover, this class of experiments has largely been restricted to aqueous solvents. But the peptide hydrogen bond in a globular protein is in a more *hydrocarbon-like medium*. We are therefore interested in the following equilibrium:

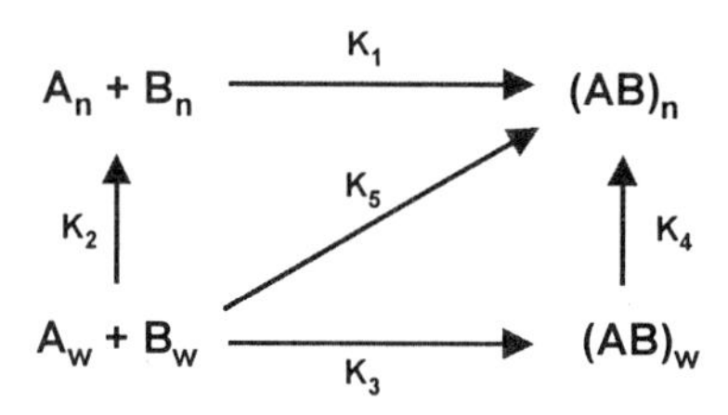

where A and B are the hydrogen-bond donor and acceptor, w is water, and n is the nonpolar solvent. We aim to determine ΔG_5, the hydrogen-bond contribution to protein stability (corresponding to equilibrium constant K_s. We obtain this by using other steps of the thermodynamic cycle. (*1*) A wide range of hydrogen-bonding species, including NMA, formamide, alcohols, carboxylic acids, and phenols, tend to associate in nonpolar solvents ($K_1 > 1$); for example, for the dimerization of NMA in CCl_4, ΔG_1 = -2.4 kcal/mol.[312,158,160,265,290] However, this free energy is solvent dependent. Hydrogen bonding strengthens in nonpolar solvents either if (*i*) the dielectric constant of the solvent is reduced, with other solvent properties held fixed,[101] as expected for Coulomb interactions, or if (*ii*) the electron- or proton-donating or -accepting capacity of the solvent is varied, with the dielectric constant held fixed.[8,161] (*2*) Transferring a hydrogen bond into a nonpolar medium is generally disfavored: ΔG_2 = +6.12 kcal/mol has been estimated for NMA from water to CCl_4.[265] (*3*) As noted above, AB dimerization is disfavored in water; for NMA, ΔG_3 = +3.1 kcal/mol.[158,265] (*4*) It follows that ΔG_4 = +0.62 kcal/mol for NMA in CCl_4;[265] ΔG_4 is also near zero for carboxylic acids in benzene[158] and is predicted to be +2.2 kcal/mol for transferring the formamide dimer from water to CCl_4.[290] (*5*) It follows from these estimates that hydrogen bonding opposes folding. For NMA in CCl_4, ΔG_5 = +3.72 kcal/mol; for formamide, ΔG_5 = +1.9 kcal/mol. However, if the transfer of A and B from water into a nonpolar medium is driven by some other force, such as hydrophobicity (see below), then hydrogen-bond formation, process 1, will be strongly favored within the folded structure. Thus whereas hydrogen bonding may not assist the *collapse* process, it would favor internal organization within the compact protein. Two additional uncertainties make it difficult to estimate the medium effect on the hydrogen-bond strength: (*i*) the nonpolar core of a protein is not a homogeneous dielectric,[136,137] and (*ii*) hydrogen-bond strength is extremely sensitive to geometric details of bond angles.[270,271] Since there are so many hydrogen bonds in native proteins, then even small errors in estimating their strength will lead to large errors in determining their effects on protein stability. Only 11% of all C=O groups and 12% of all NH groups have no hydrogen bonds.[23] Of all the hydrogen bonds to C=O groups, 43% are to water, 11% are to side chains, and 46% are to main-chain NH groups. Of all the hydrogen bonds to NH groups, 21% are to water, 11% are to side chains, and 68% are to main-chain C=O groups. To reliably estimate the stability of a protein would therefore require model studies more accurate than about $(1/5)kT$ per intrachain peptide bond. For the reasons noted above, this accuracy is not yet available from the current models.

Solvent denaturation studies indicate that hydrogen bonding is not the dominant folding force.[287,82] If it were, then solvents that form strong hydrogen bonds to the peptide backbone should compete effectively and unfold the protein. Those solvents that do not affect hydrogen bonding should not affect stability. In this light, several observations are of importance. (*1*) It would be difficult to rationalize the observation that very small concentrations of *surfactants* (1% dodecyl sulfate, for example) unfold proteins[304] since they do not destabilize helices.[180] Moreover, the effectiveness of tetraalkylammonium salts to denature proteins depends on the number of methylene groups, indicative that it is the hydrophobic interaction, rather than hydrogen bonding, which determines stability. (*2*) Since the C=O group is a strong hydrogen acceptor and the NH group is a weak donor, Singer[287] has pointed out that the solvents useful for competing with the peptide hydrogen bond would be those which are stronger donors than NH; the peptide bond will generally compete effectively against solvents that are weaker hydrogen acceptors than C=O. Dioxane is only a hydrogen-bond acceptor and therefore should not denature proteins if hydrogen bonding were the dominant folding force. However, dioxane denatures proteins.[287] (*3*) Alcohols are more hydrophobic than water, but they enhance helix formation.[61] If hydrogen bonding were the dominant folding force, alcohols should stabilize proteins. Hence, the observation that at low concentrations they destabilize proteins is inconsistent with hydrogen bonding as the principal driving force.[128,313,314,226] The caveat is that alcohols have complex effects on protein stability, depending on concentration and temperature.[39] Finally, a particularly important comparison involves the effect on protein stability of alcohols (ethanol and propanol) vs. the corresponding glycols (ethylene glycol and propylene glycol). The glycols are less hydrophobic and have more hydrogen-bond sites than the corresponding alcohols. The observation that the glycols are worse denaturants is strong evidence that hydrogen bonding is less important than the hydrophobic inter-action.[304,313]

Mutation studies show that hydrogen-bonding groups affect stability, but by an amount which can differ considerably depending on the site and nature of the mutation.[26] Fersht[95] has estimated from activation free energy measurements of tyrosyl-tRNA synthetase/substrate interact-ions that breakage of a hydrogen bond increases the free energy by 0.5-1.5 kcal/mol for unpaired uncharged donor and acceptor or about 3.5-4.5 kcal/mol if the donor or acceptor is charged. Site-directed mutagenesis experiments, in which nonhelical proline 86 in phage T4 lysozyme is replaced by other amino acids which extend the helix and add new *backbone* hydrogen bonds, show marginal *reduction,* rather than increase, in stability.[7] On the other hand, *side-chain* hydrogen-bonding groups are found to stabilize T4 lysozyme.[5,121] However, because *site-directed mutagenesis* experiments measure only the *total change in stability upon mutation*, $\Delta\Delta G$, and not the individual molecular components of that change, then these mutations may involve more than just hydrogen bonding. For example, for one of them (Thr replacing Val-157), free energy perturbation calculations show that the added stability arises from better van der

Waals interactions rather than from the difference in hydrogen-bond strength.[65]

4. Local Interactions: Intrinsic Propensities

The term "*intrinsic propensity*" does not describe any single type of force. Rather it is intended to convey the idea that there are certain conformational *preferences* of di- or tripeptides, depending on the sequence, which arise from the sum of short- and long-ranged forces that are local among connected neighboring residues. ("Local" may extend to residues three to four monomers distant and may therefore also include hydrogen bonds involved in turns or helices.) Intrinsic propensities have been studied by the measurement of helix/coil equilibria of peptides in water[297,182,181] and turn/coil equilibria.[79,80,319] The stabilities of long polypeptide helices in aqueous solution can be attributed to intrinsic propensities. Helix stability increases with chain length.[116,326,240] Therefore, although the free energy contribution from each individual residue may be small, summed over many residues, the helix can be strongly favored relative to the coil in a long chain.

The traditional view has held that the individual residue helix/coil equilibrium constants are so nearly equal to one, however, and the initiation constants so small that short helices (less than about 15-20 residues) are not stable in aqueous solution. There have been two bases for this view. First, short helices extracted from stable globular proteins have been found to be unstable in isolation in aqueous solution.[89,310,81] Second, using "guest" amino acids randomly doped into "host" copolymers of hydroxy-, propyl-, and hydroxybutyl-l-glutamine, Scheraga and his colleagues[297] showed that the intrinsic propensities of amino acids to form helices in water are small (with helix/coil equilibrium constants ranging from 0.59 to 1.39 at 20°C). (Helical propensity can be increased considerably by reducing the temperature to near 0°C.) However, these equilibrium constants will be universal, in principle, only if the host helix itself is completely inert in its effect on the helix/coil equilibrium of the guest residue. The following recent evidence with other hosts, however, suggests that the helix is not completely inert, i.e., that there are "*context" effects*. (*1*) The helix/coil constants differ in other sequences and can be as large as nearly 2 for alanine in alanine-based helices.[182,224] Although the helix/coil constant for uncharged guest residue i appears to depend on the local sequence through residues $i - 1$ and $i + 1$, it does not appear to further depend on $i - 2$ and $i + 2$ or otherwise on the position in the chain.[195] (*2*) Additional stability results if helix formation leads to *burial of nonpolar surface*.[304,56,257] (*3*) Helices can be stabilized considerably by reducing the helix *dipole moment* through reduction of the charges at the helical ends.[280,281,181] (*4*) Salt bridges and *aromatic interactions* can also affect stability.[181,283] In addition, intrinsic propensities can vary with the solvent.[255] Recent evidence[196] suggests that context effects may be at least as important as intrinsic propensities.

On the basis of these observations, considerable progress continues to be made in improving stabilities, so that higher helix/coil equilibrium constants are achieved, in shorter chains, and at increasing temperatures up to near room temperature.[181,36] Nevertheless, intrinsic propensities, in the absence of other forces, appear to be insufficient to account for the full helical stability in *globular proteins*. Helices in globular proteins are short. The average length is about 12 residues, and the most probable helix length (peak of the distribution) is less than 6 residues.[144,171,291] Yet protein helices remain 100% helical up to temperatures near the denaturation point. Other forces must therefore also be important for stabilizing helices in globular proteins. One possibility is that there may be additional "context" effects due to the environment provided by the protein interior. For example, charges are distributed in proteins so as to stabilize the helix dipole.[32,259] Helices with modified charges at the ends can affect protein stability.[199] In principle, helices could pack in pairs, antialigned, to reduce the net dipole moment; this probably contributes little to stability, however, since the ends of helices are generally found in a high-dielectric medium at protein surfaces.[261,114,243] In contrast to these effects, the protein environment may not always stabilize helices: Alber *et al.*[7] found that added hydrogen-bonded *helix-extending residues* in T4 lysozyme had little effect or destabilized the protein.

Other evidence suggests that protein architecture does not arise principally from intrinsic propensities. First, distributions of secondary structures predicted by intrinsic propensities are inconsistent with those in known protein crystal structures. Any model of protein stability based only on local interactions would predict, as is observed in helix/coil equilibria, that longer helices should be more stable, and therefore more probable, than shorter helices. In contrast, studies of the crystal structures of globular proteins[144,171,291] show just the opposite: *helix probability* decreases monotonically with length (see Figure 11). Similarly, longer sheets are observed to be less probable than shorter sheets, for both parallel and antiparallel sheets. These discrepancies are not repaired by local factors alone. For example, it is known that helices can be terminated by "*stop residues*."[155] Stop signals will not account for the data base trends, however, which are grand averages over residues, positions, and proteins, since it would then follow that most amino acids must be helix destabilizing, in contradiction to the basic premise. Moreover, even given that local interactions impart some stability to helices and turns, it is difficult to rationalize how they would give rise to sheets, which are intrinsically nonlocal. Therefore, alternative explanations for these distributions involve nonlocal factors. For example, at high densities short peptides can form stable helices in crystals,[146] suggesting the importance of *packing effects*. It is shown in section 8 that the distributions of internal architecture can be accounted for by *steric forces* of nonlocal origin.

Second, attempts to predict protein structures by use of only intrinsic propensities have had limited success.[277,145,21,311,262,251] Success rates are about 64% when averaged over many proteins.[311,262,251,135] Because this is considerably better than chance, it implies that intrinsic propensities are significant determinants of protein structure. However, according to Qian and Sejnowski,[251] who have used *neural net methods*, "no method based solely on local information is likely to produce significantly better results for non-homologous proteins." Within a given class of proteins, success rates may be higher.[159] Limitations of intrinsic propensities are also found in studies of conformations of identical pentapeptides in different proteins.[145,21] Kabsch and Sander found that in 6 of 25 cases, one pentamer could be found in a helix whereas the identical sequence in a different protein would be in a sheet. This implies that *local information alone is not sufficient to fully specify the conformation in the protein*. What does the 64% success rate tell us about the magnitude of the nonlocal factors missing from these prediction methods? If we assume that the conformation of any residue can be predicted with *a priori success rate* $p_0 = 1/3$,[55] as a rough estimate for classification as a helix, sheet, or other conformation, then a success rate p carries an amount of *information* $< I >$:

$$< I > = p \ln(\frac{p}{p_0}) + (1+p)\ln\left[\frac{(1-p)}{(1-p_0)}\right] \qquad (1)$$

Thus, a success rate of p = 0.60-0.70 implies that local factors alone account for only about 15-30% of the total information required to make a perfect prediction. This agrees with estimates[50] that the local contributions to the stability of a six-residue helix (based on a 1.05 equilibrium constant) are about 12% of the magnitude of the nonlocal steric free energy that drives helix formation, described in section 8.

5. Hydrophobic Effect

Following the elimination of electrostatics as a principal force of folding by Jacobsen and Linderstrom-Lang,[139] it was suggested that protein folding was driven by the *aversion for water of the nonpolar residues*.[175,178,148] The same aversion was known to drive micelle formation; it was then assumed to be due to van der Waals interactions.[66] In two remarkably insightful papers, Kauzmann[148,149] made the first strong case for the importance of the hydrophobic interaction in protein folding. He reasoned that the formation of one hydrophobic "bond" (which he called an "antihydrogen bond") upon folding involves the gain of a *full* hydrogen bond among water molecules, which should be more important by an order of magnitude than simply a *change of strength* of a hydrogen bond upon folding if hydrogen bonding were the dominant folding force. In support of this view, Kauzmann offered the following evidence. First, *nonpolar solvents* denature proteins.[287,313] According to a hydrophobic mechanism, the nonpolar solvent reduces the free energy of the unfolded state by solvating the exposed nonpolar amino acids. Second, experiments of Christensen[57] had shown an unusual temperature dependence in which stability not only decreases at high temperatures but also decreases at low temperatures. Kauzmann observed that this *cold destabilization* resembles nonpolar solvation: nonpolar solutes become more soluble in water at low temperatures.[247]

A considerable body of more recent evidence continues to support the *view* that *hydrophobicity is the dominant force of folding*. First, spectroscopic and high-resolution differential scanning calorimetry experiments show the resemblance of the temperature dependence of the free energy of folding and the temperature dependence of the free energy of transfer of nonpolar model compounds from water into nonpolar media.[220,244,247] Both involve large decreases in *heat capacity*. Second, a large number of crystal structures of proteins have become available since Kauzmann"s predictions. They show that a predominant feature of globular protein structures is that the nonpolar residues are *sequestered into a core* where they largely avoid contact with water.[236,53,54,317,194,123] Third, protein stability is affected by different salt species (particularly at high salt concentrations) in the same rank order as the *lyotropic (Hofmeister) series*;[313,314,19] this is generally taken as empirical evidence for hydrophobic interactions.[313,314,60,201,202] From the most stabilizing (for folded ribonuclease) to the most destabilizing, the rank order of anions is found to be SO_4^{2-}, CH_3COO^-, Cl^-, Br^-, ClO_4^-, CNS^- and the rank order of cations is NH_4^+, K^+, Na^+, Li^+, Ca^{2+}. The solubilities of benzene and acetyltetraglycyl ethyl ester in aqueous salt solutions increase in the same rank order. Fourth, accessible surface studies[256] and site-directed mutagenesis experiments, involving the replacement of a given residue by other amino acids, show that the stability of the protein is proportional to the *oil-water partitioning propensity* of the amino acid (see Figure 4).[323,324,184,185,151] Fifth, the hydrophobicities of residues in the cores of proteins appear to be more strongly conserved and correlated with structure than other types of interactions.[173,35,152,300,27] Sixth, computer simulations of incorrectly folded proteins show that the principal diagnostic of incorrect folding of proteins, apart from *inappropriate burial of charge*, is the *interior/exterior distribution of hydrophobic residues*.[214,215,29]

What are "hydrophobic" interactions? There has been some disagreement about the meaning of hydrophobic (effect, force, interaction).[130,131,209,306,124] At least three different meanings of these terms have been used. (*1*) Hydrophobic has been used to refer to the transfer of a nonpolar solute to any aqueous solution. (*2*) Alternatively, it has been used more specifically to refer to transfers of nonpolar solutes into an aqueous solution only when a particular characteristic temperature dependence, described below, is observed. These two meanings describe only experimental observations and make no reference to any particular molecular interpretation. (*3*) "Hydrophobicity" has also been used to refer to particular molecular models, generally involving the ordering of water molecules around nonpolar solutes. In this review, hydrophobicity will be defined by (*2*) for reasons discussed below.

What is unusual about the temperature dependence of the hydrophobic interaction? First it is useful to describe "*normal" solutions*. There are two driving forces relevant to the mixing of simple solutions of A with B, of *spherical particles* governed by *dispersion forces*. The tendency to mix is *driven* by an increase in the *translational entropy* since there are more *distinguishable spatial arrangements* of the A and B molecules in the mixed system than of the individual pure systems. On the other hand, mixing in simple systems is *opposed* by the enthalpies of interaction; ordinarily, dispersion

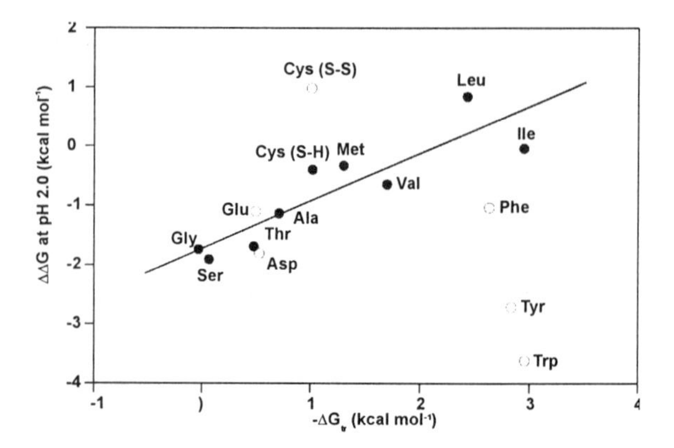

Figure 4 - Change in free energy of unfolding, $\Delta\Delta G$, of mutant T4 lysozymes at position 3 (wild type is Ile) by substitution of other amino acids, compared to the corresponding free energy of transfer from water to ethanol ΔG_{tr}. Reprinted with permission from Matsumura, M., Becktel, W. J., & Matthews, B. W. (1988) *Nature 334*, 406. Copyright (1988) Macmillan Magazines Limited.

forces leading to AB attractions are smaller than those leading to the corresponding AA and BB attractions. The latter is captured in the general rule that "like dissolves like."[132] For simple solutions, the transfer of B from pure B into A is therefore generally opposed by enthalpic interactions. When these interactions are strong, i.e., when A and B are relatively insoluble in each other, then the free energy of transfer is dominated by this opposing enthalpy, which is much larger than the mixing entropy, and transfer is disfavored. This is the ordinary form of incompatibility of two components A and B.

Oil and water are also incompatible at 25°C. Oil/water incompatibilities tend to be stronger than "normal" incompatibilities. However, the free energy alone (i.e., the solubility or the *partition coefficient*) is not the principal distinction between normal and hydrophobic processes; in both cases the transfer can be disfavored. The distinction between normal and hydrophobic processes is in the temperature dependence. What was first recognized as unusual about nonpolar transfers to water at 25°C[84,46,99] was that they are not principally opposed by the enthalpy; they are principally opposed by an *excess* (i.e., "unitary") entropy.[305,110,307,247] The excess entropy is that which remains after the *mixing entropy* (the *RT* ln *x* term in the chemical potential) is subtracted from the total measured entropy. The enthalpy of mixing oil and water is generally small at 25°C and sometimes even negative (favorable).[247] Because these conclusions derive from experiments with solutes at high dilution, they imply that the excess entropy must arise from the water solvation around the solute rather than from some possible solute-solute interaction. A molecular interpretation of these data, which is supported by computer simulations,[106,225,254] is that at 25°C nonpolar solutes are surrounded by ordered waters. Waters surrounding the nonpolar solute prefer to hydrogen bond with other waters rather than to "waste" hydrogen bonds by pointing them toward the nonpolar species (see top of Figure 6).[295,106]

The aversion of nonpolar solutes for water becomes more ordinary, and less entropy driven, at higher temperatures. This is because there is a second fundamental difference between simple incompatibility and oil/water incompatibility. For simple solutions, the *heat capacity change upon transfer* is small. *For nonpolar solutes, the heat capacity change upon transfer from the pure liquid to water is large and positive.*[99,58,247,85] This means that for simple solutions the transfer of A into B is characterized by enthalpy and entropy values which are temperature independent and a free energy which is constant or linear with temperature. However, the situation is quite different for the transfer of nonpolar solutes into water. *A large heat capacity implies that the enthalpy and entropy are strong functions of temperature, and that the free energy vs. temperature is a curved function, increasing at low temperatures and decreasing at higher temperatures.* Hence, there will be a temperature at which the solubility of nonpolar species in water is a minimum[63,30,85,247] (see Figure 5). A striking consequence follows, one which is at variance with the view that water ordering is the principal feature of the aversion of nonpolar residues for water. The free energy of transferring nonpolar solutes into water is extrapolated to be most positive in the temperature range 130-160°C.[247] Therefore, the aversion of nonpolar species for water, whatever its molecular nature, is greatest at these high temperatures. Because this maximum aversion, by definition, arises where the free energy of transfer is a maximum, and thus where the entropy (the temperature derivative of the free energy) equals zero.[247,24] In other words, at the temperature for which hydrophobicity is strongest, the entropy of transfer is zero! At 130-160°C, the aversion of nonpolar solutes for water is enthalpic, as in simple classical solutions. *It is therefore inappropriate to refer to nonpolar solvation processes in water, with large heat capacity changes, as "entropy driven" or "enthalpy driven," since either description is accurate only within a given temperature range.* At 25°C, the hydrophobic effect is entropic; at 140°C, it is extrapolated to be enthalpic.[247]

For processes with constant ΔC_p, the enthalpy and entropy of transfer are:

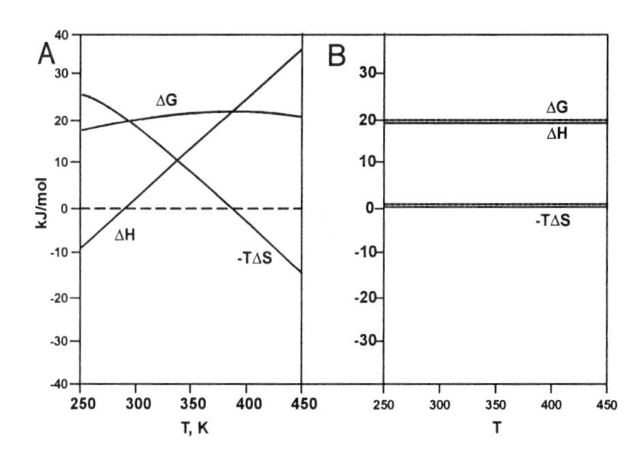

Figure 5 - Comparison of the enthalpy (ΔH), entropy (ΔS), and free energy (ΔG) of solute transfer from the pure liquid into a hypothetical regular solution and into an aqueous solution. The data in (A) are for benzene, from Privalov and Gill;[247] the entropy at high temperature is extrapolated on the basis of assumed constant heat capacity. The figure on the right represents an idealization according to regular solution theory. Reprinted with permission from Privalov, P. L., & Gill, S. J. (1988) *Adv. Protein Chem. 39*, 191. Copyright (1988) Academic Press.

$$\Delta H(T) = \Delta H(T_1) + \int_{T_1}^{T} \Delta C_p dT = \Delta H(T_1) + \Delta C_p (T - T_1) \tag{2}$$

$$\Delta S(T) = \Delta S(T_2) + \int_{T_2}^{T} \frac{\Delta C_p}{TdT} = \Delta S(T_s) + \Delta C_p \ln\left(\frac{T}{T_2}\right) \tag{3}$$

and therefore the *free energy of transfer* is

$$\Delta G(T) = \Delta H - T\Delta S = \Delta H(T_1) - T\Delta S(T_2) + \Delta C_p \left[(T - T_1) - T \ln\left(\frac{T}{T_2}\right)\right] \tag{4}$$

These quantities are defined in terms of two arbitrary *reference temperatures*: T_1 for which the enthalpy is known, and T_2, for which the entropy is known. There are three alternative ways to express this free energy in terms of two particularly convenient reference temperatures, T_h, the temperature at which the enthalpy is zero, and T_s, the temperature for which the entropy is zero:

$$T_1 = T_h \quad T_2 = T_s$$

$$\Delta G(T) = \Delta C_p \left[(T - T_h) - T \ln\left(\frac{T}{T_s}\right)\right] \tag{5}$$

$$T_1 = T_2 = T_s$$

$$\Delta G(T) = -\Delta H(T_s) + \Delta C_p \left[(T - T_s) - T \ln\left(\frac{T}{T_s}\right)\right] \tag{6}$$

$$T_1 = T_2 = T_h$$

$$\frac{\Delta G(T)}{RT} = -\frac{\Delta S(T_h)}{R} + \frac{\Delta C_p}{R}\left[\left(1 - \frac{T_h}{T}\right) - \ln\left(\frac{T}{T_h}\right)\right] \quad (7)$$

These three expressions have identical content. Which form is used depends on which set of parameters is most convenient: (ΔC_p, T_s, T_h), [ΔC_p, T_s, $\Delta H(T_s)$], or [ΔC_p, T_h, $\Delta S(T_h)$]. Ordinary thermodynamic convention is to choose a single reference temperature, rather than two, but the mixed expression is included here because it has been used for protein folding (see below). Equations 5-7 predict the type of temperature dependence shown in Figure 5A when ΔC_p is large or in Figure 5B when $\Delta C_p = 0$. What is the molecular basis for the large *heat capacity of transfer* of nonpolar solutes into water? Two observations have contributed significantly to a molecular picture. First, the entropy and heat capacity of transfer are linearly proportional to the *surface area of the nonpolar solute.*[197,111,143,141] Second, the large heat capacity of transfer for the simplest solutes[111] decreases slightly with increasing temperature. This leads to the view[99,208,111,203] that the organization of water molecules in the first shell surrounding the solute is like an "iceberg," a *clathrate*, or a "*flickering cluster*" (see Figure 6).

At room temperature, the water molecules surrounding the nonpolar solute principally populate a low-energy, low-entropy state: the waters are ordered so as to form good water-water hydrogen bonds. With increasing temperature, the waters surrounding the solute increasingly populate a higher energy, higher entropy state: they are less ordered and have weakened attractions. Hence, increased temperature causes "melting" of the surrounding water structure, insofar as the entropy and energy are increased.

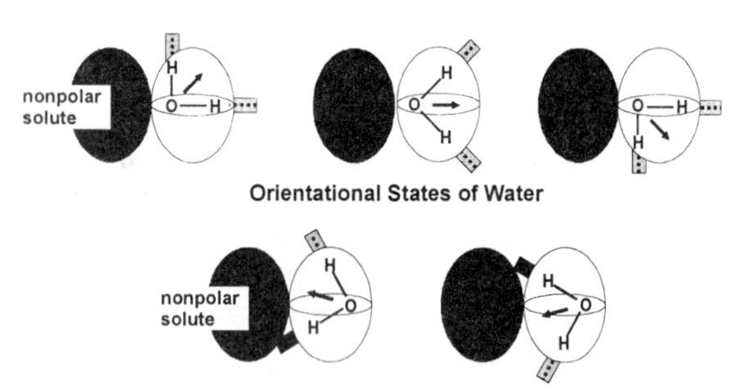

Figure 6 - "Iceberg" model for the large heat capacity of transfer of nonpolar solutes into water.[99,111] At low temperatures (near room temperature for benzene, for example), the water molecules surrounding the nonpolar solute adopt only a few orientations (low entropy), to avoid "wasting" hydrogen bonds; thus, all water configurations are fully hydrogen bonded (low energy). At higher temperatures, more conformations are accessible (higher entropy), but some of them have weaker or unformed hydrogen bonds and/or van der Waals interactions (higher energy). This contributes to the heat capacity because the system energy increases with temperature.

[This melting process is probably better represented as bent hydrogen bonds than broken bonds, since the bending energy is much smaller than the breaking energy.[170]] The reason this results in a large heat capacity is that the *two different energetic states of water provide an energy storage mechanism*. The higher energy state becomes more populated with temperature. The reason this heat capacity is so large per solute molecule is because each solute molecule is surrounded by a large number (more than 10) of first-shell water molecules, each of which can participate in this energy storage mechanism. Not yet known is the detailed breakdown of the nature of these energies, although they undoubtedly include some combination of solute-water and water-water hydrogen bonding and van der Waals and dipolar electrostatic interactions.

What is the molecular interpretation of T_s and T_h? Equations 2-7 each have temperature-dependent and temperature-independent terms. The slope of the temperature dependence in eqs. 5-7 is given by ΔC_p. T_s and T_h can be interpreted as representing a reference enthalpy or entropy at some given temperature. T_s and T_h are diagnostic for liquid-state nonpolar transfer processes.[24] Sturtevant[296] observed that several different biomolecular processes at 25°C have nearly identical values of the ratio $\Delta S/\Delta C_p$. Baldwin[24] showed that the constancy of this ratio, taken together with $T_2 = T_s$ and $T = 298$ K, substituted into eq. 3 implies that these various processes can all be characterized by a single temperature, $T_s = 114$°C. Using the data of Gill and Wadsö[110] for ΔS and ΔC_p for the transfer of liquid nonpolar compounds to water at different temperatures, T, substituted into eq. 3, Baldwin again found a single characteristic temperature, $T_s = 112.8$°C, implying that Sturtevant"s biomolecular processes resemble nonpolar solvation. Baldwin[24] preferred the use of eq. 5. For eqs. 5-7, T_s and T_h are convenient *reference quantities* because ΔH, ΔS, and ΔC_p depend linearly on solute surface area, so ratios such as $\Delta S/C_p$, and hence T_s should be independent of solute size.

On the other hand, Murphy, Gill, and Privalov[206,247] have preferred the convention given by eq. 6. They assumed that at temperature T_s nonpolar solvation is identical with classical solvation and that

$\Delta H(T_s)$ is due only to van der Waals forces. They refer to the factors other than $\Delta H(T_s)$ in eq. 6 as the "hydration effect." They note that $\Delta H(T_s)$ is a positive enthalpy disfavoring transfer and that the hydration term is always negative (or zero at $T = T_s$). It follows that the hydration effect favors nonpolar transfers into water. However, it is not clear that this separation into these molecular factors is warranted. At $T = T_s$, nonpolar solvation is not identical with simple solvation. Even though the entropy of transfer may extrapolate to zero at $T = T_s$, the heat capacity remains large.[203]

The heat capacity is probably a more fundamental characteristic of nonpolar solvation than the entropy, because the entropy depends on the choice of concentration units, as do the free energy and partition coefficient, whereas that heat capacity and enthalpy do not.

In addition, these thermodynamic models predict that for $T > T_s$, the entropy of transfer becomes positive, leading to the questionable prediction that the entropy would then favor solvation at high temperatures (see Figure 5A for benzene above 400 K). Also, since water hydrogen bonding persists to beyond 500 K,[63,100] $\Delta H(T_s)$ probably includes water-water hydrogen bonding in addition to van der Waals interactions. Thus, it is not clear that this thermodynamic separation of terms corresponds to a simple molecular picture.

Equation 7 provides yet a different view. Whereas ΔG for transferring benzene to water is most positive at $T = T_s$ (near 100°C), $\Delta G/RT$ is most positive at $T = T_h$ (near room temperature). $\Delta G/RT$ corresponds directly to a solubility, partition coefficient, or *Boltzmann population.* Benzene is least soluble in water around room temperature. In eq. 7, $\Delta S(T_h)/R$, representing *water ordering*, is the most positive contribution to $\Delta G/RT$. The remaining terms, whatever their molecular interpretation, favor *solvation*. Thus at present, T_h and T_s serve to identify nonpolar transfer processes, but their breakdown into molecular components awaits further theory and experiments.

The *thermal unfolding* of proteins shows important similarities and important differences in comparison with the solvation processes of small nonpolar solutes. The similarity is that protein unfolding involves a large increase in heat capacity, characteristic of nonpolar exposure.[220,244,24,30,247,218] Therefore, the enthalpy and entropy of folding strongly decrease with temperature, and the free energy is curved: maximum protein stabilities are in the 0-30°C range (see Figure 8). The heat capacity of protein unfolding is itself approximately independent of temperature, although it appears to decrease somewhat at higher temperatures.[30,247]

The thermal unfolding of proteins differs from nonpolar solvation in the particular values of enthalpy and entropy at any given temperature. For example, as noted above, at 25°C the transfer of small nonpolar compounds into water has approximately zero enthalpy change and a large negative excess entropy change. In contrast, at 25°C, protein unfolding has a positive enthalpy change (except for myoglobin for which the enthalpy is about zero) and a small or positive excess entropy.[24,247] There is an additional positive entropy and enthalpy of unfolding – in excess of that predicted from nonpolar solvation experiments.

What is the origin of the *residual enthalpy and entropy of folding*? As noted in section 2, the residual enthalpy of unfolding becomes more positive with increased content of polar residues.[244,247,74] If it is assumed to be approximately temperature independent,[24] then this result is expected from *classical polymer solution behavior*, in which chain monomers of type A have normal dispersion-force-driven incompatibilities with solvent B.[97] This enthalpy may arise from van der Waals or hydrogen-bonding interactions among the backbone or polar residues. *Thermodynamic models of protein unfolding have generally been based on the assumption that the large heat capacity change is fully attributable to nonpolar solvation. This is open to question.* Hydrogen bonding to water weakens with temperature, according to valerolactam dimerization and thymine dissolution experiments.[10] This would also contribute to a change in heat capacity. Also, increased temperature causes the *unfolded states* of proteins to *expand*, reducing nonpolar solvation and contributing additional temperature dependence to the enthalpy and entropy of folding.[74]

The *large residual entropy* represents an *increased disordering upon unfolding*, relative to that expected for nonpolar solvation. In sections 5 and 6, it is suggested that at least a large component of this is due to a difference in the *freedom of configurations* of the chain backbone, more severely *restricted by steric constraints* in the folded than in the unfolded states. However, there may also be an *extra* residual entropy due to additional configurational restrictions of the side chains, in the folded state. The side chains may be partially frozen if the folded protein resembles a solid-like state.[279,280] [Proteins are very compact – they're viewed as *glass-like*; stable, malleable at elevated temperatures, able to creep and deform, soluble in appropriately active solvents.]

The residual entropy of protein denaturation is $\Delta S_r(112°C)$ = 18 J/(K·mol),[24,244,206] similar to that of the dissociation of solid diketopiperazines, $\Delta S_r(112°C)$ = 16 J/(K·mol), and considerably different than that of liquid hydrocarbon dissolution for which $\Delta S_r(112°C)$ = -0.5 J/(K·mol).[206] The heat capacity change increment upon dissolution of the solid diketopiperazines is the same as for the liquid-state transfer process.[204,205]

From those thermodynamic experiments, however, it is not possible to determine how much of the entropy difference originates with the *chain expansion* and *solvation* and how much originates from *side-chain unfreezing*. Bendzko *et al.*[31] have suggested that if side-chain freezing is important, then protein denaturation should lead to an increase in partial molar volume of the protein; instead, they and others (see section 5) find a decreased volume upon unfolding. Another test of *side-chain restrictions* in the folded core is to compare crystal structure *distributions of side-chain rotamers*[241] with computer simulations of the mean position and fluctuations of side chains that are attached to spatially unconstrained backbones. Such comparisons[140,237] show that equilibrium side-chain positions are predicted relatively well by the unconstrained simulations, but they are constrained somewhat differently in different secondary structures.[237,193]

Another class of experiment bears on the issue of the solid-like vs. liquid-like nature of the protein core. Whereas the *core is solid-like* in its *density* and *compressibility*, it may behave differently insofar as the *transfer process*. The experiments involve multiple amino acid substitutions at a given site and the measurement of the change in protein stability, $\Delta\Delta G$, due to each of the replacements.[323,324,184,185,151,268] The slope of $\Delta\Delta G$ vs. the *free energy of transfer* ΔG_{tr} for the corresponding *amino acid from water to oil* (see Figure 4) provides information about similarities and differences of the protein folding process relative to the simpler process of amino acid transfer from water into liquid oil. This slope depends on a combination of factors: (*i*) the "deformability" of the native-state cavity, the energetic and entropic constraints affecting the freedom of the cavity wall residues to move to accommodate the mutated amino acid; (*ii*) the *interactions of the residue with the cavity* (w), including both its entropic restrictions (side-chain freezing) and the residue/cavity energetics; (*iii*) the degree to which the denatured-state environment of the specified residue resembles pure solvent; and (*iv*) the exposure of the residue in the native state. A slope of 1 is consistent with a process in which the amino acid is exposed in the unfolded state and is transferred into a native cavity that resembles the reference liquid oil. Simple solute transfer to a liquid involves (*a*) opening the cavity (unfavorable by approximately free energy $w/2$) and then (*b*) transferring the solute (favorable by a free energy, w). On the other hand, if a solute is transferred instead into an already opened cavity ("preformed"), then only (*b*) is involved.[151] *Thus, the free energy of transfer into a simple liquid is only half that of transfer into a preformed cavity.* Therefore, a slope of 2 is expected from these experiments if the cavity wall residues are so constrained that they do not move to accommodate the mutant residue. Protein cavities may differ. Some may have much "deformability," as in a liquid, and lead to slopes near 1. Others may have little deformability, if neighboring residues are constrained, and lead to slopes near 2. Cavity deformabilities may also depend on residue size; cavity wall residues may move to accommodate large occupants but not small ones, for example. Overall, slopes ranging from 1 to 2 are expected from factor (*i*). Slopes smaller than 1 imply that the residue may be exposed in the native state or may be significantly buried in the denatured state. Slopes considerably greater than 2 must arise at least partly from (*ii*). A negative slope could arise if the residue is more exposed at the surface of the native structure than in the denatured state. A slope of approximately 1 is observed for residue 3 in T4 lysozyme.[184] A slope of approximately 2 is observed for residues 88 and 96 in baRNase and for a pocket involving residues 35 and 47 of gene V protein of phage f1.[151,268] For two sites normally containing charged residues, Asp-80 in kanamycin nucleotidyltransferase[185] and Glu-49 in Trp synthase,[323,324] the slopes vary with pH, reaching a maximum of about 3.8. These latter experiments are more complex because the proteins have significant populations of intermediates (and do not have two-state transitions) and they have an electrostatic component for the transfer. Although it would appear that current results on noncharged residues can be explained largely by different cavity deformabilities (*i*), nothing rules out the possibility that the other factors (*ii* and *iii*) are important. This type of experiment has not yet established whether side-chain freezing is an important component of protein stability.

Which liquid "oil" best characterizes the native core? While there is some evidence that cyclohexane is good,[252] the best correlations of transfer studies with amino acid distributions in proteins appear to involve nonpolar hydrogen-bonding solvents including octanol, ethanol, and dioxane.[93,217,165,263] It may be, therefore, that some fraction of the temperature-independent enthalpy, attributed to van der Waals and hydrogen-bonding interactions, is also present in these transfer experiments and that it contributes differently for different oils.

Finally, I return to the meaning of hydrophobicity. I believe the most useful, common,[307,124] and unambiguous meaning of this concept is simply in reference to nonpolar transfers from nonaqueous media into aqueous media: (*i*) that are strongly disfavored and (*ii*) whenever there is a large associated increase in heat capacity [definition (*2*) at the beginning of this section]. It was this remarkable feature of nonpolar solvation that was first identified as unusual[84,46,99] and that merits special terminology. Definition (*1*), on the other hand, needs no special term because it otherwise describes ordinary solution processes. There are two problems with (*3*), hydrophobicity as water ordering or other molecular models: (*i*) the molecular mechanism is still not fully understood, and (*ii*) "water ordering" is an appropriate description of the entropic repulsion of nonpolar solutes near room temperature, but not over a broader temperature range. This would lead to unnecessary hairsplitting: benzene insolubility in water would be referred to as hydrophobic at 25°C but not at 100°C, where it is even more strongly expelled. This meaning is not subject to the Hildebrand objection;[130,131] he noted that *hydrophobicity is not an enthalpic disaffinity of nonpolar solutes for water but instead is due to a water-water affinity*. Also, because definition (*2*) describes hydrophobicity in terms of the full transfer process, represented by the total free energies in eqs. 5-7, rather than by a particular term in those expressions, it is not subject to difficulties of molecular interpretations.[249,72]

6. What Is Missing?

The dominant force of folding is only half the story. Nearly equal in magnitude is a large opposing force. The structures and stabilities of globular proteins result from the *balance between driving and opposing forces*. Only recently have the main opposing contributions become better understood. That hydrophobicity is not the whole story becomes immediately apparent from certain puzzles. First, Tanford[303] and Brandts[37,38] showed that hydrophobicity alone would predict protein stability an order of magnitude greater than measured values. They estimated that the free energy of unfolding would be about 100-200 kcal/mol at 25 °C. This estimate is based on free energies of transfer of hydrophobic amino acids from water into ethanol or dioxane, representative of the folded core of the protein, multiplied by the number of nonpolar residues. *However, free energies of unfolding are observed to be only about 5-20 kcal/mol.*[220,244,247] This implies that there must be a large force, of magnitude nearly equal to the hydrophobic driving force, which opposes folding. Second, while the temperature dependence of folding was an important clue for hydrophobicity, Tanford[303] suggested that it also posed a paradox. If nonpolar components associate more strongly as the temperature is increased, then proteins should fold more tightly with increasing temperature. Just the opposite is observed above room temperature: increasing temperature unfolds proteins. Proteins are typically thermally unfolded in the range of 50-100°C; the free energies of transfer of hydrocarbons into water have a maximum extrapolated to be in the range of 130-160°C. Third, if hydrogen bonding is not a dominant force of folding, then what is the origin of the considerable amounts of internal architecture, of secondary and tertiary structures, in proteins? Through what forces do the amino acid sequences so uniquely determine the native structure? The hydrophobic effect would seem to be too nonspecific and an unnatural candidate as the origin of helices and sheets.

Fourth, *the pressure dependence of protein stability* does not resemble that of model hydrophobic compounds.[41,329,150] For example, the partial molar volume of methane decreases from 60 to 37.3 ml/mol upon transfer from hexane to water.[183] The decreased volume arises because water molecules pack more efficiently surrounding a nonpolar solute molecule than in its absence. Since the unfolding of a protein leads to increased nonpolar exposure to water, then these model studies would suggest that the partial molar volume of a protein should decrease considerably upon unfolding due to similar contraction of solvating water molecules (by about 20 ml/mol multiplied by a number in the range of 10-40, representing the total hydrophobic exposure upon unfolding). While the volume change of protein unfolding is indeed generally observed to be negative, it is only in the range of -30 to -300 ml/mol (about 0.5% of the total volume),[41,329,82,256] which is somewhat smaller than the methane model would suggest. Brandts *et al.*[41] suggested that other simple factors would not account for this discrepancy; for example, if unfolding leads to exposure of charged groups, then the volume change upon unfolding would be predicted to be even more negative, increasing the discrepancy. There is an additional problem. For model compounds, increasing the pressure leads to more normal water solvation, so the volume change of transfer to water diminishes and ultimately becomes positive at about 1500-2000 atm. For proteins, on the other hand, the negative volume of unfolding does not change much with pressure. Two simple explanations, however, can account for these discrepancies between the model transfer experiments and protein unfolding. First, methane is a poor model amino acid. Better models include alcohols, ketones, and amides, which can form hydrogen bonds; these model compounds have much smaller negative volumes of transfer to water, more closely predictive of the protein experiments.[103,138] Second, the pressure dependence is at least qualitatively accounted for by recognizing (*i*) that model nonpolar solutes in water are less compressible than in the pure liquid[41] and (*ii*) that the folded state of a protein is much less compressible than the reference liquid hydrocarbon to which it is generally compared. A folded protein is typically 10-fold less compressible than organic liquids (or about half as compressible as ice)![162,105,83,156,92] This is because (*i*) the volume change in small-molecule transfers should diminish with increasing pressure and (*ii*) the volume change in protein folding should diminish much less with increasing pressure than in the small-molecule reference experiment. Therefore, it is important to recognize that a protein is not just a sum of transfers of small-molecule model side chains. Proteins are polymers. It is described below how the *chainlike* nature of proteins and the resultant *conformational freedom* lead to a strong force that opposes folding.

7. Principal Opposing Force Is Entropic

Since the 1930s it has been known that *the main force opposing protein folding is entropic.* Northrop[213] was the first to observe a sharp thermal denaturation transition; the equilibrium constant depends strongly on temperature. Only more recently has the molecular basis for the opposing entropy become clear. Just as there are *translational*, *rotational*, and *vibrational entropies* of small molecules, depending on the relevant degrees of freedom, likewise there are different possible molecular origins of the entropy gain upon protein unfolding. For example, Mirsky and Pauling[198] suggested that folding would be opposed by an *entropy arising from the proper mating of specific ion pairs*. As another example, the helix-coil transition theories[274,326,240] showed that *local degrees of freedom* could be an important source of entropy opposing helix formation.

However, it has long been known that polymers are also subject to another type of *configurational entropy*, one that is *nonlocal*.[97,68] It arises from "*excluded volume*," the impossibility that two chain segments can simultaneously occupy the same volume of space. A chain can occupy a large volume of space in any of a large number of different configurations. However, there are relatively few ways the chain can configure if it is constrained to occupy a small volume of space, simply due to severe steric constraints.

Excluded volume (steric constraints) will play a role in any process that involves a change in the *spatial density* of polymer segments. These include solution thermodynamic properties of chain molecules, *particularly their dependence on concentration, including solubilization and phase behavior, colligative properties, expansion and shrinkage, and virial coefficients and their temperature dependences.*[97] Models for predicting these effects originated with the *Flory-Huggins theory*;[97] others are based on *scaling*[68] and *renormalization group* methods.[102]

How can a local entropy be distinguished from a nonlocal entropy? By definition, the entropy is $S/Nk = -\sum_{i=1}^{s} p_i \ln p_i$, where the probabilities p_i describe the distribution of all states, $i = 1, 2, 3, \ldots s$, accessible to a system, and N is the number of particles. The issue of local vs. nonlocal is the question of what degrees of freedom change in the process of interest. Local entropies arise from the energies responsible for the conformations of connected residues:

$$p_i = z^{-1} \exp\left[\frac{-\varepsilon(\phi_i, \psi_i, \chi_i, \chi_{i+1})}{kT}\right] \qquad (8)$$

where z is the partition function and ε is the energy of the dipeptide (or tripeptide, etc.) as a function of ϕ, ψ, and χ angles. These resemble vibrational or internal rotational entropies in small molecules; their contribution to stability is probably small.[147] Local entropies are independent of global properties of the chain such as the radius of gyration, or internal/external distributions of nonpolar/polar residues. Local entropies underlie helix/coil processes. There is evidence that changes in local entropies can affect the stabilities of globular proteins. For example, Matthews *et al.*[191] have shown that replacement of a glycine for a proline in T4 lysozyme decreases protein stability by increasing the local conformational freedom of one peptide bond.

On the other hand, nonlocal entropies depend on the relative numbers of chain configurations, $N(\rho)$, as a function of the chain segment density, ρ (i.e., the number of monomers divided by the volume occupied by the chain):

$$p(\rho) = \frac{N(\rho)}{\int N(\rho)d\rho} \qquad (9)$$

Because the folding of a protein involves collapse of the chain from a large volume (the denatured state) to a small volume (the native state), it must lose a considerable amount of this nonlocal entropy in the process. Native states of globular proteins are extremely compact.[156,256] They have as little free volume as small-molecule crystals; they have *compressibilities* closer to those of *glasses* than to those of liquids; and their configurational freedom is as restricted as in glasses and crystals of polymers, as evidenced by the existence of *virtually unique native configurations* of proteins.

One test of the importance of nonlocal entropies in protein stability involves *cross-linking the chain.* Introducing a cross-link reduces the number of conformations accessible to the unfolded state, making it relatively less favorable, and in that way making the folded state relatively more stable.[149,239,47, 49,223,186] It appears that the most effective way to stabilize a protein at present, with a single amino acid change, is to add a cross-link. Increases of 29 and 25°C in denaturation temperature have been achieved in hen lysozyme and ribonuclease.[4] For the same reasons, a protein should also be stabilized if it is *constrained to be adjacent to an inert surface* or *contained within a small pore.*[73] Similarly, the restriction of configurations of the unfolded chain may also account for the observation that *carbohydrates or other chains appended to proteins* appear to *stabilize them.*[45]

8. Modeling Protein Stability

More quantitative insights into the forces of protein folding can be obtained through *theoretical models* and experimental tests of them. In principle, if the hydrophobic interaction and conformational entropy are the dominant contributions, then the free energy of folding can simply by calculated as (*i*) the difference in nonpolar exposure in native and denatured states, multiplied by the free energy of transferring nonpolar surface, and (*ii*) the difference in conformational entropies. The free energy of folding has been calculated in various studies, on the basis of several of the following *simplifying assumptions*: (*i*) in the native state the hydrophobic residues are fully buried in a nonpolar core and (*ii*) in the unfolded state the hydrophobic residues are fully exposed to solvent. The hydrophobic contributions would then be the sum of the free energies of transfer of the nonpolar residues. If in addition (*i*) there is only one configuration in the folded state, (*ii*) the *denatured state is an ideal random flight* [as in a "theta" solvent;[97] where chains liberate freely], and (*iii*) the entropy is local, then the *difference in configurational entropies* will be simply $nk \ln z$, where n is the number of rotational bonds and z is the *number of accessible conformations* per dipeptide bond. These approaches have not succeeded, perhaps for the following reasons. First, native states have nonpolar exposure. Methods now exist, however,[168,86] to determine the detailed surface exposure and the associated free energies, if the native structure is known. Second, full exposure of the unfolded state is a poor approximation. The *ensemble of denatured configurations* represents a *complex polymeric state*, with the following properties. (*1*) It is often dense, being only 1.3- to -2-fold greater in volume than the native state, rather than 10- to 100-fold as in the theta state.[248,250,119] (*2*) The net change in surface exposure to water upon unfolding can be as little as 14% of the maximum possible.[304,305,276]. (*3*) The *radius*, and therefore the free energy, of *denatured configurations* depends on the *composition* and *length of the chain*[304,119,285] and on external thermodynamic conditions, including temperature, pH, salt, and denaturants.[248] (4) In some cases, distinctly identifiable different normative states (*"folding isoforms"*) exist and exchange via configurational transitions.[90,117,119,284] (5) Some non-native states have much secondary structure.[250,28,284, 285] This is not a class of configurations that can be represented as fully exposed and independent of the solvent and thermodynamic conditions. For these reasons it is also clear that the configurational entropy of folding cannot be due principally to local factors and is not a constant independent of chain and solvent properties. Because the unfolded state contributes to stability on the same footing as does the folded state, satisfactory predictions of stability will require satisfactory *models of the unfolded states*.

Arguably, the simplest model for polymer chain collapse that treats the chain conformations more realistically is that of *homopolymers* in *poor solvents.* The nonlocal conformational entropy favors the open configurations, but when the solvent becomes poor (*incompatible*), the polymer collapses because polymer-polymer contacts are more favorable than polymer-solvent contacts.[172,67,242,267]

The collapse of a *heteropolymer* such as a protein is significantly different than that of a homopolymer. In the process of collapse to the folded state, the *ensemble of configurations of the molecule* must (*i*) decrease in radius to native compactness and (*ii*) configure so as to bury much nonpolar surface area (see Figure 7). *Hence, there are two degrees of freedom: the radius (or segment density) and the "ordering," the degree of segregation of the nonpolar residues into a core.* In contrast, for homopolymers the only degree of freedom is the density. If the free energy is known as a function of these two degrees of freedom, then the state of minimum free energy can be found. Theory has recently been developed on this basis for the collapse of heteropolymers.[71,74] The protein is represented as a chain of bead-like monomers connected by rotatable bonds. Two minima are generally found in this model. One minimum identifies a *compact state* with nonpolar core, and the other minimum identifies a *less compact ensemble* (with no nonpolar core), the distribution of which changes with solution conditions; this represents the unfolded ensemble. Because there is a *free energy barrier* between these two states, the model predicts two-state behavior, wherein *intermediate states* are less *populated than native or denatured states*. This often agrees with experiments – but not with some proteins or under certain solution conditions. Many small single-domain proteins populate predominantly two states.[179,246,244] This approach addresses the problems above. The nature of the unfolded state and its nonpolar exposure are predicted from the *balance of forces* rather than assumed. The conformational entropy difference between native and denatured states is calculated as a function of the chain and solution properties. Thus the theory simply provides a *procedure for enumerating* (*i*) the *relative amount of nonpolar surface buried* in the folded and unfolded states (some nonpolar surface is found to be exposed in the folded state, and some is buried in the unfolded state) and (*ii*) the *number of configurations accessible* in the folded and unfolded states. Three approximations are used for this counting.[71,74] First, the *Bragg-Williams mean-field approximation*[133] is used to count nonpolar contacts by assuming they are uniformly distributed within the chain volume. Second, the *Flory mean-field approximation*[97] is used to calculate how the number of conformations diminishes with density due to excluded volume; it

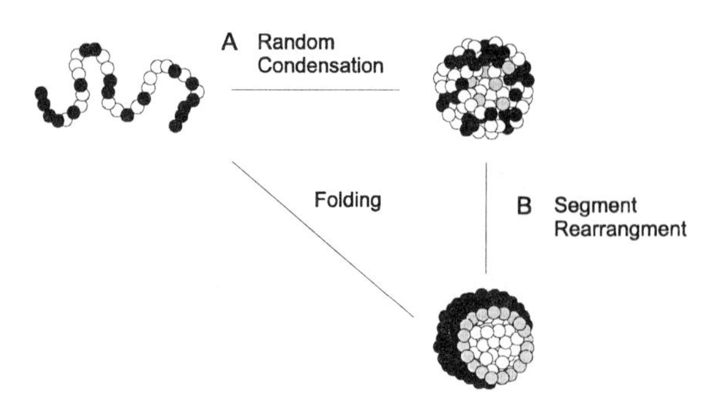

Figure 7 - Protein folding involves (A) an increase in compactness of the chain and (B) a reconfiguration of nonpolar residues into the core and polar residues to the surface. The collapse of homopolymers only involves (A).

too assumes segments are uniformly distributed. Third, *only the effects of composition are taken into account* (the number of nonpolar residues, and *not their sequence along the chain*): this is the random copolymer approximation. On this basis, the free energy of folding depends on (*i*) the chain length, (*ii*) the number of nonpolar residues, and (*iii*) the free energy of transferring an amino acid from water to a suitable "protein-like" medium.

The *general predictions of the model* are at least qualitatively consistent with known protein behavior. For example, the theory predicts that proteins should undergo a (maximally cooperative) first-order transition from the denatured to native state as the solvent becomes poor.[71,74] If there are too few hydrophobic residues, or if the chain is too short (less than several tens of residues), the molecule will not fold. The hydrophobic free energy is predicted to be about -60 kcal/mol of protein at 25°C for typical small proteins.[74] The opposing configurational free energy is predicted to be of nearly the same magnitude as the hydrophobic force: about 50 kcal/mol at 25°C. *This configurational entropy is large because there are tens of orders of magnitude fewer configurations in the folded than in the unfolded states.*[49,71] Thus, the net stability of the protein should be a small difference of these two large free energies (ca. -10 kcal/mol). This also resolves a related problem. The number of configurations of the unfolded protein is enormous. On that basis, it has been suggested that a protein would require many ages of the universe to find the native structure by random search, and thus that it could not attain its thermodynamic state of lowest free energy (the "*Levinthal" paradox*). This theory, on the other hand, predicts that simple collapse will reduce the configurational space enormously, implying that there is no inconsistency with the observation that proteins do attain their states of minimum free energy.[71]

The theory also predicts the nature of thermal and solvent stabilities of proteins, in good agreement with experiments. To predict thermal stability, the *temperature dependence of the hydrophobic interaction* must be put into the model; this is taken[74] from the nonpolar small-molecule transfer experiments of Nozaki and Tanford[217] and Gill and Wadsö.[110] The theory then predicts the existence of *two first-order transitions* vs. temperature. One, a *"cold" denaturation,* occurs at low temperatures because the dominant temperature dependence of folding at those temperatures is due to the weakening of the hydrophobic interaction with decreasing temperature. Cold denaturation has been observed or has been predicted from extrapolations of experiments in several systems.[57,127,248,51,37,221,244, 247,120] The other more familiar *thermal denaturation at higher temperatures* occurs because the gain in conformational freedom of the chain is more advantageous than the gain of interactions among nonpolar contacts. Comparison of the theoretical and experimental free energies, enthalpies, and entropies of folding is shown in Figure 8.

The theory has also been applied to the prediction of protein stabilities in *denaturing solvents*, such as urea and guanidine hydrochloride, and in stabilizing solvents.[9] In general the model can be applied to predicting stability as a function of any *external parameter* (x), provided that the oil/water partition coefficient for the representative elementary amino acid transfer is known as a function of x. In the case above, x is temperature; in this case, x is the concentration of urea or guanidine in the aqueous solvent. With use of the small-molecule transfer data of Nozaki and Tanford[216] for this purpose, the resultant protein stability theory predictions are

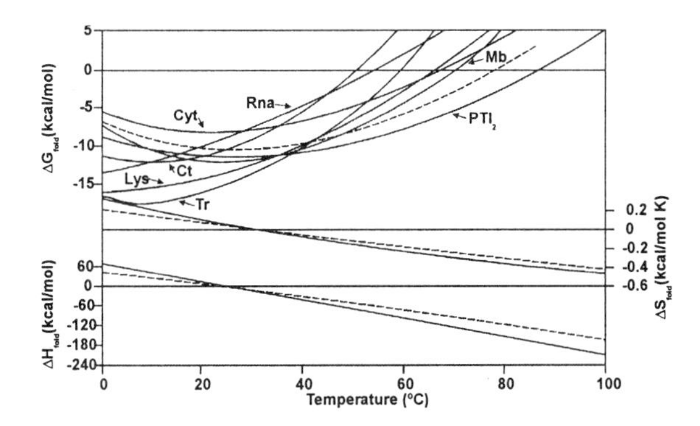

Figure 8 - Thermal stabilities of proteins. Experimental data for free energies, enthalpies, and entropies of folding taken from Privalov[244] and Privalov and Kechinashvili[246] (—). Theoretical predictions are from Dill *et al.*[74] (--).

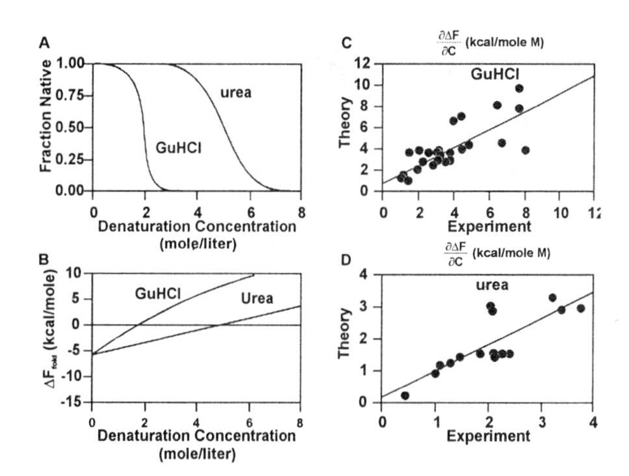

Figure 9 - Comparison of theory and experiments for denaturation of proteins by guanidine hydrochloride and urea.[9] (A) Theoretical denaturation curve. (B) Free energy vs. denaturant has approximately constant slope for small ranges of denaturant. (C and D) Comparison of theoretical and experimental slopes, $\partial \Delta F_{fold}/\partial c$, at the denaturation midpoint.

shown in Figure 9, along with experimental results. The theory predicts that the free energy of folding should be a linear function of urea concentration or a slightly curved function of guanidine concentration. These denaturants are predicted to cause unfolding by solvating nonpolar groups better than water does. Similarly, proteins should be stabilized by additives that are worse at solvating nonpolar groups than water is. Stabilizers include a wide range of sugars,[169,18,22] glycerol,[107] and polyols, poly(ethylene glycol), and some salts.[22,20]

9. Why Do Proteins Have Internal Architecture?

Globular proteins have internal organization comprised of a combination of "irregular" structures and "regular" structures such as helices and sheets. This internal organization is uniquely specified by the amino acid sequence. This organization and its uniqueness would appear to be difficult to reconcile with the picture of the nonspecific dominant forces described in the preceding sections. Secondary structures are hydrogen bonded. Therefore, hydrogen bonds must naturally play a significant role in determining *internal architecture*. Yet according to the arguments above, hydrogen bonding, although prevalent in folded proteins, is a weak driving force. Moreover, hydrogen bonding taken alone cannot readily explain sequence specificity: how the *sequence encodes only one native conformation.* Most intrachain hydrogen bonds in globular proteins are among backbone C=O and NH groups rather than among side chains.[23] From the data base of protein crystal structures, Baker and Hubbard[23] observed that of all the intrachain hydrogen bonds to C=O groups, 81.3% are with backbone NH groups and 18.7% are with side chains. Of all the intrachain hydrogen bonds to NH groups, 86.2% are with backbone C=O groups and 13.8% are with side chains. However, sequence specificity must arise from differences in side-chain properties, not from differences in backbone properties such as peptide hydrogen bonding. Any one amino acid will have essentially the same backbone interactions as any other. Therefore, if hydrogen bonding were dominant, the native structure of a protein should be essentially independent of the amino acid sequence, and native structures should be regular and periodic, either purely helix or purely sheet. Similarly, to the extent that backbone van der Waals interactions favor folding, they also are probably not very selective for one compact conformation relative to others.

Could it be that there is so much secondary structure in proteins because irregular conformations cannot form hydrogen bonds as well as helices and sheets can? The current limited evidence does not support this view. That the many irregular conformations of proteins can form good intrachain hydrogen bonds is clear from the study of Baker and Hubbard.[23] Of all the hydrogen bonds to C=O groups, 8.9% are in β-structures, 24.6% are in helices, and 5.3% are in turns. In comparison, a minimum of 18.3% of all hydrogen bonds to C=O groups are in "irregular" structures; i.e., they are intrachain (not bonded to water) and not in secondary structures. This is a minimum because the Baker and Hubbard study does not itemize bonds to water and in secondary structures as mutually exclusive categories, so only this lower bound can be obtained from their study. Similarly for NH groups, 38.3% are hydrogen bonded in helices, 13.6% in β-structures, and 9% in turns. A minimum of 17.9% of all hydrogen bonds to NH groups are in irregular structures. Hence, hydrogen bonds in irregular structures are not significantly less common than those in helices and sheets. Combined with the observation that 11% of all C=O groups and 12% of all NH groups have no hydrogen bonds,[23] this suggests that hydrogen-bonding requirements do not severely constrain the conformations accessible to the chain.

Indeed, the earliest expectations[320,232,228-231,154,153] were that internal architecture in globular proteins would be regular and periodic as in crystals, presumably with all residues in helices or sheets. For example, Pauling *et al.*[232] and Pauling and Corey[228-231] specifically sought types of structure in which every amino acid was "equivalent," i.e., interchangeable and not dependent on the sequence. Therefore, ever since the appearance of the first known structures of globular proteins,[154] the *central problem of protein architecture* has not been to *explain why proteins have so much regular structure.* It has been just the opposite: to *explain why there is so much irregular structure, and how the sequence uniquely encodes them both.* About 53% of all residues in globular proteins are in irregular structures.[144,159] Is protein architecture then a consequence of the smaller interactions, including side chain hydrogen bonding? In the following section, I suggest that sequence-specific internal architecture in globular proteins does not arise principally from these smaller forces but from the dominant forces.

Recent exhaustive simulations of all possible chain conformations have shown that protein-like internal architecture is simply a natural consequence of steric constraints in compact polymers.[47-50] Any flexible polymer, when made to be compact by *any* driving force, will have much internal architecture composed of helices and sheets. For proteins this driving force is presumably the hydrophobic interaction. *There are simply very few possible ways to configure a compact chain, and most of them involve helices and sheets.* The existence of internal organization is due to the physical impossibility of steric violation which alternative configurations would require. In other words, *consider a pearl necklace.* Squeeze it into a ball. One might expect that an ensemble of these compact necklaces would be highly disordered. This is not the case, however, it will have approximately the same distribution of internal architectures as are observed in globular proteins, comprised of helices, sheets, and irregular structures. The evidence for this is the following.

Polymer simulations and theory show that if a chain molecule is to form a single self-contact, it will prefer one that forms the smallest possible loop.[139,239,47,49] *A chain with a local (small) loop has more remaining accessible conformations (and thus greater entropy) than a chain with a nonlocal (large) loop.* [Theory and experiment also show that "stiffness," i.e., intrinsic propensities, can further favor or disfavor tight loops.[98,278,325]] Now consider *configurations that have two self-contacts; these provide the most basic description of elements of secondary structure.* A most interesting result from the theory and simulations is that if a chain forms two self-contacts, it will prefer to form a small loop and to form the second contact as close as possible in sequence to the first, simply because these, among all possible two-contact configurations, have the greatest entropy.[47,49] *The only two ways a chain molecule can form such a loop pair are either as a helix or as the beginning of an antiparallel sheet.* Hence given only the conformational freedom and steric restriction in flexible chains, there is a tendency to form helices and sheets, even in the absence of other forces, and even for chain conformations of relatively large radius of gyration. These results are obtained for chains configured in two or three dimensions, on different types of lattices, and by alternative path integral methods and therefore do not appear to be an artifact of the theoretical methods used in these predictions.

Exhaustive simulations further show that as a chain becomes increasingly compact, it develops a considerable amount of secondary structure[48] (see Figure 10). Consistent with this result, protein-like organization in compact chains has been observed by Monte Carlo methods for other lattices and chain lengths.[286,288,289]

The internal distributions of secondary structures obtained in exhaustive simulations of all possible compact conformations[50] are in good agreement with the distributions that Kabsch and Sander[144] have observed in the database of protein crystal structures (see Figure 11). For example, theory and experiment agree that the most common helices and sheets will be the shortest ones. The shortest antiparallel sheets are 3- to 4-fold more common than the shortest helices, which are about as common as the shortest parallel sheets. Among long secondary structures, in decreasing prevalence should be helices, antiparallel sheets, and then parallel sheets. It is interesting then to turn this argument around. Since the known protein architectures are distributed in the same way as in the complete ensemble of all compact conformations, this suggests that the currently known proteins are a reasonably representative sample of all the forms of internal structure that proteins could adopt [see also Finkelstein and Ptitsyn[96]].

These results provide an explanation for certain aspects of protein structure that have otherwise been puzzling. First, they show how the dominant forces due to hydrophobicity and steric constraints give rise to internal architectures. They provide a single framework for comprehending the coexistence of all the types of internal protein structure, regular and irregular, helices and sheets. Past hypotheses have tended to address a single type of architecture, often principally helices, in which the focus is on factors that are local in the sequence. *The present results suggest instead that architectures arise principally from nonlocal factors. That is, "tertiary" forces drive secondary structures rather than "secondary" forces driving tertiary structures.* It follows that helices and sheets in globular proteins are only secondarily a consequence of hydrogen bonding. In this regard, it is interesting that of the 176 known crystal structures of different synthetic polymers that are considered to be reliable, 49 are planar zigzags, and 79 are helices of 22 different types.[301] Many different crystalline polymers have helical pitches nearly the same as that

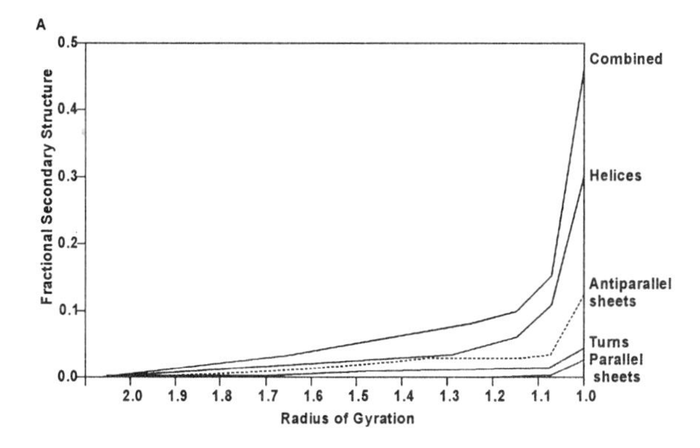

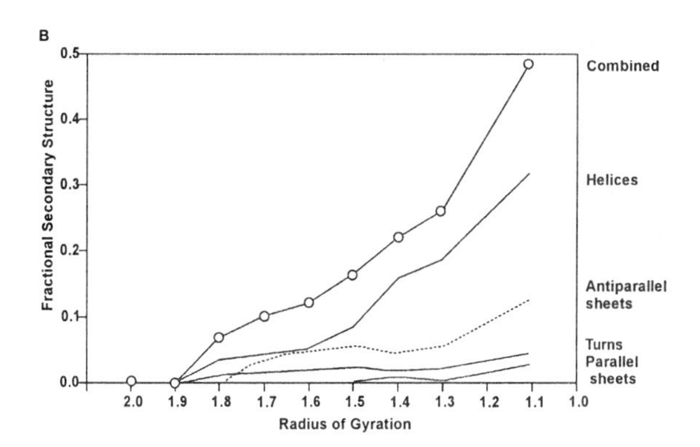

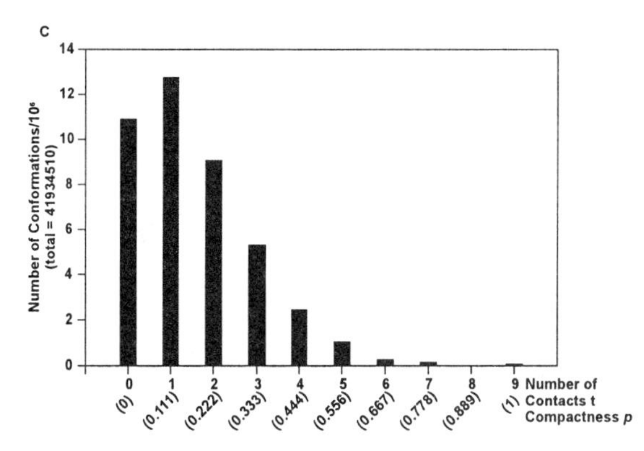

Figure 10 - Increasing compactness of a chain molecule leads to the formation of secondary structure. With increasing compactness (number of intrachain contacts), exhaustive simulations of all possible configurations of short chains on lattices show that[48,50] (C) the number of accessible configurations diminishes rapidly and (B) the amount of secondary structure increases sharply. (A) Same results as (B), except plotted against radius of gyration instead of compactness.

of the α-helix, between three and four monomers per turn, including polybutadiene, polypropylene, polyvinylnaphthalene, polybutene, and even fibrous sulfur, none of which forms hydrogen bonds.[301] This is not to say that local factors and hydrogen bonds are unimportant. This is only to say that *local factors and hydrogen bonding probably contribute to making energetic "decisions" only within an already highly restricted ensemble, one in which the chain will be forced (by the hydrophobic drive to compactness and by the highly selective steric forces) to have considerable stretches of (i, i + 3) self-contacts,* for example. In that regard, what dictates that these will be specifically α-helices, with 3.6 residues per turn, instead of any of dozens of other types of closely related helices will be local forces uniquely determined by amino acids as monomers.

One aspect of internal architecture that is *not* a consequence of packing forces alone is the spatial distribution of turns. Turns are observed to occur largely at the surfaces of proteins.[163] This does not arise from packing constraints.[48] Rather it appears to be a consequence of the polar nature of turn residues.[264] Since the middle residues of turns are

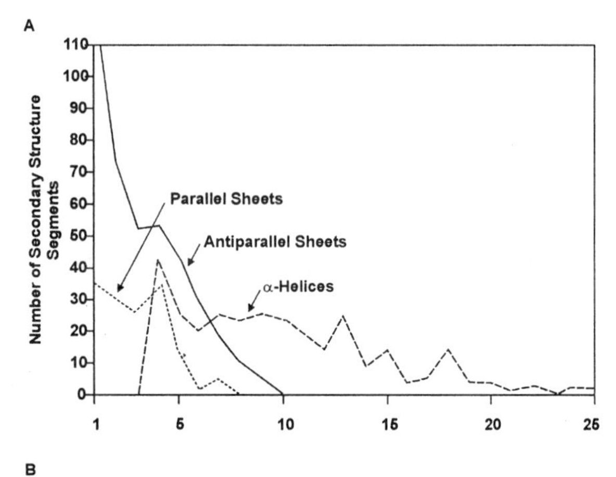

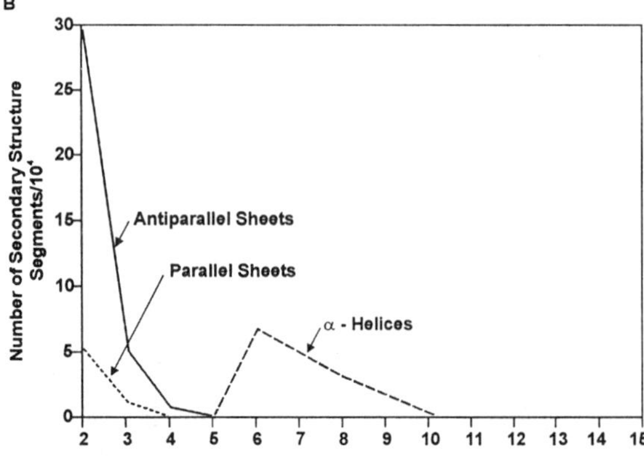

Figure 11 - Comparison of data base studies of Kabsch and Sander[144] (A) of the internal distributions of lengths of helices and sheets in globular proteins, with (B) exhaustive lattice simulations of the full compact ensemble.[50]

geometrically unable to form intrachain hydrogen bonds, then their hydrogen bonding needs can best be met by interacting with water at protein surfaces.

A second puzzle has been why some denatured states of proteins have secondary structure. According to these simulations, in the absence of other interactions, the amount of secondary structure simply depends on the radius of gyration of the chain[48,50] (see Figure 10). Therefore, *proteins that are denatured only weakly, and thus have small radii, should have some secondary structure.* This may account for secondary structures in "*molten globule*" and other compact denatured states of proteins, with radii only slightly larger than those of native molecules.[250,117,118,284,164,42,28]

Because internal architecture should therefore arise in any compact polymer, it should be possible to design other copolymers, not necessarily comprised of amino acids, which can be driven to compactness in poor solvents and which should then have protein-like architectures. Consistent with this prediction, Rao *et al.*[253] have shown that in a solution containing a large number of different sequences of amino acid copolymers roughly half the molecules are highly compact and there is 46% helix observed by circular dichroism. This implies that a large fraction of all possible sequences is capable of coding for large amounts of helix.

10. Why Is the Native Structure Unique?

A most unusual feature of globular proteins, relative to any other type of isolated polymer molecule, is that they can be found in only a single conformation, the native structure. ("Single" here means that the chain path is identifiable at relatively high resolution as an average over the ensemble of small fluctuations.) What forces, encoded in the amino acid sequence, cause this remarkable uniqueness? As noted above, there is considerable elimination of configurational possibilities simply due to the compactness. Nevertheless, even for compact chain molecules, there are still many accessible conformations; the number of maximally compact conformations increases exponentially with chain length.[48,71] Within this ensemble of physically accessible compact conformations, all the various types of interactions could play a role, including hydrogen bonding, intrinsic propensities, ion pairing, and hydrophobicity, in causing some compact conformations to be of higher free energy and others to be of lower free energy.

Is there a single one among these types of interaction, by virtue of its *encoding within the sequence, that "picks out" the one native structure?*

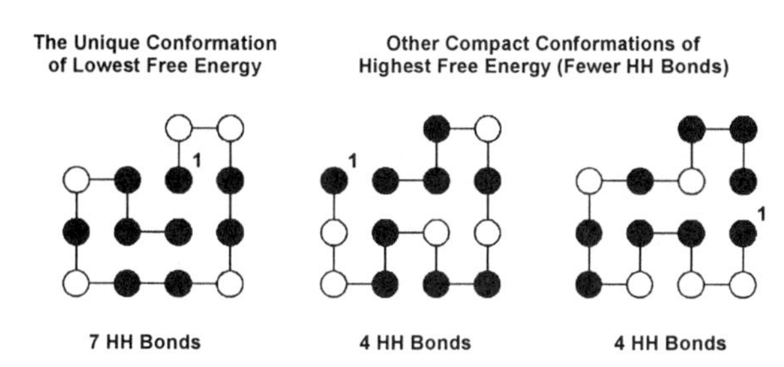

Figure 12 - Exhaustive lattice simulations of short chains show that any given sequence has many different compact conformations (such as the two on the right) but often only one native state, in which the maximum number of nonpolar contacts is formed. Nonpolar residues (H, ●); polar residues (P, O).

The most significant further restrictor of conformational space may be the hydrophobic interaction. By use of a simple model of short self-avoiding flexible chains on lattices, in which the only energetic feature of the sequence is the hydrophobic interaction, every conformation has been explored by exhaustive search in order to determine the native state(s), those at the global minimum of free energy.[166] This has been done for many different sequences. In this model, residues are either hydrophobic (H) or polar (P). The free energy of any chain conformation is determined simply by the number of HH topological contacts (topological contacts are defined in Figure 1). Therefore, "native" conformations are those with the maximum number of such HH contacts. How many native structures does any sequence have? Figure 13 shows the distribution. The most surprising result is that this is a strongly decreasing function; *far more sequences can configure into only one native structure than can configure into 10 or 100 native structures.* According to this simulation, for most folding sequences, exceedingly few ways a chain can configure to form the maximum possible number of HH contacts exist. One example is shown in Figure 12. This suggests that *hydrophobicity is strongly selective* and singles out only a very small number of candidate native structures from the compact ensemble. Other evidence also supports this view. Hydrophobicity patterns appear to be more predictive of conformational families than other types of interactions.[300,27,35] In addition, Covell and Jernigan[62] have exhaustively explored all conformations of lattice chains confined within known protein shapes using 3D lattices, weighting them using the free energies of Miyazawa and Jernigan,[200] and have compared them with known protein structures. They found the native conformation to be within the best 1.8%. Hence, within their shape-restricted ensemble, the transfer free energies (whatever their molecular bases in hydrophobicities, hydrogen bonding, van der Waals, and other interactions) considerably restrict the possible native structures. Thus, the nonlocal forces encoded within the sequence, rather than the local factors involving connected residues, would appear to be largely responsible for the uniqueness of the native structure.

Whatever role hydrophobicity may play in reducing the ensemble to a small set of possibilities, it is also clear that hydrophobicity cannot be *solely* responsible for determining the single native structure. *The hydrophobic contact free energies of the native and the next higher free energy level will, in general, differ by no more than one or two nonpolar contacts, less than 2 kcal/mol. At this level of discrimination, the nondominant interactions are important.* If they were not, then it would be impossible to account for the existence of hydrogen bonding, the partial successes of the intrinsic propensities, and the observed nonrandom distributions of charges, ion pairs, and aromatic groups.[43,44,33] Thus, hydrophobicity may "select" a relatively small number of compact conformations, from which the native structure is determined by the balance of all types of interactions.

11. Sequence Space and Origins of Proteins from Random Sequences

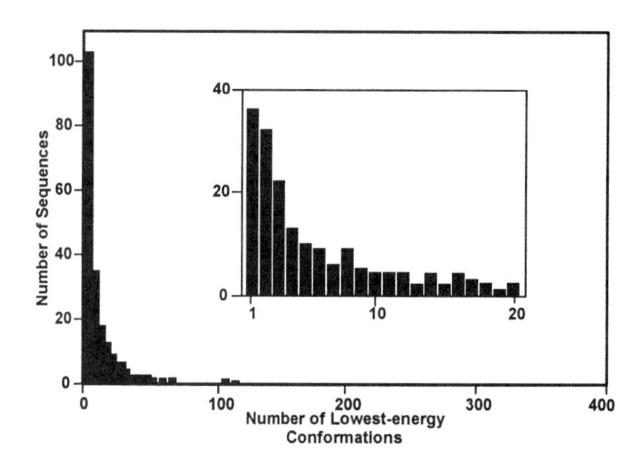

Figure 13 - How many native structures does a given sequence have? This histogram derives from a 2D lattice model of self-avoiding flexible chains (of length = 24 monomers) subject only to hydrophobic interactions.[167] Exhaustive search permits determination of the structure(s) or global minimum in free energy among the compact ensemble. These native structures are then found for many different sequences. The inset shows finer detail. This decreasing function implies that more sequences have only one native structure than have two, three, etc. Thus for a typical sequence, there are very few configurations that have the maximum possible number of nonpolar contacts.

That hydrophobicity is the dominant folding force is also consistent with expectations for the nature of mutation-induced changes in proteins. (*1*) Replacements of hydrophobic residues in the core are more disruptive of stability than other types of substitutions.[234,,6,7,173,167] (*2*) Because hydrophobicity is relatively nonspecific and orientation independent, many different core replacements are tolerated, provided only that they are hydrophobic and not significantly different in size. There is even greater tolerance for replacements of surface residues.[234, 173,35,6,167] (*3*) It follows that a large fraction of the molecules in "sequence space" (all the possible sequences) should be able to fold. Two-dimensional lattice simulations predict that more than 50% of all possible sequences will, under native conditions, fold to within 10% of the minimum possible radius of gyration and will therefore have considerable amounts of secondary structure.[166] White and Jacobs[318] have found remarkably little difference between hydrophobicity sequence patterns in more than 8000 known sequences on the one hand and random sequences on the other. (*4*) There should be an extremely large number of "convergent" sequences: i.e., a given native structure should be encodable in many different sequences.[167] This implies a significant probability that a random sequence of amino acids will encode a globular conformation in general and a particular native structure in specific. For example, although there is only one *sequence* of ribonuclease, there should be more than 10^{100} different sequences, which will all have the same native backbone *conformation* as ribonuclease (see Figure 14).[167] Indeed, there has been considerable success in designing proteins based on simple principles.[258,69,70] The nonspecific nature of the dominant folding forces may therefore be essential in explaining how functional proteins could have originated from random sequences since only a negligible fraction of sequence space could have been sampled during the origins of life.

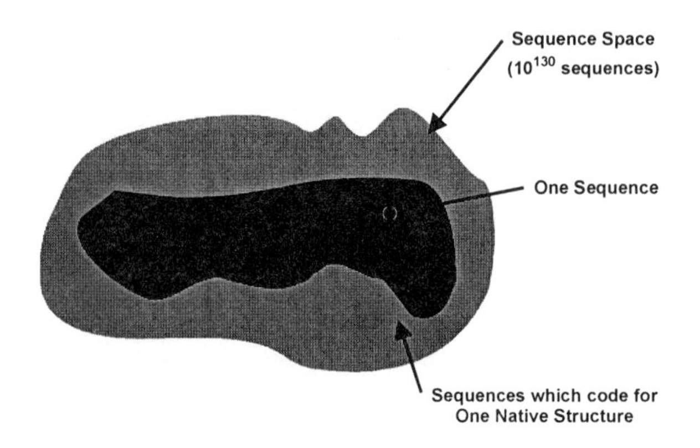

Figure 14 - For a protein of 100 residues, there are 20^{100} = 10^{130} different sequences. This is the sequence space. A given protein (such as ribonuclease) has only one sequence. Hence, the probability of drawing the *sequence* of a particular protein by random selection is 10^{-130}. On the other hand, lattice simulations[167] show that the probability of drawing any sequence which will fold to a specified *structure* (the ribonuclease configuration, for example) is estimated to be about 120 orders of magnitude larger than this.

12. Conclusions

More than 30 years after Kauzmann"s insightful hypothesis, there is now strong accumulated evidence that hydrophobicity is the dominant force of protein folding, provided that "hydrophobic" is operationally defined in terms of the transfer of nonpolar amino acids from water into a medium that is nonpolar and preferably capable of hydrogen bonding. Other forces are weaker but can affect stability. In acids and bases, electrostatic charge repulsions destabilize native proteins. Near neutral pH, ion pairing can stabilize proteins. There is evidence that hydrogen bonding or van der Waals interactions among polar amino acids may be important, but their magnitude remains poorly understood. An important contributor to protein structure and stability is the dominant opposing force, arising principally from the loss of nonlocal conformational entropy due to steric constraints in the folded state. The marginal stabilities of proteins arise from the small difference between these large driving and opposing forces. Hydrophobicity leads to compact conformations with nonpolar cores, but it is the steric constraints in compact chains that are largely responsible for their considerable internal architecture. The reason that only one native structure is encoded in the amino acid sequence may be largely attributable to the hydrophobic interaction; there are only a small number of ways to configure a chain to maximize the number of nonpolar contacts. These forces are of a nature such that proteins should be tolerant of amino acid substitution, a given native structure should be encodable in many different sequences, and a large fraction of all possible sequences should fold to compactly structured native states.

References:

1. Acampora, G., & Hermans, J. (1967) *J. Am. Chem. Soc. 89,* 1543.
2. Ahmad, F., & Bigelow, C. C. (1986) *Biopolymers 25,* 1623.
3. Alber, T. (1989a) *Annu. Rev. Biochem. 58,* 765.
4. Alber, T. (1989b) in *Prediction of Protein Structure and the Principles of Protein*

Conformation (Fasman, G. D., Ed.) p. 161, Plenum, New York.
5. Alber, T., *et al.* (1987a) *Nature 330*, 41.
6. Alber, T., *et al.* (1987b) *Biochemistry 26,* 3754.
7. Alber, T., *et al.* (1988) *Science 239*, 631.
8. Allen, G., Watkinson, J. G., & Webb, K. H. (1966) *Spectrochim. Acta 22,* 807.
9. Alonso, D. O. V., & Dill, K. A. (1990) *Biochemistry* (submitted for publication).
10. Alvarez, J., & Biltonen, R. (1973) *Biopolymers 12,* 1815.
11. Anderson, D. E., Becktel, W. J., & Dahlquist, F. W. (1990) *Biochemistry 29,* 2403.
12. Anfinsen, C. B. (1973) *Science 181*, 223.
13. Anson, M. L. (1945) *Adv. Protein Chem. 2,* 361.
14. Anson, M. L., & Mirsky, A. E. (1925) *J. Gen. Physiol. 9,* 169.
15. Anson, M. L., & Mirsky, A. E. (1931) *J. Phys. Chem. 35,* 185.
16. Anson, M. L., & Mirsky, A. E. (1934) *J. Gen. Physiol. 17,* 393.
17. Anufrieva, E. V., *et al.* (1968) *J. Polym. Sci. 16,* 3533.
18. Arakawa, T., & Timasheff, S. N. (1982) *Biochemistry 21,* 6536.
19. Arakawa, T., & Timasheff, S. N. (1984) *Biochemistry 23,* 5912.
20. Arakawa, T., Bhat, R., & Timasheff, S. N. (1990) *Biochemistry 29,* 1924.
21. Argos, P. (1987) *J. Mol. Biol. 197*, 331.
22. Back, J. F., Oakenfull, D., & Smith, M. B. (1979) *Biochemistry 18,* 5191.
23. Baker, E. N., & Hubbard, R. E. (1984) *Prog. Biophys. Mol. Biol. 44,* 97.
24. Baldwin, R. L. (1986) *Proc. Natl. Acad. Sci. U.S.A. 83*, 8069.
25. Barlow, D. J., & Thornton, J. M. (1983) *J. Mol. Biol. 168,* 867.
26. Bartlett, P. A., & Marlowe, C. K. (1987) *Science 235*, 569.
27. Bashford, D., Chothia, C., & Lesk, A. M. (1987) *J. Mol. Biol. 196*, 199.
28. Baum, J., *et al.* (1989) *Biochemistry 28,* 7.
29. Baumann, G., Frommel, C., & Sander, C. (1989) *Protein Eng. 2,* 329.
30. Becktel, W., & Schellman, J. (1987) *Biopolymers 26*, 1859.
31. Bendzko, P. I., Pfeil, W. A., Privalov, P. L., & Tiktopulo, E. I. (1988) *Biophys. Chem. 29*, 301.
32. Blagdon, D. E., & Goodman, M. (1975) *Biopolymers 14*, 241.
33. Blundell, T., et al. (1986) *Science 234*, 1005.
34. Bolen, D. W., & Santoro, M. M. (1988) *Biochemistry 27,* 8069.
35. Bowie, J. U., Reidhaar-Olsen, J. F., Lim, W. A., & Sauer, R. T. (1990) *Science 247*, 1306.
36. Bradley, E. K., Thomasen, J. F., Cohen, F. E., Kosen, P. A., & Kuntz, I. D. (1990) *J. Mol. Biol.* (submitted for publication).
37. Brandts, J. F. (1964a) *J. Am. Chem. Soc. 86,* 4291.
38. Brandts, J. F. (1964b) *J. Am. Chem. Soc. 86,* 4302.
39. Brandts, J. F. (1969) in *Structure and Stability of Biological Macromolecules* (Timasheff, S. N., & Fasman, G. D., Eds.) p. 213, Dekker, New York.
40. Brandts, J. F., & Lumry, R. (1963) *J. Phys. Chem. 67*, 1484.
41. Brandts, J. F., Oliveira, R. J., & Westort, C. (1970) *Biochemistry 9*, 1038.
42. Brems, D. N., & Havel, H. A. (1989) *Proteins 5*, 93.
43. Burley, S. K., & Petsko, G. A. (1985) *Science 229*, 23.
44. Burley, S. K., & Petsko, G. A. (1988) *Adv. Protein Chem. 39,* 125.
45. Burteau, N., Burton, S., & Crichton, R. R. (1989) *FEBS Lett. 258*, 185.
46. Butler, J. A. V. (1937) *Trans. Faraday Soc. 33*, 229.
47. Chan, H. S., & Dill, K. A. (1989a) *J. Chem. Phys. 90, 492*.
48. Chan, H. S., & Dill, K. A. (1989b) *Macromolecules 22*, 4559.
49. Chan, H. S., & Dill, K. A. (1990a) *J. Chem. Phys. 92*, 3118.
50. Chan, H. S., & Dill, K. A. (1990b) *Proc. Nat. Acad. Sci. U.S.A.* (in press).
51. Chen, B., & Scheillman, J. A. (1989) *Biochemistry 28*, 685.
52. Chick, H., & Martin, C. J. (1911) *J. Physiol. 43*, 1.
53. Chothia, C. (1974) *Nature 254*, 304.
54. Chothia, C. (1976) *J. Mol. Biol. 105*, 1.
55. Chou, P. Y., & Fasman, G. D. (1978) *Adv. Enzymol. 47*, 45.
56. Chou, P. Y., Wells, M., & Fasman, G. D. (1972) *Biochemistry 11*, 3028.
57. Christensen, J. K. (1952) *C. R. Trav. Lab. Carlsberg, Ser. Chim. 28*, 37.
58. Christian, S. D., & Tucker, E. E. (1982) *J. Solution Chem. 11*, 749.
59. Cohn, E. J., *et al.* (1933) *J. Biol. Chem. 100*, 3.
60. Collins, K. D., & Wasbabaugh, M. W. (1985) *Q. Rev. Biophys. 18*, 323.
61. Conio, G., Patrone, E., & Brighetii, S. (1970) *J. Biol. Chem. 245*, 3335.

62. Covell, D. G., & Jernigan, R. L. (1990) *Biochemistry 29*, 3287-3294.
63. Crovetto, R., Fernandez-Priri, R., & Japas, M. L. (1982) *J. Chem. Phys. 76*, 1077.
64. Cybulski, S., M., & Scheiner, S. (1989) *J. Phys. Chem. 93*, 6565.
65. Dang, L. X., Merz, K. M., & Kollman, P. A. (1989) *J. Am. Chem. Soc. 111*, 8505.
66. Debye, P. (1949) *Ann. N.Y. Acad. Sci. 51*, 575.
67. deGennes, P. G. (1975) *J. Phys. (Paris) 36*, L-55.
68. deGennes, P. G. (1979) in *Scaling Concepts in Polymers Physics,* Cornell University Press, Ithaca, NY.
69. DeGrado, W. F. (1988) *Adv. Protein Chem. 39*, 51.
70. DeGrado, W. F., Wasserman, Z. R., & Lear, J. D. (1989) *Science 243*, 622.
71. Dill, K. A. (1985) *Biochemistry 24*, 1501.
72. Dill, K. A. (1990) *Science* (in press).
73. Dill, K. A., & Alonso, D. O. V. (1988) in *Colloquium Mosbach der Gessellschaft fur Biologische Chemie: Protein Structure and Protein Engineering* (Huber, T., & Winnacker, E. L., Eds.) Vol. 39, p. 51, Springer-Verlag, Berlin.
74. Dill, K. A., Alonso, D. O. V., & Hutchinson, K. (1989) *Biochemistry 28*, 5439.
75. Doty, P., & Yang, J. T. (1956) *J. Am. Chem. Soc. 78*, 498.
76. Doty, P., *et al.* (1954) *J. Am. Chem. Soc 76*, 4493.
77. Doty, P., Bradbury, J. H., & Holtzer, A. M. (1956) *J. Am. Chem. Soc. 78*, 947.
78. Doty, P., *et al.* (1958) *Proc. Natl. Acad. Sci. U.S.A. 44*, 424.
79. Dyson, H. J., *et al.* (1985) *Nature 318*, 480.
80. Dyson, H. J., *et al.* (1988a) *J. Mol. Biol. 201*, 161.
81. Dyson, H. J., *et al.* (1988b) *J. Mol. Biol. 201*, 201.
82. Edelhoch, H., & Osborne, J. C. (1976) *Adv. Protein Chem. 30*, 183.
83. Eden, D., *et al.* (1982) *Proc. Natl. Acad. Sci. U.S.A. 79*, 815.
84. Edsall, J. T. (1935) *J. Am. Chem. Soc. 57*, 1506.
85. Edsall, J. T., & McKenzie, H. A. (1983) *Adv. Biophys. 16*, 53.
86. Eisenberg, D., & McLachlan, A. D. (1986) *Nature 319*, 199.
87. Eisenberg, M. A., & Schwert, G. W. (1951) *J. Gen. Physiol. 34*, 583.
88. Ellis, R. J. (1988) *Nature 328*, 378.
89. Epand, K. M., & Scheraga, H. A. (1968) *Biochemistry 7*, 2864.
90. Evans, P. A., *et al.* (1987) *Nature 329*, 266.
91. Eyring, H., & Stearn, A. E. (1939) *Chem. Rev. 24*, 253.
92. Fahey, P. F., Krupke, D. W., & Beams, J. W. (1969) *Proc. Natl. Acad. Sci. U.S.A. 63*, 548.
93. Fauchere, J.-L., & Pliska, V. E. (1983) *Eur. J. Med. Chem. Chem. Therm. 18*, 369.
94. Fersht, A. R. (1972) *J. Mol. Biol. 64*, 497.
95. Fersht, A. R. (1985) *Nature 314*, 235.
96. Finkelstein, A. V., & Ptitsyn, O. B. (1987) *Prog. Biophys. Mol. Biol. 50*, 171.
97. Flory, P. J. (1953) in *Principles of Polymer Chemistry*, Cornell University Press, Ithaca, NY.
98. Flory, P. J. (1969) *Statistical Mechanics of Chain Molecules*, Wiley, New York.
99. Frank, H. S., & Evans, M. W. (1945) *J. Chem. Phys. 13*, 507.
100. Franks, F. (1983) *Water*, The Royal Society of Chemistry, London.
101. Franzen, J. S., & Stephens, R. E. (1963) *Biochemistry 2*, 1321.
102. Freed, K. F. (1987) in *Renormalization Group Theory of Macromolecules*, Wiley, New York.
103. Friedman, M. E., & Scheraga, H. A. (1965) *J. Phys. Chem. 69*, 3795.
104. Friend, S. H., & Gurd, F. R. N. (1979) *Biochemistry 18*, 4612.
105. Gavish, B., Gratton, E., & Hardy, C. J. (1983) *Proc. Natl. Acad. Sci. U.S.A. 80*, 750.
106. Geiger, A., Rahman, A., & Stillinger, F. H. (1979) *J. Chem. Phys. 70*, 263-276.
107. Gekko, K., & Timasheff, S. N. (1981) *Biochemistry 20*, 4667.
108. Gibbs, J. H., & DiMarzio, E. A. (1959) *J. Chem. Phys. 30*, 271.
109. Gill, S. J., & Noll, L. (1972) *J. Phys. Chem. 76*, 3065.
110. Gill, S. J., & Wadsö, I. (1976) *Proc. Natl. Acad. Sci. U.S.A. 73*, 2955.
111. Gill, S. J., *et al.* (1985) *J. Phys. Chem. 89*, 3758.
112. Gilson, M. K., & Honig, B. H. (1988a) *Proteins 3*, 32.
113. Gilson, M. K., & Honig, B. H. (1988b) *Proteins 4*, 7.
114. Gilson, M. K., & Honig, B. H. (1989) *Proc. Natl. Acad. Sci. U.S.A. 86*, 1524.
115. Goldenberg, D. P., & Creighton, T. E. (1983) *J. Mol. Biol. 165*, 407.
116. Goodman, M., *et al.* (1969) *Proc. Natl. Acad. Sci. U.S.A. 64*, 444.
117. Goto, Y., & Fink, A. L. (1989) *Biochemistry 28*, 945.

118.Goto, Y., & Fink, A. L. (1990) *J. Mol. Biol.* (in press).
119.Goto, Y., Calciano, L. J., & Fink, A. L. (1990) *Proc. Natl. Acad. Sci. U.S.A. 87*, 573.
120.Griko, Y. V., Privalov, P. L., Sturtevant, J. M., & Venyamenov, S. Y. (1988) *Proc. Natl. Acad. Sci. U.S.A. 85*, 3343.
121.Grutter, M. G., *et al.* (1987) *J. Mol. Biol. 197*, 315.
122.Gurney, R. W. (1962) in *Ionic Processes in Solution*, Dover, New York.
123.Guy, H. R. (1985) *Biophys. J. 47*, 61.
124.Ha, J. H., Spolar, R. S., & Record, M. T., Jr. (1989) *J. Mol. Biol. 209*, 801.
125.Harrington, W. F., & Schelliman, J. A. (1956) *C. R. Lab. Carlsberg, Ser. Chm. 30*, 21.
126.Haschemeyer, R. H., & Haschemeyer, A. E. V. (1973) in *Proteins*, Wiley, New York.
127.Hawley, S. A. (1971) *Biochemistry 10*, 2436.
128.Hermans, J. (1966) *J. Am. Chem. Soc. 88*, 2418.
129.Hermans, J., & Scheraga, H. A. (1961) *J. Am. Chem. Soc. 83*, 3283.
130.Hildebrand, J. H. (1968) *J. Phys. Chem. 72*, 1841.
131.Hildebrand, J. H. (1979) *Proc. Natl. Acad. Sci. U.S.A. 76*, 194.
132.Hildebrand, J. H., & Scott, R. L. (1950) in *The Solubility of Nonelectrolytes*, Reinhold, New York.
133.Hill, T. L. (1960) in *Introduction to Statistical Thermodynamics*, Addison-Wesley, Reading, MA.
134.Hollecker, M., & Creighton, T. E. (1982) *Biochim. Biophys. Acta 701*, 395.
135.Holley, L. H., & Karplus, M. (1989) *Proc. Natl. Acad. Sci. U.S.A. 86*, 152.
136.Honig, B., & Hubbell, W. (1984) *Proc. Natl. Acad. Sci. U.S.A. 81*, 5412.
137.Honig, B., Hubbell, W., & Flewelling, R. F. (1986) *Annu. Rev. Biophys. Biophys. Chem. 15*, 163.
138.Hvidt, A. (1975) *J. Theor. Biol. 50*, 245.
139.Jacobsen, C. F., & Linderstrom-Lang, K. (1949) *Nature 164*, 411.
140.Janin, J., *et al.* (1978) *J. Mol. Bïol. 125*, 357.
141.Jolicoeur, C., *et al.* (1986) *J. Solution Chem. 15*, 109.
142.Jorgensen, W. L. (1989) *Acc. Chem. Res. 22*, 184.
143.Jorgensen, W. L., Goo, J., & Ravimohan, C. (1985) *J. Phys. Chem. 89*, 3470.
144.Kabsch, W., & Sander, C. (1983) *Biopolymers 22*, 2577.
145.Kabsch, W., & Sander, C. (1984) *Proc. Natl. Acad. Sci. U.S.A.* 1075.
146.Karle, I. L., *et al.* (1990) *Proteins 7*, 62.
147.Karplus, M., Ichiye, T., & Pettitt, B. M. (1987) *Biophys. J. 52*, 1083.
148.Kauzmann, W. (1954) in *The Mechanism of Enzyme Action* (McElroy, W. D., & Glass, B., Eds.) p. 70, Johns Hopkins Press, Baltimore, MD.
149.Kauzmann, W. (1959) *Adv. Protein Chem. 14*, 1.
150.Kauzmann, W. (1987) *Nature 325*, 763.
151.Kellis, J. T., Nyberg, K., & Fersht, A. R. (1989) *Biochemistry 28*, 4914.
152.Kelly, L., & Holladay, L. A. (1987) *Protein Eng. 1*, 137.
153.Kendrew, J. C. (1961) *Sci. Am. 205*, (6), 96.
154.Kendrew, J. C., *et al.* (1958) *Nature 181*, 662.
155.Kim, P. S., & Baldwin, R. L. (1984) *Nature 307*, 329.
156.Klapper, M. H. (1971) *Biochim. Biophys. Acta 229*, 557.
157.Klotz, I. M., & Franzen, J. S. (1962) *J. Am. Chem. Soc. 84*, 3461.
158.Klotz, I. M., & Farnham, S. B (1968) *Biochemistry 7*, 3879.
159.Kneller, D. G., Cohen, F. E., & Langridge, R. (1990) *J. Mol. Biol.* (in press).
160.Kresheck, G. C., & Klotz, I. M. (1969) *Biochemistry 8*, 8.
161.Krikorian, S. E. (1982) *J. Phys. Chem. 86*, 1875.
162.Kundrot, C. E., & Richards, F. M. (1987) *J. Mol. Biol. 193*, 157.
163.Kuntz, I. D. (1972) *J. Am. Chem. Soc. 94*, 4009.
164.Kuwajima, K. (1989) *Proteins 6*, 87.
165.Kyte, J., & Doolittle, R. F. (1982) *J. Mol. Biol. 157*, 105.
166.Lau, K. F., & Dill, K. A. (1989) *Macromolecules 22*, 3986.
167.Lau, K. F., & Dill, K. A. (1990) *Proc. Natl. Acad. Sci. U.S.A. 87*, 638.
168.Lee, B. K., & Richards F. M. (1971) *J. Mol. Biol. 55*, 379.
169.Lee, J. C., & Timasheff, S. N. (1981) *J. Biol. Chem. 256*, 7193.
170.Lennard-Jones, J., & Pople, J. A. (1951) *Proc. R. Soc. London A205*, 155.
171.Levitt, M., & Greer, J. (1977) *J. Mol. Biol. 114*, 181.
172.Lifschitz, I. M. (1968) *Zh. Eksp. Teor. Fiz. 55*, 2408.
173.Lim, W. A., & Sauer, R. T. (1989) *Nature 339*, 31.
174.Linderstrom-Lang, K. U. (1924) *C. R. Trav. Lab. Carlsberg, 15*, 70.

175. Linderstrom-Lang, K. U. (1952) *Lane Medical Lectures,* Vol. 6, p 53, Stanford University Press, Stanford, CA.
176. Lotan, N., Bixon, M., & Berger, A. (1967) *Biopolymers 5*, 69.
177. Luger, K., et al. (1989) *Science 243*, 206.
178. Lumry, R., & Eyring, H. (1954) *J. Phys. Chem. 58*, 110.
179. Lumry, R., Biltonen, R., & Brandts, J. F. (1966) *Biopolymers 4*, 917.
180. Lupu-Lotan, N., Yaron, A., Berger, A., & Sela, M. (1965) *Biopolymers 3*, 625.
181. Marqusee, S., & Baldwin, R. L. (1987) *Proc. Natl. Acad. Sci. U.S.A. 84*, 8898.
182. Marqusee, S., Robbins, V. H., & Baldwin, R. L. (1989) *Proc. Natl. Acad. Sci. U.S.A. 86*, 5286.
183. Masterson, W. L. (1954) *J. Chem. Phys. 22*, 1830.
184. Matsumura, M., Becktel, W. J., & Matthews, B. W. (1988a) *Nature 334*, 406.
185. Matsumura, M., et al. (1988b) *Eur. J. Biochem. 171*, 715.
186. Matsumura, M., Matthews, B. W., Levitt, M., & Becktel, W. J. (1989) *Proc. Natl. Acad. Sci. U.S.A. 86*, 6562.
187. Matthew, J. B., & Richards, F. M. (1982) *Biochemistry 21*, 4489.
188. Matthew, J. B., & Gurd, F. R. N. (1986) *Methods Enzymol. 130*, 413.
189. Matthews, B. W. (1987a) *Harvey Lect. 81*, 33.
190. Matthews, B. W. (1987b) *Biochemistry 26*, 6885.
191. Matthews, B. W., Nicholson, A., & Becktel, W. J. (1987) *Proc. Natl. Acad. Sci. U.S.A. 84*, 6663.
192. Mattice, W. L., & Scheraga, H. A. (1984) *Biopolymers 23*, 1701.
193. McGregor, M. J., Islam, S. A., & Sternberg, M. J. E. (1987) *J. Mol. Biol. 198*, 295.
194. Meirovitch, H., & Scheraga, H. A. (1980) *Macromolecules 13*, 1406.
195. Merutka, G., & Stellwagen, E. (1990) *Biochemistry 29*, 894.
196. Merutka, G., *et al.* (1990) *Biochemistry* (submitted for publication).
197. Miller, K. W., & Hildebrand, J. H. (1968) *J. Am. Chem. Soc. 90*, 3001.
198. Mirsky, A. E., & Pauling, L. (1936) *Proc. Natl. Acad. Sci. U.S.A. 22*, 439.
199. Mitchinson, C., & Baldwin, R. L. (1986) *Proteins 1*, 23.
200. Miyazawa, S., & Jernigan, R. L. (1985) *Macromolecules 18*, 534.
201. Morrison, T. J. (1952) *J. Chem. Soc. 3,* 3814.
202. Morrison, T. J., & Billett, F. (1952) *J. Chem. Soc. 3*, 3819.
203. Muller, N. (1990) *Acc. Chem. Res. 23*, 23.
204. Murphy, K. P., & Gill, S. J. (1989a) *J. Chem. Thermodyn. 21*, 903.
205. Murphy, K. P., & Gill, S. J. (1989b) *Thermochim. Acta 139*, 279.
206. Murphy, K. P., Privalov, P. L., & Gill, S. J. (1990) *Science 247*, 559.
207. Nelson, J. W., & Kallenbach, N. R. (1986) *Proteins 1*, 211.
208. Nemethy, G., & Scheraga, H. A. (1962) *J. Chem. Phys. 36*, 3401.
209. Nemethy, G., Scheraga, H. A., & Kauzmann, W. (1968) *J. Phys. Chem. 72*, 1842.
210. Nemethy, G., Peer, W. J., & Scheraga, H. A. (1981) *Annu. Rev. Biophys. Bioeng. 10*, 459.
211. Neurath, H., *et al.* (1944) *Chem. Rev. 34*, 157.
212. Nojima, H., *et al.* (1977) *J. Mol. Biol. 116*, 429.
213. Northrup, J. H. (1932) *J. Gen. Physiol. 16*, 323.
214. Novotny, J., Bruccoleri, R. E., & Karplus, M. (1984) *J. Mol. Biol. 177*, 787.
215. Novotny, J., Rashin, A. A., & Bruccoleri, R. E. (1988) *Proteins 4*, 19.
216. Nozaki, Y., & Tanford, C. (1970) *J. Biol. Chem. 245*, 1698.
217. Nozaki, Y., & Tanford, C. (1971) *J. Biol. Chem. 246*, 2211.
218. Ooi, T., & Oobatake, M. (1988) *J. Biochem. (Tokyo) 103*, 114.
219. Ostermann, J., *et al.* (1989) *Nature 341*, 125.
220. Pace, C. N. (1975) *CRC Crit. Rev. Biochem. 3*, 1.
221. Pace, C. N., & Tanford, C. (1968) *Biochemistry 7*, 198.
222. Pace, C. N., & Grimsley, G. R. (1988) *Biochemistry 27*, 3242.
223. Pace, C. N., Grimsley, G. R., Thomson, J. A., & Barnett, B. J. (1988) *J. Biol. Chem. 263*, 11820.
224. Padmanabhan, S., *et al.* (1990) *Nature 344*, 268.
225. Pangali, C., Rao, M., & Berne, B. J. (1982) *J. Chem. Phys. 71*, 1982.
226. Parodi, R. M., Bianchi, E., & Ciferri, A. (1973) *J. Biol. Chem. 248*, 4047.
227. Pauling, L. (1960) in *The Nature of the Chemical Bond*, 3rd ed., Cornell University Press, Ithaca, NY.
228. Pauling L., & Corey, R. B. (1951a) *Proc. Natl. Acad. Sci. U.S.A. 37*, 235.
229. Pauling, L., & Corey, R. B. (1951b) *Proc. Natl. Acad. Sci. U.S.A. 37*, 251.
230. Pauling, L., & Corey, R. B. (1951c) *Proc. Natl. Acad. Sci. U.S.A. 37*, 272.
231. Pauling, L., & Corey, R. B. (1951d) *Proc. Natl. Acad. Sci. U.S.A. 37*,729.
232. Pauling, L., Corey, R. B., & Branson, H. R. (1951) *Proc. Natl. Acad. Sci. U.S.A. 37*, 205.

233. Peller, L. (1959) *J. Phys. Chem. 63*, 1194.
234. Perutz, M. F., & Lehmann, H. (1968) *Nature 219*, 902.
235. Perutz, M. F., & Raidt, H. (1975) *Nature 255*, 256.
236. Perutz, M. F., Kendrew, J. C., & Watson, H. C. (1965) *J. Mol. Biol. 13*, 669.
237. Piela, L., Nemethy, G., & Scheraga, H. A. (1987) *Biopolymers 26*, 1273.
238. Platzer, K. E. B., *et al.* (1972) *Macromolecules 5*, 177.
239. Poland, D. C., & Scheraga, H. A. (1965) *Biopolymers 3*, 379.
240. Poland, D. C., & Scheraga, H. A. (1970) *Theory of the Helix-Coil Transition*, Academic Press, New York.
241. Ponder, J. W., & Richards, F. M. (1987) *J. Mol. Biol. 193*, 775.
242. Post, C. B., & Zimm, B. H. (1979) *Biopolymers 18*, 1487.
243. Presnell, S. R., & Cohen, F. E. (1989) *Proc. Natl. Acad. Sci. U.S.A. 86*, 6592.
244. Privalov, P. L. (1979) *Adv. Protein Chem. 33*, 167.
245. Privalov, P. L. (1989) *Annu. Rev. Biophys. Biophys. Chem. 18*, 47.
246. Privalov, P. L., & Kechinashvili, N. N. (1974) *J. Mol. Biol. 86*, 665.
247. Privalov, P. L., & Gill, S. J. (1988) *Adv. Protein Chem. 39*, 191.
248. Privalov, P. L., *et al.* (1986) *J. Mol. Biol. 190*, 487.
249. Privalov, P. L., Gill, S. J., & Murphy, K. P. (1990) *Science* (in press).
250. Ptitsyn, O. B. (1987) *J. Protein Chem. 6*, 273.
251. Qian, N., & Sejnowski, T. J. (1988) *J. Mol. Biol. 202,* 865.
252. Radzicka, A., & Wolfenden, R. (1988) *Biochemistry 27*, 1664.
253. Rao, S. P., Carlstrom, D. E., & Miller, W. G. (1974) *Biochemistry 13*, 943.
254. Ravishanker, G., Mezei, M., & Beveridge, D. L. (1982) *Faraday Symp. Chem. Soc. 17*, 79.
255. Rich, D. H., & Jasensky, R. D. (1980) *J. Am. Chem. Soc. 102*, 1112.
256. Richards, F. M. (1977) *Annu. Rev. Biophys. Bioeng. 6*, 151.
257. Richards, F. M., & Richmond, T. (1978) in *Molecular Interactions and Activity in Proteins*, (Wolstenholme, G. E., Ed.) p. 23, Ciba Foundation Symposium 60, Excerpta Medica, Amsterdam.
258. Richardson, J. S., & Richardson, D. C. (1987) in *Protein Engineering* (Oxender, D. L., & Fox, C. F., Eds.) p. 149, Alan R. Liss, New York.
259. Richardson, J. S., & Richardson, D. C. (1988) *Science 240,* 1648.
260. Robertson, T. B. (1918) *The Physical Chemistry of the Proteins*, Longmans, Green and Co., New York.
261. Rogers, N. K. (1989) in *Prediction of Protein Structure and the Principles of Protein Conformation* (Fasman, G. D., Ed.) p. 359, Plenum, New York.
262. Rooman, M. J., & Wodak, S. J. (1988) *Nature 335*, 45.
263. Rose, G. D., *et al.* (1985a) *Science 229*, 834.
264. Rose, G. D., Gierasch, L. M., & Smith, J. A. (1985b) *Adv. Protein Chem. 37*, 1.
265. Roseman, M. A. (1988) *J. Mol. Biol. 201*, 621.
266. Rothman, J. E. (1989) *Cell 59*, 591.
267. Sanchez, I. C. (1979) *Macromolecules 12*, 980.
268. Sandberg, W. S., & Terwilliger, T. C. (1989) *Science 245*, 54.
269. Santoro, M. M., & Bolen, D. W. (1988) *Biochemistry 27*, 8063.
270. Scheiner, S., & Hillenbrand, E. A. (1985) *Proc. Natl. Acad. Sci U.S.A. 82*, 2741.
271. Scheiner, S., Redfern, P., & Hillenbrand, E. A. (1986) *Int. J. Quant. Chem. 29*, 817.
272. Schellman, C., & Schellman, J. A. (1958) *C. R. Lab. Carlsberg, Ser. Chim. 30*, 463.
273. Schellman, J. A. (1955) *C. R. Trav. Lab. Carlsberg, Ser. Chim. 29*, 223.
274. Schellman, J. A. (1958a) *J. Phys. Chem. 62*, 1485.
275. Schellman, J. A. (1958b) *C. R. Trav. Lab. Carlsberg, Ser. Chim. 30*, 450.
276. Schrier, M. Y., & Schrier, E. E. (1976) *Biochemistry 15*, 2607.
277. Schulz, G. E., & Schirmer, R. H. (1979) in *Principles of Protein Structure*, Springer-Verlag, New York.
278. Semlyen, J. A. (1976) in *Advances in Polymer Science* (Cantow, H. J., *et al.*, Eds.) Vol. 21, Springer, Berlin.
279. Shakhnovich, E. I., & Finkelstein, A. V. (1989a) *Biopolymers 28*, 1667.
280. Shakhnovich, E. I., & Finkelstein, A. V. (1989b) *Biopolymers 28*, 1681.
281. Shoemaker, K. R., *et al.* (1985) *Proc. Natl. Acad. Sci. U.S.A. 82*, 2349.
282. Shoemaker, K. R., *et al.* (1987) *Nature 326*, 563.
283. Shoemaker, K. R., *et al.* (1990) *Biopolymers 29*, 1.

284. Shortle, D., & Meeker, A. K. (1989) *Biochemistry 28*, 936.
285. Shortle, D., Meeker, A. K., & Freire, E. (1988) *Biochemistry 27*, 4761.
286. Sikorski, A., & Skolnick, J. (1989) *Proc. Natl. Acad. Sci. U.S.A. 86*, 2668.
287. Singer, S. J. (1962) *Adv. Protein Chem. 17*, 1.
288. Skolnick, J., Kolinski, A., & Yaris, R. (1988) *Proc. Natl. Acad. Sci. U.S.A. 85*, 5057.
289. Skolnick, J., Kolinski, A., & Yaris, R. (1989) *Proc. Natl. Acad. Sci. U.S.A. 86*, 1229.
290. Sneddon, S. F., Tobias, D. J., & Brooks, C. L., III (1989) *J. Mol. Biol. 209*, 817.
291. Srinivasan, R. (1976) *Indian J. Biochem. Biophys. 13*, 192.
292. Stigter, D., & Dill, K. A. (1989) *J. Phys. Chem.* 93, 6737.
293. Stigter, D., & Dill, K. A. (1990) *Biochemistry 29*, 1262.
294. Stigter, D., Alonso, D. O. V., & Dill, K. A. (1990) (submitted for publication).
295. Stillinger, F. H. (1980) *Science 209*, 451.
296. Sturtevant, J. M. (1977) *Proc. Natl. Acad. Sci. U.S.A. 74*, 2236.
297. Sueki, M., et al. (1984) *Macromolecules 17*, 148.
298. Susi, H. (1969) in *Structure and Stability of Biological Macromolecules* (Timasheff, S. N., & Fasman, G. D, Eds.) Marcel Dekker, New York.
299. Susi, H., & Ard, J. S. (1969) *J. Phys. Chem. 73*, 2440.
300. Sweet, R. M., & Eisenberg, D. (1983) *J. Mol. Biol. 171*, 479.
301. Tadokoro, H. (1979) in *Structure of Crystalline Polymers*, Wiley, New York.
302. Tanford, C. (1961) in *Physical Chemistry of Macromolecules*, Chapter 7, Wiley, New York.
303. Tanford, C. (1962) *J. Phys. Chem. 84,* 4240.
304. Tanford, C. (1968) *Adv. Protein Chem. 23*, 121.
305. Tanford, C. (1970) *Adv. Protein Chem. 24*, 1.
306. Tanford, C. (1979) *Proc. Natl. Acad. Sci. U.S.A. 76*, 4175.
307. Tanford, C. (1980) in *The Hydrophobic Effect*, 2nd Ed., Wiley, New York.
308. Tanford, C., & Kirkwood, J. G. (1957a) *J. Am. Chem. Soc. 79*, 5333.
309. Tanford, C., & Kirkwood, J. G. (1957b) *J. Am. Chem. Soc. 79*, 5340.
310. Taniuchi, H., & Anfinsen, C. B. (1969) *J. Biol. Chem. 244*, 3864.
311. Thornton, J. (1988) *Nature 335*, 10.
312. Vinogradov, S. N., & Linnell, R. H. (1971) in *Hydrogen Bonding*, Van Nostrand Reinhold, New York.
313. von Hippel, P. H., & Schleich, T. (1969a) in *Structure and Stability of Biological Macromolecules* (Timasheff, S., & Fasman, G., Eds.) Vol. 11, p. 417, Dekker, New York.
314. von Hippel, P. H., & Schleich, T. (1969b) *Acc. Chem. Res. 2*, 257.
315. Wada, A., & Nakamura, H. (1981) *Nature 293*, 757.
316. Weiner, S., *et al.* (1984) *J. Am. Chem. Soc. 106*, 765.
317. Wertz, D. H., & Scheraga, H. A. (1978) *Macromolecules 11*, 9.
318. White, S. H., & Jacobs, R. E. (1990) *Biophys. J.* (in press).
319. Wright, P. E., Dyson, H. J., & Lerner, R. A. (1988) *Biochemistry 27*, 7167.
320. Wu, H. (1929) *Am. J. Physiol. 90*, 562.
321. Wu, H. (1931) *Chin. J. Physiol. 5*, 321.
322. Wu, H., & Wu, D. (1931) *Chin. J. Physiol. 5*, 369.
323. Yutani, K., et al. (1984) *J. Biol. Chem. 259*, 14076.
324. Yutani, K., et al. (1987) *Proc. Natl. Acad. Sci. U.S.A. 84*, 4441.
325. Zhang, R., & Snyder, G. H. (1989) *J. Biol. Chem. 264*, 18,472.
326. Zimm, B. H., & Bragg, J. K. (1959) *J. Chem. Phys. 31*, 526.
327. Zimm, B. H., & Rice, S. A. (1960) *Mol. Phys. 3*, 391.
328. Zimm, B. H., *et al.* (1959) *Proc. Natl. Acad. Sci. U.S.A. 45*, 1601.
329. Zipp, A., & Kauzmann, W. (1973) *Biochemistry 12*, 4217.

PROCESS RATIONALE

HYDROPHOBIC INTERACTION CHROMATOGRAPHY

Pharmacia, Piscataway, NJ, ph. (800) 526-3593. (Adapted).

1. Introduction

A classical 1948 paper "Adsorption Separation by Salting Out," by Tiselius laid the foundation for a separation method, which is now called hydrophobic interaction chromatography (HIC). He noted that proteins and other substances which are precipitated at high concentrations of neutral salts (salting out), are often adsorbed quite strongly in salt solutions of lower concentration than required for their precipitation, and some adsorbents which at increased salt concentrations become excellent adsorbents." Since then, great strides have been made in developing almost ideal stationary phases for chromatography (e.g., cellulose, cross-linked dextran, Sephadex), cross-linked agarose (Sepharose CL, Sepharose High Performance and Sepharose Fast Flow), and in developing coupling methods for immobilizing ligands of choice to such matrices. It was a combination of these two events, which, in the beginning of the 1970s, led to the synthesis of a variety of hydrophobic adsorbents for biopolymer separations based on this previously rarely exploited principle.

The first attempt at synthesizing such adsorbents was made by Yon followed by Er-el *et al.,* Hofstee and Shaltiel and Er-el. Characteristically, these early adsorbents showed a mixed ionic-hydrophobic character. Despite this, Halperin *et al.* claimed that protein binding to such adsorbents was predominantly of a hydrophobic character. Porath *et al.* and Hjertén *et al.* later synthesized charge-free hydrophobic adsorbents and demonstrated that the binding of proteins was enhanced by high concentrations of neutral salts, as previously observed by Tiselius, and that elution of the bound proteins was achieved simply by washing the column with salt-free buffer or by decreasing the polarity of the eluent. Pharmacia was first in producing commercial HIC adsorbents (Phenyl and Octyl Sepharose CL-4B) of the charge-free type and has continuously followed this up with new developments in agarose matrix design by introducing new stable HIC media based on Superose, Sepharose Fast Flow, and Sepharose High Performance, meeting various demands on chromatographic productivity, selectivity, and efficiency.

The principles that govern protein adsorption to HIC media are complementary to those of ion-exchange chromatography and gel filtration. HIC is even sensitive enough to be influenced by nonpolar groups – which are normally buried within the tertiary structure of proteins but exposed if the polypeptide chain is incorrectly folded or damaged (e.g., by proteases). This sensitivity can be useful for separating the pure native protein from other forms.

Protein binding to HIC adsorbents is promoted by moderately high concentrations of antichaotropic salts, which also have a stabilizing influence on protein structure. Elution is achieved by a linear or stepwise decrease in the concentration of salt in the adsorption buffer. Recoveries are often very satisfactory.

2. Principles of Hydrophobic Interaction Chromatography

The many theories that have been proposed for HIC are essentially based upon those derived for interactions between hydrophobic solutes and water, but none of them has enjoyed universal acceptance. What is common to all is the central role played by the structure-forming salts and the effects they exert on the individual components (i.e., solute, solvent, and adsorbent) of the chromatographic system to bring about the binding of solute to adsorbent. In view of this, Porath proposed "salt-promoted adsorption" as a general concept for HIC and other types of solute-adsorbent interactions occurring in the presence of moderately high concentrations of neutral salts.

Hofstee and later Shaltiel proposed "hydrophobic chromatography" with the implicit assumption that the mode of interaction between proteins and the immobilized hydrophobic ligands is similar to the self-association of small aliphatic organic molecules in water. Porath *et al.* suggested a salting-out effect in hydrophobic adsorption, thus extending the earlier observations of Tiselius. They also suggested that "the driving force is the entropy gain arising from structure changes in the water surrounding the interacting hydrophobic groups." This concept was later extended and formalized by Hjertén, who based his theory on the well known thermodynamic relationship:

$$\Delta G = \Delta H - T\Delta S$$

He proposed that the displacement of the ordered water molecules surrounding the hydrophobic ligands and the proteins leads to an increase in entropy (ΔS) resulting in a negative value for the change in free energy (ΔG) of the system. This implies that the hydrophobic ligand-protein interaction is thermodynamically favorable, as is illustrated in Figure 1.

An alternative theory is based on the parallelism between the effect of neutral salts in salting out

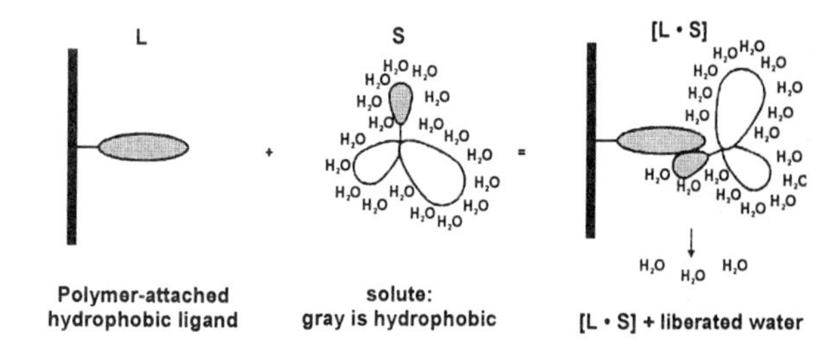

Figure 1 - Close to the surface of the hydrophobic ligand and solute (L and S), the water molecules are more highly ordered than in the bulk water and appear to "shield off" the hydrophobic ligand and solute molecules. Added salt interacts strongly with the water molecules, leaving less water available for the "shield off" effect, which is the driving force for L and S to interact with each other.

(precipitation) and HIC. According to Melander and Horvath, hydrophobic interaction is accounted for by an increase in the surface tension of water arising from the structure-forming salts dissolved in it. In fact, a combination of these two mechanisms seems to be an obvious extension and has been exploited long before HIC adsorbents were synthesized. Finally, Srinivasan and Ruckenstein have proposed that HIC is due to van der Waals attraction forces between proteins and immobilized ligands. The basis for this theory is that the van der Waals attraction forces between protein and ligand increase as the ordered structure of water increases in the presence of salting out salts.

2.1 HIC versus Reverse-Phase Chromatography. In theory, HIC and reverse-phase chromatography (RPC) are closely related LC techniques. Both are based upon interactions between solvent-accessible nonpolar groups (hydrophobic patches) on the surface of biomolecules and the hydrophobic ligands (alkyl or aryl groups) covalently attached to the gel matrix. In practice, however, they are different. Adsorbents for RPC are more highly substituted with hydrophobic ligands than HIC adsorbents. The degree of substitution of HIC adsorbents is usually in the range of 10-50 μmol/ml of gel composed of C_2-C_8 alkyl or simple aryl ligands, compared with several hundred μmol/ml of gel containing C_4-C_{18} alkyl ligands. Consequently, protein binding to RPC adsorbents is usually very strong, necessitating the use of nonpolar solvents to induce their elution. RPC has found extensive applications in analytical and preparative separations of mainly peptides and low molecular weight proteins that are stable in aqueous-organic solvents.

Compared with RPC, the polarity of the complete system of HIC is increased by decreased ligand density on the stationary phase and by adding salt to the mobile phase. The type of immobilized ligand (alkyl or aryl) determines the protein adsorption selectivity of the HIC adsorbent. In general, straight chain alkyl (hydrocarbon) ligands show "pure" hydrophobic character while aryl ligands show a mixed-mode behavior, in which both aromatic and hydrophobic interactions are possible (Figure 2). It is also established that at a constant degree of substitution, the protein binding capacities of HIC adsorbents increase with increased alkyl chain length. The charged type HIC adsorbents show an additional mode of interaction, which will not be discussed here. The choice between alkyl or aryl ligands is empirical and must be established by screening experiments for each individual separation problem.

HIC media from Pharmacia are all based on the glycidyl ether coupling procedure, which produces gels that are charge-free and that should thus only have hydrophobic interactions with proteins. The phenyl group also has a potential for π-π interactions. The glycidyl ether coupling technique will introduce a short spacer but the effect of this will be very limited since the short hydrophobic chain is "neutralized" with the hydrophilic OH-group.

2.2 Degree of Substitution. The protein binding capacities of HIC adsorbents increase with increased degree of substitution of immobilized ligand (Figure 2.) At a sufficiently high degree of ligand substitution, the apparent binding capacity of the adsorbent remains constant (plateau is reached) but the strength of the interaction increases. Solutes bound under such circumstances are difficult to elute due to multipoint attachment.[34]

2.3 Type of Base Matrix. It is important not to overlook the contribution of the base matrix. The two most widely used types of support are strongly hydrophilic carbohydrates, e.g., cross-linked agarose,

A

$$-O-CH_2-\underset{}{\overset{OH}{CH}}-CH_2-O-(CH_2)_3-CH_3$$

B

$$-O-CH_2-\overset{OH}{CH}-CH_2-O-(CH_2)_7-CH_3$$

C

$$-O-CH_2-\overset{OH}{CH}-CH_2-O-C_6H_5$$

D

$$-O-CH_2-\overset{OH}{CH}-CH_2-O-CH_2-C(CH_3)$$

Figure 2 - Different hydrophobic ligands coupled to cross-linked agarose matrices.

or synthetic copolymer materials. The selectivity of a copolymer support will not be exactly the same as for an agarose-based support substituted with the same type of ligand. To achieve the same type of results on an agarose-based matrix as on a copolymer support, it may be necessary to modify adsorption and elution conditions.

2.4 Type and Concentration of Salt. The addition of various structure-forming ("salting out") salts to the equilibration buffer and sample solution promotes ligand-protein interactions in HIC. As the concentration of such salts is increased, the amount of proteins bound also increases almost linearly up to a specific salt concentration and continues to increase in an exponential manner at still higher concentrations. This latter phenomenon, where total binding capacity of Phenyl Sepharose High Performance, for α-chymotrypsinogen and RNase was examined at gradually increasing salt concentrations.

In this experiment, the column was first equilibrated with buffer containing varying concentrations of salt. The sample was dissolved in buffer including this initial salt concentration prior to application to the column. However, in those experiments where the protein begins to precipitate at high salt concentration (1.3 M and 2.3 M ammonium sulfate for α-chymotrypsinogen and RNase, respectively) the sample was dissolved at a slightly lower salt concentration.

The samples were loaded on the column until breakthrough could be observed at the column outlet. Then start buffer at the initial concentration was run through the column until UV-absorption in the eluent returned to the baseline. Finally, the bound proteins were eluted with a decreasing salt gradient. A significant increase in adsorption capacity can be seen when the salt concentration is increased above the precipitation point. This phenomenon is probably due to the precipitation of proteins on the column. It has a concomitant negative effect on the selectivity of the HIC adsorbent.

The effects of salts in HIC can be accounted for by reference to the *Hofmeister series* for the precipitation of proteins or for their positive influence in increasing the molal surface tension of water. These effects are summarized in Tables 1 and 2.

Table 1 - Hofmeister Series on the Effect of Some Anions and Cation in Precipitating Proteins

← Increasing precipitation ("salting out") effect

Cations: $NH4^+$, Rb^+, K^+, Na^+, Cs^+, Li^+, Mg^{2+}, Ca^{2+}, Ba^{2+}

Anions: PO_4^{3-}, SO_4^{2-}, CH_3COO^-, Cl^-, Br^-, NO_3^-, ClO_4^-, I^-, SCN^-

Increasing chaotropic effect ("salting in") →

Table 2 - Relative Effects of Some Salts on the Molal surface Tension of Water

$Na_2SO_4 > K_2SO_4 > (NH_4)_2SO_4 > Na_2HPO_4 > NaCl > LiCl \ldots > KSCN$

In both instances, sodium, potassium, or ammonium sulfates produce relatively higher "salting-out" (precipitation) or molal surface tension increment effects. It is also these salts that effectively promote ligand-protein interactions in HIC. Most of the bound proteins are effectively desorbed by simply washing the HIC adsorbent with water or dilute buffer solutions at near neutral pH.

The effect of pH in HIC is also not straightforward. In general, an increase in pH weakens hydrophobic interactions. probably as a result of increased titration of charged groups, thereby leading to an increase in the hydrophilicity of the proteins. On the other hand, a decrease in pH results in an apparent increase in hydrophobic interactions. Thus, proteins that do not bind to a HIC adsorbent at neutral pH bind at acidic pH. Hjertén *et al.* found that the retention of proteins changed more drastically at pH values above 8.5 and/or below 5 than in the range pH 5-8.5.

2.5 Effect of Temperature. Based on theories developed for the interaction of hydrophobic solutes in water. Hjertén proposed that the binding of proteins to HIC adsorbents is entropy driven:

$$\Delta G = \Delta H - T\Delta\Delta \approx - T\Delta S$$

which implies that the interaction increases with an increase in temperature. Experimental evidence for this effect has been presented by Hjertén and Jennissen. It is interesting to note that the van der Waals attraction forces, which operate in hydrophobic interactions, also increase with increase in temperature. However, an opposite effect was reported by Visser and Strating, indicating that the role of temperature in HIC is of a complex nature. This apparent discrepancy is probably due to the differential effects exerted by temperature on the conformational state of different proteins and their solubilities in aqueous solutions. In practical terms, one should thus be aware that a downstream purification process developed at room temperature might not be reproduced in the cold room, or *vice versa.*

2.6 Additives. Low concentrations of water-miscible alcohols, detergents, and aqueous solutions of chaotropic ("salting-in") salts result in a weakening of the protein-ligand interactions in HIC leading to desorption of bound solutes. The nonpolar parts of alcohols and detergents compete effectively with the bound proteins for the adsorption sites on the HIC media resulting in displacement of the latter. Chaotropic salts affect the ordered structure of water and/or that of the bound proteins. Both types of additives also decrease the surface tension of water (see Table 3), thus weakening the hydrophobic interaction to give a subsequent dissociation of the ligand-solute complex.

Although additives can be used in the elution buffer to affect selectivity during desorption, there is a risk that proteins could be denatured or inactivated by exposure to high concentrations of such chemicals. However, additives can be very effective in cleaning up HIC columns that have strongly hydrophobic proteins bound to the gel medium.

Table 3 - Physical Properties of Some Solvents Used in HIC (Data at 25°C).

Solvent	Viscosity (centripoise)	Dielectric Constant	Surface Tension (dynes/cm)
Water	0.89	78.03	72.00
Ethylene glycol	16.90	40.70	46.70
Dimethyl sulfoxide	1.96	46.70	43.54
Dimethyl formamide	0.79	36.71	36.76
n-propanol	2.00	20.33	23.71

3. Optimizing a HIC Step

The main purpose of optimizing a chromatographic step is to reach the predefined purity level with highest possible recovery by choosing the most suitable combination of the critical chromatographic parameters. In process applications, there is also a need to reach the highest possible throughput. The screening experiments outlined previously will mainly help in establishing the most suitable medium to use. The sections below will deal with some important guidelines for optimizing the critical operational parameters, which affect the maximum utilization of the HIC step. These parameters include type of buffer salt, salt concentration, buffer pH, temperature, bed height, flow rate, gradient shape, and gradient slope.

In conclusion, when purifying large molecules such as proteins, relatively short columns can be used if the selectivity of the adsorbent is exploited in an optimal way. The linear flow rate should, if required, be sufficiently reduced in order to optimize the kinetics of the adsorption and desorption processes. Also, this can be further enhanced by choosing a smaller bead size. Smaller beads will also provide the necessary increase in efficiency when more difficult separation problems are encountered.

3.1 The solvent. This is one of the most important parameters to have a significant influence on the binding capacity and selectivity of an HIC medium. In general, the adsorption process is often more selective than the desorption process and it is therefore important to optimize the starting "binding" buffer conditions with respect to critical parameters such as pH, type of salt, concentration of salt, and temperature. The combination of salt and pH can be manipulated to give optimum selectivity during purification by HIC. Optimal conditions differ from application to application and are best established by running linear gradients and varying the parameters in a controlled way (for example by using "factorial" design). Changes of temperature and pH are sometimes restricted by the stability of the substance of interest or by system constraints, etc., but may often be of interest to evaluate. The Hofmeister series (Table 1) gives important guidelines in choosing the type of salt to use. The most efficient salts are normally ammonium sulfate and sodium sulfate but also "weaker" salts such as sodium chloride should be considered. In an ideal situation, the correct choice of salt and salt concentration will result in the selective binding of the protein of interest while the majority of the impurities pass through the column unretarded. If the protein of interest binds weakly to the column, an alternative approach is to choose the starting buffer conditions, which will result in the maximum binding of a large proportion of the contaminating proteins but will allow the protein of interest to pass through unretarded. An extension of this strategy is to increase the salt concentration in the unbound fraction to such an extent that the protein of interest binds to the same column in a second run while most of the impurities pass through the column unretarded. The column of Alkyl Superose was equilibrated with varying concentrations of ammonium sulfate (2 M to 0.8 M) and its selectivity for the IgG_1 investigated. The results show that high selectivity for IgG_1 is obtained using 1 M ammonium sulfate in the binding buffer.

It should be pointed out that the higher the salt concentration in the equilibration buffer, the greater the risk that some of the proteins in the sample will precipitate. Since such precipitates can clog tubing and column filters, the sample must be filtered or centrifuged. This extra step can be avoided by equilibrating the sample in a lower salt concentration than is required for its precipitation and then applying it to a column that is equilibrated with a higher salt concentration. Some of the proteins will precipitate on the column (zone precipitation) but they redissolve upon reduction of the salt concentration during stepwise or gradient elution.

References:
(*Note*: Only references retained in the adapted version are given, using the original numbering.)
8. On the mode of adsorption of proteins to "hydrophobic columns." Biochem. Biophys. Res. Commun. 72 (1976) 108-113, Wilchek, M., Miron, T.
10. Salting-out in amphiphilic gels as a new approach to hydrophobic adsorption. Nature 245

(1973) 465-466, Porath, J., Sundberg, L., Fornstedt, N., Olson, I.

11. Hydrophobic interaction chromatography. The synthesis and use of some alkyl and aryl derivatives of agarose. J. Chromatogr. 101 (1974) 281-288, Hjertén, S., Rosengren, J., Påhlman, S.
12. Hydrophobic interaction chromatography on Phenyl- and Octyl-Sepharose CL-4B. in *Chromatography of Synthetic and Biological Macromolecules*. Roger, E., Ed., Ellis Horwood Ltd., Chichester, England, 1978. Janson, J. C., Låås, T.
13. Hydrophobic interaction chromatography of serum proteins on Phenyl-Sepharose Cl-4B. J. Chromatog. 242 (1982) 385-388, Hrkal, Z., Rejnkova, J.
15. Nuclear proteins. VI. Fractionation of chromosomal non-histone proteins using hydrophobic chromatography. Biochem. Biophys. Acta 563 (1979) 253-260, Comings, D. E., Miguel, A. G., Lesser, H. H.
17. Hydrophobic interaction chromatography of proteins, nucleic acids, viruses and cells on non-charged amphiphilic gels, in: *Methods of Biochemical Analysis* (D. Glick, Ed.), John Wiley & Sons, Inc., 1981, pp. 89-108. Hjertén, S.
20. Utilization of Hydrophobic interaction for the formation of an enzyme reactor bed. Biotechnology & Bioengineering 17 (1975) 613-616, Caldwell, K. D., Axén, R., Porath, J.
24. Salt-promoted adsorption: recent developments. J. Chromatog. 376 (1986) 331-341, Porath, J.
39. Temperature-dependent van der Waals forces. Biophys. J. 10 (1970) 664-674, Parsegian, V. A., Ninham, B. W.
41. Some general aspects of hydrophobic interaction chromatography. J. Chromatog. 87 (1973, 325-331, Hjertén, S.
46. Characterization of hydrophobic interaction and Hydrophobic Interaction Chromatography media by multivariate analysis. J. Chromatog. 599 (1992) 131-136, Kårsnås, P., Lindblom, T.
47. Differences in retention behavior between small and large molecules in Ion Exchange Chromatography and Reversed Phase Chromatography. Anal. Biochem. 142 (1984) 134-139, Ekström, B., Jacobson, G.
50. Expression, Purification and Crystallization of the HIV-1 Reverse Transcriptase (RT). AIDS Res. Hum. Retrovir. 6 (1990) 1297-1303, Unge, T., Ahola, H., Bhikhabhai, R., Bäckbro, K., Lövgren, S., Fenyö, E. M., Honigman, A., Panet, A., Gronowitz, G. S., Strandberg, B.
51. Factors involved in specific transcription by mammalian RNA polymerase II. Identification and characterization of Factor II. H., J. Biol. Chem. 267 (1992) 2786-2793, Flores, O., Lu, H., Reinberg, D.
54. Phenyl-Sepharose-mediated Detergent-Exchange Chromatography: Its application to exchange of detergents bound to membrane proteins. Biochemistry 23 (1984) 6121-6126, Robinson, N. C., Wiginton, D., Talbert, L.
55. Characterization of the dermatan sulfate proteoglycans, DS-PGI and DS-PGII, from bovine articular cartilage and skin isolated by Octyl Sepharose chromatography. J. Biol. Chem. 264 (1989) 2876-2884, Choi, H. U., Johnson, T. L., Pal, S., Tang, L. H., Rosenberg L., Neame, P. J.
59. Production of clinical grade recombinant Exotoxin A from *E. coli.* in *Recovery of Biological Products VI, An Engineering Foundation Conference,* Interlachen, Switzerland, September, (1992) Tsai, A. M., Kaufman, J. B., Shiloach, J., Gallo, M., Fass, S.
60. Recombinant toxins for cancer treatment. Science 254 (1991) 1173-1177, Pastan, I., FitzGerald, D.
61. Structure of exotoxin A of *Pseudomonas aeruginosa* at 3.0-Ångström resolution. Proc. Natl. Acad. Sci. USA 83 (1986) 1320-1324, Allured, V. S., Collier, R. J., Carroll, S. F., McKay, D. B.
62. Functional domains of *Pseudomonas* exotoxin identified by deletion analysis of the gene expressed in *E. coli.* Cell 48 (1987) 129-136, Hwang, J., FitzGerald, D. J., Adhya, S., Pastan, I.
63. Functional analysis of domains II, Ib, and III of *Pseudomonas* exotoxin. J. Biol. Chem. 264 (1989) 14256-14261, Siegall, C. B., Chaudhary, V. K., FitzGerald, D. J., Pastan, I.
64. Anti-tumor activities of immunotoxins made of monoclonal antibody B3 and various forms of *Pseudomonas* exotoxin. Proc. Natl. Acad. Sci. USA 88 (1991) 3358-3362, Pai, L. H., Batra, J. K., FitzGerald, D. J., Willingham, M. C., Pastan, I.
65. Solvent modulation in Hydrophobic Interaction Chromatography. Biotechnol. Appl. Biochem. 13 (1991) 151-172, Arakawa, T., Narhi, L. O.
66. Hydrophobic Interaction Chromatography in alkaline pH. Anal. Biochem. 182 (1989) 266-270, Narhi, L. O., Kita, Y., Arakawa, T.

Alternative Approach

Centriprep Microconcentrators for Small Volume Concentration; Centricon-3 and Centricon-100

Amicon™, Beverly, MA, ph. (508) 777-3622. (Adapted)

1. Introduction

1.1 Product Description. Centriprep concentrators are disposable ultrafiltration devices for purifying, concentrating, and desalting biological samples in the 5-15 ml volume range. These complete, ready-to-use ultrafiltration devices are designed for operation in most centrifuges that can accommodate 50-ml centrifuge tubes. Concentrator components include a sample container with a twist-lock cap, a filtrate collector containing an Amicon YM membrane, plus an air-seal cap for sample isolation.

The main features are ease of use, high flow rate, and low adsorption. Using the Centriprep-30 device, a 15-ml sample can be concentrated to 0.7 ml in just 30 minutes. With smaller volumes, concentrating proteins takes less time. The YM membrane"s low adsorptivity leads to high concentration factors. Greater than 90% recovery of retained macromolecular solutes is typical (see Table 2). To prevent filtration to complete dryness, a built-in deadstop provides a final concentrated sample volume of 0.6-0.7 ml.

1.2 Molecular Weight Cutoffs. Centriprep-3, 3,000 Daltons; Centriprep-10, 10,000 Daltons; Centriprep-30, 30,000 Daltons; Centriprep-100, 100,000 Daltons.

1.3 Applications. Centriprep-3 concentrators can be used to concentrate, and desalt oligonucleotides, peptides, growth factors and small proteins. Some applications for the Centriprep-10 and -30 concentrators are (*1*) concentrating and desalting column eluates and gradient fractions; (*2*) recovering biomolecules from cell culture supernatants, lysates, extracts, or other biological samples; and (*3*) purifying low-molecular-weight components (e.g., amino acids and antibiotics) from physiological fluids, cell culture media, or fermentation broths. The Centriprep-100 device can be used for separating free ligand from antibody conjugates and to concentrate virus preparations.

1.4 Product Advantages. Centriprep concentrators offer many advantages over concentration techniques, such as chemical precipitation, evaporation and lyophilization. Set-up time with the Centriprep devices is minimal; the device is simply spun until the desired concentration is achieved. In addition, the filtration process itself is gentle, avoiding potential problems such as sample denaturation and concentration of buffer salts. Since concentration and desalting occur simultaneously within the Centriprep device, sample losses due to material transfer are eliminated.[1] The Centriprep concentrators all-in-one design also provides added safety when working with hazardous samples; since all materials are contained within the Centriprep device, operator exposure to biohazards is minimized.

2. Principle of Operation

To begin operation, the twist-lock cap of the sample container is loosened and the filtrate collector removed. Sample is added, then the filtrate collector is carefully reinserted back into the sample container (displacing solution) and the cap locked, sealing the device. The assembled concentrator is centrifuged at 500-3000 x *g* to the concentration factor desired. Centrifuges with swinging bucket or fixed-angle rotors may be used.

Immersing the filtrate collector in the sample solution creates a slight hydrostatic pressure differential, which exerts an upward buoyancy force on the membrane at the base of the filtrate collector. By itself, the hydrostatic pressure exerted by the displaced solution is too weak to produce ultrafiltration. Centrifugation increases this pressure, forcing low-molecular-weight materials and solvent through the membrane into the filtrate collector.[2] Solutes with molecular weights above the membrane cutoff remain in the sample container (retentate) and become increasingly concentrated as the operation continues. Ultrafiltration occurs in the direction opposite the centrifugal force vector (see Figure 1).

The pressure differential created by the raised solution level is important in concentrator operation. During centrifugation, the sample solution meniscus falls as the filtrate meniscus rises. In the process, the filtrate collector loses its buoyancy and sinks to the bottom of the sample container, permitting maximum filtrate collection. Eventually, an equilibrium is reached where the menisci reach equal height. Filtration then stops, since the hydrostatic pressure difference is now zero. For *g*-force and centrifugation time guidelines, see Table 1.

To resume concentration, filtrate is decanted before spinning the device for a second time. Decanting reestablishes the pressure differential between the sample meniscus and the filtrate meniscus, allowing filtration to resume. Filtration continues until a new equilibrium point is achieved. Note that the pressure differential is constantly changing during operation as the menisci rise and fall.

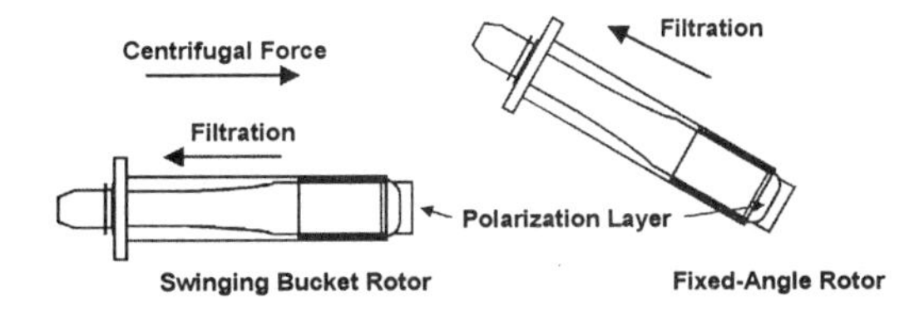

Figure 1 - Polarization control of protein concentration procedure (see text).

Centriprep concentrators are designed to maintain high flow rates by minimizing solute build-up on the membrane during operation. Centrifugal force causes dense materials to sink away from the membrane to the sample container bottom.[3] Because the Centriprep membrane "floats" above this polarization layer, the full surface of the membrane is utilized at all times and filtrate flow is unrestricted. This is particularly beneficial when working with suspensions, which can easily foul a membrane.

3. Operation

1. Turn twist-lock cap on sample container counterclockwise, then slide filtrate collector assembly out and set it on its side. Add solution to the sample container (the line on the side of the container shows the maximum volume, 15 ml).

CAUTION: While filling the sample container, the membrane surface at the bottom of the filtrate collector is exposed. Be careful not to touch, scratch, or damage the membrane when handling.

NOTE: Solutions with High Solids Content. For solutions such as cell suspensions, a starting volume over 5 ml will result in reduced flow rates (see *"Performance")*. For best results when working with a sample having over 10% solids, limit the initial sample volume to 5 ml.

2. Make sure the twist-lock cap seats fully onto the shoulder of the filtrate collector. If necessary, slide the cap downward until it stops at the shoulder. *Be careful not to scratch or damage the membrane surface!*
3. Carefully insert the capped filtrate collector into the sample container, gently pushing it down so the collector displaces solution. Turn the twist-lock cap clockwise to seal the sample container. Finally, make sure the air-seal cap is snug on the twist-lock cap.
4. Insert assembled concentrator into the centrifuge. The carrier should swing freely to a horizontal position and not touch the rotor or the trunnion ring. (For added clearance, Centriprep concentrators may be used in adaptors without rubber cushions.)

CAUTION: Inspect swinging-bucket rotors for proper clearance before centrifuging. Any obstruction may result in damage to the concentrator and possible loss of sample.

5. Spin concentrator at appropriate g-force until the fluid levels inside and outside the filtrate collector equilibrate. See Table 1 for guidelines on achieving various concen-trations.
6. To continue concentrating after equi-libration, remove the device from the centrifuge and snap off the air-seal cap. With vent groove oriented upward, decant the filtrate; replace cap and spin device a second time.
7. After the second spin, decant the remaining filtrate. If further concentration is desired, spin the device again; otherwise, proceed to step 8.
8. Loosen the twist-lock cap (turn counter-clockwise) and remove the filtrate collector. Using a pipette, withdraw the sample or simply pour the concentrate into a suitable container.

3.1 Samples. Centriprep-3: cytochrome *c*, 0.25 mg/ml; Centriprep-10 and -30: bovine serum albumin, 1 mg/ml; Centriprep-100: gamma globulin, 1 mg/ml.

3.2 g-Forces. Centriprep-3 and -10: 3000 x *g*; Centriprep-30: 1500 x *g*; Centriprep-100: 500 x *g*. Swinging-bucket or fixed-angle rotor, equilibrated at 25°C.

3.3 Limitations. Maximum Centrifugal Force: Centriprep-3 and -10 devices, 3000 x *g*; Centriprep-30 device, 1500 x *g*; Centriprep-100 devices, 500 x *g*.

Operation at relative centrifugal forces above these limits is not recommended.

Table 1 - Centrifugation Guidelines for Dilute Protein Solutions.

Starting Volume (ml)	Spin Time (minutes) and Retentate Volume (ml)		
Centriprep-3	Spin #1	Spin #2	Spin #3
5	50 1.00	10 0.65	10 0.60
10	65 1.90	10 0.90	10 0.65
15	95 3.00	35 0.90	10 0.65
Centriprep-10	Spin #1	Spin #2	Spin #3
5	15 0.95	10 0.60	N/A
10	30 1.70	10 0.70	5 0.60
15	40 3.00	10 0.90	5 0.65
Centriprep-30	Spin #1	Spin #2	Spin #3
5	5 1.10	5 0.65	5 0.60
10	10 1.85	5 0.70	5 0.60
15	15 3.00	10 0.85	5 0.65

Table 2 - Typical Retentate Recovery.

		% Recovery per Model			
Solute and Starting Concentration	Nominal MW	3	10	30	50
Vitamin B_{12} (0.2 mg/ml)	1,355	15	5	5	5
Protamine sulfate (1 mg/ml)	5,000-10,000	80	45	5	5
Cytochrome *c* (0.25 mg/ml)	12,400	90	85	15	5
β-Lactoglobin A (1 mg/ml)	18,700	95	90	80	5
α-Chymotrypsin-ogen (1 mg/ml)	24,500	90	90	75	5
Carbonic anhydrase (1 mg/ml)	30,000	95	90	75	5
Ovalbumin (1 mg/ml)	46,000	95	95	90	5
Bovine serum albumin (1 mg/ml)	67,000	95	90	90	50
Gamma globulin	160,000-	95	90	90	85

[With fixed-angle rotors, polarization control may be adversely affected at low g-forces (below 2000 x *g*). Depending upon sample composition and solute concentration, filtrate flow may be reduced.]

3.4 Prerinsing. The membranes in Centriprep-3, -10, and -30 concentrators contain trace amounts of glycerin. The membrane in the Centriprep-100 concentrator contains glycerin with low levels of sodium azide. If these materials interfere with analysis, rinse the device before use by centrifuging approximately 15 ml of buffer or deionized water until the first equilbration point is reached. Decant the filtrate and retentate (repeat rinsing if necessary), then begin operation with the sample solution (membranes must be kept wet to remain functional).

3.5 Leakage. Should leakage occur as a result of either membrane damage or a mechanical defect in the device seal, recombine the filtrate and concentrate, then repeat operation using a new filtrate collector.

3.6 Membrane Expansion. The dry YMT membrane used in the Centriprep-3, -10, and -30 devices swells upon first contact with aqueous solutions. After centrifugation, the concentrator membrane may therefore appear to be slightly swollen. In addition, a small amount of filtrate (approximately 0.5 ml) may remain in the area between the membrane and the support. This is normal and does not affect the operation or performance of the device.

3.7 Concentrator Reuse. Centriprep concentrators are intended for one-time use only! Attempts to clean the device could damage the membrane surface or its seal, possibly compromising the integrity of the unit and thereby affecting performance. For this reason, be sure to dispose of all Centriprep concentrators immediately after use.

4. Performance

Flow rates and recovery characteristics of Centriprep concentrators have been tested with several well-known solutes. These results, which are summarized in the following tables, may be used to estimate performance with other solutes. Relative centrifugal force (RCF) is measured at the filtrate collector base and is calculated thus:

$$\mathrm{RCF} = 1.118 \times 10^{-5} \times (r)\,\mathrm{RPM}^2$$

where r is the radius, the distance in centimeters measured from the center of rotation to the base of the filtrate collector. RCF and RPM are not the same; 3000 x *g* is not equivalent to 3000 rpm. Check your centrifuge operating guide for instructions on converting g-force to RPM, or use the above formula.

4.1 Flow Rate. The filtration rate is affected by several operating parameters, including sample concentration, starting volume, relative centrifugal force, type of rotor used, membrane cut-off, and temperature. When concentrating dilute solutions, swinging-bucket and fixed-angle rotors yield comparable flow rates. For solutions containing over 10% solids, a starting volume greater than 5 ml typically results in reduced flow rates. A volume of 5 ml or less produces the fastest flow rates because this limits the amount of solids that can pack onto and bind the membrane.

Spin times must be lengthened when working at low temperatures. At 4°C, flow rates are approximately 1.5 times slower than at 25°C. Viscous solutions may also require longer spin times. For example, using a Centriprep-10 device, a 1 mg/ml BSA solution in 50% glycerin would take 5 times longer to concentrate than the identical protein in an aqueous buffer solution.

4.2 Retention and Recovery. The anisotropic, hydrophilic YM membranes in Centriprep concentrators are characterized by a nominal molecular weight cut-off, i.e., their ability to reject molecules above a specified molecular weight. Generally, solutes below the cut-off will pass through the membrane and into the filtrate; solutes above the cut-off will be retained.

Because it is an easy parameter to work with, molecular weight is used in rating "membrane perm-selectivity." Although based on results achieved with well-characterized proteins, this procedure is an arbitrary convenience and may not be valid for all solutes. For instance, retentivity may be greater for a globular molecule than for a linear molecule having the same nominal molecular weight. The effects of pH and ionic strength on a molecule's Stokes radius and behavior in a particular solution must also be considered when assessing solute retentivity. Low solute recovery in the retentate may indicate possible adsorptive losses and/or solute passage through the membrane. Adsorptive losses depend upon solute concentration, nature of solute (hydrophobic or hydrophilic), temperature, time of contact with component surfaces, sample composition and pH (see Tables 2 and 3).

15 ml starting volume; temperature, 25°C; swinging-bucket rotor.

Centriprep-3: *g*-force 3000 x *g*, 95 minute first spin, 35 minute second spin.

Centriprep-10: g-force 3000 x *g*, 40 minute first spin, 10 minute second spin.

Centriprep-30: g-force 1500 x *g*, 15 minute first spin, 10 minute second spin.

Centriprep-100: g-force 500 x *g*, 45 minute first spin, 15 minute second spin.

Bovine serum albumin; 15 ml starting volume 25°C; g-force: 1500 x *g*; swinging-bucket rotor, two 25-minute spins.

4.3 Desalting. For desalting or solvent exchange, the sample is concentrated until the equilibration point is reached. Filtrate is discarded, then the sample is reconstituted to the original volume by adding an appropriate solvent, then vortex mixing. The sample is then concentrated and reconstituted once more. This process is repeated until the concentration of the contaminating microsolute is sufficiently reduced. To demonstrate, 15 ml of 1 mg/ml gamma globulin in 0.5 M NaCl was concentrated down to approximately 3 ml in a Centriprep-30 under the conditions listed in Table 4. The resulting concentrate was then reconstituted back to 15 ml by adding 10 mM Tris-HCl; the solution was then centrifuged a second time. After three desalting spins, the NaCl concentration dropped to 5 mM for a salt removal of 99%. For a 15 ml starting volume. g-force: 1500 x *g* swinging-bucket rotor; 25°C, 20 minute spin.

4.4 Biological Integrity. To demonstrate the enzyme activity remaining in the retentate after centrifugation, 15 ml of glucose-6-phosphate dehydrogenase containing 1 unit of enzyme activity/ml and 1 mg/ml bovine serum albumin (to minimize nonspecific adsorption) were spun in a Centriprep-10 concentrator for 60 minutes at 3000 x *g* (two spins, with filtrate decanted between spins).

Table 3 - Typical Protein Recovery, Centriprep-30

Concentration	% Retentate Recovery
10 μg/ml	77.7 +/- 5.8
100 μg/ml	80.3 +/- 0.7
500 μg/ml	92.1 +/- 2.5
1000 μg/ml	89.9 +/- 2.4

Table 4 - Desalting Gamma Globulin (1 mg/ml) with Centriprep-30 Concentrator

Spin #	Protein Recovery	NaCl Concentration
Initial sample	-	500 mM
1	95.1%	140 mM
2	94.2%	25 mM
3	94.0%	5 mM

Approximately 88% of the original activity was recovered. A *unit* is the amount of activity that will oxidize 1 μmole of glucose-6-phosphate to 6-phosphogluconate per minute in the presence of NADP at pH 7.8, 30°C. [Note that the conditions must be stipulated and adhered to if one is to ensure reproducibility.]

5. Quantitating Recoveries

The density of most dilute protein solutions is nearly equal to that of water (1 g/ml). Using this property, the retentate and filtrate recoveries can be quantified by weighing them, then converting the units from grams to milliliters. This provides a close approximation of the recovery volumes for protein solutions with concentrations up to roughly 20 mg/ml. Total recovery, percent retentate, and percent filtrate may be calculated mathematically using the formulas provided. These formulas may then be entered into a data reduction program.

5.1 Weighing Procedure.

1. Before use, weigh the sample container and an empty test tube (15 ml capacity).
2. Fill the sample container with solution and reweigh. The starting volume equals the difference in container weight after adding the solution.
3. Assemble the concentrator and spin per instructions. Decant filtrate into preweighed test tube. If further concentration is desired, spin a second time, decanting the filtrate into a test tube containing the first filtrate.
4. To calculate the filtrate volume, reweigh the test tube plus filtrate, then subtract the weight of the empty test tube. To obtain the concentrate volume, reweigh the sample container with retentate, then subtract its dry weight.

5.2 Calculations.

1. Using an appropriate assay procedure (spectrophotometry, radioimmunoassay, refractive index, or conductivity) assay filtrate, retentate, and sample of the starting material to determine solute concentrations.
2. With the volume and concentration data from above, determine relative quantities of retentate and filtrate. Using the weights (and corresponding volumes) and measured concentrations, calculate relative quantities of retentate and filtrate recovered as follows:

$$\% \text{ filtrate recovery} = 100\, W_f C_f / W_0 C_0$$

$$\% \text{ retentate recovery} = 100\, W_r C_r / W_0 C_0$$

$$\% \text{ total recovery} = \% \text{ filtrate recovery} + \% \text{ retentate recovery}$$

Definition of Variables:
W_r = total retentate weight before the assay, W_0 = starting material weight, W_f = filtrate weight, C_r = retentate concentration, C_0 = starting material concentration, C_f = filtrate concentration.

References:

1. Blatt, W. F., Robinson, S. M., Bixler, H. J., Membrane Ultrafiltration: the Diafiltration Technique and Its Application to Microsolute Exchange and Binding Phenomena. Anal. Biochem. 26:151 (1968).
2. Bowers, W. F. and Tiffany, D. B., U.S. patent 4,832,851.
3. Rigopulos, P. N., U.S. patent 3,488,768.

INNOVATION/INSIGHT

SUBCELLULAR FRACTIONATION

Finean, J. B., Coleman, R. and Michell, R. H. (1974) *Membranes and Their Cellular Function*, pp. 12-13, John Wiley and Sons, New York. (Adapted)

1. Introduction

Subcellular fractionation can be used in two interrelated ways, either as an analytical method for defining the intracellular location of a particular component or process, or as a preparative method for isolating a particular entity for detailed study. Before the isolated fractions, which are always mixtures, can be used for either purpose, they must be thoroughly characterized. This assessment may include optical and electron microscopy – in experiments designed to monitor concentrations of a variety of "markers" relative to the homogenate. The *markers* are chemical, enzymatic, or antigenic constituents that have previously been shown to be localized or concentrated at known cellular sites – and can, therefore, be used to define their presence and distributions in the isolated fractions of material from these sites. The choice of markers is simplified somewhat by the fact that particular constituents usually, but not always, occur in the same sites in different types of cells. When distributions of markers are measured, it is essential to ensure that all the material present in the original homogenate is taken into account. Enzymes located on the inner faces of isolated membrane vesicles, or in solution within them, are often not freely accessible to substrates present in the incubation medium bathing the sealed vesicles. When the *membranes are disrupted*, e.g., by *hypotonicity*, *detergents*, or *ultrasonic treatment*, this "latency" disappears and the enzymes can be quantitatively assayed. Such latency, which may be a valuable pointer to intracellular compartmentation and was crucial in the discovery of lysosomes, must always be considered during the design of assays for intracellular enzymes.

In preparative experiments, the monitoring of markers acts as a constant check that the material being studied is not severely contaminated with unwanted materials. It can also suggest which contaminating materials might be significant – both in terms of mass and of enzyme activities. *Enzymatic reactions can often be satisfactorily studied in material of only moderate purity*, but detailed chemical or physical analysis of an organelle isolating the enzyme in a highly purified state.

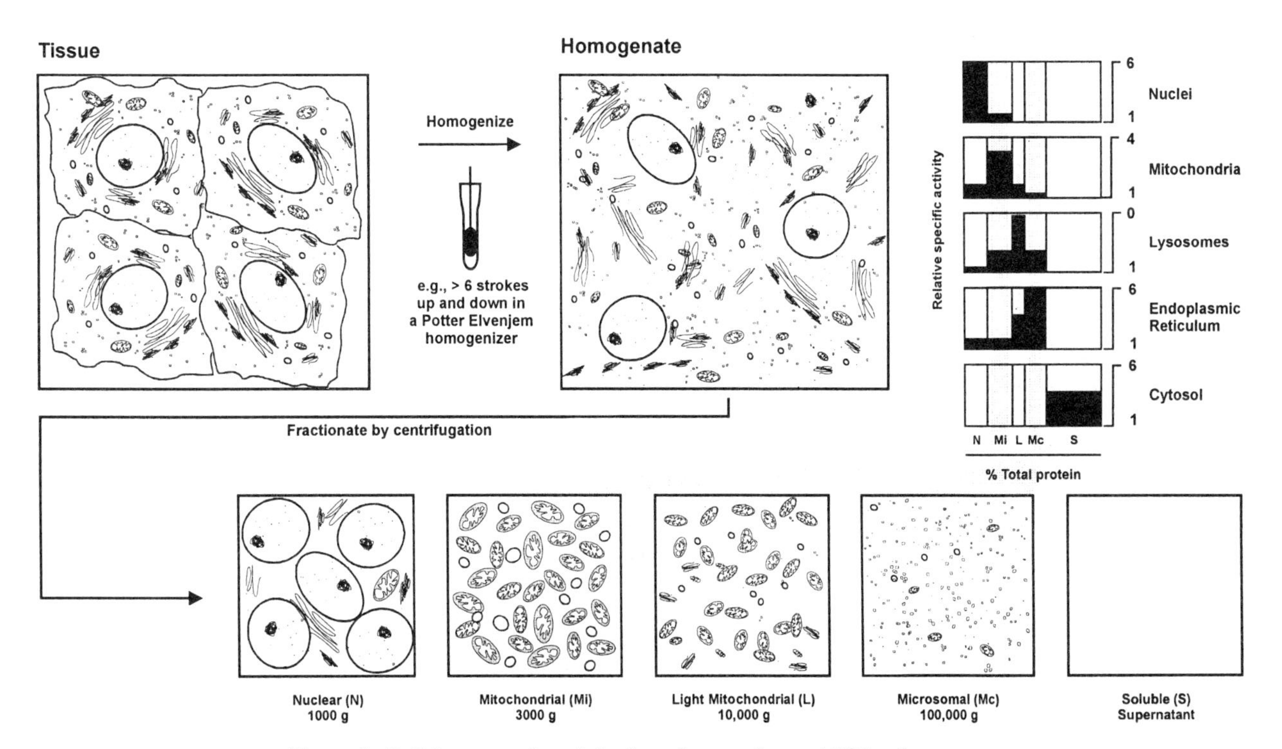

Figure 1 - Cellular compartmentalization of organelles and RSAs of enzymes.

2. Commonly Used Compartmental Markers

The *relative specific activity* (RSA) of a marker expresses the degree to which each isolated fraction contains an enriched or diluted amount of particular organelles (Figure 1). RSA is the ratio of specific activities (SA) of the (impure) fraction of interest and homogeneous enzyme ("pure" homogenate)..

$$RSA = \frac{SA_{impure}}{SA_{pure}}$$

The SA refers to the activity per mg of protein. Note that one does not directly stipulate whether the protein is correctly folded or fully functional from molecule to molecule. The researcher is seeking a means to track activities in complex materials. It is crucial that one keep in mind the real and potential limitations implicit in trying to interpret such population-averaged information in terms of a single molecule. The width of each bar in Figure 1 expresses the proportion of the cell protein in the fraction and its area represents the proportion of the marker and organelle (Table 1.) The "theoretical" maximum RSA of any marker is inversely proportional to the contribution of that organelle to the cell mass (Table 1).

Table 1 - Commonly Used Compartmental Markers.

Compartment	Marker
Nucleus	DNA
Mitochondria	Succinate dehydrogenase, cytochrome oxidase
Lysosome	Acid phosphatase, acid DNases
Endoplasmic Reticulum	Glucose-6-phosphatase, esterase (of napthyl acetate ester)
Plasma Membrane	5'nucleotidase, Na^+/K^+-dependent ATPases
Golgi Complex	Thymine pyrophosphatase, UDP-galactose:*N*-acetylglucosamine galactosyltransferase
Peroxisomes	Catalase, D-amino acid oxidase
Chloroplasts	Chlorophyll, ribulose diphosphate carboxykinase
Cytosol	Lactate dehydrogenase, 6-phosphogluconate

LAB 7.1:
THE PROTEIN PURIFIER: A LEARNING AID FROM PHARMACIA

Overview:

This software learning aid from Pharmacia simulates the experience of working through a protein purification problem using a range of separation and analytical techniques. You will gain insight into the strategies commonly employed in a purification, without actually performing the experiments, thus, saving time and materials. Practical experience is still important, but these skills can be learned far more thoroughly with an initial grounding in the fundamental principles.

The software is intended to run on all PCs with Enhanced Color Graphics Monitors. The software is not copy protected. Please make copies only for your own use and for the purpose of security. After an initial series of screens where you"ll be given some basic information, you"ll be asked to choose which protein you want to purify.

After these initial screens, a menu of separation methods is displayed. Depending on the method chosen, you must enter the appropriate experimental conditions, from which the computer calculates the experimental results. These appear either as a "statement of recovery," in the case of ammonium sulfate precipitation, or in the form of a chromatogram for the chromatographic methods.

Having obtained your first result, you are presented with an "analysis menu" from which you can determine where the activity lies. You may also choose to check the purity of your protein by electrophoresis. From the results of the electrophoresis, you"ll be able to decide

1. which fractions to take for the next separation step and
2. what the next separation step should be.

After each separation method, you"ll be able to see your results on the purification record – which shows the ever-improving purity of your protein. The progress is expressed in terms of International Units/man-hour to give an idea of your efficiency as well as the amount of pure protein obtained so far. You may repeat a purification using different strategies to see which is the most appropriate for a given protein.

Experimental Outline:

1. Program start-up.
2. Instructions.

Procedures:

1. Program Start-up

1. Switch on the computer.
2. Wait for the prompt and insert the disk into the disk drive.
3. Type the command a:\proteins and ENTER. The program will automatically load.

2. Short Instructions

The program follows a simple logical pathway. You may choose from 20 different proteins or, alternatively, you may continue with an uncompleted purification file stored on disk. Sample volume and composition are assumed to be appropriate when you select a method. In reality, buffer salt, pH and concentration changes often have to be made. Sample dilution or concentration is also sometimes required. In this program, we assume that such manipulations are carried out automatically. This is reasonable for the purposes of learning the general principles; we should point out that sample state, volume, pH, and concentration often play a major role in the decision as to which, and in what order, techniques are used. (A discussion of such considerations can be found in "A strategy for protein purification," Separation News, Vol. 13, No. 6, Pharmacia-LKB.)

The separation techniques have been chosen for their general applicability. Many more specific techniques, e.g., affinity chromatography, have been omitted since their proper use requires detailed knowledge of the protein to be purified. These techniques are frequently extremely valuable and should be considered carefully in practical purification schemes. Further information on these techniques is available from Pharmacia.

3. Full Instructions

This section gives a more detailed screen-by-screen description of the program. After loading the program, the first screen to appear after the title and copyright screens is Screen 1. This is the starting point of the program; a solution containing 20 different proteins. Note that in reality, sample volume, concentration and clarity are important considerations. Within this program, such parameters are considered to be optimal for a chosen methodology.

Screen 1

```
Protein Purification
```

You have a solution containing 511 mg of protein. You must purify 20 enzymes from this solution using a variety of separation techniques. You must try to obtain as high a yield as possible, but keep an eye on the cost in man-hours. Time is money! Now press the space bar to start. Pressing the space bar gives Screen 2.

Screen 2

```
Protein Purification
```

You may choose to start from the beginning or to continue a purification that has been stored on disk. To choose between these two alternatives, press the space bar to highlight the appropriate line and then ENTER. If you choose to continue from stored data, Screen 3 appears.

Screen 3

```
Protein Purification
```

Enter the file name of this experiment specifying drive and path if necessary. A wrong file name will result in a return to Screen 2: Protein Purification.

Screen 4

```
Which enzyme do you want to
purify?
```

Enter no. (1-20). If we assume that you choose protein no. 4, the next screen to appear is Screen 5.

Screen 5

```
Enzyme 4

This enzyme is stable for
several hours at
temperatures up to 40°C and
at pH values between 4.5
and 10.
Press the space bar to
continue.
```

Screen 5 provides basic information concerning the stability of the enzyme. Physicochemical stability commonly plays a deciding role in the choice of conditions that can be employed in a separation step. Disregarding these constraints may lead to loss of activity! Pressing the space bar gives Screen 6: Separation Methods. Screen 6 is a menu of the separation methods available. These may be chosen with the space bar and ENTER key.

Screen 6

```
Separation Methods
ammonium sulfate
precipitation
gel filtration
ion exchange chromatography
chromatofocusing
hydrophobic interaction
chromatography (HIC)
       start again
```

If you choose ammonium sulfate precipitation, you will be asked for the following information on Screen 7.

Screen 7

```
Ammonium Sulfate
Fractionation
```

```
Initial % saturation?
```

Enter the value of the starting concentration of ammonium sulfate. Let us assume that you chose 10%. You are then asked for the final concentration on Screen 8.

Screen 8

```
Ammonium Sulfate
Fractionation
Initial % saturation? 10
Final % saturation?
```

Supplying the value for the final % saturation leads to a result expressed in terms of precipitated activity and precipitated protein (Screen 9).

Screen 9

```
Ammonium Sulfate
Fractionation
Initial % saturation? 10
Final % saturation? 50

You have precipitated
89.4% of the enzyme
activity
and 49.0% of the protein.

Press the space bar to
continue or
ESC to repeat the
fractionation.
```

If the result is unacceptable, the ability to repeat the fractionation allows you to investigate various "fractionation windows" (just as in real life) until you attain an optimum trade-off between recovery and enrichment. As soon as you have an acceptable result, press the space bar:

Screen 10

```
...and now?
SDS-PAGE
continue
```

Selecting SDS-PAGE results in a diagrammatic representation of the results from an SDS-PAGE separation of the precipitated and soluble fractions from the ammonium sulfate fractionation. The results give information concerning the molecular weight distribution of the SDS-treated fractions.

If you select continue, you must then choose to work with the precipitated material or the soluble material. After selecting the appropriate fraction, a screen such as the following will appear (Screen 11):

Screen 11

```
Situation after step 1
Total protein:  250.4 mg
Total enzyme:   6225 units
Enrichment:     1.8
Yield of enzyme:     89.4%
Cost so far:    0.08 man-hours/100 U

Press the space bar to
continue.
```

The table in Screen 11 summarizes the result of the individual separation step just performed in terms of total protein and total enzyme activity remaining after the step. The cumulative values for enrichment, yield, and cost are also shown.

Pressing the space bar allows you to choose a detailed summary of the purification to date or to continue with the next separation technique (Screen 12).

Screen 12

and now?
examine record of purification
Continue

Screen 13 - Purification of Enzyme 4

Step	Procedure	Protein (mg)	Enzyme (units)	Enzyme yield (%)	Enrichment (m.h./100 U)	Cost
Initial		511.0	7000	100.0	1.0	
1	Amm. 250.4 sulfate	6225	89.4	1.8	0.8	

Selecting Continue returns you to the Separation Methods menu Screen 6.

Selecting Gel filtration from the Separation Methods menu (Screen 6) calls up the Gel Filtration menu (Screen 14).

Screen 14

Gel Filtration Media
Sephadex G-50
Sephadex G-75
Sephadex G-100
Sephacryl S-200 HR
Sephacryl S-300 HR

Choose the gel you think most appropriate for your sample. From your choice, the program will produce a simulated chromatogram on the screen, which, when complete, will be supplemented with a table asking for instructions for the next step (Screen 15).

Screen 15

What next?
dilute fractions and re-assay
assay for enzyme activity
1D SDS-PAGE 2D SDS-PAGE
pool fractions
repeat this step
go to next step
start again
go home

Selecting Ion exchange gives Screen l6.

Screen 16

Ion Exchange Media
Q-Sepharose Fast Flow
S-Sepharose Fast Flow

Q-Sepharose Fast Flow is a strong anion exchange gel, i.e., it retains its positive charge over a wide pH range. S-Sepharose Fast Flow is similarly a strong cation exchanger.

Selecting chromatofocusing gives Screen 17.

Screen 17

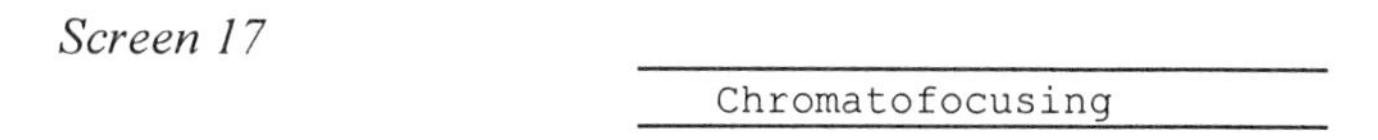

You must establish a descending pH gradient with Polybuffer which spans no more than three and no less than two pH units.

```
pH at start of gradient?
```

Chromatofocusing is a chromatographic technique that separates proteins based on their ability to bind to an anion exchange gel in an ever-decreasing pH gradient. As proteins are often eluted close to their p*I*, the technique has often been regarded as the chromatographic equivalent of IEF. This is an oversimplification but is, nevertheless, a useful analogy. Selecting Hydrophobic Interaction yields Screen 18.

Screen 18

Hydrophobic Interaction Media
Phenyl-Sepharose CL-4B
Octyl-Sepharose CL-4B

Phenyl Sepharose CL-4B and Octyl Sepharose CL-4B are both hydrophobic gels. The former gel is less hydrophobic than the latter and is used for separations of strongly hydrophobic proteins. Conversely, Octyl-Sepharose CL-4B is used for separations of less strongly hydrophobic proteins.

The choices available are:

1. Dilute fractions and reassay. This is useful if the peaks in the chromatogram are so concentrated that they are off-scale, i.e., it is impossible to see the peak tops. This function requests a dilution factor and then replots the chromatogram accordingly.
2. Assay for enzyme activity. Selecting this function locates the first and last active fractions in the chromatogram.
3. 1D SDS-PAGE. Up to 15 fractions may be analyzed by SDS-PAGE. The information gained from such an analysis may be useful in determining which fractions to pool before continuing. SDS-PAGE may also be selected after pooling the appropriate fractions, and may then provide useful information as to which technique to use in the next step.
4. 2D SDS-PAGE. Individual fractions can be analyzed prior to pooling, or alternatively the pooled fractions can be analyzed, often yielding useful information prior to the next step.
5. Pool fractions. Collect specific fractions from the chromatogram.
6. Repeat this step. If the result is disappointing, you may repeat the separation using different parameters. This function may not be used after fractions have been pooled.
7. Start again.
8. Go home. Progress to date with a particular separation problem may be stored on disk for continuation later.

The program continues in this interactive fashion cyclically moving from separation to analysis and result reporting and back to separation again until you judge that the enzyme is pure.

Note:

1. Sephadex, Sephacryl, Sepharose, and Polybuffer are registered trademarks of Pharmacia-LKB Biotechnology AB, Uppsala, Sweden.

Reference:

1. Booth, A. (1987) Pharmacia, Piscataway, NJ, ph. (800) 526-3593.

LAB 7.2:
INDUCTION AND PURIFICATION OF β-GALACTOSIDASE FUSION PROTEIN FROM BACTERIA

Overview:

In this segment of the course (Unit 7), you will purify the *C. elegans* β-galactosidase-*myo*-3 fusion protein from *E. coli.* The objective here is to obtain a substantially purified fusion protein preparation using several different types of column chromatography. Bacterial cells harboring the β-galactosidase-*myo*-3 fusion protein will be physically disrupted by a combination of *lysozyme treatment, freeze-thaw*, and *sonication.* The soluble extract (homogenate) of your clone contains only a small fraction of the total amount of recombinant protein as well as many other contaminating bacterial proteins. Most of the recombinant protein will remain as insoluble inclusion bodies. These *inclusion bodies* are one component of the total insoluble debris resulting from cell lysis. Overexpression of the plasmid-encoded β-gal-*myo*-3 fusion gene leads to accumulation of large amounts of these insoluble β-gal-*myo*-3 fusion gene products. Inclusion bodies are aggregated or crystallized forms of the abundant fusion protein. You will remove this insoluble debris in order to retain only soluble protein that is easily purified from contaminants. The protein has to be obtained in solution prior to its purification.

In future labs, the fusion protein will be selectively removed from the homogenate by adsorption to various types of chromatography columns. These purification steps will be monitored by measuring protein concentrations and performing β-galactosidase enzyme activity assays. The purity of the preparations will be analyzed by SDS-PAGE. Throughout it"s isolation and purification, steps must be taken to ensure that your protein is not inactivated or denatured by physical or biological factors. Sample solutions must be carefully buffered at a pH that is conducive to protein stability and functional integrity. Avoid temperature above 25°C to minimize thermal denaturation and protease activity.

Experimental Outline:

1. Grow bacterial cultures that have been transformed with recombinant plasmid containing the *myo*-3 gene.
2. Harvest IPTG-induced expression vector cultures.
3. Lyse the bacterial cells.
4. Remove insoluble debris.

Required Materials:

Buffer A (recipe below); bacterial suspension cultures; IPTG; 2XYT medium; 1-L Erlenmeyer flask; 50 mg/ml ampicillin; shaker/incubator; centrifuge; PMSF (phenylmethylsufonyl fluoride); 10 mg/ml lysozyme; vortex; dry ice; Oakridge tube; sonicator; protective ear guards

Prelab Preparations:

1. Liquid Bacterial Culture Growth Containing the Cloned Recombinant Plasmid

1. Prepare this culture on the day before protein induction in the recombinant plasmid.
2. Inoculate a tube containing 5 ml of 2XYT/amp medium (containing 250 μg of ampicillin) with bacterial cells containing the recombinant plasmid. These cells should be picked from a single isolated bacterial colony grown on agar plates.

2. Prepare Buffer A

1. Mix together first, then add H_2O to 99 ml.

1 M	Tris, pH 7.4	20	ml
0.1 M	EDTA, pH 7	20	ml
	NaCl	1.46	g
	β-mercaptoethanol	70	μl
	glycerol	5	ml

2. Add last:

0.1 M PMSF in ethanol	1	ml

Store in a cold room.

3. Grow Bacterial Suspension Cultures for Fusion Protein Purification

The fusion protein encoded by your clone will be purified from a 250 ml culture grown in the presence of IPTG for 3-5 hours. The night before, you should have inoculated a tube containing 5 ml of 2XYT/amp medium with cells picked from a single colony of your clone.

1. Obtain a 1-L Erlenmeyer flask containing 250 ml of sterile 2XYT medium.
2. Add 250 μl of 100 mg/ml ampicillin to the flask.
3. Inoculate the flask of 2XYT/amp/IPTG medium with 250 μl of the overnight culture of your clone.
4. Incubate the culture in a 37°C shaker/incubator at 300 rpm, until A_{600} = 0.5. Then, add IPTG to a final concentration of 1 mM. Incubate an additional 2 hours.

Procedures:

1. Extraction of Recombinant Protein from *E. coli.*

1. Harvest the bacteria containing the IPTG-induced expression vectors:
   a. Pour all of the culture into a 250-ml centrifuge bottle. Balance your bottle with a bottle prepared by another student.
   b. Centrifuge for 5 minutes at 7,000 rpm to pellet the bacteria.
   c. Remove and discard the clear supernatant. Invert the tube over a paper towel for 5 minutes to remove the residual liquid.

2. Lyse the Bacterial Cells

CAUTION: Buffer A contains a potent protease inhibitor (PMSF) that is TOXIC if ingested.
DON"T PIPET BY MOUTH!

1. Add 3 ml of Buffer A to the pellet.
2. Vortex to suspend the pellet.
3. Transfer the solution to a 50-ml Oakridge tube.
4. Add 75 μl of 10 mg/ml lysozyme.
5. Vortex and incubate on wet ice for 15 minutes.
6. Place the microfuge tube on bed of dry ice to freeze.
7. Remove the microfuge tube from the ice to thaw.

NOTE: Put on protective ear guards before activating the sonicator.

8. Sonicate the culture. Sonication uses high-frequency sound waves to help release more of the fusion peptide from the bacterial cell and reduces the viscosity of the solution by shearing chromosomal DNA.
   a. Immerse the tip of the sonicator in the bacterial solution. Make sure the tip does not contact the sides of the tube.
   b. Sonicate the ice cold solution 5-6 times at maximum power for 15 seconds, returning the tube to ice in between sonications. This fraction represents the "crude homogenate."
   c. Record the approximate volume of the crude homogenate: _________.
9. Remove samples of crude homogenate for the Bradford protein assay (20 μl) and β-gal enzyme assay (20 μl). Label the tubes and freeze at -20°C.
10. Remove an additional 20 μl of the crude homogenate and make a sample for SDS-PAGE (20 μl sample + 20 μl 2x sample buffer). Label tubes and freeze at -20°C.

3. Removal of Insoluble Debris

1. Centrifuge the crude homogenate in a 4°C refrigerated centrifuge for 20 minutes at 12,000 rpm to remove insoluble debris. DO NOT DISCARD THE PELLET.
2. Transfer all of the supernatant to a fresh centrifuge tube and STORE IT ON ICE. Save three 20 μl samples. Label these aliquots as the soluble fraction. As a precautionary measure, also retain and freeze the remaining soluble fraction (≈3 μl) as three 1-ml aliquots.
3. Add 1 ml of Buffer A to the pellet from step 1.
4. Vortex the tube to resuspend the pellet.
5. Remove 20 μl samples for SDS-PAGE, protein determination, and enzyme assays as above and freeze the remainder at -20°C. These samples will represent the "insoluble fraction."
6. Record the approximate volume of the insoluble fraction: __________.

LAB 7.3:
GEL FILTRATION OF MOLECULAR WEIGHT STANDARDS AND PROTEIN FRACTIONATION

Overview:

During this lab, you will prepare a gel filtration chromatography column, and calibrate it with molecular weight markers. This calibration determines how well the column was packed and at what point in the elution process proteins of a particular size elute. After calibration, you will fractionate your β-gal-*myo*-3 fusion protein. Fractions from both the molecular weight markers and fusion protein will be collected for analysis in the next lab.

Gel filtration separates proteins on the basis of their size and shape using porous beads packed on a column. Large elongated proteins cannot enter the pores in the beads and, as a result, elute from the bottom of the column first. Smaller proteins enter the beads, so they require more solvent before they elute from the column. Thus, the *myo*-3 protein can be separated from contaminating proteins of different sizes and from other contaminants such as small salt molecules.

Experimental Outline:

1. Prepare gel filtration column.
2. Calibrate gel filtration column with molecular weight markers.
3. Perform gel filtration column chromatography to purify the β-gal-*myo*-3 fusion protein.

Required Materials:

Sephacryl 300HR; desiccator; dry Ottawa sand; 170-ml BioRad Econocolumn with buffer reservoir; Buffer A (recipe in Lab 7.2); molecular weight marker mixture; β-mercaptoethanol; β-gal-*myo*-3 fusion protein; BioRad Gel Filtration Standard; column chromatography fraction collector

Procedures:

1. Prepare Gel Filtration Column

1. Warm a 50% slurry (1:1 buffer A/matrix) of Sephacryl 300HR to room temperature.
2. Deaerate the slurry in a desiccator for 15 minutes to remove air bubbles.
3. Add 1 inch dry Ottawa sand (20-30 mesh) to the bottom of a 170 ml BioRad Econocolumn to prevent the matrix from obstructing the outlet.
4. Turn off the stopcock and add the resin slurry to the column.
5. Continue adding the resin slurry until the packed gel reaches about 4 inches below the top of the column.
6. Attach a buffer reservoir filled with Buffer A to the top of the column.
7. Wash the column with buffer for 20 minutes.

2. Calibrate Gel Filtration Column with Molecular Weight Markers

1. Add molecular weight markers containing Blue Dextran to the top of the column (e.g., BioRad Standard, described below).
2. Add Buffer A throughout the run to avoid allowing the column to run dry.
3. During the column run, collect fractions of eluate. (Fraction collection can be accomplished using an automated device overnight.)
4. After the run, wash the column with 750 ml of Buffer A to remove residual molecular weight marker proteins and prequilibriate it for the next use. Reequilibriate it again if there is a time lag since strongly "included" materials may elute upon standing. If this lag period is >1 day, the resin must be preserved in 20% ethanol or by including 0.1% sodium azide in the buffer.

BioRad Gel Filtration Standard Proteins:

Component	Molecular Weight	Amount per Vial (mg)
Thyroglobulin (bovine)	670,000	5.0
Gamma globulin (bovine)	158,000	5.0
Ovalbumin (chicken)	44,000	5.0
Myoglobin (horse)	17,000	2.5
Vitamin B_{12}	1,350	0.5

3. Run Gel Filtration Chromatography of your Protein Sample

1. Wash the column with Buffer A containing β-mercaptoethanol before adding the fusion protein to remove any residual proteins and contaminants from the previous column run..
2. Run the soluble protein sample containing the β-gal-*myo*-3 fusion protein (extracted from Lab 7.2) through the column. Use the same procedure as for the molecular weight marker.
3. Collect the column eluate fractions; this can be done overnight. The fractions will be saved until the next lab.

Notes:

1. Be certain your column doesn"t leak. Replace faulty joint fixtures if you detect leaks.
2. The column should be continuously eluted with buffer when not in use; this helps avoid matrix cracking, which occurs upon drying.
3. To save a column overnight for later use, run 20% ethanol through it. Before adding protein to the reused column, remove excess ethanol from the column.

Reference:

1. BioRad Gel Filtration Standard Instruction Manual.

PROCESS RATIONALE

GEL FILTRATION CHROMATOGRAPHY

Pharmacia, Piscataway, NJ, ph. (800) 526-3593. (Adapted)

1. Characterization of Solute Behavior.

Results of gel filtration experiments are typically expressed in the form of an elution diagram showing the variation of solute concentration in the eluent with the volume of eluent passed through the column. For protein and nucleic acid work (and in many other applications), continuous detection using a UV-monitor and recorder produces an immediate and permanent record. From this diagram, we can determine the *elution volume* (V_e) of a given solute protein, nucleic acid, etc.

- When very small samples are applied (small enough to be neglected compared with V_e), the position of the peak maximum in the elution diagram should be taken as V_e.
- When samples whose volumes cannot be neglected compared with V_e are applied, the V_e values of the macromolecules or standards are measured as the volume when half of the component has eluted to the position of the maximum of its elution peak.
- When very large samples are used (giving a plateau region in the elution curve), the volume eluted from the start of sample application to the inflection point (or half height) of the rising part of the elution peak should be taken as V_e. This criterion is often incorrectly applied to samples not giving plateau regions.

Normally, in gel filtration with Sephadex, Sepharose, or Sephacryl, solutes have linear *partition isotherms* and give symmetrical peaks. Elution volumes are, therefore, easily determined by these methods. More sophisticated criteria are seldom useful in practice. V_e is not in itself sufficient to define the behavior of the sample substance, since this parameter varies with the *total volume of the packed bed* (V_t) and with the way the column has been packed. V_e must be expressed as either of the following ratios in order to define the behavior of the sample substance.

$$\frac{V_e}{V_t} \text{ or } \frac{V_e}{V_0}$$

By analogy with other types of partition chromatography, the elution of a solute is best characterized by a *distribution coefficient* (K_d).

However, for a given gel there is a constant ratio of K_{av}/K_d, which is independent of the nature of the solute or its concentration. K_{av} is easily determined and, like K_d, defines solute behavior independently of the bed dimensions and packing. Other methods of normalizing data give values that vary depending upon how well the column is packed.

$$K_d = \frac{V_e - V_0}{V_s} \quad (1)$$

$$K_{av} = \frac{V_e - V_0}{V_t - V_0} \quad (2)$$

The *void volume* (V_o) is the elution volume of molecules that never enter the matrix from the mobile phase; they are larger than the largest pores in the gel. The *volume of the stationary phase* (V_s) in gel filtration is equal to V_i the volume of the solvent inside the gel – which is available to very small molecules, i.e., the elution volume of a solute that will distribute freely between the mobile and stationary solvent phases minus the void volume. Thus, K_d represents the fraction of the stationary phase that is available for diffusion of a given solute species. In practice, the volume of the stationary phase defined in this way is rather difficult to determine. Methods involve measurement of the elution volumes of radioactive ions such as ^{23}Na. It is much more convenient to substitute the term $V_t - V_o$ for V_s, when we obtain for K_{av} (see above). K_{av} represents the fraction of the stationary gel volume, which is available for diffusion of a given solute species. Since $V_t - V_o$ includes the volume of the gel forming substance, which is inaccessible to all solute molecules, K_{av} is not a true partition coefficient (Figure 1).

2. Theoretical Considerations

The use of gel filtration for the determination of well documented. In practice, it is found that for a series of compounds of similar molecular shape and

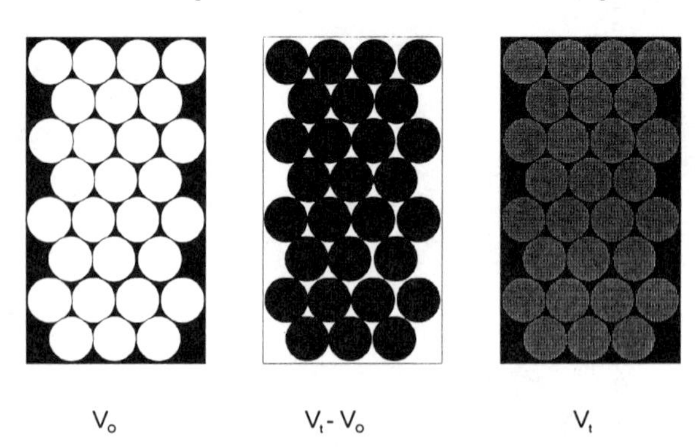

Figure 1 - Diagrammatic representation of V_t and V_o. Note that $V_t - V_o$ includes the volume of the actual fibers forming the matrix of each bead. (Fisher, L., *Laboratory Techniques in Biochemistry and Molecular Biology*. Vol. 1, part II. *An Introduction to Gel Chromatography*. North-Holland Publishing Company, Amsterdam.)

density a sigmoidal relationship exists between their K_{av} values and the logarithms of their molecular weights (MW). *Calibration curves* constructed in this way for a particular gel type are often termed "*selectivity curves*," and over a considerable range there is a conveniently linear relationship between K_{av} and log MW. In ideal gel filtration behavior, no molecules can be eluted with a K_{av} greater than 1 or less than 0.1. If the K_{av} is greater than 1, some kind of adsorption is indicated. If the K_{av} is less than 0, *channeling* in the chromatographic bed is indicated and the column must be repacked. Several models have been proposed to describe the behavior of solutes during filtration. Most have regarded the partition of solute molecules between the gel particles and surrounding fluid as an entirely steric effect.

3. Experimental Considerations

Figures 2 and 3 are diagrams of column chromato-graphy apparatus setups.

3.1 Resolution. The first objective of the gel filtration experiment is to achieve adequate resolution of the components of interest.

Resolutions (R_s) greater than about 1.2 indicate baseline separation (peaks do not overlap). In many experiments, adequate resolution is not difficult to achieve. In these cases, consideration can then be given to other objectives of well-designed experiments. Since most separations are preparative, these *objectives* are *to achieve maximum recovery with the least sample dilution in the shortest possible time*.

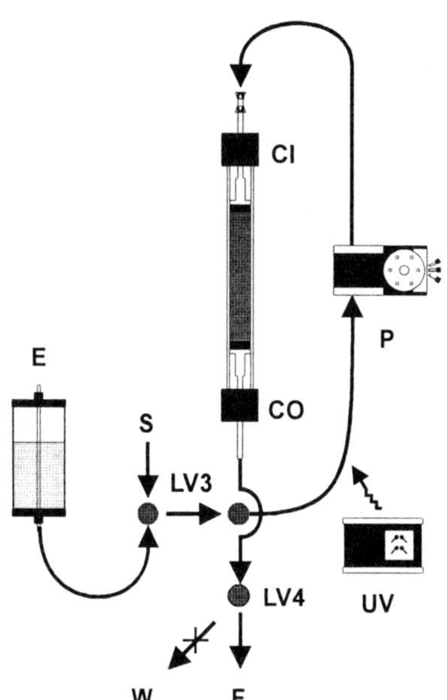

Figure 2 - Increasing the effective column height by recycling. Eluent (E) and sample (S) are connected to the 3-way valve LV4 that can be closed during recycling. The 4-way valve connects the column outlet (CO) to the inlet (CI), or the sample/eluent (S/E) to CI, and CO to the fraction collector (F). Placing a UV monitor inline prior to F allows one to record the final output (UV out.) Placing a UV monitor between LV4 and CI allows one to monitor the initial loading of sample and the progress of a recycled batch over the course of > 1 round of fractionation (UV recycle). The bypass from E to P allows one to wash the column bed then divert the effluent to F or waste (W). (Flow variations are shown in Figure 3.)

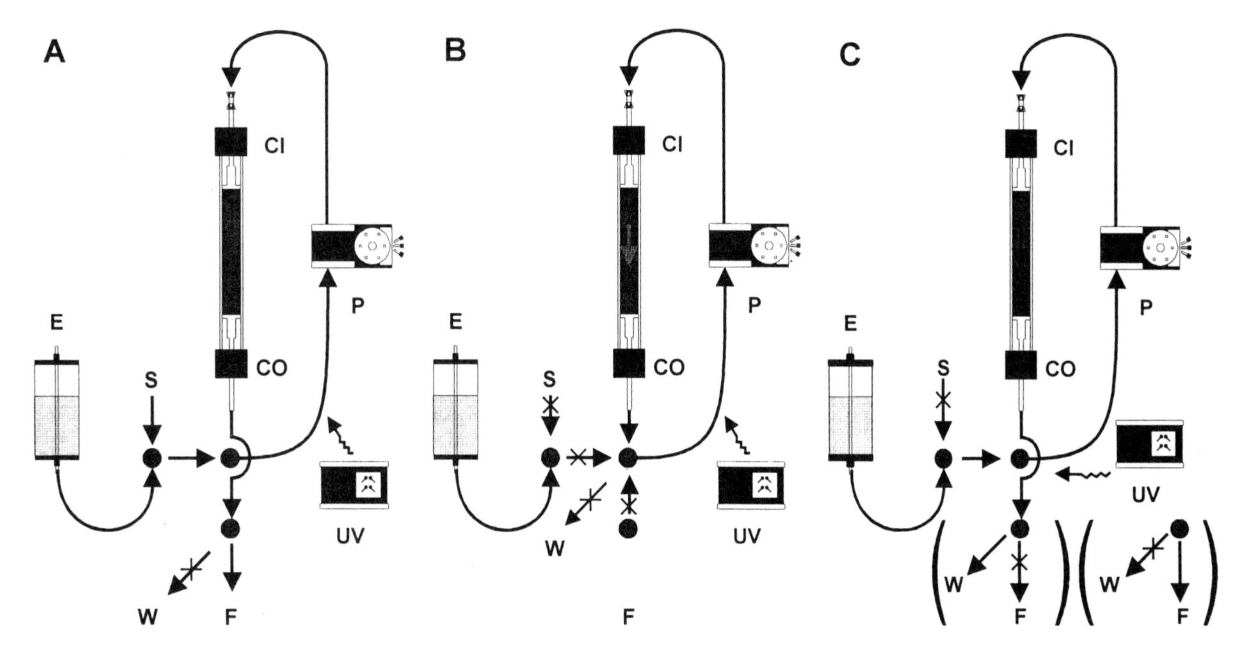

Figure 3 - Connection variations achieved using different merge and bypass possibilities open using 3- and 4-way values in assembling column chromatography systems. (A) Sample loading. Pump (P) drives loading of S, plus E if desired, through and onto the column, with monitoring of the input (CI) by absorbance at UV1. (B) Purification by sample recycling; UV1 monitors the progress of cycling steps. (C) Sample elution/column bed rinsing. P drives E through the column, which is either collected by F or shunted to a waste receptacle (W). Other abbreviations are defined in Figure 2.

3.2 Choice of Gel Type. The choice is easiest with group separations such as *desalting* and *buffer exchange* since there is a large difference in molecular weight of the two component groups and complete resolution is not difficult to achieve. A gel should be chosen to ensure that the high MW substances elute at V_o (K_{av} = 0). This will give minimum zone broadening and dilution, and reduce the time the components of interest are on the column. The low molecular weight substances should be eluted near V_t (K_{av} = 1). In the majority of desalting applications, Sephadex G-25 medium is the recommended gel. It combines good rigidity, for easy handling and good flow characteristics, with more than adequate resolving power to desalt molecules down to about 5,000 Da.

For fractionation, where a number of substances of similar molecular weight are to be separated, careful choice of gel is important. Special care should be taken when choosing the gel if dissociating media or detergents are to be used, to allow for their influence upon the conformation of the molecules to be separated.

For MW determinations, a gel should be chosen so that the sample's expected value falls on the linear part of the selectivity curve and in the middle of a suitable range of calibration standards. The use of *detergents* as solubilizing agents has also become popular due to the interest in proteins with *low* aqueous solubility, e.g., membrane components. Once again, the effect of these agents on protein conformation usually requires a more open gel for gel filtration. Detergents do not appear to influence the pore structure of the gel. Since the selectivity curves of the various gels do overlap, more than one gel type may seem appropriate for a particular separation. In this case, it is normally best to use the gel with the lower exclusion limit, because the substances of interest will be eluted sooner, possibly improving the recovery of active molecules. Furthermore, considering gels with the same chemical composition, one with a lower exclusion limit will have greater rigidity – allowing greater flexibility in the choice of flow rate.

Important characteristics of Pharmacia columns include:

- Dead space at the outlet of less than 0.1% of the column volume. Minimizes dilution and prevents remixing of separated zones.
- Advanced design bed supports that give uniform flow without clogging.
- Most of the columns can be fitted with flow adaptors for upward flow experiments and easy sample application.

3.3 Pressure and Solvent Resistance. Pharmacia Fine Chemicals" columns and accessories are designed in several resistance classes. If organic solvents are to be used, it is important to choose a column and adaptor of the SR series. These are constructed from materials that will resist organic solvents. The SR 10/50 column can also be used at higher pressures than columns of the K series – to 140 psi. – and are autoclavable. Columns of the K series are designed for routine laboratory use with aqueous solvents at pressures up to 14 psi.

3.4 Sample Size. For analytical purposes and difficult fractionation experiments where maximum resolution is required, the starting zone must be narrow relative to the length of the column. A sample volume of 1-5 % of the bed volume is recommended. Smaller volumes do not normally improve resolution.

In group separations and some fractionation experiments, where peaks are well resolved, it is often appropriate to improve the experimental design by increasing the sample size. In order to minimize sample dilution, which is an inevitable consequence of gel filtration, a maximum sample volume should be used within the limits set by the separation distance. If no *zone broadening* were to occur during passage down a column, the maximum *sample size* could be as great as the *separation volume* (V_{sep}).

$$V_{sep} = V_{eB} - V_{eA} \tag{3}$$

V_{eA} is the elution volume of one substance and V_{eB} is the elution volume of a second substance. As a result of microturbulence, nonequilibria between the stationary phase and the mobile phase and longitudinal diffusion in the bed, the zones will always be broadened. The sample size must, therefore, always be smaller than the separation volume. In desalting and buffer exchange, volumes up to approximately 30% of the *total bed volume* (V_t) can be used to minimize dilution and still retain good separation. Where complete recovery of desalted sample is the major requirement, sample volumes of between 15 and 20% of V_t are recommended. For routine small scale desalting with the PD-10 columns, use of the maximum recommended sample volume of 2.5 ml results in effectively complete recovery of desalted material in a volume of 3.5 ml, with a dilution factor of only 1.4.

When designing an optimized desalting system, the volume of gel required should be packed in a short wide column rather than a long narrow one. This allows more rapid recovery of desalted materials since higher volume flow rates can be achieved with the wider column (Table 1).

3.5 Sample Composition. Due to the linear partition isotherms, gel filtration is largely independent of sample concentration. However, in addition to the solubility of the solutes, the viscosity of the sample often limits the concentration that can be used. A high sample *viscosity* causes instability of the zone (the migrating and partitioning peak material), producing an irregular flow pattern. This leads to very broad and skewed zones. The critical variable is the viscosity of the sample relative to the eluent.

In practice, the relative viscosities of sample and eluent should not differ by more than a factor of about 2, corresponding to a protein concentration in the sample of about 70 mg/ml, when a dilute aqueous buffer is used as the eluent.

3.6 Eluent. If the product is to be lyophilized, *volatile buffers* such as ammonium acetate, ammonium bicarbonate, or ethylenediamine acetate can be used. In desalting, the separation volume is so large that, in general, charged substances can also be treated with distilled water as eluent. Complete removal of salt is not possible, but the amount of ions excluded and, therefore, eluted with the high MW fraction is so small it can be neglected in most cases.

The most important consideration in the choice of eluent in gel filtration is its effect on the sample molecules. The pH and ionic composition of the buffer, and the presence of dissociating media or detergents can cause conformational changes, dissociation of proteins into subunits, and dissociation of enzymes from cofactors and of hormones from carrier molecules.

3.7 Flow Rate. Resolution decreases with increasing flow rate under conditions usually encountered in gel filtration. The optimum flow rate for maximum resolution of proteins is of the order of 2 ml cm^{-2}/h^{-1} in laboratory columns; although flows up to 5 times faster can often be used without much deterioration, maximum resolution is obtained with a long column and a low flow rate. The fastest run is obtained with a short column and a high flow rate. Good resolution and short running times are not compatible.

If peaks are well separated at a low flow rate using a long column, the excess *resolution may be traded off for speed.* The flow rate can be increased and a shorter column can be used or, alternatively, more sample can be applied. For preparative purposes, the advantage of a higher flow rate (and consequently a faster separation) often outweighs the loss of resolution in the chromatographic run. However, a prerequisite for this kind of optimization is the use of a rigid gel that will tolerate the higher operating pressures needed to produce higher flow rates.

The second kind of separation, fractionation, involves molecules of similar sizes and, consequently, much smaller differences in K_{av}. Here, the correct choice of gel and operating conditions is critical. Practical simplicity, excellent recovery, free choice of elution conditions, and straightforward interpretation of results make fractionation by gel filtration an invaluable part of any purification scheme (Table 1).

Table 1 - Comparison of Methods for Protein Desalting.

Criterion	Method		
	Gel Filtration with Sephadex G-25 or G-50	Conventional Dialysis in Cellulose Bags	Hollow Fiber Method
Simplicity	Easy	Easy	Easy
Efficiency of desalting	Usually approaching 100%	May reach 100% if taken to completion	About 95%
Reliability	Completely reliable	Bags may split or leak	Fibers may break or leak
Cost	Inexpensive	Inexpensive	Expensive
Speed	Typically nearly 100% desalting in 15 minutes	Very slow, can take several days	95% desalting in 20 minutes at best
Recovery	Usually approaching 100% at low protein concentrations	Absorption troublesome	May be 70% to 100% at lower protein at low concentrations
Recovery of small molecules	Usually approaching 100%	Not usually possible	Not usually possible
Dilution	Depends on bed loading	Small	Small
Lifetime	Usually 100 runs minimum. Columns may frequently be used for 1-2 years without repacking		Up to 50 runs or 4 months

VENDOR LITERATURE

SEPHADEX AND SEPHACRYL

Gel Filtration: Principles and Methods, 6th Edition; Pharmacia Biotech Inc., Piscataway, NJ, ph. (800) 526-3593. (Adapted)

1. Sephadex

1.1 Chemical and Physical Properties. Sephadex is a bead-formed gel prepared by cross-linking dextran with epichlorohydrin. A partial structure is shown in Figure 1. The large number of hydroxyl groups renders the gel extremely hydrophilic. Consequently, Sephadex swells readily in water and electrolyte solutions. The G-types of Sephadex differ in their degree of cross-linking and hence in their degree of swelling and fractionation range. The degree of swelling of Sephadex is substantially independent of the presence of salts and detergents. However, the effective fractionation range will be altered if the salts or detergents change the conformation of the substances being fractionated. Sephadex also swells in dimethyl sulfoxide and formamide. Dimethylformamide may be used with Sephadex G-10 and G-15 and mixtures of water with the lower alcohols may be used with Sephadex G-10, G-15, G-25, and G-50. It should be noted that the degree of swelling in organic solvents or their mixtures will not be the same as in water alone. Sephadex LH-20 and LH-60 (described in separate booklets), Sephacryl, and Sepharose CL are recommended for gel filtration in organic solvents.

Sephadex is available in different particle size grades. The Superfine grade is intended for column chromatography requiring very high resolution and for thin-layer chromatography. The Fine grade is recommended for preparative purposes where the extremely good resolution that can be achieved with the Superfine grade is not required, but where the flow rate is of greater importance. The Coarse and Medium grades are intended for preparative chromatographic processes where a high flow rate at a low operating pressure is essential. In addition, the Coarse grade is suitable for batch procedures.

Figure 1 - Partial structure of Sephadex.

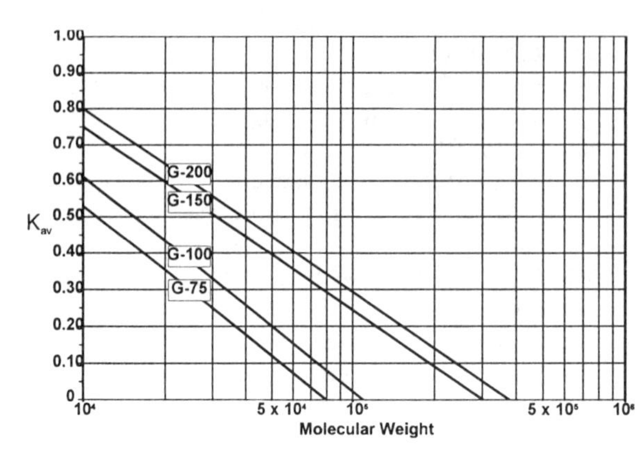

Figure 2 - Ranges of molecular weights (MW) retained by Sephadex G-type resins ("selectivity curves").

1.2 Chemical Stability. Sephadex is insoluble in all solvents (unless it is chemically degraded). It is stable in water, salt solutions, organic solvents, alkaline and weakly acidic solutions. In strong acids, the glycosidic linkages in the gel matrix are hydrolyzed. However, Sephadex can be exposed to 0.1 M HCl for 1-2 hours without noticeable effects, and in 0.02 M HCl Sephadex is still unaffected after 6 months. Prolonged exposure to oxidizing agents will affect the gel and should be avoided.

1.3 Physical Stability. Sephadex does not melt and may be sterilized in the wet state at neutral pH or dried by autoclaving for 30 minutes at 120°C without affecting its chromatographic properties. If dry Sephadex is heated to more than 120°C, it will start to caramelize. The mechanical strength of Sephadex depends on the degree of cross-linking. Highly cross-linked gels such as Sephadex G-10 and G-15 are rigid and high flow rates may be used. Gels with a small degree of cross-linking, e.g., Sephadex C-200 and G-150, may be compressed and the maximum flow rate that can be used varies with the type of Sephadex (Figure 2).

2. Sephacryl

2.1 Chemical and Physical Properties. Sephacryl is prepared by covalently cross-linking all dextran with *N,N'*-methylene bisacrylamide to give a rigid gel with a carefully controlled range of pore sizes (Figure 3). A small number of carboxyl groups may be present. The wet bead diameter is 40-105 mm, the average bead diameter approximately 70 mm. By varying the production conditions, gels with different fractionation ranges are obtained. Sephacryl is usually used with aqueous eluents. However, its gel structure allows the water to be replaced with organic solvents with a much smaller effector pore size than is observed with Sephadex (Table 1).

2.2 Chemical Stability. Sephacryl is insoluble in all solvents unless it is chemically degraded. It may be used in the range from pH 3 to 11. At lower pH, limited hydrolysis of the dextran chains may occur. Sephacryl has been treated with 0.2 M NaOH for 100 hours at room temperature without significant effect on flow rates or porosity. Sephacryl may be used with eluents containing detergents, e.g., SDS, and with structure-breaking media such as 6 M guanidine hydrochloride (Figure 4).

2.3 Molecular Weight Determination. A calibrated column of Sephacryl, Sephadex, or Sepharose provides a simple and well-documented way of determining the MW (defined in Figure 2) of proteins as a natural stage in their purification. Unlike electrophoretic techniques, gel filtration provides a means of determining the MW or size of native or denatured globular proteins under a wide variety of conditions (pH, ionic strength, temperature, etc.) freeing the researcher from the constraints imposed by the charge state of the molecules. Gel filtration in the presence of urea and guanidine hydrochloride, dissociating agents that transform polypeptides and proteins to a random coil configuration and thereby reduce structural differences, has proved particularly useful for MW determinations.

Figure 3 - Partial structure of Sephacryl resin.

Table 1 - Bed Volumes of Sephacryl in Organic Solvents.[a]

Organic Solvent	Approx. Volume (ml) Using S-200.	Approx. Volume (ml) Using S-300.
Dimethylsulfoxide	100	100
Dimethyl formamide	100	100
Ethyl acetate	70	90
Chloroform	70	85
n-Heptane	65	70
Tetrahydrofuran	65	75
Acetone	65	65
Toluene	60	85

[a]Using 100 ml of Sephacryl, sedimented in H_2O.

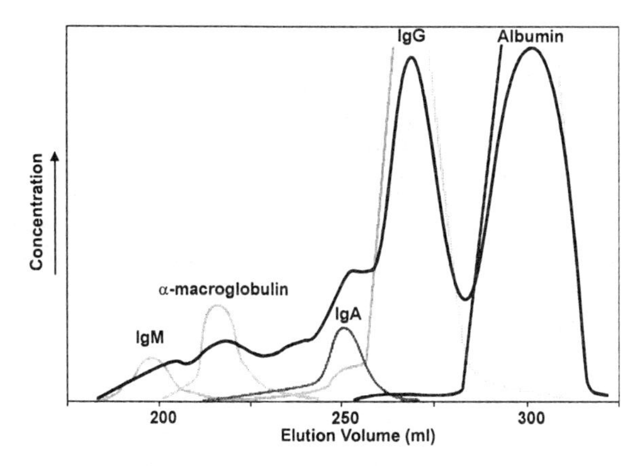

Figure 4 - Gel filtration of serum protein on Sephacryl S-300 Superfine. Pharmacia Column K 26/100; bed height 94 cm; eluent Tris-HCl buffer solution (I = 0.1, pH 8) containing NaCl (0.5 M); flow rate: 2.4 ml/cm^{-2} h^{-1}. I = ionic stregenth.

VENDOR LITERATURE

SIGMA™ GEL FILTRATION MOLECULAR WEIGHT MARKERS

Sigma Chemical Company, St. Louis, MO, ph. (800) 325-3010. (Adapted)

1. Introduction

Gel filtration chromatography is an established method or determining the size and molecular weight (MW) of proteins. Fractionation is based on the differential diffusion of various molecules into the gel pores. Proteins having high MWs do not enter the gel pores. Instead, they pass through the fluid volume of a column of porous gel particles faster than those having low MWs. Proteins elute from the gel in order of decreasing MW.

MW size determinations of unknown proteins are made by comparing the ratio of V_e/V_0 for the protein in question to the V_e/V_0 of protein standards of known MW (V_e is the elution volume and V_0 is the void volume). The V_0 of a given column is based on the volume of effluent required to elute a large molecule such as Blue Dextran (MW ~2 x 10^6 Da), which is excluded from entering the particles. As a result, it runs right through the column and elutes in the minimal volume. A calibration curve can then be prepared by plotting the logarithms of known MW of protein standards versus their V_e/V_0 values.

The ratio V_e/V_0 is essentially independent of column size and protein concentration but may be temperature dependent for some proteins. Unreliable MW may be obtained if the protein forms a complex with the gel, contains a large amount of carbohydrate, aggregates to larger complexes, or dissociates into subunits under the conditions used.[1] The MW of an impure protein may be determined using this procedure if a specific detection test is available.

2. Column Packing and Equilibration

To prepare the column, pour a slurry of approximately equal volumes of buffer and gel into a vertical column that is one-third filled with buffer. Let this settle about 15 minutes without flow, then allow the excess buffer to drain through the growing gel bed. Continuously add gel slurry until a bed height of at least 90 cm is obtained. Connect a buffer reservoir to the top of the column and let 2-3 column volumes of buffer pass through the column. At this time the bed height may be adjusted by addition or removal of preequilibrated gel.

NOTES:

1. The column should be continuously eluted with buffer when not in use.
2. For our testing, a resin bed 90 cm in height and 1.6 cm in diameter was used. This column was built and run at 0-5°C.
3. Gel preparation and column packing *must* be done carefully.

MW-GF 1000 KIT (For use with proteins in the MW 29 to 700 kDa range). Prepare cross-linked Sepharose 6B in an appropriate equilibration buffer. Suspend the resin in about 2 volumes of buffer. Let the gel settle (approx. 15-30 minutes) and then remove the smallest particles by decantation. Repeat last step. Degas the gel suspension before use.

3. V_0 Determination

Dissolve the Blue Dextran in equilibration buffer containing 5% glycerol. Glycerol is added to increase the density of the solution but its use is optional. The sample volume should be less than 2% of the total gel bed volume. For each resin, the concentration of Blue Dextran that will give an A_{260} of approximately 1.0 in the peak fraction is listed in Table 1. Carefully apply the Blue Dextran sample to the column to determine V_0 and to check the continuity of the column packing (avoid disturbing the gel bed surface). Immediately after applying the sample, begin collecting fractions of 0.5-1.5% of the total gel bed volume. Monitor the progress of the Blue Dextran sample down the column. Skewing of the blue band during chromatography reveals a fault in the column, although some tailing is normal. The Blue Dextran (D 4772) included in this kit contains some material of very high MW, which is excluded from Sepharose 6B, and a smaller amount of material of lower MW

Table 1 - Sephadex, Sephacryl, and Sepharose Kit Information[a]

Kit	Resin	Blue Dextran Concentration (mg/ml)	Flow Rate (ml/hr)
MW-GF-70	Sephadex G-75-50	1	12
MW-GF-200	Sephacryl S-200	2	20
MW-GF-1000	Sepharose CL-6B-200	2	20

[a]Column size: 90 cm x 1.6 cm; sample volume: 2.0 ml; fraction volume: 25 ml.

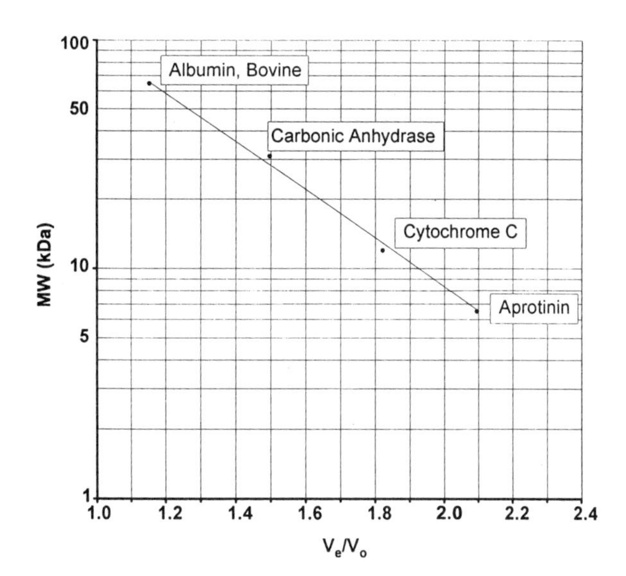

Figure 1 - Calibration curve typically obtained with MW-GF-7D Kit proteins on Sephadex G-75-50.

that is partially retained in the resin. The leading peak indicates V_0.

Elution of Blue Dextran may be followed by absorbance readings at either 280 or 610 nm. Determine the V_e for Blue Dextran (column V_0) by measuring the volume of effluent collected from the point of sample application to the blue peak center.

NOTES:

1. We do not advise mixing Blue Dextran with kit standards or sample proteins – many proteins bind to Blue Dextran.
2. It is recommended that protein standards and Blue Dextran solutions be prepared fresh. Occasionally, some aggregated protein may appear at V_0.

4. V_e Determinations for Protein Standards

Dissolve individual protein standards in equilibration buffer containing 5% glycerol. For a 90 x 1.6 cm column, apply 2 ml containing the following concentrations, as mg solid per ml. This will produce an A_{280} of ~1.0 in the peak fraction.

Apply protein standards to the column using a volume of sample equal to that used for the Blue Dextran sample and collect fractions at the flow rate used for the Blue Dextran sample. The elution of the standard proteins may be followed by readings at 280 nm. Determine the V_e for the protein standards spectrophotometrically by measuring the volume of effluent collected from the point of sample application to the center of the protein peak.

5. Standard Curve

Plot: MW vs. V_e/V_0 for each respective protein standard on semilog paper (see Figures 1, 2, and 3).

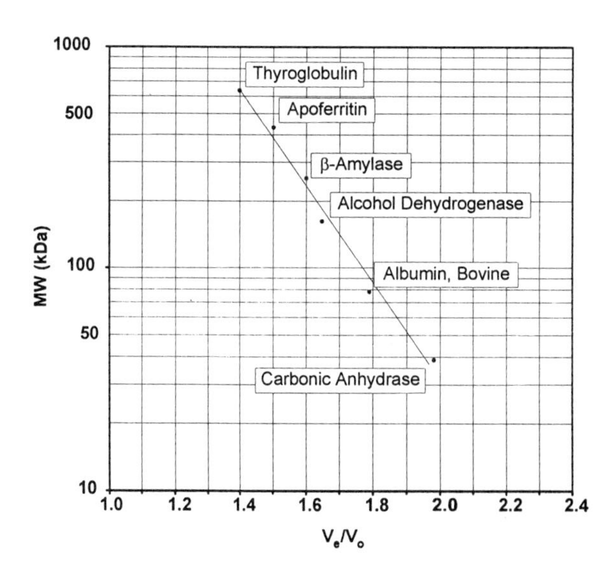

Figure 2 - Calibration curve typically obtained with MW-GF-200 Kit proteins Sephacryl S-200.

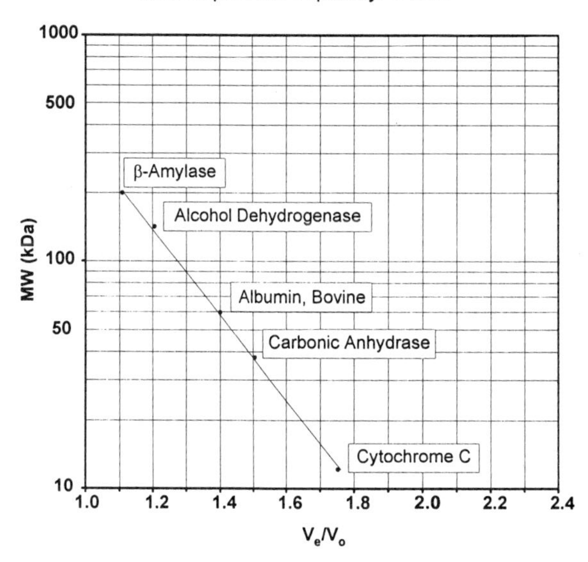

Figure 3 - Calibration curve typically obtained with MW-GF-1000 Kit proteins on Sepharose 6B.

6. Elution Volume Determination for Unknown Protein

Apply unknown sample to the column at an appropriate concentration (volume of sample applied should be the same as that used for the Blue Dextran and the protein standards). Determine the V_e of the unknown using the same methods applied to the standards. Calculate the V_e/V_0 for the unknown and determine its molecular weight from your standard curve.

References:

1. Whitaker, J. R., Anal. Chem. 35, 1950 (1963).
2. Andrews, P., Biochem J. 91, 222 (1964).
3. Marshall, J. J., J. Chromatogr. 53, 379 (1970).

LAB 7.4:

MICROPLATE β-GALACTOSIDASE ASSAY TO DETERMINE FRACTIONS CONTAINING FUSION PROTEIN; MW DETERMINATION

Overview:

In this lab, microplate β-galactosidase assays will be performed with the fractions from the β-gal-*myo*-3 fusion protein collected in Lab 7.3. During the β-galactosidase assay *o*-nitrophenol-para-galactoside is cleaved by β-galactosidase to form *o*-nitrophenol and β-D-galactoside. *o*-Nitrophenol is a highly colored reaction product that absorbs light at 410 nm. β-Galactosidase activity is directly related to the increase in absorbance at 410 nm. In otherwords, if the enzyme is present, the substrate ONPG is converted to galactose and the yellow colored *o*-nitrophenol. The chemical reaction is as follows:

ONPG + H_2O —(β-galactosidase)→ β-D-galactoside + *o*-nitrophenol + H^+

The gel filtration chromatography fractions containing the protein of interest will be pooled. Then, the molecular weight (MW) marker fractions from the gel filtration column will be used to construct an elution profile. Next, a standard curve will be constructed from the MW markers of log MW versus V_e/V_0 (where V_e is the elution volume of a given protein and V_0 is the void volume of the column – that is the volume of the solvent space surrounding the beads). There is a linear relationship between the relative elution volume of a protein V_e/V_0 and the log MW. Thus, a standard curve can be constructed using results obtained with proteins of known mass. The molecular weight of the *myo*-3 fusion protein will be extrapolated from this standard curve. In future labs, you will perform protein assays and SDS-PAGE with the pooled β-gal-*myo*-3 fusion protein to determine the effectiveness of the gel filtration column for β-gal-*myo*-3 fusion protein isolation.

Experimental Outline:

1. Perform microplate β-galactosidase assays of gel filtration fractions to determine which ones contain the β-gal-*myo*-3 fusion protein.
2. Pool fractions containing the β-gal-*myo*-3 fusion protein.
3. Construct the elution profile using results from absorbance measurements at 280 nm.
4. Construct a standard curve using results obtained with MW markers: log MW vs. V_e/V_0.
5. Interpolate to obtain the MW of the fusion protein from the standard curve.

Required Materials:

Substrate Mix and Stop Buffer (recipes below); microtiter plate; microtiter plate reader; graph paper; spectrophotometer

Procedures:

1. Microtiter Plate β-Galactosidase Assay

1. Prepare solutions
   a. *Substrate Mix*:

4 mg/ml	ONPG (*o*-nitrophenol-β-D-galactopyranoside)
1 mM	$MgCl_2$
50 mM	β-mercaptoethanol

0.1 M Na_2HPO_4 (pH 7.3)

b. *Stop Buffer*:

1 M $NaHCO_3$

c. *Buffer B*:

		NaCl	29	g
1	M	Tris-HCl, pH 7.4	10	ml
1	M	$MgCl_2$	10	ml
0.1	M	EDTA	10	ml
		β-mercaptoethanol	10	ml
		Triton X-100	1	ml

Add H_2O to 1000 ml. Store at room temperature.

2. Procedure
   a. Add 2 μl of each enzyme sample to 18 μl of Buffer B in separate wells of a microtiter plate. Mix well by shaking. Make sure to include wells for the blanks.
   b. Aliquot 2 μl of the diluted sample above (1/10 dilution) into 98 μl of Buffer B.
   c. To the samples in step b add 100 μl of the substrate mix and mix well to start the reaction. Time the addition of enzyme so that there is 5-15 seconds between the addition of each sample. This will ensure that all of the samples are incubated for equivalent times. Exact times must be recorded for each assay.
   d. Incubate at room temperature until a yellow color develops (time may vary).
   e. Add 100 ml of the stop buffer; timed as in step c. The stop buffer ends the reaction by increasing the pH to 11, thus inactivating the β-galactosidase enzyme.
   f. Read the plates at 405 nm in a microtiter plate reader.
   g. Pool the fractions containing the β-gal-*myo*-3 fusion protein.

2. Construct the Molecular Weight Elution Profile

1. Determine the absorbance at 280 nm for all the MW marker fractions.
2. Graph these absorbances versus the elution volume.

3. Determine the Molecular Weight of the *myo*-3-β-galactosidase Fusion Protein

1. Calculate the elution volume (V_e) of each MW marker fraction.
2. Calculate the void volume (V_0) of the gel filtration column.
3. Construct a standard curve of log MW (provided in the literature) versus V_e/V_0.
4. Determine the ratio V_e/V_0 for the β-gal-*myo*-3 fusion protein.
5. Interpolate the MW of the β-gal-*myo*-3 fusion protein.

Note:

1. ONPG (orthonitrophenol β-D-galactoside) is a β-galactoside analog that produces a yellow color when cleaved by β-galactosidase.

Additional Report Requirements:

1. Microtiter plate β-galactosidase absorbance readings.
2. Absorbances at 280 nm of the MW markers.
3. Elution profile of molecular weight marker.
4. Standard curve of MW of markers versus V_e/V_0 showing MW of β-gal-*myo*-3 fusion protein extrapolation.
5. Records of which fractions were pooled.

References:

1. Miller, J. M. (1973) *Experiments in Molecular Genetics,* Cold Spring Harbor Laboratory, Cold Spring Harbor, NY.
2. Pardee, A. B., Jocob F. and Monod., J. (1959) The genetic control and cytoplasmic expression of "inducibility" in the synthesis of β-galactosidase by *E. coli.* J. Mol. Biol. 1:165.
3. Zabin, I. and Fowler, A. (1970) β-Galactosidase and thiogalactoside transacetylase. *The Lactose Operon*, p. 27, Cold Spring Harbor Laboratory.

PROCESS RATIONALE

TIME COURSE ASSAY OF β-GALACTOSIDASE

Miller, J. M. (1973) *Experiments in Molecular Genetics*, Cold Spring Harbor Laboratory, Cold Spring Harbor, NY. (Adapted)

1. Introduction

The enzyme β-Galactosidase hydrolyzes β-D-galactosides. The reaction can be measured easily with chromogenic substrates, colorless substrates which are hydrolyzed to yield colored products. An example is *o*-nitrophenyl-β-D-galactoside. This compound is colorless, but in the presence of β-galactosidase it is converted to galactose and o-nitrophenol. The *o*-nitrophenol is yellow and can be measured by its absorption at 420 nm. If the o-nitrophenyl-β-D-galactoside (ONPG) concentration is high enough, the amount of *o*-nitrophenol produced is proportional to the amount of enzyme present and to the time the enzyme reacts with the ONPG. In order for the assay to be linear, the ONPG must be in excess. For best results, the amount of enzyme should be such that it takes between 15 minutes and 6 hours for a faint yellow color to develop. The reaction is stopped by adding a concentrated Na_2CO_3 solution, which shifts the pH to 11 and inactivates β-galactosidase.

2. Materials

2.1 Day 1. Overnight cultures of strains CSH23 and CSH36, 10 ml of 1x A medium supplemented with 20 μg/ml B1, 10 ml of 3 M $MgSO_4$, and 0.4% glucose, 10 ml of the same medium + 1 mM IPTG, 10 ml of each of the media used in Day 1.

2.2 Day 2. Spectrophotometer, ice bucket, 6 ml of Z buffer for β-galactosidase assay: 4 test tubes with 0.9 ml of Z buffer each; 4 test tubes with 0.5 ml of Z buffer each; 1 ml toluene; 2 ml ONPG, 4 mg/ml in 0.1 M phosphate buffer, pH 7 (1x A); 5 ml 1 M Na_2CO_3; vortex, stop watch, water bath at 28°C.

3. Methods

3.1 Day 1. In order to obtain precision in assaying β-galactosidase, it is necessary to use exponentially growing cells; with the following experiment, we will assay two different strains. These have the following genotypes: $i^+z^+y^+$ and $i^-z^+y^+$. Prepare an overnight culture of each of the two strains in minimal medium 1x A supplemented with B1, $MgSO_4$, glucose, and IPTG. Also prepare an overnight culture in the same medium without IPTG for each of the two strains.

3.2 Day 2. Subculture each of the four overnight cultures into fresh media of the exact types used in Day 1. Add 4 drops of the overnight culture to 5 ml of fresh media. Continue to aerate at 37°C until the cultures are at 2-5 x 10^8 cells/ml (A_{600} of 0.28-0.70). Cool the cultures to prevent further growth by immersing in an ice bucket containing a mixture of ice and water. After 20 minutes β-galactosidase activities of the cultures can be assayed and the bacterial densities measured.

Record the cell density by measuring the absorbance at 600 nm. Withdraw 1 ml of the culture for this measurement. Immediately add aliquots of the cultures to the assay medium (Table 1). The final volume will always be 1 ml. If high levels of β-galactosidase are being assayed, add 0.1 ml of culture to 0.9 ml of Z buffer. If low levels are being determined, then add 0.5 ml of culture to 0.5 ml of Z buffer. In this experiment, use both of these dilutions for each of the four cultures. Add one drop of toluene with a Pasteur pipette to each tube and immediately vortex for 10 seconds. Toluene partially disrupts the cell membrane, allowing small molecules, such as ONPG, to diffuse into the cell. Allow the toluene to evaporate by placing the tubes on a rotor at 37°C for 40 minutes with the top open. (Mild shaking at 37°C in a shaker bath is also sufficient.) Place the tubes in a water bath at 28°C for 5 minutes.

Start the reaction by adding 0.2 ml of ONPG (4 mg/ml) to each tube and shake for a few seconds. Monitor the time progress of the reaction with a stopwatch. Stop it by adding 0.5 ml of a 1 M Na_2CO_3 solution after sufficient yellow color has developed.

Record the absorbances at both 420 nm and 550 nm for each tube. Ideally, the A_{420} reading (yellow color) should be 0.6-0.9. The reading at 420 nm is actually a combination of absorbance by the *o*-nitrophenol and light scattering by the cell debris. Errors due to this latter component can be corrected by obtaining the absorbance at another wavelength where only light scattering occurs (e.g., there is no contribution from *o*-nitrophenol at 550 nm). The light scattering at 420 nm is proportional to that at 550 nm.

Table 1 - Preparation of Z Buffer[a]

Z Buffer (per liter)	Amount	Concentration (M)
$Na_2HPO_4 \cdot 7H_2O$	16.1 g	0.06 M
$NaH_2PO_4 \cdot H_2O$	5.5 g	0.04 M
KCl	0.75 g	1 mM
$MgSO_4 \cdot 7H_2O$	0.246 g	0.05 M
β-Mercaptoethanol	2.7 ml	0.05 M

[a]Do not autoclave. Adjust pH to 7.

For *E. coli*, the A_{420} (light scattering) = 1.75 A_{500}. Using this correction factor we can compensate for the light scattering and compute the true absorbance of the *o*-nitrophenol. For our activity units, we use the following formula:

$$N_{BGU} = \frac{1000\left[A_{420} - (1.75 A_{550})\right]}{V \Delta t A_{600}}$$

where, N_{BGU} is the number of β-galactosidase units, A_{280} and A_{550} are read from the reaction mixture, A_{600} reflects the cell density just before assay, *t* is the time of the reaction in minutes and *V* is the volume of culture used in the assay in ml. These units are proportional to the increase in *o*-nitrophenol per minute per bacterium. They are convenient because a fully induced culture grown on glucose has approximately 1000 units, and an uninduced culture approximately 1 unit. Instead of correcting for the cell debris interference, it is also possible to spin down the debris in a small centrifuge, thus eliminating the 550 nm reading.

4. Sample Calculation

Suppose we assay an IPTG-induced culture with an A_{600} of 0.60. We add 0.1 ml of this culture to 0.9 ml of Z buffer + toluene + ONPG, and after 15 minutes stop the reaction. The A_{420} is 0.900, and A_{550} 0.050. Since 1.75 x 0.050 = 0.088 we have

$$N_{BGU} = \frac{1000\,(0.900 - 0.088)}{(15)(0.60)(0.1)} = \frac{812}{0.9} = 902 \text{ Units}$$

If after several hours there is still no yellow color in the basal level assays, cover them with aluminum foil and allow the reaction to continue overnight. In this case, prepare a control sample containing 1 ml of Z buffer, but no cells, and incubate along with the basal level cultures, after addition of ONPG. This will serve as a control for the spontaneous splitting of ONPG.

The specific activity of β-galactosidase is defined in terms of units/mg protein. Many investigators define a unit of β-galactosidase (U) as the amount of enzyme that produces 1 nmol *o*-nitrophenol/min at 28°C, pH 7. Under the above conditions 1 nmol/ml *o*-nitrophenol has an A_{420} of 0.0045 using a 10 mm light path. Therefore, the above units can easily be converted into these units. Although protein is usually determined directly from extracts or estimated from the dry weight of cells, we can estimate the amount of protein by assuming that 109 cells yield approximately 150 μg of protein and that an A_{600} of 1.4 corresponds to 109 cells/ml. Using this information, calculate the specific activity of β-galactosidase in terms of nmol of *o*-nitrophenol/ min/mg protein.

5. Additional Methods

As an alternative to the toluene method, 2 drops of chloroform and 1 drop of 0.1% SDS solution are added to each ml of assay mix. The tubes are then vortexed for 10 seconds. With this procedure, more cells are opened and the evaporation step is eliminated.

References:

1. Pardee, A. B., Jacob, J. and Monod, J. (1959) The genetic control and cytoplasmic expression of "inducibility" in the synthesis of β-galactosidase by *E. coli.*, J. Mol. Biol. 1:165.
2. Zabin, I. and Fowler, A. (1970) β-Galactosidase and thiogalactoside transacetylase. *The Lactose Operon*, p. 27, Cold Spring Harbor Laboratory.

VENDOR LITERATURE

β-GALACTOSIDASE SUBSTRATES

Fluka Chemie AG, Buchs, Switzerland, ph. (USA): (800) 358-5287. (Adapted)

1. Isopropyl-β-D-1-thiogalactopyranoside, IPTG

$C_9H_{18}O_5S$ MW: 238.31 [367-93-1]
59740 **BioChemika > 99%** (TLC)
250 mg - $24
1 g - $70

1.1 Applications/Literature. IPTG is an inducer of β-galactosidase activity in bacteria and is used in conjunction with X-Gal to detect *lac* gene activity during cloning experiments: S. Cho *et al.*, Biochem. Biophys. Res. Commun. 128, 1268 (1985); D. G. Yansura, D. J. Henner, *Genet. Biotechnol. Bacilli* (A. T. Ganesan, J. A. Hoch, eds.), 2nd ed., p. 249, Academic Press, Orlando, Fla. (1983).

1.2 Storage Handling/Safety. Store at -18°C. Keep under argon.

2. 2-Nitrophenyl-β-D-galactopyranoside

$C_{12}H_{15}NO_8$ MW: 301.26 [369-07-3]
73660 **BioChemika ~ 99%** (HPLC)
Appearance: off-white powder
1 g – $15
5 g – $60
25 g – $230
Fluka-Specification
mp 185-190°C dec.; $[\alpha]^{20}$-86±1⁺; $[\alpha]^{20}$-69±1⁺

2.1 Applications/Literature. Substrate for the determination of β-galactosidases: H. Jagota *et al.*, J. Food Sci. 46, 161 (1981). Beil. 17, 7*N*, 52

2.2 Storage Handling/Safety. Store at 0-4°C. Hygroscopic. Keep under argon. Sensitive to humidity.

3. 4-Nitrophenyl-α-D-galactopyranoside

$C_{12}H_{15}NO_8$ MW: 301.26 [7493-95-0]
73653 **BioChemika > 99%**
250 mg – $14
1 g – $40

3.1 Applications/Literature. Substrate for α-galactosidase: C. A. Dangelmaier, H. Holmsen, Anal. Biochem. 104, 182 (1980). Binds to the lactose carrier protein of *E. coli*: P. Overath *et al.*, Biochemistry 18, 1 (1979); G. J. Kaczorowski *et al.*, Proc. Natl. Acad. Sci. U.S.A 77, 6319 (1980). Beil. 17, 7*N*, 55

3.2 Storage Handling/Safety. Store at 0-4°C. Hygroscopic. Keep under argon. Sensitive to humidity.

4. 4-Nitrophenyl-β-D-glucopyranoside, ONPG

$C_{12}H_{15}NO_8$ MW: 301.26 [3150-24-1]
73760 **BioChemika > 99%** (HPLC)
1 g – $20
5 g – $85

4.1 Applications/Literature. Substrate for the determination of β-galactosidase: V. Buonocore *et al.*, J. Appl. Biochem. 2, 390 (1980). Beil. 17, 7*N*, 55

4.2 Storage Handling/Safety. Store at 0-4°C. Hygroscopic. Keep under argon. Sensitive to humidity.

5. 2-Nitrophenyl-β-D-glucopyranoside

$C_{12}H_{15}NO_8$ MW: 301.26 [2816-24-2]
73674 **BioChemika > 99%** (HPLC)
1 g – $32
5 g – $135

5.1 Applications/Literature. Substrate for the detection of β-glucosidase: H. Verachtert *et al.*, J. Chromatogr. 147, 443 (1978). Beil. 17, 7*N*, 52

5.2 Storage Handling/Safety. Store at 0-4°C. Hygroscopic. Keep under argon. Sensitive to humidity.

INNOVATION/INSIGHT

LUMINESCENT REPORTER GENE ASSAYS FOR LUCIFERASE AND β-GALACTOSIDASE USING A LIQUID SCINTILLATION COUNTER

Fulton, R. and Van New, B. (1993) Biotechniques 14: 762. (Adapted)

1. Introduction

The increased sensitivity and ease of detection of firefly luciferase may appeal to anyone doing reporter gene assays. We report here that chemiluminescent assays can be performed in a liquid scintillation counter using both luciferase and β-galactosidase reporter genes. These assays are conveniently performed from the same cell extract. The increased sensitivity over *chloramphenicol acetyltransferase* (CAT) assays allows the transfection of significantly fewer cells, using less material with a shorter turn-around time. Its application is demonstrated here using a kappa immunoglobulin reporter gene construct that shows inducibility in pre-B cells. Mouse B-cell lines are transfected by a modification of the DEAE-dextran method.

We have applied this system to the study of the developmental control of mouse kappa immunoglobulin light chain expression. We constructed a series of luciferase reporter gene vectors containing a kappa V-region promoter, which was paired with either the kappa intron enhancer or 3" enhancer. Transfection of these constructs into a mouse pre-B cell line shows that both enhancers drive a higher level of expression than the promoter alone. The intron enhancer is induced to still higher expression levels with the mitogen, lipopolysaccharide. The luminescent substrate, *AMP-GD*, makes cotransfection with a β-galactosidase expressing vector, such as pCFI110, a very convenient internal control for transfection efficiency. This assay can be performed using the same lysate as the luciferase assay and can also be measured on a liquid scintillation counter. Thus, (activity) values are corrected for transfection efficiency using pCH110 as the cotransfectant control. It was previously reported that the sensitivity of the luciferase assay allows transfections using viral transcriptional control elements to be scaled down significantly, to 1/100 of the size of a normal CAT assay. The high sensitivity reported for the luminescent β-galactosidase assay may allow its use in microtransfection as well. Indeed, we have found that luciferase microtransfection is suitable for use with vectors containing immunoglobulin control regions (data not shown). In this system, however, more consistent results are achieved using the mid-size transfection described above. This still amounts to a sizable savings of time and materials compared with CAT assays.

In summary, we have demonstrated rapid, reliable, and sensitive chemiluminescent detection techniques that make luciferase and β-galactosidase transfections accessible to most laboratories without the purchase of additional equipment.

2. CAT Assay

Chloramphenicol acetyltransferase catalyzes the transfer of an acetyl group from acetyl-Coenyme A (CoA) to the 3'-hydroxy position of chloroamphenicol.

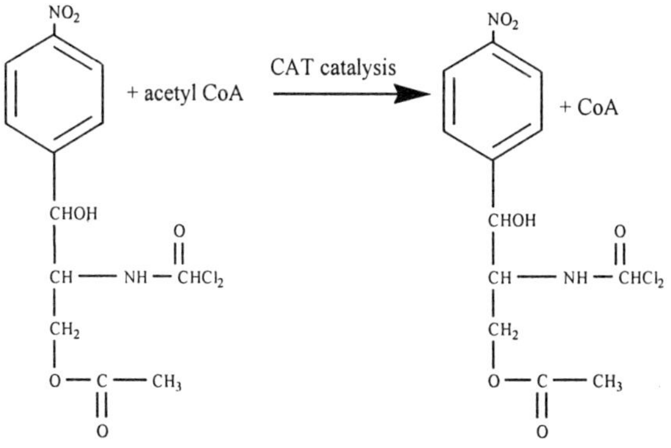

NO_2 + acetyl CoA —CAT catalysis→ NO_2 + CoA

CHOH, CH — NH — $CHCl_2$ (C=O), CH_2, O — C(=O) — CH_3

Chloramphenicol

Reference:

1. Schwartz, O., J. L. Virelizier, L. Montagnier and U. Hazan. 1990. A microtransfection method using luciferase-encoding reporter gene for the assay of human immunodeficiency virus LTR promoter activity. Gene 88:1, 97-205.

LAB 7.5:
ION EXCHANGE COLUMN CHROMATOGRAPHY

Overview:

Anion exchange chromatography will be used to purify the β-galactosidase-*myo*-3 fusion protein. Ion exchange chromatography separates proteins on the basis of their net charges. A positively charged gel matrix separates proteins based on difference in net negative charge. You will use the Macro-Prep High Q Strong Anion Exchange Support (BioRad) – a macroporous ion exchange hydrophilic matrix with high chemical, mechanical, and thermal stability.

The negatively charged fusion protein will be eluted from the positively charged resin by increasing the concentration of NaCl in the elution buffer. Elution occurs because the chloride ions compete with the negatively charged groups on the protein for binding sites on the matrix. Thus, proteins of differing charges will elute from the column at characteristic salt concentrations. Proteins with a low density of negative charge elute first, followed by those with more negative charge densities. Following purification, you will perform β-galactosidase enzyme assays to determine which fractions contain the majority of the fusion protein. Those containing the highest concentration of fusion protein will be pooled. In future sessions, you will perform protein assays and SDS-PAGE to determine the effectiveness of this purification step.

Experimental Outline:

1. Prepare and run anion exchange chromatography.
2. Perform microtiter plate β-galactosidase assay.

Required Materials:

Macro-Prep High Q Strong Anion Exchange Support (BioRad); dry Ottawa sand; chromatography column (2.5 x 20 cm); Buffer A (50 mM $NaPO_4$, pH 6.5); β-gal-*myo*-3 fusion protein; fraction collector; substrate mix (recipe below); stop buffer (recipe below); microtiter plate; microtiter plate reader; graph paper; spectrophotometer

Prelab Preparations:

1. Prepare buffers A, A2, A3, A4, A5, A6, A7, A8, A9, and A10. Components of these buffers are the same as Buffer A, except that the NaCl concentration is varied. NaCl concentrations are as follows: A: 0.1 M; A2: 0.2 M; A3: 0.3 M; A4: 0.4 M; A5: 0.5 M; A6: 0.6 M; A7: 0.7 M; A8: 0.8 M; A9: 0.9 M; A10: 1.0 M.

Procedures:

1. Prepare Anion Exchange Chromatography Column

1. Assemble the chromatography column (2.5 x 20 cm) on a ring stand. Attach the stopcock to the bottom.
2. Add dry Ottawa sand (20-30 mesh) to about 0.5 to 1 inch deep in the bottom of the column.
3. Gently pour 100 ml of a 50% slurry of Macro-Prep High Q Strong Anion Exchange Support into the column.
4. Turn on the stopcock and allow the column to pack.
5. Attach a buffer reservoir filled with Buffer A to the top of the column.
6. Equilibrate the column with 100 ml of Buffer A. Allow the buffer to flow down until it is almost even with the gel bed. Turn off the stopcock.
7. Attach small diameter tubing to the bottom of the stopcock and make connections to the fraction collector.
8. Program the fraction collector to collect a volume of 150 drops.
9. Start the fraction collector.
10. Gently apply 1 ml of the crude soluble extract collected in Lab 7.2.
11. Turn on the stopcock and allow the sample to move into the gel until it is even with the gel surface. Turn off the stopcock. Be careful to not let the matrix run dry.
12. Gently layer a few milliliters of Buffer A onto the top of the gel bed. Turn on the stopcock and allow the buffer to move into the gel until it is even with the gel surface.
13. Repeat step 12 several times until the sample is safely into the gel.
14. Add 20 ml of Buffer A. Turn on the stopcock and allow the buffer to move into the gel until it is even with the gel surface.

15. Repeat step 14 successively with Buffers A2 through A10.
16. Add an additional 100 ml of Buffer A to the column. Allow the column to run until the buffer is just above the gel bed. Your teaching assistants will regenerate the gel for future use. (Regeneration instructions given in the Macro-Prep instructions.)

2. Microtiter Plate β-Galactosidase Assay

1. Prepare solutions

a. *Substrate Mix*:

4	mg/ml	ONPG (o-nitrophenol-β-D-galactopyranoside)
1	mM	$MgCl_2$
50	mM	β-mercaptoethanol
0.1	M	Na_2HPO_4 (pH 7.3)

b. *Stop Buffer*:

1	M	$NaHCO_3$

c. *Buffer B*:

		NaCl	29	g
1	M	Tris-HCl, pH 7.4	10	ml
1	M	$MgCl_2$	10	ml
0.1	M	EDTA	10	ml
		β-mercaptoethanol	10	ml
		Triton X-100	1	ml

Add H_2O to 1000 ml. Store at room temperature.

3. Procedure

a. Add 2 μl of each enzyme sample to 18 μl of Buffer B in separate wells of a microtiter plate. Mix well by shaking. Make sure to include wells for the blanks.

b. Aliquot 2 μl of the diluted sample above (1/10 dilution) into 98 μl of Buffer B.

c. To the samples in step b add 100 μl of the substrate mix and mix well to start the reaction. Time the addition of enzyme so that there is 5-15 seconds between the addition of each sample. This will ensure that all of the samples are incubated for equivalent times. Exact times must be recorded for each assay.

d. Incubate at room temperature until a yellow color develops (time may vary).

e. Add 100 ml of the stop buffer; timed as in step c. The stop buffer ends the reaction by increasing the pH to 11, thus inactivating the β-galactosidase enzyme.

f. Read the plates at 405 nm in a microtiter plate reader.

g. Pool the fractions containing the β-gal-*myo*-3 fusion protein.

Notes:

1. The column should be continuously eluted with buffer when not in use.
2. Alternate procedure: substitute 0 to 1 M NaOAc for NaCl.

Additional Report Requirements:

1. Microtiter plate β-galactosidase absorbance readings.
2. Records of which fractions were pooled.

Reference:

1. BioRad Ion-Exchange Standard Instruction Manual.

PROCESS RATIONALE

ION EXCHANGE CHROMATOGRAPHY

Ion Exchange Chromatography: Principles and Methods, pp. 4-7 (1987) Pharmacia, Piscataway, NJ. (Adapted)
Wankat, P. C. (1990) *Rate Controlled Separations,* Elsevier Science Publishers, New York, NY. (Adapted)

1. Introduction

Separation in ion exchange chromatography depends upon the reversible adsorption of charged solute molecules to an immobilized ion exchange group of opposite charge. Most ion exchange experiments are performed in five main stages. These steps are illustrated schematically in Figure 1.

The first stage (A) is equilibration, in which the ion exchanger is brought to a starting state, in terms of pH and ionic strength, which allows the binding of the desired solute molecules. The exchanger groups are associated at this time with exchangeable counterions (usually simple anions or cations, such as chloride or sodium).

The second stage (B) is sample application and adsorption, in which solute molecules carrying the appropriate charge displacc counterions and bind reversibly to the gel. Unbound substances can be washed out from the exchanger bed using starting buffer.

In the third stage (C), substances are removed from the column by changing to elution conditions unfavorable for ionic bonding of the solute molecules. This normally involves increasing the ionic strength of the eluting buffer or changing its pH. In Figure 1, desorption is achieved by introducing an increased salt concentration gradient. Solute molecules are released from the column in the order of their strengths of binding, the most weakly bound substances are eluted first.

The fourth (D) and fifth stages (E) are the removal from the column of substances not released by the previous experimental conditions and reequilibration at the starting conditions for the next purification.

Separation is obtained because different substances have different degrees of interaction with the ion exchanger due to differences in their charges and charge densities. These interactions can be controlled by varying conditions such as ionic strength and pH. The differences in charge properties of biological compounds are often considerable. Since ion exchange chromatography is capable of separating species with very minor difference in properties, e.g., two proteins differing by only one charged amino acid, it is a very powerful separation technique.

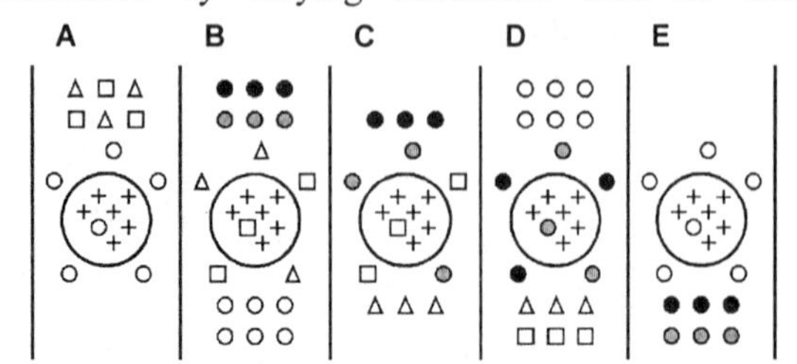

Figure 1 - Ion exchange chromatography: salt gradient elution. ○ - initial buffer counterions. Δ, □ - sample (□ - included, Δ - excluded). ●, ◉ - gradient ions (◉ - included, ● - excluded).

In ion exchange chromatography, one can choose whether to bind the substances of interest and allow the contaminants to pass through the column, or to bind the contaminants and allow the substance of interest pass through. Generally, the first method is more useful since it allows a greater degree of fractionation and concentrates the substance(s) of interest.

The conditions under which substances are bound (or become free) are discussed in detail in the sections dealing with choice of experimental conditions. In addition to the ion exchange effect, other types of binding may occur. These effects are small and are mainly due to van der Waals forces and nonpolar interactions.

Ion exchange separations may be carried out in a column or by a batch procedure. Both methodologies are performed using the definite stages of equilibration, sample adsorption etc.

2. Matrix

An ion exchanger consists of an insoluble porous matrix to which charged groups have been covalently bound. The charged groups are associated with mobile counterions. These counterions can be reversibly exchanged with other ions of the same charge without altering the matrix.

It is possible to have both positively and negatively charged exchangers (Figure 2). Positively charged exchangers have negatively charged counterions (anions) available for exchange and so are termed anion exchangers. Negatively charged exchangers have positively charged counterions (cations) and are termed cation exchangers.

The matrix may be based on inorganic compounds, synthetic resins, polysaccharides, etc. The characteristics of the matrix determine its chromatographic properties, such as resolution, capacity, recovery, and physical properties such as its mechanical strength and flow properties. The nature of the matrix will also affect its behavior towards biological substances and the maintenance of biological activity.

The first ion exchangers were synthetic resins designed for applications such as demineralization,

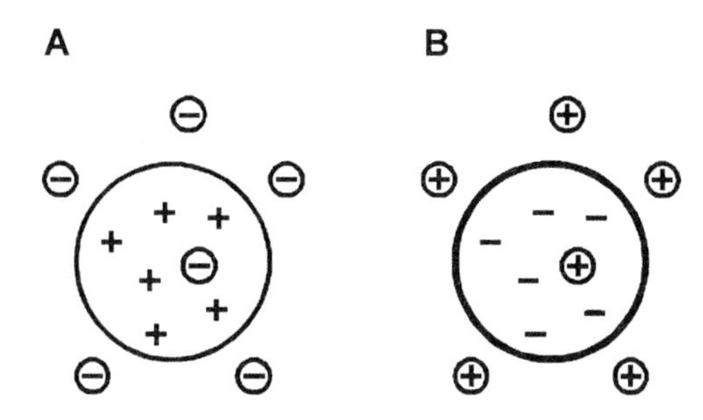

Figure 2 - Ion exchanger types. (A) Anion exchanger with exchangeable counterions. (B) Cation exchanger with exchangeable counterions.

water treatment, and recovery of ions from wastes. Such ion exchangers consist of hydrophobic polymer matrices highly substituted with ionic groups, and have very high capacities for small ions. Due to their low permeability, these matrices have low capacities for proteins and other macromolecules. In addition, the extremely high charge density gives very strong binding and the hydrophobic matrix tends to denature labile biological materials. Thus, despite their excellent flow properties and capacities for small ions, these types of ion exchanger are unsuitable for use with biological samples.

The first ion exchangers designed for use with biological substances were the cellulose ion exchangers. Because of the hydrophilic nature of cellulose, these exchangers had little tendency to denature proteins. Unfortunately, many cellulose ion exchangers had low capacities (or the cellulose became soluble in water) and poor flow properties (due to their irregular shapes).

Ion exchangers based on dextran (Sephadex), followed by those based on agarose (Sepharose CL-6B) and cross-linked cellulose (DEAE Sephacel) were the first ion exchange matrices to combine the correct spherical form with high porosity, leading to improved flow properties and high capacities for macromolecules.

Subsequently, developments in gel technology have enabled this macroporosity to be extended to the highly cross-linked agarose-based media such as Sepharose Fast Flow and Sepharose High Performance, and the synthetic polymer matrix, MonoBeads. These modern media offer the possibility of fast high-capacity high-resolution ion exchange chromatography at both analytical and preparative scales.

3. Charged Groups

The presence of charged groups is a fundamental property of an ion exchanger. The type of group determines the type and strength of the ion exchanger; their total number and availability determine the capacity. There are a variety of groups that have been chosen for use in ion exchangers; some of these are shown in Tables 1a and 1b.

Table 1a - Functional Groups Used on Anion Exchangers.

Anion Exchangers	Functional Group
Diethylaminoethyl (DEAE)	$OCH_2CH_2N^+H(CH_2CH_3)_2$
Quaternary aminoethyl (QAE)	$OCH_2CH_2N^+(C_2H_5)_2CH_2CH(OH)CH_3$
Quaternary ammonium (Q)	$CH_2N^+(CH_3)_3$

Table 1b - Functional Groups Used on Cation Exchangers.

Cation Exchangers	Functional Group
Carboxymethyl (CM)	OCH_2COO^-
Sulfopropyl (SP)	$CH_2CH_2CH_2SO_3^-$
Methyl sulfonate (S)	$CH_2SO_3^-$

Sulfonic and quaternary amino groups are used to form strong ion exchangers; the other groups form weak ion exchangers. The terms *strong* and *weak* refer to the extent of the variation of ionization with pH and not the strength of binding. Strong ion exchangers are completely ionized over a wide pH range, whereas with weak ion exchangers, the degree of dissociation and thus exchange capacity varies much more markedly with pH.

Some properties of strong ion exchangers are:

1. Sample loading capacity does not decrease at high or low pH values due to loss of charge from the ion exchanger.
2. A very simple mechanism of interaction exists between the ion exchanger and the solute.
3. Ion exchange experiments are more controllable since the charge characteristics of the media do not change with changes in pH. This makes strong exchangers ideal for working with data derived from electrophoretic titration curves.

4. Resolution

The *resolution* (R_s) is defined as the distance between peak maxima compared with the average base width of the two peaks. Elution volumes and peak widths should be measured with the same units to give a dimensionless value to the resolution. R_s is a measure of the relative separation between two peaks and can be used to determine if further optimization of the chromatographic procedure is necessary. If R_s = 1.0 (Figure 4), 98% purity has been achieved at 98% of peak recovery, provided the peaks are Gaussian and approximately equal in size. Baseline resolution requires that R_s = 1.5. At this value, the purity of the peak material is 100%.

A completely resolved peak is not equivalent to a pure substance. This peak frequently represents a series of components that are not resolvable using the selected separation parameter. The resolution

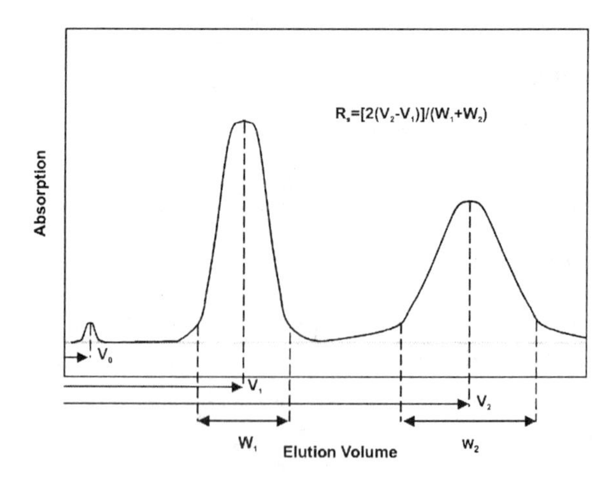

Figure 3 - Determination of the resolution (R) between two peaks. V_o = void volume, V_1 = elution volume for peak 1, V_2 = elution volume for peak 2, V_t = total volume, W_1 = peak width for peak 1, W_2 = peak width for peak 2.

achievable in a system is proportional to the product of the *selectivity* (α, section 7), the *efficiency* (H, section 6), and the *capacity* (k', section 5) of the system, the three most important parameters to control in column chromatography. The analytical expression for R_s is:

$$R_s = \frac{1}{2}\left(\frac{\alpha - 1}{1+\alpha}\right) N^{\frac{1}{2}} \left(\frac{k'}{1+k'}\right)$$

This equation is referred to as the "fundamental equation" of linear chromatography.

5. Capacity Factor

The *capacity* or *retention factor* (k') is a measure of the retention of a component and should not be confused with loading capacity (mg sample/ml) or ionic capacity (mmol/ml). The capacity factor is calculated for each individual peak. For example, k' for peak 1 in Figure 3 is derived from the equation:

$$k'_1 = V_1 - V_o / V_o$$

In the equation for R_s, k' is the average of k'_1 and k'_2 (calculated for peak 2).

Adsorption techniques such as ion exchange chromatography can have high capacity factors since experimental conditions can be chosen which lead to peak retention volumes greatly in excess of V_t. This contrasts with the technique of gel filtration, where capacity is low, since all peaks must elute within the volume (V_t - V_o)

6. Efficiency

The *efficiency factor* (N) is a measure of the zone broadening which occurs on the column (peak width) and can be calculated from the expression:

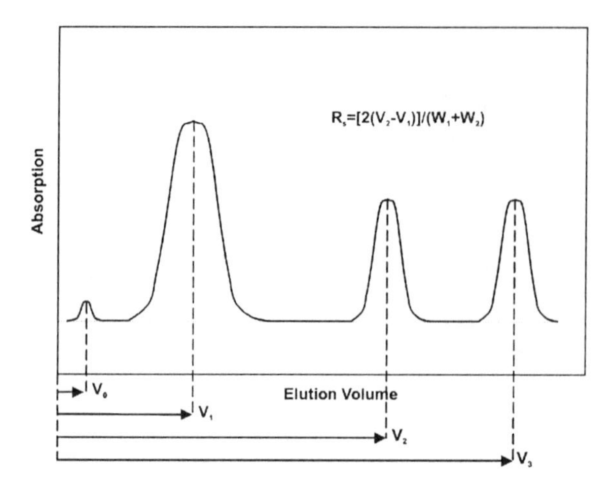

Figure 4 - Hypothetical chromatogram. Definitions are as in Figure 3.

$$N = L^2/\sigma^2$$

where the bed length (L) is divided by the standard deviation of the peak (σ).

It is expressed as the number (N) of *theoretical plates* for the column under specified experimental conditions. Efficiency (H) is frequently stated in the number of theoretical plates per meter chromatographic bed, or height equivalent to a theoretical plate, which is the *bed length* (L) divided by the plate number (N).

$$H = L/N$$

Since the observed value for N is dependent on experimental factors such as flow rate and sample loading, it is important that comparisons are done under identical conditions. In the case of ion exchange chromatography, efficiency is measured under *isocratic conditions*, using a substance that does not interact with the matrix, e.g., acetone.

One of the main causes of zone broadening in a chromatography bed is longitudinal diffusion of the solute molecules. The diffusion effect is minimized if the distances available for diffusion, in both the liquid phase and the gel beads, are minimized. In practice, this is achieved by using small uniform bead sizes. Important developments in ion exchange chromatography have been the introduction of 10-μm and 34-μm diameter particles such as MonoBeads and Sepharose High Performance, to give high efficiency media.

After bead size, the second major contributory factor to efficiency is good experimental technique. Badly, unevenly packed chromatography beds and air bubbles will lead to *channeling*, *zone broadening*, and loss of resolution. Good separations require well packed columns and the importance of column

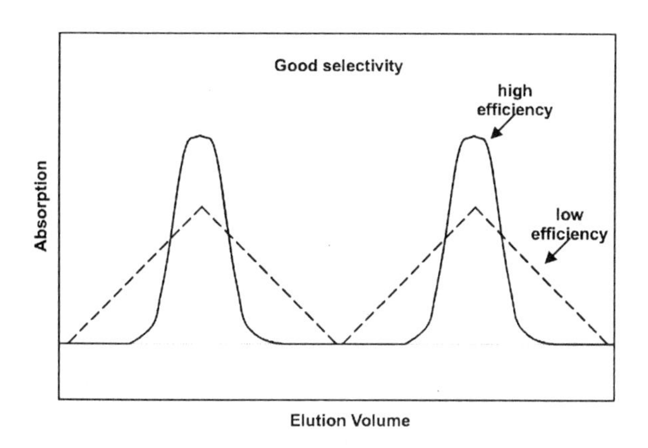

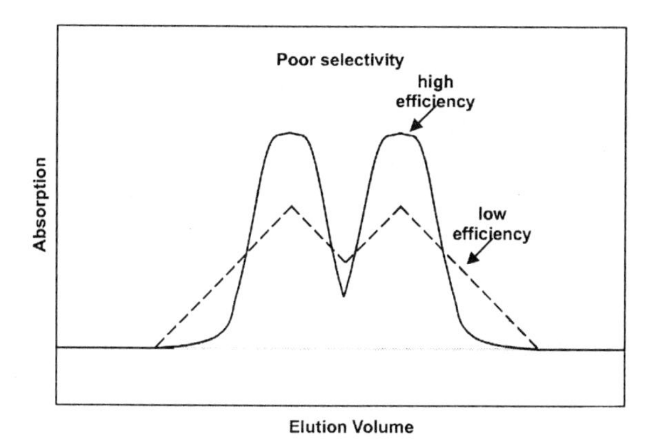

Figure 5 - The effects of selectivity and efficiency on resolution. Good (left) and poor (right) selectivity.

packing increases in direct proportion to the performance required.

7. Selectivity

The *selectivity* (α) defines the ability of the system to separate peaks, i.e., the distance between two peaks. The selectivity factor can be calculated from the chromatogram (Figure 5) using the expression:

$$\alpha = V_2 - V_0/(V_1 - V_0)$$

Good selectivity is a more important factor than high efficiency in determining resolution (Figure 5) since R_s is linearly related to selectivity but quadratically related to efficiency. This means that a four-fold increase in efficiency is required to double the resolution under isocratic conditions.

Selectivity in ion exchange chromatography is not only dependent on the nature and number of the ionic groups on the matrix but also on the experimental conditions, such as pH and ionic strength. It is the ease and predictability with which these experimental conditions, and thus the selectivity, can be manipulated. This gives ion exchange chromatography the potential of extremely high resolution.

THEORY/PRINCIPLES

THE ISOELECTRIC POINT: PROTEIN CHARGE NEUTRALITY AT A PARTICULAR PH

Mahler, H. R. and Cordes, E. H. (1971) *Biological Chemistry*, 2nd Ed., Harper & Row, New York, pp. 92-96. (Adapted)

1. Protein pH Titrations

Titration curves of proteins as a function of pH are complex and difficult to interpret theoretically as a result of two factors. First, there are ordinarily a large number of titratable groups present in each protein molecule; 50 to 60 such groups per 100 kDa M_r are typical figures. Second, the value of pK_a for each of the titratable groups of proteins may differ by one or more pH units from that for the simple amino acid. Thus, if a protein contains 10 aspartic acid residues, each β-carboxyl is, in general, characterized by a distinct value of pK_a – ranging from 3 to 6. Such variation in pK_a for groups present in proteins, or other charged macromolecules, may be ascribed to three factors: (1) electrostatic effects resulting from the ionization of other groups on the protein, (2) medium effects due to the proximity of hydrophobic residues, and (3) hydrogen bonding.

2. Isoelectric Point

In the course of titration of a protein from the completely acidic to the completely basic form, at some pH the mean charge on the protein must be zero. This pH is termed the isoelectric point, p*I*. Experimentally, the pI is defined as that pH at which the protein does not migrate in an electric field. It may be approximately calculated from amino acid composition data. The condition of electroneutrality of proteins is affected by the presence of salts, as a result of the capacity of proteins to bind ions and the alteration in pK_a due to ionic-strength effects. The p*I* is a function of the nature and concentration of the buffer, and, in general, of any other solutes when the p*I* is measured. On the other hand, p*I*s are not a function of protein concentration. For most proteins, they are close to neutrality, reflecting approximately equal numbers of acidic and basic residues as constituents of the protein. In some cases, however, distinctly different behavior is noted. For example, pepsin, an enzyme found in the stomach where conditions are normally very acidic, has a p*I* near 1. Protamines typically have p*I*s near 12.

Reference:

1. Young, E., Gordon, I. N., and Florkin, M. and Stotz, E. H., (Eds.), *Comprehensive Biochemistry*, Elsevier, New York. 1963, Vol. 7, p. 25.

Alternative Approach

Ion-Pair Chromatography

Forum, Mac Mod Analytical Inc., Chadds Ford, PA, ph. (800) 441-7508. (Adapted)

Ion-pair chromatography (IPC) is a useful high-pressure/performance liquid chromatography (HPLC) technique to separate acids and bases. It is similar to reverse-phase chromatography in that solutes elute from typically either C_8 or C_{18} silica-based columns – with a mobile phase comprised of an organic modifier and an aqueous buffer. Unlike reverse-phase, the buffered, aqueous mobile-phase contains an "ion-pair (IP) reagent." Commonly used ion-pair reagents are short-chain ions, such as hexane sulfonate or tetraethylammonium. Methanol, in lieu of acetonitrile, is recommended as the organic modifier for this separation technique, as it provides greater solubility for buffer and ion-pair reagent solutions. Several theories have described the mechanism of separating acids and bases by this technique. One theory holds that the alkyl portion of the IP-reagent partitions into the bonded phase, leaving the ionized portion directed away from the bonded phase (Figure 1). At high concentrations of IP-reagent, this creates a charged stationary phase that behaves similar to an ion-exchange support.

Although potentially similar from a mechanistic view to ion-exchange, IPC offers several advantages. Tailing bands are rarely observed in IPC. Silica-based C_8- and C_{18}-bonded columns are more reproducible and more stable than typical ion-exchange supports. IPC can provide more selectivity options, as the separation can be varied continuously from reverse-phase (zero IP reagent) to ion-exchange (high concentration of IP reagent).

When optimizing an IP separation, there are five major variables to consider: (*1*) % organic, (*2*) ion-pair size, (*3*) concentration, (*4*) pH, and (*5*) buffer and/or salt concentration. The % organic (methanol) is adjusted to give the needed *retention* (k').

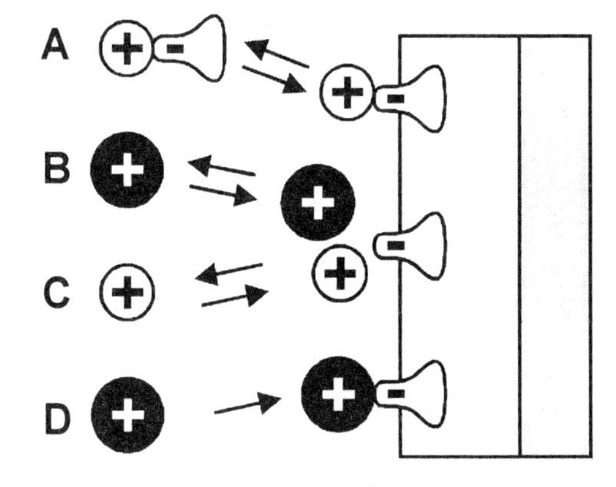

Figure 1 - Ion-pair chromatography. The solute is a protonated base (black circle) and the IP reagent is an alkyl anion (white anion). The purified material (white circle) is bound to the resin as an ion-pair with the IP reagent.

Varying the pH and IP concentration can change k' and/or *relative peak position*. Typically, peaks should elute with $1 < k' < 20$. Buffers concentration should be between 25 and 30 mM and should have adequate capacity at the needed pH.

A wide variety of IP reagents are available. Researchers have shown that if the IP reagent is adjusted to give the same run-time, the band spacing provided by different reagents is usually similar. Instead of trying different IP reagents, it is more productive to cause a change in band spacing than to vary the concentration of the IP reagent itself. The pH of the mobile phase affects the retention of an acid or base, whether using reverse-phase or ion-pair chromatography, but not in the same way. Under reverse-phase conditions, the retention of an acid will be maximized when the pH is significantly below its pK_a (2 pH units), i.e., when the acid is protonated or not ionized. With a large concentration of IP reagent in the mobile phase, the surface of the stationary phase is now charged and the IP reagent will cover the bonded phase. Ideally, the retention of an acid will be maximized when the acid is ionized or when the pH is significantly greater than its pK_a. In many cases, a mixed-mode separation is observed; both reverse-phase and IP mechanisms are operational. Retention is larger still at pH 5 in 1 mM TBA, as reverse-phase partitioning is more significant when these acids are not ionized. The IP k' of these acids at pH 8 increases when the TBA concentration is increased.

These examples illustrate that varying the IP reagent concentration and pH of the buffered mobile-phase causes tremendous changes in k' and α. This is a very real advantage of IPC.

Problems that may arise during the course of IPC method development are mentioned. Baseline artifacts are more common in IPC than in reverse-phase separations. These artifacts can be minimized by (*1*) dissolving the samples in the mobile phase, (*2*) injecting less than 50 μl samples, and (*3*) using purified buffers and IP reagents. It may take longer for the column to reach equilibrium; therefore, ensure that replicate injections are made to verify equilibration. IP reagents must be carefully removed from the bonded phase. Short-chain IP reagents can be washed from the column bonded phase more easily than long-chain reagents.

Table 1 - Suggested Ion-pair Reagents.

Cationic Samples (Protonated Bases)	Anionic Samples (Ionized Acids)
Pentane Sulfonate	Tetraethylammonium
Hexane Sulfonate	Tetraethylammonium

Although removal is achievable, it is probably best to use a column exposed to IP reagents exclusively for IP applications (see Table 1 for recommended reagents). Ionizable compounds are particularly sensitive to temperature changes, so thermostating the column is strongly advised. Good reviews of IPC are given.[3,4]

References:

1. S. O. Jansson and S. Johanson, J. Chromatogr. 242, 41 (1982).
2. D. S. Lu, J. Vialle, H. Tralongo, and R. Longeray, J. Chromatogr. 268, 1 (1983).
3. M. T. W. Hearn, Ed., *Ion-Pair Chromatography*, Marcel Dekker, New York, 1985.
4. J. H. Knox and R. A. Hartwick, J. Chromatogr. 204, 3 (1981).

Alternative Approach

HPLC: Ion Exchange and Reverse-Phase Methods: Literature Sources

1. Introduction

Chromatography takes advantage of the fact that different molecules have different affinities for certain stationary or mobile phases and can therefore be separated due to differences in rates of partitioning between these phases. A variety of stationary and mobile phase combinations can be used to effect molecular separations based on this fundamental partitioning mechanism.

1.1 Ion Exchange Chromatography. The stationary phase used in ion exchange chromatography is composed of residues that are charged either positively (anion exchange) or negatively (cation exchange) under the conditions of the experiment. Anionic samples bind to the cationic stationary phase. When a sufficiently chaotropic or concentrated mobile-phase anion exchanges with the sample molecule and occupies the cationic stationary phase binding site, the sample partitions off of the stationary phase and into the mobile phase – thus eluting from the column. Proteins elute from the column as their charge interactions with the column are disrupted. This disruption occurs when another soluble anion from the solvent competes for the cationic binding site. Salt anions such as chloride and acetate are added to the mobile phase as a gradient to gradually disrupt protein interactions with the column.

1.2 Reverse-Phase Chromatography. The stationary phase in reverse-phase chromatography is coated with a relatively hydrophobic hydrocarbon and the mobile phase is typically a polar liquid (e.g., H_2O), which is gradually made more hydrophobic by gradient addition of a less polar miscible liquid (e.g., methanol, acetonitrile). In this technique, sample molecules bind to the column via hydrophobic interactions. More polar molecules elute first by partitioning preferentially into the polar mobile phase. As the concentration of less polar (more hydrophobic) solvent in the mobile phase increases, more hydrophobic molecules partition into the mobile phase and elute from the column. Molecules, which are retained on the stationary phase longest, are most hydrophobic and only partition into the mobile phase when it becomes sufficiently nonpolar.

Reverse-phase chromatography (RPC) separates molecules via differences in hydrophobic affinities for the hydrocarbon column packing (in this case, C_8). Differences in mobile-phase pH can drastically change the degree of ionization of a particular protein and thus greatly alter how much it partitions into the mobile phase. For example, if a protein is protonated at a certain pH in such a way that it has a positive surface charge, it will be less likely to partition onto the hydrophobic stationary phase and more likely to partition into the polar mobile phase where the surface charge can be solvated. At a higher pH, this same protein might deprotonate to establish a relatively uncharged surface. In this case, the protein will favor hydrophobic interactions, such as can be established with the stationary phase, and partitions preferentially into that phase, unless or until the mobile phase becomes sufficiently nonpolar to accommodate the nonpolar protein.

RPC methods have some important limitations. Because they rely on hydrophobic interactions and typically use denaturing solvents, RPC methods frequently denature proteins. Thus, *reverse-phase* separations may produce much lower enzyme activities relative to ion exchange separations. Another limitation is that the pH of reverse-phase HPLC cannot go above about pH 8 due to the sensitivity of the column packing (SiO_2 matrix) to base hydrolysis.

1.3 High Performance Liquid Chromatography. Chromatographic techniques have become much more sensitive with the advent of HPLC (high-performance liquid chromatography) technology. HPLC stepping pumps were designed to produce even liquid flow rates and reproducible solvent gradients under high and sometimes variable pressure conditions. High pressures are generated due to back pressure from the mobile liquid phase as it passes through the tightly packed (small bead containing) chromatography column. Smaller beads result in larger surface areas and increase the number of theoretical plates. This increases the separation capacity of the column and permits separation of very similar molecules by taking advantage of slight differences in affinity for the column matrix.

1.4 FPLC. Pharmacia™ distributes a widely used multicomponent chromatography system with the trade name "Fast Pressure Liquid Chromatography" (FPLC), which operates at medium pressures (200 to 1000 psi) and has the advantage of much larger separation capacities than typical HPLC setups.

2. Instrumental Aspects

2.1 Reverse-Phase HPLC.

1. C_8 column and a "guard column"
2. 20 µl sample volume
3. UV detector reading at 254 nm
4. Flow rate = 0.75 ml/minute
5. Column pressure = 2000 psi
6. Solvent buffers:
   A = 50 mM sodium phosphate (pH 6.5)

$B_{pH6.5}$ = 50 mM sodium phosphate (pH 6.5), containing 70% methanol, or

B_{pH4} = 50 mM sodium phosphate (pH 4), containing 70% methanol

7. Solvent gradient: 0 to 100% B in 30 minutes; hold for 5 minutes; return to 0% B
8. Fraction collector; acquires 0.75 ml fractions every minute
9. Chart speed = 0.5 cm/minute

2.2 Ion Exchange HPLC.

1. DEAE-Cellulose column
2. 50 µl and 200 µl sample volume
3. UV detector reading at 280 nm
4. Flow rate = 1 ml/minute
5. Column pressure = 2000 psi
6. Solvent buffers:
   A = 50 mM sodium phosphate (pH 6.5)
   B_{Cl} = 50 mM sodium phosphate (pH 6.5) containing 1 M NaCl, or
   B_{Ac} = 50 mM sodium phosphate (pH 6.5) containing 1 M sodium acetate
7. Solvent gradient:
   0% to 100% B in 30 minutes; hold for 10 minutes; return to 0% B
8. Fraction collector collects 1 ml/minute
9. Chart speed = 1.25 cm/minute

3. Specific Activity as a Function of Purification Stage

Following the increase in protein purification provided by each subsequent technique monitors the course of a purification protocol. This requires calculating the specific activity of each pooled protein fraction (see equation 1). Percent purification is determined by the percentage change in specific activity of the purified protein fraction relative to that of the crude cell homogenate (see equation 2). To calculate each specific activity, one requires the total protein concentration in the fraction and the specific activity of the protein of interest in that fraction. Total protein in each fraction is measured using the BCA assay.

$$\text{specific acitivity} = \frac{\text{activity (units/ml)}}{\text{total protein (mg/ml)}} \quad (1)$$

$$\text{\% purification} = \frac{\text{specific activity of fraction}}{\text{specific activity of crude homogenate}} \quad (2)$$

4. Literature

4.1 Ion-Exchange.

1. Kopaciewicz, W., Rounds, M.A., Fausnaugh, J. and Regnier, R. F. (1983) Retention Model for High-Performance Ion-Exchange Chromatography, J. Chromatogr. 266, 3-21.
2. Kopaciewicz, W. and Regnier, R. F. (1983) Mobile Phase Selection for the High-Performance Ion-Exchange Chromatography of Proteins, Anal. Biochem. 133, 251-259.
3. Regnier, F. E. (1984) High Performance Ion-Exchange Chromatography, Methods Enzymol. 104, 170-189.

4.2 Reverse-Phase HPLC Chromatography.

1. Szepesi, G. (1992) How to Use Reverse-Phase HPLC, VCH Publishers, NY.
2. Lough, W. J. and Wainer, I. W. (Eds.) (1995) High Performance Liquid Chromato-graphy, Blackie Academic & Professional, Glasgow, UK.
4. Ahuja, S. (1090) Selectivity and Detect-ability Optimizations in HPLC, John Wiley & Sons, NY.
5. McMaster, M. C. (1994) HPLC – A Practical User's Guide, VCH Publishers, NY.
6. Bidlingmeyer, B. A. (1992) Practical HPLC Systems, John Wiley & Sons, NY.
7. Sadek, P. C. (2000) Troubleshooting HPLC Systems, John Wiley & Sons, NY.

4.3 General.

1. Rossomando, E. F. (1990) Ion-Exchange Chromatography, Methods Enzymol. 182, 309-317.
2. Wang, N. -H. L. (1990) Ion Exchange in Purification, in Separation Processes in Biotechnology, Asenjo, J. A. (Ed.), Nature Pub. Co., pp. 359-400.
3. Richey, J. S. (1984) Optimal pH Conditions for Ion Exchangers on Macroporous Supports, Methods Enzymol. 104, 223-233.
4. Schwab, H., Rieman, W. III and Vaughan, P. A. (1957) Theory of Gradient Elution through Ion Exchangers, Anal. Chem. 29, 1357-1361.

LAB 7.6:

AFFINITY CHROMATOGRAPHY AND MICROPLATE β-GALACTOSIDASE ASSAYS TO DETERMINE FRACTIONS CONTAINING FUSION PROTEIN

Overview:

Affinity chromatography will be used to purify the β-galactosidase fusion peptide encoded by your clone. Affinity chromatography is based on the fact that a protein can specifically bind to another molecule, its ligand. The high specificity of this protein purification step depends upon the ligand that is immobilized on the insoluble matrix. When the solution of proteins to be purified is applied to the column, ideally only the protein of interest binds to the ligand and all other proteins pass through. You will use the affinity matrix *p*-aminobenzyl-1-thio-β-D-galactopyranoside-agarose. This material is composed of a substrate analog covalently coupled to the agarose beads which binds β-galactosidase.

After washing the column to remove nonspecifically bound proteins, high pH buffer is applied to release and elute bound β-gal-fusion protein from the immobilized ligand in a highly purified form.[1] You will then perform β-galactosidase enzyme assays to determine which protein fractions to pool. In future sessions, you will perform protein assays and SDS-PAGE to determine the effectiveness of this procedure.

Experimental Outline:

1. Prepare affinity chromatography column.
2. Microtiter plate β-galactosidase assay.

Required Materials:

saturated ammonium sulfate solution (recipe below); 0.1 M phenylmethylsulfonyl fluoride (PMSF); 100% ethanol. Store in dark at room temperature; 10 mM Tris, pH 7.4; microfuge tubes; Buffer B (recipe below); Elution buffer (EB: 0.1 NaBorate, pH 10); Tris-EDTA-PMSF (TEP) buffer; 1 M Tris-HCl, pH 7.4; *p*-aminobenzyl-1-β-galactopyranoside-agarose (Sigma); chromatography column (2.5 x 20 cm); fraction collector; substrate mix (recipe below); stop buffer (recipe below); microtiter plate; microtiter plate reader; graph paper; spectrophotometer

Prelab Preparations:

1. Prepare Solutions (for 40 students)

1. *Saturated ammonium sulfate solution*: Dissolve 75 g of ammonium sulfate in 100 ml of dH_2O. Store at 4°C.
2. 0.1 M *phenylmethylsulfonyl fluoride* (PMSF, a protease inhibitor): CAUTION PMSF IS HIGLY TOXIC! WEAR GLOVES.
   Dissolve 0.435 g of PMSF in 25 ml of 100% ethanol. Store in the dark at room temperature.
3. *Buffer B*:

	NaCl	29	g
1 M	Tris-HCl, pH 7.4	10	ml
1 M	$MgCl_2$	10	ml
0.1 M	EDTA	10	ml
	β-mercaptoethanol	10	ml
	Triton X-100	1	ml

   Add H_2O to 1000 ml. Store at room temperature.
4. *Elution buffer* (EB: 0.1 M NaBorate, pH 10):
   Dissolve 3.1 g of boric acid in 490 ml H_2O. Adjust pH to 10 with NaOH. Add H_2O to 500 ml.
5. *Tris-EDTA-PMSF (TEP) buffer*:

First, add	0.1 M	PMSF in ethanol	1 ml
Then add	1 M	Tris-HCl, pH 7.4	10 ml
	0.1 M	EDTA, pH 7	10 ml
		dH_2O	to 100 ml

6. 1 M Tris-HCl, pH 7.4. — 50 ml
7. p-aminobenzyl-1-β-galactopyranoside agarose — 40 ml

Procedures:

1. Prepare the Affinity Chromatography Column

1. The instructor will provide 4 ml of a 2:1 slurry of *p*-aminobenzyl-1-β-galactosidase-agarose affinity matrix in Buffer B. Invert and allow the beads to settle to about the 2 ml mark. Be gentle, the beads are fragile.
2. Remove the supernatant with a pasteur pipet. Add 5 ml of Buffer B and resuspend the matrix by gentle inversion.
3. Assemble the chromatography column (2.5 x 20 cm) by attaching a stopcock to the bottom and a reservoir collar to the top. Mount the assembled column on a ring stand.
4. Make sure the stopcock is closed. Add a uniform slurry of matrix material to the column. Open the stopcock to allow the buffer to drip through. Do not allow the buffer to drop below the top of the settling beads.
5. Wash the column with 15 ml of Buffer B.
6. Open the column stopcock and allow the liquid level to drop until just above bed. Close the stopcock.
7. Add 1 ml of the crude soluble extract collected in Lab 7.2 to the column.
8. Open the stopcock for a slow flow rate (5-10 ml/hour or 1-2 drop/minute) and collect 1 ml of buffer in a microfuge tube. Label as "wash 1" and freeze.
9. Close the stopcock and allow the column to stand for 15 minutes.
10. Open the stopcock entirely for this step. Add 4 successive 1 ml volumes of Buffer B and collect each in a separate microfuge tube. Label these tubes as "wash 2-5" and freeze.
11. Wash the column with an additional 10 ml of Buffer B to remove any non-β-gal proteins left behind. Discard these into the flowthrough solution.
12. Add 50 μl of 1 M Tris-HCl, pH 7.4 to each of 12 separate 1.5-ml microfuge tubes before eluting the column. This is done in order to lower the pH, which will renature the fusion protein immediately after elution from the column.
13. Elute the column with 12 successive additions of 0.35 ml of elution buffer ("EB": 0.1 M NaBorate, pH 10). At pH 10, β-galactosidase is partially denatured, thereby releasing the fusion peptide from the column matrix.
14. Collect each fraction (about 0.35 ml) in a separate microfuge tube from step 7 above and label as "eluate 1-12."
14. Chill the eluates on ice immediately after collection and add 1.2 ml of cold, saturated ammonium sulfate. Vortex to mix. Let stand on ice for 20 minutes.
15. Spin the ammonium sulfate precipitated elution samples in a cold microfuge for 10 minutes. Remove and discard the supernatant.
16. Resuspend the invisible pellets in 100 μl of Tris-EDTA-PMSF (TEP) buffer.

2. Microtiter Plate β-Galactosidase Assay

1. Prepare solutions

a. *Substrate Mix*:

4	mg/ml	ONPG (*o*-nitrophenol-β-D-galactopyranoside)
1	mM	$MgCl_2$
50	mM	β-mercaptoethanol
0.1	M	Na_2HPO_4 (pH 7.3)

b. *Stop Buffer*:

1	M	$NaHCO_3$

c. *Buffer B*:

		NaCl	29	g
1	M	Tris-HCl, pH 7.4	10	ml
1	M	$MgCl_2$	10	ml
0.1	M	EDTA	10	ml
		β-mercaptoethanol	10	ml
		Triton X-100	1	ml

Add H_2O to 1000 ml. Store at room temperature.

2. Procedure

a. Add 2 μl of each enzyme sample to 18 μl of Buffer B in separate wells of a microtiter plate. Mix well by shaking. Make sure to include wells for the blanks.

b. Aliquot 2 μl of the diluted sample above (1/10 dilution) into 98 μl of Buffer B.

c. To the samples in step b add 100 μl of the substrate mix and mix well to start the reaction. Time the addition of enzyme so that there is 5-15 seconds between the addition of each sample. This will ensure that all of the samples are incubated for equivalent times. Exact times must be recorded for each assay.
d. Incubate at room temperature until a yellow color develops (time may vary).
e. Add 100 ml of the stop buffer; timed as in step c. The stop buffer ends the reaction by increasing the pH to 11, thus inactivating the β-galactosidase enzyme.
f. Read the plates at 405 nm in a microtiter plate reader.
g. Pool the fractions containing the β-gal-*myo*-3 fusion protein.

Additional Report Requirements:

1. Microtiter plate β-galactosidase absorbance readings.
2. Records of which fractions were pooled.

References:

1. Germino, J., Gray, J. B., Charbonneau, H., Vanaman, T. and Bastia, D. (1983) Use of gene fusions and protein-protein interaction in the isolation of a biologically active regulatory protein: the replication inhibitor protein of plasmid R6K Proc. Natl. Acad. Sci. U.S.A 80: 6848-6852.
2. Bradford, M. M. (1976) A rapid and sensitive method of measuring microgram quantities of proteins utilizing the principle of protein-dye binding. Anal. Biochem. 72: 248-264.
3. Pardee, A. B. Jacob, F. and Monod, J. (1959). The genetic control and cytoplasmic expression of inducibility in the synthesis of β-galactosidase in *Escherichia coli*. J. Mol. Biol. 1: 165-168.

Process Rationale

Affinity Chromatography

Pharmacia, Piscataway, NJ, ph. (800) 526-3593. (Adapted)

1. Introduction

Affinity chromatography is a type of adsorption chromatography in which the molecule to be purified is specifically and reversibly adsorbed by a complementary *binding substance* (*ligand*, L) immobilized on an *insoluble support* (*matrix*, M). Purification is often of the order of several thousand-fold and recoveries of active material are generally very high. Many spectacular separations have been achieved in a single step, allowing immense time-saving over less selective multistage procedures. Affinity chromatography has a concentrating effect, enabling large convenient processes. The high selectivities of the separations derive from the natural specificities of the interacting molecules. For this reason, affinity chromatography can be used (*1*) to purify substances from complex biological mixtures, (*2*) separating native from denatured forms of the same substance, and (*3*) removing small amounts of biological material from large amounts of contaminating substances.

The first application of affinity chromatography was the selective adsorption of amylase onto insoluble starch in 1910. The complex organic chemistry required, synthesizing a reliable matrix and covalently attaching ligands, prevented the technique from becoming generally established in biological laboratories. However, in 1967, Axén, Porath and Ernback reported that molecules containing primary amino groups could be coupled to polysaccharide matrices that have been activated by cyanogen bromide. This represented the beginning of affinity chromatography as a routine separation technique; cyanogen bromide activation has now become the most generally used method for coupling ligands.

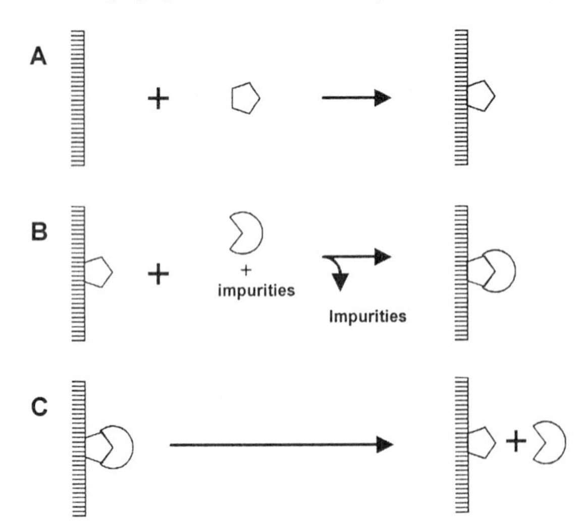

Figure 1 - Steps of affinity chromatography. (A) Immobilize the ligand; (B) Absorb the substrate and separate the impurities; (C) Desorb the pure substrate.

The principle of affinity chromatography is shown in Figure 1. A successful separation requires that a biospecific L be available and that it can be covalently attached to a chromatographic bed material M. It is important that immobilized L retains its specific binding affinity for the *substance of interest* (S) and that methods are available to selectively desorb the bound substances in an active form, after washing away unbound material. Biological systems that use affinity chromatography frequently used are listed below.

Any component can be used as a ligand for purifying its respective binding substance; see Table 1 for examples.

2. The Matrix, M

Sepharose™ is a bead-formed agarose gel (Figure 2) which displays virtually of all the features required of a successful M to immobilize biologically active molecules. The hydroxyl groups on the sugar residues can be easily derivatized for covalent attachment of a L. Sepharose 4B is the most favored and widely used M.

The open pore structure makes the interior of the M available for L attachment and ensures good binding capacities, even for large molecules. The exclusion limit of Sepharose 4B in gel filtration is MW 20 x 10^6. Sepharose 4B exhibits extremely low nonspecific adsorption; this is essential because the power of affinity chromatography relies on specific L-S binding interactions. The bead form of the gel provides good flow properties with minimal channeling in the bed ensuring rapid separation. Adsorbents based on Sepharose are stable under a wide range of experimental conditions such as at high and low pH, and in the presence of detergents and dissociating agents. However, in situations that require organic

Table 1 - Ligand • Substance Pairs.

Binding Molecule	Ligand
Enzyme	Substrate analogue, inhibitor, cofactor
Antibody	Antigen, virus, cell
Lectin	Polysaccharide, glyco-protein, cell surface receptor, cell
Nucleic acid	Complementary base sequence, histones, nucleic acid polymerase, binding protein
Hormone, vitamin	Receptor carrier protein
Cell	Cell surface-specific protein, lectin

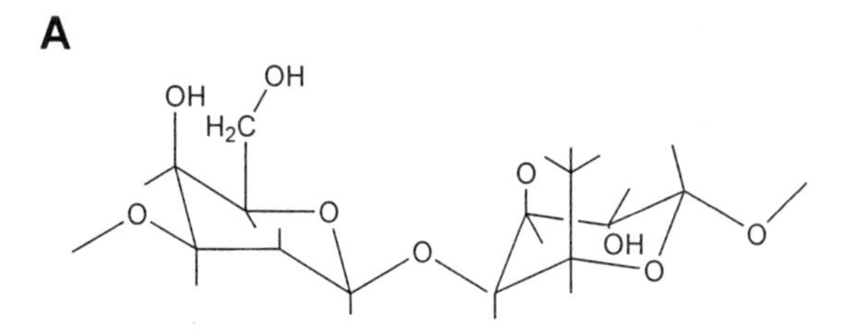

Figure 2 - Partial structure of agarose. (A) Chemical linkage, and (B) macrostructural branching.

solvents (e.g., chemical modification), high temperatures (e.g., autoclaving), or disruptive eluents (e.g., guanidine hydrochloride), covalently cross-linked Sepharose is ideal (Sepharose CL, Sepharose Fast Flow and Sepharose High Performance).

3. The Ligand, L

The selection of L for affinity chromatography is influenced by two factors. First, it should exhibit specific and reversible binding affinity for the substance to be purified. Second, it should have chemically modifiable groups that allow it to be attached to M without destroying its L-binding activity. The dissociation constant (K_d) for the [L•S] complex should ideally be in the range 10^{-4} to 10^{-8} M in free solution. Interactions involving K_d greater than 10^{-4} M, for example, the binding reaction between an enzyme and a weak inhibitor, is likely to be too weak for successful affinity chromatography. Conversely, K_d is lower than approximately 10^{-8} M, for example, the affinity between a hormone and hormone receptor, elution of the S without inactivation is likely to be difficult. If no information is available on the strength of the binding complex, a trial and error approach should be used.

4. Spacer Arms

The active site of a biological substance is often located deep within the molecule. Adsorbents prepared by coupling small ligands (e.g., enzyme cofactors) directly to Sepharose can exhibit low capacities due to steric interference between the M and substances binding to L. In these circumstances, a "spacer arm" is interposed between M and L to facilitate effective binding (Figure 3).

The length of the spacer arm is critical. If it is too short, the arm is ineffective and the ligand fails to bind substances in the sample. If it is too long, non-specific effects become pronounced and reduce the selectivity of the separation.

5. Coupling Gel

Methods are available for ligand immobilization which are quick, easy and, safe. The correct choice of coupling method depends on the L material to be immobilized and purified. Many of the reagents used to activate M prior to coupling are extremely hazardous and should only be handled by qualified personnel. CNBr-activated Sepharose 4B enables ligands containing primary amino groups to be safely, easily, and rapidly immobilized by a spontaneous reaction.

6. Group Specific Adsorbents

Group specific adsorbents have affinity for a group of related substances rather than for a single substance. The same general L can therefore be used to purify several substances (e.g., a class of enzymes) without the requirement that a new adsorbent be prepared for each different S to be purified. Each group of adsorbed substances is either structurally or functionally similar. The specificity in group specific (general L) affinity chromatography derives both from the selectivity of L and the use of selective elution conditions. Group-specific adsorbents based on the following L materials are available ready coupled (see Table 2).

7. Coupling Gels for L Immobilization

The correct choice of coupling gel is dictated by both the type of groups available on the L molecule for coupling and the nature of the binding reaction with the S to be purified. Groups that are frequently used for immobilizing L are listed in Table 3; coupling gels are available for attachment via each of these groups.

Immobilization should be attempted through the least critical region of the molecule, to ensure minimal interference with the normal binding reaction. For example, an enzyme inhibitor containing amino groups

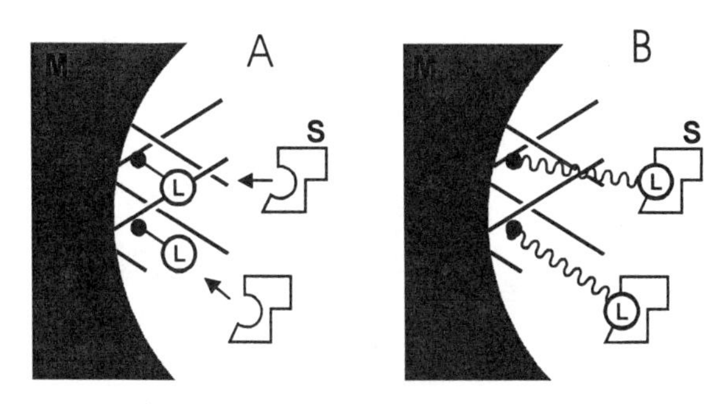

Figure 3 - The principle of spacer arms. (A) Ligand attached directly to M; L binding is obstructed; (B) ligand attached to M via a spacer arm, facilitating L binding.

could be attached to the matrix through its amino groups, providing that the specific binding activity with the enzyme is retained.

Table 2 – Group Specific Adsorbents

Ligand	Specificity
2", 5" ADP	Enzymes which have $NADP^+$ as cofactor
5" AMP	Enzymes which have NAD^+ as cofactor and ATP-dependent kinases
Arginine	Proteases and zymogens including prothrombin, prekallikrein, clostripain
Benzamidine	Proteases including trypsin, urokinase, kallikrein, prekallikrein
Blue (Cibacron Blue)	Broad range of enzymes which have nucleotide cofactors; serum albumin, etc.
Calmodulin	Calmodulin-regulated kinases, phosphodiesterases, cyclases, and other enzymes
Con A (concanavalin A)	Terminal β-D-gluco-pyranosyl, β-D-mannopyranosyl or sterically similar residues
GammaBind G, type 2	F_c region of IgG and related molecules similar to protein G
GammaBind G, type 3	F_c region of IgG and related molecules similar to protein G
Gelatin	Fibronectin
Helix pomatia lectin	*N*-acetyl-β-D-galactosaminyl residues
Heparin	Many growth factors, coagulation factors, restriction endo-nucleases, and other nucleic acid binding proteins
Lentil lectin	Similar to that of Con A, but lower binding affinity for simple sugars
Lysine	Plasminogen, ribosomal RNA
Poly(A)	Nucleic acids and oligonucleotides which contain poly(U) sequences; RNA-specific proteins
Poly(U)	Nucleic acids, especially mRNA, which contain poly(A) sequences; poly(U)-binding proteins
Protein A	F_c region of IgG and related molecules
Protein G	Similar to that of protein A but different affinities for IgG's from different species
Red (Procion Red)	Broad range of enzymes which have nucleotide cofactors, etc.
Wheat germ lectin	*N*-acetyl-D -glucosamine

8. Cyanogen Bromide Activation

The reaction of cyanogen bromide with Sepharose 4B[1,3] results in a reactive product – to which proteins, nucleic acids, or other biopolymers can be coupled, under mild conditions, via primary amino groups or similar nucleophilic groups. The structure of Sepharose is shown in Figure 2. Cyanogen bromide reacts with hydroxyl groups on Sepharose. The dominating active groups in CNBr-activated Sepharose 4B are cyanate esters.[4,5] The activated groups react with primary amino groups of L to form isourea linkages (Figure 4). Multipoint attachment ensures that proteins and other biopolymers do not hydrolyze from M. The activation procedure also cross-links Sepharose and thus enhances its chemical stability. This offers much greater flexibility in the choice of elution conditions and derivatization.

Table 3 - Coupling Gels for Immobilizing Ligands via Different Chemical Groups.

Ligand	Chemical Group	Coupling Gel
Protein, peptide, amino acid	Amino	CNBr-activated Sepharose 4B HiTrap NHS ECH Sepharose 4B Activated CH Sepharose 4B Epoxy-activated Sepharose 6B
	Carbonyl	EAH Sepharose 4B
	Thiol	Thiopropyl Sepharose 6B Activated Thiol Sepharose 4B Epoxy-activated Sepharose 6B
Sugar	Hydroxyl	Epoxy-activated Sepharose 6B
	Amino	ECH Sepharose 4B Activated CH Sepharose 4B HiTrap NHS Epoxy-activated Sepharose 6B EAH Sepharose 4B
polynucleotide	Amino, carboxyl	CNBr-activated Sepharose 4B
	Mercurated base	Thiopropyl Sepharose 6B
Coenzyme, cofactor, antibiotic, steroid	Amino, carboxyl, thiol or hydroxyl	Use gel with spacer arm (see above)

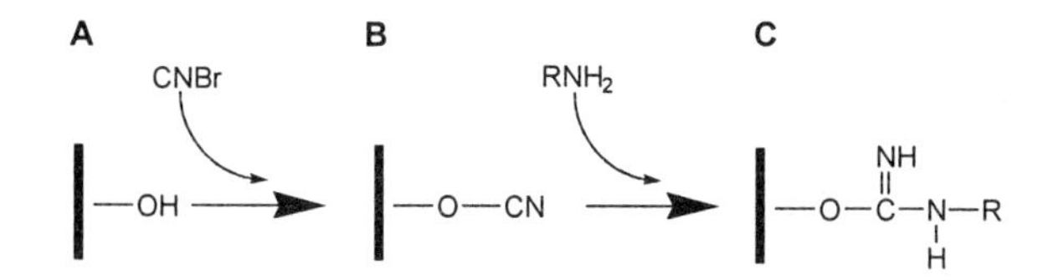

Figure 4 - Activation by cyanogen bromide and coupling to the activated gel. (A) Hydroxyl is converted to (B) cyano then to (C) isourea.

CNBr-activated Sepharose 4B is used for coupling proteins and nucleic acid ligands. For example, human hemoglobin has been coupled to CNBr-activated Sepharose 4B and this adsorbent was used to fractionate antihemoglobin antibodies by using an elute containing a gradient of decreasing pH and increasing acetic acid concentration.[6] CNBr-activated Sepharose 4B is widely used to prepare adsorbents for purifying specific antibodies which are subsequently radiolabeled for use in immunoassay.[7] Adsorbents prepared using CNBr-activated Sepharose 4B and containing transferrin or antitransferrin have been used repeatedly for more than two years without loss of activity.[8] DNA has also been coupled to Sepharose 4B using the cyanogen bromide method and the DNA Sepharose was used for both isolation of nucleic acids by hybridization and purification of enzymes which bind DNA.[9]

9. Immobilized Protein A

Protein A supplied by Pharmacia LKB is isolated from the culture medium of a strain of *Staphylococcus aureus* which does not incorporate the protein into its cell wall. Protein A interacts with the F_c part of IgG molecules from many species. In humans, it binds IgG of subclasses 1, 2, and 4.

Protein A binds IgG from most mammalian species. Goudswaad *et al.* made a detailed study of the interaction of protein A with immunoglobulins from different species. Some polyclonal IgM and IgA from pig, dog, and cat were fractionated on protein A Sepharose CL-4B, which shows different binding affinity for the various subclasses of mouse IgG.

Although IgG is the major reactive human immunoglobulin, some other types have also been demonstrated to bind with protein A. Interaction takes place with human IgA as well as human myeloma IgA_2 but not IgA_1. Some human monoclonal IgM and some IgM from normal and macroglobulinemic sera bind to protein A.

10. Specificity of Protein G

Like protein A, protein G binds specifically to the F_c region of IgG. However, a much wider range of polyclonal IgGs bind strongly to protein G and adsorption takes place over a wide pH range. Among the important differences in binding to IgGs of different subclasses, that protein G binds to all human subclasses (including IgG_3, which constitutes approximately 8% of total IgG) and all mouse IgG subclasses (including mouse IgG_1) bind equally well under standard buffer conditions. IgGs from sheep, cow and horse, as well as rat IgG_{2a} and IgG_{2b}, which either do not bind or bind only weakly to protein A, can be purified on immobilized protein G. This matrix binds human myeloma IgM, IgA, and IgE molecules, although some do bind to protein A. The fast and efficient purification of monoclonal IgG from rat hybridoma cell culture fluid has been described. In the method, a sample volume of 20 ml was applied to a HiTrap Protein G column in a volume of only 1 ml.

References:

1. Selective enzyme purification by affinity chromatography. Proc. Natl. Acad. Sci. U.S.A 61 (1968) 636-643, Cuatrecasas, P., Wilchek, M., Anfinsen, C.B.
2. Covalent attachment of DNA to agarose. Improved synthesis and use in affinity chromatography. Eur. J. Biochem. 54 (1975) 411-418, Arndt-Jovin, D. J., Jovin, T. M., Bahr, E. *et al.*
3. Purification of recombinant chimeric B72.3 F_{ab}, and $F_{(ab")2}$ using streptococcal protein G. Prot. Express. Purific. 3 (1992) 368-373, Proudfoot, K.A., Torrance, C., Lawson, A.D.G. *et al.*
4. Protein A reactivity of various mammalian immunoglobulins. Scand. J. Immunol. 8 (1978) 21-28, Goudswaard. J., van der Donk, J.A., Noordzij, A. *et al.*
5. Isolation of pure IgG1 IgG2a and IgG2b, immunoglobulins from mouse serum using Protein A-Sepharose. Immunochemistry 15 (1978) 429-436, Ey, P. L., Prowse, S. J., Jenkin, C. R.
6. Human colostral IgA interacting with staphylococcal protein A. Acta Pathol. Microbiol. Scand. Sect. C 84 (1976) 71-72, Grov, A.
7. Differential binding of IgA proteins of different subclasses and allotypes to staphylococcal protein A. Immun. Forsch. 153 (1977) 466-469, Patrick C. C., Virella, G., Koistinen, J. *et al.*
8. Protein A reactivity of two distinct groups of human monoclonal IgM. Scand. J. Immunol. 4 (1975) 843-848, Lind. I., Harboe, M., Follinf, I.
9. Human IgM interacting with staphylococcal protein A. Acta Pathol. Microbiol. Scand. Sect. C 83 (1975) 173-176, Grov, A.
10. Metal chelate affinity chromatography, a new approach to protein fractionation. Nature 258 (1975) 598-599, Porath, J., Carlsson, J., Olsson, I. *et al.*

PROCESS RATIONALE

AFFINITY CHROMATOGRAPHY: ONE STEP PURIFICATION OF HYBRID PROTEINS CARRYING FUSED β-GALACTOSIDASE ACTIVITY

Ullmann, A. (1984) Gene 29: 27-31. (Adapted)

1. Introduction

1.1 Terms. Recombinant DNA; lac; gene fusions; affinity chromatography; salt effect; antibodies.

1.2 Summary. A one-step purification method of hybrid protein exhibiting β-galactosidase activity, based on *affinity chromatography* in the presence of high salt concentration, is described. Starting from crude bacterial extracts, several milligrams of near-homogeneous proteins can be obtained in a few hours with an overall yield of 85 to 95%. The purified hybrid proteins can be used to obtain antibodies against the foreign portion of the protein fusion.

1.3 Abbreviations. SDS, sodium dodecyl sulfate; TPEG, *p*-amino-phenyl-β-D-thiogalactoside.

1.4 General. The increasing possibilities offered by genetic engineering to construct hybrid genes opened a new field to study gene expression and protein structure and allowed the identification of proteins which are otherwise difficult to detect in cell extracts.[1]

Since the first description of gene fusion in *Escherichia coli*[8] several methods were developed to fuse the *lac* operon to other genes. Muller-Hill and Kania[11] first described a hybrid protein obtained by genetic fusion of the *lac* repressor to β-galactosidase. The hybrid protein had full repressor and β-galactosidase activities in spite of the fact that 23 amino acids from the amino-terminal part of β-galactosidase were lacking. Several general methods were subsequently developed to fuse the *lac* operon to other genes yielding hybrid proteins. The *in vivo* gene fusion method developed by Casadaban[3] using Mu phage was extended by *in vitro* methods using various cloning vectors[4,14] in which foreign DNA was cloned and expressed in the promoter proximal part of the *lac Z* gene of *E. coli.* The resulting hybrid proteins retained β-galactosidase activity and this facilitates their purification. Several hybrid proteins, obtained by the *in vivo* method of Casadaban,[3] were purified and characterized.[6] The purification involved several steps and the overall yields were 20-40%. Hybrid proteins, especially when they are large, are subject to proteolytic degradation or form aggregates. Therefore, classical, multistep purification procedures lead to low yields of the native protein. In this paper, a general, one-step method for the purification of enzymatically active hybrid β-galactosidase proteins is presented.

Table 1 - Bacterial Strains and Corresponding Hybrid Proteins

Strain	Hybrid Proteins	Reference
2E01[a]		Ullmann *et al.* (1968)[17]
BMH71-I8 [pUK230LyD3][b]	Lysozyme-β-galactosidase	Ruther *et al.* (1982)[14]
CSH50-3 [pMK38-62]	gal repressor-β-galactosidase	Wilcken-Bergmann *et al.* (1983)[18]
pop3891	Maltodextran phosphorylase-β-galactosidase	Gutierrez and Raibaud (1984)[7]
TP2339 [pDIA11861]	Adenylate cyclase-β-galactosidase	Roy *et al.* (1983)[13]
CAG1139 [pDIA4016]	Enzyme III ^{Glc}-β-galactosidase[c]	Huttner, E., De Reuse, H. and Danchin, A.

[a]Strain constitutive for the wild-type β-galactosidase.
[b]Indicates plasmid-carrier state.
[c]Enzyme III^{Glc} (encoded by the *crr* gene of *E. coli*) is a component of the phosphoenolpyruvate sugar transport system (PTS).

2. Materials and Methods

2.1 Bacterial Strains and Growth Conditions. All bacterial strains producing hybrid proteins were kindly provided by A. Danchin, E. Huttner, B. Muller-Hill, and M. Schwartz and are listed in Table 1. The construction of the gene fusions and the relevant genotypes of the strains are described in the corresponding references. Cultures were grown in LB-rich medium[10] at 37°C (in the presence of 25 μg/ml ampicillin for strains carrying plasmids). Strain pop3891 was grown in LB medium in the presence of 5 mM maltose at 30°C.

2.2 Preparation of Bacterial Extracts. Bacteria were harvested by centrifugation and disrupted by sonication in 20 mM Tris-HCl, 10 mM $MgCl_2$ (pH 7.4). Cell debris were removed by centrifugation. The extracts could be stored for several weeks at -20°C without significant loss of activity.

2.3 Preparation of Affinity Column. CH-Sepharose 4B (Pharmacia Fine Chemicals) was thoroughly washed according to the supplier"s

Table 2 - Binding Capacity of TPEG-Sepharose Column.

NaCl Concentration (M)	β-Galactosidase Retained (units)
-	70,000
1.25	320,000
1.4	700,000
1.6	1,200,000

recommendation. One mmol of TPEP (Sigma) was added to 100 ml of gel in 10 ml water. The pH was adjusted to 4.7, and 7.5 mmol of *N*-acetyl-*N*'-(3-dimethylaminopropyl)-carbodiimide (Sigma) in 8 ml of water was added dropwise to the mixture. The pH was maintained at 4.7 for 1 hour and the reaction was allowed to proceed for 24 hours at room temperature by gentle stirring. The gel was thoroughly washed with distilled water and stored at 4°C.

2.4 Analytical Procedures. β-Galactosidase activities were measured according to Pardee *et al.*[12] One unit was defined as the amount of enzyme that converted 1 nmol of substrate per minute at 28°C. SDS-polyacrylamide gel electrophoresis was performed as described Laemmli.[9] Proteins were assayed according to Bradford.[2]

3. Results

3.1 Purification of Hybrid Proteins. The purification method is based on the affinity chromatography procedure described by Steers *et al.*,[16] who used TPEG affinity column as the last step of β-galactosidase purification. The major feature of the one-step purification method is the use of high salt concentration. NaCl increases the capacity of the affinity-column for β-galactosidase and thus allows the elution of enzyme concentrations around 1 mg/ml. Data presented in Table 2 show the capacity of TPEG-Sepharose columns for β-galactosidase contained in crude extracts as a function of NaCl concentration. As can be seen, 1 ml of affinity matrix retains more than 2 mg of β-galactosidase (500,000 units represent 1 mg of enzyme) at 1.6 M NaCl. This should be compared to 1,400,000 units retained in the absence of NaCl when pure β-galactosidase is applied (data not shown). It appears, therefore, that high salt concentration prevents nonspecific adsorption of foreign proteins on the affinity column.

Columns of 1 ml affinity matrix were loaded with crude extracts containing adenylate cyclase-β-galactosidase (80,000 units/ml β-galactosidase, 18 mg/ml protein).

3.2 Purification Procedure. Routinely, the purification procedure was carried out at 20°C as follows: Bacterial extract (15-25 mg of protein/ml) prepared in 20 mM Tris-HCl, 10 mM $MgCl_2$ (pH 7.4) was mixed with 5 M NaCl in order to obtain a final concentration of 1.6 M NaCl; β-mercaptoethanol was present at 10 mM. The solution was applied to the TPEG-Sepharose column equilibrated with the same buffer (20 mM Tris-HCl, 10 mM $MgCl_2$, 1.6 M NaCl, 10 mM β-mercaptoethanol, pH 7.4).

The bed volume of the column was 1 ml; 800,000 units of β-galactosidase were adsorbed. The column was washed with the buffer used for equilibration (about 100 bed volumes) until no more material absorbing at 280 nm appeared in the flow through. The protein was eluted from the column with 100 mM sodium carbonate, 10 mM β-mercaptoethanol, pH 10.95.

Table 3 - Purification of Hybrid Proteins.
β-Galactosidase Specific Activity (units/mg Protein)

Protein	Crude Extract	Pure Protein
β-Galactosidase	15,000	490,000
Lysozyme-β-galactosidase	6,700	360,000
Gal repressor-β-galactosidase	7,600	550,000
Maltodextran phosphorylase-β-galactosidase	2,800	220,000
Adenylate cyclase-β-galactosidase	4,800	480,000
Enzyme III^{Glc}-β-galactosidase	14,500	580,000

Starting from crude bacterial extracts, several milligrams of pure protein can be obtained in 3-4 hours with overall yields of 85-95%. Table 3 shows the specific activities of wild-type β-galactosidase and different hybrid proteins, in crude extracts and after the one-step purification. As can be seen, depending upon the specific activities of crude proteins, 33- to 100-fold purification could be achieved. The proteins are in general 85 to 95% pure and are suitable for immunization or amino acid sequence analysis.

4. Discussion

The one-step procedure described here can be used to purify different types of hybrid proteins exhibiting β-galactosidase activity, obtained by various methods of gene fusion. Several milligrams of near homogeneous protein can be obtained in a few hours. Subunits of the purified hybrid proteins contain between 11 and 500 amino acids of the foreign sequence; they are nevertheless able to fold into a conformation leading to tetrameric protein endowed with β-galactosidase activity. The purified hybrid proteins can be used to obtain antibodies against the foreign portion of the protein fusion.[14] Indeed, antibodies have already been obtained against the adenylate cyclase-β-galactosidase hybrid protein which recognize wild type adenylate cyclase.[5] The antibodies could be used as a tool to isolate the native protein from the organism of interest.

The proteins isolated by the procedure described above seem sufficiently homogeneous to be submitted to amino acid sequence analysis (Danchin *et. al.*, 1984; Strosberg, A.D., personal communication). The sequence information could then be used to produce synthetic polypeptide antigens.

References:

1. Bassford, P., Beckwith, J., Berman, M., Brickman, E., Casadaban. M., Guarente, L., Saint- Girons, I., Sarthy, A., Schwartz, M., Shuman, H. and Silhavy, T.: Genetic fusions of the *lac* operon: A new approach to the study of biological processes. in Miller, J.H. and Reznikofi, W.S. (Eds.), *The Operon*, Cold Spring Harbor Laboratory, Cold Spring Harbor, NY, 1978, p. 245-261.
2. Bradford, M. M.: A rapid and sensitive method of microgram quantities of proteins utilizing the principle of protein-dye binding. Anal. Biochem. 72 (1976) 248-264.
3. Casadaban, M.: Transposition and fusion of the *lac* genes to selected promoters in *Escherichia coli* using bacteriophage lambda and Mu. J. Mol. Biol. 104 (1976) 541-555.
4. Casadaban, M. J., Chou, J. and Cohen, S. N.: *In vitro* gene fusion that join an enzymatically active β-galactosidase segment amino-terminal fragments of exogenous proteins: *Escherichia coli* plasmid vectors for the detection and cloning of translational initiation signals. J. Bacteriol. 143 (1980) 971-980.
5. Danchin, A., Guiso, N., Roy, A. and Ullmann, A.: Identification of the *Escherichia coli cya* gene product as authentic adenylate cyclase. J. Mol. Biol. (1984) in press.
6. Fowler, A. V. and Zabin, I.: Purification, structure and properties of hybrid β-galactosidase proteins. J. Biol. Chem. 258 (1983) 14354-14358.
7. Gutierrez, C. and Raibaud, O.: Point mutations which reduce the expression of *malPQ*, a positively controlled operon of *Escherichia coli.* J. Mol. Biol. (1984) (in press).
8. Jacob, F., Ullmann, A. and Monod, J.: Deletions fusionnaant l"operon lactose et un operon purine chez *Escherichia coli.* J. Mol. Biol. 24 (1965) 704-719.
9. Laemmli, V. K.: Cleavage of structural proteins during the assembly of the head of bacteriophage T4. Nature 227 (1970) 680-685.
10. Miller, J. H.: *Experiments in Molecular Genetics*, Cold Spring Harbor Laboratory, Cold Spring Harbor, NY, 1972, p. 433.
11. Muller-Hill, B. and Kania, J.: Lac repressor can be fused to β-galactosidase. Nature 249 (1974) 561-563.
12. Pardee, A. B., Jacob, F. and Monod, J.: The genetic control and cytoplasmic expression of inducibility in the synthesis of galactosidase of *Escherichia coli.* J. Mol. Biol. 1 (1959) 165-168.
13. Roy, A., Haziza, C., and Danchin, A.: Regulation of adenylate cyclase synthesis in *Escherichia coli*: nucleotide sequence of the control region. EMBO J. 2 (1983) 791-797.
14. Ruther, U., Koenen, M., Sippel, A. E. and Muller-Hill, B.: Exon cloning: immuno-enzymatic identification of exons of the chicken lysozyme gene. Proc. Natl. Acad. Sci. U.S.A 79 (1982) 6852-6855.
15. Shuman, H. A., Silhavy, T. J. and Beckwith, J. R.: Labeling of proteins with β-galactosidase by gene fusion. J. Biol. Chem. 255 (1980) 168-174.
16. Steers Jr., E., Cuatrecasas, P. and Pollard, H. B.: The purification of β-galactosidase from *Escherichia coli* by affinity chromatography. J. Biol. Chem. 246 (1971) 196-200.
17. Ullmann, A., Jacob, F. and Monod, J.: On the subunit structure of wild-type vs complemented β-galactosidase of *Escherichia coli.* J. Mol. Biol. 32 (1968) 1-13.
18. Wilcken-Bergmann, B., Koenen, M., Griesser, H. W. and Muller-Hill, B.: 72 residues of gal repressor fused to β-galactosidase repress the gal operon of *E. coli.* EMBO J. 2 (1983) 1271-1274.

Lab 7.7:
BCA Protein Concentration Assays and β-Galactosidase Assays to Construct an Enzyme Purification Table

Overview:

In this lab, you will perform the BCA (bicinchonic acid) protein concentration assay, construct a bovine serum albumin standard curve, and determine the β-galactosidase-*myo*-3 fusion protein concentrations of all pooled protein samples. You will determine the concentrations of pooled proteins in the crude homogenate and eluates obtained from the gel filtration, ion exchange, and affinity chromatography steps. Microplate β-galactosidase assays will then be performed with the pooled β-galactosidase-*myo*-3 fusion protein. Enzyme activities will be calculated from the β-galactosidase assays absorbance at 405 nm. With these two calculations, an enzyme purification table will be constructed. The table will show the protein concentration, enzyme concentration, specific activity, enrichment, total protein and enzyme contents, and enzyme yields in the respective eluates.

Experimental Outline:

1. BCA protein assays to determine protein concentrations (BCA™ method).
2. β-Galactosidase assays to determine enzyme activities.
3. Construct enzyme purification table.

Required Materials:

UV-visible spectrophotometer; 10 mM sodium phosphate (pH 7); bovine serum albumin standard solution (2 mg/ml); BCA (Pierce) reagents A and B (see Lab 7.3); 60°C water bath; substrate mix (recipe below); stop buffer (recipe below); microtiter plate; microtiter plate reader; graph paper

Procedures:

1. Protein Concentration Determination: BCA Method

1. Prepare a stock protein, then dilute it with sodium phosphate buffer.
   a. Prepare a 10 mM sodium phosphate buffer (pH 7) for protein dilutions:
      i. Weigh 0.069 g sodium phosphate, monobasic.
      ii. Adjust to pH 7 and raise the volume to 50 ml with H_2O.
   b. Prepare 5 ml of bovine serum albumin (BSA) standard at a concentration of 2 mg/ml:
      i. Weigh 0.01 g of BSA.
      ii. Dilute the BSA to 5 ml.
   c. Prepare the "working reagent" (2 ml per sample): For 17 ml of "working reagent": 50 parts reagent A (16.667 ml) per 1 part reagent B (333 μl).
      i. 1 blank (working reagent only).
      ii. 5 standards (BSA and working reagent).
   d. Prepare the following set of protein standards in duplicate:

Volume of Albumin Standard	Volume of Sodium Phosphate	Final Concentration
50 μl	1.95 ml	50 μg/ml
100 μl	1.90 ml	100 μg/ml
150 μl	1.85 ml	150 μg/ml
200 μl	1.80 ml	200 μg/ml
250 μl	1.75 ml	250 μg/ml

   e. Prepare the unknown dilutions:

Sample	Source Experiment	Dilution
Crude hmogenate	Lab 7.2	20 μl protein/980 μl 10 mM sodium phosphate buffer (1:50)
Isoluble faction	Lab 7.2	20 μl protein/980 μl 10 mM sodium phosphate buffer (1:50)
Sluble fraction	Lab 7.2	20 μl protein/980 μl 10 mM sodium phosphate buffer (1:50)
Poled gel filtration chromatography (GFC) fractions	Lab 7.3	20 μl protein/380 μl 10 mM sodium phosphate buffer (1:20)

Pooled ion exchange chromatography (IEC) fractions	Lab 7.5	20 μl protein/380 μl 10 mM sodium phosphate buffer (1:20)
Pooled affinity chromatography (AC) fractions	Lab 7.6	20 μl protein/380 μl 10 mM sodium phosphate buffer (1:20)

With the above dilutions, use four volumes for each protein sample to run the BCA assay (24 reactions total).

f. Pipet 100 μl of each standard or unknown protein sample into labeled tubes. For a blank use 100 μl of 10 mM sodium phosphate buffer.
g. Add 2.0 ml of "working reagent" to each tube, then mix well.
h. Incubate all the tubes at 60°C for 30 minutes.
i. Cool all the tubes to room temperature.
j. Quickly measure the absorbance at 562 nm for each tube. Use the "reagent blank' to set the blank on the spectrophotometer and read the sample values.

Standard Curve Values	#1	#2
50 μg/ml		
100 μg/ml		
150 μg/ml		
200 μg/ml		
250 μg/ml		

Unknown Protein Values	10 μl	50 μl	100 μl	200 μl
Crude homogenate				
Insoluble fraction				
Soluble fraction				
Pooled GFC fractions				
Pooled IEC fractions				
Pooled AC fractions				

To find the concentration (μg/ml) for each unknown protein dilution, find the concentration on the standard curve corresponding to its absorbance reading. Then, adjust this concentration value with the dilution factor so that all values are comparable to using 100 μl in each assay. (For example, if 50 μl unknown protein gives a concentration of 0.35, multiply times two to get a value for a 100 μl sample = 0.70 μg/ml.) Be sure to take into account the original dilution. Average all of the readings for each fraction to achieve a more accurate concentration for each fraction (only average values that fall on the standard curve). Record the protein concentrations on the enzyme purification table.

2. Microtiter Plate β-Galactosidase Assay

1. Prepare solutions

a. *Substrate Mix*:

4 mg/ml ONPG (*o*-nitrophenol-β-D-galactopyranoside)
1 mM $MgCl_2$
50 mM β-mercaptoethanol
0.1 M Na_2HPO_4 (pH 7.3)

b. *Stop Buffer*:

1 M $NaHCO_3$

c. *Buffer B*:

	NaCl	29	g
1 M	Tris-HCl, pH 7.4	10	ml

1 M	$MgCl_2$	10	ml
0.1 M	EDTA	10	ml
	β-mercaptoethanol	10	ml
	Triton X-100	1	ml

Add H_2O to 1000 ml. Store at room temperature.

2. Procedure

a. Add 2 μl of each enzyme sample to 18 μl of Buffer B in separate wells of a microtiter plate. Mix well by shaking. Make sure to include wells for the blanks.

a. Aliquot 2 μl of the diluted sample above (1/10 dilution) into 98 μl of Buffer B.

b. To the samples in step b add 100 μl of the substrate mix and mix well to start the reaction. Time the addition of enzyme so that there is 5-15 seconds between the addition of each sample. This will ensure that all of the samples are incubated for equivalent times. Exact times must be recorded for each assay.

d. Incubate at room temperature until a yellow color develops (time may vary).

e. Add 100 ml of the stop buffer; timed as in step c. The stop buffer ends the reaction by increasing the pH to 11, thus inactivating the β-galactosidase enzyme.

f. Read the plates at 405 nm in a microtiter plate reader.

g. Calculate the enzyme activity as follows:

$$[\text{Enzyme}] = \frac{\Delta A_{405}}{\Delta t\ (\text{min}) \times \text{volume of undiluted sample (ml)}}$$

The volume of the undiluted sample will be 0.0002 ml if diluted as indicated above.

3. Quantitate the Enzyme Activity

a. Fill in the following chart:

Protein fraction	Time Change (minutes)	Absorbance at 405 nm	Enzyme Activity (U/ml)
Crude homogenate			
Insoluble fraction			
Soluble fraction			
Pooled GFC fractions			
Pooled IEC fractions			
Pooled AC fractions			

3. Construct Enzyme Purification Table

a. Part I

Protein Fraction	Protein Conc. (μg/ml)	Enzyme Activity (U/ml)	Specific Activity (U/mg)
Crude homogenate			
Insoluble fraction			
Soluble fraction			
Pooled GFC fractions			
Pooled IEC fractions			
Pooled AC fractions			

Record any important observations (be observant).

b. Part II

Protein Fraction	Enrichment[a]	Total Protein Amount[b]	Total Enzyme Amount[c]	Enzyme Yield[d]
Crude homogenate				
Insoluble fraction				
Soluble fraction				
Pooled GFC fractions				
Pooled IEC fractions				
Pooled AC fractions				

a. $\text{Enrichment} = \dfrac{\text{Specific activity of fraction studied}}{\text{Specific activity of crude homogenate}}$

b. Total Protein = Protein Concentration × sample volume

c. Total Enzyme = Enzyme Concentration × sample volume

d. $\text{Enzyme yield} = \dfrac{\text{Total enzyme amount of fraction studied}}{\text{Total enzyme amount of crude homogenate}} \times 100$

Record any useful observations (e.g., Were there any unusual increases in enzyme activity?).

Additional Report Requirements:

1. BCA assay standard curve
2. Extrapolated protein concentrations from BCA standard curve
3. Microtiter plate β-galactosidase absorbance readings
4. Enzyme purification table and all corresponding calculations

References:

1. Smith *et al.* (1985) Anal. Biochem. 150: 76-85; Pierce Biochemicals, ph. (800) 874-3723.
2. Miller, J. M. (1973) *Experiments in Molecular Genetics*, Cold Spring Harbor Laboratory, Cold Spring Harbor, NY.
3. Pardee, A. B., Jocob, F. and Monod, J. (1959) The genetic control and cytoplasmic expression of "inducibility" in the synthesis of β-galactosidase by *E. coli.* J. Mol. Biol. 1: 165.
4. Zabin, I. and Fowler, A. (1970) β-Galactosidase and thiogalactoside transacetylase, *The Lactose Operon*, p. 27, Cold Spring Harbor Laboratory.

UNIT 8: DISCONTINUOUS GEL ELECTROPHORESIS, PROTEIN MOBILITIES, AND APPARENT SIZE DETERMINATION

1. Disc Gel Electrophoresis

Discontinuous (disc) gel electrophoresis and its simpler forerunner *sodium dodecyl sulfate polyacrylamide gel electrophoresis (SDS-PAGE)* are used to separate and characterize proteins in partially purified and homogeneous preparations. Disc gel electrophoresis has the particular advantage of focusing proteins into a narrow ("tight") population ("band") before they enter the connected gel for fractionation by SDS-PAGE. Disc gel electrophoresis refers to the fact that *three different pH-buffer systems and two different polyacrylamide gels* are polymerized, stacked one on the other, in one gel.[1] The first step is to prepare a "running gel." Following electrophoresis, the running gel is stained to observe fractionated proteins and is usually polymerized from an 8% to 12% polyacrylamide pregel solution – prepared in a buffer containing 5 to 10% SDS and Tris-HCl buffer at pH 8.7. Under these circumstances, proteins are denatured to approximately linear forms and "coated" with a monolayer of negatively charged SDS molecules. As a result, they migrate toward the positive electrode at the bottom of the gel. Separation of proteins occurs in the running gel; electrophoretic mobilities depend on both the size and charge of individual proteins. The charges and mobilities of the SDS-coated polypeptides are usually proportional to their molecular sizes.

A "*stacking gel*" that is strikingly different from the "*running gel*" is placed on top of the latter. The 1 to 5% stacking gel serves as a solid matrix; proteins are focused into a narrow band by the combined activities of the buffer chemicals, the functional groups of the proteins, the gel matrix components, and migration in an electrical field. The stacking gel contains SDS and Tris-HCl at pH 6.5, not 8.7. This 2 log unit difference between the upper *stacking* and lower *running* buffers is critical for protein focusing. It is important that the sample buffer pH also be 6.5. Both pH and polyacrylamide content are lower in the stacking gel compared to the running gel. As a result, the boundary below the stacking gel provides a clear-cut line of discontinuity.

There is a second discontinuity boundary above the stacking gel. The buffer in the upper reservoir of the electrophoresis chamber contains SDS and a buffer at pH 8.7 – the same as in the running gel but composed predominantly of glycine, not Tris-HCl. Substituting glycine is critical for protein focusing in the stacking gel.

When all gels and buffers are in place, the protein sample is applied to the slots (lanes), and the current is begun by applying a voltage between the upper and lower buffer reservoirs, using the gel as an in-line resistor. At first, negatively charged ions begin moving toward the positively charged electrode in the reservoir in which the bottom of the gel is submerged. The following electrophoretic mobilities are observed: Cl^- > proteins$^-$ > glycinate$^-$. Almost immediately, glycinate ions cross the discontinuity between the upper reservoir buffer (pH 8.7) and the stacking gel/sample buffer (pH 6.5). At this point, the population of negatively charged glycine molecules acquires a predominantly neutral (zwitterion) charge: the amino group becomes NH_3^+, neutralizing COO^-. As a result, molecules do not move in the electric field that spans the length of the gel. A drop in mobile negative ions occurs rapidly at the very top of the protein sample well. Since the overall voltage has not changed and the overall current must remain constant throughout the gel, the protein migration rates increase, balancing the sudden change in glycinate migration. While proteins in the upper portion of the sample migrate faster, proteins in the lower edge of the sample cannot move past the chloride ions accumulated below them. The chloride ions provide sufficient numbers of mobile negative charges to support the current, and the larger, less mobile proteins align at the boundary of the running and stacking gels. Proteins move rapidly only above the chloride ions. This compresses the proteins into a narrow band above the running gel. When the proteins, chloride and glycine enter the pH 8.7 running gel, the mobilities of the ions allow the "focused proteins" to separate based on their size-dependent migration through the polyacrylamide.

2. Protein Separation by Size and Use of Disulfide Destabilizing Reagents

Protein secondary and tertiary structures are covalently stabilized by disulfide bonds, reinforcing the tendency to bury less polar residues in the "hydrophobic" core. β-Mercaptoethanol (β-ME) and dithiothreitol (DTT) are mild reducing agents that can reduce disulfide bonds and thereby disrupt an important set of contributors to protein stability. SDS can disrupt hydrophobic and ionic interactions in proteins very effectively. Treatment of proteins with β-ME (or DTT) and SDS denatures most proteins, producing a "random coil" state. Surprisingly, a relatively uniform charge-to-mass ratio is observed for most proteins, permitting their separation by SDS-PAGE according to differences in molecular mass. An exception to this trend occurs with heavily glycosylated proteins, which bind anomalous amounts of SDS, resulting in unusual electrophoretic mobilities compared with unglycosylated proteins of the same mass.

A collection of nonglycosylated proteins spanning a range of molecular masses are boiled for several minutes in 10% SDS and 0.5 M DTT or 1 M β-ME. Analysis by SDS-PAGE yields relative mobilities (1.0 is typically assigned to that of Coomassie Brilliant Blue). This routine procedure is used almost uniformly to determine apparent molecular weights of proteins; accuracies are typically in the +/- 20% range.[2]

References:

1. Bruening, G., Criddle, R., Preiss, J. and Rudert, F. (1970) *Biochemical Experiments*, Wiley, New York.
2. Weber, K. and Osborn, M. (1969) The reliability of molecular weight determinations by dodecyl sulfate-polyacrylamide gel electrophoresis. J. Biol. Chem. *244*: 4406-4412.

PROCESS RATIONALE

DISCONTINUOUS GEL ELECTROPHORESIS AND PROTEIN SIZE DETERMINATION

Cooper, T. G. (1977) *The Tools of Biochemistry*, pp. 204-208, Wiley-Interscience, New York. (Adapted)

Discontinuous pH electrophoresis, or simply *disc electrophoresis,* is a modification of the zone electrophoretic techniques. The significant differences are (1) the use of a two gel system and (2) the unique buffer systems used in the gel matrix and in the buffer reservoirs. The two gel system is illustrated in Figure 1. The lower-separating or running gel is prepared using about the same amount of acrylamide (5-10%) as would be used for an analogous zone electrophoretic experiment. It is in this gel that the macromolecules subsequently separate. The buffer used in this gel is usually an amine such as Tris, which is adjusted to the proper pH (e.g., 8.7) using hydrochloric acid. After the separating gel has polymerized, a second, small (1 cm) layer of gel is polymerized on top of it. This is called the upper or stacking gel and is prepared using much less acrylamide than was used in preparation of the running gel. A total acrylamide content of 2 to 3% is common. The buffer used in the stacking gel is also an amine such as Tris, but in this case the pH is adjusted with hydrochloric acid to a value about 2 pH units lower than that of the running gel (pH 6.5). The buffer used in the protein sample should be identical to that used in the stacking gel. The buffer used in the lower reservoir is of the same composition and pH as in the running gel, although it may be somewhat more dilute. Buffer for the upper reservoir is also an amine. However, it is adjusted to a pH the same as or slightly above that used in the running gel with a weak acid whose pK_a is at the desired pH. *Glycine* is commonly used for this adjustment; it also serves another important function (see below).

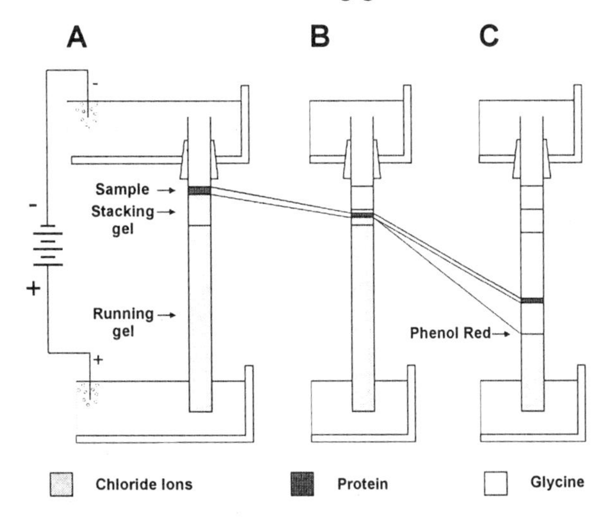

Figure 1 - Schematic diagram of a two gel system and the movement of the various ionic species during electrophoresis [from Bruening, G., R. Criddle, J. Preiss and F. Rudert. (1970) *Biochemical Experiments*, Wiley, N.Y.].

The electrophoretic behavior observed with this method is best understood by following the events occurring immediately after the power is turned on. Glycine in the upper buffer reservoir exists as both a zwitterion with a net charge of zero and glycinate ion with a charge of minus one:

$$NH_3CH_2COOH + H^+ \rightleftharpoons {}^+NH_3CH_2COO^- + H^+ \rightleftharpoons NH_2CH_2COO^- + H^+$$

When the electric field is established, chloride, protein, bromophenol blue, and glycinate anions all begin to migrate toward the anode. However, as glycinate ions enter the sample buffer and stacking gel they encounter a condition of *low pH*, shifting the equilibrium toward formation of *zwitterions*, which are *immobile*. Failure of glycine zwitterions to move into the sample and stacking gel creates a deficiency of mobile ions, which in turn decreases current flow. However, a constant current must be maintained throughout the entire electrical system. This is accomplished in the area between the leading chloride ions and the trailing glycinate ions by an increase in voltage. The result is a very highly localized voltage gradient occurring between the chloride ions and the glycinate ions. In this condition the relative ion mobilities are

glycinate < protein < bromophenol blue < Cl

In this strong local electric field, the anionic proteins all migrate rapidly – the stacking gel has large pores, so their progress is not impeded very much. If any of the proteins overtake the leading chloride ions, they slow down. Wherever chloride ions reside there is no ion deficiency, and hence the large field strength disappears. Therefore, the proteins migrate quickly until they reach the area containing chloride ions and then slow down drastically. Rapid movement of the proteins behind the chloride front and their decreased rate of migration as they approach the front result in a piling up (concentrating) of the proteins in a tight disc between the glycinate and chloride ions. As the disc of protein encounters the running gel, its migration is slowed by the smaller pores of the gel. This permits the small glycinate ions to catch up with the proteins.

Crossing the interface between the stacking and running gels, glycinate ions become fully charged once again, eliminating the ion deficiency. Hence, from this point on there is a constant field strength throughout the gel, and separation of the proteins proceeds just as with zone electrophoresis. The advantage of this modification is that the protein sample enters the separating gel as a narrow zone. The resulting protein bands are therefore much more compact, and this increases the resolution of the technique. A large number of buffer systems have been developed which cover the range of pH values between 3.5 and 9.5.[1-4]

SDS acrylamide gel electrophoresis can be used for protein size determination. Equations 1, 2, and 3 state that the mobility of a protein in acrylamide gels is a function of both its net charge and size:

$$v = \frac{Eq}{d6\pi r\eta} \quad (1)$$

where v is the velocity of the molecule, E/d is the field strength, q is the molecular charge, r is the effective sphere-like (isotropic) radius of the molecule, and η is the solution viscosity:

$$\log M = \log M_o - K_R T \quad (2)$$

where M is the electrophoretic mobility, M_o is the free mobility in a sucrose solution, and K_R is the retardation coefficient, which is defined as:

$$K_R = C(R + r) \quad (3)$$

where C is a constant, R is the geometric mean radius, and r is the radius of the gel fibers.

Hypothetically, two proteins of different molecular weights may migrate toward the anode at the same rate if their size differences are balanced by compensating charge differences. For this reason, acrylamide gel electrophoresis using gels of only one pore size, as discussed above, may not be used to gain information about the molecular weight of a molecule. Techniques employing varying pore size are subsequently discussed. A second restriction placed on electrophoretic techniques concerns the number of species observed on the gel. Molecules which are tightly, but not covalently, bound together do not usually separate from one another during electrophoresis. For example, core RNA polymerase (composed of three nonidentical subunits) appears as a single band if subjected to either zone or disc electrophoresis. Therefore, the number of protein or RNA bands observed on a gel represents only the minimum number that may in fact be present.

In an effort to surmount these problems, Shapiro *et al.*[5] attempted to separate a mixture of proteins in the presence of sodium dodecyl sulfate (SDS), an anionic detergent. The chemical formula of SDS is as follows:

$$[CH_3 - (CH_2)_{10} - CH_2 - O - SO_3^-]\, Na^+$$

The results of their preliminary efforts prompted Weber and Osborn[6] to determine the mobilities of about 40 proteins in the presence of SDS. These investigators observed empirically that the mobilities of these proteins were a linear function of the logarithms of their molecular weights (see Figure 2). Sodium dodecyl sulfate has been shown to bind to the hydrophobic regions of proteins and to separate most of them into their component subunits. SDS binding also imparts a large negative charge to the denatured, randomly coiled polypeptides. This charge largely masks any charge normally present in the absence of SDS. The precise reasons for the success of this technique are obscure, but it is widely used empirically as an assay of the molecular weight and subunit composition of purified proteins. It should not, however, be considered universal because cases have been reported where incorrect information has been obtained with this method. Very large or structural proteins (such as collagen) are particularly troublesome.

If component subunits of a protein are held together by disulfide bonds, these bonds may be broken before electrophoresis by heating the preparation in the presence of SDS and β-mercaptoethanol, which reduces them to sulfhydryl groups. These groups are then blocked with an appropriate alkylating agent to prevent reformation of the disulfide bonds. Although the mobilities of proteins with molecular weights between 12 and 7 kDa behave in the expected manner, some curvature was observed in a plot of mobility versus log molecular weight for proteins in the 40 to 200 kDa size range. This, however, may be a result of the investigators using a decreased amount of cross-linker "bis" [*N*',*N*'-methylene-bis(acrylamide)]. Whenever this method is used for molecular weight determination, it is advisable to include at least three to four standard proteins with molecular weights spread above and below the unknown. Bracketing the unknown protein in this manner yields a linear plot from the protein standards on which the unknown sample is appropriately positioned to determine its molecular weight. An example of this approach appears in the molecular weight determination of HMG-CoA synthase shown in Figure 3.

A few practical notes should be emphasized regarding SDS electrophoresis. Potassium salts should be avoided and sodium salts used instead

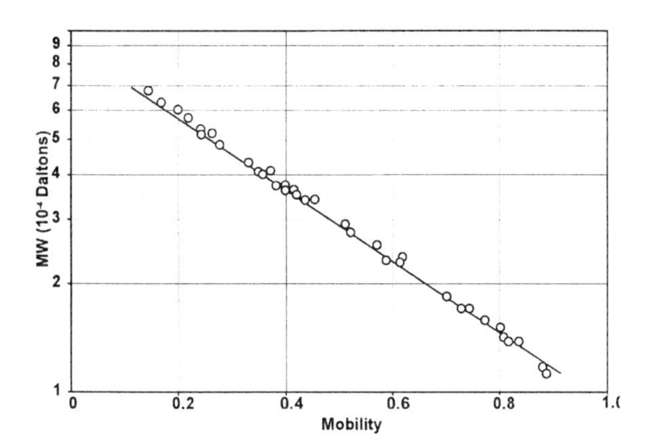

Figure 2 - Comparisons of the molecular weights of 37 different polypeptide chains in the MW range from 11 to 70 kDa with their electrophoretic mobilities.[6]

because potassium dodecyl sulfate is quite insoluble. In addition, even sodium dodecyl sulfate is insoluble below about 10°C. Although not specifically cited above, disc gel electrophoresis may also be used in the presence of SDS. These techniques, described by Studier[7] and Ames,[8] are of great advantage when the sample volume is in excess of 10 to 20 µl per gel. A widely used buffer system for SDS acrylamide gel electrophoresis is that developed by Laemmli.[9]

A second method for the determination of molecular weights has been described by Hedrick and Smith.[3,10] Their procedures are based on measuring the extent to which a protein migrates into acrylamide gels containing different total amounts of acrylamide and hence possessing different pore sizes. Although this method is more cumbersome than the SDS procedure described above, it has the advantage that molecular weights of "native," undissociated proteins may be obtained. These methods may also be used to distinguish charge isomers of proteins such as lactate dehydrogenase.

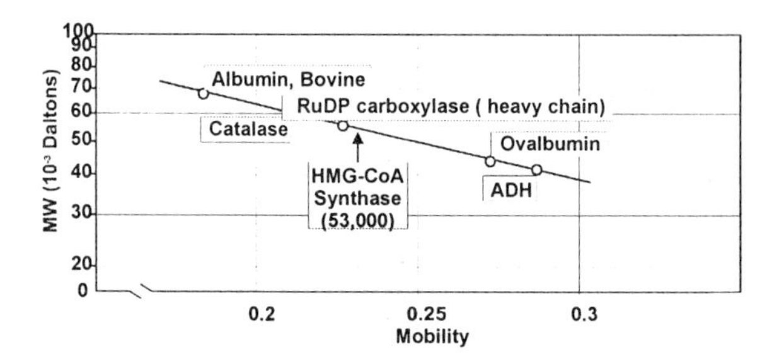

Figure 3 - Estimation of the subunit mass of mitochondrial HMG-CoA synthase by SDS gel electrophoresis [Reed, W. D., K. D. Clinkenbeard and M. D. Lane (1975) J. Biol. Chem. 250, 3120].

References:

1. Rodbard, B. and A. Chrambach (1971) Anal. Biochem. 40, 95-134.
2. Gabriel, O. (1971) Methods Enzymol. 222, 565-577.
3. Hendrick, J. L. and A. J. Smith (1968) Arch. Biochem. Biophys. 126, 155-164.
4. Paterson, B. and R. C. Strohman (1970) Biochemistry 9, 4094-4105.
5. Shapiro, A. L., E. Vinuekla and V. J. Maizel (1967) Biochem. Biophys. Res. Commun. 28, 815-820.
6. Weber, K. and Osborn, M. (1969) J. Biol. Chem. 244, 4406-4412.
7. Studier, W. F. (1973) J. Mol. Biol. 79, 237-248.
8. Ames, G. F. L. (1974) J. Biol. Chem. 249, 634-644.
9. Laemmli, U. K. (1970) Nature 227, 680-685.
10. Kempe, T. L., D. M. Gee, G. M. Hathaway and E. A. Noltmann (1974) J. Biol. Chem. 249.

LAB 8.1: DISCONTINUOUS SDS GEL ELECTROPHORESIS

Overview:

In this lab, SDS discontinuous gel electrophoresis will be performed with crude homogenate, insoluble and soluble fractions, and gel filtration, ion exchange, and affinity chromatography fractions containing pooled β-galactosidase-*myo*-3 fusion proteins. SDS-PAGE uses sodium dodecyl sulfate to denature proteins and saturate them with an overall negative charge. As a result, each protein has nearly identical charge-to-mass ratio and is separated primarily on the basis of its mass. The smallest proteins move farthest. You will prepare SDS "running" and "stacking" gels, which will be poured and polymerized for electrophoresis. After loading samples onto the gel, the β-galactosidase-*myo*-3 fusion proteins will be electrophoresed, stained, and destained. Proteins are visualized directly in gels by staining with Coomassie Brilliant Blue. This analysis will allow you to determine the effectiveness of your protein purification procedures, step by step, the degree of purity of the protein sample, and its molecular mass (MW). The molecular mass can be calculated because the mobilities of most polypeptide chains are directly proportional to log MW. Thus, the MW of the *myo*-3 protein can be determined by comparing its electrophoretic mobility with those proteins (markers) of comprising a range known MWs.

Experimental Outline:

1. Running gel preparation.
2. Stacking gel preparation.
3. Electrophoresis preparation.
4. Protein sample preparation.
5. Electrophoresis.
6. Protein staining and destaining.

Required Materials:

running gel; *N,N,N',N'*-tetramethylethylenediamine (TEMED); ammonium persulfate (APS); acrylamide; sodium dodecyl sulfate (SDS); stacking gel (recipe listed below); running gel (recipe below); 2x sample buffer (recipe below); 5x running buffer (recipe below); Pasteur pipet; Coomassie blue stain (recipe below); destaining solution (recipe below); electrophoresis apparatus; water bath; vacuum; fine-tipped syringe

Procedures:

1. Prepare Electrophoresis Solutions

a. SDS-PAGE Reagents and Buffers

i. Running Gel (10%)

40%	acrylamide	10 ml
10%	SDS	0.4 ml
1 M	Tris, pH 8.8	15 ml
	H_2O	14.4 ml
15%	APS	0.25 ml
	TEMED	32 µl

ii. Stacking Gel (5%)

40%	acrylamide	2.5 ml
10%	SDS	200 µl
1 M	Tris, pH 6.8	2.5 ml
	H_2O	14.7 ml
15%	APS	125 µl
	TEMED	16 µl

iii. 2x Sample Buffer:

0.5 M	Tris, pH 6.8	2 ml
	glycerol	2 ml
20%	SDS	2 ml
1%	bromophenol blue (1:3)	2 ml
1 M	DTT (add before use)	2 ml

vi. 5x Running Buffer:

Tris base	15.1	g
glycine	94	g
10% SDS	50	ml
dH_2O	to 1000	ml

vii. Coomassie Blue Stain:

Coomassie Brilliant Blue R-250	2.5	g
methanol	450	ml
glacial acetic acid	100	ml
H_2O	450	ml

viii. Destaining Solution:
10% acetic acid
10% methanol (may be increased)

2. Preparation of Running Gel

CAUTION: Acrylamide is a cumulative neurotoxin, but can be worked with safely if proper precautions are taken.

1. Prepare a 10% acrylamide solution as shown above (do not add TEMED or APS).
2. Degas the acrylamide stock for 5 to 10 minutes by vacuum. Degassing appears to be beneficial to remove polymerization inhibitors and to ensure optimal polymerization conditions. SDS is a detergent and can cause troublesome foaming if you swirl the solution too vigorously.
2. Before pouring the running gel, add 32 μl of TEMED and 250 μl of 15% APS. Make the APS 10 minutes before use (weigh this out by group: 50 mg/0.5 ml).
3. Swirl the solution to make it homogeneous and quickly pour it between two glass plates. Fill to 3 cm below the top of the glass plate and very gently (but quickly) layer 1/4 inch of buffer-saturated butanol on top of the running gel. This will keep the gel from drying and provide a flat top for the running gel. Allow the running gel to polymerize for about 1 hour.

3. Preparation of Stacking Gels

1. Prepare a 5% acrylamide solution as shown above (do not add TEMED or APS).
2. Degas the acrylamide stock for 5 to 10 minutes by vacuum.
3. Pour off the 1/4 inch of buffer-saturated butanol from the top of the running gel.
4. Before pouring the stacking gel, add 16 μl TEMED and 125 μl of 15% APS (made freshly).
5. Swirl the solution to make it homogeneous and quickly pour it on top of the running gel, adjusting the level to about 1 cm from the top of the cut out plate.
6. Insert the well comb at a 45° angle to avoid trapping bubbles. Allow the stacking gel to polymerize for approximately 1 hour.

4. Electrophoresis Preparation (Hoeffer apparatus)

1. After the stacking gel is polymerized, gently remove the comb.
2. Rinse the sample wells with distilled water and drain by inverting the unit over a sink.
3. Fill the sample wells and the upper and lower buffer chambers with 1x Running buffer.
4. Bubbles must be removed from between the glass plates in the lower buffer chamber. Failure to do so will result in an extreme drop in current and smearing of the front. The best way to do this is to use a 10-ml disposable syringe, filled with running buffer, equipped with an 18-gauge needle bent at a 45° angle. For best results, work at eye level with the bottom of the gel.

5. Sample Preparation

Next, apply your protein samples to the SDS polyacrylamide gel. Each group will apply the samples from their protein purifications, plus the appropriate molecular weight standards. Each protein sample to be loaded will be mixed with the appropriate amount of 2x sample buffer. The instructor will tell you which samples to use and how much to load. The amount of protein and sample buffer to load will depend on the protein concentration.

1.
2.
3.
4.
5.
6.
7.
8.
9.
10.

6. Electrophoresis

1. Add each sample to the appropriate well. BE CERTAIN THAT THE POWER IS NOT HOOKED UP TO YOUR ELECTROPHORESIS APPARATUS.
2. Turn on the electrophoresis apparatus. The current per gel should be about 20 mA. Electrophoresis will take about 1.5 hours. DO NOT TOUCH THE APPARATUS WHILE THE POWER IS ON!
3. After electrophoresis, remove the gel from the plates and place it in a staining tray filled with Coomassie Brilliant Blue R-250 for about 1/2 hour.
4. Place the gel in a tray containing destaining solution for 0.5 to 3 hours, as required, based on the marker dye migration position.
5. Take a picture of the gel.

Notes:

1. APS is used as a source of free radicals, which initiates the polymerization of acrylamide monomers.
2. TEMED catalyzes the formation of free radicals produced by APS.

Additional Report Requirement:

1. Picture of your gel.

References:

1. Gordon, A. H. (1975) in *Laboratory Techniques in Biochemistry and Molecular Biology*, 2nd Edition, pp. 103-108, (T. S. Work and E. Work, Eds.), North Holland Publishing Co., Amsterdam.
2. SE 250 Hoefer Mighty Small II Instruction Manual.

UNIT 9: IMMUNOCHEMICAL TECHNIQUES

1. Introduction

This discussion is kept to a minimum because the appendices provide a wealth of details about the use of antibodies in biotechnology. Antibodies are used to detect antigens in the following scenarios: (*i*) in solutions, (*ii*) in gels, (*iii*) on blotting membranes, (*iv*) during or after attachment to affinity chromatography matrices, and (*v*) *in situ* and *in vivo* within cells, tissues cells, systems, and organs. An "immunoassay" involves using an antibody to detect an antigen. In the "simplest" situation, the *antigen* is detected by preferential retention of antibody-antigen complexes to solvent filters or blotting membranes; antibody-free antigen is not retained. If the antibody is recognized, the protein binds the antigen and the complex adheres to the filter or "blot" matrix.

2. Antibody and Antigen Labeling; Detection Strategies

One "labels" either the antigen or the antibody. "Labeling" involves several types of processes; a range of examples is given to illustrate some typical applications.

In one approach, the antigen is labeled directly with a radioisotope, fluorophore, or ligand that can be "grabbed" by a subsequent reagent in the assay protocol. The isotope is detected by exposure to a film or image storage plate. Fluorophores can be imaged by exciting them with light at the appropriate absorption wavelength and measuring the fluorescent pattern at the emission wavelength. Such images have been obtained with aqueous samples in Eppendorf tubes, using antigens that have been separated by electrophoresis or chromatography, and with antigens in complex biological structures (*in situ* and *in vivo*). It's not surprising that this methodology has produced a huge amount of information. These images have allowed us to reconstruct the organizations of many biological materials and, as a result, better understand the mechanisms of the respective functions.

Radioisotopes are affixed to both substrates and immunoglobulins. Antibodies have been labeled for years with radioactive iodine, which reacts with and binds tyrosine covalently, then is detected by autoradiography, imaging plates, or with liquid samples using a "liquid scintillation cocktail." This "cocktail" contains the sample and dissolved scintillation chemicals ("fluors") that produce light in proportion to the amount of absorbed α, β, or γ radiation produced by the isotope. This light is quantitated within a "liquid scintillation counter" to determine the amount of antigen attached to labeled antibody or vice versa.

Another major approach to detecting antigen-antibody complex formation involves attaching a "secondary antibody" to the antigen-bound "primary antibody" then assessing the presence of bound secondary antibody using its attached label. The second antibody binds the "constant region" of the primary antibody. Since different species have antibodies with different constant region sequences and structures, one can study patterns produced by more than one antibody, directed against different antigens, within the same sample. Many applications can be accomplished using commercially available secondary antibodies. In many applications, the label is replaced by a label-generating appendage called a "conjugated ligand" (or simply "conjugate"). Many types of conjugates have been developed. General categories include (*i*) coenzymes (e.g., biotin), (*ii*) gold beads, (*iii*) magnetically attractive beads, and (*iv*) peptide fragments. Examples of the latter include coenzyme mimics (e.g., the "strep-tag") and fragments that bind specific cations (e.g., nickel by the "his-tag") or resemble a domain of the parent protein. Two other important ligands that have been conjugated to immunoglobulins are (*v*) enzymes and (*vi*) nucleic acids.

Enzyme conjugates amplify signals by generating multiple labels in the vicinity of the enzyme by converting colorless reactants into colored products. Nucleic acid conjugates amplify the signal by producing nucleic acid duplexes via PCR ("immuno PCR"). In case (*v*), the secondary antibody and its tethered enzyme bind the primary antibody, which has bound the antigen. The amount of enzyme product is proportional to the initial amount of antigen-primary antibody complex formed. In case (*vi*), the secondary antibody and its tethered nucleic acid (and PCR primer sites) bind the primary antibody, which has bound the antigen. The amount of DNA produced by PCR is proportional to the initial amount of antigen-primary antibody complex formed.

3. Recombinant Antibodies

Another area of immunochemical research involves efforts to isolate and manipulate immunoglobulin genes in expression vectors. The goal is to produce "*single-chain antibodies*" (scF_v) for use in biotechnology and medicine. ScF_v

constructs are produced from plasmids that contain the heavy- and light-chain genes, connected by a linker-encoding fragment, along with control sequences to allow protein expression. As a result, one can produce scF_v protein in *E. coli*, allowing one to study the detailed binding characteristics of a specific antigen binding domain.

This leads us to "*combinatorial biochemistry*." To pursue a combinatorial approach, one makes sets of biomolecules that differ at one or a few positions while the rest of the sequence is preserved. A key advantage of this approach is that a range of different variants can be tested in one reaction mixture. Selecting one or a small set of molecules with superior functional capability involves discriminating and separating them from less active molecules and other variants.

An inherent assumption is that different protein sequence variants (or RNA, DNA in those applications) behave similarly and do not interfere with each other's functional capability in the same selection "pot." Tests must be incorporated into protocols to avoid "missing and losing" these "victims" as a result of interferences or nonanalogous behaviors.

A "panel" of expressed antibodies can be probed to test their ability to bind a desired resin- or bead-bound ligand or substrate. Materials generated by a panel of separated colonies can be probed in the same mixture. One only need investigate subspecies in the "combinatorial sample" if the antigenicity or other activity is found. The advantage is that many samples are tested simultaneously, allowing the investigator to avoid substantial effort by eliminating a broad sweep of less capable molecules from further investigation. If the materials are generated by a library of cells, it also becomes unnecessary to propagate the colonies that generate inferior competitor products. These subjects are addressed in detail in Unit 10.

4. Two-Dimensional Gradient Gel Electrophoresis

Two-dimensional gel procedures have been developed to probe binding characteristics of ligand, e.g., nucleic acid binding proteins. In this technique one varies the pH or the concentration of substrate, protein, or coeffectors by forming a gradient of the varied material across the first dimension in a tube gel. One then adheres the first dimension tube gel to the top of a slab gel containing SDS, a "sizing" gel. This second dimension of electrophoresis allows one to separate the species and assess their differential behaviors with respect to the gradient material incorporated into the first dimension of the gel. Studies that combine the use of these procedures, combinatorial genetics, and protein expression are yielding important new insights into the relation between sequences, domain structures, and functions.

5. Western "Blotting"

The goal of "*Western immunoblotting*" and related techniques (e.g., "Northwestern" and "Southwestern" experiments) is to separate proteins as a function of MW, to blot or electroblot them to a membrane, then to "probe" them with an antibody. When this antibody recognizes an "antigenic determinant" on the blotted protein pattern, the corresponding band (or bands) binds a secondary antibody that carries an attached conjugate moiety (an "immunoconjugate"). The conjugate is then supplied with a substrate that reacts to yield a colored (or otherwise detectable) product. Color only develops in the vicinity of the blot-bound antigen and attached primary antibody.

6. Molecular Mass Determination of β-Gal Fusion Protein from Western Blot

To estimate the MW of the antigenic β-gal fusion protein, the colored bands in the immunoblot are superimposed on their locations on the original SDS gel pattern. Next one compares its mobility with those of coelectrophoresed MW standards. Since the MW of the β-gal fragment attached to the fusion protein can be determined from the DNA sequence, the MW of the fusion protein itself can be calculated. Plotting the electrophoretic mobilities versus log MW produces a calibration line that can be used to estimate log MW corresponding to the mobility of the fusion protein.

This approach to determining MW assumes that the fragment is the expected one. If so, the β-gal promoter and protein-encoding DNA sequences located upstream and downstream from the inserted fusion protein fragment DNA are known. As a result, PCR can be used to determine the DNA sequence using the dideoxy DNA-sequencing technique. PCR should also indicate the placement and orientation of the inserted fusion fragment.

INNOVATION/INSIGHT

IMMUNOCHEMICAL TECHNIQUES

Cooper, T. G. (1977) *Tools of Biochemistry*, Chapter 8, Wiley-Interscience, New York. (Adapted)

1. Introduction

Immunoglobulins are classified at three successively finer levels of resolution. The first subdivision separates these proteins on the basis of structural and immunochemical differences into five groups called *isotypes* (Table 1). Each of the isotypes is divided into subgroups called *allotypes*. Allotypic differences are minor amino acid variations observed from one individual to another; that is, they are allelic forms within an isotype. The third level of classification, *idiotypes*, depends on differences that are clone-specific for clones of immunoglobulin-producing cells. Idiotypes are variations of structure occurring close to the ligand binding site of the γ-globulin.

The most detailed structural studies have been performed using IgGs because this isotype accounts for 85% or more of the serum immunoglobulins. These molecules are Y-shaped (see Figure 1) with a M_r of *ca.* 150 kDa. They contain *two heavy chains* (MW 50 kDa) and *two light chains* (MW 25 kDa) held together by three to seven of the 20 to 25 kDa disulfide bonds occurring per molecule. The heavy chains are bent at their midpoints to permit formation of the Y shape. Wide variation of the bend angle has been observed, but it is unclear whether a single molecule can bend to multiple angles. Cleavage of purified antibodies with *papain* yields three fragments as shown in Figure 1. Two fragments forming the arms of the Y (denoted F_{ab}, *antigen binding fragments*) retain the specific ligand binding sites but are univalent, that is, contain only one binding site per fragment. The remaining fragment (denoted F_c, *crystallizable fragment*) appears to be uniform in all rabbit IgG molecules.

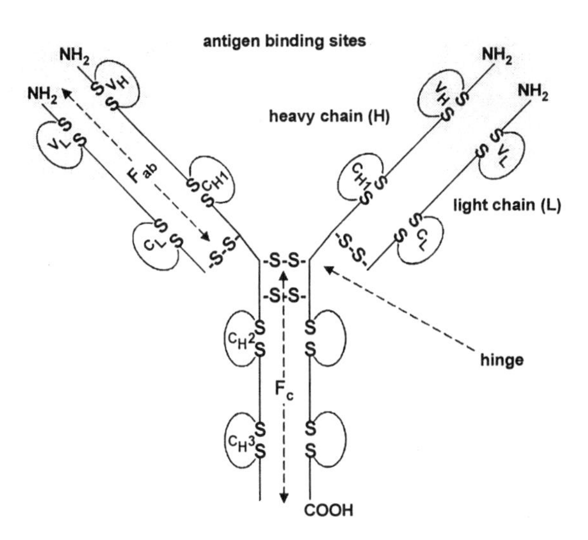

Figure 1 - Schematic representation of polypeptide chains in the basic immunoglobulin structure.

2. Antibody Formation

An *antigen* is defined as any compound that (*1*) can stimulate production of antibodies when injected into a test animal, and (*2*) reacts specifically with the antibodies produced. Both parts of this definition are necessary to distinguish antigens from *haptens* (see Figure 2), which are small molecules that can react with specific antibodies, but do not elicit specific antibody production unless injected in a *conjugated form*, that is, *bound to an antigen.*

The most important parts of the antigen are its *antigenic determinants*, the specific regions of the macromolecule directly involved in determining specificity of the immunochemical reaction. Each antigen may have many different determinants, each of which elicits a specific immune response and reacts with specific antibodies produced as a part of

Table 1 - Characterization of Immunoglobin Isotypes

Iso-type	MW (kDa)	Carbohydrate Content (%)	Serum Concentrations[b]	Biological Properties
IgG	160	2.9	750-2200	Main serum immunoglobulin
IgA	160[b]	7.5	51-380	Main immunoglobulin of secretions
IgM	1,000[c]	10.9	21-279	Increase in initial immunological response
IgD	160		1-56	
IgE	160		$1\text{-}14 \times 10^{-4}$	Allergic antibodies

[a]From J. Clausen, Immunochemical Techniques for the Identification and Estimation of Macromolecules, in *Laboratory Techniques in Biochemistry and Molecular Biology* (T. S. Work and E. Work, Eds.), American Elsevier, New York, 1969, p. 413.
[b]Concentration in normal human serum expressed as mg/100 ml of serum.
[c]These immunoglobulins polymerize.

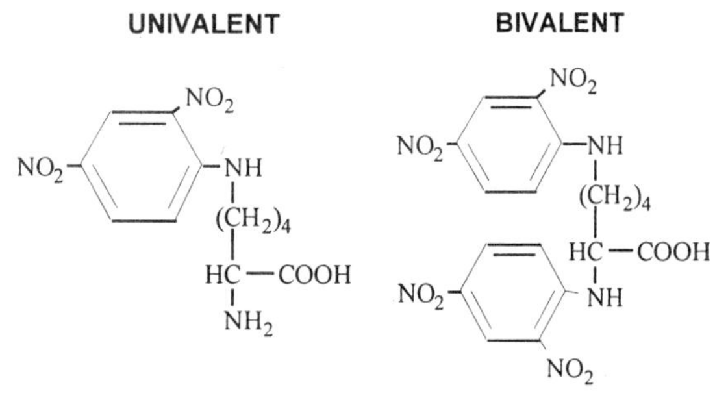

Figure 2 - Examples of a univalent hapten, ε-DNP-lysine, and a bivalent hapten, α, ε-bis-DNP-lysine.

that response. Antibodies, like enzymes, possess very specific three-dimensional requirements for binding to an antigenic determinant. Departure from the required shape of determinant, or the antibody binding site, markedly decreases or totally prevents binding. Although maximum binding requires a perfect fit, it must be emphasized that significant binding may be possible even in the absence of total complementarity.

An antigen injected into a test animal elicits an immune response by binding to determinant-specific receptor sites on two types of small lymphocytes denoted B and T. These *receptor sites* have been shown to be immunoglobulins, integrated into the lymphocyte cell membrane, which accounts for their *determinant specificity*. In response to the binding of an antigen, both lymphocyte varieties divide and increase in size, becoming large lymphocytes (Figure 3). Large lymphocytes begin producing limited quantities of antibody, but more importantly *continue differentiating into plasma cells* containing an extensive rough endoplasmic reticulum. It is the specialized plasma cells that secrete most of the immunoglobulins. Although T lymphocytes do not secrete immunoglobulins, they do play a critical accessory role in stimulating and regulating proliferation and differentiation of B lymphocytes.

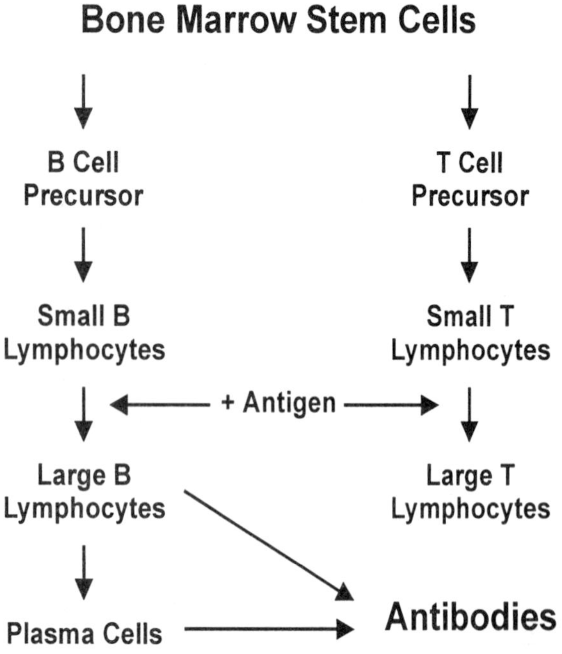

Bone Marrow Stem Cells

B Cell Precursor → Small B Lymphocytes → (+ Antigen) → Large B Lymphocytes → Plasma Cells → Antibodies

T Cell Precursor → Small T Lymphocytes → (+ Antigen) → Large T Lymphocytes

Large B Lymphocytes → Antibodies

Figure 3 - Development routes of the principal components involved in the immune response. Macrophages, which play an important accessory role in the immune response, have been omitted.

One of the puzzles concerning antibody formation is the observation that a wide variety of antibody molecules are produced against each type of antigenic determinant injected into the test animal. Of the several attempts to explain these observations, Burnet's clonal selection hypothesis currently receives greatest support. This view suggests that there are many individual diversified B lymphocytes, which may each respond to only one or a few antigenic determinants. An antigen injected into an animal interacts with only those cells carrying a receptor site that is complementary, or nearly so, to the injected determinant (Figure 4). On dividing, each of the responding lymphocytes gives rise to a clone of identical progeny, which give rise to clones of plasma cells, each producing their own specific type of antibody. All the antibodies produced by the member cells of one clone are identical, but those produced by different clones are each different, reflecting the structure of the immunoglobulin receptor to which the antigen was originally bound.

3. Practical Aspects of Antibody Production

3.1 The Antigen. Success or failure of immunochemical methods is largely determined by the initial step, preparation of the antigen. The immune systems of most animals are intensely sensitive, detecting even the smallest traces of contaminant antigens. B lymphocyte proliferation, leading to plasma cell secretion of antibody, is

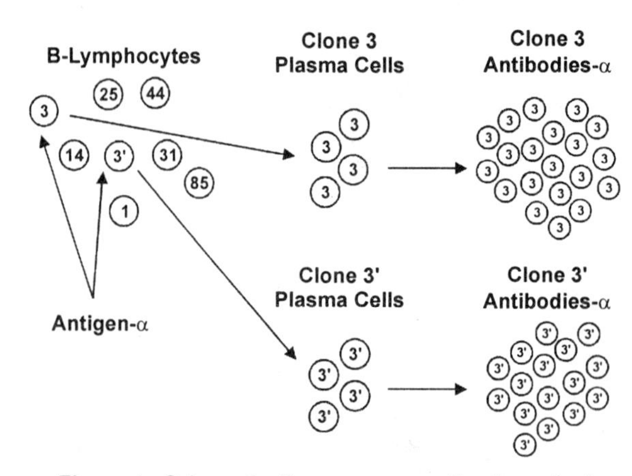

Figure 4 - Schematic diagram representing the salient features of the clonal selection theory.

primarily responsible for the amplified sensitivity. Therefore, every effort should be made to ensure that the antigen to be injected is homogeneous. At minimum, the preparation should yield only one band, with no visible contamination, when subjected to acrylamide gel electrophoresis. At best, however, this ostensibly ensures only that antibody is produced against one protein. If the protein is made of subunits, each subunit is likely to have multiple antigenic determinants and result in production of a set of antibodies in varying proportions. In view of this, the specificity of antibody preparations may be increased by first separating the protein into subunits and injecting only one type of polypeptide. This is not absolutely necessary unless maximum specificity is required. It is also important to note that injecting only one subunit of a protein requires that the isolated subunit possess a set of determinants similar to those of the subunit in its native position (microscopic environment) within the complete protein.

3.2 Adjuvants. The immunogenic potency of soluble antigens may be considerably increased if they persist within the tissues of the test animal. This may be accomplished by giving multiple small injections of antigen over a period of time. It has been shown with Diphtheria toxoid for, example, that several small doses – given over a period – evoke a much greater response than an identical quantity given all at once. Another method of obtaining antigen persistence is through use of an *adjuvant.* This term may be used broadly to denote *any substance that increases the immune response toward an injected immunogen.* Inorganic gels such as aluminum hydroxide or aluminum phosphate served as the earliest adjuvants. Immunogens were adsorbed on their surfaces and the mixture injected. These compounds have, however, been largely replaced by the oil-water emulsion adjuvants developed by Freund (Table 2). Immunogen in aqueous solution is mixed with one of Freund's adjuvants and a stable emulsion is produced. This may be done by very rapid mixing, sonication, or drawing the mixture repeatedly into a syringe (20-40 gauge needle) and squirting it out again forcibly. Stability of the emulsion may be ascertained by placing a single drop in a beaker of cold water. If it is stable, the first droplet spreads evenly across the liquid surface, but

Table 2 - Composition of Freund's Complete and Incomplete Adjuvants

Components	Complete	Incomplete
Mannoside mono-oleate, 1.5 ml	+	+
Paraffin oil, 85 ml	+	+
Mycobacterium butyricum, 5 mg (killed, dried)	+	-

subsequent drops remain at (or below) the surface and do not break up. Injecting an unstable emulsion into a test animal results in a loss of most advantages incurred by using adjuvants. After either subcutaneous or intramolecular injection, the stable emission forms small droplets that spread widely from the site of injection. The stability of emulsion results in a very slow release of immunogen, lasting sometimes up to several months or longer.

The presence of heat-killed mycobacteria in Freund's complete adjuvant serves to generally stimulate the entire reticuloendothelial system of the test animal. In addition, there is some evidence these dead cells also stimulate T lymphocyte proliferation and their production of B cell-stimulating factors. Use of the complete adjuvant is recommended for the first dose of immunogen, but incomplete adjuvant is adequate for subsequent injections.

4. Animals, Dose, and Route of Inoculation

Animals commonly used for antibody production are rabbits, goats, guinea pigs, chickens, rats, donkeys, and horses. Of these sources, young New Zealand White rabbits are most often employed when limited quantities of serum are needed; goats are used to produce larger quantities. If antibody preparations are to be used for low or medium specificity studies, sera may be collected from a number of rabbits which were immunized with the same preparation of immunogen. However, if fine specificity is required, it is advisable to use a larger animal such as a goat, so that the entire study can be performed using well characterized antibody preparations – obtained from a single or a few closely spaced bleedings. This is because, as discussed below, antibody preparations differ from animal to animal and even within the same animal over a period of time.

The correct *dose* of immunogen depends on its nature, whether adjuvant is used, and whether the animal has received previous injections. For example, 100 mg to 1 mg serum albumin injected into a rabbit in the absence of adjuvant evokes only a minimal response, whereas in combination with Freund's adjuvant it is highly effective. A second injection at a later time would require only 10 to 50 mg for a threshold response. After the threshold dose is exceeded, increasing quantities of immunogen lead to increasing responses. The degree of increase, however, is not proportional to dose. In fact, there is a very real danger of using an excessive amount of immunogen and establishing a state of unresponsiveness or immunological tolerance. Such tolerance is also evoked when certain immunogens are given in amounts just below threshold levels. Some trial and error is necessary, therefore, to establish the best dosage for a specific antigen and the conditions of injection.

For most primary injections using Freund's complete adjuvant, *intramuscular injection* provides rapid access to the lymphatic circulatory system. This route is much better than using foot pad injections, which usually produce severe swelling, ulceration, and necrosis. Secondary injections in the presence of incomplete adjuvant may be given either intramuscularly or subcutaneously. The latter route is advantageous in that the antigen is absorbed slowly, minimizing the possibility of *anaphylactic shock*. This route should not be used with complete adjuvant because an abscess will form.

4.1 Response to Inoculation. The rate of antibody production following inoculation depends on whether it is the first (primary) or a subsequent (secondary) injection. The two patterns of response are shown in Figure 5. After the first injection of immunogen there is a lag period of 1 to 30 days before appearance of serum antibodies; 5 to 7 days is quite common for many soluble proteins. After this, the concentration of serum antibodies increases exponentially, reaching a maximum usually at around 9 to 11 days when soluble proteins are used. The duration of maximum antibody production and the rate of its decline are largely influenced by the effectiveness of adjuvant and stability of the adjuvant-antigen emulsion. These facts are illustrated in Figure 6. In the absence of adjuvant, serum antibody concentration increases only a small amount and then falls rapidly. Alum adjuvant increases the maximum level and slows the rate of decline, but not nearly to the degree observed with Freund's complete adjuvant (Table 3).

If a second injection of immunogen is made after the primary response has declined, a secondary (amnestic or memory) response ensues. The nature of this response is shown in Figure 5. By comparison,

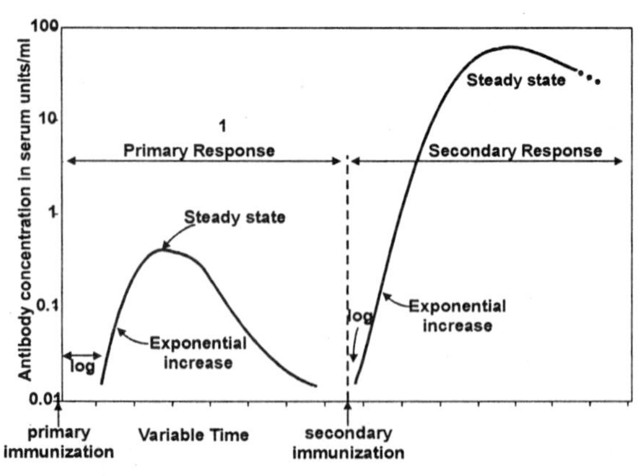

Figure 5 - Comparison of serum antibody concentrations following the first and second injections of immunogen (indicated as arrows beneath the abscissa). Note the logarithmic scale for antibody concentration. Time units are unspecified to indicate the great variability encountered with different immunogens.

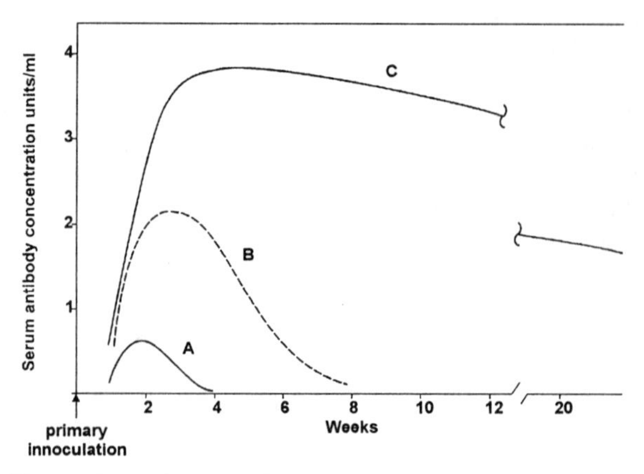

Figure 6 - Influence of adjuvants on the production of antibodies following injection of an immunogen. Schematic view of antibody produced by rabbits in response to one injection (arrow) of a soluble protein (A) in a dilute salt solution, (B) adsorbed on precipitated alum, or (C) incorporated into a stable water-in-oil emulsion containing mycobacteria (Freund's complete adjuvant).

the secondary response (*1*) occurs after a much shorter lag time, (*2*) results in significantly higher levels of circulating antibodies, and (*3*) persists longer. Of practical significance is the much smaller amount of antigen needed to evoke a secondary response than for the primary inoculation. Therefore, care should be taken to decrease the dose rate. In addition, complete adjuvant is no longer needed since the reticuloendothelial system is still stimulated from the primary injection; incomplete adjuvant is usually used for booster injections.

The immunoglobulin fraction may be isolated and concentrated by ammonium sulfate fractionation or column chromatography. Specific antibodies may be isolated using affinity chromatography with the original antigen as the bound ligand. In early purifications of this type, antibody was eluted with dilute acid or base, often resulting in antibody inactivation. These procedures are now replaced by elution with 4 M magnesium chloride, with much less inactivation.

5. Reaction of Macromolecular Antigens and Antibodies in Solution

Most, if not all, contemporary applications of immunochemical techniques are based on the reaction of antibodies with the antigen used to induce their production, to yield a very stable complex. If both antigen and antibody are present in solution, a precipitate forms as long as the antibody is in molar excess. This is known as a *precipitin reaction*. For example, a highly purified preparation of avidin was injected into a rabbit, which then produced avidin-specific antibodies. Since biotin forms a very stable complex (dissociation constant $\approx 10^{-15}$ M^{-1}) with avidin, complexation of avidin with ^{14}C-biotin provides a convenient means of detecting the protein's whereabouts. Following 24 hours of incubation at 4°C, the immunoprecipitate was separated from soluble components by centrifugation and the amount of radioactive biotin in the precipitate and supernatant was determined. The curve describing the amount of radioactivity found in the precipitate may be divided into three portions: an ascending limb termed the region of antibody excess, a region of *equivalence*, and a descending limb termed the region of *antigen excess*. In the regions of antibody excess and equivalence, all of the antigen is found in the precipitate. In the region of antigen excess, however, less and less ^{14}C-biotin precipitates and more can be found in solution. Note that none of the ^{14}C-biotin is precipitated when nonimmune control serum is used in place of immune serum. This is essentially a titration of the serum's capacity to precipitate avidin. Heidelberger and Kendall performed a similar experiment but also isolated the precipitates and determined that the antibodies had two ligand binding sites per molecule. Antigens also have a varying number of determinant sites, which increases with increasing molecular weight. Antibody bivalency and the lattice theory explain the failure of immunoprecipitates to form at low antibody-antigen molar ratios; there is an insufficient number of binding sites on the antibodies compared to the number of available determinant sites to form the required lattice (Table 3).

6. Use of Antibodies for Specific, High Resolution Assay of Proteins

Past studies concerning enzyme production and its control depended heavily on measurement of enzymatic activity to detect the presence of a protein. Although major advances in our understanding of molecular biology attest to the success of such approaches, they are nonetheless limited. Their principal deficiency is the requirement for activity before a given enzyme may be detected. It is becoming clear, however, that many interesting and important events occur prior to the point at which an enzyme becomes active. For example, some enzymes are synthesized in an inactive form and are subsequently activated by proteolytic cleavage or other chemical modification. Protein aggregates such as phage structural proteins, ribosomes, and multienzyme complexes undergo elaborate assembly procedures prior to becoming functional. An increasing number of cases have been found in which an enzyme is inactivated by chemical modification rather than being proteolytically degraded. In all of these situations, enzyme activity is absolutely inadequate as an experimental probe.

Reliance on enzyme activity as the monitoring device also limits study to enzyme-type proteins; however, many interesting proteins do not have catalytically assayable functions. Some proteins are regulatory elements, which modulate synthesis or activity of enzymatically active proteins; others are structural elements or proteins, performing their function only when integrated into a membrane, that is, permeases (transporters). The functions and modes of operation for many of these proteins are being elucidated with the help of immunochemical assay procedures.

These procedures may be divided into two classes, direct and indirect. *Direct methods* involve synthesis or modification of the molecule to be measured in the presence of a radioactively labeled precursor followed by its isolation using immunochemical precipitation. *Indirect methods* involve competition of a nonradioactive protein, obtained from an experimental sample, with a radioactively labeled standard protein, for a limited amount of antibody. The quantity of desired protein in the experimental sample is determined by the degree to which it successfully competes with radioactive standard for binding to antibody. The latter method is known as radioimmunoassay (RIA).

Table 3 - Composition of Immunoprecipitates Obtained with Various Sets of Antigens and Corresponding Antibodies

Antigen	Molecular Weight	Mole Ratio of Antibody to Antigen[a]
Bovine ribonuclease	13,400	3
Egg albumin	42,000	5
Horse serum albumin	67,000	6
Human γ-globulin	160,000	7
Horse apoferritin	465,000	26
Thyroglobulin	700,000	40
Tomato bushy stunt virus	8,000,000	90
Tobacco mosaic virus	40,700,000	650

[a]Precipitates were prepared under conditions of extreme antibody excess. These values represent a minimal estimate of the antigen valence.

From E. A. Kabat and M. M. Mayer, *Experimental Immunochemistry,* 2nd Ed., C. C. Thomas, Springfield, IL, 1961.

7. Radioimmunoassay

Radioimmunoassay procedures are capable of measuring trace amounts of any substance that can serve as an antigen or hapten. Resolution is limited principally by the specific activity of the radioactively labeled standard and by the specificity of the antibody preparation. At present the practical limit is on the order of 10^{-11} to 10^{-9} g antigen per sample. In the past, many biologically important molecules such as drugs and hormones could be assayed only at low resolution using predominantly bioassay procedures. Today an ever-increasing number of these enzymatically inactive molecules are being reliably quantitated in samples of biological fluids such as blood, urine, and spinal fluid.

In RIA, an experimental sample of unlabeled antigen is mixed with a constant, known amount of the same antigen radioactively labeled. To this mixture is added a limited amount of rabbit antibody prepared against the antigen being measured. *It is critical to add sufficiently small amounts of antibody* so that the system is in a condition of *antigen excess*, even when unlabeled antigen is absent. Since the system is antibody limited, the amount of labeled antigen bound to the rabbit antibodies is inversely proportional to the quantity of nonradioactive antigen present. This is basically an *isotope dilution* type of *experiment*. The final step of the procedure is separation of antigen-antibody complexes from free antigen, a prerequisite for measuring the amount of antigen bound. Although many methods are available the one most widely used experimentally is immunoprecipitation. This method of separation can be used successfully because (*1*) the antigenic determinants of an antibody molecule are not located very close to its antigen binding sites and (*2*) immunoprecipitation of an antibody molecule does not adversely affect its ability to bind antigen. In practice, a second antibody preparation is added to precipitate the antigen-rabbit antibody complexes. A good candidate in this example would be goat antibody obtained following immunization of the animal with the γ-globulin fraction of rabbit serum. Some caution is needed in selecting the pair of antibody preparations to be used because some do not work well together, as is the case for donkey antibodies prepared against guinea pig γ-globulins.

A number of investigators have reported that the F_c portion of IgG binds tightly to the A protein of *Staphylococcus aureus* cell walls (*J. Immunol.* l04:140 and 105:1116, l970). Therefore, these organisms may be heat-inactivated, fixed with formaldehyde, and used to precipitate antigen-antibody complexes (*Eur. J. Immunol.* 4:29, 1974; *J. Immunol.* 115: 1617, 1975). Heat-treated organisms may be substituted for the second antibody preparation normally used in radioimmunoassays. They are also gaining popularity for direct immunoprecipitation, because with these methods the antigen-antibody complex is bound to a bacterium and, as such, can be more easily and thoroughly washed. This results in a significant lowering of nonspecific precipitation of antigens unrelated to those being studied.

References:

1. B. D. Davis, R. Dulbecco, H. N. Eisen, H. S. Ginsberg, and W. B. Wood, Jr., *Microbiology*, 2nd Ed., Harper and Row, New York, 1973.
2. J. F. Miller, A. P. Mitchell, G. F. Davies, and R. B. Taylor, *Transplant. Rev.*, 1:3 (1969). Antigen-Sensitive Cells, Their Source and Differentiation.
3. J. Clausen, Immunochemical Techniques for the Identification and Estimation of Macro-molecules, in *Laboratory Techniques in Biochemistry and Molecular Biology* (T. S. Work and E. Work, Eds.), American Elsevier, New York, 1969.
4. D. H. Campbell, J. S. Garvey, N. E. Gremer, and D. H. Sussdorf, *Methods in Immunology*, 2nd Ed., Benjamin, New York, 1970.
5. D. M. Weir, Ed., *Handbook of Experimental Immunology*, F. A. Davis, Philadelphia, 1967.
6. D. J. Shapiro, J. M. Taylor, G. S. McKnight, R. Palacios, C. Gonzalez, M. L. Kiely and R. T. Schimke, *J. Biol. Chem.*, 249:3665 (1974). Isolation of Hen Oviduct Ovalbumin and Rat Liver Albumin Polysomes by Indirect Immunoprecipitation.
7. N. H. Axelsen, J. Kroll, and B. Weeke, Eds., *A Manual of Quantitative Immunoelectrophoresis*, BioRad Laboratories, Richmond, CA.
8. H. G. Minchin and T. Freeman, *Clin. Sci.*, 35:403 (1968). Quantitative Immunoelectro-phoresis of Human Serum Proteins.
9. J. G. Feinberg, *Int. Arch. Allergy,* 11:129 (1957). Identification, Discrimination and Quantification in Ouchterlony Gel Plates.
10. E. A. Kabat and M. M. Mayer, *Experimental Immunochemistry,* 2nd Ed., C. C. Thomas, Springfield, IL, 1967.
11. R. D. Palmitter, *J. Biol. Chem.*, 248:2095 (1973). Ovalbumin Messenger Ribonucleic Acid Translation.
12. J. I. Thorell and B. G. Johansson, *Biochim. Biophys. Acta*, 251:363 (1971). Enzymatic Iodination of Polypeptides with ^{125}I to High Specific Activity.
13. J. Roth, Methods for Assessing Immunologic and Biologic Properties of Iodinated Peptide Hormones, in *Methods in Enzymology*, Vol. 37 (B. W. O"Malley and J. G. Hardman, Eds.), Academic Press, New York, 1975, p. 223.
14. W. M. Hunter, The Preparation of Radio-iodinated Proteins of High Activity, Their Reaction With Antibody *In Vitro*: the Radioimmunoassay, in *Handbook of Experimental Immunology* (D. M. Weir, Ed.), F. A. Davis, Philadelphia, 1967.
15. R. Graham and W. M. Stanley, *Anal. Biochem.*, 47:505 (1972). An Economical Procedure for the

Preparation of L-(^{35}S)Methionine of High Specific Activity.
16. F. T. Wood, M. M. Wu and J. C. Gerhart, *Anal. Biochem.*, 69:339 (1975). The Radioactive Labeling of Proteins with an Iodinated Amidination Reagent.
17. A. E. Bolton and W. M. Hunter, *Biochem. J.*, 133:529 (1973). The Labeling of Proteins to High Specific Radioactivities by Conjugation to a ^{125}I-Containing Acylating Agent. Application to the Radioimmunoassay.

INNOVATION/INSIGHT

THE ENZYME-LINKED IMMUNOSORBENT ASSAY (ELISA)

Voller, A., Bidwell, D. E. and Bartlett, A. (1979) Dynatech Laboratories Inc., Alexandria, VA, ph. (800) 336-4543. (Adapted)

1. Introduction

Immunological reactions are used for assays or detection procedures because they can give high levels of specificity and sensitivity. Antibodies or antigens labeled with various markers have been found to be particularly useful for assays of biological substances. Fluorescent dyes have been used for this purpose for many years and immunofluorescence is now one of the best established diagnostic methods in microbiological and clinical immunology laboratories. However, immunofluorescence is time consuming, is not easy to perform in an automated manner, and reading of results is usually subjective. The other labels used on antigens or antibodies have been isotopes and it is with these that the science and industry of radioimmunoassay has been developed. These radioimmunoassay techniques have a very high level of sensitivity and reproducibility and are amenable to automation for large scale processing. However, the use of isotopes poses some problems: the costly reagents have a short shelf life, complex equipment is required to read the results, and special safety measures must be observed in the handling and disposal of reagents. These factors have meant that isotopic assays have been restricted to more sophisticated centers of the affluent world. Increasing public awareness and concern about the dangers of isotopes may well lead to more stringent regulations, which could further limit their use.

In the present article, attention will be focused on heterogeneous enzyme-immunoassays. These all involve a separation step where the reacted enzyme-labeled component is separated by washing from the unreacted enzyme-labeled material. These assays are particularly suitable for the detection and measurement of large molecular weight substances (MW over 10,000). The homogeneous enzyme-immunoassays, which are at present restricted to the assay of small molecular weight substances and which are exemplified by the EMIT system, will not be considered in detail here, however, a brief explanation of their basis is necessary.

In a homogeneous enzyme-immunoassay, a hapten is linked to an enzyme in such a manner that the enzyme activity is altered when the hapten combines with antibody. The assay system consists of enzyme-labeled hapten + antibody + enzyme substrate + test sample (Figure 1). If the test sample contains any of the hapten, it will combine with the antibody, so leaving the enzyme-labeled hapten free to degrade the substrate, which is read in a spectrophotometer. These assays, which do not require separation steps, can be performed in a minute or two. The assays have been used widely for assaying drugs of abuse (e.g., opiate, barbiturates, amphetamines, benzodiapine, etc.) and for measuring therapeutic drug levels (e.g., digoxin, carbamazepine, lidocaine, phenytoin, phenobarbital, primidone, and theophylline) and more recently assays of this type have been established for T4.

Heterogeneous *enzyme-immunoassays* combine the advantages of *immunofluorescence* and *radioimmunoassay* and overcome many of the disadvantages of the other two methods. Enzyme-labeled reagents are cheap to prepare, and

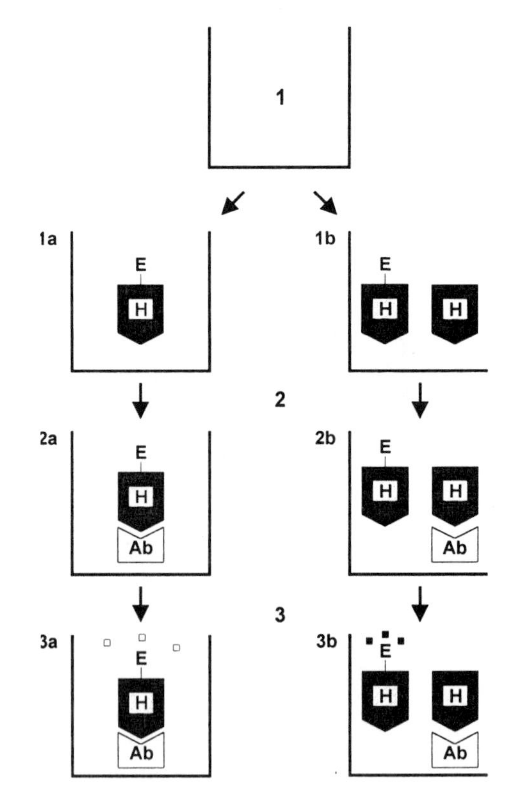

Figure 1 - The homogeneous enzyme-immunoassay method. 1. Unknown sample. 1a. Add sample without hapten and enzyme hapten complex (unknown sample – negative). 1b. Add sample containing hapten and enzyme hapten complex (unknown sample – positive). 2. Add antibody. 2a. When enzyme and antibody bind, enzyme activity is altered. 2b. Unknown sample hapten binds antibody preferentially, leaving the enzyme hapten complex free and the enzyme activity remains unaltered. 3. Add enzyme substrate. 3a. Substrate not degraded by enzyme, indicating a negative test sample. 3b. Substrate degraded, indicating a positive test sample.

are highly stable (giving a long shelf life), yet yield assays which approach the sensitivity of radioimmunoassay and give objective results that can be determined, either visually, or with rather simple equipment.

Heterogeneous enzyme-immunoassays depend on the assumption that either an antigen or antibody can be linked to an enzyme while retaining both immunological and enzymatic activity in the resultant conjugate. Chemical linkage was shown to be possible a decade ago and satisfactory immunological linkage of enzyme and antibody followed later.

The first workers in this field were primarily interested in the visual detection and localization of antigens, but even in 1967 Nakane and Pierce[5] indicated that an indirect immunoenzyme method, analogous to indirect immunofluorescence, was feasible. In the first quantitative assays of antibody by indirect enzyme-immunoassays, the results were assessed visually, in a manner analogous to that of immunofluorescence. The only advantage was that the immunoenzyme staining technique results could be read with an ordinary microscope.

The next stage in the development of enzyme-immunoassays was the linkage of soluble antigens or antibody to an insoluble solid phase in a way in which the reactivity of the immunological component was retained. This was the basis for the techniques known as ELISA (*Enzyme-Linked Immunosorbent Assay*) pioneered by Engvall and colleagues and by Van Weemen and Schuurs.

2. Competitive ELISA

This system (Figure 2) for detection and measurement of antigen is carried out as follows:

1. Specific antibody (or immunoglobulin containing specific antibody) is attached to the solid phase, which is then washed.

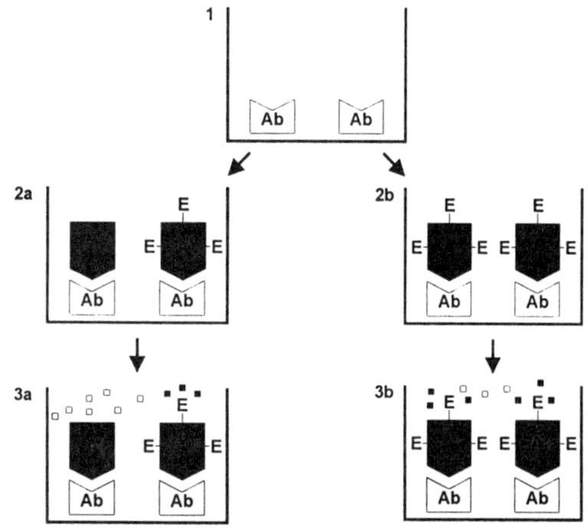

Figure 2 - Competitive ELISA for Assaying Antigen. 1. Absorb antibody to surface. 2a. Add enzyme labeled antigen and unknown antigen sample. 2b. Add enzyme labeled antigen. 3a. Enzyme substrate on right altered. 3b. Enzyme substrate on both sides altered.

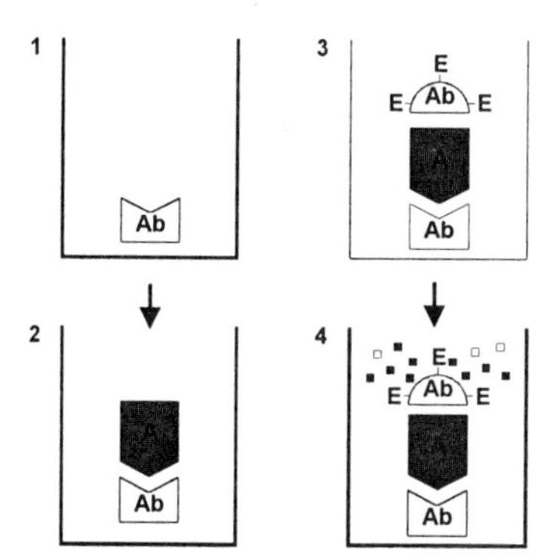

Figure 3 - The double antibody sandwich ELISA for measuring antigen. 1. Antibody adsorbed to plate. Wash. 2. Test solution containing antigen added. Wash. 3. Add enzyme-labeled specific antibody. Wash. 4. Add enzyme substrate. Amount of hydrolysis ≡ amount of antigen present.

2. Test solution, thought to contain antigen, is then added. It is either mixed with enzyme-labeled antigen or enzyme-labeled antigen is added after a short time. Plates are then incubated and washed.

Enzyme substrate is added. Reference wells containing only enzyme-labeled antigen show coloration. The inhibition of that color change in the wells with test samples is proportional to the amount of antigen in the test samples.

3. Double Antibody Sandwich ELISA

This system (Figure 3), for the detection and measurement of antigen, is carried out as follows:

1. Immunoglobulin containing specific antibody or purified specific antibody (for antigen to be measured) is attached to the solid and washed.
2. The test solution thought to contain antigen is then incubated with the sensitized solid phase, which is then washed.
3. Enzyme-labeled specific antibody to the antigen is then incubated with the solid phase followed by washing.
4. Enzyme substrate is added. The color change is proportional to the amount of antigen in the test solution in step 2.

4. Modified Double Antibody Sandwich ELISA

This system can be modified (Figure 4) for the detection and measurement of antigen and is carried out as follows:

1. Immunoglobulin containing specific antibody (e.g., from antiserum) is attached to the solid phase, which is then washed.
2. The test sample containing antigen is then added and incubated with the sensitized solid phase, which is then washed.

3. Immunoglobulin solution containing specific antibody is then added, and then incubated and washed. This immunoglobulin is from antisera produced in a species different from that in step 1 (e.g., it could be from goat or sheep antisera).
4. Enzyme-labeled antiimmunoglobulin is then added and incubated, followed by washing. This antiimmunoglobulin must be reactive with the immunoglobulin used in step 3 (e.g., goat or sheep) but must not react with the immunoglobulin used to coat the plate (e.g., rabbit in this instance).
5. The enzyme substrate is added. The color change is proportional to the amount of antigen in the test solution in step 2.

5. Modification of Indirect ELISA for Assay of Antigen

This is another system (Figure 5) that can be used for small molecular weight substances.

1. The relevant antigen is attached to the solid phase, which is then washed. For small antigens the attachment may have to be made via a larger carrier (e.g., bovine serum albumin).
2. The test solution, which is thought to contain antigen, is then mixed with a reference antiserum containing specific antibody and this mixture is incubated with the sensitized solid phase, which is then washed. This antibody can be enzyme-labeled, so step 3 is not required.

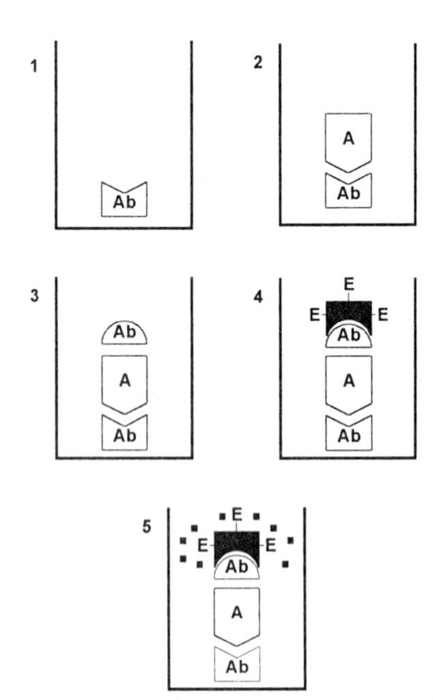

Figure 4 - Modified double antibody sandwich ELISA for measuring antigen. 1. Adsorb antibody A to surface (e.g., rabbit). Wash. 2. Add sample containing antigen. Wash. 3. Add antibody B (of a different species, e.g., goat). Wash. 4. Add enzyme-labeled anti-β-globulin (e.g., anti-goat IgG).

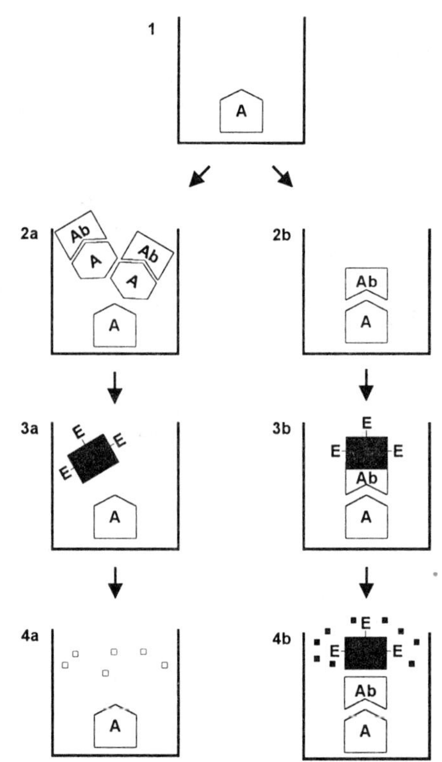

Figure 5 - Modification of the indirect ELISA for assay of antigen of antigen. 1. Adsorb antigen to surface. 2a. Add test sample thought to thought to contain antigen mixed with reference antibody.. 2b. Add control sample containing no antigen. 3a. Add enzyme antiglobulin conjugate and wash. Conjugate does not bind to an immobilized antibody. 3b. Add enzyme antiglobulin conjugate and wash. Conjugate does not bind to an immobilized antibody. 4a. Add enzyme substrate. No substrate degradation indicates test sample contains antigen. 4b. Add enzyme substrate. Substrate degradation indicates test sample contains no antigen.

3. Enzyme-labeled antiglobulin (reactive with the antibody used in step 2) is then added, incubated and washed.
4. Enzyme substrate is added. The difference in color change between a reference sample containing no antigen and the test solution in step 2 indicates the amount of antigen in the test samples. This is a competitive assay; high antigen concentrations result in less color at the end of the test.

6. Solid Phase Anti-IgM ELISA

1. The solid phase is coated with a solution of immunoglobulin from a specific antiserum produced against IgM. The solid phase is then washed.
2. The test sample (usually serum) being assayed for antibody is then added and incubated and the plate is then washed.
3. A solution of the viral antigen or of whole viruses is then incubated and the plate washed.

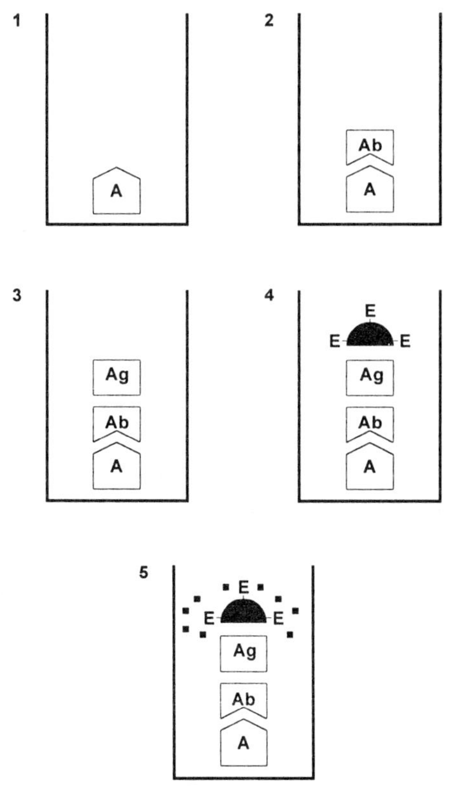

Figure 6 – Indirect Method for Assay of Antibody. 1. Absorb antigen to plate. 2. Add serum: any specific antibody attaches to antigen. 3. Add antigen that attaches to the antibody. 4. Add enzyme labeled antiglobulin which attaches to antigen. 5. Add substrate. Amount of hydrolysis ≡ amount of antibody present.

4. Enzyme-labeled specific antibody (to the virus) is then incubated in the plate, followed by washing.
5. The enzyme substrate is then added. The color change is proportional to the amount of virus-specific IgM in the serum tested in step 2.

7. The Solid Phase

Either antigen or antibody is attached to the solid phase. This permits the separation of immunologically reacted from unreacted material during the test. It is imperative that the solid phase should take up an adequate amount of the reactant in a reproducible manner. Variability at this stage is probably still the major factor in determining the precision of all solid phase immunoassays.

Various carriers have been used such as particles of cellulose, polyacrylamide, cross-linked dextrans, silicone rubber, microcrystalline glass, and plastic. All of these are satisfactory but they must be centrifuged in the washing steps, which is inconvenient and time consuming. The use of preformed materials such as tubes, beads, discs, and microplates overcomes the need for centrifugation as they can be washed easily. Various workers have used all of these formats, molded from polystyrene, polypropylene, polyvinyl, and other plastics. In our experience the specially designated microplates made of polystyrene are easy to use and give adequate, reproducible uptake of most antigens merely by passive adsorption in alkaline solution. Such plates must be checked by the manufacturer or user to ensure that the antigen uptake on the surface is uniform across the plate. The individual polystyrene wells, which can be fit into plate holders, are convenient if only a few tests are to be done at a time. For some purposes, where a high uptake of the immunological reagent is needed, polyvinyl plates may be preferred. These have been found to be especially suitable for coating with immunoglobulin for use in assays for antigens and they can, of course, be easily cut up and counted individually in isotopic assays. In general the polyvinyl plates yield higher backgrounds, so are less suitable for visually read ELISA tests.

7.1 Materials for Coating Solid Phase. The optimal conditions for coating the solid phase (e.g., concentration of reactant, time of coating, temperature, and pH) are best determined by checkerboard titrations using reference reagents. These conditions should be strictly adhered to in subsequent tests if reproducible results are to be obtained. For most proteins and lipoproteins satisfactory sensitization is achieved with solutions at 1-10 μg/ml in carbonate/ bicarbonate buffer (pH 9.6). The adsorption occurs rapidly, most being completed within one to two hours at 20-25°C. However, for convenience, overnight coating at 4°C is often used and is satisfactory.

There is no satisfactory way to predict whether a particular antigen will be suitable for use in the indirect ELISA method. It is best to test the material at various dilutions using reference sera and reference conjugates. Because antigens are suitable for other serological tests does not mean that they will give acceptable results in ELISA. Many microbiological antigen preparations are crude extracts and it is especially important to determine whether these are reacting nonspecifically with test sera. For this purpose, it may be necessary to prepare and use control antigen (e.g., from uninfected tissue cultures or uninfected animals) extracts. Values given by these "control" materials should be subtracted from those given by the antigen preparations (Table 1).

In the double antibody sandwich method, specific antibody can be used for coating the solid phase but this is an expensive procedure and for most purposes it has been found that the immunoglobulin fraction containing the specific antibody can be used. Hermann and Collins showed that the immunoglobulin adsorption is dependent on the initial immunoglobulin concentration and on the time

Table 1 - ELISA Test for Antibody to *Varicella* Illustrating the Importance of Using Control Antigens in Viral ELISA Tests

	+Ve			-Ve		
	1/ 100	1/ 200	1/ 400	1/ 100	1/ 200	1/ 400
Varicella antigen	1.14	0.79	0.50	0.36	0.31	0.26
Control antigen	0.37	0.35	0.25	0.32	0.31	0.27
Correct ELISA	0.77	0.44	0.25	0.04	0.00	0.00

allowed for adsorption.

Once the solid phase is coated adequately, excess coating reagent can be removed by washing in PBS Tween, then the sensitized materials are ready for immediate use or they can be stored. If the sensitized microplates are to be stored they are finally given an additional wash in distilled water then they are thoroughly dried (over silica gel or $CaCl_2$). These plates must then be sealed into air-tight, waterproof packs (e.g., foil/polyester/paper containing desiccants). When prepared in this manner the sensitized plates can be shipped at ambient temperature and, until opened, show virtually no loss of activity even after storage periods of one year or more at 4°C. This long shelf life is a particularly valuable asset for assays to be used in small laboratories doing few tests at unpredictable times and for the developing countries of the world. The possibility of preparing large batches of sensitized plates also reduces the manufacturer's quality assurance costs and should eventually result in lower prices for the user.

8. Washing Steps

The heterogeneous enzyme-immunoassays of the ELISA type being considered here all consist of a series of incubations of different reagents separated by washing steps. These washing steps must be carried out in such a manner that there is no carryover from one step to the next. For many purposes, washing can be carried out manually merely by emptying the tube or plate, refilling with phosphate-buffered saline containing Tween 20, and leaving for a few moments. Three washes of this type are usual, after which the plate is shaken dry and the next reagent is added.

The microplate format is very convenient for the washing steps since 96 wells can be washed simultaneously while the sequential washing of individual tubes, beads, or discs is very cumbersome by comparison.

Fully automatic ELISA processing apparatus including washing arrangements have been described but these are only suitable for very large scale operations. However, if rigid control of the washing process is required or if highly infectious agents are involved (e.g., HB Ag) it is advisable to use a washing/aspiration device, which collects the aspirated material in a closed container for subsequent disposal.

9. Test Sample

In the ELISA tests for large molecular weight substances there is often a tendency for materials in the samples being assayed to attach in a nonspecific manner to the solid phase. This is especially so in the case of indirect ELISA tests where the sample is often serum, plasma, CSF, saliva, etc. This problem can be largely avoided by making a solution of the test sample in a buffer (e.g., PBS) containing a detergent (e.g., Tween 20), which prevents nonspecific attachment to the solid phase. It may be necessary to further dilute the test sample and/or include additional protein (e.g., albumin) to reduce the background nonspecific uptake to an acceptable level.

The ratio of the absorbance reading of a highly positive sample to that of a negative sample in indirect ELISA should be at least 5:1 and preferably 10:1, so it is important to keep the "nonspecific" background as low as possible.

In indirect tests, the serum or plasma sample may contain substances (e.g., rheumatoid factor) that may react with other serum components (e.g., specific IgG antibody) which then become fixed to the sensitized solid phase. This poses a special problem for assays of IgM antibody. It can be avoided by chromatographic or centrifugal separation of the IgM and IgG or by use of absorbants specific for one of the immunoglobulins (e.g., Rivanol, DEAE, or Protein A).

Quantitation of the material in the test sample being assayed can usually be made on a single dilution of that test sample. However, serial dilutions of the test sample can be used, especially where a visually read end-point is required. For antigen assays a standard curve is normally set up and the values of "unknowns" read from it. In some instances high antigen levels might be missed due to the "high dose hook effect" (that is the situation that exists when antigen levels are extremely high and give falsely low absorbance values). In these situations, several sample dilutions must be employed.

10. Conjugates

A wide variety of enzymes have been used at one time or another as labels for antigens and antibodies. The enzymes must be stable, highly reactive, available in a highly purified form, yield stable conjugates, and be cheap and safe to use. In addition, a convenient substrate detector system must be available. Enzymes that have been tried for this purpose include acetylcholinesterase, cytochrome *c*, β-D-galactosidase, glucoamylase, glucose oxidase, β-D-glucuronidase, lactate dehydrogenase, lacto-peroxidase, ribonuclease and tyrosinase. However ,alkaline phosphatase and horseradish peroxidase have been most favored and the latter is now a clear favorite for most workers because of its low cost and easy conjugation and because it has such a wide variety of substrates.

The objective is to link the enzyme with the antibody or antigen in such a way that each retains the maximum amount of their reactivity. This is achieved by use of cross-linking agents with at least two active groups. The cross-linking can be accomplished in a one-step procedure where enzyme antibody and cross-linking reagent are mixed together.

Alternatively a two-step method can be used. The enzyme is first reacted with the cross-linking agent then the activated enzyme is reacted with the antibody or antigen.

Glutaraldehyde is one of the most popular cross-linking agents and has been used in both one- and two-step procedures. Conjugates made by the one-step reaction are usually heterogeneous (i.e., they contain high and low molecular weight polymers of enzyme and antibody). However, these conjugates contain little or no free antibody or enzyme, and most of the enzyme and immunological activity is retained. Unfortunately peroxidase does not couple well by the one-step glutaraldehyde method. The two-step glutaraldehyde method is suitable for peroxidase and it yields homogeneous conjugates (i.e., one molecule peroxidase is coupled to one molecule of antibody).

The other popular coupling reagent is sodium periodate. This can be used in a two-step method to activate peroxidase, forming aldehyde groups which react in the second step with the amino groups of the antibody being labeled. These conjugates are heterogeneous. A later modification of the periodate method is easier to perform and also gives good results.

In our experience, the periodate method yields about 4 times as much conjugate (per mg antibody) as the one-step glutaraldehyde and 3 times as much as the two-step glutaraldehyde method when peroxidase is used as the label.

Other coupling agents, which have been used with success, include N^1O^1-phenylenedimaleimide and *p*-benzoquinone. A comprehensive review of all these conjugation methods has recently been published by Avrameas and colleagues.

The level of purity required of the material to be labeled depends on the particular type of assay to be performed. In the case of antibodies, it is not always necessary to use antibody purified by immunosorbent methods; for many purposes the total immuno-globulin fraction of the sera can be labeled. It is not, however, advisable to label whole serum as this results in unacceptably high background levels in ELISA tests, and is costly in terms of enzyme. Much of our experience has been gained with sheep or rabbit antihuman immunoglobulin conjugates. For these, we have found that ammonium or sodium sulfate precipitation of the serum provides adequate material for the enzyme labeling. However, the use of affinity chromatography purified antibody gave slightly lower background levels in indirect ELISA tests and the conjugates could be used at a higher dilution. F_{ab} fragments derived from the immuno-globulin part of the antiserum and labeled with the enzyme gave high activity with a marked reduction in background staining levels when they were used in indirect ELISA systems.

The conjugates, when prepared as described above, may contain a high level of antibody-enzyme complexes but there may be residual contaminating unlabeled antibody and free enzyme. Purification by means of salt precipitation $(NH_4)_2SO_2$, Na_2SO_4, or gel filtration (Sephacryl 200) is to be recommended.

Usually, in indirect tests, an enzyme-labeled antiglobulin is used, but there may be some instances where indirect ELISA tests need to be carried out on sera from several species and then a universal conjugate is more convenient. To date the best candidate is enzyme-labeled protein A. This can be used instead of the antispecies conjugates as it will react with the IgG of most species. Atanasiu found these protein A conjugates were very satisfactory for detecting rabies antibody in many different animals. However, in other viral systems, using crude antigens, our experience has been that the protein A conjugates can give high background values. A second labeled antibody can also be used.

For most diagnostic purposes in infectious diseases, enzyme conjugated to the IgG fraction of an antiserum will be used, and we have found that these are remarkably stable. Alkaline phosphatase-labeled antihuman IgG preparations, made by the glutar-aldehyde methods and stored in the concentrated form with a preservative, have retained full activity for at least two years. Peroxidase conjugates stored in the lyophilized state have kept their activity for a similar time span. Storage in the liquid form reduces wastage of the conjugates when only small amounts are used daily, however, lyophilized stock material might be better for very long-term storage.

The conjugates are best stored in the concentrated form, and diluted immediately before use, in PBS Tween. The working dilution is determined according to the conjugate incubation time and substrate reaction time. Highly diluted conjugate will need a long incubation time, whereas if rapid results are needed, concentrated conjugate is used with a short incubation time.

The use of high temperature incubations for

sample, conjugate, and substrate will reduce the incubation times needed. With the small volumes used in the microplate system, sample incubation times of 2-3 hours and conjugate incubations of 2-3 hours at ambient temperature or overnight at 4°C are usually employed and the concentration of reactants is arranged to give substrate reaction times of 15-60 minutes. If more dilute conjugates are used the substrate reaction times can be lengthened to compensate.

11. Substrates

The chief requirement of a substrate is to provide a sensitive detection method for the enzyme in the conjugate. Most authors have used chromogenic substrates, which are initially colorless and give a strong color following degradation. Ideally the substrate should give completely soluble products with a high extinction coefficient (i.e., dense color per unit degraded). The substrate should also be cheap, safe, and easy to use. For some purposes a fluorogenic substrate may be preferred (e.g., methylumbelliferin).

In the case of alkaline phosphatase, a very convenient substrate is *p*-nitrophenyl phosphate. This is available commercially as tablets from which a working solution (which is not photosensitive) can be quickly and easily prepared. The enzyme reaction of alkaline phosphatase can be stopped with concentrated NaOH and the yellow product is quite stable. It can be assessed visually or photometrically at 405 nm.

A wide variety of peroxidase substrates oxidized by H_2O_2 are available but care must be taken to choose one with adequate solubility. Diaminobenzidine is insoluble and 5-aminosalicylic acid (5AS) and *o*-dianisidine are partially insoluble, although some workers have used them in ELISA. ABTS (diammonium salt) [2-2'-azino-di-(3-ethylbenzothiazolin sulfone-6] also gives a poor dose-esponse curve; *o*-tolidine is better but there is little doubt that *o*-phenylenediamine (OPD) is by far the best peroxidase substrate.

The product is completely soluble and it has a high extinction coefficient at 492 nm. This substrate is excellent for visual reading, yielding an orange color after stopping with H_2SO_4. OPD is photosensitive so that reasonable care is required when it is used. OPD, along with some other peroxidase substrates, has been reported as being mutagenic. However, there are no indications that it is carcinogenic and it should not present problems when used under normal laboratory conditions. It is not subject to restrictive legislation.

All substrates should be used with care and direct contact with skin or eyes should be avoided.

12. Visual Readings

These can be very consistent and accurate, especially where a "yes" or "no" answer is required. Positive results are indicated by a colored reaction product in the microplate wells. Solutions with absorbance of 0.1-0.2 in the yellow-brown range (400-500 nm) can be distinguished from the colorless negative wells. If a titer is required, serial dilutions of the test sample are made; the last dilution giving a visible colored reaction product is the titer.

13. Photometric Readings

For assays of high precision, the results must be read photometrically. This can be done by removing the substrate reaction solution from the microplates or tubes and transferring to the cuvette of a spectrophotometer and reading its absorbance at the maximal absorption wavelength (405 nm for *p*-nitrophenyl phosphate, 492 for OPD). It should be possible to retain the sample, thus enabling those reading "off the scale" to be diluted for rereading. Suitable spectrophotometers must have microcuvettes to enable readings to be made on sample volumes of 200-300 μl. At present, various photometers with manual and automatic sampling devices are being introduced for ELISA.

The alternative is photometric measurement of the absorbance of the substrate solution *in situ* in the reaction tube or microplate. Rapid microplate readers, capable of reading a whole plate in less than one minute, are just becoming available. Automatic reading devices have also been incorporated into the complete ELISA processing machinery of Ruitenberg and Brosi. Kinetic measurements could be made but this is not necessary and should be considered only if the equipment is already available.

1. As "positive" or "negative." All test samples giving an absorbance value above a threshold level (e.g., 0.2 or 0.4) may be considered as positive. This threshold level is predetermined by testing a large number of "negative" samples and the threshold value is at the upper limit of these "negatives."
2. As the "absorbance value," e.g., 0.2, 0.7, 1.2. These are values obtained at the end of ELISA tests, which have been carried out under defined conditions or which have been controlled by inclusion of reference samples. The absorbance values can be used as a direct measure of the samples" reactivity.
3. As a "ratio." The ratio of the absorbance value of the sample to the mean of a group of known negatives. A ratio value above a certain figure (e.g., 2 x negative, 3 x negative) is considered positive.
4. As an end-point "titer." The samples are serially diluted and all dilutions are tested by ELISA.

The titer is the last dilution that yields a value above that of a group of known negative samples.

5. As a "unit." At the same time as the samples are tested, a series of known content samples are also included. A standard curve is constructed from the absorbance of these and the unknown samples" content is calculated from that curve. This method is used for well-defined antigens or haptens, but is less suitable for antibody assay due to variation of antibody affinity in test sera.

In practice, we find that for indirect ELISA tests for antibody it is essential to include a reference positive and negative sample on each plate. The substrate reaction is stopped when the reference positive absorbance reaches a predetermined value. This enables valid comparisons to be made between tests on different occasions and in different laboratories.

In general, the results of indirect ELISA tests for antibody, and of double antibody sandwich assays for antigens which are now well defined, are expressed according to methods 1-4 above. For antibody assays methods 1-3 above tend to minimize the real differences in concentration of antibody between samples, and method 4 involves the tiresome procedure of serial dilution and increased cost. In addition the small absorbance differences towards the end point leads to inaccuracies. In an attempt to solve this problem, Felgner described a method of expressing results as a *m*ultiple *of* *n*ormal *a*ctivity (MONA) of negative sera. This method, based on the *dose-response curve* of each system, enables values more representative of the serum titer to be obtained from tests on a single serum dilution. It should be borne in mind that the end result in antibody assays reflects the combined effects of antibody concentration and its affinity.

14. Setting Up

14.1 Testing of Plates for Acceptability. Polystyrene microplates are checked by adding 200 µl of a solution of human IgG (100 ng/ml) made up in coating buffer (see below) to each well and incubating overnight at 4°C. Plates are washed in PBS-Tween then enzyme-labeled anti-human globulin solution is added and incubated for 18 hours at 4°C. After washing, the substrate is added and the reaction is allowed to proceed until the absorbance of the contents of the wells is about 1. The reaction is then stopped. The absorbance of contents of each well is then read individually. New plates should give results similar to those of previous satisfactory batches and the total variation of values in duplicate pairs of wells should be within ±10% of the mean (set at about 1) of the whole plate. Plates that are acceptable will need to be tested later in a similar manner by indirect ELISA using the actual test system intended for use. This will confirm that the respective antigens adsorb uniformly at their working dilution.

14.2 Determination of Optimal Test Conditions for ELISA for Antibody to Measles.

1. Serial dilutions of the antigen (e.g., measles antigen in this case) are made up in coating buffer. Of each dilution 200 µl is added to a horizontal row of the wells in a microplate and are incubated, with lids, overnight in a humid box at 4°C. This allows the antigen to absorb to the plastic surface. The plates are washed by emptying, filling with PBS Tween, then leaving for a few moments. This process is repeated 3 times and plates are then shaken dry.
2. Of three reference sera (positive, weak positive, and negative for antibodies to measles antigen) serially diluted in PBS Tween 200 µl is added to vertical rows of wells so that each serum dilution is reacted with each antigen dilution. A row containing only PBS Tween is included. Plates are then incubated 2 hours at room temperature in a humid box and washed as before.
3. The stock conjugate (e.g., sheep anti-human immunoglobulin labeled with an enzyme) is diluted in PBS Tween and 200 µl amounts are added to each well. The plate is then incubated for 3 hours at room temperature. Plates are washed as before.
4. Of substrate solution (see below) 200 µl amounts is added to each well and incubation is allowed to proceed at room temperature. The reaction is stopped with strong acid or alkali after a given time (about 30 minutes). The samples are removed from each well and are read in a spectrophotometer. That combination of the antigen/serum dilutions, which gives maximal separation between positive and negative test samples but still gives low negative values (under 0.2), is chosen for the subsequent tests.

From this checkerboard titration it was decided to use the antigen diluted 1/200 and the serum 1/200. This not only gave a good distinction between the positive and negative reference sera but would also be economical with antigen and produce a reasonable substrate reaction time of about 30 minutes for the indirect ELISA.

14.3 The Indirect ELISA for Antibody to Measles Antigen.

1. The above measles antigen and its control antigen (prepared from uninfected cells) are diluted 1/200 in coating buffer. The measles antigen is added in 200 µl amounts to duplicate rows of wells on a polystyrene plate and the control antigen to single rows of wells. The plate is incubated overnight at 4°C in a humid box.

The plate is washed by emptying, filling with PBS Tween from a wash bottle and leaving for a few moments. This process is repeated three times.

2. Test sera, reference positive and negative sera are diluted l/200 in PBS Tween and 200 μl amounts added to duplicate wells of antigen and to a single well of control antigen in the plate. The plate is incubated 2 hours at room temperature in a humid box then washed as before.
3. Freshly diluted conjugate (200 μl) is added to each well and incubated 3 hours at room temperature, after which plates are washed.
4. Substrate solution (p-nitrophenyl phosphate for alkaline phosphatase conjugates, OPD for peroxidase conjugates) (200 μl) is added to each well. The time of addition is carefully noted.
5. Samples are taken from the antigen wells with the positive reference sample and the absorbances read. When an absorbance of 1 is reached, the time interval is noted. To all the wells in the plate, 50 μl of the stopping solution (NaOH for *p*-nitrophenyl phosphate or H_2SO_4 for OPD) is added after the same interval. The absorbance of the contents of each well is then read (at 405 nm for *p*-nitrophenyl phosphate, at 492 nm for OPD). The substrate solution containing the appropriate amount of NaOH or H_2SO_4, added after the same time intervals, is used as the spectrophotometer blank. (A defined substrate incubation time, e.g., 30 minutes, can be used.)

Due to difficulties in obtaining precise reproducibility of incubation times and temperatures and of reagents, it is usual to include a reference positive serum on each plate. This reference positive is chosen in advance to give an absorbance reading of 1.0 with the specified conditions of the test. Correction of the sample readings back to the reference positive compensates for slight day-to-day variations in the test conditions.

15. Materials Needed

15.1 Coating Buffer. Carbonate-bicarbonate (pH 9.6): 1.59 g of Na_2CO_3, 2.93 g of $NaHCO_3$, and 0.2 g of NaN_3 made up to 1 liter with distilled water. Store at room temperature for not more than 2 weeks.

15.2 PBS Tween. Phosphate-buffered saline plus 0.05% Tween consists of 8.0 g of NaCl, 0.2 g of KH_2PO_4, 2.9 g of $Na_2HPO_4 \cdot 12\ H_2O$, 0.2 g of KCl, 0.5 ml of Tween 20, and 0.2 g of NaN_3 in 1 liter of distilled water at pH 7.4. Sodium azide should not be used in solutions when peroxidase is used as enzyme.

15.3 Conjugates.

1. Alkaline phosphatase-labeled sheep antihuman immunoglobulin. Store in concentrated form at 4°C with sodium azide as preservative. Dilute stock solution in PBS Tween immediately before use.
2. Horseradish peroxidase (HRP)-labeled rabbit anti-human IgG. Store in lyophilized state or in solution at -20°C or at 4°C with 50% glycerol. Make up amount needed and dilute in PBS Tween immediately before use.

15.4 Substrates.

15.4.1 For Alkaline Phosphatase Conjugates. Diethanolamine buffer (10%): 97 ml of diethanolamine, 800 ml of water, 0.2 g of NaN_3, 100 mg of $MgCl_2 \cdot 6\ H_2O$; 1 M HCl is added until the pH is 9.8. The total volume is made up to 1 liter with water. Store at room temperature in an amber bottle.

Substrate solution is *p*-nitrophenyl phosphate (1 mg/ml). Tablets (5 mg) are stored at -20°C in the dark until used. Immediately before use, one (5 mg) tablet is dissolved in each 5 ml of 10% diethanol-amine buffer at room temperature. It must be used the same day. The reaction stopping solution is 3 M NaOH. Read in a spectrophotometer at 405 nm.

15.4.2 Peroxidase conjugates. Phosphate-citrate buffer pH 5 consists of 24.3 ml of 0.1 M citric acid (19.2 g/1000 ml), 25.7 ml of 0.2 M phosphate (28.4 g Na_2HPO_4), 50 ml of H_2O. Substrate solution to be made up freshly *immediately* before OPD use. Dissolve 40 mg OPD in 100 ml of above buffer and add 40 μl of 30% H_2O_2. This substrate is light sensitive and must be used *at once.* Stopping solution: 2.5 M H_2SO_4. Read in a spectrophotometer at 492 nm.

16. Conjugation Methods

The following descriptions will apply to antibody labeling. Before using any of the procedures, the immunoglobulin fraction of the antisera must first be obtained by salt precipitation or by chromatography. It should then be made up in PBS at pH 7.2, dialyzed extensively against PBS and adjusted to 2 mg protein/ml in PBS.

16.1 One-Step Glutaraldehyde Method.[24] Suitable for alkaline phosphatase, peroxidase, and many other enzymes.

1. Alkaline phosphatase is usually supplied as a precipitate in ammonium sulfate. Measure the appropriate volume to contain 5 mg enzyme (at least 1000 units/mg protein). Centrifuge, discard supernatant.
2. Add 0.85 ml of the immunoglobulin solution (2 mg/ml) to the enzyme pellet. Dialyze against PBS at 4°C overnight to remove ammonium sulfate.
3. Make volume up to 1 ml with PBS. Add 10 μl of 20% glutaraldehyde solution. Mix well and allow to react for 2 hours at room temperature. Dialyze

against PBS at 4°C overnight with 2 changes of 2 liters PBS.

4. Transfer dialysis tube to fresh Tris buffer, pH 8. Dialyze extensively.
5. Dilute conjugate to 4.0 ml with Tris buffer containing 1.0% bovine serum albumin and 0.02% sodium azide. Filter through millipore filter 0.22 µm. Store in the dark at 4°C.

16.2 Two Step Glutaraldehyde Method. Suitable for Peroxidase.

1. HRP (10 mg) is dissolved in 0.2 ml of 0.1 M PBS at pH 6.8 containing 1.25% glutaraldehyde. This mixture is left overnight at room temperature.
2. The mixture is dialyzed against normal saline or passed down a Sephadex G-25 column to remove free glutaraldehyde. The solution is made up to 1 ml in normal saline.
3. The globulin solution of the antiserum is prepared as previously and is adjusted to 5 mg/ml in normal saline.
4. Of the globulin solution 1 ml is mixed with 1 ml of the activated HRP solution and 0.1 ml of 1 M carbonate bicarbonate buffer, pH 9.5 is added and the mixture is allowed to stand at 4°C for 24 hours.
5. Of 0.2 M lysine solution 0.1 ml is added and the mixture is kept at room temperature for 2 hours.
6. The mixture is dialyzed against PBS (pH 7.2) overnight.
7. The enzyme-labeled globulin is precipitated by addition of an equal volume of saturated ammonium sulfate solution. The precipitate is washed twice in half-saturated ammonium sulfate, and is suspended in 1 ml of PBS.
8. The conjugate is dialyzed against PBS extensively (or passed down a Sephadex G-25 column).
9. Conjugate is centrifuged at 10,000 x *g* for 30 minutes and sediment is discarded. Bovine serum albumin is added to 1%.
10. Conjugate is filtered through a Millipore filter (0.22 µm). It can be stored at -20°C or at 4°C if an equal amount of glycerol is added.

INNOVATION/INSIGHT

HOW THE IMMUNE SYSTEM LEARNS ABOUT SELF

von Boehmer, H. and Kisielow, P. (April, 1993) Scientific American, p. 82. (Adapted)

The immune system can be our salvation and our ruin. It protects us from bacteria, viruses, and other harmful microorganisms, but it can also reject lifesaving transplants of kidneys, hearts, and bone marrow. It will accept organ grafts from an identical twin but reject organs from a complete stranger or even another family member. In such autoimmune diseases as multiple sclerosis, it will attack a healthy tissue as though it were a pathogenic invader. From this behavior we know that the immune system can ordinarily distinguish "*self*," tissue that is genetically identical to that of the body, from "*nonself*," genetically foreign material.

Despite 100 years of debate and speculation, the principles and mechanisms of this discrimination process have until recently been obscure. Experimental analyses were hindered by the vast diversity of immunologic cells and receptor molecules, which enable the immune system to recognize the wide variety of self and nonself substances collectively called antigens.

One idea in particular, *clonal deletion*, has been the center of much controversy. According to this hypothesis, the immune system initially contains cells that have the potential to attack the body's tissues, but these cells are somehow eliminated before they can do any harm. Although the concept is easy to outline, determining whether such cellular deletions actually occur and probing the specifics of how they take place were difficult until the advent of modern genetic technologies.

Using those techniques, we created mice whose immune systems produce only one type of antigen receptor instead of 100 million. Then, by following the development of cells bearing this receptor in various animals, we were able to prove the existence of clonal deletion and to describe, in detail previously impossible, the discrimination between self and nonself. Our improved understanding of the cellular and molecular mechanisms of self-nonself discrimination could eventually lead to more rational medical strategies for correcting such immune disorders as immunodeficiency and autoimmunity and for preventing transplant rejection.

To explain what we and other investigators have learned about self-nonself discrimination and the experiments that led us to that knowledge, we must first introduce a few facts about the components and development of the immune system. The immune system of humans and other animals is in large part composed of millions of white blood cells, the lymphocytes. Morphologically, one lymphocyte looks much like another. In reality, each clone, or genetically identical set of cells, differs from all others because it carries several thousand copies of a unique receptor protein on its surface. A receptor can fit an antigen as a lock does a key. Structurally the *receptors* consist of a constant part, which is the same on many lymphocytes, and a variable part, which is specific to each lymphocyte and allows the receptor to bind to its antigen.

Aside from the differences in their antigen receptors, lymphocytes can also be divided into groups according to their origin and functional role in the immune system. Lymphocytes that mature in the thymus (a spongy gland under the breastbone) are known as *T cells*; those that develop in the bone marrow are *B cells*. Unlike T cells; B cells are able to secrete their receptors, which circulate in the blood as *antibodies*. T cells respond to antigen in other ways; and on that basis they can be further subdivided into two classes.

The most aggressive lymphocytes are the *cytotoxic or "killer" T cells*. Their main task is to screen other cells for signs of viral infection and other abnormalities, such as development into cancer cells. Viruses hide and multiply inside a host cell until it bursts, releasing thousands of new virus particles to infect other cells. The immune system can usually interfere with this vicious circle, even though the viruses are hidden, because cells constantly degrade proteins, including viral proteins, into fragments called peptides. The peptides are transported to the cell surface and *presented* to the immune system. The antigen receptors of killer T cells allow them to recognize the viral peptides, which signal that a cell must be destroyed as unhealthy.

The astonishing mechanism that transports and *presents* peptides has been revealed in pioneering studies by Howard M. Grey (National Jewish Center for Immunology and Respiratory Medicine in Denver) and Emil R. Unanue (Washington University School of Medicine, St. Louis), as well as in more recent crystallographic work by Pamela J. Bjorkman and Don C. Wiley (Harvard University). They have shown that most peptides bind to major histocompatibility complex (MHC) molecules inside a cell. The molecules are of two types: *class I MHC*, which displays peptides from proteins made inside the cell, and *class II MHC*, which displays peptides from proteins that have entered the cell from the outside (such as bacterial toxins).

Both kinds of MHC molecules carry peptides to the cell surface and present them to prekiller (unactivated) T lymphocytes. When a prekiller T cell bearing a correctly fitting receptor encounters an antigen-MHC complex, the T cell divides repeatedly; all the daughter cells eventually become active killer T lymphocytes with the same receptor and the power to destroy infected cells. Killing the cells deprives a virus of life support and makes it accessible to antibodies, which can finally eliminate it (Figure 1).

We still do not know exactly how killer T cells recognize peptides, but we have a good idea which molecules are involved. Antigen recognition by killer T lymphocytes is peculiar because the cells are specific for both the peptide and the peptide-presenting MHC molecule. This dual specificity was first recognized in killer T cells by Rolf M. Zinkernagel and Peter Doherty in 1974 (then at the Australian National University). At first, it was not clear whether the dual specificity was a property of a single receptor or of two, or whether it might be the result of two different lymphocytes working in concert. Then, in 1978, Hans Hengartner, Werner Haas, and one of us (von Boehmer, Basel Institute for Immunology) isolated a single killer T cell and propagated it alone in tissue culture. The resulting clone of identical cells still exhibited the original specificity for peptide and MHC molecules, indicating that a single T cell had dual specificity. For several years, that was all we knew about the problem.

During the mid-1980s, the long, controversial search for the elusive receptor of killer T cells culminated in the discovery of the genes and proteins of the alpha-beta *T cell receptor* (TCR). Many laboratories contributed to that work, including those of Mark M. Davis (then at the NIH in Bethesda), Tak W. Mak (Ontario Cancer Institute in Toronto), James P. Allison (then at the University of Texas, Dallas), Ellis L. Reinherz (Harvard Medical School), and Philippa C. Marrack and John W. Kappler (National Jewish Center for Immunology and Respiratory Medicine). The newly discovered receptor molecule consisted of one alpha and one beta polypeptide chain, each encoded by a separate gene.

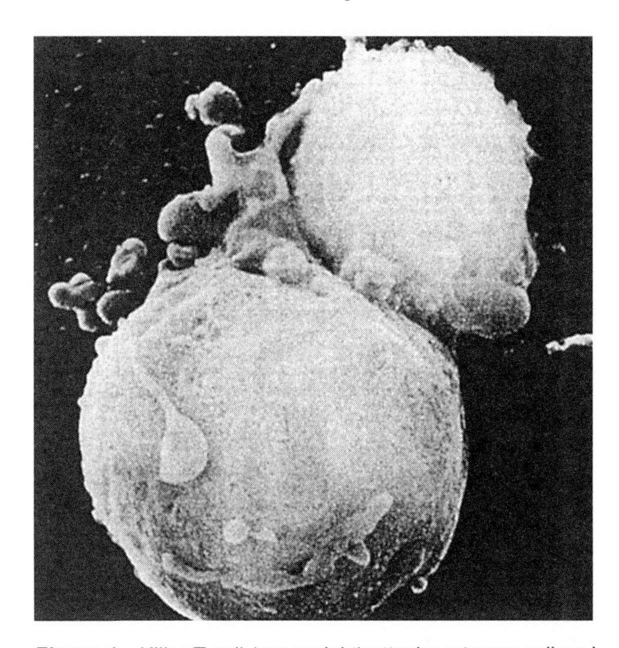

Figure 1 - Killer T cell (upper right) attacks a tumor cell and destroys it. To fend off disease, T cells and other parts of the immune system must distinguish the "self" of the body from the abnormal cells. Investigators have recently discovered how the immune system selects for receptor-bearing T cells that can discriminate between types of cells.

In 1986, Zlatko Dembic, Michael Steinmetz, Haas, and one of us (von Boehmer) transferred the alpha and beta TCR genes from one T cell clone into a second clone with a different specificity. The resulting T cells had the specificities of both the donor and the recipient cells. A single molecule – the alpha-beta TCR – therefore determined a cell's specificity for both one peptide and one MHC molecule.

By itself, the binding of the alpha-beta TCR to a peptide and MHC molecule is usually insufficient to change prekiller T cell into a killer T cell. Full activation requires the binding of another molecule, the *CD8 receptor*. That receptor, which is the same on all killer T cells, also binds to MHC molecules but at a site different from that of the alpha-beta TCR. The CD8 protein was discovered in 1968 by Edward A. Boyse and his colleagues (Sloan-Kettering Institute for Cancer Research, New York City). In 1974, while at Sloan-Kettering, Hiroshi Shiku and one of us (Kisielow) discovered that the CD8 protein was unique to killer T cells and could be used to distinguish them from other lymphocytes (Figure 2).

The function of the CD8 coreceptor became apparent in 1987. By transferring a CD8 gene into a CD8-negative killer cell line, Dembic and one of us (von Boehmer) showed that the CD8 coreceptor is actively involved in antigen recognition by killer T cells. At about the same time, Frank Emmrich and Klaus Eichmann (Max Planck Institute for Immunology Freiburg) discovered that a killer T cell is activated most effectively when an alpha-beta TCR and a CD8 coreceptor are bound by the same molecule.

A second class of T lymphocytes, the helper T cells, also has an alpha-beta TCR and an invariant coreceptor that work together to activate the immunologic defense. The TCR on the helper cells is encoded by the same alpha and beta genes that make those of the killer T cells. The invariant coreceptor of the helper cells is the CD4 protein, however, not CD8. Whereas alpha-beta TCRs and CD8 coreceptors on killer T cells bind to class I MHC molecules and peptides from proteins made inside cells, alpha-beta TCR and CD4 coreceptors on helper T cells bind to

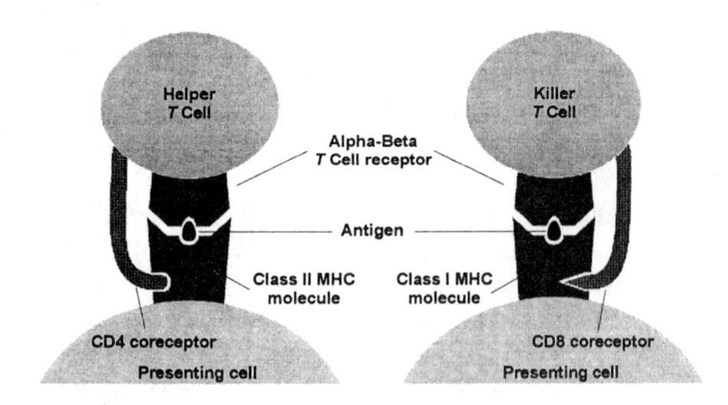

Figure 2 - Receptor complexes on T cells govern their activity. Both killer and helper cells have alpha-beta T cell receptors (TCRs) that can recognize antigens displayed by major histocompatibility (MHC) proteins on cell surfaces. Coreceptor molecules on killer and helper cells bind to different classes of MHC molecules.

class II MHC molecules and peptides from proteins ingested by cells.

Helper T cells also play a different role: they cooperate with B cells in the antibody response to antigens such as bacterial toxins. After a toxic protein is released into the bloodstream, it is taken up by *macrophages*, cells that nonspecifically scavenge and ingest various substances. B cells, too, ingest toxin molecules that have bound to the unique antibody-type receptors on their surfaces.

Inside the macrophages and B cells, the toxin is degraded, and its peptides are then transported to the surface by class II MHC molecules. Once the alpha beta TCRs and the CD4 coreceptors of a prehelper T cell are bound by the same MHC molecules on a presenting cell, the T cell starts to divide and produces active helper T cells. (Macrophages are particularly potent at activating the cells.) The active helper T cells produce factors called *interleukins* that further stimulate the B cells to divide and to secrete large amounts of their specific antibodies, which circulate freely in the blood, bind to the toxin, and neutralize it.

The tremendous diversity of alpha-beta TCRs and MHC molecules is the key to the selective activity of the T cells; the patterns and causes of that diversity, therefore, bear directly on self-nonself discrimination. The genes encoding the variable part

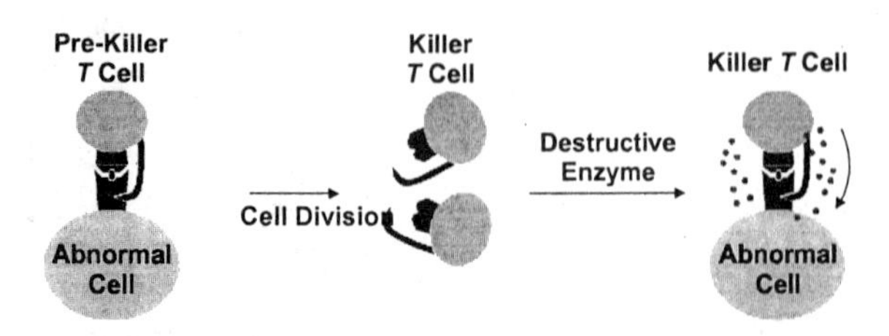

Figure 3 - How killer T cells act. Killer cells recognize foreign antigens displayed by abnormal cells, which they destroy with enzymes (*).

of the TCR are inherited from our parents – not as a continuous stretch of genetic information but instead as little pieces that combine randomly in the developing lymphocytes. That recombination mechanism creates the genes for 100 million or more different TCRs in an individual. The variability seen in MHC molecules is different; although many different MHC genes are present in the population, one individual will have only two genes for each type of MHC molecule.

Because TCRs are generated purely by chance, without regard for which MHC molecules and peptides are in the body, one can imagine that, for any individual, certain receptors will be useful, some will be useless, and still others harmful. *Useful receptors* are those that can help defend the body by binding to nonself-peptides from viral or bacterial proteins presented by self-MHC molecules. Useless receptors are ones that cannot recognize any peptides when presented by self-MHC molecules. *Useless receptors* would bind to self-peptides presented by self-MHC molecules. *Harmful receptors* would bind to self-peptides presented by self-MHC molecules; lymphocytes with such receptors might attack the body's own tissue. Self-nonself discrimination by the immune system is therefore a question not only of how harmful T cells are prevented from destroying the body but also of how the wasteful accumulation of useless T cells is prevented (Figures 3 and 4).

An answer to the problem of harmful T cells was proposed in 1948 by Frank J. Fenner and Sir F. Macfarlane Burnet (Walter and Eliza Hall Institute of Medical Research, Melbourne) and in 1963 by Joshua Lederberg (then at the University of Wisconsin, Madison), that harmful cells were deleted or removed early.

The assumption underlying their *clonal deletion hypothesis* was simple: *lymphocytes with antigen receptors* pass through *two phases* of *development* characterized by radically different responses to antigen binding. In the first phase, binding with an antigen would kill an immature lymphocyte; during the second phase, antigen binding would activate the cell rather than kill it. Because self-peptides are always present, *lymphocytes that have receptors for self-antigens would be exposed to them and deleted* early in development, thus leaving only the cells that have receptors for nonself antigens. The latter cells could mature further and could be stimulated when foreign antigens enter the body.

The immune systems selective accumulation of useful T cells has been studied by several investigators [including Jonathan Sprent and Michael J. Bevan (Research Institute of Scripps Clinic), Zinkernagel and one of us (von Boehmer)]. They have raised the possibility that T cells with receptors capable of binding to self-MHC molecules proliferate

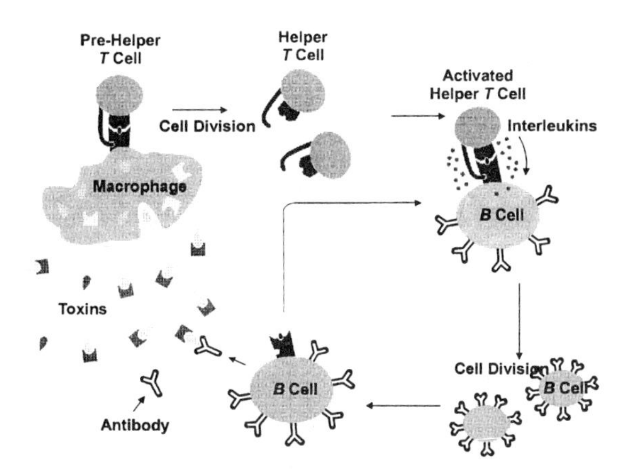

Figure 4 - How helper T cells act. (from top left) Helper T cells respond to antigens presented by macrophages and B cells, which ingest bacterial toxins and other foreign antigens in the blood. The helper cells then multiply and secrete compounds called interleukins that encourage B cells to multiply and release disease-fighting antibodies.

and accumulate preferentially in the lymphoid organs. Cells that could not recognize self-MHC molecules would not expand their numbers.

Both hypotheses initiated many experiments and even more arguments. The clonal deletion hypothesis was favored and disfavored by approximately equal numbers of scientists. In the 1960s and 1970s, Sir Gustav J. V. Nossal (Walter and Eliza Hall Institute), Melvin Cohn (Salk Institute), and others stood on the "pro-deletion" side, whereas Richard K. Gershon (Yale University School of Medicine), Niels Jerne (Basel Institute for Immunology), and others took the "contra-deletion" side. Experiments that could conclusively settle the matter could not be done. At

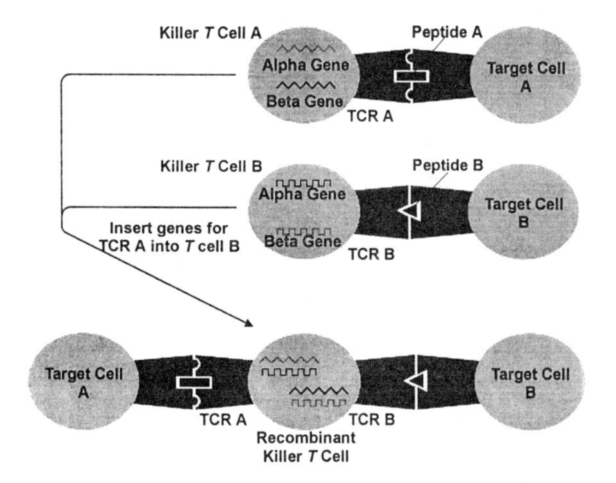

Figure 5 - Transfer of the genes for the antigen receptor will alter the reactivity of the recipient cell (Recombinant killer T cell). If the genes for a unique TCR are extracted from one T cell and inserted into a T cell with a different TCR, the resulting cell will produce both receptors and respond to both antigens (peptide A and B), which will be recognized by both donor and recipient cells (target cells A and B).

that time, the nature of TCRs was unknown and reagents for identifying the specificity of TCRs (such as *monoclonal antibodies*) were not available. Investigators could examine only whether antigen-specific T cells could be activated under certain experimental conditions. Such tests could not distinguish T cells that were silent (not induced) from those that might be absent (physically eliminated) (Figure 5).

The same problem dogged experiments for testing whether some mechanism preferentially expanded or selected clones of useful T cells. Sprent, Zinkernagel, one of us (von Boehmer), and others insisted on such a mechanism, whereas Polly Matzinger (University of California, San Diego), Leroy E. Hood (California Institute of Technology), Philippe Kourilsky (Pasteur Institute, Paris), and others contested it – again on the basis of inconclusive experiments. As a result, we had decades of hot debate.

By the mid-1980s, the nature of TCRs was disclosed, and it became possible to raise TCR-specific antibodies. Before investigators could perform conclusive experiments on self-nonself discrimination, however, they had to cope with another obstacle: it was impossible to follow the development of the few T cells bearing any one particular receptor because they represented only a tiny fraction of all the lymphocytes in a normal animal. The few cells of interest would become lost in the crowd.

For that reason, in 1985 one of us (von Boehmer) decided to study self-nonself discrimination in TCR transgenic mice. *Transgenic mice* carry genetic material that has been introduced artificially. The fundamental technique was developed in the mid-1970s by Ralph L. Brinster and Rudolph Jaenisch. We transferred the TCR genes from one T cell clone into fertilized mouse eggs. The mice that developed from those eggs had integrated the added genes into their own genome and expressed them by making the encoded TCR.

Because we hoped to create a mouse that made the transgenic TCR exclusively, we used mice that had recently been discovered by Melvin Bosma (Fox Chase Cancer Center, Philadelphia). Those mice had a genetic defect, *severe combined immune deficiency* (SCID), which resulted from the inability to combine antigen-receptor gene segments properly. Consequently, they could not produce any antigen receptors of their own. By introducing one functional alpha and one beta TCR gene into a SCID mouse, we could obtain an animal that expressed only the transgenic TCR.

With the help of Yasushi Uematsu of Steinmetz's laboratory (Basel Institute for Immunology), Anton Berns (Netherlands Cancer Institute, Amsterdam),

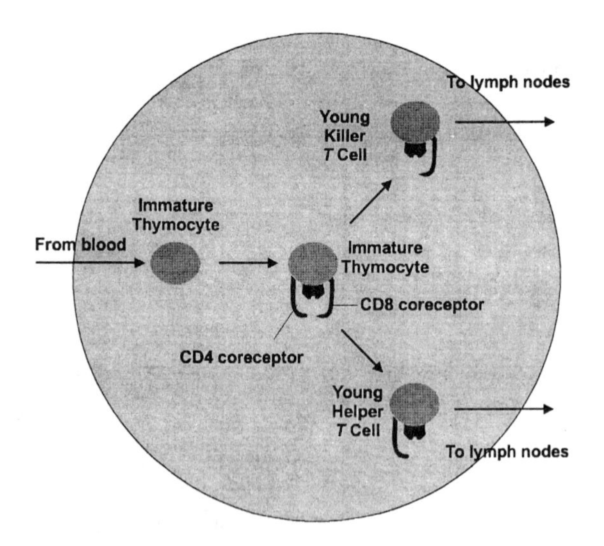

Figure 6 - T cell development occurs principally in the thymus gland. Immature cells entering the thymus do not initially bear a T cell receptor or either type of coreceptor. Later, the slightly more mature thymocytes express all three molecules. Depending on its experiences in the thymus, a T lymphocyte eventually stops making one of the coreceptors and becomes either a killer cell or a helper cell.

and Horst Bluthmann (Central Research Units, F. Hoffmann-La Roche & Co.), we produced such a mouse strain. With these mice we could conclusively address many crucial questions about self-nonself discrimination by the immune system (Figure 6).

We had chosen a TCR specific for the HY peptide, an intracellular peptide that is present in male mice but absent in females. This peptide, we knew, was presented by class I MHC molecules called D^b. The alpha and beta genes encoding the receptor were isolated from a clone of killer T cells and were introduced into the SCID mice. By breeding our transgenic SCID mice in various permutations, we could produce animals in which all the developing lymphocytes would carry the transgenic TCR and would be predictably harmful, useless, or potentially useful. In male mice that had D^b MHC molecules, the transgenic TCR would be harmful (it would bind with the self-peptide HY and the self-MHC). In females that lacked D^b, the TCR would be useless because it could not bind to the MHC molecules. The TCR would be useful in females that had D^b and could present HY as a nonself-peptide.

We set out to analyze the fates of the harmful, useless, and useful T cells in each of these cases. Our analysis was crucially dependent on an antibody against the transgenic TCR that was made by Hung Sia Teh (University of British Columbia, Vancouver). Without that antibody we could never have been sure that we were analyzing the development of cells expressing the transgenic TCR.

The results showed that when the TCR was harmful, that is, in male transgenic SCID mice carrying both HY peptides and D^b MHC molecules, the *thymus* contained few immature *thymocytes* (the precursors of T cells) and no mature helper or killer T cells at all. Because any mature T cells would have been harmfully self-reactive, they were deleted before they could become dangerous. That result was entirely consistent with the clonal deletion hypothesis. Follow-up experiments by our student Wojciech Swat (Institute of Immunology and Experimental Therapy Wroclaw, Poland) have shown that physical deletion, rather than arrested development, causes the absence of the immature thymocytes.

When the TCR was useless, as it was in female transgenic SCID mice lacking D^b MHC molecules, immature thymocytes were present, but mature helper and killer T cells were absent. That observation indicated that useless cells incapable of recognizing and interacting with the self-MHC molecules do not mature; they die after a short life span (Figure 7).

Finally, when the TCR was potentially useful, as in the female mice carrying class I D^b MHC molecules, we found immature thymocytes and mature killer T cells but no mature helper T cells. That result holds two implications. First, the binding of the alpha-beta TCR to MHC molecules in the absence of the antigen peptide rescues immature cells from death and induces maturation. Second, the specificity of the alpha-beta TCR for class I or class II MHC molecules determines whether a developing T cell will become a helper cell or a killer cell. Because the genes for the TCR that we had inserted came from a killer T cell clone, all the T cells in our transgenic SCID mice became killer cells.

By extrapolating our results to receptor selection in normal, nontransgenic animals, we arrived at the following picture of *self-nonself discrimination* by the immune system. Immature thymocytes express a wide variety of antigen receptors, which are generated through random combinations of TCR gene segments and through random pairings of various alpha and beta TCR chains. If the receptor on one of these cells

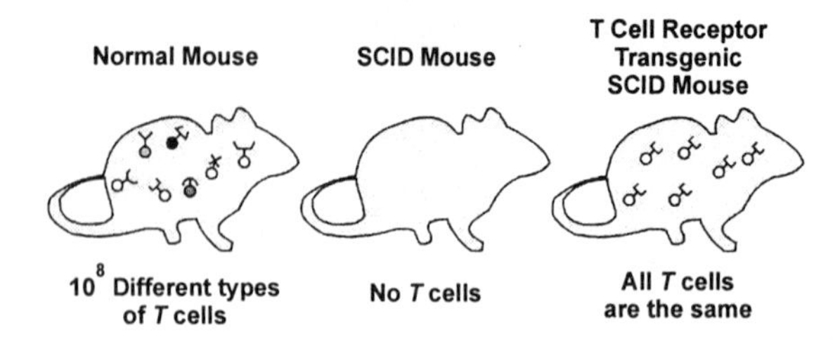

Figure 7 - Transgenic mice are useful for studying the fates of T cells bearing specific receptors. Normal mice produce T cells with so many different receptors that it is difficult to follow one set of cells. Mice that have the SCID mutation, however, do not produce any T cells. If the genes for one T cell receptor are inserted into a SCID mouse egg, all of the T cells in the resulting mouse will have the same receptor and meet the same developmental end.

does not bind to any molecules in the thymus, the cell is useless and dies after about three days. If the receptor binds in the thymus to both the peptide and the MHC molecule, the cell is destroyed because it is harmful. If the receptor binds to MHC molecules in the absence of its antigen peptide, the cell is potentially useful and is therefore selected for further maturation. Depending on whether the receptor binds to a class I or class II MHC molecule, the selected cell will become either a killer or a helper T lymphocyte.

Helper and killer cells patrol the body's lymph nodes, spleen, and other peripheral lymphoid organs. After they leave the thymus, they react differently when their receptor binds to a peptide and an MHC molecule. Binding of a receptor to both molecules results in T cell activation and the generation of effector cells. Thus, the immune system learns to distinguish self from nonself by screening lymphocytes: the useful are selected, the useless are neglected, and the harmful are rejected.

There is now good evidence that the deletion of immature harmful cells also occurs in normal mice (Figure 8). As we have already mentioned, it is difficult to follow the fate of cells with one specific receptor in normal animals. Yet Kappler and Marrack have found that certain molecules called super antigens will bind many different TCRs and class II MHC molecules. Super antigens do not bind to MHC molecules and TCRs at the same sites as peptide antigens. Nevertheless, using super antigens in normal mice, Kappler and Marrack, Zinkernagel and H. Robson (Ludwig Institute of Cancer Research, Lausanne) have obtained data that are compatible with our studies of TCR transgenic mice. Dennis Loh (Stanford University), Stephen Hedrick (University of California, San Diego), and Zinkernagel used different TCR transgenic mice to reach similar conclusions and have extended our results.

Most investigators agree that clonal deletion is probably not the only way to silence harmful T cells. In our minds, the question never was whether clonal deletion was the only mechanism, but whether it existed at all. We now know that deletion is most easily induced in immature thymocytes, which implies that only those peptides produced in the thymus, or carried there, can induce deletion. If all self-peptides were present in sufficient quantity to induce deletion, autoimmunity would not exist. The fact that it does indicates that the deletion mechanism of self-nonself discrimination is not perfect and that other processes probably do exist.

Although ample circumstantial evidence supports the existence of *"suppressor" T cells*, which prevent other T cells from becoming active, as yet we have no conclusive evidence on how these cells operate (Figure 9). Marc K. Jenkins, Ronald H. Schwartz, and B. J. Fowlkes (NIH), David Lo (Scripps Clinic), and Jacques Miller (Walter and Eliza Hall Institute) have demonstrated that certain peptides, when presented on cells other than B cells and macrophages, can induce anergy in T cells: *anergic T cells* do not die,

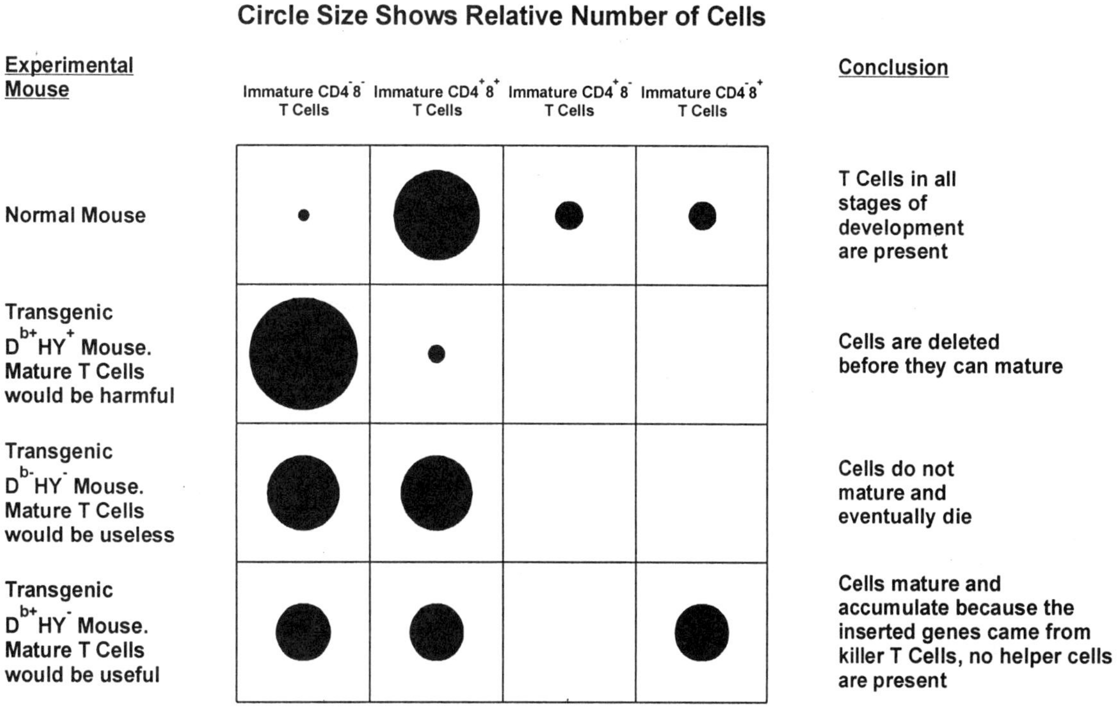

Circle Size Shows Relative Number of Cells

Experimental Mouse
Immature CD4$^-$8$^-$ T Cells
Immature CD4$^+$8$^+$ T Cells
Immature CD4$^+$8$^-$ T Cells
Immature CD4$^-$8$^+$ T Cells
Conclusion
Normal Mouse
T Cells in all stages of development are present
Transgenic D^{b+}HY$^+$ Mouse. Mature T Cells would be harmful
Cells are deleted before they can mature
Transgenic D^{b-}HY$^-$ Mouse. Mature T Cells would be useless
Cells do not mature and eventually die
Transgenic D^{b+}HY$^-$ Mouse. Mature T Cells would be useful
Cells mature and accumulate because the inserted genes came from killer T Cells, no helper cells are present

Figure 8 - A test of clonal deletion.

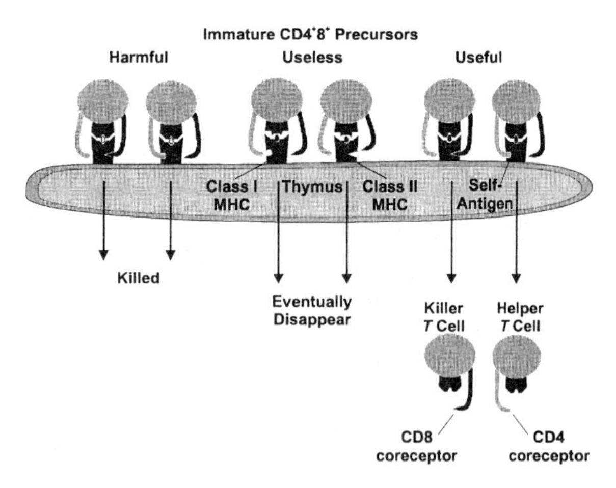

Figure 9 - Fate of maturing T cells is determined in the thymus. Harmful cells, with receptors that would recognize the body"s antigens and MHC proteins, are destroyed. Useless cells that cannot recognize the body's MHC proteins die of neglect. Useful cells that can recognize antigens on the MHC proteins, but not the self-antigens, mature and accumulate.

but they are unresponsive to antigenic stimulation. Despite some progress in understanding self-nonself discrimination by mature T cells, we are still far from achieving the clinical goal of specifically silencing them. Achieving that goal would make organ transplantation far more successful than it is today.

Our experiments were designed to probe the mechanisms of self-nonself discrimination and not to solve the problem of *organ transplantation* or *autoimmunity*. Yet we may ponder how the results might bear on those problems. Some forms of autoimmunity are almost certainly caused by the recognition of antigens that are not present in the thymus and that therefore do not delete T cells. That is seemingly the case with myelin basic protein, which is normally sequestered in the nervous system; injections of myelin basic protein into some animals can induce experimental *allergic encephalomyelitis* (EAE), a disease akin to *multiple sclerosis* in humans. Many investigators also suspect that antigens found on insulin-producing pancreatic islet cells do not reach the thymus. If these antigens become the targets of autoimmune T cells, the result could be *diabetes*.

Of course, it would be tempting to try to prevent autoimmune diseases by identifying the organ-specific antigens involved and introducing them into the thymus to delete the offensive T cells. Hartmut Wekerle (Max Planck Institute for Psychiatry, Munich) and Eli E. Sercarz (University of California, Los Angeles) have performed such experiments by injecting relatively high doses of myelin basic protein into neonatal mice. The procedure did protect the animals from developing EAE after subsequent injections.

Ali Naji and his co-workers (Hospital of the University of Pennsylvania, Philadelphia) have explored the application of thymic tolerance mechanisms to organ transplantation. They implanted foreign pancreatic islets in the thymuses of rats; subsequent grafts of the islet tissue were permanently accepted by the animals.

It remains to be shown whether the tissue-specific antigens in the thymus caused the deletion of the developing, potentially harmful T cells, but at present that is the most logical explanation. These experiments do not provide an immediate solution to the problem of autoimmunity and transplantation. They do suggest that by presenting certain antigens from normal tissues, or from future tissue grafts to young T cells in the thymus, we will eventually be able to prevent unwanted immune reactions efficiently.

Further Reading:

1. P. Kisielow *et al.*, Tolerance in T-cell receptor transgenic mice involves deletion of non-mature $CD4^+8^+$ thymocytes. Nature, 333, 6175, 742-746; June 23, 1988.
2. H. S. Teh *et al.*, Thymic major histocompatibility complex antigens and the alpha-beta T-cell receptor determine the CD4/CD8 phenotype of T-cells. Nature, 335, 6187, 229-233; September 15, 1988.
3. H. von Boehmer, Developmental biology of T-cells in T-cell receptor transgenic mice. Annu. Rev. Immunol. 8, 531-556; 1990.

INNOVATION/INSIGHT

MAKING MONOCLONAL ANTIBODIES THAT WON'T FIGHT BACK

Rhein, R. (1993) Journal of National Institute of Health Research 5: 40-48. (Adapted)

The Holy Grail of monoclonal antibody research, creating therapeutic immunoglobulins that are not recognized as foreign by the human immune system, now seems within reach. Methods of producing antibodies *in hybrid mouse-human cells* are quickly being supplanted by new techniques that bypass the mouse component altogether.

One strategy is to create human antibodies targeted to specific antigens without first immunizing any animal, reported Richard Lerner (Scripps Research Institute) and colleagues in *Science* (Nov. 20, 1992). *Fragments of human monoclonal antibodies* (F_{ab}) can be produced by genetically engineered bacteriophages growing in *Escherichia coli* bacteria. The phages carry genes that code for the heavy- and light-antibody chains; *antibody diversity* is achieved by *random association and expression of large numbers of these genes*.

Lerner and his colleagues have created completely human F_{ab} molecules that neutralize *human immunodeficiency virus-type 1* and *respiratory syncytial virus*, which causes severe lower respiratory tract illness in young children, *in vitro*. They have produced others that bind *hepatitis B surface antigen*. Competing research institutions and biotechnology companies are adding to the list. Rebecca Mullinax and her colleagues at Stratagene Inc. (La Jolla) have created F_{ab} molecules that bind to *tetanus toxoid*, for example.

So far, none of these F_{ab} preparations has been clinically tested in humans. Nevertheless, "it is likely that before the turn of the century, machines will be available to produce any antibody on demand without recourse to animal immunization," says Carlos Barbas, one of the key monoclonal antibody researchers at Scripps (Figure 1). [*Note*: This has not come to pass. Ed.]

Another strategy for reducing immunogenicity of synthetic antibodies is to "humanize" monoclonal antibodies or antibody fragments. Sherie Morrison at Columbia University and her colleagues at Stanford, and Greg Winter and his colleagues at the Medical Research Council, have pioneered the creation of antibodies that are primarily, but not completely, "human." They immunize a mouse with the target antigen to produce a typical monoclonal antibody; then, they isolate the genes that code for the variable region and combine them with human genes that code for the rest of the antibody, to create a *chimeric mouse-human antibody*.

Winter has gone a step further and isolated just those genes that code for the *complementarity-determining region* (CDR) of the variable portion of the mouse antibody – the region that actually *binds* the *target antigen*. By combining these genes with human genes for the rest of the antibody, Winter's group produced "humanized" antibodies.

In the mid-1970s, Cesar Milstein and Georges Kohler of MRC hybridized mouse lymphoma and spleen cells to form a *"hybridoma" cell line* that perpetually manufactures mouse monoclonal antibodies. Ever since, the medical promise of these disease-fighting protein structures has been obvious. The vexing problem of their immunogenicity became apparent when, in early clinical trials, many human recipients of the monoclonals mounted immune responses against them.

Such *human anti-mouse antibody* (HAMA) responses can be dangerous, causing anaphylactic shock in rare instances. Even mild HAMA responses can prevent the mouse-derived monoclonal antibodies from achieving their therapeutic goals, after their first use, because the immune system of the patient has been primed to clear future mouse antibodies as soon as they appear in the bloodstream. Worse, from a clinical perspective, no other murine-based antibody therapy will work in patients who develop the HAMA response.

The logical strategy has been to make the

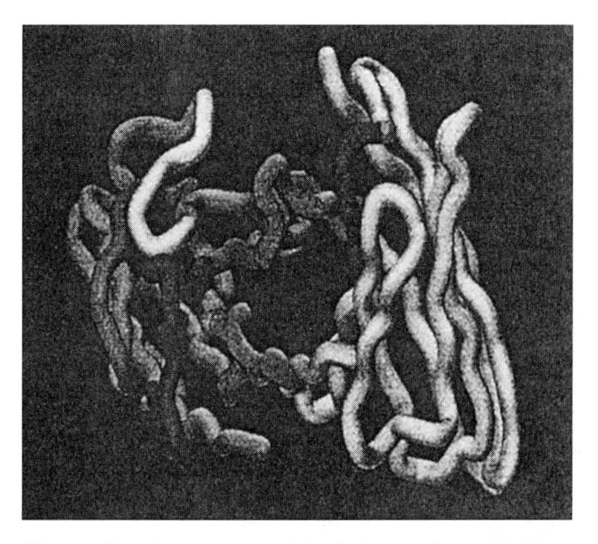

Figure 1 - Computer model of the binding site of an antibody molecule, showing the heavy chain (far left) and the light chain (far right). Sequence variability within the complementarity determining regions (shorter center sections) allows antibodies to evolve to bind any antigen. [Courtesy Carlos Barbas, Scripps Research Institute, La Jolla, CA.]

potentially therapeutic antibodies more "human." Indeed, natural human antibodies – immunoglobulins taken from the sera of people previously exposed to an infectious agent – have been used for many years to combat hepatitis and other diseases. But the proportion of antibodies in the sera that recognize and bind tightly to specific antigens – that is, antibodies with high specificity and binding affinity for their target antigens – is often low. Deliberately and repeatedly exposing human donors to a disease-causing antigen in order to boost the specificity and binding affinity of their antibodies would certainly be unethical. Human antibodies, derived either from natural sera or from hybridomas, have thus become impractical.

In 1984, Morrison and her colleagues synthesized "chimeric" mouse-human antibodies to greatly reduce the "mouse" component. An immunoglobulin G (IgG) molecule, the class of antibodies most commonly produced in hybridoma systems, consists of *four polypeptide chains: two identical "heavy" chains and two identical "light" chains*. Together, they form a *Y-shaped molecule*, the arms of which contain variable polypeptide regions that bind antigen. The *trunk of the Y is a constant polypeptide region, essentially the same for all IgG antibodies* (Figure 2). In Morrison's mouse-human chimeric antibody, 75 percent of the molecule consists of human constant regions, and 25 percent consisted of murine variable regions.

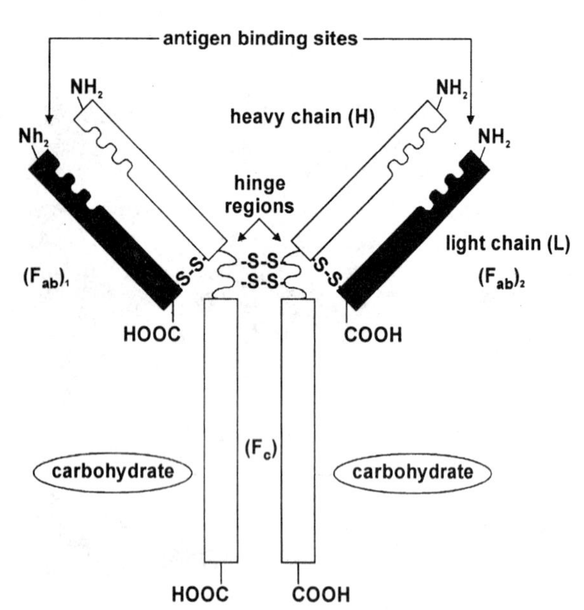

Figure 2 - The typical Y-shaped structure of an antibody molecule. The molecule is formed from two identical heavy chains (H) and two identical light chains (L). The amino terminal regions of both heavy and light chains form the antigen binding sites. The heavy chains form the tail and hinge regions. [Reprinted with permission from the Molecular Biology of the Cell, Garland Publishing Inc., New York and London, (1989).]

Two years later, Winter and his colleagues "humanized" monoclonal antibodies even further. Winter's group incorporated into human immunoglobulins only 10 percent of the mouse antibody – the antigen-recognition part, or CDR.

Since the mid 1980s, the hope of monoclonal antibody researchers has been that the body's immune system would not "see" these reduced regions of mouse protein in chimeric antibodies. In fact, different researchers report different results. Robert Jordan and his colleagues (Centocor) reported that their mouse-human chimeric monoclonal antibody for inhibiting platelet aggregation caused no HAMA response in 64 patients. In contrast, of 86 patients who received *murine* monoclonal antibody, 17 patients, or 20 percent, showed a HAMA response. But Albert LoBuglio (of the University of Alabama at Birmingham Cancer Center) reported immune responses in as many as 67 percent of patients who were administered chimeric antibodies to treat gastrointestinal adenocarcinomas.

Bacterial immunoprotein constructs. In an effort to bypass animals altogether, researchers turned to genetically engineered microorganisms to express human antibodies. Early researchers used plasmids as vectors to transport human antibody genes into mammalian cells or bacteria. But bacteriophages are now the most common vectors used to insert the desired genes into bacteria.

Phages have been used to insert foreign DNA into bacteria for many years. However, it was not until 1989, when Winter's group reported in *Nature* that it had produced a single, heavy-chain F_{ab}, and that phages successfully carried antibody genes into *E. coli.* A few months later, William Huse, then at Scripps, and his colleagues described in *Science* their own success in generating large numbers of heavy- and light-chain, variable-region antibody fragments. Like Winter's group, the Scripps team was working with phage lambda and blood from immunized mice.

"*Combinatorial libraries*" of these synthetic F_{ab} constructs mimic a natural process in B lymphocytes – in which hypermutation of genes encoding the variable regions of IgGs produces combinations of antibodies with varying binding specificities and affinities to particular antigens. The new technique offers the prospect of creating *highly selective, high-affinity human antibodies* without first immunizing anyone with the target antigens.

Lerner and his colleagues have written numerous articles describing methods of producing human antibody fragments without immunizing animals or humans. Winter says his own group at Cambridge was the first to achieve the goal, however. "We did it, and we did it first," he says, citing an article he co-authored for the *Journal of Molecular Biology* in 1991 describing the MRC group's success.

There are similarities and differences in the MRC and Scripps teams, approaches to creating synthetic human antibodies. Both groups sequence the genes for variable regions of "naive" human antibodies – that is, antibodies derived from donors who have not been exposed to the target antigen – from the B lymphocytes or hybridomas that produce them. They use the polymerase chain reaction (PCR) to amplify the mRNA. Then, they create strands of DNA complementary to the mRNA, which they insert into the genomes of bacteriophage.

To achieve antibody diversity, Lerner's group uses *error-prone PCR during amplification*, thus creating millions of slightly different F_{ab} genes – and millions of slightly different antibody fragments. This "chemical solution to the diversity problem ... has the potential to create approximately 10^{20} complementarity-determining regions for antibody heavy chains," Lerner and his colleagues reported (in the May 1992 issue of the *Proceedings of the National Academy of Sciences*). "When combined with light chains and expressed on phage surfaces, high-affinity antibodies could be selected from 5.0 x 10^7 *Escherichia coli* transformants," they said.

Winter's group achieves a comparably large combinatorial library by cloning the genes for different parts of the antibodies" variable regions and randomly assembling these CDR genes when they are inserted into the phage genomes. This results in "a library of millions of different phages," they reported in the Aug. 15, 1992 issue of the *Journal of Biological Chemistry.*

With their different methods, both groups are able to produce enough different antibody fragments that, statistically, at least some will be specific to the target antigens and will bind tightly to them. Just creating millions of different antibody fragments is not enough. The researchers must efficiently locate the few that fit best into the folded amino acid structure of the target antigen, like a key in a lock. These F_{ab}s are the ones that will be multiplied to neutralize the target antigen.

Both teams of researchers, as well as others in private industry, have taken advantage of a relatively new method of accelerating the "maturation" of their synthetically created antibody fragments to produce F_{ab}s that are highly specific against target antigens and attain the highest binding affinities. They use threadlike "filamentous" phages to insert the genes into *E. coli.*

Unlike phage lambda, which simply carries F_{ab} genes into *E. coli, filamentous phages display the expressed F_{ab} proteins* on their surfaces as they bud from the bacteria. Winter describes this process as "mimicking the strategy of the immune system." In living animals, antibodies are produced and displayed on the surfaces of B lymphocytes. In the body, antigen binding to these antibodies triggers the expression of B-cell genes to synthesize heavy and light IgG protein chains, each of which contains three hypervariable regions. Hypermutation of the variable region genes produces a large variety of antibodies. Cells that produce the "best" antibodies – those with the highest affinity and binding specificity to the antigen – are cloned by the body in great quantities, whereas those producing antibodies that do not bind well are not replicated.

In the laboratory, combinatorial antibody libraries mimic the display of antibody on the surface of the B cell, antigen-driven selection, and affinity maturation, according to the MRC researchers. One method of accomplishing this, used by the Scripps team, is to attach the human genes coding for the variable regions of antibody heavy chains, derived from PCR-amplified messenger RNA coding for the different regions of the antibody protein, to viral genes encoding phage surface proteins. Genes encoding the light chains, also inserted into the phage genome, produce free antibody light-chain polypeptides in the periplasmic space of *E. coli* – where the heavy and light chains are assembled. Because the heavy chains are anchored in the phage coat protein, the resulting antibody fragment is attached to it when the phage buds from the bacterium. At this point, researchers can harvest the F_{ab} molecules from the culture media, or they can go on to the next step to increase their specificity and affinity to particular antigens by mimicking nature's method.

In one technique, called "*panning*," a plastic surface is coated with antigen and exposed to the antibody-bearing phage particles. Phages with the highest-affinity antibody will bind tightly to antigen on the plastic; the low-affinity antibody can be washed off. *E. coli* are then reinfected with the phages that contain the high-affinity antibodies. The cycle is repeated several times (*in vitro/in vivo* evolution) to increase antibody binding constants to at least 10^{-10} M^{-1}, which is the reciprocal concentration of antibody that will be half-saturated by antigen (0.1 nM).

The idea of having phages not only carry genes into bacteria but also display the express proteins on their surfaces was conceived in 1985 by George Smith of the University of Missouri in Columbia. Winter's group reported using filamentous phages in creating antibody libraries in 1990. The Scripps group published its work with filamentous phages in 1991.

Former Scripps researcher Huse, who, in 1990, started his own company – Ixsys Inc. in San Diego – to make peptides and human monoclonal antibodies, has gone a step further to reduce the size of the combinatorial library. He points out that most of the changes in antibody structure do not increase binding affinity or selectivity. "The fundamental problem with the panning technique is that it's a molecular event," he notes. "All molecular events are random." Creating just those antibodies that are potentially useful would save time and effort," he says.

Ixsys has developed a novel *codon-based mutagenesis procedure* for increasing the useful diversity of antibody fragments produced by the filamentous phage-vector system. It involves replacing entire DNA-nucleotide triplets rather than individual nucleotides during mutagenesis of the CDR.

In Ixsys' procedure, random alterations are introduced in specific target codons to synthesize only a predefined region of the CDR, while a bias toward the parental sequence of amino acids is maintained to keep the mutation from veering too far from the original. "We ought to be 1,000 times more efficient in identifying high-affinity antibodies," Huse says. "Within 100 mutations, we get a five-fold increase in affinity."

One problem with the filamentous phage-vector system is that, unlike mammalian cells, *E. coli* cannot express whole antibodies. However, for many therapeutic purposes, antibody fragments may actually work better than whole antibodies because they clear from the bloodstream faster and, in cancer patients, penetrate tumors more efficiently. Of course, heavy (H)- and light (L)-chain, variable-region antibody fragments with high-binding affinity produced by phage libraries can always be attached to conserved L- and H-chain fragments to form whole IgG antibodies.

This is the therapeutic approach at Stratacyte Corp., in La Jolla, a subsidiary of Stratagene. Stratacyte's director of research, Bob Shopes, says that his company has screened F_{ab} fragments for specificity, "but once we get into therapeutics, we'll use whole antibodies." These may not be normal Y-shaped IgG antibodies, he says, but rather *dimeric antibodies* – resembling IgA antibodies, where the two Ys are linked together to resemble the spokes of a wheel. "Dimeric antibodies kill [target] cells 100 times better than normal IgG antibodies," he says. This is because antibodies are concentrated at a particular site to give the *complement cascade* a "kick start," Shopes says.

Researchers at several other biotechnology and pharmaceutical companies are also working in this field (including Avi Ashkenazi and Mark O'Connell at Genentech Inc., Robert Ames at SmithKline Beecham, and Dale Yelton of Bristol-Myers Squibb, who has worked with Ixsys's Huse). JoAnn Suzich (at MedImmune Inc.), is developing combinatorial antibody libraries to protect against respiratory syncytial virus. Suzanne Zebedee (at R. W. Johnson Pharmaceutical Research Institute) and Hermann Gram, (at Sandoz Pharma) have worked closely with the Scripps team.

In other areas of monoclonal-antibody research, scientists are exploring different methods of reducing the antigenicity of antibodies produced in animals by immunization. Xoma Corp. was one of the first companies to manufacture classical *monoclonal antibodies* – antibodies produced from clones of single mouse-human hybridoma cells – for therapeutic use in humans to combat *septic shock*. A Xoma team led by the company's president and science director, Patrick Scannon, is now developing second-generation "humanized" antibodies against autoimmune diseases by reducing the amount of "mouse" in the peptide chains that make up the antibody.

To further reduce the immunogenicity of monoclonal antibodies, Roland Newman and his colleagues (at IDEC Pharmaceuticals Corp.) have created *"primatized" antibodies*, in which the antigen-recognition molecules are derived from macaque monkeys instead of mice. IDEC researchers use the same techniques that are used to make mouse human chimeric antibodies. They immunize cynomologous macaques with antigen, *immortalize* the B cells that produce the monkey antibodies, then use the cell lines to isolate and sequence the genes that encode the heavy- and light-chain variable regions. Newman and his collaborators insert the monkey genes – along with human genes for encoding the constant regions of antibody – into the genomes of Chinese hamster ovary cells, which express the resulting monkey-human chimeric antibodies *in vitro*. The constant region is human, and the variable region is macaque.

"For all intents and purposes, the primatized antibody should be recognized as "human" by the patient's immune system," says Newman, IDEC's director of gene cloning. Gene sequences that encode antibodies in humans and macaques are quite similar – between 84 percent and 96 percent amino acid homology for the variable regions. At the same time, Newman says, there is enough difference for the primatized antibody to recognize a human antigen such as the CD4 protein on T helper lymphocytes, which may allow the chimeric antibody to block autoimmune reactions in diseases such as rheumatoid arthritis.

Although Newman and his colleagues have not yet tested the primatized antibodies in humans, he notes that the reverse – putting human antibodies into macaques – elicited no adverse immune responses in the monkeys during the course of a year.

A more direct route to creating human antibodies may be to make them in transgenic mice. Jonathan MacQuitty (chief executive officer of GenPharm

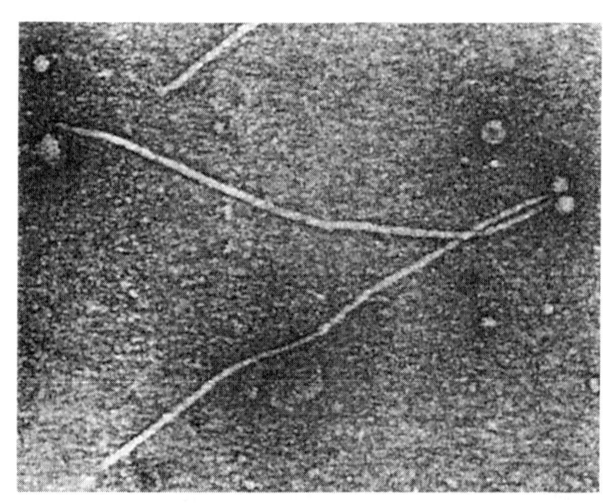

Figure 3 - Electron micrograph of hepatitis B antigen particles (round), and the filamentous phages that are genetically engineered to produce F_{ab} antibody fragments against the antigen. The phages display the F_{ab} fragments on their surfaces, thus binding the hepatitis B antigen particles. [Courtesy Carlos Barbas, Scripps Research Institute, La Jolla, CA.]

International) says researchers at his company have microinjected "unrearranged" human antibody genes into fertilized mice ova. When the mice from those eggs grow up, they are challenged with an appropriate antigen to produce human antibodies. Then, their spleen cells are fused with mouse myeloma cells to make hybridomas, which secrete human monoclonal antibodies against the antigen (Figure 3).

One of the first therapeutic goals will be to generate antibodies that bind to and disable the human T cells that cause rheumatoid arthritis, MacQuitty says. No matter how the antibody is created, the goal of each of these efforts is to produce an antibody that will not cause an immune reaction in the patient when used as therapy. "Even when you have human antibodies, don"t be surprised if you trigger an immune response," says Scannon.

Additional Reading:

1. Barbas, C. F., E. Bjorling, F. Chiodi, N. Dunlop, D. Cababa, T. M. Jones *et al.*, "Recombinant human F_{ab} fragments neutralize human type 1 immunodeficiency virus *in vitro*," Proc. Natl. Acad. Sci. USA 89, 9339 (1992).
2. Barbas, C. F., J. E. Crowe, D. Cababa, T. M. Jones, S. L. Zebedee, B. R. Murphy *et al.*, "Human monoclonal F_{ab} fragments derived from a combinatorial library bind to respiratory syncytial virus F glycoprotein and neutralize infectivity," Proc. Natl. Acad. Sci. USA 89, 10164 (1992).
3. Zebedee, S. L., C. F. Barbas, Y. L. Hom, R. H. Caothien, R. Graf, J. DeGraw, J. Pyati *et al.*, "Human combinatorial antibody libraries to hepatitis B surface antigen," Proc. Natl. Acad. Sci. USA 89, 3175 (1992).
4. Mullinax, R. L., E. A. Gross, J. R. Amberg, B. N. Hay, H. H. Hogrefe, M. M. Kubitz *et al.*, "Identification of human antibody fragment clones specific for tetanus toxoid in a bacteriophage lambda immunoexpression library," Proc. Natl. Acad. Sci. USA 87, 8095 (1990).
5. Larrick, J. W., L. Danielsson, C. A. Brenner, M. Abrahamson, K. E. Fry, and C. A. Borrebaeck *et al.*, "Rapid cloning of rearranged immunoglobulin genes from human hybridoma cells using mixed primers and the polymerase chain reaction," Biochem. Biophys. Res. Commun. 160, 1250 (1989).
6. Kohler, G., and C. Milstein, "Continuous cultures of fused cells secreting antibody of predefined specificity," Nature 256, 495 (1975).
7. Morrison, S. L., M. J. Johnson, L. A. Herzenberg, and V. T. Oi, "Chimeric human, antibody molecules:mouse antigen-binding domains with human constant region domains," Proc. Natl. Acad. Sci. USA 81, 6851 (1984).
8. Jones, P. T., P. H. Dear, J. Foote, M. S. Neuberger, and G. Winter, "Replacing the complementarity-determining regions in a human antibody with those from a mouse," Nature 321, 522 (1986).
9. Winter, G., and C. Milstein, "Man-made antibodies," Nature 349, 293 (1991).
10. Ward, E. S., D. Gussow, A. D. Griffiths, P. T. Jones, and G. Winter, "Binding activities of a repertoire of single immunoglobulin variable domains secreted from *Escherichia coli,"* Nature 341, 544 (1989).
11. Huse, W. D., L. Sastry, S. A. Iverson, A. S. Kang, M. Alting-Mees, D. R. Burton *et al.*, "Generation of a large combinatorial library of the immunoglobulin repertoire in phage lambda," Science 246, 1275 (1989).
12. Lerner, R. A., A. S. Kang, J. D. Bain, D. R. Burton, and C. F. Barbas, "Antibodies without immunization," Science 258, 1313 (1992).
13. Marks, J. D., H. R. Hoogenboom, T P Bonnert, J. McCafferty, A. D. Griffiths, and G. Winter, "Bypassing immunization: human antibodies from V-gene libraries displayed on phage," J. Mol. Biol. 222, 581 (1991).
14. Barbas III, C. F., J. D. Bain, D. M. Hoekstra, and R. A. Lerner, "Semisynthetic combinatorial antibody libraries: a chemical solution to the diversity problem," Proc. Natl. Acad. Sci. USA 89, 4457 (1992).
15. Marks, J. D., H. R. Hoogenboom, A. D. Griffiths and G. Winter, "Molecular evolution of proteins on filamentous phage," J. Biol. Chem. 23, 16007 (1992).
16. Smith, G. P., "Filamentous fusion phage: novel

expression vectors that display cloned antigens on the virion surface," Science 228, 1315 (1985).
17. McCafferty, J., A. D. Griffiths, G. Winter, and D. J. Chiswell, "Phage antibodies: filamentous phage displaying antibody variable domains," Nature 348, 552 (1990).
18. Kang, A. S., C. F. Barbas, K. D. Janda, B. J. Benkovic, and R. A. Lerner, "Linkage of recognition and replication functions by assembling combinatorial antibody F_{ab} libraries along phage surfaces," Proc. Natl. Acad. Sci. USA 88, 4363 (1991).
19. Newman, R., J. Alberts, D. Anderson, K. Carner, C. Heard, F. Norton *et al.*, "“Primatization” of recombinant antibodies for immunotherapy of human diseases: a macaque/human chimeric antibody against human CD4," Bio/Technology 10, 1455 (1992).

LAB 9.1:
WESTERN BLOTTING

Overview:

In this lab, we will first electrophorese β-galactosidase-*myo*-3 fusion protein on polyacrylamide gels then transfer the protein to nitrocellulose and perform immunodetection with monoclonal antibodies specific for the fusion protein. The use of antibodies to visualize a specific protein of interest is called an immunoblot (Western blot). The nitrocellulose retains an exact negative imprint image of the pattern of proteins that were blotted from the gel. A sensitive immunoassay method will be used to identify the fusion protein based on recognition by the monoclonal antibody. Excess binding sites on the nitrocellulose are blocked with nonspecific proteins before placing the nitrocellulose in a solution containing the (primary) antibody – which recognizes the protein of interest. Then excess unbound antibody is removed and the primary antibody – now specifically bound to the protein of interest – is detected with an enzyme-coupled secondary antibody. Finally, the nitrocellulose is immersed in a solution of substrate – which is converted into a colored insoluble product by the enzyme that is coupled to the secondary antibody. This procedure verifies the enzymatic integrity of your isolated protein.

Experimental Outline:

1. Prepare solutions.
2. Immunoblotting.
3. Electrophoretic transfer of proteins to nitrocellulose.

Required Materials:

Tris buffered saline (TBS) (recipe below); BSA blocking solution; Genius blocking solution; Tris-buffered saline with Tween (TTBS); 4-chloro-1-napthol solution (recipe below); PAGE gel (recipe Lab 8.1); water bath; loading dye containing bromophenol blue; molecular weight markers; nitrocellulose paper; Transphor unit (Hoeffer); gloves; goat anti-mouse peroxidase IgG; Hoeffer Western Blotting apparatus

Procedures:

1. Prepare Solutions

1. *10x Tris-buffered saline* (TBS):

1 M Tris, pH 7.4	100	ml
NaCl	43.8	g

Add H_2O to 500 ml and store at room temperature.

2. *1x TBS*: Mix 200 ml of 10x TBS with 1800 ml of H_2O to make 2 L of 1x buffer. Store at room temperature.
3. *BSA blocking solution*: 1x TBS with 3% BSA.
4. *1x TTBS*: Mix 100 ml of 10x TBS with 899 ml of H_2O. Add 1 ml of Tween 20 slowly while stirring. Store at room temperature.
5. *Goat anti-mouse peroxidase IgG working solution*: dilute to greater than 1/2000 with BSA blocking solution.
6. *4-Chloro-1-napthol*: WEAR GLOVES, SUSPECTED CARCINOGEN.
   Dissolve 60 mg of 4-chloro-1-napthol in 20 ml of methanol. Just before use, add 100 ml of TBS and 60 μl of 30% H_2O_2.
7. *Transfer buffer*: Add 18.2 g of Tris base and 86.5 g of glycine to 4 liters of H_2O (optional: 0.1% SDS may be added). Add 1.2 liters of methanol and adjust to 6 liters with H_2O. The pH of the solution should be 8.2-8.4.

2. Immunoblotting (Genius Procedure)

1. Prepare 2 polyacrylamide gels (recipe Lab 8.1).
2. Load matching samples onto the 2 gels as directed by your instructor; record their specifications below.
   i.
   ii.
   (sample specifics, continued)

iii.
iv.
v.
vi.
vii.
viii.

3. Mix one part protein solution and one part loading dye containing bromophenol blue in a microfuge tube for each gel sample. Heat these tubes and the molecular weight markers for 10 minutes at 90-100°C then place them on ice until you load them.
4. Electrophorese each gel at 20 mA. Stop the gel when the bromophenol blue is almost at the bottom.
5. Transfer the proteins from one of the gels to nitrocellulose using the Transphor unit (Hoeffer). See below for Transphor unit instructions.
6. After the proteins have been transferred, label the blot so the proteins can be visualized in the correct orientation.
7. Use the autoclaved pipet tip box top as a tray for the immunoblot. Place the tray on the shaker at room temperature.
8. Block the blot by immersing it in Genius blocking solution for 30 minutes.
9. Incubate the blot in 20 ml of BSA blocking solution containing 0.2 ml of mAb 5-6 (diluted 100-fold) for 1 hour. Remove the blot and place it in a new tray.
10. Rinse the blot 10 minutes in 1x TTBS, then for 10 minutes in 1x TBS.
11. Incubate in 20 ml of BSA blocking solution containing 10 μl of 2000-fold diluted anti-mouse peroxidase IgG for 1 hour. Remove the blot and place it in a new tray.
12. Repeat step 10.
13. Stain the blot by immersing it in the 4-chloro-1-napthol solution.
14. Stop the staining reaction by rinsing the blot with distilled water. Dry the blot between filter papers.
15. Compare the immunoblot to the bromophenol blue-stained protein gel.

3. Electrophoretic Transfer of Proteins to Nitrocellulose (from # 5 above)
Instructions for the Hoeffer TE 22 Mighty Small Transphor Unit

1. Attach the tubing to the ports of the heat exchanger in the base of the unit for circulation of cold water or 50% ethylene glycol-water. Circulate coolant at a pressure no greater than 18 liters/minute, or 10 psi. A red pressure relief valve is located between the water inlet and outlet ports at the base of the buffer chamber. The purpose of the relief valve is to prevent breakage of the alumina cooling plate in the event that pressure in the cooling chamber builds due to obstruction of the water outlet port. The relief valve is set to relieve internal pressure at 15 psi. Attach a piece of tubing to the barbed fitting on the valve and run the tubing to a drain. It is most unlikely that the pressure relief valve will be needed. It is, however, an additional safety precaution to protect the instrument and prevent an accident in the lab.
2. Carefully place a magnetic stirring bar in the buffer chamber. Do not drop the bar into the buffer chamber. This may crack the aluminum plate. Set the unit on top of a magnetic stirring motor. Even if cooling is not critical for your system, stirring is important to prevent buffer depletion at the electrodes.
3. Add just under 1 liter of transfer buffer to the chamber. You may want to precool the buffer. The level of buffer should match a pair of nicks in the red side panels, as viewed through the clear front of the chamber. After the cassettes have been inserted, the buffer should cover both the electrodes and cassettes without touching the banana plugs.
4. Pour transfer buffer to a depth of at least one inch into a tray big enough to hold the cassette. The cassette should be loaded under buffer, as follows:
   a. Place one half of the opened cassette in the tray of buffer. For negatively charged macromolecules (i.e., nucleic acids and most proteins), submerse the light-colored half of the cassette.
   b. Place one foam sponge on the submersed cassette half and press down gently to force out trapped air bubbles.
   c. Place one sheet of blotter paper on the sponge.
   d. Place a 7 x 8-cm transfer membrane on top of the blotter paper, fully within the open grid of the cassette half.
   e. Place the gel, which has been subjected to electrophoresis and then equilibrated with transfer buffer, on top of the membrane, within the open grid. Roll a glass pipet or test tube gently over the gel to expel trapped air bubbles between the membrane and gel; no transfer occurs in areas where air bubbles are trapped.

f. On top of the gel, add another piece of blotting paper and another sponge. Expel any trapped air bubbles and close the cassette.

5. Lift the loaded cassette out of the buffer tray and insert it into one of the sets of vertical slots in the red end plates of the Mighty Small Transphor buffer chamber. For only one or two gels, use the central cassette positions. Each cassette should be inserted with the hinges on top, and with the light-colored cassette half facing the positive electrode, and the dark-colored half facing the negative electrode. When so positioned, with the gel in each cassette on the same side of the membrane, all macromolecular species of the same net charge will thus migrate from the gel toward the membrane when the electrical field is applied. (In most situations, the macromolecules are negatively charged and migrate toward the positive electrode.)
6. The cassette with its packing should hold the gel in firm contact with the transfer membrane without squeezing the gel. If the packing seems loose, add an additional sheet of blotter paper in step 4; if the packing seems too tight, replace the upper sponge with a sheet of blotter paper. However, the thickness of at least two sheets of blotter paper between the gel and the cassette is necessary to ensure an even current flux. Otherwise, you may see a grid pattern from the cassette on the transfer membrane after it is stained. For most gels, the symmetrical arrangement of sponges and blotter paper should produce the right pressure without current distortion.
7. After inserting a cassette, tap it a few times to shake out air bubbles introduced in passage from the buffer tray to the tank. A few bubbles left in the sponges should not interfere with the transfer.
8. Check the buffer level to make sure it is at least a centimeter below the banana plugs, as these will corrode if exposed to buffer.
9. Place the lid on the chamber so that the red lead connects with the positive electrode, and the black lead with the negative electrode, as determined by the transfer conditions and the way you have oriented the gels and the membranes. Remember that the direction of migration depends on both the sample being transferred and the pH of the transfer buffer. If the species of interest are negatively charged in this buffer, the anode (+) should be on the membrane side of the cassette (light-colored cassette half), and the cathode (-) on the gel side (the dark-colored cassette half), so that the sample will migrate from the gel toward the membrane. Most proteins, whether or not they are denatured with SDS, migrate toward the anode in the Towbin Tris/glycine/methanol buffer system.
10. Connect these leads to the corresponding terminals of your power supply. Cooling is always recommended. The power conditions you use depend in part on the power supply available. Transfers are frequently carried out in the constant voltage mode; as heat is released, however, the temperature of the buffer tends to rise, and the current increases. If the current exceeds 500 mA, the voltage should be turned down. (Do not carry out long runs, even at lower voltages, without supervision.)
    We recommend that you carry out transfers in constant current mode. With the Towbin buffer system, and with watercooling to about 15 °C, the temperature should not rise significantly when carrying out a protein transfer at 200 mA constant current.

NOTE: Transfer of nucleic acids requires more attention than the transfer of proteins. The higher conductivity of the buffer requires a higher constant current setting to obtain a comparable voltage gradient. To keep the voltage nearly constant throughout the run, it may be desirable to run the transfer at a lower temperature (4°C). A constant current setting of about 500 mA will give good transfer in two hours.

NOTE: Electrophoretic transfers are performed at high currents, with considerable heat generated during a run. Because the plastics used to make the Mighty Small Transphor can warp at high temperatures, do not allow the buffer temperature to exceed 60°C.

11. When the transfer is complete, turn off the power and remove the lid from the unit. Use the plastic hook provided with the unit to lift out the cassettes. Open them carefully, remove the gels and transfer membranes, and discard the blotter paper. The foam sponges may be reused indefinitely.
    Although transfer buffer may be used several times, it should not be stored in the unit. If you are not doing another run immediately, pour out the buffer and rinse the chamber, with electrode panels in place, in distilled water. If you plan to do a subsequent transfer in the same buffer, allow the buffer to cool to room temperature between runs. Otherwise, the buffer may overheat and damage the unit.

Additional Report Requirement:

1. Immunoblot.

References:

1. Towbin, H., Staehelin, T. and Gordon, J. (1979) Electrophoretic transfer of proteins from polyacrylamide gels to nitrocellulose sheets: procedure and some applications. Proc. Natl. Acad. Sci. U.S.A 76: 4350-4354.
2. Gershoni, J. M. and Palade, G. E. (1983) Protein blotting: principles and applications. Anal. Biochem. 131: 1-15.
3. Miller, D. M., Ortiz, I, Berliner, G. C. and Epstein, H. F. (1983) Differential localization of two myosins within nematode thick filaments. Cell 34: 477-490.
4. Hoeffer TE 22 Mighty Small Transphor Instructions.

PROCESS RATIONALE

IMMUNOBLOTTING

Molecular Probes, Eugene, OR, ph. (503) 465-8300. (Adapted)

1. ELF and Alkaline Phosphatase Immunoblotting

Molecular Probes has made a major breakthrough in the development of substrates that yield fluorescent precipitates at the site of enzymatic activity – a process we call Enzyme-Labeled Fluorescence (ELF™). This *substrate* is a derivative of a highly *insoluble heterocyclic ring molecule containing a hydroxyl group* that forms an intramolecular hydrogen bond (Figure 1). Molecular Probes scientists recognized that if they *attached a phosphate* to this hydroxyl group, they could effectively disrupt the intramolecular hydrogen bond, *creating a soluble phosphatase substrate* with spectral properties that are significantly different from the unmodified molecule. Indeed, the *substrate* contained in our ELF kits *fluoresces only weakly in the blue range*, but *once its phosphate is enzymatically removed, the resulting product exhibits an intense yellow-green fluorescent precipitate*, at the site of phosphatase activity.

1.1 ELF-AP Immunohistochemistry Kit. Molecular Probes has developed and optimized a method for using the ELF alkaline phosphatase substrate to detect antigens in tissue sections. Our new ELF-AP Immunohistochemistry Kit contains the key reagents and protocols for immunohistochemical applications.

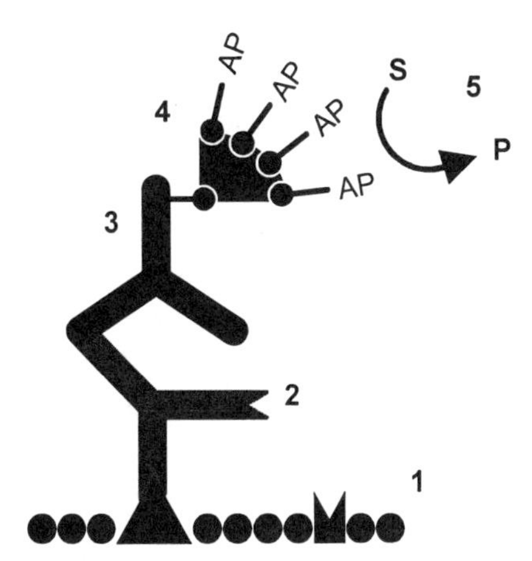

Figure 1 - Amplified detection of membrane bound antigens. **1.** Nonfat dry milk blocks unoccupied sites on the membranes. **2.** Primary antibody to a specific antigen is incubated with the membrane. **3.** Biotinylated antibody is added to bind to the primary antibody. **4.** Streptavidin-biotinylated AP complex is incubated with the membrane. **5.** color development reagent is then added to the blot; amplified signal at the site of the complex is observed by formation of a colored precipitate.

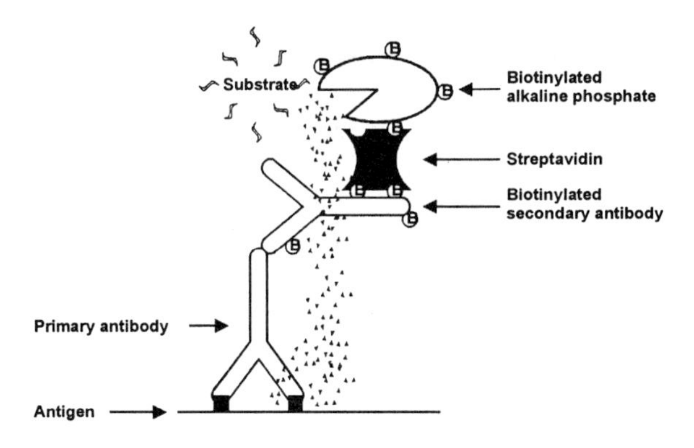

Figure 2 - Schematic diagram of the methods employed in alkaline phosphatase-mediated immunohistochemical techniques using Molecular Probes ELF-AP Immunohistochemical Kit.

A schematic of the method recommended in this kit is shown in Figure 2. Streptavidin is used to link a biotinylated secondary antibody to the biotinylated alkaline phosphatase – a common immunohistochemical technique for optimizing tissue penetration.

2. Amplified Alkaline Phosphatase Immuno-Blot Kits

BioRad, Hercules, CA, ph. (510) 741-1060. (Adapted)

Increased detection sensitivity is achieved with the new Amplified Alkaline Phosphatase Immuno-Blot assay kit for Western blotting. This kit employs specific streptavidin-biotin binding properties to detect as little as 10 pg of membrane-bound protein, a 5-fold increase in sensitivity over other blotting systems.

1. *Superior antibody* is adsorbed to remove cross-reacting antibodies to human immunoglobulins and human serum; results in improved signal-to-noise ratios.
2. Streptavidin enables *more specific binding* than avidin, reducing nonspecific binding and eliminating high backgrounds on your blots. (For further information, request bulletin 1600.)

Process Rationale

Western Blots Using Stained Protein Gels

Thompson, D. and Larson, G. (1992) BioTechniques 12: 656-658. (Adapted)

1. Abstract

A general method is described for the electrophoretic transfer of proteins from stained gels to membranes and subsequent Western detection of specific proteins on the stained membranes. Proteins are separated by sodium dodecyl sulfate (SDS) polyacrylamide gel electrophoresis (PAGE), and the gels are stained using either of two different methods followed by electrophoretic transfer to nitrocellulose or Immobilon-P™ membranes. The transferred proteins remain stained during immunodetection, providing a set of background markers for protein location and size determination.

2. Introduction

SDS-PAGE is a standard technique for the separation and analysis of proteins. Following electrophoresis, gels are typically stained with dyes such as Coomassie Brilliant Blue in order to visualize the proteins bands. Gels can also be stained during electrophoresis using Chroma-PhorT™ stain. Although protein staining is usually the first choice for locating protein bands, it is not the most sensitive method available. Western blotting, using specific antibodies, is a more sensitive method for detecting proteins of interest.[2,5] Western blots can be used to confirm the identity of stained protein bands or to visualize proteins present in amounts not detectable by protein staining. The procedure for Western blotting often involves running duplicate gels in order to have a separate gel for staining and one for blotting. We describe a simple method for the electrophoretic transfer of proteins from stained gels to membranes that are then used for Western detection of specific proteins, eliminating the need for duplicate gels. The transferred proteins remain stained during immunodetection and serve as background markers for protein location and size determination.

3. Materials and Methods

3.1 Protein Separation and Staining. β-Galactosidase and luciferase proteins were synthesized in standard *in vitro* transcription/ translation reactions using *E. coli* S30 extracts (Promega, Madison, WI). For controls, increasing amounts of pure β–galactosidase and luciferase protein were added to S30 extract, and the proteins were separated on 10% SDS-PAGE, 0.75-mm mini-gels according to Laemmli.[3] Electrophoresis was carried out in a Mini-PROTEAN™II (Bio-Rad, Richmond, CA) apparatus in buffer containing 25 mM Tris-HCl, pH 8.3, 192 mM glycine, and 0.1% SDS. Gels were run at 180 V for 45 minutes to 1 hour. The gels were stained after electrophoresis with Coomassie Blue (0.25% Coomassie Brilliant Blue R-250 in 25% isopropanol, 10% acetic acid), or during electrophoresis using the ChromaPhor visualization system (Promega),[4] followed by destaining in 10% acetic acid overnight.

3.2 Electroblotting. Prior to transfer, the stained gels were soaked in 50 mM Tris-HCl, pH 7.5, 1% SDS for 1 hour followed by 15 minutes in transfer buffer (25 mM Tris-HC1, 192 mM glycine and 20% [v/v] methanol). The stained proteins were transferred from the gels to either nitrocellulose (BA85, Schleicher & Schuell, Keene, NH) or Immobilon-P™ (Millipore, Bedford, MA) membranes using a Mini Trans-Blot (Bio-Rad) apparatus. Transfer was carried out at 100 V for 1.5 hours at 4°C.

3.3 Immunodetection. After transfer, the membranes with the stained proteins present were blocked in 5% bovine serum albumin (BSA)/TBS (100 mM Tris-HCl, pH 7.4, 0.9% NaCl) for at least 1 hour and subsequently washed (3 x 5 minutes) in 0.1% BSA/TBS. The membranes were then incubated for 1 hour in a 1:5000 dilution of β-galactosidase monoclonal antibody (Promega) or luciferase polyclonal antibody in Ab buffer (1% BSA/0.05% Tween 20/TBS). Membranes were again washed (3 x 5 minutes) in 0.1% BSA/TBS, followed by incubation for 45 minutes to 1 hour with a 1:5000 dilution (in Ab buffer) of alkaline phosphatase-conjugated anti-mouse or anti-rabbit antibody (Promega). After a final wash (3 x 5 minutes) in 0.l% Tween 20/TBS, color was visualized with NBT/BCIP (nitro blue tetrazolium/5-bromo-4-chloro-3-indolyl-phosphate). (Promega or Vector Labs, Burlingame, CA).

4. Results and Discussion

Upon drying, the nitrocellulose membrane retains more of the color from the stain and provides sharper contrast between the immunodetected band and the background, while the stain that remains on the Immobilon-P™ membrane, though still visible, tends to be less intense. This difference is a result of the membrane used, and not the staining method, because Coomassie stain is also more intense on nitrocellulose. Although the intensity of the background stain is different on the membranes, the immunodetection sensitivity levels are the same. Under either set of conditions, we were able to immunodetect less than 25 ng of protein. Although the limits of detection have not been fully explored, it is possible that the stain may interfere with antibody binding. Other factors that affect detection sensitivity are the efficiency of electrophoretic transfer and binding of proteins to the membrane.

Transfer efficiency will vary depending on a number of different factors such as protein size, buffer conditions, time of transfer, and nature of the specific proteins being transferred.[1] We routinely separate and blot proteins, from *E. coli* S30 *in vitro* transcription/translation reactions, and see relatively efficient transfer over a broad size range based on staining after transfer. We can detect less than 25 ng of starting protein in a background of S30 proteins after Western blotting. We are also able to detect synthesized protein in 5 ml of a standard 50 ml S30 *in vitro* transcription/translation reaction.

Although only two proteins were presented as examples in this report, we have found this method of electrophoretic transfer of proteins after gel staining to be a general method regardless of whether the gel was stained with Coomassie Brilliant Blue R-250 or Chroma-Phor™ stain and whether the stained gels were transferred to Immobilon-P™ membranes or nitrocellulose. This technique has been used to detect a variety of proteins, synthesized in S30 *in vitro* transcription/ translation reactions. Examples include β-galactosidase, luciferase, chloramphenicol acetyl transferase (CAT), and proteins from crude extracts of *Salmonella* outer membranes. Peptide fragments and proteins, cross-linked in the gel with formaldehyde, have also been detected by Westerns after transfer from stained gels. We have found this method useful for confirming the identity of stained protein bands as well as identifying the specific proteins in a heavy background of total protein – as we see in S30 *in vitro* transcription/translation reactions.

Although proteins are thought to be fixed in the gel after the staining and destaining process, we have found that stained proteins can be transferred to membranes. The membranes are then used for Western detection of specific proteins, eliminating the need to run a separate unstained gel for Western blotting. In addition, the proteins remain stained during immunodetection and provide a set of background markers for protein localization and size determination.

References:

1. Bers, G. and D. Garfin. 1985. Protein and nucleic acid blotting and immunobiochemical detection. BioTechniques 3: 276-288.
2. Burnette, W. M. 1981. Western blotting: Electrophoretic transfer of proteins from sodium dodecyl sulfate-polyacrylamide gels to un-modified nitrocellulose and autoradiographic detection with antibody and radioiodinated protein. Anal. Biochem. 112: 195-203.
3. Laemmli, U. K. 1970. Cleavage of structural proteins during the assembly of the head of bacteriophage T4. Nature 227: 680-685.
4. Shultz, J. and G. Larson. 1991. ChromaPhor protein visualization system. Promega Notes, No. 29.
5. Towbin, H., T. Staehelin and J. Gordon. 1979. Electrophoretic transfer of proteins from polyacrylamide gels to nitrocellulose sheets: Procedure and some applications. Proc. Natl. Acad. Sci. U.S.A. 76: 4350-4354.

UNIT 10: COMBINATORIAL BIOCHEMICAL TECHNOLOGY

1. Overview

Most of the literature passages in this section were abstracted from *Methods in Enzymology*, Volume 276: *Combinatorial Chemistry* (Abelson, J. N., Ed.), Academic Press, San Diego, CA, 1996, which addresses molecular biological approaches to combinatorial selection strategies. More traditional chemical problems have also been solved using analogous approaches, as described in other articles from the volume that were not included here.

To illustrate the breadth of this emerging field, consider two recent *Chemical and Engineering News* articles. The first (Borman, 1999) describes problems that have both molecular biological and chemical aspects. The second (Dagani, 1999) describes chemical and materials science problems that have been solved using combinatorial approaches. These applications proceed via steps analogous to those used in the molecular biological selection scenarios. Comparing these analogies illustrates generalized features of all combinatorial programs and highlights key areas requiring improvement. This is necessary to fully appreciate the capabilities promised by such approaches in drug and materials discovery.

The following passages from the seminal article by Gold and colleagues (Irvine *et al.*, 1991) describe the general steps in the Systematic EvoLution by EXponential (SELEX) enrichment method with RNAs targeted by RNA-binding proteins. Survivor RNAs are selectively bound in competitive reactions using sequence pools containing varying degrees of sequence randomization.

"After mixing a protein with a pool of RNA molecules in the appropriate buffer solution, protein-bound RNA molecules are separated from unbound RNA molecules by passing the solution through a nitrocellulose filter. Most protein sticks to the filter, including protein-RNA complexes, while most unbound RNA molecules wash through. RNA molecules collected on the filter are then eluted and cDNA copies are amplified by PCR, followed by *in vitro* transcription to restore the RNA pool to a concentration high enough for the next round. These selection and amplification steps are repeated until the desired level of enrichment is attained for RNA sequences that bind well to the protein."

"Any method of partitioning RNA, DNA or peptide sequences using any target can be utilized (in principle) in an enrichment process like SELEX. In general, the three main steps involved in each round are: (*1*) selection of ligand sequences that bind to a target, (*2*) partitioning of bound and unbound ligand sequences; and (*3*) PCR amplification of ligand sequences in the desired fraction."

"Particular variations in experimental conditions can dramatically alter the outcome of SELEX experiments. Since each round of SELEX with available methods is labor-intensive, and a typical SELEX experiment requires several rounds to complete, employing the current laboratory techniques to test an extensive range of conditions would require an excessive amount of time and effort. Alternatively, by first applying mathematical analysis and computer simulation in order to understand enrichment processes like SELEX better, available laboratory results can be used to test predictions made from first principles critically."

"The correlations (between experiment and simulations) ... predict that the equilibrium mechanism proposed for SELEX is sufficient to explain the high levels of enrichment attained in the laboratory after just a few rounds. ... Further analysis ... should provide accurate predictions for the levels of enrichment to expect under different conditions."

2. Affinity Maturation of Phage-Displayed Peptide Ligands (Yu, J. and Smith, G. P., pp. 3-27).(extracted)

Many experiments in this volume start with large libraries of random amino acid or nucleotide sequences of a certain length from which a tiny subset is selected according to some criterion of "fitness" – most often, affinity for a chosen target receptor. In most cases, the library represents sequences of the same length exceedingly sparsely. Even the very best (fittest) sequence in a sparse initial library (limited permutations) may be much inferior to the best sequence of the same length (of all possible permutations).

If the sequences are capable of heritable mutation – phage display and random RNA and DNA libraries fall into this category – the problem of sparseness might be addressed by encouraging fitter sequences to "evolve" from parent sequences in the initial library. This sort of artificial evolution is exemplified by the "greedy" strategy: Step A, from the initial library select the very best sequence; call this the "initial champion." Step B, mutagenize the initial champion randomly, producing a "clan" of closely related mutants. Step C, from that clan select the mutant with the very best fitness. Step D, repeat steps B and D as needed until an optimal ligand is found. Each round of selection thus selects "greedily" for the very best sequence available in the current population.

In order thus to broaden the search for fitter sequences, the stringency (fitness threshold) can be reduced in the early rounds of selection, so as to include sequences somewhat inferior to the initial champion: Step A', from the initial library select a mixture of sequences with diverse fitnesses (ideally, above a certain threshold). Step B', mutagenize the entire population of selected sequences to produce many clans of mutants. Step C', from those clans select a mixture of sequences with diverse finesses (ideally, above a slightly higher threshold than in Step A'). Step D', repeat steps B' and C' as often as desired, possibly increasing the stringency of selection with succeeding rounds. Step E', after the final round of mutagenesis, stringently select the very best sequence in the current population.

Alternating nonstringent selection with mutagenesis in this way makes it possible to discover "dark horses": sequences in the initial library that are inferior to the initial champion, yet can be mutated to even higher fitness than the champion. A dark horse will usually lie in a different neighborhood than the initial champion, since in most cases two sequences in the same small neighborhood will be able to mutate to the same local optimum.

3. Phage Display of Proteases and Macromolecular Inhibitors (Wang, C.-I., Yang, Q. and Craik, C. S., pp. 52-57) (extracted)

The challenge for locating the primary determinants of protein function has stimulated two innovative advances in the field: (*1*) the generation of molecular repertoires with sufficient diversity for altered function to be represented; and (*2*) the development of screens or selections to effectively sort through the large repertoires for desired properties. Phage display is an *in vitro* selection technology that meets these criteria. This technology allows for a foreign protein or peptide to be displayed on the surface of filamentous phage, linking the phenotype of the phage to its genotype. This is accomplished by fusing the foreign gene to the gene coding for either the major (pVIII) or minor (pIII) coat proteins of the filamentous phage. Once expressed, the fusion proteins are incorporated onto the surface of the phage particles during phage assembly. The gene fusion can then be used to introduce mutations into the coding sequences of the target protein or peptide. This results in a library of variants, each displayed on the surface of a phage particle that contains the DNA encoding the variant.

The phage particles in the library displaying the variants of interest must be identified. The method most commonly used for this purpose is referred to as "biopanning," analogous to the process of "panning for gold." A protein or ligand that will preferentially bind to the phage displayed variant is immobilized on an insoluble matrix. The library of variants is then passed over this matrix and the phage encoding the protein of interest is enriched because it binds to the matrix through the foreign protein domain. The phage particles are stable in a wide range of pH and salt concentrations, allowing flexible conditions for washing and elution procedures. The bacteriophage eluted from the matrix are then used to infect bacteria for subsequent amplification of the enriched variants.

The ability to carry out multiple biopanning cycles provides a powerful enrichment procedure for phage display. Even if the enrichment of a single round of biopanning is not very significant, strong binders may still be identified through repetitive rounds. When productive, this cycle greatly increases the sensitivity of the *in vitro* selection and enables rare individual clones with desired binding properties to be isolated from large, random libraries with up to $10^{8\text{-}9}$ individual clones. This approach was initially applied to libraries of short peptides to isolate antigenic epitopes. As it became clear that proteins could be expressed on phage without seriously altering their activity, the technology entered the realm of protein engineering.

4. *In Vitro* Genetic Analysis of RNA Binding Proteins Using Phage Display Libraries (Laird-Offringa, I. and Belasco, J. G., pp.149-170).(extracted)

In vitro selection of RNA-BP variants that bind tightly to a target RNA. (*1*) A combinatorial library of mutagenized cDNA fragments encoding the RNA-binding domain of the protein of interest is inserted into a phagemid vector such that the resulting proteins become fused to a signal peptide, a $(His)_6$ tag, a Myo epitope tag, and the gene III coat protein of filamentous bacteriophage fd. The library is transformed into an *E. coli* host. (*2*) A pool of phage displaying a library of RNA-BP variants is generated by infecting the host with helper phage. (*3*) This phage pool is allowed to bind to the RNA target annealed to a biotinylated oligo-deoxynucleotide. RNA-bound phage are captured on the streptavidin-coated beads. (*4*) After washing, phage bound to the beads are used to infect *E. coli*, allowing the cycle of amplification and selection to be repeated.

In our genetic system depicted in Figure 1, the selective step takes place *in vitro* and involves binding the RNA-BP to an RNA target immobilized on beads. To amplify the bound protein molecules, it is necessary to link each protein to its corresponding gene. We chose to do this by phage display, a technique in which a foreign protein is displayed on the exterior of filamentous bacteriophage that carry the corresponding gene in their interior. Using such fusion phage, we have been able to physically couple RNA-BPs to their genes, thereby allowing RNA-BP variants with desired binding characteristics to be alternatively selected and amplified from a combinatorial phage display library.

5. Cell-Free Synthesis of Peptide Libraries Displayed on Polysomes (Mattheakis, L. C., Dias, J. M. and Dower, W. J., pp. 195-210)

We have developed an *in vitro* peptide expression system which displays the peptide library on polysomes. This system, which avoids bacterial transformation, can generate libraries that are several orders of magnitude larger than those of cell-based systems. In addition, the diversity of sequences synthesized on polysomes may also be greater since secretion, phage assembly, and other cellular processes are not required for peptide display *in vitro*.

A summary of the method begins with the construction of a DNA. The library consists of random peptide-coding sequences that are fused in-frame to the 5' end of a spacer sequence. The DNA library is incubated in a DNA-dependent *in vitro* transcription/translation system, and polysomes are isolated by high-speed centrifugation. The polysomes, consisting of nascent peptides linked to their encoding mRNAs, are screened by affinity selection of the nascent peptides on an immobilized target. The polysome-bound mRNA is recovered, copied onto cDNA, and amplified by polymerase chain reaction (PCR) to produce template for the next round of *in vitro* synthesis and selection. After each round, the enriched peptides are identified by cloning and sequencing a portion of the amplified template and their binding specificities are determined using various assays. The following describes these steps in detail and discusses applications of the technology to ligand discovery.

6. A SELEX Primer (Fitzwater, T. and Polisky, B., pp. 275-301)

The SELEX (System Evolution of Ligands by EXponential enrichment) combinatorial chemistry process is a drug and diagnostic discovery technology that permits isolation of oligonucleotide ligands from large (10^{15} members) random-sequence single-stranded oligonucleotide populations that bind molecular targets with high affinity and specificity. For a review of SELEX technology and its applications, see Irvine *et al.* (1991). An initial DNA oligonucleotide population is generated synthetically and can serve as a transcription template to generate RNA pools for the RNA SELEX process. The SELEX procedure can also be carried out with DNA derived directly from solid-phase synthesis (DNA SELEX). A key step in the SELEX process is "partitioning," the separation of oligonucleotides with affinity for the target from those without substantial affinity. For many protein targets, partitioning can be carried out using nitrocellulose filtration to capture all protein and associated oligonucleotides. Techniques using this partition method are described in this chapter. Other partitioning processes used to separate bound from free oligonucleotides include immobilization of the target on a solid matrix or immunoprecipitation of the target. Immobilization of targets in the SELEX process is often an attractive strategy as it can facilitate "counter-SELEX technique" elution schemes using soluble compounds that are structurally related to but distinct from the desired target to improve recognition specificity of the "winning" oligonucleotide family.

The RNA pool can be "front-loaded" with modified nucleoside triphosphates such as 2'-fluoro- or 2'-amino-CTP and -UTP resulting in oligonucleotide pools that are resistant to endonuclease digestion or with 5-iodo- or 5-bromo-UTP permitting a search for oligonucleotides potentially capable of covalently cross-linking to the target. (For a review of nucleoside modifications and their potential role in SELEX process, see Eaton and Picken). Iterative cycling of selection and amplification allows isolation of a small number of oligonucleotide sequences that interact specifically with the target.

The range of targets against which the SELEX technique has been successful is large. Targets include proteins, peptides, amino acids, small organic molecules, and polysaccharides. It is clear that high-affinity ligands can be isolated against proteins not known to interact "naturally" with nucleic acids, including proteases, phospholipases, growth factors, antibodies, and hormones. Finally, oligonucleotides selected from random pools can have catalytic properties similar to classical enzymes.

6.1 Evaluating Affinity of Random Pool for Target: Theoretical Aspects. Irvine *et al.* have developed a quantitative analysis of the parameters that have the greatest impact on the outcome of the SELEX process. The aim of this study was to establish guidelines for optimizing enrichment of the highest affinity members of the initial population even when they were present in a single copy, in a minimum number of rounds. Key parameters affecting the number of rounds needed to complete the SELEX process include target and nucleic acid concentrations, the K_d of the initial random pool for the target, the efficiency of the partition system, the background nonspecific binding, and the ratio of high- and low-affinity oligonucleotides in the initial population. As pointed out by Irvine *et al.* (1991), the probability of recovering the highest affinity oligonucleotides in a population increases with the number of such oligonucleotides in the population, with the ratio of their K_d to bulk K_d, and with the total amount of target used in the round. However, the use of high target concentrations throughout selection has the effect of reducing binding competition among the oligonucleotide population, thus reducing the enrichment per round and dramatically increasing the number of rounds required. The optimum strategy utilized high target levels in initial rounds to ensure the capture of high-affinity binders that my be rare in the population, followed by rounds in which target levels are reduced to ensure competition and rapid enrichment of the best binders.

Obviously, in most cases one does not know the K_d of the highest affinity oligonucleotide in the starting population. Here one desires a strategy that ensures adequate enrichment regardless of the difference in K_d between the bulk population and the best binder. Irvine *et al.* (1991) show that the "near optimal" target concentration can be determined by the following relationship:

$$T_{no} = [(K_d) + (\text{oligo})](0.7)\left(\frac{BG}{CP}\right)^{2/3}$$

where T_{no} is the "near optimal" target concentration, K_d is the bulk oligonucleotide K_d for the target, BG is the fraction of free RNA that partitions as nonspecific background, *oligo* is the concentration of total oligonucleotide nucleotide, and CP is the oligonucleotide fraction that partitions due to binding to the target. When the total oligonucleotide concentration is substantially greater that the bulk K_d, the K_d term in this equation can be ignored.

An alternative rule of thumb for determining target concentrations is to use the target level that binds 5-10% of the total ligand pool. Computer simulations indicate that in this situation the number of rounds required to enrich the best binding oligonucleotide from one copy in 10^{15} to at least half the pool is within seven rounds of that required when the target concentration was theoretically optimal. In practice, SELEX process experiments are typically initiated at oligonucleotide:target ratios ranging from 10 to 100. In subsequent rounds, the ratio is increased to promote competition.

6.2 Determination of Dissociation Constant. To determine the dissociation constant between an oligonucleotide and the target, a model describing the stoichiometry of the reaction being observed is required. Initially this is not known so a model using a 1:1 stoichiometry is used as shown:

$$P:R = P + R$$

The dissociation constant (K_d) for this reaction is defined as

$$K_d = \frac{[P][R]}{[P:R]} \quad (1)$$

where $[P]$ is the free protein concentration, $[R]$ is the free oligonucleotide concentration, and $[P : R]$ is the concentration of the complex. The mass balance equations for the protein and the RNA are defined as

$$[P]_t = [P] + [P:R] \quad (2)$$

$$[R]_t = [R] + [P:R] \quad (3)$$

where the subscript t refers to the total input concentration of either oligonucleotide or protein. Using eqs. (1)-(3) and solving for $[P : R]$ yields eq. (4):

$$[P:R] = [P]_t + [R]_t + K_d - \left[\left(-[P]_t - [R]_t - K_d\right)^2 - \left(4[P]_t[R]_t\right)^{1/2}\right]^2 \quad (4)$$

To determine % bound, eq. (4) is divided by $[R]_t$ and normalized by the % bound maximum (max) and % bound minimum (min) as shown in eq. (5):

$$\%\,\text{bound} = \frac{[P:R]}{[R]_t}(\%\,\text{max} - \%\,\text{min}) + \%\,\text{min} \quad (5)$$

Nonlinear least-squares analysis of eqs. (4) and (5) yield best-fit parameters for the K_d, % max, and % min to experimental data. These equation can be input into commercially available programs: **KaleidaGraph**, Synergy Software, 2457 Perkiomen Avenue, Reading, PA 19606; **Prism GraphPad**, 10855 Sorrento Valley Road, Suite 203, San Diego, CA 92121.

7. Affinity Selection-Amplification from Randomized Ribooligonucleotide Pools (Ciesiolka, J., Illangasekare, M., Majerfeld, I., Nickles, T., Welch, M., Yarus, M. and Zinnen, S., pp. 315-335)

7.1 Random Region. When probing RNA function using a random region, it may prove useful to understand the sequence space that can be explored. In particular, the number and the distribution of randomized positions must be decided. This is complicated by the emergent state of the field, in which the size and the nature of the RNA needed for most biochemical functions are still unknown. Typically, the random region consists of a stretch of 25-100 contiguous randomized nucleotides between constant regions to provide primer sites and promoters. It may prove useful for some selections, however, to punctuate the random region with fixed sequence, to link regions into longer randomized sequences, or to insert the randomized positions into a fixed structural context.

7.2 Selections. The choice of an appropriate selection procedure is crucial to the success of a selection-amplification experiment. Although the specific selection protocol must be tailored to the particular task, the procedure should always be selective, specific, and simple. The method should be selective so that molecules that do not meet the selection criteria are eliminated as efficiently as possible and are selected exclusively according to clearly specified criteria. Because the selection process is repeated in every round, sometimes more than once (see below), simplicity is very desirable.

Several different separation techniques have been used in selection-amplification experiments. Affinity chromatography, filter binding, gel mobility shift, and gel electrophoresis are the most commonly used. For selection of RNAs that bind small molecular weight ligands like derivatized amino acids, free amino acids, small organic dyes, GDP ATP, GDP and arginine simultaneously, antibiotics, the alkaloid theophylline, cyanocobalamin, and the transition state analog of a bridged biphenyl, affinity chromatography was particularly useful. In the case of RNAs that bind proteins or protein cofactors, filter binding has been the technique of choice.

8. *In Vitro* Evolution of Randomized Ribozymes
(Tsang, J. and Joyce, G. F., pp. 410-425)

Darwinian evolution can be performed *in vitro* to obtain ribozymes with optimized function or novel catalytic abilities. Molecules with the desired characteristics are selected from a pool of randomized sequences, amplified and mutated. This process is repeated until the ribozymes of interest are obtained.

The key difference between *in vitro* evolution and *in vitro* selection is that the former involves the repeated introduction of new mutations. Thus, with *in vitro* evolution, even if the desired ribozyme does not exist in the initial pool, suboptimal variants can be selected and, over the course of evolution, gain new mutations that result in the behavior of interest. As a consequence, the number of possible sequences that can be surveyed is greatly increased compared to what can be represented in an initial pool.

There are two types of initial pools. One is derived from a preexisting ribozyme sequence and the other is composed of completely random sequences. The former allows for a thorough search of the sequences that are closely related to a prototype and is particularly useful when optimizing or subtly altering the function of an existing ribozyme. The latter allows for a broader but less comprehensive search of RNA sequences, which may be necessary when attempting to develop ribozymes with entirely new catalytic abilities.

8.1 Creating Pools of Randomized Ribozymes. There are two basic strategies for creating a pool of randomized ribozymes. One strategy employs degenerate, synthetic oligodeoxynucleotides, which can be either directly transcribed to RNA or incorporated into a DNA template that encodes the ribozyme. Such degenerate oligonucleotides can be produced on an automated DNA synthesizer using nucleoside 3'-phosphoramidite solutions that have been doped with a small percentage of the three incorrect monomers, correcting for the differential coupling efficiency of the various phosphoramidites. Another strategy employs an error-prone polymerase to introduce mutations into the ribozyme-encoding DNA template. One such method, mutagenic polymerase chain reaction (PCR), exploits the highly error-prone activity of *Taq* DNA polymerase.

8.2 Randomization Strategies. Our basic strategy for randomizing a ribozyme is to generate a population of variants that is both diverse and comprehensive, i.e., a population containing variants with the greatest possible diversity yet including all possible variants that are closely related to the prototype sequence. For a sequence of length n, mutagenized at a degeneracy d, the average number of mutations per sequence is nd. The mean number of copies per sequence for a sequence that has k mutations is given by $S[P(k, n, d)]/N(k)$, where S is the size of the population, P is the proportion of the population having k mutations, and N is the number of distinct sequences having k mutations. P and N are given by the equations

$$P(k,n,d)=\left[\frac{n!}{(n-k)!k!}\right]d^k(1-d)^{n-k} \quad (1)$$

$$N(k)=[n!(n-k)!k!]3^k \quad (2)$$

In one study, for example, we wished to randomize a total of 140 positions within an existing ribozyme (n = 140). For a population size of 20 pmol ($S = 1.2 \times 10^{13}$ molecules), we chose a degeneracy of 5% (d = 0.05). In this way, more than half of the mutagenized population contained seven or more mutations, yet all possible 1-, 2-, 3-, 4-, and 5-mutation variants were represented. If we had introduced mutations at a lower frequency, say 2.5% (d = 0.025), then variants with 4 or fewer mutations would have been represented comprehensively, but the vast majority of individuals would have remained very closely related to the original ribozyme. If we had introduced mutations at a higher frequency, say 10% (d = 0.10), then the variants would have contained, on average, a high number of mutations, but sequences that are closely related to the original ribozyme would not have been well represented. The degree of randomization should be based on the level of confidence one has that the starting ribozyme is a good lead compound in the search for the desired catalytic function.

References:

1. Borman, S. (1999) Reducing time to drug discovery, Chem. and Eng. News, Mar. 8: 33-48.
2. Dagani, R. (1999) A faster route to new materials, Chem. and Eng. News, Mar. 8: 51-60.
3. Irvine, D., Tuerk, C. and Gold, L. (1991) Selexion: Systematic evolution of ligands by exponential enrichment with integrated optimization by non-linear analysis. J. Mol. Biol. 222: 739-761.

INNOVATION/INSIGHT

EXAMPLES OF COMBINATORIAL TECHNIQUES

These Science and Technology Concentrates from *Chemical and Engineering News* (C&EN) provide examples of results obtained by combinatorial analysis of RNA populations. The third article illustrates a biotechnological application of antisense RNA.

1. Better Ribozyme Developed by *in vitro* Selection

C&EN, Aug. 3, 1992, p. 24.

An *in vitro* selection technique has been used to obtain variants of a *ribozyme* (catalytic RNA) that use a different metal ion cofactor than the naturally occurring ribozyme [*Nature*, 361: 182 (1993)]. Coordination of Mg^{2+} or Mn^{2+} to the *Tetrahymena* ribozyme is normally required for the ribozyme to catalyze sequence-specific phosphodiester cleavage of an external RNA oligonucleotide substrate. *In vitro evolution* is an iterative process of selecting molecules with the desired activity from a large randomized pool of RNA, amplifying the selected molecules, and mutating them. Niles Lehman and Gerald F. Joyce (Scripps Research Institute) used this technique to obtain a population of RNAs that use sCa^{2+}, in addition to Mg^{2+} or Mn^{2+}, as a cofactor. Joyce's group had earlier used the same technique to find variant forms of *Tetrahymena* ribozyme with altered substrate specificity.

The new findings "extended the range of different chemical environments available to RNA enzymes and illustrate the power of *in vitro* evolution in generating macromolecular catalysts with desired properties," the researchers say.

2. Isolation of RNA Catalysts Supports "RNA-World" Hypothesis

C&EN, Sept. 13, 1993, p. 33.

In a study that sheds light on the way biocatalysts may have evolved, molecular biologists David P. Bartel and Jack W. Szostak of Massachusetts General Hospital, Boston, have isolated a new class of catalytic RNAs from a large pool of random-sequence RNAs [*Science,* 261: 1411 (1993)]. The work buttresses the hypothesis of an "RNA world" in which DNA and proteins did not yet exist. The hypothesis was proposed to solve a quandary about what evolved first – DNA molecules, which are needed to code for proteins, or proteins (such as enzymes), which are needed to catalyze DNA replication. The RNA-world hypothesis suggests that early organisms used RNA both for coding and catalysis, and that DNA and proteins developed later. Bartel and Szostak used a technique called *in vitro* selection to isolate specific RNAs from an initial pool of 10^{15} random-sequence RNAs. The RNAs were selected for their ability to form terminal 3', 5'-phosphodiester bonds with a short piece of RNA in a manner analogous to single-nucleotide additions catalyzed by RNA polymerases. The technique yielded RNAs capable of catalyzing self-ligation reactions. The study supports the RNA-world hypothesis because it demonstrates that RNA is capable of catalyzing this type of RNA ligation. It casts doubt on another proposal of the hypothesis – that a replicase (self-replicating RNA) may have arisen spontaneously from a primitive random pool of RNA – because it shows that the probability of this occurring is very small. But this doesn't doom the hypothesis because replicases could have been generated in some other way.

3. Technique Targets, Cleaves Specific RNA Sequences

C&EN, Feb. 15, 1993, p. 22.

A new technique for targeting and cleaving specific RNA sequences may provide a way to inhibit gene expression for the treatment of human diseases. In the technique, 2'-5' adenylate(A)-dependent RNase, an enzyme activated by oligo(A), is the agent that cleaves RNA. To direct the enzyme to a unique RNA sequence, an oligo(A) is covalently linked to an oligonucleotide whose sequence is antisense to complementary in Watson-Crick base pairing to that of the RNA target. The *antisense oligonucleotide* binds to the RNA target, and the oligo(A) activates the RNase, inducing RNA cleavage. Paul F. Torrence of NIH and co-workers developed the technique and used it to cleave a human immunodeficiency virus messenger RNA *in vitro* [Proc. Natl. Acad. Sci. U.S.A. 90, 1300 (1993)]. "Because 2'-5' A-dependent RNase is present in most mammalian cells," they say, "the control of gene expression based on this technology including therapies for cancer, viral infections and certain genetic diseases can be envisioned."

INNOVATION/INSIGHT

MAKING ANTIBODY FRAGMENTS USING PHAGE DISPLAY LIBRARIES

Clackson, T., Hoogenboom, H. R., Griffiths, A. D. and Winter, G. (1991) Nature 352: 624-628. (Adapted)

1. Introduction

To by-pass hybridoma technology and animal immunization, we are trying to build antibodies in bacteria by mimicking features of immune selection.[1] We recently used fd phage to display antibody fragments fused to a minor coat protein,[3,4] allowing enrichment of phage with antigen.[3] Using a random combinatorial library of the *rearranged heavy* (V_H) *and kappa* (V_κ) *light chains*[5-8] from mice immune to the *hapten* 2-phenyloxazol-5-one (phOx), we have now *displayed diverse libraries of antibody fragments on the surface of fd phage.* After a single pass over a *hapten affinity column*, fd phage with a range of phOx binding activities were detected, at least one with a high affinity dissociation constant, $K_d = 10^{-8}$ M^{-1}.

For the random combinatorial libraries, fdDOG1 RF was extensively digested with *Not*I and *Apa*LI, purified by electroelution[24] and 1 μg was ligated to 0.5 μg (5 μg for the hierarchical libraries) of the assembled single-chain antibody (*scF*$_v$) genes in 1 ml with 8,000 units of T4 DNA ligase (New England Biolabs) overnight at 16°C. Purified ligation mix was electroporated in six aliquots into MC1061 cells[25] and plated on NZY medium[24] with 15 μg ml^{-1} tetracycline, in 243 x 243 mm dishes (Nunc); 90-95% of clones contained *scFv* genes by PCR screening (see legend to Figure 1).

Table 1 - Affinity Selection of Hapten-Binding Phage Clones Binding to phOx

	Pre-column	After First Round	After Second Round	After Third Round
Random combinatorial libraries phOx-immunized mice	0/568 (0%)	48/376 (13%)	175/188 (93%)	-
Unimmunized	-	-	0/388 (0%)	-
Hierarchical libraries				
V_{H-B}/V_k reP library	6/190 (3%)	348/380 (92%)	-	-
V_H-rep/V_{k-d} library	0/190 (O%)	23/380 (7%)	-	-
Fractionation of V_{H-B}/V_{k-d} and V_{H-B}/V_{k-b} phage[a]	88/1,896 (4.6%)	55/95 (57.9%)	1,152/ 1,156 (99.7%)	1,296/ 1,299 (99.8%)
Mixture of clones 44/1,740 (2.5%)[b]				-

[a] Numbers refer to V_{H-B}/V_{k-d}.

[b] Numbers after three reinfections and cycles of growth. This control, omitting the column steps, confirms that a spurious growth or infectivity advantage was not responsible for the enrichment of clone V_{H-B}/V_{k-d}.

Colonies were scraped into 50 ml 2x TY medium[26] and shaken at 37°C for 30 minutes. Liberated phage were precipitated twice with polyethylene glycol and resuspended to 10^{17} *transducing units* (TU) ml^{-1} in water (titrated as in ref. 3). For *affinity selection*, a 1-ml column of phOx-BSA-Sepharose[27] (M. Dreher and C. Milstein, unpublished results) was washed with 300 ml PBS, and 20 ml PBS containing 2% skim milk powder (MPBS). Phage (10^{12} TU) were loaded in 10 ml MPBS, washed with l0 ml MPBS and finally 200 ml PBS. The bound phages were eluted with 5 ml of 1 mM 4-ε-aminocaproic acid methylene 2-phenyl-oxazol-5-one (phOx-CAP). About 10^6 TU of eluted phage were amplified by infecting 1 ml of log phase *E. coli* TG1 and plating as above. For a further round of selection, colonies were scraped into 10 ml of 2xTY medium and then processed as above. For the hierarchical libraries, V_{H-B} and V_{k-d} genes were individually recloned, then assembled with the V_H or V_k repertoires. For the fractionation of clone V_{H-B}/V_{k-d}, 7 x10^{10} TU of phage in the ratio 20 V_{H-B}/V_{k-b} to 1 V_{H-B} /V_{k-d} were loaded onto a phOx-BSA-Sepharose column in 10 ml MPBS and eluted as above. Eluted phage were used to reinfect *E. coli* TG1, and phage produced and harvested as before. About 10^{11} TU of phage were loaded onto a second affinity column and the process repeated to give a total of three column passes. Dilutions of eluted phage at each stage were plated in duplicate and probed separately[24] with oligonucleotides specific for V_{k-b} (5'-GAGCGGGTA ACCACTGTACT) or V_{k-d} (5'-GAATGGTATAGTA CTACCCT). *A second pass enriched for the strong binders at the expense of the weak.* The binders were encoded by V genes similar to those found in anti-phOx hybridomas, but in promiscuous combinations (where the same V gene is found with several different partners). By combining a promiscuous V_H or V_κ gene with diverse repertoires of partners to create hierarchical libraries, we elicited many more pairings with strong binding activities. Phage display offers new ways of making antibodies from V-gene libraries, altering V-domain pairings and selecting for antibodies with good affinities.

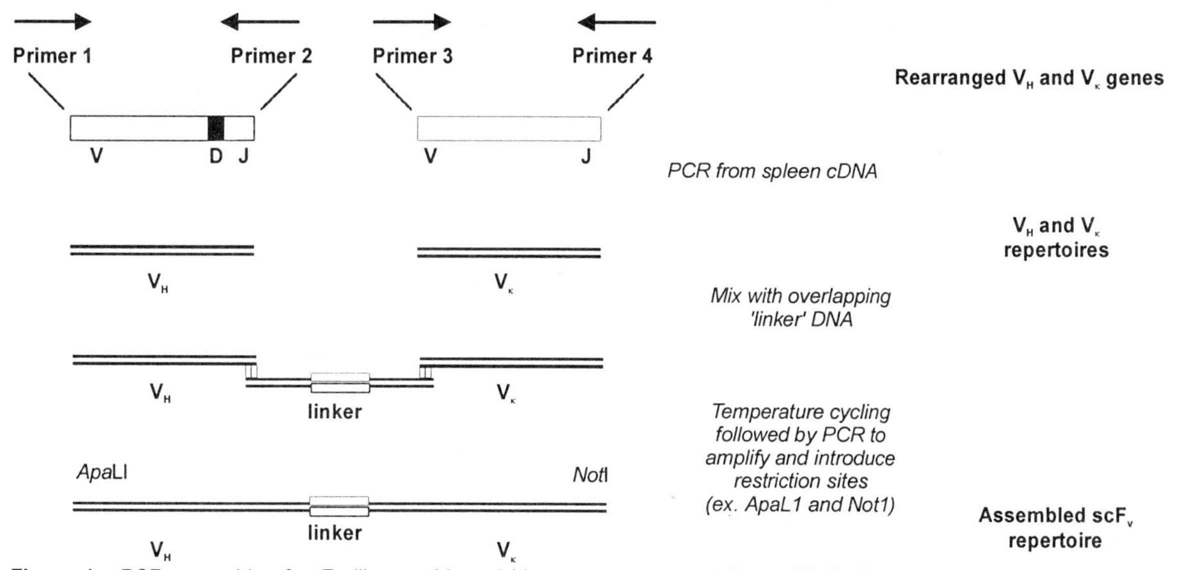

Figure 1 - PCR assembly of scF_V library. V_H and V_κ genes are separately amplified, then mixed with a linker fragment that overlaps them both. The linker (93 base pairs) encodes the short peptide, (Gly, Ser), which links V_H and V_κ in scF_Vs. Cycles of annealing-denaturation, followed by reamplification of the mixture, generate a random combinatorial cassette of V_H and V_κ genes joined in-frame for expression.

We used the polymerase chain reaction (PCR)[9] to amplify the V_H and V_κ genes from the spleen messenger RNA of mice immunized with phOx, and also developed a *"PCR assembly" process* to link these genes together randomly for expression as single-chain F_v (scF_v) fragments[11,12] (Figure 1). The assembled genes were cloned in a single step into the vector fdDOG1 for display as a fusion with the fd gene III coat protein. This initial library of 2 x 10^5 clones seemed to be diverse, and sequencing revealed the presence of most V_H groups[13] and V_κ subgroups[14] (data not shown). None of the 568 clones tested bound to phOx as detected by enzyme-linked immunosorbent assay (ELISA).

The library of phages was passed down a phOx affinity column (Table l), and eluted with hapten. Of the eluted clones, 13% bound to phOx, and ranged from poor to strong binding in ELISA. We *sequenced* 23 of these hapten-binding clones and found eight different V_H genes in a variety of pairings with seven different V_κ genes. Most of the domains, such as V_{H-B} andV_{k-d}, were able to bind hapten with any of several partners.[15] The probability of finding multiple partners for a given chain should depend mainly on the inherent promiscuity of the chain and on the number of available partners and competing chains. Two other examples of promiscuous pairings have been noted in random combinatorial libraries made in λ phage,[6-8] so this may prove to be a feature of small combinatorial libraries from immunized animals.

We *screened for binding of the phage to hapten by ELISA*: 96-well plates were coated with 10 μg ml^{-1} phOx-BSA[27] or 10 μg ml^{-1} BSA in PBS overnight at room temperature. Colonies of phage-transduced bacteria were inoculated into 200 μl 2xTY medium[26] with 12.5 μg ml^{-1} tetracycline in 96-well plates ("cell wells," Nucleon) and grown with shaking (300 rpm) for 24 hours at 37°C. At this stage, cultures were saturated and phage titers were reproducible (10^{10} TU ml^{-1}). Phage supernatant (50 μl), mixed with 50 μl PBS containing 4% skimmed milk powder, was then added to the coated plates.

2. Methods

For the random combinatorial libraries, cytoplasmic RNA was isolated[29] from the pooled spleens of either 5 male BALB/c mice boosted 8 weeks after primary immunization with phOx coupled to chicken serum albumin,[27] or of 5 unimmunized mice. The cDNA was made with avian myoblastosis virus reverse transcriptase (Anglian Biotech)[30] and primers that straddle the junction between the variable and constant regions of γ heavy chains and κ light chains (C. Marks, unpublished data). V_H and V_κ repertoires were amplified from the cDNA with 25 cycles of PCR (94°C for 1 minute, 60°C for 1 minute, 72°C for 2 minutes) using Vent polymerase (New England Biolabs) and the primers VHIBACK[19] and VHIFOR-2.[31]

References:

1. Winter, G. & Milstein, C., Nature 349, 293-299 (1991).
2. Smith, G. P., Science 228,1315-1317 (1985).
3. McCafferty, J., Griffiths, A. D., Winter, G. & Chiswell, D. J., Nature 348, 552-554 (1990).

4. Hoogenboom, H. R. *et al.,* Nucleic Acids Res. 19, 4133-4137 (1991).
5. Huse, W. D. *et al.,* Science 248, 1275-81 (1939).
6. Caton, A. J. & Koprowski, H., Proc. Natl. Acad. Sci. U.S.A. 87, 6450-6454 (1990): correction on 88, 1590 (1991).
7. Mullinax, R. L. *et al.,* Proc. Natl. Acad. Sci. U.S.A. 87, 8095-8099 (1990).
8. Persson, M. A. A., Caothien, R. H. & Burton, D. R., Proc. Natl. Acad. Sci. U.S.A. 88, 2432-2436 (1991).
9. Saiki, R. K. *et al.,* Science 239, 487-491 (1988).
10. Horton, R. M, Hunt, H. D., Ho, S. N., Pullen, J. K. & Pease, L. R., Gene 77, 61-68 (1989).
11. Huston, J. S. *et al.*, Proc. Natl. Acad. Sci. U.S.A. 85, 5879-5883 (1988).
12. Glockshuber, R., Malia, M., Pfitzinger, I. & Pluckthun, A., Biochemistry 29, 1362-1367 (1990).
13. Dildrop, R., Immun. Today 5, 85-86 (1984).
14. Kabat, E. A., Wu, T. T., Reid-Miller, M., Perry, H. M. & Gottesman, K. S., *Sequences of Proteins of Immunological Interest* (Department of Health and Human Sciences, Government Printing).
15. Hudson, N. W., Muogett-Hunter, M., Panka, D. J. & Margolies, M. N., Immunology 139, 2715-2723 (1987).
16. Berek, C., Griffiths, G. M. & Milstein, C., Nature 316, 412-418 (1985).
17. Alzari, P. M. *et al.,* EMBO J. 1 9, 3807-3814 (1990).
18. Perelson, A. S., Immunol. Rev. 110, 5-33 (1989).
19. Orlandi, R., Gussow. D. H., Jones. P. T. & Winter, G., Proc. Natl. Acad. Sci. U.S.A. 86, 3833-3837 (1989).
20. Bass, S., Greene, R. & Wells, J. A., Proteins 8, 309-314 (1990).
21. Cwirla, S. E., Peters, E. A., Barrett. R. W. & Dower, W. J., Proc. Natl. Acad. Sci. U.S.A. 87, 6378-6382 (1990).
22. Devlin, J. J., Panganiban, L. C. & Devlin. P. E., Science 249, 404-406 (1990).
23. Scott. J. K. & Smith, G. P., Science 249, 336-390 (1990).
24. Sambrook, J., Fritsch, E. F. & Maniatis, T., *Molecular Cloning: A Laboratory Manual* (Cold Spring Harbor Laboratory, New York, 1989).
25. Dower, W. J., Miller, J. F. & Ragsdale, C. W., Nucleic Acids Res. 16, 6127-6145 (1988).
26. Nagai, K. & Thogersen, H. C., Methods Enzymol. 153, 461-481(1987).
27. Makela, O., Kaartinen, M., Pelkonen, J. L. T. & Karjalainen, K., J., Exp. Med. 148, 1644-1660 (1978).
28. Gibson. T., Thesis, Univ. Cambridge (1984).
29. Gherardi, E.. Pannell, R. & Milstein, C. J., Immunol. Methods 126, 61-68 (1990).
30. Clackson. T., Gussow, D. & Jones, P. T. in *PCR a Practical Approach* (Eds., McPherson, M. J., Taylor G. R. & Quirke, P.) (IRL, Oxford, in press).
31. Ward, E. S., Gussow, D., Griffiths, A. D., Jones, P. T. & Winter, G., Nature 341, 544-546 (1989).
32. Gussow., D. & Clackson, T., Nucleic Acids Res. 17, 4000 (1989).
33. Sanger, F., Nicklen, S. & Coulson, H. R., Proc. Natl. Acad. Sci. U.S.A. 74, 5463-5467 (1977).
34. Eisen, H. N., Methods Med. Res. 10, 115-121(1964).
35. Segal, I. *Enzyme Kinetics* 73-74 (Wiley-Interscience, New York, 1975).
36. Foote, J. & Milstein. C., Nature (in press).
37. Better, M., Chang, C. P., Robinson, R. R. & Horwitz, A. H., Science 240, 1041-1043 (1988).
38. Munro, S. & Pelham, H. R. B., Cell 46, 291-300 (1986).
39. Shimatake, H. & Rosenberg, M., Nature 292, 128-132 (1981).
40. Gottesman, M. E., Adhya, S. & Das, A. J., Molc. Biol. 140, 57-75 (1980).
41. McManus, S. & Riechmann, L., Biochemistry 30, 5851-5857 (1991).
42. Vieira, J. & Messing, J., Methods Enzymol. 153, 3-11(1987).
43. De Bellis, D. & Schwartz, I., Nucleic Acids Res. 18, 1311 (1990).
44. Harlow, E. & Lane, D., *Antibodies: A Laboratory Manual* (Cold Spring Harbor Laboratory, New York, 1988).

INNOVATION/INSIGHT

BUILDING A BETTER ENZYME

Roberts, S. S. (1993) Journal of National Institute of Health Research 5: 42. (Adapted)

1. Introduction

If imitation is the sincerest form of flattery, enzymes must rival presidents and movie stars in number of psychopaths. These catalytic proteins have many uses, but some scientists have sought to go beyond what nature provides. These researchers are seeking ways to make enzymes more valuable by tailoring them to specific applications. Some researchers are even modern-day alchemists, transmuting RNAs and antibodies into enzymes.

Speed and specificity, the qualities that make *enzymes* essential to living organisms, also make them valuable in the lab. Enzymes speed up biochemical reactions in the test tube, just as they do in the body. They also allow researchers to quickly and accurately assay for a particular protein or chemical. With the right enzyme, researchers can copy a molecule or cut it into pieces. Techniques such as DNA sequencing and nucleic acid amplification are unthinkable without enzymes. But sometimes no enzymes are available to catalyze the reaction a scientist wants to perform. Sometimes an enzyme is available, but it produces an unwanted by-product.

To get around these problems, scientists and companies are constantly looking for new enzymes. Although some scientists look for new naturally occurring enzymes, a newer approach is enzyme engineering. Not only can enzymes – such as the restriction endonucleases described in this article – be changed to make them more useful, but scientists can now alter molecules such as RNA and antibodies to give them catalytic properties, a process that is also described below.

Restriction endonucleases are enzymes made by bacteria that recognize specific sequences in DNA strands and cut them apart. Researchers use restriction endonucleases for many purposes, including manipulating *recombinant DNA*, analyzing DNA (and RNA) occurrence and concentrations by DNA *hybridization* (Southern blots, Northern blots), and mapping genes by *pulsed-field gel electrophoresis*. Although small nucleic acid fragments are useful for many experiments, sometimes scientists prefer to have fewer – but longer – strands of DNA. Then, they need so-called *rare cutters*, restriction endonucleases that cut DNA in only a few places and so produce fewer fragments of greater length. Although some restriction enzymes are naturally rare cutters, researchers have developed techniques for turning ordinary restriction enzymes into rare cutters.

Like the "Saturday Night Live" parody of a commercial in which two people argue whether a new product is a floor polish or a dessert topping, catalytic RNAs (ribozymes) belong to two seemingly mutually exclusive categories. They are nucleic acids, yet they have catalytic activity like enzymes. Some splice their own sequence out of RNA, whereas others play a role in replication or tRNA maturation. The discovery of these dual-personality molecules won Thomas Cech (University of Colorado, Boulder) and Sidney Altman (Yale) the 1989 Nobel Prize in chemistry. Researchers have since been looking for other naturally occurring ribozymes and are also trying to design their own by methods such as *directed evolution*.

Like ribozymes, some antibodies with catalytic ability – called catalytic antibodies or *abzymes* – occur naturally. Artificial abzymes were developed first (independently by Richard Lerner of the Scripps Research Institute in La Jolla, California, and Peter Schultz of the University of California at Berkeley); natural abzymes were discovered later. A newly developed model abzyme will allow scientists to tinker with its binding region, possibly increasing its catalytic ability or letting it perform new reactions.

Engineered and ersatz enzymes (inferior substitutes) are likely to be useful in many laboratory experiments. Altered restriction enzymes might cut nucleic acid sequences at places no known natural enzymes cut. The new catalytic molecules are also shedding light on the workings of enzymes in general.

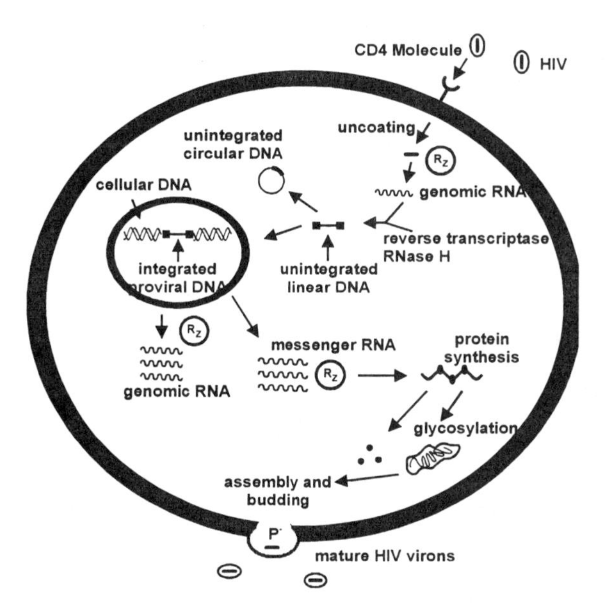

Figure 1 - How HIV enters, infects, and replicates within a cell (beginning at upper right).

One day, designer enzymes may be valuable clinically. For example, researchers suggest that ribozymes that could clip apart viral nucleic acids, and so block viral replication, might be able to ameliorate viral diseases, including AIDS (Figure 1). Specifically designed catalytic antibodies might be able to destroy viral coat proteins.

2. Restriction Enzymes: Making A Rare Cutter

With the twentieth century's emphasis on efficiency and productivity, restriction enzymes are an anomaly, a product that does its job too well. For some important applications, restriction enzymes are far too enthusiastic in cutting DNA. For example, most restriction enzymes recognize a DNA sequence six bases long, resulting in cuts as often as every 4096 bases or so. Short DNA pieces (approximately 4000 base pairs long) are impractical for mapping and sequencing.

There are such things as *rare cutters* – restriction enzymes that recognize longer DNA sequences – but not many have yet been found. Most rare cutters that do exist have an eight base-pair recognition sequence.

Rare cutter *Not*I, for instance, cuts human DNA – which contains 3 billion base pairs – every million bases or so, on average. Researchers are devising ways to make ordinary restriction enzymes less efficient, in effect turning them into rare cutters.

One of these methods is called *Achilles' heel cleavage.* A description of this technique was first published by Michael Koob and Waclaw Szybalski in 1988 and was made applicable to more situations in 1992. "Now I can cut at basically any sequence," says Koob, a postdoc in Szybalski's lab at the McArdle Laboratory for Cancer Research at the University of Wisconsin at Madison. "What I'm working on is to include cloning in the process," which would make Achilles' cleavage more practical for sequencing DNA.

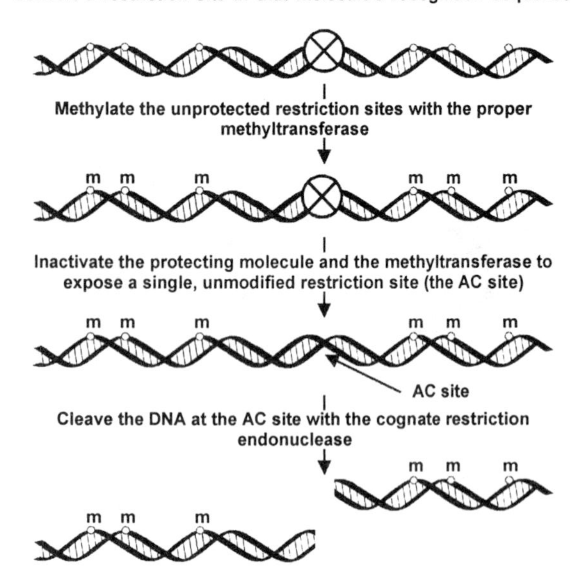
Add a DNA binding molecule to purified plasmid or genomic DNA to convert a restriction site in that molecule's recognition sequence
Methylate the unprotected restriction sites with the proper methyltransferase
m m m m m m
Inactivate the protecting molecule and the methyltransferase to expose a single, unmodified restriction site (the AC site)
m m m m m m
AC site
Cleave the DNA at the AC site with the cognate restriction endonuclease
m m m
m m m

Figure 2 - Achilles' Heel cleavage (AC) allows any restriction enzyme to act as a rare cutter, snipping apart DNA at only one site. (Reproduced with permission from M. Koob, "Conferring new cleavage specificities of restriction endonucleases," Methods Enzymol. 216, 321, 1992.)

Achilles' cleavage takes advantage of the ability of methylases (enzymes that add methyl groups to nucleic acids) to block the action of restriction enzymes (Figure 2). When a DNA sequence is methylated, it is protected from being cut. In Koob and Syzbalski's method, before the DNA is cut with any restriction enzyme, all cutting sites but one are methylated with a methyltransferase specific for the same sequence as the restriction enzyme. The result? The restriction enzyme cuts in only one place instead of in dozens or hundreds. Researchers can, if they desire, protect more than one cutting site from methylation by experimenting with different reagent concentrations, temperatures, and other reaction conditions.

Because so many restriction enzymes and methylases exist, one of the trickiest parts of Achilles' cleavage has been finding proteins to carry out the crucial step in the technique, binding tightly and specifically to one and only one cutting site to protect it from methylation. The protein Koob and Szybalski used first was a repressor molecule, either the *lac* repressor or the phage cI operator. Last year, they made the method more general by using instead an *Escherichia coli* protein, either the *integration-host-factor protein* or the *RecA protein,* to prevent methylation. (*In vivo*, RecA is best known for its role in homologous recombination, whereas integration host factor is thought to be involved in DNA replication, recombination, and transcription.)

"The RecA system allows us to cleave at any specificity," says Koob. "In that respect, it is very general." But, he says, "RecA has some limitations. There's probably a better enzyme that will bind at any [desired DNA] sequence without some of the manipulations I have to go through now."

Koob would also like to combine cutting and cloning to make the technique convenient for gene mapping. "I'd like to cleave at the sequence-tagged sites along the genome and then directly clone out that intervening DNA," he says. So far, Achilles' cleavage has been used to map only a few yeast and bacterial genes.

Koob says that Achilles' cleavage may offer a way around a recent problem with *yeast artificial chromosomes* (YACs). Researchers use YACs to clone large stretches of DNA by incorporating them into yeast chromosomes. When the chromosome duplicates, so does the inserted DNA. But now, gene mappers are discovering that yeast cells do not just copy the extra DNA, they also subtract or scramble pieces, resulting in clones that differ greatly from the original DNA.

"If you cut something directly from the [human] chromosome [by Achilles' cleavage] and then clone it directly [instead of in YACs] ... you would have a high degree of confidence that it came from the portion of the genome you thought it was coming from," says Koob.

Several research groups adapted the Achilles' cleavage technique to study *triplex DNA*. They attached a synthetic DNA strand (instead of a protein) to double-stranded DNA to protect a restriction site from methylation (see January 1992 issue, page 40). Koob also suggests that Achilles' cleavage could be used as an assay for the DNA binding specificity of some enzymes: "You can tell the relative strength of binding between two different sites by comparing the extent of cleavage between sites."

Additional Reading:

1. M. Koob, "Conferring new cleavage specificities of restriction endonucleases," Methods Enzymol. 216, 321 (1992).
2. M. Koob, A. Burkiewicz, J. Kur, and W. Szybalski, "RecA-AC: single-site cleavage of plasmids and chromosomes at any predetermined restriction site," Nucleic Acids Res. 20, 5831 (1992).
3. M. Koob, E. Grimes, and W. Szybalski, "Conferring operator specificity on restriction endonucleases," Science 241, 1084 (1988).
4. J. Kur, M. Koob, A. Burkiewicz, and W Szybalski, "A novel method for converting common restriction enzymes into rare cutters: integration host factor mediated Achilles' cleavage (IHF-AC)," Gene 110, 1 (1992).

3. Abzymes: Antibody Alchemy

Physicists study the debris left behind by subatomic particles whizzing through atoms to learn about the particles themselves. Geologists look at the many barren wastelands orbiting the sun to understand better why the Earth is verdant and teeming with life. So perhaps it is not surprising that Stephen Benkovic of the Department of Chemistry at Pennsylvania State University in University Park studies the mechanisms of enzyme catalysis by looking at antibodies.

His antibodies, however, are no ordinary antibodies, but ones endowed with catalytic powers, called abzymes or catalytic antibodies. Such antibodies hint at what primitive enzymes might have been like and provide a fruitful avenue for studying the process of catalysis. Although his lab is working on many projects, "the general theme for all of the antibody projects is trying to understand the chemical mechanism that these antibodies use for their catalytic activity," says Jon Stewart, a postdoc in Benkovic's lab. He says that as they study the mechanisms of abzymes, they constantly keep in mind enzymes that catalyze analogous reactions. Among the recent efforts of scientists in Benkovic's lab are characterizing a model catalytic antibody and designing an abzyme that catalyzes a reaction for which no enzyme is known to exist.

Usually, antibodies do not have catalytic properties. However, like enzymes, antibodies bind tightly to proteins and small molecules.

Enzymes are believed to catalyze reactions by binding to a transient structure (called a transition state) that the reactant passes through during the reaction. By producing monoclonal antibodies to transition states, scientists can sometimes create an antibody capable of catalyzing a reaction, although it is more sluggish than an enzyme fine-tuned by eons of evolution to perform the same function (see July 1990 issue, page 77).

The Benkovic lab's model abzyme, a monoclonal antibody called NPN43C9, catalyzes amide and ester hydrolysis. This abzyme, first described in 1988, was originally isolated from a combinatorial library of mouse immunoglobulin fragments in collaboration with Richard Lerner and Kim Janda of the Scripps Research Institute in La Jolla, California. "We view this as a model for antibodies that have the ability to cleave the amide bond of proteins," says Stewart. In as-yet-unpublished research, Benkovic's lab collaborated with Victoria Roberts and Elizabeth Getzoff, both of Scripps, to build a three-dimensional computer model of NPN43C9.

Once the researchers knew the antibody's likely shape, they introduced mutations into selected amino acid side chains to see how these alterations would affect the molecule's catalytic abilities. They determined that "there are two amino acid residues of 43C9 that are important for catalysis," says Stewart, a histidine residue at position 91 and an arginine residue at position 96, both in the light chain of the antibody molecule. "We're now looking at amino acid changes to improve the catalytic efficiency of 43C9," Stewart says, and to allow it to catalyze reactions with the help of a metal cofactor, such as zinc or copper. Once they can attach metal ion cofactors and make other changes to the binding pocket of NPN43C9, the antibody may be able to catalyze new reactions, such as oxidation-reductions. Modifying an existing model antibody should be a more efficient way of producing new catalytic abilities than starting from scratch.

Stewart says that NPN43C9 is in some ways more primitive than enzymes that catalyze similar reactions. "The closest analogues to 43C9 are the enzymes chymotrypsin and trypsin," he says. Those enzymes each have three amino acid residues important in cleavage of the amide bond, whereas in 43C9, only the histidine residue plays a role. Also, NPN43C9 is somewhat slower; It accelerates hydrolysis a million-fold over spontaneous rates, whereas trypsin and chymotrypsin accelerate the reaction 25 million-fold.

In a separate project, scientists in Benkovic's lab generated an antibody with a new catalytic specificity. Benkovic, Diane Schloder at Scripps, and Penn State collaborators Louis Liotta, Scott Taylor, Patricia Benkovic, and Grover Miller created a catalytic antibody called RG2-23C7. It *catalyzes the deamination of an asparagine-glycine dipeptide.* No known enzyme catalyzes this reaction.

Stewart says that when this deamination reaction occurs in a protein, the body breaks down that protein. "If we could do this for a protein of interest," says Stewart, "we might be able to target it for degradation. If an enzyme were operating out of control, this [antibody] might be a way to modulate its activity indirectly." The antibody would thus function something *like a restriction enzyme for proteins.* Although such an antibody could have important therapeutic and research applications, Stewart points out that not all enzymes have the correct dipeptide. Those that do sometimes have this dipeptide buried within their structures, where the antibody would not be able to get at it.

Additional Reading:

1. S. J. Benkovic, "Catalytic antibodies," Annu. Rev. Biochem. 61, 29 (1993).
2. R. A. Gibbs, S. Taylor, and S. J. Benkovic, "Antibody catalyzed rearrangement of the peptide bond," Science 258, 803 (1992).
3. L. J. Liotta, P. A. Benkovic, G. P Miller, and S. J. Benkovic, "A catalytic antibody for imide hydrolysis featuring a bifunctional transition-state mimic," J. Am. Chem. Soc. 115, 350 (1993).

4. Ribozymes: Evolution Revolution

4.1 Unnatural Selection and Irrational Drug Design. Although Gerald Joyce (Dept. of Chem. and Mol. Biol., Scripps Research Institute) prefers to describe his work as "directed evolution," he doesn't reject the more colorful labels people have applied to his research on making new and better ribozymes (RNA molecules with catalytic capabilities). After all, it's a rather delicious irony that Joyce's fumbling-around-in-the-dark approach endows ribozymes with new properties in circumstances in which the logical procedures of rational drug design are useless – for example, when too little is known about the ribozyme's structure. And Joyce wouldn't have it any other way. "If I knew the answer to a problem, I really wouldn't be as interested in doing the experiment," he says.

Joyce's approach may be random, but it is based on a method proved effective over the past 4 billion years on earth. To design a better ribozyme, Joyce lets evolution take its course in a test tube. He starts with trillions of ribozymes adapted for one function, and he *pressures the ribozyme population to evolve a new function.* Just like evolution, Joyce puts the ribozyme population through *repeated cycles* of *mutagenesis*, *amplification* (the equivalent of reproduction in organisms), and *selection.* When the technique is successful, the population of ribozymes changes, generation by generation, in the desired direction.

Ribozymes are valuable because they can break and form covalent bonds in RNA. Most known ribozymes are self-effacing in the most literal sense; their function is to cut themselves out of RNA strands. In his first successful evolution experiment, Joyce and Scripp's technician Amber Beaudry wanted to design a ribozyme that would cleave single-stranded DNA under physiological conditions. However, without knowing the crystal structure of the ribozyme, the researchers could not use conventional methods of drug design.

So, they evolved the new ribozyme. They started with a naturally occurring ribozyme known as the *Tetrahymena* ribozyme, which is derived from an RNA precursor produced by the protozoan *Tetrahymena thermophila.*

Directed evolution is much like a race – in which successive hurdles change the average characteristics of the runners, in discrete stages, by sieving out the least-proficient ones. For example, very tall hurdles might eliminate the shortest runners, whereas many randomly scattered short hurdles might force the clumsiest runners from the race. The population of runners at the end of the race will thus differ greatly from that at the start. Similarly, directed molecular evolution – artificial selection consisting of cycles of amplification, mutation, and selection – can change the catalytic abilities of a population of ribozymes in a desired direction. Each hurdle (selection) acts as an evolutionary bottleneck (sieve). For example, if the only RNA molecules permitted to "breed" (be amplified) are those that can cleave DNA, the DNA-cutting abilities of the population improve with each generation.

They generated trillions of copies of the ribozyme, deliberately making errors that resulted in some copies having mutations. Next, they exposed the molecules to DNA, giving them the opportunity to cleave DNA. A few could. The researchers then amplified and diversified the most efficient DNA cleavers by an *error-prone* polymerase chain reaction method that introduced mutations into many ribozyme copies. After 10 generations, Joyce and Beaudry had blindly evolved a new ribozyme that could cut DNA 100 times as efficiently as the wild-type version could – although it was much sloppier about cutting in the correct place. In as-yet-unpublished results, Joyce and Scripp's graduate student Joyce Tsang continued the experiment another 18 generations with different selection pressures and were able to sharpen the specificity greatly.

In a second set of experiments, Joyce and Scripps postdoc Niles Lehman changed the cofactor of the *Tetrahymena* ribozyme. The wild-type ribozyme requires a magnesium or manganese ion before it can cut RNA. After eight generations of directed selection, the population had evolved two different complexes of mutations that allowed molecules to cut RNA with only calcium ions present.

Directed-evolution experiments have another benefit. They make evolutionary biology truly an experimental science as well as a theoretical one. "You see evolutionary history play out, but at a very detailed molecular level," says Joyce. "Evolution is happening in nature, but on a time scale that's just achingly slow." With the trillions of individuals in each generation and the short generation time, directed evolution makes evolution proceed at warp speed, giving scientists their first detailed view of evolution to compare with theoretical predictions and *post hoc* explanations.

Although previous *in vitro* selection methods mimicked evolution in some respects, Joyce's directed evolution is the first to make use of new mutations incorporated during every cycle. He says the technique still has some disadvantages. First, because the technique depends on the random appearance of useful mutations, "you don't know something is going to work until it does," he says. Second, "there are some things that RNA, by virtue of its functional groups, just isn't going to be able to do. RNA's playbook is rather limited." For example, he suspects that it would be almost impossible to evolve a ribozyme that performed a complex series of reactions like glycolysis.

Additional Reading:

1. J. M. Burke and A. Berzal-Herranz, "*In vitro* selection and evolution of RNA: Applications for catalytic RNA, molecular recognition, and drug discovery," FASEB J. 7, 106 (1993).
2. A. A. Beaudry and G. F. Joyce, "Directed evolution of an RNA enzyme," Science 257, 635 (1992).
3. G. F. Joyce, "Directed molecular evolution," Sci. Am. 267(6), 90 (1992).
4. N. Lehman and G. F. Joyce, "Evolution *in vitro* of an RNA enzyme with altered metal dependence," Nature 361, 182 (1993).

INNOVATION/INSIGHT

THE IMMUNOZAP™ CLONING AND EXPRESSION SYSTEM

Stratagene Corporation, La Jolla, CA, ph. (619) 535-0180. (Adapted)

To produce an ImmunoZAP™ library, cDNA is first synthesized from RNA isolated from human or murine B cells derived from peripheral blood, lymph node, or spleen tissue. To enrich this population of cDNA for immunoglobulin sequences, heavy and light chain polymerase chain reactions (PCR) are performed using *primers* supplied with the ImmunoZAP™ cloning kit. These PCR products are then *directionally cloned* into the ImmunoZAP™ vectors.

The expressed antibody molecules (or subfragments) can be detected with various screening methods. Typically, the antibody proteins are "lifted" from *phage plaques* grown on *E. coli lawns* using *nitrocellulose filters* and the replica filters are subsequently incubated with labeled antigen. Detection of bound antigen at certain locations on the filter identifies the positions of phage plaques expressing antibody molecules having affinity for the antigen.

The heavy and light chain libraries may be screened separately, or the heavy and light chains may be coexpressed to screen for assembled *antigen binding fragments*. Alternatively, one may wish to identify a heavy chain with useful binding properties and coexpress it with a light chain library. The *diversity generated* using this strategy may reflect the diversity generated by the association of heavy chain and light chain pairs in lymphocytes, although combinations that do not exist in nature will also occur. It is possible that these combinations may exhibit desirable characteristics that would be difficult or impossible to obtain with *conventional hybridoma technology*.

Classically, *monoclonal antibodies* (MAbs) have been generated with hybridoma technology by the fusion of myeloma cells with B-lymphocytes.[1] Selection of the hybridoma cells is accomplished by culturing the fusion products in several 96-well microtiter plates in the presence of HAT medium. This process generally yields a handful of immortal hybridoma cells that are then screened for the desired MAb.

While this methodology has often been successful, production and screening are both time consuming and costly. The production of human MAbs has been further limited by persistent problems, such as *low clonal frequency*, low antibody production, and *cell-line instability*.[2-4] Due to the constraints implicit in current hybridoma technology, it is impossible to efficiently survey the induced antibody response to a given antigen. Desirable MAbs for some antigens may take years to isolate or indeed may never be isolated. In addition, once a MAb has been isolated, altering its affinity or specificity to create a designer MAb requires additional resources with complex manipulations.[5-8]

To circumvent some of the difficulties associated with producing MAbs, Stratacyte has applied the powerful tools of molecular biology to develop the ImmunoZAP™ cloning and expression system. This novel technology involves the creation of *immunoglobulin-specific bacteriophage libraries* in *E. coli*.[9-12] These libraries allow for the screening and expression of immunoglobulin light, heavy, or light-heavy proteins. Libraries may be constructed to express murine, human, or murine-human immunoglobulin for the identification of the MAb of interest. The genes defining a desired MAb are easily isolated using the ImmunoZAP™ cloning system, and the affinity or specificity can then be altered, or the effector or reporter function modified to produce a designer MAb.

This exciting new technology expands the utility of using Mabs as tools to facilitate basic research and may potentially produce reagents that have promise for clinical analysis and therapeutic use.

References:

1. Kohler, G. and Milstein, C. (1975) Nature 256:495-497.
2. Glassey, M. C. and Dillman, R. O. (1988) Mol. Biother. 1:7-12.
3. Dillman, R. O. (1990) Antibody Immuno-conjugates Radiopharmaceut. 3, 1-15.
4. Boyd, J. and James, K. (1989) *Advances in Biotechnological Processes. Monoclonal Antibodies: Production and Application*, A. Mizrahi, Ed. (Alan R. Liss; N.Y.), Vol. 11, pp. 2-43.
5. Morrison, S. L., Johnson, M. J., Herzenberg, L. A., and Oi, V. T. (1984) Proc. Natl. Acad. Sci. U.S.A. 81:6851-6855.
6. Jones, P. T., Dear, P. H., Foote, J., Neuberger, M. S. and Winter, G. (1986) Nature 321:522-525.
7. Verhoeyen, M., Milstein, C. and Winter, G. (1988) Science 239:1534-1536.
8. Queen, C. *et al.* (1989) Proc. Natl. Acad. Sci. U.S.A. 86:10029-10033.
9. Sastry, L. *et al.* (1989) Proc. Natl. Acad. Sci. U.S.A. 86:5728-5732.
10. Iverson, S. A. *et al.* (1989) Cold Spring Harbor Symp. Quant. Biol. 54:273-282.

11. Huse, W. D. *et al.* (1989) Science 246:1275-1281.

12. Mullinax, R. L. *et al.* (1990), Proc. Natl. Acad. Sci. U.S.A. 87:8095-8099.

APPENDICES

Part 1 Terms List

alkaline phosphatase - enzyme that catalyzes the removal of 5'-phosphates from the ends of nucleic acids. Prevents self-ligation of cloning vectors.

aptamer - a combinatorially-produced nucleic acid selected by participation in a biochemical recognition process. For example, quadruplex-forming DNAs can inhibit thrombin-mediated blood clotting by competitively inhibiting essential thrombin/fibrinogen interactions.

artificial chromosomes - Yeast artificial chromosomes (YACs), contain the following essential elements: terminal telomeres, internal centromeric sequences (CENS), and one or more autonomous replication sequences (ARS).

assay sensitivity - involves two considerations, minimal and maximal detection ranges and standard errors associated with replicate determinations.

cDNA - duplex DNA sequence, produced by reverse transcription from an RNA template carried in a cloning vector. Baltimore and Temin showed that RNA can be created by reverse transcription of RNAs, then used by retroviruses to make cDNAs that can be integrated into the genome of the host.

central dogma of molecular biology - statement that DNA is replicated to make daughter generation DNA, RNA is made by transcription from the DNA template, and proteins are synthesized based on the sequence of specific mRNAs.

clone - one of a group of genetically identical cells or organisms derived from a common ancestor.

cloning vector - a self-replicating DNA or RNA carrying a covalently attached genetic element that can be propagated and amplified within a host cell.

cohesive ("sticky") ends - complementary single-stranded DNAs that protrude from opposite ends (both 5' or both 3') of a duplex fragment or from the ends of different duplex molecules; generated by staggered cutting by a restriction endonuclease.

combinatorial approaches - using *chemical DNA synthesis* to make DNAs that contain two defined sequences (e.g., polylinkers) that bracket a randomized sequence. *Randomization* of the DNA sequence is accomplished in chemical synthesis by injecting an equimolar mixture of all four of the nucleoside phosphoramidites when inserting nucleotides corresponding to intended randomized positions. *Selection procedures*, followed by *cloning* and *purification* of single clones from the library, produce purified selected sequences that survived a selection procedure from a randomly mixed population. Selected nucleic acids or proteins bind or perform some task better than others from the investigated population. Steps can be incorporated to *evolve a composite population* across several generations with selection occurring at each.

competent - cells capable of taking up DNA from their surroundings. In some cases, this is a transient physiological state that occurs at some time during the division cycle. Other cells must be chemically treated (e.g., in $CaCl_2$ or LiCl solutions) in order to induce competence.

cultured cells - bacterial colonies grown on agar plates or in liquid media. Examples of eukaryotic cell cultures are Jar cells (a carcinoma of the chorion); yeast (fairly easy to grow, manipulate, and predict using Mendelian genetics); *Tetrahymena* (protozoan; easy to grow and produce amplified numbers of specialized chromosomes); mitogen-transformed Chinese hamster ovary (CHO) cells (used for cell cycling studies); transformed mouse fibroblasts (3T3); and primary sheep pituitary cells (first generation; not transformed).

digoxigenin - a modified deoxythymidine triphosphate containing a fluorescent steroid-like reporter group attached at C5 of the base. Used to label DNA by random priming with Klenow DNA polymermase for Southern or Northern blot experiments. Detected by immune enzyme conjugate binding then reaction with colorimetric or chemiluminescent substrates.

directional cloning - both plasmid and foreign DNA are cut with two different enzymes creating two different sets of cohesive termini (plasmid and insert fragments). This reduces self-ligation of vector and forces ligation in only one orientation.

DNA sequencing - Two major classes of DNA sequencing methods are now in use. Both generate a series of radiolabeled oligonucleotides starting from a single point and terminating randomly at specific nucleotide residues. When these reaction mixtures are electrophoresed on high-resolution polyacrylamide gels, sequences of about 250-350 nucleotides can be determined accurately.

episomes - an extragenomic genetic element. Generalized examples of episomes include viruses, transposons, and viroids.

ethidium bromide - planar aromatic fluorescent compound commonly used to stain nucleic acids.

ethidium intercalation - stacking of the aromatic carcinogen ethidium (Etd^+ bromide salt) between base pairs. The high fluorescence yield of DNA-bound Etd^+ makes bound DNA appear orange above a purple background upon viewing UV illuminated Etd^+-stained gels.

expression vector - a vector designed with the proper signals to drive transcription of the gene or genes of interest and translation of the mRNA product. Used to produce a protein product from the gene of interest, or a suitable fusion protein from a fused gene.

fluorescence enhancement - increase in fluorescence yield from a chromophore when transferred into a less polar environment. Used to study binding, surface exposure, and other types of structural transitions.

fusion protein - fused product of an exchangeable fragment from a protein of interest and part or all of a second well-characterized marker protein such as β-galactosidase. The latter can be used for enzymatic selection, Western blotting, or affinity chromatography.

genome - the apparatus that protects and propagates the nucleic acid-encoded blueprint from parent to offspring generation. Fundamental processes are replication of the parental genome to give the product offspring genome, segregation of parental and offspring genomes after replication, and regulation of the timing and fidelity of the cycle. When gametes fuse to form the fertilized ovum, the developmentally fully capable cell is said to carry a "pluripotent genome."

genotype - the nucleic acid/protein code that is physically segregated during cell division and carries the functional characteristics known as phenotypes.

growth-selectable genetic markers - a genetic element that encodes a protein that typically degrades an antibiotic. When the element is present in the cell, it grows; when not, it won't.

high-resolution gel electrophoresis - polyacrylamide gels used for techniques such as sequencing. The technique is able to resolve DNA fragments that differ in length by a single nucleotide.

hybridization - noncovalent complex formation between complementary nucleic acids mediated by hydrogen bonding and stabilized primarily by stacking forces. Also called duplex formation and annealing.

insertional inactivation - usually used when cloning with vectors containing two antibiotic resistance markers. A foreign DNA fragment is cloned into a restriction site within one of the two antibiotic genes. The other antibiotic resistance gene is used to select for transformants. Transformants are screened for resistance to the second antibiotic. Plasmids containing inserts will be sensitive to the second antibiotic because insertion of the foreign DNA inactivates the antibiotic resistance gene.

intercalate - stacking of a planar aromatic compound between two base pairs in a nucleic acid. This process decreases the exposure of fluorophores such as ethidium, producing a high fluorescence yield.

ligate - to form a phosphodiester bond linking two adjacent nucleotides; sealing a nick in one strand of a double helix. Also refers to blunt-ended ligation of two duplexes to form a linked DNA and covalently linking two RNA strands.

Low-resolution gel electrophoresis - method using agarose gels that can resolve only DNA fragments that differ significantly in size. These gels may be used to analyze plasmids, restriction mapping, Southern and Northern hybridizations, and other moderate to large-sized DNAs.

nick translation - method to label DNA in which a preexisting nick in a duplex DNA is used as a starting point by DNA polymerase I to extend and replace the nucleotides of one strand with labeled nucleotides.

Northern blot hybridization - method used to probe for the specific RNA species in the presence of a heterogeneous population. The specific DNA is labeled, the RNA is electrophoresed then transferred laterally to a filter to preserve the gel pattern, the DNA is hybridized to the filter-bound RNA, and excess DNA is removed. The resulting autoradiogram will correspond only to specific RNA sequences complementary to the labeled DNA probe.

operon - model of gene expression regulation. Operator sites in promoter regions control the expression of downstream genes. Developed from the classical experiments of Jacob and Monod with the lactose operon.

phages - bacterial viruses; examples in *E. coli* include the T-even and -odd phages (T2, T3, T7), ϕx174 and phage M13, whose life cycle has dsDNA, circular ssDNA and circular ssRNA forms. The M13 DNA system is used in ssDNA subcloning and is important for dideoxy (dd) DNA-sequencing strategies.

phenotype - Morphogenic or enzymatic result of gene expression activity.

plasmid - an autonomous self-replicating extrachromosomal circular DNA; pBR322 was a popular early colecin E1-based vector; the more recent pUC- and pET-based vectors are popular and useful in a variety of diverse scenarios. The key tool of molecular biology.

pluripotent - When gametes fuse to form the fertilized ovum, the developmentally fully capable cell is said to carry a "pluripotent genome," all of the genetic information necessary to produce all cells of the organism.

poly(A)⁺ mRNA - a subpopulation of eukaryotic messenger RNAs that contains more than 15 adenylic acid residues at the 3' terminus. The poly(A) tail and attached mRNA can be isolated and purified by taking advantage of their affinity to oligo(dT) bound to chromatography supports.

polylinker - a sequence of DNA containing many different types of restriction sites, which enables a number of possible routes into a clone construction strategy.

polymerase chain reaction (PCR) - a process in which two sets of single-stranded primer DNAs are hybridized to either (*i*) ssDNA or (*ii*) ssRNA, then the DNA sequence bracketed by the primer binding sites is produced using either (*i*) *Taq* DNA polymerase or (*ii*) *Taq* reverse transcriptase plus DNA polymerase. In the second type of reaction, cDNA is produced in the first round of amplification, then DNA polymerase produces DNA in subsequent rounds. By cycling through steps involving duplex melting then reannealing with primers, one can get large-scale synthesis of any sequence located between two known primer DNA sequences. High-temperature-stabilized *Taq* enzymes allow stringency selection against nonspecific primer-template complexes, favoring production of DNAs only from correctly primed DNAs.

radioisotopes - high-energy atoms that emit subatomic particles, including alphas (a helium nucleus), betas (e^-), neutrinos and gamma rays. Examples: ^{32}P, ^{35}S, ^{14}C, and ^{3}H.

random priming - a method for labeling DNA involving duplex denaturation, hybridization to oligonucleotides, and extension using Klenow DNA polymerase and labeled dNTPs.

receptors - soluble or membrane-bound proteins that bind chemical effectors then transduce the "signal" to produce a biochemical effect. For example, several of the steroid-based hormones bind to soluble receptors, which then attain an increased affinity for a set of DNA regulatory sites on growth control genes, resulting in altered gene expression.

replica plating - a filter or sterile velvet cloth is wrapped around a round wooden or plastic block slightly smaller than a petri dish and gently pressed down on the surface of an agar plate containing many colonies (the "master plate"). This is pressed down on the surface of a plate identical to the first plate and/or onto the surface of a plate containing selective media (such as an antibiotic). Clonal colonies which grow on the newly inoculated plates are positioned in the exact same position as the colonies on the master plate. Descendent colonies with particular phenotypic characteristics can be identified, yet preserved as uncontaminated master colonies.

repressors - proteins that bind the promoter of a gene and inhibit RNA polymerase binding, thereby preventing gene expression. For example, galactose corepresses the *lac* operon by binding *lac* repressor protein. When yeast are grown with glucose, the operon is activated. Growing cells in galactose turns the operon off.

restriction endonucleases - type I: recognize a sequence, but then cut elsewhere (not useful for site-specific cloning procedures). These enzymes mediate differences in strain compatibility. Type II: recognize and hydrolyze a specific DNA sequence in the 4-8 base pair size range.

restriction mapping – method used to document the order of specific restriction sites along the length of a DNA fragment by analyzing the sizes of restriction fragments in an agarose gel. Electrophoretic patterns obtained following digestion with one enzyme or combinations of single, double, triple, or sequential digestions are analyzed to reconstitute the overall map.

RFLP mapping – "restriction fragment length polymorphism"; a method to analyze complex preparations, such as human chromosome extracts, for individual-specific sequence heterogeneity in regions between defined genetic loci. The techniques uses rare-cutting restriction endonucleases to look for alterations in fragment patterns (relative to suitable controls) to determine which genotype pattern is present. This is the basis of genetic fingerprinting and the large increase in our knowledge of the molecular basis of genetic diseases.

sequence complexity - the population size of a given nucleic acid sequence, the copy number. Determined by the rate of hybridization to a probe nucleic acid. In the case of DNA, this analysis is done using a plot called a "c_0t curve," referring to the initial concentration multiplied by time, which gives the rate of hybridization. The RNA plot is called an "r_0t curve." More highly represented sequences will hybridize faster because they can "find a partner" quicker if there's more of them. Some nucleic acids carry multiple copies of a sequence on one strand, which must be factored into the analysis. Hybridization kinetics measurements allow determination of quantitative RNA synthesis levels on a per-gene copy basis, an important step in characterizing expression patterns.

shuttle vector - vector with control elements that allow gene expression in either prokaryotic or eukaryotic cells. Used to propagate eukaryotic genes in bacteria then transfer them to the eukaryote for protein-mediated activity or regulation studies.

signal transduction - methods used to transfer chemical signals in organisms between intracellular compartments, cells, or systems. Examples of second messengers and pathways include cyclic AMP, cyclic GMP, Ca^{2+}, cell cycle gating factors, various kinase/phosphatase pathways, prostaglandins, and cytokines.

Southern blot hybridization - detection of a DNA in a heterogeneous mixture by sequence-specific hybridization to a labeled DNA probe. (Also see Northern blot hybridization.)

stacking forces - forces between aromatic bases in adjacent nucleotides in nucleic acids, or between intercalated aromatic drugs (e.g., ethidium) and the flanking bases. Stabilizing forces involve entropy due to the release of surface-bound solvent, the hydrophobic effect, and dipolar interactions between electronic systems in the adjacent aromatic rings (London dispersion forces).

stringency - increasing the fidelity in nucleic acid hybridization reactions by decreasing the stability of mismatched duplexes. For example, decreasing the NaCl concentration decreases phosphodiester charge shielding, enhancing strand-strand repulsion. The result is more serious destabilization of mismatched complexes relative to correctly paired duplexes. This ensures more faithful probe recognition.

transformation - acquisition of new genetic material (sequence markers) by foreign DNA uptake. Cells that can absorb DNA are typically produced by chemical treatment (e.g., with $CaCl_2$); this condition is called *competent.*

vector - plasmid, phage, or chromosomal DNA carrier. Foreign DNA may be inserted, cloned, and manipulated. RNAs and proteins, produced by transcription and translation, respectively, may be studied either *in vitro*, *in situ*, or *in vivo* depending on the context of the problem. Plasmids are used to transport genes, their products, and their associated biochemical functions between different cells, tissues, systems, organisms, and crops. The genetic signals contained in the DNA control site sequences confer means to control how much biochemical effector is made, thereby allowing control of the biological fate of the metabolism mediated by the RNA or protein being expressed.

PART 2 TERMS LIST

abzymes - enzymes elicited against a transition-state analogue that can catalyze a selected chemical transformation.

adjuvant - preparation of immunoactivating material (e.g., killed viral components) – mixed with an antigen to enhance the likelihood of antibody production.

adsorbent - chromatographic materials that cause separation as a result of physical and chemical interactions with the sample components.

ammonium sulfate precipitation - salting out a particular protein, or all of the other proteins, from a mixture of proteins as a means to fractionate them. The precipitates can usually be redissolved, so the method can be used to obtain the protein of interest in either the solubilized or precipitated fraction following resolubilization.

antibodies - usually divalent immunoglobulin G molecules (IgGs); glycoproteins that bind to specific antigens. Higher numbers of binding sites are present in IgA (4 Ig subunit) and IgM (10 subunit) complexes.

antigen - recognized substrate of an antibody; recognized structural subcomponents are called "determinants"; the structural domains are called "epitopes."

band broadening - widening of a chromatographic peak due to either extra- or intracolumn effects.

bed volume - for a given mass of sorbent packed in a cylindrical bed, the volume of solvent required to fill all pores of the sorbent bed and all interstitial spaces between the sorbent particles.

capacity - ability of a column to tolerate a certain sample concentration before overloading. It is related to the amount of liquid phase loading (or film thickness) and the relative polarities of the sample components and the stationary phase.

capillary column - capillary columns are divided into two classes, high and low resolution. Low-resolution columns have internal diameters of 0.4 to 0.75 mm; high-resolution columns have internal diameters of 0.1 to 0.4 mm.

capillary electrophoresis - electrophoresis of molecules through a functionalized capillary column equipped with an optical quartz window for spectroscopic visualization of the separated electrophoresing molecules.

carbohydrate cell surface antigen - linear and/or branched chains of monosaccharide units, such as *N*-acetylgalactosamine, *N*-acetylnuraminic acid, fucose, and galactose, that determine the blood type of a patient. Interactions between hydroxyl-rich carbohydrates (CHO) and H_2O usually slow and often smear the patterns of glycoproteins on gels. More CHO-enriched protein structures are called "proteoglycans."

"carrier free" radioisotope - 100% specific activity. When no non-radioactive material is added, the undiluted potency is called "carrier free."

Cerenkov radiation - when a solid radioactive substance emits a β^- particle, the particle must transfer from the atom nucleus (solid state) into the surrounding medium. When that happens, the velocity of the β^- particle decreases (the relative speed of light changes) because it's in the new medium. In order to conserve energy, it must give off some energy. This excites fluor molecules in the medium, which emit the energy as light. When the radioactive particle changes its medium from a solid to liquid, the "extra" relativistic energy is given off as light, called Cerenkov radiation.

chaperonins - a set of catalytic protein-assembling complexes, typified by the *groEL* gene product of *E. coli*, which folds other proteins properly in a relatively nonspecific ATP-driven process.

chromatography - separation of compounds by exchanging (or "plating") on and off of a solid phase into a moving liquid (LC) or gaseous (GC) phase. Liquid methods are typically used for biomolecular chromatography.

column chromatography - separation of a solution of molecules by differential partitioning between a liquid carrier phase and a solid matrix phase. The chemical nature of the bound modifications (charge state, polarity, pK_a, biomolecular shape) determines the affinity of the sample components for the column matrix (resin, "packing").

complement - a complex proteolytic cascade that triggers the cellular immune response, activated by cytokines and cell-cell interactions, which results in the destruction of antigens and their engulfment by macrophages.

counterion - an ionic species associated with the ionic functional group of opposite charge on the sorbent surface. Isolate retention occurs through exchange of the counterion for the isolate. Counterions are important because the degree of interaction between a sorbent functional group and different counterions varies widely, with implications for isolate retention. This difference is referred to as the counterion selectivity of a sorbent.

cross-linked - a polymerized stationary phase that has improved durability and allows rinsing of the column with solvents without the loss of stationary phase film.

curie (Ci) – measure of radioactivity. 2.22×10^{12} disintegrations per minute (dpm); a mCi is typically sufficient for tens of experiments.

dead volume - the void volume of a column. Can be measured by the time it takes to elute an unretained component.

denaturation - cooperative loss of noncovalent interdomain connections in a protein or nucleic acid. Strong denaturants are classified as "chaotropic" agents (e.g., urea, guanidium, SDS).

Donnan layer - layer of hydration and bound electrolytes condensed around a charged macromolecule. This layer increases the effective radius of the macromolecule and is detectable using X-ray scattering and ion NMR techniques.

efficiency - a measure of the ability of a column to produce narrow peaks. The unit of measure is plates/length.

electronic energies - the energies of molecules participating in chemistry reactions. They can be represented and characterized by the Jablonski diagram, which is a map of the status of the valence electrons as a function of their energy.

elution - removal of a chemical species from a sorbent by changing the solvent or matrix chemistry to disrupt the isolate/sorbent interaction.

enzyme-linked immunosorbant assay (ELISA) - immunoassay in which the antibody is used to probe the sample of interest while bound to a bare or derivatized multiwell polystyrene ELISA plate. This is followed by incubation with a second antibody that binds to the constant region of the primary antibody. The second antibody has an enzyme (typically alkaline phosphatase, horseradish peroxidase) covalently linked (conjugated) to it. After thorough washes, the color-forming enzyme substrate is added. Binding of the specific primary antibody is monitored by the rate of appearance of the colored conjugate-enzyme product.

excited state - high energy state that occurs in a molecule during the typically short span of time after light or chemical energy is absorbed. The energy is reemitted to some sink interaction and the molecule goes on to form a new, less energetic excited state or a ground state (e.g., ground state + energy ∀ excited state ∀ ground state or lower energy excited state).

expression vector - a plasmid that carries signals sufficient to drive both transcription and translation from a gene and the product mRNA, resulting in production of the corresponding protein. *Overexpression* occurs when a "strong" promoter is used to drive transcription.

extraction - transfer of a chemical species from one phase into another. In sorbent extraction, transfer of the species of interest from the matrix environment onto the sorbent (retention) or transfer from the sorbent into the solvent (elution). Isolate purification is achieved via a series of selective extraction steps.

F_{ab} – antibody-binding immunoglobulin fragment; structurally diverse antigen-binding IgG region.

F_c – constant immunoglobulin fragment; structurally constant IgG region.

film thickness - actual thickness of the stationary phase coating the supporting material.

fractionation - physical separation of biomolecules based upon differences in solubility, pH, isoelectric point, hydrophobicity, substrate affinity, size, and other properties.

functional group - a group of atoms on a chemical species having properties that can be exploited for retention on sorbents. Functional groups important to sorbent extraction include ionic groups, hydrogen bonding groups, dipoles, aromatic rings, and virtually any other regions on molecules exhibiting nonpolar, polar ion exchange or other potentials for interaction with sorbents.

fused silica - synthetic material made of extremely pure silicone dioxide. The high purity of the material gives very inert surfaces for improved chromatography of active compounds such as amines or acids.

fusion protein - a protein produced by fusing an exchangeable fragment from a protein of interest, and usually a fragment or all of a second well-characterized marker protein such as β-galactosidase, which can be used for enzymatic selection, Western blotting, or affinity chromatography.

gas chromatography - gas phase separation method accomplished by injecting the sample into heated columns made of heat-stable materials such as silica gel. Samples must be volatile under the mobile carrier phase conditions and can be derivatized to decrease the solubility or increase the propensity to thermal breakdown. Low molecular weight organic compounds are separated most easily, however lipids, nucleic acids, carbohydrates, and other biomolecules can be analyzed.

Geiger-Mueller counter - the "Geiger counter" contains an easily excited gas in an enclosed chamber with a pair of electrodes carrying a large interelectrode voltage. When these gas molecules are ionized by nuclear radiation, they gather around the electrode of opposite charge and induce a current in the circuit, which is registered in the deflection of a meter needle or digital readout.

gene expression - gene transcription and mRNA translation.

HETP - (height equivalent of a theoretical plate). The length of column occupied by one theoretical plate.

high-performance liquid chromatography (HPLC) - column chromatography separation method in which liquids are pumped at pressures ranging up to > 2000 psi at rapid rates over reinforced columns containing suitably derivatized matrices. Components include sample injectors, two pumps under computer control to program changes between different conditions (via solution gradients), the "solvent programmer" interface between the computer and pumps, and fraction collection and output viewing/printing devices.

humoral immune response - immune response in which an antigen binds to B or T cells and induces clonal proliferation of the antibody-producing cell. A memory response can be raised by rapid mobilization of previously developed cells.

immunoglobulin - Both serum-soluble and cell surface-bound forms occur. Immunoreceptor proteins (see antibody).

immunological detection - accomplished by labeling either an antibody, or second antibody that binds to the constant region of the first antibody. Binding is monitored via reporter functionalities such as ^{125}I, ^{35}S, fluorophores, enzymes, or nucleic acid probes.

inclusion bodies - vacuole-compartmentalized overexpressed proteins found in clone-expressing cells. Proteins cannot be exported and may be either in native, denatured, or aggregated conformations.

interaction - attraction or repulsion between two chemical species in a specific chemical environment. In sorbent extraction, the three principal interactions are isolate/sorbent, matrix/sorbent, and isolate/matrix. Specific interactions include nonpolar, polar, ion exchange, and covalent.

interferences - undesired components in the sample matrix that obfuscate the matrix/isolate binding interaction.

Ion exchange - interaction of an ionic isolate functional group with an ionic functional group of opposite charge on the sorbent surface. Ion-exchange interactions are manipulated using solvent pH, ionic strength, and selectivity of counterions. Ion exchange interactions are referred to as strong or weak depending on the energy of the isolate/sorbent interaction. Strong ion exchange resins include the Dowex products. Weak ion-exchange resins are divided into the anion and cation exchangers. An example of an anion-exchanger is quaternary-amine-derivatized diethylaminoethyl (DEAE) cellulose. An example of a cation exchanger is carboxylate-modified QAE Sepharose.

isoelectric point - pH at which the net charge of a protein is zero; determined by the protonation state of the pH-titratable functional groups.

isotopes – atoms that differ by the number of neutrons.

labeling – for example, use of polynucleotide kinase to affix ^{32}P-phosphate to the 5' termini of oligonucleotides, allowing the researcher to monitor the fate of the molecule, reactions affecting the ends, or hybridization. Many types of isotopic and molecular (fluorescent, affinity binding) labels and labeling schemes have been devised for all classes of biomolecules.

Meselson/Stahl experiment - proof that DNA replication is semiconservative. Growth of *E. coli* cells with ^{1}H medium then "chase" it with ^{2}H medium in the second generation. The DNA has the same buoyant density in both daughter cells, intermediate between those obtained with DNA preparations from ^{1}H- and ^{2}H-adapted cells, respectively. This is because the ^{2}H goes equally into both daughter chromosomes (replication is *semi-conservative*).

methyloctylsilyl - C_8 reverse-phase packing.

mobile phase - the agent used to induce the sample to move through the column.

monoclonal antibodies - single epitope-specific immunoglobulin produced by hybridoma cells, the result of fusion of a spleen cell with a sarcoma. The hybridoma is transplanted into the peritoneal cavity of a mouse and proliferates to produce large amounts of ascites culture, a rich source of the antibody.

native folded conformation - biologically active form of a macromolecule, containing intact hydrogen bonds, salt bridges, hydrophobic "bonds," and disulfide bonds, when a coenzyme normally binds to an enzyme (e.g., the porphyrin in cytochromes); the coenzyme-free protein is called an apoprotein.

nonpolar - a commonly used mechanism in sorbent extraction. Nonpolar interactions occur between nonpolar isolate functional groups and nonpolar functional groups on the surfaces of a sorbent. Also refers to functional groups that exhibit predominantly van der Waals interactions. Thus, C_{18} is referred to as a nonpolar sorbent because of its hydrocarbon chain.

overload - saturation of the stationary phase by the solute evidenced by band broadening, tailing, and flat edged chromatographic peaks.

partition ratio - (also capacity ratio, k). The ratio of the amount of time the solute spends in the stationary phase to that spent in the mobile phase.

pH - negative log of the hydrogen ion concentration of an aqueous solution.

pK_a - for an acidic or basic functional group, the pH at which half of the functional groups in solution are charged and half are neutral. The pK_a of a specific functional group is related to the chemistry of the species in which the group is contained. The pK_a value for a group is important in sorbent extraction, especially when using ion-exchange, because isolate or interference retention requires that both the isolate and sorbent be charged.

polar - polar interactions occur between isolate functional groups exhibiting dipole moments, and similar groups on the sorbent. Also refers to the functional groups themselves. For example, unbonded silica is a very polar sorbent because of the high content of silanols.

polyacrylamide gel electrophoresis - analytical and preparative method to characterize a mixture of proteins or nucleic acids based on differences in particle size and charge. The gel is cast between two glass plates by polymerizing acrylamide and bis-acrylamide using ammonium persulfate and TEMED catalysts.

porous polymer - polystyrene beads (cross-linked with divinylbenzene) that are particularly good for aqueous samples. Other modifications are available with unique separation properties.

primary isotope - the undiluted isotope, straight out of the bottle.

protein detection assays - methods used to determine the presence of protein. Examples include the Lowry, Bradford, BCA, and silver-stain techniques. Each relies on the presence of certain types of reagent-binding sites on the protein.

protein purification chart - indicates the total yield and the increase in biological activity per unit of protein determined by specific activity assay over the course of the purification procedure. Typical steps are cell growth, preparation of crude extract, ammonium sulfate precipitation (ppt, supernate), gel filtration chromatography, and other chromatographic separations.

radioisotope – high-energy particle-emitting atomic species. (e.g., ^{32}P, ^{35}S, ^{14}C, and ^{3}H). High-energy particles include alpha and beta particles, neutrinos, and gamma rays.

resolution - a measure of the ability of a column to separate two peaks through a combination of column efficiency and selectivity.

retention - attraction of a chemical species for a sorbent such that the species is immobilized on the sorbent. The degree of this attraction is referred to as the strength of the retention.

retention time - time between the moment of injection until a compound elutes and is detected at the peak maximum.

reverse-phase - separation based upon the polarity (hydrophobicity/hydrophilicity) of the biomolecule. The matrix is typically composed of unmodified and derivatized alkyl chains containing 8 to 18 carbons (C_{8-18}).

reverse-phase chromatography - the gradient principle is applied by increasing the concentration of a less polar solvent (e.g., ethanol) relative to the H_2O concentration, which decreases the strength of hydrophobic interactions between the matrix and biomolecule and leads to elution.

salt-gradient elution - by steadily increasing the NaCl or other salt concentration, charge-charge interactions between an ion-exchange matrix and the biomolecule can be outcompeted, resulting in its elution.

scintillation - when a radioactively excited molecule induces a "fluor" solvent to emit light.

selectivity - the degree to which discrimination of one chemical species relative to others occurs. Selectivity of a sorbent for a species is the ability of that sorbent to retain only that species. Selectivity of an extraction is the degree of isolate purity achieved as a result of the extraction. Selectivity of a counterion is the degree to which an ionic sorbent prefers that counterion relative to other counterions. Selectivity is important in sorbent extraction since the more selective each step in the extraction is, the fewer steps are required. Also, the more selective each step, the smaller the mass of sorbent required.

shuttle vector - contains control elements sufficient to allow expression of the gene in either prokaryotic or eukaryotic systems. Used to propagate eukaryotic genes in bacteria then transfer them to eukaryotic hosts for activity or regulation studies.

silica gels - silicon chains, networks that are commonly used in strongly organic systems, e.g., lipids, carbohydrates. Separations based on derivatives include all of those described here. Since silica is degraded under acidic conditions, pH-dependent experiments must be done using derivatized silicas, or other replacement polymer matrices.

size exclusion - a chromatographic method that utilizes porous packings of a certain size. Molecules that are too large are excluded from the pores while small molecules may penetrate easily, resulting in fractionation based on particle size.

sodium dodecyl sulfate (SDS) - C_{12} alkyl chain connected to a sulfate group, which binds relatively nonspecifically to the peptide backbone of a protein. This coats the molecule as a linear entity and depends essentially upon the length, and thus normalizes the linear shape of the molecule so that protein sizes can be compared directly during electrophoresis.

solid support - a variety of high temperature materials upon which stationary phases are deposited.

solute - nonsolvent components in a sample.

solvation - process that prepares a sorbent for sample application. In most cases, solvation consists of 2-3 steps: first, wetting of the sorbent with an organic solvent (this solvent should be miscible with the solvent used in the next step), followed by removal of excess wetting solvent from the sorbent using a solvent similar to the sample matrix to be applied.

solvent - the main nonsolvent component of a sample.

sorbent - porous, chemically modified silica used for selective extraction of chemical species from liquids. Typical sorbents are 40-μm particle size, 60-Å porosity bonded silica gels.

sorbent bed - sorbent packed into a configuration such that solvents and liquid samples can be passed through the sorbent. Sorbents are more efficient at extraction and concentration when used in a bed configuration since all of the sample will be exposed to sorbent (compared to adding sorbent directly to a sample then filtering the sorbent out), and manipulation of the sorbent is simpler.

specific activity - biological activity determined by assay per unit mass of protein. This definition introduces the somewhat arbitrary nature of enzymatic activity units. The amount of radioactivity per total amount of mass; moles of radioactive material/total number of moles of all isotopes (e.g., 1000 dpm [^{14}C]alanine/mmole alanine).

stacking gel theory - use of a separately polymerized 'stacking gel' above the main running gel, which is adjusted to a different pH. Due to a rather complicated set of interactions, the proteins compress at the interface between the first and second gel, aligning them so they begin to electrophorese at the same time. This helps to compress the bands prior to resolution by size on the gel, giving a cleaner separation and final pattern.

staining and destaining - protein gels are typically stained by soaking in either Coomassie brilliant blue or silver, which is then reduced to Ag^0 in a process similar to black-and-white photography to give dark brown to black bands. After staining, the excess unbound dye is removed by soaking the gel in destaining solution (e.g., dilute acetic acid for Coomassie blue).

stationary phase - chemical that is a liquid or a solid at operating temperatures and which is the main factor in causing separation of a component blend to occur.

substrate (ligand) affinity - separation by preferential binding to a substrate-derivatized matrix. Variations include substrates such as ATP, antibodies and protein A. The latter is from *Staphylococcus aureus* and binds the constant region of immunoglobulins; they are released at acidic pH.

tailing - a skewing occurring on the "following edge" (eluting after the maximum) of a chromatographic peak resulting from a higher affinity population of active sites in the chromatographic resin.

theoretical plate - a parameter measured from the bandwidth obtained using a given set of operating conditions that characterizes the resolving capacity of the chromatography resin.

titer - the amount of specific antibody in a mixture (e.g., serum) determined by serial dilutions of the solution of interest and subsequent immunoassay.

Western blotting - used to detect the presence of a specific protein in a complex mixture using an antibody for specific detection. The mixture is first resolved according to size on a polyacrylamide gel, then the proteins are transferred laterally onto and attached to a filter in the same pattern. The filter is then probed by soaking with the antibody preparation, rinsed thoroughly and tested for adhered probe antibody by binding a second antibody peroxidase conjugate. When the filter is exposed to 4-chloronapthol (or 3,3'-diaminobenzidene) and H_2O_2, the dye is oxidized, turns black, and reveals the position of the antigen protein in the electrophoretic pattern on the blot.

Laboratory Reagents

The following list of reagents is included to provide detailed functions of each component used to construct the reactions described in this manual. It is imperative that the student use this information in tandem with the experiments describe in the units. Student can easily consider several concepts at a time to address a complex problem. Reaction planning and problem troubleshooting will benefit from this information. Listing our own facts is not intended to replace those of a teacher who disagrees or has more to add. We merely intend to provide a starting basis for individual student reflection, student-student and student-teacher discussions, and classroom roundtables.

Send us tips for the next edition (hardin@bchserver.bch.ncsu.edu).

ϕx174 - a single-stranded DNA bacteriophage with a genome of 5375 bases; the first genome to be completely sequenced, by F. Sanger and colleagues in 1977. Restriction digests of this bacteriophage are commonly used as molecular weight markers.

4-chloro-1-napthol - chromogenic substrate for visualizing horseradish peroxidase conjugates. Cleavage of 4-chloro-1-napthol yields a blue precipitate that is soluble in alcohol and xylene but not in aqueous solutions. This substrate is commonly used in Western blotting protocols.

agarose - linear polysaccharide polymer extracted from agar extracted from a red seaweed. Commonly used to prepare electrophoresis gels or culture plates. When a solution of agarose, which has been chemically modified to be structurally weakened, is heated to its boiling point and then allowed to cool, it hardens, forming a matrix. The density of the hardened agarose gel is determined by the agarose concentration. Pore size is determined by adjusting the agarose concentration; the higher the agarose concentration, the smaller the pore size. Agarose gels are fragile; they are held together by weak hydrogen and hydrophobic bonds. DNA fragments in the 0.2-50 kb range can be separated by agarose gel electrophoresis.

alcohol/buffer solution - buffer containing a hydroxylated alkane; used in the Elu-Quick DNA purification kit to remove salts and undesired macromolecules from samples.

alkaline phosphatase - (e.g., bacterial alkaline phosphatase, calf alkaline phosphatase) – hydrolytic enzymes that dephosphorylate the 3' and 5' ends of nucleic acids (e.g., ends of plasmids). Commonly used to dephosphorylate the 5' phosphorylated termini of linearized double-stranded DNA to prevent self-ligation. The enzymes are active at 65°C for at least one hour and inactivated by phenol extraction.

ammonium sulfate - salt commonly used to precipitate proteins; binds charged residues and thereby reduces and neutralizes charge repulsion, thus inducing aggregation and precipitating the protein from solution. Ammonium sulfate ions are nonchaotropes, therefore, proteins treated with this salt typically do not denature. Ammonium sulfate is commonly used to purify enzymes and proteins when added to solution at a high concentration. High ionic strengths can be achieved with ammonium sulfate due to its high solubility in water.

ammonium persulfate (APS) - initiator of acrylamide polymerization. Ammonium persulfate donates the free radicals required to catalyze the polymerization process.

ampicillin - a β-lactam antibiotic that arrests *E. coli* cell growth by inhibiting the number of enzymes in the cell membrane and thereby disrupting cell wall synthesis. Ampicillin is commonly used to select for *E. coli* cells that have been transformed with a plasmid carrying the *bla* gene (encoding β-lactamase).

bactotryptone - pancreatic digest of casein used to prepare microbiological cell culture media.

bactoyeast extract - extract from autolysed yeast cells used to prepare microbiological cell culture media.

BCIP/NBT solution - nitroblue tetrazolium chloride and 5-bromo-4-chloro-3-indolylphosphate *p*-toluidine; chromatogenic substrate used to visualize alkaline phosphatase conjugates in immunoblotting procedures. These compounds are used together in a 1:1 molar ratio under alkaline conditions to obtain optimal sensitivity. The alkaline phosphatases mediated cleavage of the BCIP phosphate produces a blue color and the NBT is reduced to yield an insoluble purple precipitate.

bicinchonic acid (BCA) - sodium salt used at alkaline pH in protein concentration determination assays to assist the reduction of Cu^{2+} to Cu^{1+}. Chromogenic reagents react with Cu^{1+} to form a purple complex with a characteristic absorbance maximum; the absorbance is directly proportional to protein concentration, which is determined by comparing protein samples of unknown concentrations to protein standards at known concentrations. The BCA assay is preferable to other protein concentration determination assays because of low levels of interference from reagents used in protein preparation procedures.

bovine serum albumin (BSA) - widely available protein extract. Bovine serum albumin is commonly used for nonspecific blocking reactions on Western Blotting protocols. BSA is used to competitively reduce nonspecific binding of antibodies to proteins. Bovine serum albumin can be acetylated to inactivate contaminating nucleases and proteases. It is suitable for use as an enzyme stabilizer during purification or for dilution of restriction endonucleases and a broad range of other enzymes.

bromophenol blue - a dye exhibiting a blue color at pH > 1.6. In polyacrylamide or agarose gels, bromophenol blue migrates in an electric current at the same rate as double-stranded DNA fragments of about 300 bp. Commonly used to monitor the migration of DNA, taking advantage of its electrophoretic characteristics.

BstEII digest - digest of lambda phage DNA yielding 14 fragments suitable for use as molecular weight standards in agarose gel electrophoresis.

*Bst*EII, *Hind*III - type two restriction endonucleases that recognize and cleave 4-8 base pair duplex DNA sequences with two-fold axes of symmetry. *Bst*EII is derived from *Bacillus stearothermophilus*; *Hind*III is derived from *Haemophilus influenzae*.

buffers - molecules that form both conjugate acid and base forms as a result of a protonation/deprotonation equilibrium and are used to resist changes in pH when H^+ or OH^- are added to a solution. A high buffer capacity with a low buffer concentration is needed to avoid an undesirable fluctuation of pH. Growth of cell cultures depends on both the pH of the media and the species of buffer present. Buffer species vary in toxicity in different types of cells so care must be taken in choosing buffers and their concentrations for use with new cell lines, e.g., Tris buffers are cytotoxic in certain cases. HEPES is the most generally used zwitterionic buffer. Improved pH control between pH 6.7 and 8.4 may be obtained when 20 to 50 mM HEPES is incorporated into a culture medium.

calcium chloride ($CaCl_2$) - divalent cation salt. Calcium chloride is commonly used in transformation procedures to neutralize negatively charged DNA, thus allowing the DNA to bind and enter a competent cell. Heat shock weakens specific proteins in the cell wall, inducing DNA uptake.

calf-intestine alkaline phosphatase (CAP) - a phosphomonoesterase that removes the 3' and 5' phosphates from DNA and RNA. Commonly used to dephosphorylate 5' termini of vector DNA to prevent self-ligation. It is inactivated by heating in the presence of the Zn^{2+} chelator nitrilotriacetic acid (NTA).

chloroform - organic solvent used in extraction procedures to separate nucleic acids from proteins (see "solvents").

Coomassie blue stain - blue dye used to detect proteins on polyacrylamide gels and in colorimetric assays. It binds proteins in a specific mass ratio making them intensely colored. Within a specific range of protein concentrations, these reagents give rise to absorption whose intensity is linearly proportional to protein concentration.

DEAE (diethylaminoethyl) Sepharose - ion exchange chromatography matrix used for protein purification. This strong, high-capacity anion exchange support is composed of cross-linked agarose and binds proteins via positively charged DEAE groups.

3,3'-diaminobenzidine tetrahydrochloride dihydrate (DAB) - chromogenic substrate that reacts with horseradish peroxidase to form a brown reaction product.

diethyl pyrocarbonate (DEPC) - modification reagent for His and Tyr residues at the surface of proteins. A *ca.* 0.1% solution inactivates RNase by reacting with and blocking their catalytic sites, thus protecting RNA against degradation. Moisture hydrolyzes DEPC to ethanol and carbon dioxide.

DNA ligase - hydrolytic enzyme that links adjacent nucleotides from two different DNA strands through the formation of a phosphodiester bond. Requires Mg^{2+} and ATP.

DNase I - deoxyribonuclease that digests single- and double-stranded DNA nonspecifically.

dithiothreitol (DTT) - a.k.a. "Cleland's reagent"; a chemical reagent commonly used to reduce disulfides to the corresponding thiols. DTT is commonly used at low concentrations to stabilize enzymes containing free sulfhydryl groups. Higher concentrations of DTT are used to cleave disulfide linkages in polypeptides and to facilitate protein denaturation by detergents and chaotropic agents.

Econocolumn™ chromatography column - a commercial, high-quality, affordable low-pressure chromatography column. A porous polymer bed support at the bottom of the column retains fine particulate matrices while providing quantitative sample elution.

ethanol - an organic solvent used to precipitate nucleic acids. The precipitate can be redissolved to a defined volume to produce a solution having a desired macromolecule concentration, a useful step in a broad range of protocols. Commercial 95% ethanol is preferred over 100% preparations because the latter contains benzene. Potassium or sodium acetate are also typically added to increase the amount of precipitant; the former is more soluble in aqueous ethanol.

ethidium bromide - nucleic acid intercalating agent that binds by sliding between every other pair of base pairs. Ethidium bromide is used to stain and visualize nucleic acids in agarose gel electrophoresis because it fluoresces when bound to them. Upon intercalation, the buoyant density of DNA increases, allowing nucleic acid samples to sediment into the wells of agarose gels. It can be used to detect both single- and double-stranded nucleic acids.

ethylenediaminetetraacetate (EDTA) - a metal-chelating agent. Commonly used as an inhibitor of enzymes that require metal cofactors (e.g., endonucleases, metalloproteases). EDTA decreases the amount of free metals in solution. If metals are specifically required for enzyme activity, EDTA can reduce the efficiency of target DNA binding. EDTA is also used to weaken bacterial cell walls in DNA extraction procedures by chelating Mg^{2+}, which normally help maintain cell wall integrity. EDTA also chelates cations that precipitate nucleic acids or enhance disulfide bond formation via redox chemistry.

formamide - organic molecule used to denature nucleic acids and proteins. It competes with the structural hydrogen bonding substituents of nucleic acids and proteins for the hydrogen bonds of water molecules, thereby destabilizing their secondary, tertiary, and quaternary structures.

glass beads - negatively charged beads of glass used in the Elu Quick procedure to extract DNA from agarose. Sodium perchlorate coats the negatively charged beads with positively charged sodium ions, which in turn bind the negatively charged DNA. Low salt buffer is used to elute the DNA from the glass beads after removing impurities using buffer/alcohol solution.

glycerol - an organic solvent commonly used to concentrate and store proteins at low temperatures. A 50% w/v glycerol solution will not freeze at -20°C, therefore, proteins will not denature as a result of ice formation. Glycerol has a density of 1.26 g/ml and is frequently used as a component in electrophoresis loading buffers.

glycine - amino acid that exists as either a protonated monocation, a zwitterion, or as glycinate with a single negative charge. Glycine is used in buffers for discontinuous gel electrophoresis; the glycinate ion changes to the uncharged and immobile zwitterion in low pH buffer (< 2.3). This causes a deficiency in mobile ions, which decreases the current flow. However, current must be maintained, so the voltage is increased. Therefore, there is a zone of leading chloride ions and tailing glycinate ions with an increased voltage gradient. This phenomenon results in compression of all proteins into a compact zone, producing a better separation. With pK_{a1} of 2.3 and pK_{a2} of 9.6, glycine is suitable for buffering at the upper and lower extremes of the physiologically relevant pH range.

goat anti-mouse peroxidase (GAMP) - antibody conjugate containing goat serum as a protein stabilizer.

guanidinium thiocyanate - a strong protein denaturant that is used at high concentrations to induce the unfolding of many proteins and separate unfolded proteins into their constituent polypeptide chain(s).

hydrogen bond competitors - reagents such as formamide and guanidinium salts that compete at high concentrations with the hydrogen bonds of water and the intramolecular hydrogen bonds of biomolecules and their ligands. The result is molecular denaturation.

hydrophobic additives - molecules that interact directly with hydrophobic portions of macromolecules and alter their surrounding solvent structure. They are often added to solution to decrease the dielectric constant.

isopropanol - an organic solvent used to precipitate nucleic acids from solution. When compared to ethanol, 50% less is required, thus minimizing the total volume of solution necessary for precipitation. Isopropanol forms hydrogen bonds with water. When isopropanol competes with nucleic acids and proteins for hydrogen bonds, their solubility decreases and they precipitate.

isopropyl-β-D-thiogalactopyranoside (IPTG) - a nonmetabolized gratuitous inducer of the *E. coli lac* operon. IPTG achieves this by binding lacI repressor proteins. It is widely used as an inducer in connection with cloning and expression vectors that utilize the lac promoter (operator DNA).

Klenow DNA polymerase - an enzyme that catalyzes filling-in reactions that can produce blunt or recessed (staggered) ends. This tool is used to make randomly-primed DNA. A set of combinatorially selected or randomized sequenced oligonucleotides is employed as primers, allowing one to incorporate nonradioactive or ^{32}P-labeled nucleotides into DNA fragments. Klenow DNA polymerase is derived from the large fragment of *E. coli* DNA polymerase I; it is genetically engineered to eliminate the 3' to 5'-directed exonuclease activity. As a result, it retains the polymerization fidelity of the holoenzyme but does not degrade 5' termini.

$lacI^q$ - a biological marker that imparts and indicates overproduction of lac repressor protein.

Luria-Bertani (LB) broth - a liquid medium used to maintain and propagate *E. coli* cells.

lysozyme - small (M_r 14,600) hydrolytic enzyme of ubiquitous cellular distribution that hydrolyzes 1,4-β linkages between *N*-acetylmuramic acid and *N*-acetyl-*D*-glucosamine in peptidoglycans of bacteria cell walls.

magnesium salt ($MgCl_2$, Mg^{2+}) - divalent cation salt that is usually essential as an enzyme cofactor. Divalent cations such as Mg^{2+} also stabilize charged biological macromolecules.

media - solutions used to culture cells. The commonly used LB medium provides good growth for overnight cultures. Richer media such as NZCYM produce faster growth. Media are optimized by adding salts and stabilizers to enhance cell propagation and maintenance. Generally, no more than 10% of a flask volume should be occupied by culture medium to optimize cell growth and aeration.

N,N'-methylenebisacrylamide (bisacrylamide) - most frequently used cross-linking agent for polyacrylamide gels.

mineral oil - overlay in PCR reactions. PCR reaction overlays promote temperature homogeneity and reduce sample evaporation during thermocycling.

monoclonal antibody - antibodies of a single specificity (i.e., originating from a single cloned antibody-producing cell). Monoclonal antibodies are useful as highly specific markers (detector molecules) *in vitro* in cellular ultrastructure studies.

n-butanol - an alcohol that is layered on top of freshly poured polyacrylamide gels to reduce the size of the meniscus, which can distort the banding pattern. *N*-Butanol interacts with the water in the gel through hydrogen bonding. The *n*-butanol also excludes oxygen from the gel, which can inhibit its polymerization.

nitrocellulose paper - sheet medium used in molecular biology procedures to bind nucleic acids and proteins, a.k.a. blotting (typically as 0.45-μm pore-sized sheets). Solid support for the immobilization of proteins and other macromolecular antigens, either by transfer from gels after electrophoresis (Western blots) or by blotting directly from pipetted reaction solutions (dot blots).

NP40 - detergent commonly used to lyse bacterial cell walls.

organic solvents - molecules that decrease the dielectric constant of solutions, make them less polar, and consequently alter various molecular interactions (e.g., intramolecular, solute-solute and solvent-solute forces, ionic and hydrogen bonding).

o-nitrophenol-β-D-galactopyranoside (ONPG) - an artificial β-galactosidase substrate. When ONPG is cleaved, a yellow product (*o*-nitrophenol) forms and easily assayed spectrophotometrically at 420 nm. ONPG is commonly used in conjunction with a β-galactosidase reporter gene in gene expression studies.

peroxidase - an oxidoreductase that catalyzes reactions of the general type: donor + $H_2O_2 \rightarrow$ oxidized donor + H_2O; oxidation with reduction of hydrogen peroxide to water.

phenol - organic solvent used to separate and purify nucleic acids from proteins and other cellular components. In buffered aqueous phenol solution, proteins denature and collect at the interphase while most nucleic acids remain in the aqueous phase. Nucleic acids should not be stored in phenol since it can catalyze alkaline hydrolysis.

Phenylmethylsulfonyl flouride (PMSF) - chemical reagent used as an inhibitor of serine proteases (e.g., trypsin and chymotrypsin); commonly used to ensure protein integrity during protein isolation procedures.

phosphate buffer - common buffer; disadvantages are that it is relatively ineffective above pH 7.5. It can bind and sometimes precipitate polyvalent cations, and it frequently stimulates or inhibits enzymes. Solutions are prepared by mixing precalculated amounts of monobasic and dibasic sodium or potassium phosphate stocks to produce the appropriate pH.

polyacrylamide gel - a polymerized and cross-linked gel formed from acrylamide and bisacrylamide. Used to electrophorese proteins and nucleic acids. Proteins can be separated by a variety of techniques, including denaturing gels using SDS or urea, isoelectric focusing gels, and native nondenaturing gels, in a wide variety of buffers.

polyethylene glycol (PEG) - organic solvent used to precipitate soluble proteins. Polyethylene glycol competes with the polar groups of proteins for the hydrogen bonding sites of water, thus modifying the surrounding solvent structure. PEG separates protein solutions into two immiscible liquid phases, containing either protein or polyethylene glycol. Most proteins precipitate in 30% w/v PEG.

proteinase K - proteolytic enzyme used to hydrolyze internal peptide bonds of proteins. Proteinase K is a general broad-spectrum, nonspecific, degradative serine protease that is active over a wide pH range. It is compatible with SDS, EDTA, and urea, and stable at high temperatures.

pUR288 - fusion protein expression vector containing the *lacZ* gene, which is transcribed under control of the *lac* UV5 promoter. The *lacZ* gene is normally repressed, but is induced (derepressed) when IPTG is added to the growth medium. Protein-encoding DNA fragments are inserted into the polylinker site at the end of the *lacZ* gene and expressed (transcribed, translated) to produce fusion proteins.

ribonuclease (RNase) - hydrolytic endoribonuclease that cleaves RNA with either limited or no specificity. RNases are commonly used to remove contaminating RNA from solutions (e.g., when preparing RNA-free DNA samples).

salts - used to modify solutions so they selectively solubilize and stabilize polyelectrolytic macromolecules. Salts can act by: (a) increasing the ionic strength and inhibiting electrostatic interactions by shielding charged functional groups. This decreases both repulsions between polyelectrolytes and direct electrostatic interactions between charged residues. (b) Nonpolar interactions can occur between solvent-exposed hydrophobic regions and the hydrophobic portion of organic salts (e.g., carboxylates, sulfonates, or ammonium). (c) A competition between solvent-solute and intramolecular interactions produces the net stability of a given biomolecular structure. These effects follow the order of cations and anions in the Hofmeister (lyotropic) series. For example, sulfate ions are lyotropic (stabilize the native form of macromolecules), while ClO_3^- and SCN^- are chaotropic (disrupt water structure). Chaotropes decrease the stability of native conformations of macromolecules, promoting denaturation at high concentrations. (d) Nucleic acids and proteins can "salt in" or "salt out of" solution depending on their composition, the solution conditions, and their resulting solubility. The solubility is increased at low salt concentrations due to constructive interactions between the charged macromolecule and the ionic species of the salt. "Salting out" is achieved at high salt concentrations because the macromolecules behave as neutral dipoles and solubility is governed by the hydrophobic effect. Proteins aggregate in a process called hydrophobic collapse.

Sephacryl S-300-HR - gel filtration chromatography resin; an insoluble but highly hydrated polymer of polyacrylamide forms the porous bead matrix.

SOB/SOC media - complex cell growth media that ensure maximum transformation efficiency. SOB and SOC media are identical except that SOC contains glucose as an additional carbon source.

sodium azide - bacteriocidal agent commonly used to protect stored proteins.

sodium dodecyl sulfate (SDS) - (or sodium laurel sulfate) - amphipathic protein denaturant that binds the hydrophobic regions of proteins, unfolds them, and imparts a net negative charge. SDS binds and produces a nearly constant charge/mass ratio (essentially overcoming differences in protein charges); it is used in polyacrylamide gel electrophoresis to help separate proteins by mass; and denatures proteins, typically by producing rod-like structures as a result of mutual repulsions between the negatively charged complexes. It is commonly used to disrupt cell walls and dissociate nucleic acid-protein complexes in DNA and RNA extraction protocols.

standard sodium citrate (SSC) - common medical buffer solution containing NaCl and sodium citrate. SSC is typically used in nucleic acid hybridization and blot transfer applications to stabilize DNA charges. SSC is commonly used in medical applications because it simulates the ionic stregenth and buffering capacity of blood serum.

sucrose - organic additive used in electrophoresis loading dyes to increase the sample density, ensuring that they layer down into the wells and do not diffuse out into the buffer reservoir.

TAE buffer - common buffer containing Tris, acetate, and EDTA.

Taq polymerase - a 94 kDa thermostable DNA-dependent DNA polymerase, containing a 5' to 3' exonuclease activity, which catalyzes DNA replication. *Taq* polymerase is purified from *Thermus aquaticus* YTI and can withstand prolonged incubation at temperatures up to 95°C without significant loss of activity. *Taq* catalyzes the template-dependent synthesis of polynucleotide chains from deoxyribonucleoside 5'-triphosphates by directing sequential phosphodiester bond formation between the 3' OH of the terminal nucleotide residue and the α phosphate of the added nucleotide triphosphate, forming the covalently linked sugar-phosphate backbone of nucleic acids, and eliminating pyrophosphate.

TBE buffer - common buffer containing Tris, borate, and EDTA; commonly used in polyacrylamide and agarose gel electrophoresis primarily because borate resists overheating.

TENS buffer - commonly used buffer containing Tris, EDTA, NaCl, and SDS.

TEP buffer - commonly used buffer containing Tris, EDTA, and PMSF.

tetramethyl ethylenediamine (TEMED) - chemical used to catalyze the polymerization of acrylamide gels by stabilizing free radicals.

tris(hydroxylmethyl)aminomethane (Tris) - base commonly used in buffers (e.g., Tris-HCl, Tris-acetate). Tris buffers effectively in the pH 7 to 9 range and is compatible with a variety of enzymes, including many restriction endonucleases and DNA-modifying enzymes.

tris-buffered saline (TBS) - common buffer used to optimize the experimental ionic strength and pH; commonly used to wash unbound antibodies and contaminants from nitrocellulose during immunoblotting.

Triton X-100 (polyethylene glycol p-isooctyl phenyl ether) - nonionic detergent commonly used to solubilize biological membranes.

TTBS - Tris-buffered saline containing the detergent Tween 20; commonly used in molecular biology protocols to wash unbound antibody and contaminants from nitrocellulose during immunoblotting.

Tween 20 - a nonionic detergent that is commonly added to TBS (Tris-buffered saline); this wash solution is used in Western blotting protocols and cell lysis buffers.

XL1-Blue E. coli - *E. coli* strain whose engineered genotype is optimized for selected cloning protocols (consult product literature).

xylene cyanol - dark blue/black dye (MW 554.6 g/mol) that is used in protein MW marker mixtures in electrophoresis.

ABBREVIATIONS LIST

NOTE: This listing does not include acronyms for plasmids, restriction enzymes, most genes, or most buffers.

%C, percent crosslinker;

η, solution viscosity;

-NHCO-, amide (peptide) group;

2D SDS-PAGE, two-dimensional SDS-PAGE;

2XYT medium, bacteria growth media

2XYT/amp medium, bacteria growth media with the antibiotic ampicillin added

2XYT/amp/IPTG medium, bacteria growth media with the antibiotic ampicillin and a gratuitous inducer, IPTG added

3T3, transformed mouse fibroblast cell line;

5AS, 5-aminosalicylic acid;

a, chemical activity (= γ C);

α, fraction of risk (student t test);

α, selectivity;

$[\alpha]^{20}$, specific rotation at 20°C;

A_{280} or A_{λ}, absorbance at 280 nm or wavelength λ;

ABTS, 2-2'-azino-di-3-ethyl-benzothiazolin sulfone-6, diammonium salt;

AC, Achilles' heel cleavage;

AC, affinity chromatography;

$Ac(Gly)_4Et$, acetyltetraglycyl ethyl ester;

ADH, alcohol dehydrogenase;

AMP, adenosine-5'-monophosphate;

AMPGD, disodium 3-(4-methoxyspiro-[1,2-dioxetane-3-2'-tricyclo[3.3.1.1^{3,7}]decane)-4-yl)phenyl-β-D-galactoside ;

AMPPD, disodium 3-(4-methoxyspiro-[1,2-dioxetane-3-2'-tricyclo[3.3.1.1^{3,7}]decane)-4-yl)phenyl phosphate;

amp^r, ampicillin resistance genotype/phenotype (gene);

amp^s, ampicillin sensitive genotype/phenotype;

AMV, avian myeloblastosis virus;

AP, alkaline phosphatase;

APS, ammonium persulfate

ARS, autonomous replication sequence;

b, cuvette pathlength;

β-gal, β-galactosidase gene (or protein);

β-gal-*myo*-3, β-galactosidase-myosin-3 gene (or fusion protein);

β-ME, β-mercaptoethanol;

BAP, bacterial alkaline phosphatase;

BBR, Boehringer blocking reagent;

BCA, bicinchonic acid

BCIP, 5-bromo-4-chloroindoyl phosphate;

BCIP/NBT, nitroblue tetrazolium chloride/ 5-bromo-4-chloro-3-indolylphosphate *p*-toluidine

BD-Sepharose, benzoylated DEAE Sepharose;

BG, fraction of free RNA partitioning as nonspecific background;

BiP, immunoglobulin binding protein;

bis, *N'*,*N'*-methylene-bis(acrylamide);

BL21 (DE3), *E. coli* strain containing T7 RNA polymerase;

BME, β-mercaptoethanol;

BSA, bovine serum albumin;

bz-DEAE, benzoylated DEAE;

C, concentration;

C, perturbant;

C_8, octyl;

C_{18}, octadecyl;

CAP, calf intestinal alkaline phosphatase;

CAT, chloramphenicol acetyltransferase;

CD, circular dichroism;
CD4, coreceptor of helper cells;
CD8, killer T cell-specific coreceptor protein;
cDNA, complementary DNA;
CDR, complementarity determining region (immunoglobulin);
CENS, centromeric DNA sequences;
cfu, colony-orming units;
CHO, Chinese hamster ovary cell line;
Ci, Curie;
CIP, calf intestinal phosphatase;
CM, carboxymethyl;
CMC, critical micelle concentration;
CMP, cytosine-5'-monophosphate;
CNBr, cyanogen bromide;
C_o, starting material concentration;
COSY, correlation (two-dimensional) spectroscopy;
c_0t, DNA concentration x time;
CP, fraction of free RNA partitioning due to target binding;
C_{rr}, retentate concentration;
C_s, perturbant (salt) concentration;
CSF, cerebrospinal fluid;
CSPD, disodium 3-(4-methoxyspiro-(1,2-dioxetane-3-2'-(5'-chloro)tricyclo[3.3.1.1^{3,7}]decane)-4-yl)phenyl phosphate;
C_t, filtrate concentration;
d, average deviation;
d, degeneracy;
Da, Dalton (= molecular weight of 1);
DAB, 3,3'-diaminobenzidine tetrahydrochloride dihydrate;
D^b MHC molecule, self protein;
dCMP, 2'-deoxycytosine-5'-monophosphate;
ΔC_p, change in heat capacity;
ΔG, Gibbs free energy;
ΔH, enthalpy change;
$\Delta\Delta G$, change in free energy (e.g., change in stability upon mutation);
ddNTP, 2',3'-dideoxynucleoside-5'-triphosphate;
DEPC, diethyl pyrocarbamate;
DHFR, dihydrofolate reductase;
ΔH_{res}, enthalpy of melting per residue of polymer;
DIFP, diisopropyl fluorophosphate
DIG, digoxygenin;
DIG-ddUTP, DIG-dideoxyUTP;
DIG-dUTP, DIG-deoxyUTP;
DIG-NHS ester, digoxigenin-3-*O*-methylcarbonyl-ε-aminocaproic acid-N-hydroxysuccinimide ester
DMD, Duchenne muscular dystrophy;
DNA, 2'-deoxyribonucleic acid;
DNA^{n-}, DNA with a net anionic charge of *n* units;
DNase, deoxyribonuclease;
DNP, dinitrophenol;
DP, discriminating power;
$\Delta pK_a/\Delta T$ (°C), pK_a temperature dependence;
ΔS, entropy change;
dsDNA, double-stranded DNA;
DTE, dithioerythritol;
DTT, dithiothreitol;
ε, energy of a dipeptide (etc.) as a function of torsion angles ϕ, φ, χ;
ε_{280} or ε_λ, molar extinction coefficient at 280 nm or wavelength λ;
E, voltage;
E/*d*, electrophoretic field strength at charge separation *d*;
EAE, allergic encephalomyelitis;

EDTA, ethylenediaminetetraacetic acid

EGTA, ethylene-bis(oxyethylenenitrilo)tetraacetic acid;

ELF, enzyme-labeled fluorescence;

ELISA, enzyme-linked immunosorbant assay;

ETBS, Tris-buffered saline with EDTA;

Etd^+, ethidium;

EtOH, ethanol;

eu, entropy units ($Kelvin^{-1}$);

F_{ab}, antibody binding fragment (immunoglobulin);

F_c, constant fragment (immunoglobulin)

FDA, Food and Drug Administration;

FPLC, fast pressure liquid chromatography;

F.W., formula weight;

γ, activity coefficient;

γ_o, activity coefficient for a model compound in H_2O;

GAMP, goat-anti-mouse peroxidase antibody conjugate;

GAPDH, glyceraldehyde-3-phosphate dehydrogenase;

GC, gas chromatography;

GFC, gel filtration chromatography;

GLP, good laboratory practice;

GM-CSF, granulocyte-macrophage colony stimulating factor;

GMP, good manufacturing practice;

GMP, guanosine-5'-monophosphate;

groES, groEL, *E. coli* stress proteins (folding chaperonins);

GTC, guanidine thiocyanate;

Gu^+, guanidium;

H, efficiency;

H, hydrophobic;

HAMA, human anti-mouse antibody;

HB Ag, hepatitis B antigen

HEPES, (see buffers);

HETP, height equivalent of a theoretical plate;

HIC, hydrocarbon interaction chromatography;

His-Tag, $(histidine)_n$ (n = ca. 6) peptide fragment used for affinity purification;

HMG-CoA, hydroxymethylglutamyl-coenzyme A;

HRP, horseradish peroxidase;

hsp70, hsp90, stress proteins;

hspI, heat shock protein gene;

HY, self peptide;

I, current;

I, ionic strength;

<I>, information;

IB, inclusion bodies;

IEC, ion exchange chromatography;

IEF, isoelectric focusing;

IgG, immunoglobulin G;

IL-1a, interleukin 1a;

IMAC, immobilized-metal affinity chromatography;

IP, ion-pair(ing);

IPC, ion-pairing chromatography;

IPTG, isopropyl-β-D-thiogalactoside;

IRS-PCR, interspersed repeated sequence PCR;

k', capacity (retention factor);

k, deviation factor;

k, number of mutations;

K_{av}, fraction of stationary gel volume available for diffusion of a solute species;

kb, kilobase;

kbp, kilobase pairs;

K_d, distribution coefficient;

K_d, dissociation equilibrium constant;

k_i, salting-out coefficient for group i;

K_m, Michaelis-Menten constant;

K_m, slope of T_m vs. C_s;

KOAc, potassium acetate;

K_R, retardation coefficient;

kT, Boltzmann's constant x T (Kelvin);

L, bed length;

L, ligand;

l, liter;

LB, Luria-Bertani medium;

LC, liquid chromatography;

LINE, long interspersed repeat sequence;

LMP, low melting point;

Lumi-Phos 530, 4-methoxy-4-(3-phosphate-phenyl)-spiro(1,2-dioxetane-3,2'-adamantane) disodium

M, electrophoretic mobility;

M, insoluble support matrix;

m, mass;

M, median of a set of measurements;

M, molar;

μ, true value of a sample measurement;

M-CSF, macrophage colony-stimulating factor;

mAb, monoclonal antibody;

MCS, multiple cloning site;

MHC, major histocompatibility complex;

MHC (class I), displays peptides from inside the cell;

MHC (class II), displays peptides from proteins that enter the cell;

MMLV, Moloney murine leukemia virus;

M_o, free mobility in sucrose solution;

MONA, multiple of norman activity;

MPBS, PBS containing 2% skim milk powder;

MPD, methyl-2,4-pentanediol

M_r, molecular weight,

MSDS, Material Safety Data Sheets

MW, molecular weight;

myo-3, myosin-3 gene (or protein);

N(ρ), relative number of chain configurations;

N, number of distinct sequences having k mutations;

N, number of particles;

n, number of rotational bonds;

N, number of theoretical plates;

n, sequence length;

NAD^+, nicotinamide adenine dinucleotide;

N_{BGU}, number of β-galactoside units;

NBT, nitroblue tetrazolium salt;

nd, average number of mutations per sequence;

NH_4OAc, ammonium acetate;

NMA, *N*-methylacetamide;

NMP, *N*-methylpropionamide;

NMR, nuclear magnetic resonance;

NOESY, nuclear Overhauser effect (two-dimensional) spectroscopy;

NP40, Nonidet P-40 detergent;

NTA, nitrilotriacetic acid;

NZCYM,

OD_{280} or OD_λ, A_{280} (or A_λ) (older nomenclature);

oligo(dT), oligonucleotide composed of up to ca. 30 2'-deoxythymidines;

ONPG, o-nitrophenol-β-D-galactopyranoside;

OPD, o-phenylenediamine;

OSHA, Occupational Safety and Health Act

P, polar;

P, power (= I^2R);

P, proportion of the population having k mutations;

p, success rate;

p_0, a priori success rate;

PAGE, polyacrylamide gel electrophoresis;

PBMC, peripheral blood mononuclear cells;

PCR, polymerase chain reaction;

PDGF, platlet-derived growth factor A;

Pefabloc SC, [4-(2-aminoethyl)-benzene-sulfonyl fluoride];

PEG, polyethylene glycol;
pelB, signal peptide targeting the periplasmic space;
Pfu, *Pyrococcus furiosus*;
pH, -log $[H^+]$;
phe, phenylalanine;
phOx, 2-phenyloxazol-5-one;
p*I*, isoelectric point (pH);
p_i, probability describing the distribution of states;
pIII, minor coat protein of filimentous phage;
pVIII, major coat protein of filamentous phage;
pK_a, -log K_a (acidity equilibrium constant);
PMSF, phenylmethylsulfonyl fluoride;
poly(A), poly(adenylic acid)
poly(dC), poly(deoxycytidine);
$pp60^{src}$, cell growth regulatory protein (tumor growth);
pph, parts per hundred;
ppm, parts per million;
ppt, parts per thousand;
PTS, phosphoenolpyruvate sugar transport system;
q, molecular charge;
Q, partition coefficient;
Q, quaternary ammonium;
$Q_{0.9}$, rejection quotient at subscripted confidence level (Q test);
QAE, quaternary aminoethyl;
R & D, research and development;
ρ, chain segment density;
r, effective sphere-like (isotropic) radius;
R, gas constant;
R, geometric mean radius;
r, radius of the gel fibers;
r, molecular radius;
R, resistance;
r^{-p}, distance-dependent molecular interations (long-range, $p \leq 3$; short-range $p > 3$);
r^2, linear regression goodness-of-fit parameter;
RCF, relative centrifugal force;
$r_{exterior}$, exterior residue of a macromolecule
RFLP, restriction fragment length polymorphism;
RIA, radioimmunoassay;
$r_{interior}$, interior residue of a macromolecule
RNA, ribonucleic acid;
RNA-BP, RNA binding protein;
RNase, ribonuclease;
r_0t, RNA concentration x time;
RPA, RNase protection analysis;
RPM, revolutions per minute;
R_s, resolution;
RSA, relative specific activity;
r_{Stokes}, Stokes radius;
RT, reverse transcriptase;
RuDP, ribulose diphosphate;
S, methyl sulfonate;
S, population size;
S, solubility in aqueous solution containing perturbant;
S, solute;
σ, standard deviation;
S, substance of interest;
$[S]_i$, substrate concentration for sample i;
S/N, signal-to-noise ratio;
SA, specific activity;
SA-PMP, Streptavidin paramagnetic particle;
scF_v, single-chain antibody;
SCID, severe combined immune deficiency;
SD, standard deviation;
SDS, sodium dodecyl sulfate;
SDS-PAGE, sodium dodecyl sulfate-polyacrylamide gel electrophoresis;

SELEX, systematic evolution by exponential enrichment;
SINE; short interspersed repeat sequences;
S_o, solubility of a model compound in H_2O;
SOB, *E. coli* growth medium;
SOC, SOB medium plus glucose;
SP, sulfopropyl;
SSC, standard saline citrate;
ssDNA, single-stranded DNA;
ssRNA, single-stranded RNA;
STET, sodium, Tris, EDTA, Triton X-100 buffer;
STS, sequence tagged sites;
t, constant determined in Student's T test;
t-PA, tissue plasminogen activator;
TA, teaching assistant;
TAE, Tris, acetate, EDTA buffer;
TAF, TBP-associated factors;
Taq, *Thermus aquaticus*;
TBE, Tris, borate, EDTA buffer;
TBP, TATA binding protein;
TBS, Tris-buffered saline;
T_c, critical temperature;
TCR, T cell receptor;
TdT, terminal deoxynucleotidyl transferase;
TE, Tris, EDTA buffer;
TENS, Tris, EDTA, NaCl, SDS buffer;
TEP, Tris, EDTA, PMSF buffer;
TETBS, Tris-buffered saline with EDTA and Tween 20;
tet^r, tetracycline resistance genotype/phenotype (gene);
tet^s, ampicillin sensitive genotype/phenotype;
TF, Transcription factor;
T_h, reference T at which $\Delta H = 0$;
TIBS, *Trends in Biochemical Sciences*;
TLC, thin-layer chromatography;
T_m, melting temperature;
TMAO, trimethylamine *N*-oxide;
T_m^o, melting temperature of a molecule in H_2O;
T_{no}, "near optimal" target concentration;
TPEG, *p*-amino-phenyl-β-D-thiogalactoside;
Tris, tris(hydroxymethyl)aminomethane;
tRNA, transfer RNA;
trp, tryptophan;
T_S, reference T at which $\Delta S = 0$;
TTBS, Tris-buffered saline containing Tween 20 detergent;
TU, transducing units;
tyr, tyrosine;
UMP, uridine-5'-monophosphate;
UV, ultraviolet
V_κ, immunoglobulin κ light chain;
v, velocity;
v/v, volume to volume content;
V_e, elution volume;
V_H, immunoglobulin heavy chain;
v_{max}, maximum velocity;
V_o, void volume;
$v_{o,i}$, initial velocity for sample i;
V_s, column stationary phase volume;
V_s, separation volume;
V_t, total packed column bed volume
w/v, weight to volume content;
W_1, peak 1 width (2 for peak 2 ...);
WB, Western blot;
W_o, starting material weight;
W_r, total retentate weight before the assay;
W_t, filtrate weight;
x g, times the force of gravity;
$\bar{X}$, arithmetic mean;
X-phosphate, 5-bromo-4-chloro-3-indolyl phosphate;

XL1-Blue, *E. coli* strain not containing the T7 RNA polymerase gene;

X_t, result t;

YAC, yeast artificial chromosome;

z, number of accessible conformations;

z, partition function;

Literature Sources for Biochemical Analyses, Methods, and Preparations

Biochemical Techniques: Theory and Practice. Robyt, J. and White, B. Waveland Press, Inc. P.O. Box 400. Prospect Heights, Illinois 60070, 1987.

1. Overview

Literature sources consist of "series works" and journals. Sources of the first type contain reviews and summaries of methods in the research literature. They are collections of articles on general but limited topics contributed by experts in the specialties contained in the series. Sources of the second type, the various biochemical and related research journals, can be further divided into those that are exclusively devoted to methods and those that contain general biochemical studies, offering in the experimental sections of their papers many methods, some of which are new. Following is an annotated bibliography literature sources for methods and techniques used in biochemical laboratory studies.

2. Series Sources

2.1 *Methods in Enzymology.* A 120-volume series introduced in 1955 and published by Academic Press, New York. There are various volume editors and many contributors (one or more different authors for each article). The series, which is still being published, covers a very wide range of subjects relating to enzymology (e.g., biomembranes; nucleic acids; recombinant DNA; action of hormones, drugs, and enzyme targeting).

2.2 *Methods of Biochemical Analysis.* A 31-volume series introduced in 1954, edited by David Glick, and published by Interscience Publications of John Wiley & Sons, New York. This series, which is still being published, covers a wide variety of analytical methods involved in the qualitative and quantitative determination of compounds of interest to biochemistry. Topics include chemical, physical, and microbiological methods of analysis, and basic techniques and instrumentation for the determination of enzymes, vitamins, coenzymes, hormones, lipids, carbohydrates, proteins, nucleic acids, and products of metabolism.

2.3 *Methods of Carbohydrate Chemistry.* An 8-volume series introduced in 1962, edited by R. J. Whistler, R. J. Smith, and J. N. BeMiller, and published by Academic Press. This series, which is still being published, covers a wide range of general and specific methods involving all aspects of carbohydrate chemistry including analysis, isolation, synthesis, and physical chemical methods for monosaccharides, oligosaccharides, polysaccharides, and glycoproteins.

2.4 Laboratory Techniques in Biochemistry and Molecular Biology. Several volumes on biochemical laboratory techniques, edited by T. S. Work) covers chromatography, electrophoresis, centrifugation, isoelectric focusing, sequencing, cell culture, immunochemistry, liquid scintillation spectrometry, an related topics.

2.5 Biochemical Preparations. A series published in 13 volumes from 1949 to 1971, containing the details for the preparation of many biochemical compounds, metabolites, enzymes, and coenzymes. Each preparation was reported by one laboratory and independently checked by a second laboratory. This series was particularly useful before the common biochemicals were available commercially and each laboratory had to prepare its own materials. Biochemical Preparations still serves as an important source for the exact methods involved in the preparation of important biochemicals.

3. Methods Journals

3.1 Analytical Biochemistry [*Anal. Biochem.*] Published by Academic Press; serves as an outlet for the publishing of new analytical and preparative methods in biochemistry.

3.2 Analytical Chemistry [*Anal. Chem.*] Published by the American Chemical Society, Washington, D. C., serves as an outlet for the publishing of analytical methods of all types, some of them applicable to biochemistry.

3.3 Biotechniques. Published five times a year by Eaton Publishing Co., Natick, MA; features reviews and new techniques, especially in the areas of molecular biology and biotechnology.

***3.4 Preparative Biochemistry* [*Prep. Biochem.*]** Published by Marcel Dekker, New York, started in 1971 as a substitute for the discontinued Biochemical Preparations series. It is devoted to the rapid dissemination of information on new preparative methods and procedures in biochemistry, immunology, pharmacy, clinical chemistry, biophysics, and molecular biology.

3.5 Journal of Chromatography [*J. Chromatogr.*] Published by Elsevier, Amsterdam; serves as an outlet for new methods and procedures in all aspects of chromatographic separations.

3.6 Carbohydrate Research [*Carbohydr. Res.*] Published by Elsevier, Amsterdam. Although devoted to publishing all aspects of research on carbohydrate chemistry and biochemistry, the journal publishes a respectable number of analytical and preparative methods.

3.7 Clinical Chemistry [*Clin. Chem.*] Published by American Association for Clinical Chemistry, Washington, D. C. Has a high percentage of bioanalytical papers, most of which are devoted to methods of analysis of medically important compounds.

3.8 Trends in Biochemical Sciences [*TIBS*] A monthly review journal, published by the Internation Union of Biochemistry and Elsevier Science Publishers. The journal features a special section on emerging techniques that presents an up-to-date review of specialized techniques used in biochemistry and molecular biology.

4. General Journals

Archives of Biochemistry and Biophysics [*Arch. Biochem. Biophys'*]
Biochemical and Biophysical Research Communications [*Biochem. Biophys. Res. Commun.*]
Biochemical Journal [*Biochim. Biophys. Acta*]
Biochemistry
Biochimica et Biophysica Acta [*Biochim. Biophys. Acta*]
Bioinorganic Chemistry [*Bioinorg. Chem.*
Bioorganic Chemistry [*Bioorg. Chem.*]
Biopolymers
Canadian Journal of Biochemistry [*Can. J. Biochem.*]
Comparative Biochemistry and Physiology [*Comp. Biochem. Physiol.*]
European Journal of Biochemistry [*Eur. J. Biochem.*]
Federation of European Biochemical Societies Letters [*FEBS Lett.*]
International Journal of Biochemistry [*Int. J. Biochem.*]
International Journal of Peptide and Protein Research [*Int. J. Peptide Protein Res.*]
Journal of Applied Biochemistry [*J. Appl. Biochem.*]
Journal of Bacteriology [*J. Bacteriol.*]
Journal of Biochemistry (Tokyo) [*J. Biochem. (Tokyo)*
Journal of Biological Chemistry [*J. Biol. Chem.*
Journal of Chemical Education [*J. Chem. Educ.*]
Journal of Inorganic Biochemistry [*J. Inorg. Biochem.*]
Journal of Lipid Research [*J. Lipid Res.*]
Journal of Molecular Biology [*J. Mol. Biol.*]
Lipids
Macromolecules
Molecular and Cellular Biochemistry [*Mol. Cell. Biochem.*]
Naturwissenschaften
Nucleic Acid Research [*Nucleic Acid Res.*]
Proceedings of the National Academy of Sciences (U.S.A) [*Proc. Natl. Acad. Sci. U.S.A.*]
Proceedings of the Society of Experimental Biology and Medicine [*Proc. Soc. Exp. Biol. Med.*]
Science
*The abbreviation sanctioned by the International Standards Organization is given after the full name of the journal. Journal names that consist of a single word are not abbreviated.

COPYRIGHT ACKNOWLEDGMENTS

Articles are classified into four categories. *Original literature* describes a single project or the solution to a limited problem, *review articles* summarize the results in an area of study and examine the details of mechanisms and interpretations, *commercial literature* can accomplish either original or review objectives or both, and *books* typically summarize fundamental concepts. Grateful acknowledgment is made to the following sources for permission to reprint material copyrighted or controlled by them.

1. Original

"Measurement of Protein in 20 Seconds," by R. E. Akins and R. S. Tuan, reprinted by permission from *BioTechniques*, Volume 12, 1992.

"Immuno-PCR: Very Sensitive Antigen Detection by Means of Specific Antibody-DNA Conjugates," by T. Sano, C. L. Smith, and C. R. Cantor, reprinted from *Science*, Volume 258, 1992.

"Reduction of Background Problems in Nonradioactive Northern and Southern Blot Analyses Enables Higher Sensitivity than 32-P Based Hybridizations," by G. Engler-Blum, M. Meier, J. Frank, and G. A. Muller, reprinted from *Analytical Biochemistry*, Volume 210, 1993.

"One-step Purification of Hybrid Proteins Which Have β-galactosidase Activity," by A. Ullmann, reprinted from *Gene*, Volume 29, Elsevier Science, 1984.

The following articles are published by *Chemical and Engineering News*:
"Better Ribozyme Developed by *in Vitro* Selection," reprinted from Volume 71, Number 4, January 25, 1993. "Isolation of RNA Catalysts Supports 'RNA-World Hypothesis," reprinted from Volume 71, Number 37, September 13, 1993. "Technique Targets, Cleaves Specific RNA Sequences," reprinted from Volume 71, Number 37, February 15, 1993.

"Making Antibody Fragments Using Phage Libraries," by T. Clackson, H. R. Hoogenboom, A. D. Griffiths, and G. Winter, reprinted from *Nature*, Volume 352, August 15, 1991.

"Luminescent Reporter Gene Assays for Luciferase and β-Galactosidase Using a Liquid Scintillation Counter," by R. Fullton and B. Van Ness, reprinted by permission from *BioTechniques*, Volume 14, 1993.

"Western Blots Using Stained Protein Gels," by D. Thompson and G. Larson, reprinted by permission from *BioTechniques*, Volume 12, Number 5, 1992.

2. Review

"Yeast of Burden – Yoking the YAC," by S. S. Roberts, reprinted from the *Journal of National Institute of Health Research*, Volume 2, 1990.

"How Cells Respond to Stress," by W. J. Welch, reprinted from *The Scientific American*, May 1993.

"The Polymerase Chain Reaction," by W. C. Timmer and J. M. Villalobos, reprinted from *Journal of Chemical Education: Concepts in Biochemistry*, Volume 70, Number 4, John Wiley and Sons, 1993.

"Polymerase Chain Reaction Used for Antigen Detection," reprinted from *Chemical and Engineering News*, Volume 70, Number 40, October 5, 1992.

"Dominant Forces in Protein Folding," by K. A. Dill, reprinted from *Biochemistry*, Volume 29, Number 31, August 7, 1990.

"Ion Effects on the Solution Structure of Biological Macromolecules," by P. H. Von Hippel and T. Schleich, reprinted from *Accounts of Chemical Research*, Volume 2, Number 9, 1969.

The following articles are published by the *Journal of National Institute of Health Research,* Volume 5,

June 1993:
"Making Monoclonal Antibodies That Won't Fight Back," by R. Rhein. "Building a Better Enzyme (Achilles Cleavage, Ribozymes, Abzymes)," by S. S. Roberts.

"Solubility as a Function of Protein Structure and Solvent Components," by C. H. Schein, reprinted from *Nature Bio/Technology*, Volume 8, 1990.

"The Enzyme Linked Immunosorbent Assay (ELISA)," by A. Voller, D. E. Bidwell, and A. Bartlett, Dynatech Laboratories, Chantilly, Virginia, 1979.

"How the Immune System Learns About Self" by H. Von Boehmer and P. Kisielow, reprinted from *The Scientific American*, April 1993.

3. Commercial

"Oxford® Pipettes," reprinted from *Oxford® Benchmate® Continuously Adjustable Pipette Instruction Manual*, Oxford Labware, Sherwood Medical, Bridgeton, Missouri.

"Concentration and Temperature Effects on pK," edited by D. E. Gueffroy, reprinted from *Buffers: A Guide for the Preparation and Use of Buffers in Biological Systems*, CalBiochem Corporation, San Diego, California, 1975.

"DNA Digests," reprinted from *New England Biolabs Annual Catalog*, New England Biolabs, Beverly, Massachusetts, 1996.

"Restriction Enzymes *Hin*dIII and *Bst*EII," reprinted from *Promega Corporation Annual Catalog*, Promega Corporation, Madison, Wisconsin, 1996.

"Elu-quick® Cleanup of Plasmid Digests," reprinted from *Schleicher and Schuell Elu-quick® DNA Purification Kit*, Schleicher and Schuell, Keene, New Hampshire, 1992.

"The Use of β-Agarase® to Recover DNA from Gel Slices," by J. A. Krall, reprinted from *Comments*, Volume 20, Number 1, United States Biochemical/Amersham Life Science, Cleveland, Ohio, 1993.

"Gibco BRL® T4 DNA Ligase," reprinted from *Focus on Applications, T4 DNA Ligase, Technical Bulletin*, Life Technologies, Gaithersburg, Maryland, 1994.

"GELase®," reprinted from *Epicentre Technologies Annual Catalog*, Epicentre Technologies, Madison, Wisconsin, 1995.

"The pET® Bacterial Expression Plasmid System," reprinted from *Novagen Annual Catalog*, Novagen, Inc., Madison, Wisconsin, 1993.

"The Qiagen® QIA*express* System," reprinted from *The Qiaexpressionist*, Qiagen® Inc., Chatsworth, California, 1992.

"Ultracomp® Transformation Kit," reprinted from *Ultracomp® E. coli Cells*, Version A, Invitrogen Corporation, San Diego, California.

"Isolation of Total RNA from *E. coli*," by P. Chomczynski, reprinted from *TRI Reagent®-RNA, DNA, Protein Isolation Reagent, Manufacturer Protocol,* Molecular Research Center, Inc., Cincinnati, Ohio, 1993.

PolyATtract® System 1000, reprinted from *PolyATtract™ System 1000, Technical Bulletin 228,* Promega Corporation, Madison, Wisconsin, 1995.

"Digoxigenin Labeling of DNA by Random Priming with Klenow DNA Polymerase with Chemiluminescence Detection," reprinted from *The Genius® System User's Guide for Filter Hybridization*, Version 2.0, Boehringer-Mannheim, Indianapolis, Indiana, 1992.

"The Protein Purifier®: A Learning Aid from Pharmacia" by A. Booth, Pharmacia Biotech, Inc., Piscataway, New Jersey, 1987.

"Fractionation of Proteins: Gel Filtration Chromatography," reprinted from *Gel Filtration: Principles and Methods*, Sixth Edition, Pharmacia Biotech, Inc., Piscataway, New Jersey.

"Sephadex®," reprinted from *Gel Filtration: Principles and Methods,* Sixth Edition, Pharmacia Biotech, Inc., Piscataway. New Jersey.

"Gel Filtration Molecular Weight Markers," reprinted from *Sigma Chemical Company Technical Bulletin*, October 1987, Sigma Chemical Company, St. Louis, Missouri.

"Centriprep Microconcentrators for Small Volume Concentration," reprinted from *Centriprep Concentrators, Operating Instructions*, publication #I-320H, Amicon, Inc., Beverly, Massachusetts, 1993.

"β-Galactosidase Substrates," reprinted from *Fluka Chemie AG Annual Catalog*, Fluka Chemie AG, Buchs, Switzerland and Ronkonkoma, New York, 1995.

"Ion Exchange Chromatography," reprinted from *Ion Exchange Chromatography: Principles and Methods*, Edition AC, Pharmacia Biotech, Piscataway, New Jersey.

"Ion-pair Chromatography," reprinted from *Forum*, Volume 14, 1992, Mac Mod Analytical, loc., Chadds Ford, Pennsylvania.

The following articles are published by Pharmacia Biotech, Inc., Piscataway, New Jersey:
"Affinity Chromatography," reprinted from *Affinity Chromatography: Principles and Methods.*
"Hydrophobic Interaction Chromatography," reprinted from *Principles and Methods.*

"ELF AP (Alkaline Phosphatase Immunohistochemistry Kit)," reprinted from *BIOProbes, New Products and Applications*, Volume 18, Molecular Probes, Inc., Eugene, Oregon, 1993.

"Amplified Alkaline Phosphatase Immune-blot Kits," reprinted from *BioRad Annual Catalog*, BioRad, Inc., Hercules, California, 1996.

"The ImmunoZAP™ Cloning and Expression System," reprinted from *Stratagene Annual Catalog*, Stratagene Corporation, La Jolla, California, 1995.

4. Books

"Introduction to the Biotechnology Laboratory," by D. W. Burden and D. B. Whitney, reprinted from *Biotechnology: Proteins to PCR*, Birkhauser Boston, 1995.

"Treatment of Analytical Data," by J. S. Fritz and G. H. Shenk, reprinted from *Quantitative Analytical Chemistry*, Allyn & Bacon, 1974.

"Preparation and Handling of Biological Macromolecules for Crystallization," by B. Lorber and R. Giege, reprinted from *Crystallization of Nucleic Acids and Proteins: A Practical Approach*, edited by A. Ducruix and R. Giege, IRL Press, 1992.

"Membranes and Their Cellular Function," by J. B. Finean, R. Coleman, and R. H. Mitchell, John Wiley and Sons, 1974.

"Time Course Assay of β-Galactosidase, " by J. M. Miller, Cold Spring Harbor, reprinted from *Experiments in Molecular Genetics*, 1973.

"Protein at a Particular pH: Isoelectric Point," by H. R. Mahler and E. H. Cordes, reprinted from *Biological Chemistry*, Second edition, Harper and Row, 1971.

"Discontinuous Gel Electrophoresis and Protein Size Determination," by T. G. Cooper, reprinted from *The Tools of Biochemistry*, Wiley-Interscience, 1977.

"SDS Gel Electrophoresis," by A. H. Gordon, reprinted from *Laboratory Techniques in Biochemistry and Molecular Biology*, North Holland Publishing Company, 1975.

"Immunochemical Techniques," by T. G. Cooper, reprinted from *The Tools of Biochemistry*, Wiley-Interscience, 1977.

SUGGESTED SCHEDULE

PART 1 - NUCLEIC ACIDS & CLONING

PART 2 - PROTEIN PURIFICATION

SUGGESTED INSTRUCTIONS FOR LAB REPORTS

Lab reports will be due one week after the last day of the lab. All calculations in your notebooks will be checked and the text of the report will be graded for completeness, precision and terseness. Include the following subjects:

1. Title

A description of the work that summarizes the important results. Include a list of authors (your byline and name of this course and addresses).

2. Abstract

A brief summary of the work. This section will be limited to 100 words. Don't waste them!

3. Introduction

What are you doing and why? Be brief. State the biological problem and necessary background regarding the problem being studied.

4. Materials and Descriptions

Materials used and descriptions of the techniques performed in the research. Do not copy the Methods procedures into your lab manual. Simply summarize them.

5. Results

Present the results of your experiments in logical order. Do not discuss the significance of the results or the logic that led you from one step to the next, just the results. The logical progression of activities and significance should be covered in the Discussion.

6. Discussion

What did your results show you? What logic led you to reach this or these conclusion(s)? What is the significance? Discuss the sensitivity of the technique and how the conclusions might be affected.

7. Tables

Include a title, clear column headings and define any abbreviations and other information in footnotes.

8. Figures

Include clearly worded legends. All symbols should be defined in the legend; avoid writing on the figure. Make symbols and fit lines easy to read; maximize the information size within the plotted field. It is not always necessary to show zero on the axes.

Write out the report in draft form somewhere else. The reports you hand in should be polished and finalized. You can write them out by hand unless we cannot read your handwriting; however, we prefer that the reports be typed/word processed. Your fellow students will help us with judgments regarding readability. In general, these reports should range in length between approximately 2 and 4 pages.

SUPPLIES REQUIRED (FOR 40 STUDENTS)

ITEM	QUANTITY	SIZE
λ *Hin*dIII DNA marker	250 mg	
β-galactosidase	1,000 U	
β-mercaptoethanol (β-ME)	100 ml	
ϕx174 *Hae*III DNA marker	250 mg	
"The Protein Purifier" software (Pharmacia)		
10x TBS	2 liter	
4-chloro-1-napthol	5 g	
70% ethanol (EtOH)	1 liter	
95% ethanol (EtOH)	5 gal	
96-well microtiter plastes (no lids)	200	
acrylamide/bis (C_2H_5NO) (40% - 29:1)	1 liter	
adenosine-5'-monophosphate (sodium salt)	10 g	
agar	1 kg	
agarose	500 g	
agarose minigel apparatus	20	
alumnium foil	1 box	
ammonium acetate (NH_4OAc)	500 g	
ammonium persulfate [$(NH_4)_2S_2O_8$]	25 g	
ammonium sulfate [$(NH_4)_2SO_4$]	3 kg	
ampicillin (amp)	25 g	
autoclave bags	20	
autoclave tape	1 roll	
bacteria culture flasks	20 of each size	250 ml, 500 ml, 1 liter
BCA reagent, A and B (Pierce)	8 kits	
beakers (glass or plastic)	10 of each size	50 ml, 100 ml, 1 liter, 2 liter, 4 liter
blotting paper	1 box	3 MM
bovine serum albumin (BSA)	100 g	
bromophenol blue	25 g	
calcium chloride ($CaCl_2$)	1 kg	
calf intestinal alkaline phosphatase	1000 U	
camera film (Polaroid 667)	15 packs	
centrifuge	1	
Centriprep 30 ultrafiltration concentrators (Amicon)	2 packs (24/pk)	
chloroform ($CHCl_3$)	500 ml	
Clorox	1 bottle	
chromatography column	40	50-100 ml (2.5 x 20 cm)
conical tubes (sterile)	1 case per student	15 ml, 50 ml
Coomassie blue	25 g	
cuvettes	40	
desiccator	1	
dialysis clips	40	
diaminobenzidene (DAB)	5 g	
diethylpyrocarbonate (DEPC)	100 ml	
digoxygenin hybridization buffer	500 ml	
dishwashing soap	1 bottle	
dithiothreitol (DTT)	1 g	
deoxyribonuclease (DNase)	15,000 U	
dry ice		
E. coli XL-1 Blue cells		

Econo columns (BioRad)	20	1.5 x 75 cm
Eluquik™ kit (Schleicher & Schuell™)	1 kit	
Eppendorf tubes (screw cap)	5,000	2 ml
erlenmeyer flasks	10 of each size	100 ml, 250 ml, 500 ml, 1 liter, 2 liter
ethidium bromide (EtBr)	1 g	
ethylenediaminetetraacetate - disodium or tetrasodium salt - (EDTA)	1 kg	
filter forceps (flat ended)	40	
fine-tipped syringe	20	
fraction collector	1 for each column	
fraction collector tubes	5,000	
gel filtration molecular weight standards	2-4 sets	
Genius Labeling Kit (Boehringer-Mannheim)	1 kit	
Genius nylon membrane	40 sheets	10 x 15 cm
glacial acetic acid (CH_3COOH)	4 liter	
gloves (latex or nitrile)	1,000 of each size	small, medium, large
glucose ($C_6H_{12}O_6$)	1 kg	
glycerol ($C_3H_8O_3$)	1 L	
glycine	1 kg	
goat anti-mouse peroxidase conjugate	1 ml	
goggles	40	
graph paper	several packets	
*Hin*dIII and buffer to accompany	10,000 U	
hybridization bags (if a heat sealer is available, if not use Zip-Loc bags)	200	
hydrochloric acid – concentrated (HCl)	2 liter	
hydrogen peroxide (H_2O_2) – 30%	500 ml	
incubator	1	
isopropanol	1liter	
isopropyl-β-D-thiogalactodase (IPTG)	5 g	
l-tryptophan (sodium salt)	10 g	
Lumiphos 530 (Boehringer-Mannheim)	40 ml	
lysozyme	2 g	
Macro-Prep High Q Strong Anion Exchange Support (BioRad)	1 liter	
magnesium chloride ($MgCl_2$)	500 g	
magnesium sulfate ($MgSO_4 \bullet 7\ H_2O$)	1 kg	
media bottle filters	10	large
media bottles (autoclavable)	50	250 ml, 500 ml, 1 liter
methanol (CH_3OH)	4 liter	
microfuge tubes	10,000	1.5 ml
microfuge tubes	5,000	0.6 ml
microtiter plate reader	1	
mineral oil	1 bottle	
mouse monoclonal antibody (for *myo*-3)	1 ml	
N,*N*,*N*',*N*'-tetramethyl ethylenediamine (TEMED)	25 ml	
nitrocellulose or nylon circles	100	to fit 100-mm petri dish
nitrocellulose	several large sheets	0.45 mm
N-lauryl sarcosine	500 g	
Nonidet P-40 (NP-40)	100 ml	
NZ amine (Sigma)	1 kg	
Oak Ridge centrifuge tubes (polypropylene or other chemical resistance type)	100	50 ml
o-nitro-pyrogalactopyranoside (ONPG)	10 g	
Ottawa sand		

p-aminobenzyl-1-β-galactosidase-agarose (Sigma)	50 ml	
PCR primers:		
1: AAGCTTTGCGACAGCTTC	50 nmol	
2: GGATCCGTCGACTCTAGA	50 nmol	
petri dishes	1 case	100 mm
pH 5 buffer	500 ml	
pH 7 buffer	1 liter	
pH 10 buffer	500 ml	
pH meter	2	
phenol/chloroform/isoamyl alcohol	100 ml	
phenylmethylsulfonyl fluoride (PMSF)	5 g	
pipet tips in boxes	100 of each size	20 μl, 50 μl, 200 μl, 1 ml
pipet tips in bulk	10,000 of each size	20 μl, 50 μl, 200 μl, 1 ml
plastic wrap	1 box	
potassium chloride (KCl)	500 g	
potassium hydroxide (KOH)	1 kg	
potassium phosphate dibasic (K_2HPO_4)	1 kg	
potassium phosphate monobasic (KH_2PO_4)	1 kg	
protective ear guards	2	
proteinase K	100 mg	
RNase A	100 mg	
rocking platform/shaker	1	
scapel	1	
SDS-PAGE molecular weight standards (prestained and unstained)	1 vial of each	
Sephacryl HR 300	3 liter	
small glass beads	several boxes	3-6 mm diameter
sodium acetate ($C_2H_3NaO_2$)	1 kg	
sodium borate ($Na_2B_4O_7$)	500 g	
sodium carbonate (Na_2CO_3)	500 g	
sodium chloride (NaCl)	5 kg	
sodium citrate ($C_6H_6Na_2O_7$)	1 kg	
sodium dodecyl sulfate (SDS)	1 kg	
sodium hydroxide (NaOH)	5 kg	
sodium phosphate dibasic (Na_2HPO_4)	2 kg	
sodium phosphate monobasic (NaH_2PO_4)	2 kg	
spatulas	50	
stir bars	several kits	
stir-plate	20	
sucrose ($C_{12}H_{22}O_{11}$)	1 kg	
T4 DNA ligase	100 U	
Taq DNA polymerase and assay buffer to accompany	300 U	
thermocycler	1	
toothpicks (flat)	10 boxes	
top-loading balance	2	
Transphor unit (Hoeffer)	10	
trays (for blotting, staining, etc.)	5-10 per student	small, medium, large
Tris base	5 kg	
Triton X-100	100 ml	
tryptone	1 kg	
Tween 20	100 ml	
tygon tubing	50 ft each	0.8 mm, 1.6 mm, 2.4 mm, 3.2 mm
UV face shield	2	
UV gel illuminator box	1	

UV-visible spectrophotometer	5	
vacuum concentrator	1	
volumetric flasks	20 of each size	250 ml, 500 ml
vortex	20	
water bath	1	
weigh boats		
weighing paper		
X-ray cassette	20	
X-ray film	50	
xylene cyanol	25 g	
yeast extract	1 kg	

INDEX

Bold entries in longer list indicate more significant occurrences.
Unit 1 Terms: pp. 388-391; Unit 2 Terms: pp. 392-396.

B

C

D

E

F

G

H

I

K

L

M

N

O

P

Q

R

S

T

U

V

W

X

Y

Z